TechConnect Briefs 2015

BIOTECH, BIOMATERIALS AND BIOMEDICAL

Edited by
Matthew Laudon   Bart Romanowicz

TechConnect

CRC Press
Taylor & Francis Group
Boca Raton London New York

CRC Press is an imprint of the
Taylor & Francis Group, an **informa** business

First published 2015 by TechConnect and CRC Press
Taylor & Francis Group
6000 Broken Sound Parkway NW, Suite 300
Boca Raton, FL 33487-2742

ISBN 13: 978-1-4987-4729-5 (pbk)
DOI: 10.1201/9781315138435

Visit the Taylor & Francis Web site at
http://www.taylorandfrancis.com

and the CRC Press Web site at
http://www.crcpress.com

CHAPTER EDITORS - 2015 TechConnect Briefs

Biofuels & Bioproducts

Scott Twary, *Los Alamos National Laboratory, USA*
Srinivas Iyer, *Los Alamos National Laboratory, USA*

ELECTRONICS & MICROSYSTEMS

MEMS & NEMS Devices and Applications

William (Cy) Wilson, *NASA Langley Research Center, USA*

Flexible Electronics

Mandakini Kanungo, *Xerox Innovation Group, USA*

Photonic Materials & Devices

Loucas Tsakalakos, *GE Global Research, USA*

Nano Electronics for 2020 and Beyond

Aaron D. Franklin, *Duke University, USA*
Gernot Pomrenke, *USAF/AFOSR/Optoelectronics and Photonics, USA*
Curt Richter, *NIST, USA*
Charles Ying, *NSF, USA*

Micro, Nano & Bio Fluidics

Edward Furlani, *University at Buffalo SUNY, USA*
Kwang Oh, *University at Buffalo SUNY, USA*

Inkjet Design Materials and Fabrication

Chris Menzel, *Fujifilm Dimatix, Inc., USA*

ENERGY & SUSTAINABILITY

Energy Storage

Imre Gyuk, *U.S. Department of Energy, USA*

Water Technologies

Armin Volkel, *Palo Alto Research Center, USA*

Advanced Materials for Oil & Gas

Willard C. (Bill) Capdevielle, *Oil Patch Engineering, PLLC, USA*
Rushan Sakhibgareev, *Chevron, USA*

Solar Power Technologies

Brian Korgel, *University of Texas at Austin, USA*
Heath Naquin, *University of Texas at Austin, USA*

NanoNuclear Materials, Fuels & Applications

Brent Segal, *Lockheed Martin, USA*
Thomas McKrell, *MIT, USA*
Steven Winston, *Steven Winston & Assoc., USA*

Carbon Capture

Robert Quinn, *Air Products and Chemicals (retired), USA*
Elizabeth Burton, *Lawrence Berkeley National Laboratory, USA*

Materials for Green Building

Elena Enache-Pommer, *Dow Building Solutions R&D, USA*

Key Elements for Successful Commercialization: Safety and Sustainability

Charles Geraci, Jr., *NIOSH, USA*
Phillip Demokritou, *Harvard University, USA*
Sally Tinkle, *Science and Technology Policy Management, USA*

PERSONAL & HOME CARE, COSMETICS & FOOD

Food Materials

Pat Spicer, *University of New South Wales, AU*
Krassimir Velikov, *Unilever R&D, NE*

Materials for Personal and Home Care

Fiona Case, *Nano Science and Technology Institute, USA*

TABLE OF CONTENTS

Cancer Nanotechnology

Sensors, Diagnostics & Imaging

Micro, Nano & Bio Fluidics

Biomaterials

Materials for Personal/Home Care & Cosmetics

Soft Matter & Colloids

PREFACE

TechConnect provides a multi-disciplinary and multi-sector forum focused on accelerating applied research and early-stage innovation towards commercialization. The technologies described in the TechConnect Briefs are provided by innovators working in industry, academic and government laboratories around the world. A review committee comprising more than 300 internationally recognized experts reviewed the research and innovations described and selected them for inclusion in this volume, and for presentation at the 10th annual TechConnect World Innovation Conference & Expo, held jointly with the 18th annual Nanotech Conference & Expo, and the 2015 National SBIR/STTR Conference, June 14-17, 2015 Washington DC.

Through TechConnect prospecting programs, transformational innovations are identified in Advanced Materials, Advanced Manufacturing, Materials for Energy & Sustainability, Electronics & Microsystems, Biotech, Medical & Pharma, Personal & Home Care, Cosmetics, and Food industries. The 2015 TechConnect Briefs are also aligned with the National Nanotechnology Initiative and the National Security Technology Accelerator interest areas.

This volume (one of 4) is focused on research and innovations in the Biotech, Biomaterials and Biomedical track, including industrially viable new materials that will transform industry and new platform technologies that enable their discovery and development. The following chapters are included:

- Materials for Drug & Gene Delivery
- Cancer Nanotechnology
- Sensors, Diagnostics & Imaging
- Micro, Nano & Bio Fluidics
- Biomaterials
- Materials for Personal/Home Care & Cosmetics
- Soft Matter & Colloids
- Food Materials

It is our distinct honor to edit the 18th annual TechConnect publication of applied research technologies at the forefront of innovation change for industry and society. The TechConnect Program Committees make every effort to provide a scientifically and commercially relevant publication. The grassroots review and ranking process is a source of pride for these joint committees, and provides a yearly evaluation and technical validation of emerging technologies addressing our world's most pressing needs. We would like to take this opportunity to thank the many individuals who have worked so hard to make this annual publication possible. We hope that you enjoy these TechConnect Briefs, and we hope that the presented technologies are significant steps towards addressing large global challenges.

Matthew Laudon
Austin, Texas

Bart Romanowicz
Austin, Texas

Fiona Case
San Diego, California

Program Review Committee

Wade Adams, *Rice University, USA*
Susan Addo Ntim, *U.S. Food & Drug Administration, USA*
I. Yucel Akkutlu, *Texas A&M University, USA*
Evangelyn Alocilja, *Michigan State University, USA*
Jay Amarasekera, *Sabic, USA*
Mansoor Amiji, *Northeastern University, USA*
Thomas Anchordoquy, *University of Colorado, USA*
Mohamed Ansari, *Universiti Tenaga Nasional Kajang, Malaysia*
Sivaram Arepalli, *Sungkyunkwan University, Korea*
Wolfgang Bacsa, *University of Toulouse, France*
Stephen Bart, *Analog Devices, Inc., USA*
S Rahima Benhabbour, *University of North Carolina at Chapel Hill, USA*
Gary Bernstein, *University of Notre Dame, USA*
Jean Berthier, *CEA Grenoble, France*
Mousumi Mani Biswas, *Western Digital, USA*
Gregory R. Bogart, *Sandia National Laboratories, USA*
Philip Bradford, *North Carolina State University, USA*
Jakob Bredsguard, *Biosyntehtic Technologies, USA*
Elizabeth Burton, *Lawrence Berkeley National Laboratory, USA*
Ahmed Busnaina, *Northeastern University, USA*
Willard C. (Bill) Capdevielle, *Oil Patch Engineering, PLLC, USA*
Fiona Case, *Case Scientific, USA*
Martin Case, *Janssen, Pharmaceutial Companies of Johnson & Johnson, USA*
Mansun Chan, *Hong Kong University of Science & Technology, China*
Alessandro Chiolerio, *Politecnico di Torino, Italy*
Bum-Kyoo Choi, *Sogang University, Korea*
Schuyler Corry, *Life Technologies, USA*
Lance Criscuolo, *Zyvex Technologies, USA*
Jaime Jose Cubillo Capuz, *Acciona, Spain*
Zheng Cui, *Suzhou Institute of Nanotech, Chinese Academy of Sciences, China*
Phil Demokritou, *Harvard University, USA*
Guy DeRose, *California Institute of Technology, USA*
George Elvin, *Green Technology Forum, USA*
Dorothy Farrell, *NCI Alliance for Nanotechnology in Cancer, USA*
Aaron Franklin, *Duke University, USA*
Nicolas Feltin, *Laboratoire national de métrologie et d'essais, France*
Thomas Fengyuan *Li, Freescale Semiconductor, USA*
Lisa Friedersdorf, *National Nanotechnology Coordination Office, USA*
Ian Fuller, *Nanotek Instruments, USA*
Edward Furlani, *University at Buffalo SUNY, USA*
Charles Geraci, *Jr., NIOSH, USA*
Harris Goldberg, *InMat, Inc., USA*
Imre Gyuk, *U.S. Department of Energy, USA*
Greg Haugstad, *University of Minnesota, USA*
Tim Hebrink, *3M Corporate Research Process Laboratory, USA*
Michael Helmus, *Consultant / Synogen, Inc., USA*
Andreas Hieke, *Stanford University, USA*
Bin Hu, *University of Tennessee, USA*
Zhaohua Huang, *University of California, San Francisco, USA*
Kiki Ikossi, *Defense Threat Reduction Agency, USA*
Srinivas Iyer, *Los Alamos National Laboratory, USA*
Mark Jack, *Florida A&M University, USA*
Oana Jurchescu, *Wake Forest University, USA*
Mandakini Kanungo, *Xerox, USA*
Barbara Karn, *National Science Foundation, USA*
Brian Korgel, *University of Texas at Austin, USA*

Jana Kukutschova, *Technical University Ostrava, Czech Republic*
James Lamb III, *Brewer Science, USA*
William A Lambert,i *ExxonMobil, USA*
David Lashmore, *Nanocomp Technologies, Inc., USA*
Matthew Laudon, *TechConnect, USA*
Kock-Yee Law, *Research and Innovative Solutions, USA*
Stephen Lehrman, *National Nanotechnology Coordination Office, USA*
James Alexander Liddle, *National Institute of Standards & Technology, USA*
Igor Linkov, *U.S. Army Engineer Research and Development Center, USA*
Yuri Lyubchenko, *University of Nebraska, USA*
Katrin Mackenzie, *Helmholtz Centre for Environmental Research - UFZ, Germany*
Michael Makowski, *PPG Industries R&D, USA*
Stuart Maloy, *Los Alamos National Laboratory, USA*
Anil Mane, *Argonne National Laboratory, USA*
Larry Mann, *Shoei Electronic Materials, Inc., USA*
Thomas McKrell, *Massachusetts Institute of Technology, USA*
Scott McNeil, *National Cancer Institute, USA*
Joey Mead, *University of Massachusetts Lowell, USA*
Michael Meador, *NASA Glenn Research, USA*
Chris Menzel, *Dimatix, Inc., USA*
Meyya Meyyappan, *NASA Ames Research, USA*
Stephen Miller, *Northwestern University Medical School, USA*
Moein Moghimi, *University of Copenhagen, Denmark*
Ron Munson, *Global CCS Institute, Australia*
Chang-Yong Nam, *Brookhaven National Laboratory, USA*
Andrzej Napieralski, *Technical University of Lodz, Poland*
Heath Naquin, *University of Texas at Austin, USA*
Tinh Nguyen, *National Institute of Standards & Technology, USA*
Patricia Nieva, *University of Waterloo, CA*
Greg Nordin, *Brigham Young University, USA*
Kwang Oh, *University at Buffalo SUNY, USA*
Pierre Panine, *Xenocs SA, France*
Hyung Gyu Park, *ETH Zurich, Switzerland*
Anil Patri, *National Institutes of Health, USA*
Joachim Piprek, *NUSOD Institute, USA*
Elena Polyakova, *Graphene 3D Lab, USA*
Gernot Pomrenke, *U.S. Air Force, USA*
Sotiris Pratsinis, *Swiss Federal Institute of Technology, Switzerland*
Serge Prudhomme, *University of Texas at Austin, USA*
Robert Quinn, *Air Products and Chemicals, USA*
YuanQiao Rao, *The Dow Chemical Company, USA*
Robin Ras, *Aalto University School of Science, Finland*
Somilkumar Rathi, *Intel Corporation, USA*
Mark Reed, *Yale University, USA*
Erik Reimhult, *Universität für Bodenkultur Wien, Austria*
Curt Richter, *National Institute of Standards & Technology, USA*
Bart Romanowicz, *TechConnect, USA*
Robert Rudd, *Lawrence Livermore National Laboratory, USA*
Richard Sadus, *Swinburne University of Technology, Australia*
Rushan Sakhibgareev, *Chevron, USA*
Robert Savinell, *Case Western Reserve University, USA*
Derek Schorzman *Liquidia Technologies, Inc., USA*
David Schut, *Shoei Electronic Materials, Inc,. USA*
Keana Scott, *National Institute of Standards and Technology, USA*
Brent Segal, *Lockheed Martin, USA*
Philippe Serp, *Institute National Polytechnique de Toulouse, France*
Neeraj Sharma, *3M, USA*
Emilie Siochi, *NASA Langley Research Center, USA*

CONFERENCE OPERATIONS MANAGER

TechConnect Briefs 2015 Volumes

ADVANCED MATERIALS:

TechConnect Briefs 2015, Vol. 1, ISBN: 978-1-4987-4727-1:

1. Nanoscale Materials Characterization
2. Informatics, Modeling & Simulation
3. Carbon Nano Structures & Devices
4. Graphene
5. Nanoparticle Synthesis & Applications
6. Nanocomposite Materials
7. Advanced Materials for Engineering Applications
8. Coatings, Surfaces & Membranes
9. Polymers
10. Soft Matter & Colloids
11. Nanotech Commercialization: Safety & Sustainability

MATERIALS FOR ENERGY, EFFICIENCY AND SUSTAINABILITY:

TechConnect Briefs 2015, Vol. 2, ISBN: 978-1-4987-4728-8:

1. Energy Storage
2. Fuel Cells & Hydrogen
3. Nanomaterials for Catalysis
4. Materials for Oil, Gas & Biofuels
5. Carbon Capture & Utilization
6. Solar Power Technologies
7. Materials for Green Building
8. Water Technologies
9. Materials for Sustainability & Efficiency

BIOTECH, BIOMATERIALS AND BIOMEDICAL:

TechConnect Briefs 2015, Vol. 3, ISBN: 978-1-4987-4729-5:

1. Materials for Drug & Gene Delivery
2. Cancer Nanotechnology
3. Sensors, Diagnostics & Imaging
4. Micro, Nano & Bio Fluidics
5. Biomaterials
6. Materials for Personal/Home Care & Cosmetics
7. Soft Matter & Colloids
8. Food Materials

ADVANCED MANUFACTURING, ELECTRONICS AND MICROSYSTEMS:

TechConnect Briefs 2015, Vol. 3, ISBN: 978-1-4987-4730-1:

1. Advanced Manufacturing
2. Sustainable Nanomanufacturing
3. NanoFab: Manufacturing & Instrumentation
4. Flexible Electronics
5. Photonic Materials & Devices
6. Nanoelectronics, Materials & Devices
7. Modeling & Simulation of Microsystems
8. MEMS & NEMS Devices & Applications
9. Micro, Nano & Bio Fluidics
10. Inkjet Design, Materials & Fabrication

TECHCONNECT BRIEFS 2015: FOUR VOLUME SET

ISBN: 978-1-4987-4731-8 (hardcopy)

TECHCONNECT BRIEFS 2015: FOUR VOLUME SET

ISBN: 978-1-4987-4733-2 (DVD)

Engineering of Recombinant Spider Silk Proteins Allows Defined Drug Uptake and Release

T. Scheibel*

*Chair of Biomaterials, Faculty of Engineering Science, University of Bayreuth, Universitätsstraße 30, 95440 Bayreuth, Germany, thomas.scheibel@bm.uni-bayreuth.de
Bayreuther Zentrum für Kolloide und Grenzflächen (BZKG)
Institut für Bio-Makromoleküle (bio-mac)
Bayreuther Zentrum für Molekulare Biowissenschaften (BZMB)
Bayreuther Materialzentrum (BayMAT)

ABSTRACT

Drug delivery systems allow tissue / cell specific targeting of drugs in order to reduce total drug amounts administered to an organism and potential side effects upon systemic drug delivery. Most drug delivery systems are polymer-based, but the number of possible materials is limited since many commercially available polymers induce allergic or inflammatory responses or lack either biodegradability or the necessary stability in vivo. Spider silk proteins represent a new class of (bio)polymers that can be used as drug depots or drug delivery systems. The recombinant poly-anionic spider silk protein eADF4(C16), which can be processed into different morphologies such as particles, films, or hydrogels, has been shown to fulfil most criteria necessary for its use as biomaterial. Further, eADF4(C16) particles have been shown to be well-suited as drug carriers for poly-cationic or neutral drugs, but cellular uptake of such particles is low. Novel variants of eADF4(C16) with inversed net charge or incorporated cell penetrating peptides and receptor interacting motifs show an increased cellular uptake. Further, poly-cationic variants allow incorporation of negatively charged drugs including high molecular weight substances, like nucleic acids.

Keywords: spider silk, particles, drug delivery, cellular uptake, biodegradation

1 INTRODUCTION

Conventional drug administration such as intravenous injection or oral administration often requires the application of high drug dosages to attain the biologically effective concentration at the target tissue, since this form of administration lacks selectivity and furthermore suffers from a poor biodistribution of the drug. These disadvantages can be overcome using drug delivery vehicles to which drugs can be bound covalently or non-covalently [1].

The vast majority of drugs have their primary target within cells and tissues [2]. It is thus preferable to deliver drugs into the cytosol especially if significant amounts of the drug are exported out of the cell by ATP-dependent efflux transporters such as the permeability glycoprotein 1 [3]. Drug delivery vehicles which are taken up into cells can act as drug depots and lead to a sustained drug release in order to achieve constant drug levels thereby lowering undesired side effects caused by over- or under-dosage. Moreover, nanoscale drug delivery platforms offer the possibility to protect labile drugs from rapid degradation in the bloodstream and prevent premature renal or hepatic clearance. Thus, the dosage of the drug can be reduced. One major problem during pharmacotherapy is the delivery of a drug to its target site [4]. There, drug delivery vehicles can help to direct the drug to its desired site of action. Cell or tissue specific targeting can be realised if the carrier's surface is modified with biological signalling peptides, antibodies or aptamers to achieve accumulation of the carrier together with its payload in the respective tissues. Thereby, local drug concentrations can be increased several folds within the target tissue and help to reduce the necessary drug concentration. Besides active targeting mechanisms, particles of a size of around 100 nm can be used for the passive targeting of inflamed and cancerous tissues via the EPR (enhanced permeability and retention) effect. Within these tissues, the envelope of the vascular system by endothelial cells is more loose and leaky and therefore supports particle extravasation into the interstitial space, while in healthy tissues particles are retained in the blood stream.

In addition to the tissue specific targeting, drug release from the carrier should be controllable. Many drugs exert their positive effect only within a certain range of concentrations, the so-called therapeutic window. If the drug concentration falls below the lower boundary, no therapeutic effect will be seen while exceeding the upper boundary results in massive toxic side effects. It is therefore desirable to achieve drug delivery systems with defined control properties.

Concerning the application of drug delivery vehicles in humans, the vehicles should fulfil some requirements. For example, they should be non-allergenic, non-immunogenic, non-toxic as well as biocompatible. Furthermore, control over their structure and morphology is prerequisite [5-7]. A broad range of drug delivery vehicles has been developed including liposomes, micelles, dendrimers, and inorganic or polymeric nanoparticles with a size in the range of a few to a few hundreds of nanometres. In general, polymeric nanoparticles are considered as the preferred form of drug delivery vehicles as they are very stable compared to e.g. liposomes and a variety of different materials can be used [4]. Among natural polymers, polysaccharides such as chitosan or alginate and proteins such as albumin or collagen are widely used [4, 8]. Besides their biocompatibility and biodegradability, most natural polymers can be processed without any need for harsh processing conditions, like high temperature, high pressure, or toxic (organic) solvents which is furthermore beneficial for their biocompatibility. Silk as a member of natural polymers combines the required features for its use in biomedical applications; it is non-toxic, non-allergenic, biodegradable and exhibits a good biocompatibility. Silks are therefore a suitable material for the use as drug delivery vehicles.

Dragline silk which builds the frame and the radii of an orb web is one of the best studied types of silk but the availability of spider silk is limited as the cannibalistic nature of spiders obstructs farming of spiders in large scale [9]. A recombinantly producible spider silk protein has been established to generate the desired protein in big amounts and constant quality [0]. The engineered spider silk protein eADF4(C16) is derived from ADF4, one of the dragline proteins of the European garden spider *Araneus diadematus*. A consensus sequence (GSSAAAAAAAAS-GPGGYGPENQGPSGPGGYGPGGP), called C module, has been identified and is repeated 16 times in the recombinant spider silk protein eADF4(C16) to design a protein of a molecular mass of 48 kDa (Figure 1) [1]. In every C module, there is one glutamate residue which is negatively charged under physiological conditions. Hence, eADF4(C16) exhibits a negative net charge.

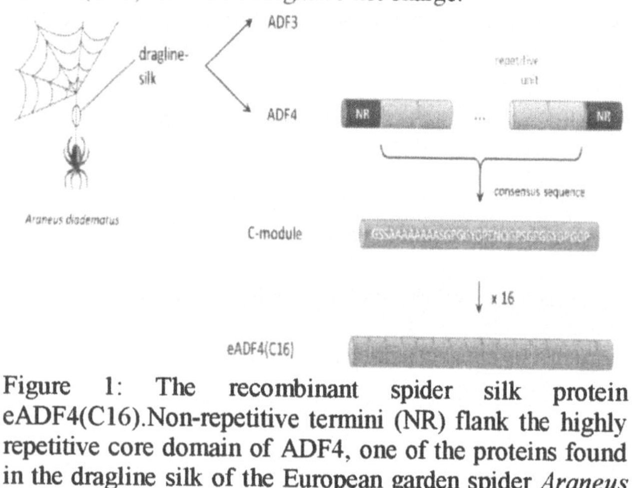

Figure 1: The recombinant spider silk protein eADF4(C16).Non-repetitive termini (NR) flank the highly repetitive core domain of ADF4, one of the proteins found in the dragline silk of the European garden spider *Araneus diadematus*. A consensus sequence called C-module is derived from ADF4, and repeated 16 times in the engineered protein eADF4(C16).

In addition to eADF4(C16), a genetically modified variant of eADF4(C16), eADF4(K16) has been engineered. In this protein, the negatively charged amino acid glutamate is exchanged to the positively charged amino acid lysine to yield the so-called K module. This K module is repeated 16 times in the protein eADF4(K16). Under physiological conditions, eADF4(K16) therefore possesses a positive net charge. Furthermore, genetic engineering allows the modification of eADF4(C16) with functional peptides, e.g. cell penetrating peptides which makes theses protein constructs highly interesting for biomedical applications [12].

One feature of recombinant spider silk proteins is their processability. It has been shown, that the recombinant spider silk protein can be processed into different morphologies, for example into fibers, films, hydrogels or particles [13]. Owing to the amino acid composition the recombinant spider silk protein is unstructured when dissolved in aqueous solutions. The addition of potassium phosphate in concentrations higher than 400 mM to spider silk protein solutions leads to a salting out process in which solid particles with no obvious submicrostructure are formed [14]. During the particle formation process, β-sheet rich structures are formed which are responsible for the stability of the particles in aqueous solutions. The size of the produced particles can be controlled by adjusting the protein concentration used during the salting out process and the mixing intensity of the spider silk protein solution during the addition of potassium phosphate (Figure 2). While increasing protein concentration yields bigger particles, the particle diameter decreases with increasing mixing intensity. Using this production method, particles of a size of 250 nm up to 3 μm can be produced [15].

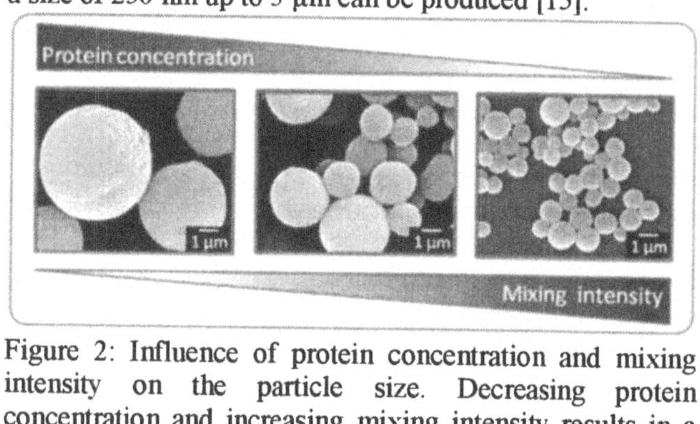

Figure 2: Influence of protein concentration and mixing intensity on the particle size. Decreasing protein concentration and increasing mixing intensity results in a decrease in particle diameter.

2 RESULTS

2.1 Cellular uptake of spider silk particles

First, the properties concerning cellular uptake of the recombinant spider silk particles have been assessed. For these experiments, particles of a size from 220 nm to 360 nm have been used. It could be shown that the uptake of particles produced from eADF4(C16) is rather poor which is mainly the result of electrostatic repulsion effects between the negative charge of eADF4(C16) particles and the negatively charged cellular surface. In contrast, the positively charged eADF4(K16) is taken up into cells much more efficiently. Furthermore, hybridizing the recombinant spider silk protein eADF4(C16) with the artificial cell penetrating peptide R_8 (eight arginines) significantly enhanced particle uptake in comparison to eADF4(C16) (Figure 3) [16].

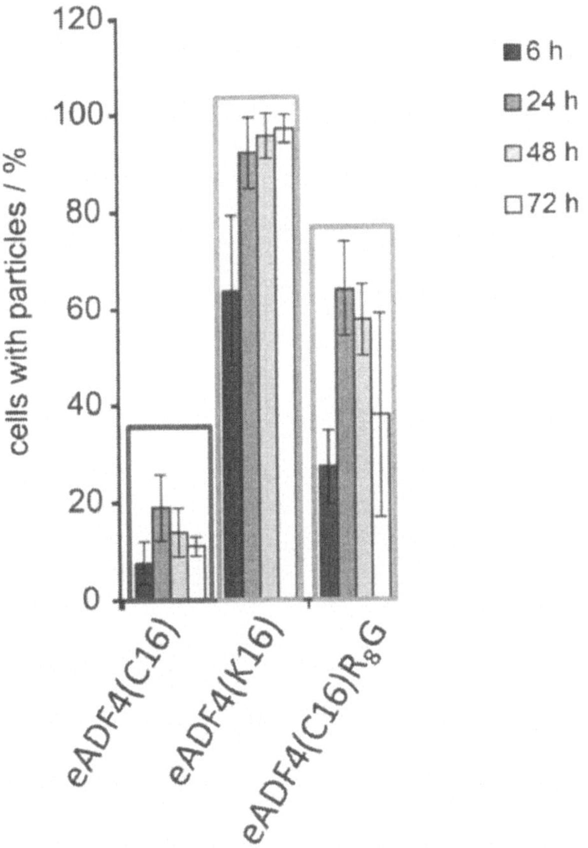

Figure 3: Uptake of recombinant spider silk particles by HeLa cells. Uptake of the respective spider silk particles after the indicated incubation period has been analysed by flow cytometry.

2.2 Drug loading and release of spider silk particles

eADF4(C16) particles are e.g. suitable transport vehicles for low molecular weight molecules and proteins [17, 18]. Uptake and release of water-soluble substances such as rhodamine B serving as a model can be, to some extent, controlled by processing conditions. Co-precipitation of rhodamine B with eADF4(C16) increased the loading efficiency compared to loading of particles by diffusion processes. Crosslinking of eADF4(C16) reduced the rate of rhodamine B release, depending on the production route. The influence of salt on release was little due to the twitterionic behavior of rhodamine B and differed depending on the loading process as well as on crosslinking of eADF4(C16). For non-crosslinked particles no influence of salt on the release of rhodamine B could be detected, while for crosslinked spheres a significant difference in the rhodamine B release was observed in presence of salt. These findings can be explained by changes in protein hydrophobicity due to the formation of di-tyrosines, the osmotic gradients between the particles and the surrounding media, as well as the partially electrostatically shielded rhodamine B molecules.

Particles of the polycationic spider silk protein eADF4(κ16) can be efficiently used to encapsulate negatively charged substances [12]. Nucleic acids are bound to eADF4(κ16) particles by strong electrostatic interactions due to the high charge density of the DNA. Nucleic acids are slowly released in contrast to small molecular weight substances which show a burst release within the first 30 min of incubation. Using Layer-by-Layer-coating of eADF4(κ16) particles with eADF4(C16) to slow down release of small molecular substances was not successful.

REFERENCES

[1] Panyam, J. and V. Labhasetwar, *Targeting intracellular targets*. Current Drug Delivery, 2004. **1**(3): p. 235-47.

[2] Farokhzad, O.C. and R. Langer, *Impact of nanotechnology on drug delivery*. ACS Nano, 2009. **3**(1): p. 16-20.

[3] Nicolas, J., et al., *Design, functionalization strategies and biomedical applications of targeted biodegradable / biocompatible polymer-based nanocarriers for drug delivery*. Chemical Society Reviews, 2013. **42**(3): p. 1147-235.

[4] Faraji, A.H. and P. Wipf, *Nanoparticles in cellular drug delivery*. Bioorganic & Medicinal Chemistry, 2009. **17**(8): p. 2950-62.

[5] Freiberg, S. and X.X. Zhu, *Polymer microspheres for controlled drug release*. International Journal of Pharmaceutics, 2004. **282**(1-2): p. 1-18.

[6] Langer, R. and N.A. Peppas, *Advances in biomaterials, drug delivery, and bionanotechnology*. AIChE Journal, 2003. **49**(12): p. 2990-3006.

[7] George, M. and T.E. Abraham, *Polyionic hydrocolloids for the intestinal delivery of protein drugs: alginate and chitosan--a review.* Journal of controlled release : official journal of the Controlled Release Society, 2006. **114**(1): p. 1-14.

[8] Orive, G., et al., *Biomaterials for promoting brain protection, repair and regeneration.* Nature Reviews. Neuroscience, 2009. **10**(9): p. 682-92.

[9] Hardy, J.G., L.M. Romer, and T.R. Scheibel, *Polymeric materials based on silk proteins.* Polymer, 2008. **49**(20): p. 4309-4327.

[10] Heidebrecht, A. and T. Scheibel, *Recombinant production of spider silk proteins.* Advances in Applied Microbiology, 2013. **82**: p. 115-53.

[11] Huemmerich, D., et al., *Primary structure elements of spider dragline silks and their contribution to protein solubility.* Biochemistry, 2004. **43**(42): p. 13604-13612.

[12] Doblhofer, E. and T. Scheibel, *Engineering of recombinant spider silk proteins allows defined uptake and release of substances.* J. Pharm. Sci., 2015. **104**: p. 988-994.

[13] Leal-Egana, A. and T. Scheibel, *Silk-based materials for biomedical applications.* Biotechnol Appl Biochem, 2010. **55**(3): p. 155-67.

[14] Slotta, U.K., et al., *An engineered spider silk protein forms microspheres.* Angewandte Chemie-International Edition, 2008. **47**(24): p. 4592-4594.

[15] Lammel, A., et al., *Processing conditions for the formation of spider silk microspheres.* ChemSusChem, 2008. **1**(5): p. 413-416.

[16] Elsner, M., Herold, H., Müller-Herrmann, S., Bargel, H. and T. Scheibel, *Enhanced cellular uptake of engineered spider silk particles.* Biomaterials Science, 2015. **3**: p. 543-551.

[17] Lammel A., Schwab, M., Hofer, M., Winter, G. and T. Scheibel, *Spider silk submicro-particles for controlled drug delivery.* Biomaterials, 2010. **32**: p. 2233-2240.

[18] Blüm, C. and T. Scheibel, *Control of drug loading and release properties of spider silk sub-microparticles.* BioNanoScience, 2012. **2**: p. 67-74.

Nitric oxide releasing PLGA microspheres for biomedical applications

Gergely Lautner *,**, Steven P. Schwendeman **,***, Mark E. Meyerhoff *

* Department of Chemistry, The University of Michigan, 930 N. University, Ann Arbor, Michigan 48109, USA, glautner@umich.edu

** Department of Pharmaceutical Sciences, The Biointerfaces Institute, University of Michigan, 2800 Plymouth Rd., Ann Arbor, MI 48109, USA

*** Department of Biomedical Engineering, University of Michigan, 2800 Plymouth Rd., Ann Arbor, MI 48109, USA

ABSTRACT

We encapsulated an *S*-nitroso species (*S*-nitroso-*N*-acetylpenicillamine (SNAP)) in biodegradable PLGA microspheres that release nitric oxide (NO) for weeks in a continuous and controlled manner. The rate and duration of NO release are tunable by varying the formulation (e.g., SNAP loading, acid terminated or ester-capped PLGA and lactic-glycolic acid ratio). The NO release rate can also be tuned to desired levels by use of copper ions or light. The encapsulation of the SNAP crystals in the polymer leads to highly stable microspheres when stored at room temperature in the dark. The PLGA-SNAP microspheres can be used for several biomedical applications where extended and controlled NO release is desired (e.g. in wound healing), due to its substantial antimicrobial effects and ability to promote angiogenesis. To prepare new types of NO-releasing wound dressings, we incorporated the PLGA-SNAP microspheres into ointments and alginate films (PLGA-SNAP/alginate) that could potentially be used for treatment of chronic wounds (e.g., diabetic ulcers) and skin burns.

Keywords: nitric oxide, PLGA microspheres, controlled release, wound healing

1 INTRODUCTION

Nitric oxide (NO) is a well known endogenous gaseous molecule with potent antimicrobial [1][2][3], vasodilating, smooth muscle relaxant [4], and growth factor stimulating effects [5]. The efficiency of gaseous NO as a therapeutic agent (e.g., inhaled for pulmonary treatments) is already well established [6]; however, its wider biomedical utility is hindered by its cumbersome administration, very short lifetime (seconds) and short diffusional distance (some hundred µm) [7]. To overcome the drawbacks of using gaseous NO, many NO donor molecules have been developed and studied. The most frequently used NO donors are low-molecular-weight molecules, such, *S*-nitroso, *N*-nitroso, *C*-nitroso, metal ion-NO complexes, heterocycles and diazeniumdiolates [8][9]. The mechanism of the NO release from these donors can be based on spontaneous decomposition, redox or catalytic processes. Upon exposure to aqueous media, NO donors typically yield an initial NO burst release behavior, and many of the NO donors and/or their decomposition products are toxic or carcinogenic. Very high NO release is undesirable, as it can react with other reactive oxygen species to produce secondary reactive nitrogen species such as peroxynitrite. A major disadvantage of prevalent diazeniumdiolate (NONOate) donor molecules is that they are precursors of carcinogenic nitrosamines [10]. The most critical challenge in the development of new NO delivery systems is the formation of a local, long-term NO release at an optimal concentration to achieve a given therapeutic effect while minimizing toxic effects. Thus, NO donors are usually attached to, or encapsulated within, polymers in order to mimic endogenous NO release for biomedical applications [11][12][13].

Endogenously produced NO plays a crucial role in wound helaing, mediates vasodilation, inhibits platelet aggregation and protects against pathogens. In the proliferative phase, NO stimulates fibroblasts, keratinocytes, endothelial cells and vascular endothelial growth factor (VEGF) production. NO can also accelerate angiogenesis, augmented interfollicular stem cell recruitment, enhance hair follicular reconstruction and increase collagen deposition in burn wounds [14].

The topical supplementation of NO during wound healing can be enormous, especially in burn wound healing and in diabetic chronic wound healing, where the immune system or the endogenous NO production is impaired. Toward this goal, we now encapsulate a *S*-nitrosothiol (RSNO) NO donor in biodegradable poly(lactic-co-glycolic acid) (PLGA) microspheres in order to provide extended NO release to support wound healing and prevent infections. PLGA is the major excipient in numerous controlled release products [15] approved by the Food and Drug Administration (FDA) and the applied NO donor (*S*-nitroso-*N*-acetylpenicillamine - SNAP) and its decomposition products are expected to exhibit excellent biocompatibility as the precursor of SNAP, penicillamine, is an FDA approved chelating agent [16].

Herein, we investigate different ways to apply PLGA encapsulated SNAP (PLGA-SNAP) microspheres, i.e. incorporate them in ointments and hydrogels, for potential wound healing applications.

2 MATERIALS AND METHODS

2.1 Synthesis of SNAP

SNAP was synthetized as previously described. [17] Briefly 2 g of *N*-acetylpenicillamine (NAP) (Fluka) was dissolved in a mixture of 50 mL methanol, 16.7 mL H_2O, 8.3 mL HCl and 2.5 mL H_2SO_4. Twenty mL of 1 M $NaNO_2$ was then slowly added to the mixture. The color of the solution turned to dark red. The solution was placed onto ice, and by blowing N_2 over the solution, the methanol was evaporated until crystals appeared (~3 h). The green crystals were collected by solution filtration, washed with ice cold water and vacuum dried. The whole process was performed with protection from light to avoid the decomposition of SNAP. Before encapsulation in PLGA, SNAP was micronized by a CryoMill (Retsch, Düsseldorf, Germany). Five hundred mg SNAP was cryomilled in two 5 mL stainless steel jars, with 16 stainless steel 3 mm balls in each. After 7:50 min long precooling with liquid nitrogen at 5 Hz, SNAP was ground for 40 min at 20 Hz, then freeze-dried (-41 °C, 0.160 mbar) for 1 day. Prior to freeze-drying the cryomilled SNAP was flash frozen in liquid nitrogen.

2.2 SNAP Loaded PLGA Microsphere Preparation

SNAP loaded microspheres were prepared by a solid-in-oil-in-water emulsion water evaporation technique. 300-400 mg PLGA 50/50 (free carboxylic acid and ester endcapped, M_w = 24,000-54,000) polymer (Evonik Röhm GmbH, Darmstadt, Germany) was dissolved in 1 mL of methylene-chloride. Then, 60-100 mg of cryomilled SNAP was added and homogenized at 10 000 rpm for 1 min with a Tempest IQ^2 homogenizer (The VirTis Co., Gardiner, NY, USA) equipped with a 10 mm shaft. Four mL of 5% (w/V) PVA solution was immediately pipetted into the solution and vortexed (Genie 2, Fisher-Scientific Industries, Inc., Bohemia, NY, USA) at the highest speed for 1 min, then poured into 100 mL of 0.5% (w/V) PVA solution under rapid stirring with a magnetic stirbar and hardened for 3 h. The final microspheres were then sieved and washed with 500 mL DI water. The fraction with diameter between 20 μm and 125 μm was collected and freeze-dried (-41 °C, 0.160 mbar) for 2 days. Prior to freeze-drying, the microspheres were flash frozen in liquid nitrogen.

2.3 Measurement of SNAP Loading of PLGA Microspheres

To determine the SNAP loading in the microspheres, 5 mg of microspheres were dissolved in 1 mL acetonitrile and the concentration of SNAP was measured by absorbance at 340 nm using a Synergy Neo Microplate Reader (BioTek U.S., Winooski, VT, USA).

2.4 PLGA-SNAP/alginate Hydrogel Preparation

Sodium alginate (Sigma) was dissolved in DI water with continuous stirring overnight. PLGA-SNAP microspheres were then suspended in the sodium alginate solution, poured into a Petri dish and degassed with vacuum. Alginate gels were crosslinked with $CaCl_2$ or $CuCl_2$ solution. PLGA-SNAP/alginate gels were characterized right after cross-linking or stored dried, then rehydrated with DI water before use.

2.5 Scanning electron microscopy

For scanning electron microscopy (SEM) imaging, PLGA-SNAP samples were coated with 40 nm gold using a sputter coater (Desk II, Denton Vacuum Inc., Hill, NJ, USA) for 120 s. Scanning electron micrographs were taken by a Hitachi S-3200N Variable Pressure SEM (Hitachi High-Technologies Corp., Tokyo, Japan). The applied accelerating voltage was 15 keV. For cross-sectional images, samples were cut using a razor-blade on glass slide prior to coating.

2.6 Nitric oxide release measurement by Nitric Oxide Analyzer

For NO release experiments, samples were incubated in 5-10 mL release media in amber 50 mL centrifuge vials with gently shaking at 37°C. Nitric oxide release experiments were performed by using an ozone-chemiluminescence technology-based Sievers Nitric Oxide Analyzer (NOA, Model 280i, GE Analytical Instruments, Boulder, CO, USA). During NO measurement, NO was purged out from the release media via a N_2 purge gas.

Light modulated NO release measurements were performed in a transparent vial exposed to a tungsten halogen lamp equipped with a Fostec DCR II EKE cold light source (Schott-Fostec, LLC, New York, USA) set at its highest intensity.

3 PLGA-SNAP MICROSPHERES FOR NO DELIVERY

SNAP was encapsulated in biodegradable PLGA microspheres with diameter of 20-125 μm using a solid in oil in water emulsion and solvent evaporation method (Figure 1). The encapsulation efficiency of SNAP was > 40% in all cases. The loadings were between 6-13%, increasing with the amount of added SNAP.

The NO release of PLGA-SNAP microspheres was tested in PBS buffer, pH 7.4, containing 100 μM EDTA (PBSE), 1 mM ascorbic acid, and variable amounts of Cu^{2+} (PBSACu). The testing was conducted at 37°C, in the dark (Figure 2 and Figure 3). In PBSE buffer, the EDTA can coordinate Cu^{2+} ions, inhibiting the decomposition of SNAP. However, the release of NO is controlled without an NO burst even in the presence of decomposing agents like Cu^{2+} and a high concentration of ascorbic acid. The NO release was continuous for 2 weeks in the carboxylic acid terminated PLGA was used and for over 4 weeks in the case of the ester terminated PLGA. The amount of NO was also tunable by varying the SNAP loading of PLGA microspheres (data not shown).

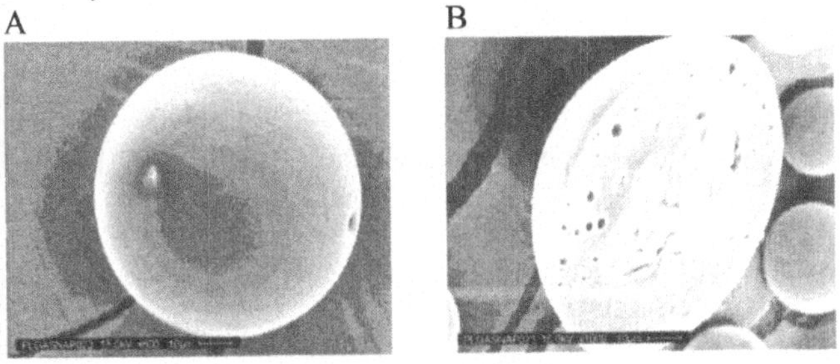

Figure 1. Scanning electron micrograph of PLGA-SNAP microsphere (A) and its cross-section (B) with the SNAP crystals within.

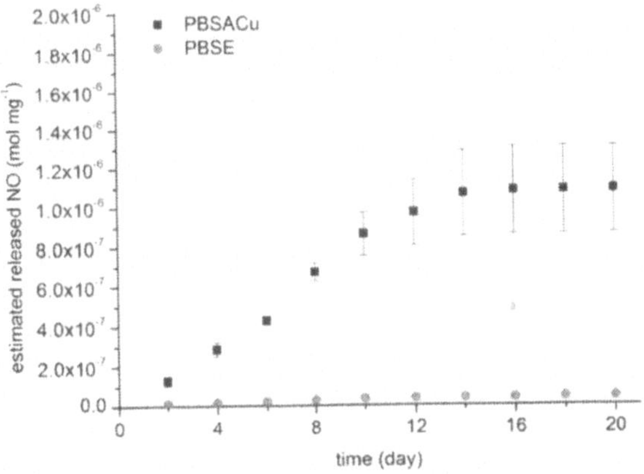

Figure 2. NO release of free carboxylic acid terminated PLGA-SNAP microspheres in PBSE and PBSACu release media changed at every timepoint

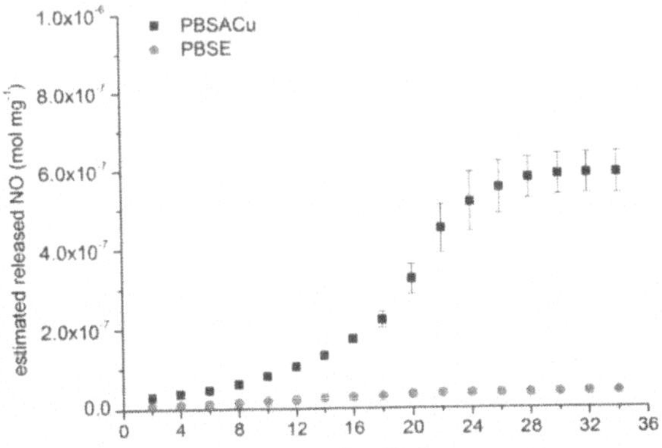

Figure 3. NO release from ester terminated PLGA-SNAP microspheres in PBSE and PBSACu release media changed at every timepoint

The NO release also can be induced and kept at desired levels by shining light on the microspheres. Dry particles released 75% of their loading in the form of NO in 20 h (Figure 4), and the level of released NO was tunable by the intensity of the light (data not shown).

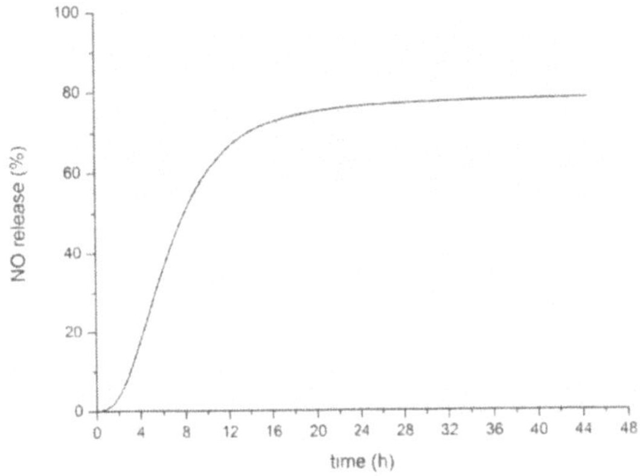

Figure 4. Light modulated NO release from ester terminated PLGA-SNAP microspheres at room temperature

The encapsulation of the SNAP crystals in the polymer led to highly stable microparticles that retained > 90% of their NO release properties over a month period when stored at room or elevated (37°C) temperature in the dark.

4 PLGA OINTMENT AND PLGA-SNAP/ALGINATE HYDROGEL FOR WOUND HEALING

PLGA-SNAP microspheres can be used for several biomedical applications where extended and controlled NO release is desired (e.g. in wound healing) since NO has substantial antimicrobial effects and promotes growth of new blood vessels.

We incorporated the PLGA-SNAP particles in hydrophobic ointment, and we were able to induce the NO release by using a simple light source. The PLGA-SNAP particles are highly stable at room temperature in the absence of humidity; thus, we expect appropriate stability of the microspheres in the ointment, which could be administered directly onto the wound and activated by a light source.

To prepare a new type of NO releasing wound dressing, we also incorporated our PLGA-SNAP microspheres into alginate hydrogels (PLGA-SNAP/alginate) (Figure 5).

Figure 5. Cross-sectional scanning electron micrograph of PLGA-SNAP/alginate film

Alginate hydrogel dressings have great potential in wound healing applications due to their smoothness, flexibility and elasticity. They also can be removed without much trauma. Since such dressings can absorb high amount of water, it can protect the wound from drying out or can be used as an occlusive dressing. Since trace amounts of copper ions are desired for efficient NO release of the PLGA-SNAP microspheres in the absence of light, the alginate hydrogel can be used as the source of copper ions by crosslinking the hydrogel with Cu^{2+}. Indeed, Cu^{2+} cross-linked alginate hydrogels were recently reported as effective wound dressing [18]. The PLGA-SNAP incorporated Cu^{2+}-alginate film continuously released NO at endothelial NO release levels for over a one week period (Figure 6). Hence, such materials can potentially be employed in chronic and burn wound management.

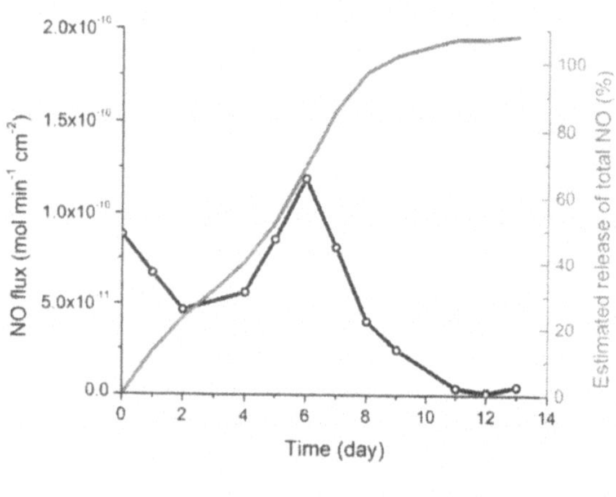

Figure 6. Estimated NO release of PLGA-SNAP/alginate film

ACKNOWLEDGEMENT

We thank the University of Michigan MCubed program and the NIH (EB-000783) for partially supporting this work.

REFERENCES

[1] C. F. Nathan, J. B. Hibbs Jr, Current Opinion in Immunology, 3, 65, 1991.
[2] M. A. De Groote, F. C. Fang, Clinical Infectious Diseases, 21 , S162, 1995.
[3] C. Bogdan, Nature Immunology, 2, 907, 2001.
[4] M. Tare, H. C. Parkington, H. A. Coleman, T. O. Neild, G. J. Dusting, Nature, 346, 69, 1990.
[5] A. Papapetropoulos, G. García-Cardeña, J. A. Madri, W. C. Sessa, Journal of Clinical Investigation, 100, 3131, 1997.
[6] M. J. D. Griffiths, T. W. Evans, The New England journal of medicine, 353, 2683, 2005.
[7] D. D. Thomas, X. Liu, S. P. Kantrow, J. R. Lancaster, Proceedings of the National Academy of Sciences of the United States of America, 98, 355, 2001.
[8] P. G. Wang, T. B. Cai, N. Taniguchi, Nitric Oxide Donors: For Pharmaceutical and Biological Applications, John Wiley & Sons, 2005.
[9] A. W. Carpenter, M. H. Schoenfisch, Chemical Society reviews, 41, 3742, 2012.
[10] L. K. Keefer, ACS Chemical Biology, 6, 1147, 2011.
[11] E. J. Brisbois, H. Handa, T. C. Major, R. H. Bartlett, M. E. Meyerhoff, Biomaterials, 34, 6957, 2013.
[12] M. C. Frost, M. M. Reynolds, M. E. Meyerhoff, Biomaterials, 26, 1685, 2005.
[13] D. a Riccio, M. H. Schoenfisch, Chemical Society reviews, 41, 3731, 2012.
[14] H. Zhu, X. Wei, K. Bian, F. Murad, Journal of burn care & research : official publication of the American Burn Association, 29, 804, 2008.
[15] C. Wischke, S. P. S. Schwendeman, International journal of pharmaceutics, 364, 298, 2008.
[16] Z. Cui, P. R. Lockman, C. S. Atwood, C. H. Hsu, A. Gupte, D. D. Allen, R. J. Mumper, European Journal of Pharmaceutics and Biopharmaceutics, 59, 263, 2005.
[17] I. Chipinda, R. H. Simoyi, The journal of physical chemistry. B, 110, 5052, 2006
[18] W. Klinkajon, P. Supaphol, Biomedical materials (Bristol, England), 9, 045008, 2014

Electrospun nanofibers for Sustained Delivery of Cetylpyridinium Chloride for Buccal Topical Application

MHA. Zanin*, VA. Santos**, AM. Oliveira*, NNP Cerize* and MA. Borsatti**

* Institute for Technological Research of the State of São Paulo – IPT, Laboratory of Chemical Processes and Particle Technology - Bionanomanufacturing, São Paulo- SP, Av. Prof. Almeida Prado 532, Cidade Universitária, CEP 05508-901, Brazil, mhzanin@ipt.br, ncerize@ipt.br, amarim@ipt.br
** University of São Paulo, São Paulo, Brazil
Department of Stomatology, School of Dentistry, Av. Professor Lineu Prestes, 2227 Cidade Universitária 05508-000, São Paulo, Brazil, valalves@usp.br, maborsat@usp.br

ABSTRACT

Electrospun polymeric nanofibers were developed using polyvinylpyrrolidone (PVP-Kollidon® 90F), ammonio methacrylate copolymer type B (PMMA-Eudragit® RS100) and cetylpyridinium chloride (CPC), 10:10:1 (wt/wt%) for buccal topical application. The morphology of mats with and without the CPC was analyzed and also the the thermal behaviors and the physicochemistry characteristics of the nanofiber mats. The SEM images revealed that all of the fibers were smooth and the average diameter of fibers was from 200 nm to 500 nm. The thermal analysis showed that the incorporation of cetylpyridinium chloride in the nanofiber through electrospinning process changed the cristallinity of cetylpyridinium chloride and/or conferred to this drug a new amorphous state. The *in vitro* release of CPC from the fiber in PBS buffer (pH 7.4) showed a sustained release over the experimental time of 24 h and the drug did not permeated all the skin achieving the medium receptor.

Keywords: Eletrospinning, nanofiber, cetylpyridinium chloride, Drug Delivery System

1 INTRODUCTION

Cetylpyridinium chloride is a drug known for its antifungal activity *in vitro* and *in vivo* against *Candida* albicans [1]. The Candida albicans is a comensal organism that can be isolated from the gastrointestinal tract, oral and vaginal mucosa of up to 80% of healthy individuals. Normally it doesn´t represent a problem, however it can cause symptomatic infections of mucosal membranes[2].

Currently, there are a variety of topical and systemic antifungal agents available for treating oral candidiasis. The topical antifungal agents most commonly used are nystatin and miconazole (MCZ), which have undesirable side effects and the risk of developing fungal resistance. The main adverse effects observed with nystatin are unpleasant taste, nausea and motion sickness. Miconazole may cause drug interactions, even in the case of topical use. Mouthwashes containing antiseptics, such as chlorhexidine, cetylpyridinium chloride, triclosan and others, have been studied as alternatives for traditional antifungal agents. It has been shown that CPC has proven antifungal activity in vitro and in vivo, with advantages over triclosan, and chlorhexidine[3]. A CPC delivery-controlled system, transported in polymer nanofibers (PVP/PMMA) was developed as an altenative to protect the drug and improve the drug bioavailability in contact with the oral mucosa.

2 MATERIAL AND METHOD

2.1 Materials

Cetylpyridinium chloride (Sigma-Aldricht Corp., Saint Louis, USA); Polyvinylpyrrolidone (PVP-Kollidon® 90F, BASF, Ludwigshafen, Germany); Ammonio methacrylate copolymer type B (PMMA, Eudragit® RS100, Evonik, Germany); Acetone and Ethyl alcohol (Synth).

2.2 Production of CPC nanofiber by electrospinning process

A 10.0% (wt/wt) PVP:PMMA (1:1) blend was solubilized in acetone and ethyl alcohol solution (50:50 v/v) and the CPC was added at 5% (wt/wt). The mixed solution was pumped (infusion pump PHD2000 – Harvard Apparatus) at a constant rate of 1 mL/h, forming a bead of solution at the tip of syringe. A high voltage (ES30P-10W Gamma High Voltage Research, Florida, USA) of 18 kV was applied between the nozzle needle, a negative potential, and the grounded collector. As the jet breaks up into fibers from the Taylor cone, the liquid was evaporated and gave rise to dry fibers which were subsequently spun on the wrapped collector with aluminum foil. The operation occurred until that a multilayer fiber mat was obtained. The distance between the tip of syringe and the collector was of 6 cm. The velocity of rotating collector used was 30 rpm. The Figure 1 shows the set up of the electrospinning equipment in lab scale where the CPC nanofiber was developed.

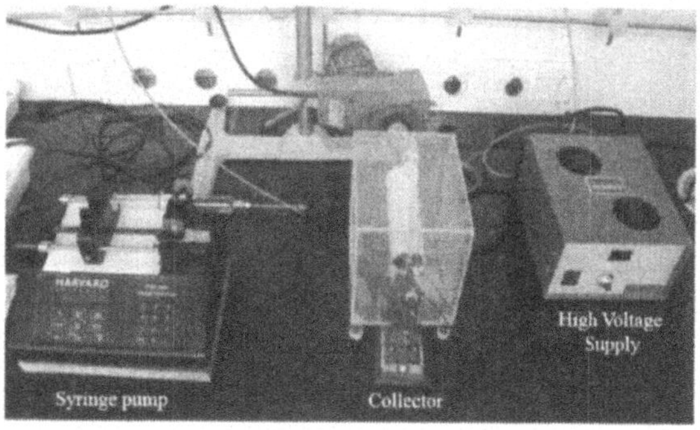

Figure 1: Picture of the Lab-built device employed in all of the electrospinning experirments.

2.3 Morphology analysis

The Scanning electron microscopy (SEM) was done using a Scanning Electron Microscopy of High Resolution - FEG - SEM (FEI Quanta 3D).

2.4 Thermal analysis

The techniques used was Differential Scanning Calorimetry (DSC). The DSC curves were obtained with a Mettler Toledo 822e 202 System. Aliquots of about 5 mg of each sample were placed in an aluminum pan without pin. Convencional DSC measurements were performed by heating the sample from 25 ^{0}C to 700 ^{0}C at a rate of 5^{0}C/min under a nitrogen flow of 50 mL/min.

2.5 FT-IR analyses

FT-IR spectrum was recorded on NICOLET 6700, Thermo Scientific FT-IR spectrometer. The samples were scanned from 600 to 4000cm^{-1}.

2.6 Permeation studies using esophageal mucosa as membran model

In vitro permeation study was done in Franz cell diffusion using a pig esophageal mucosa as model for buccal tissue [4] and phosphate-buffered-saline (PBS) (pH 7.4) as receptor medium.

Briefly, the Franz cells with a diffusion area of 1.77 cm2 were used. The connective side of tissue (Pig Esophageal Mucosa) was spread over the PVP/PMMA nanofiber and the ensemble was mounted in the diffusion cell. It was used a frozen tissue that was first placed in PBS pH 7.4 for 15 min before assembly on the diffusion cells. The donor and receptor compartments were filled with the same buffer and the cells were allowed to stabilize for 30 min in a water bath at 37^{0}C.

The receiver compartment volume was 6.0 mL and 1 mL samples were collected sequentially and replaced by fresh buffer. The CPC concentrations were determined by UV-vis spectrophotometty. The diffusion experiments were conducted for 24 h with a agitation velocity of 300 rpm.

2.7 Quantification of Cetylpyridinium chloride of the nanofiber

The amount of CPC pure and the remaining in the PVP/ PMMA nanofiber mats was determined using UV-vis spectrophotometer, GBC10 MODEL, Cintra, based on the literatura[5].

3 RESULTS

The Figure 2 illustrates a picture of the PVP/PMMA blend nanofiber developed and the Figure 3 presents the SEM micrographs of nanofiber with incorporated CPC. The PVP/PMMA blend nanofibers with CPC presented a average diameter from 200 to 500 nm and showed a round-shaped and smooth morphology.

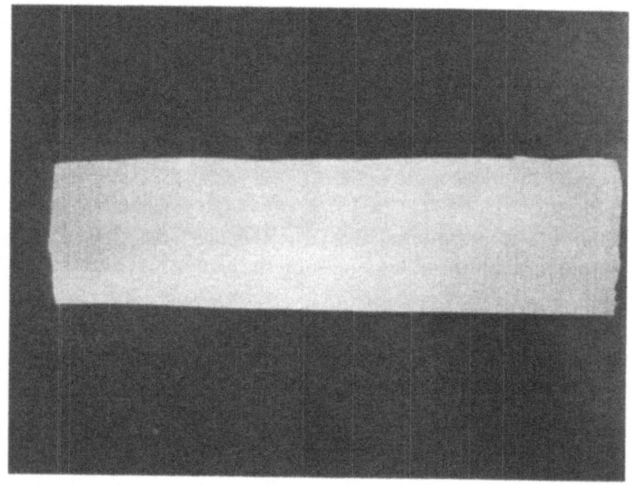

Figure 2: Picture of an electrospun PVP/PMMA fiber mat with CPC (5% wt/wt).

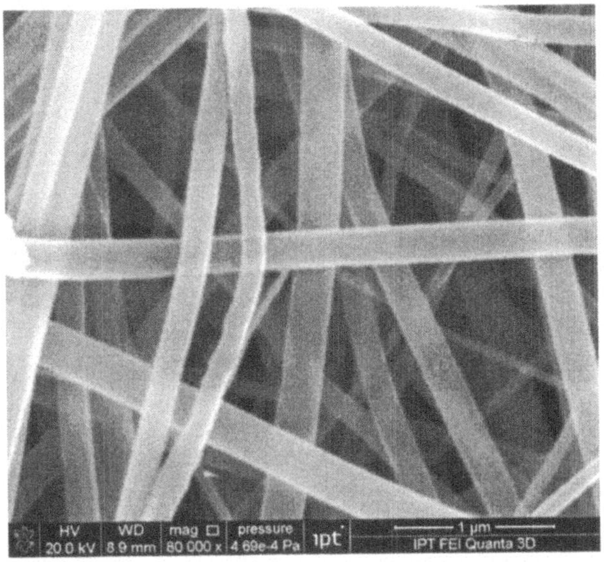

Figure 3: SEM photograph of an electrospun PVP/PMMA fiber mat with CPC (5% wt/wt). 80000x

The DSC profiles showed a single endothermic peak at 85^{0}C [6] for the CPC profile that was not detected in the nanofiber with CPC. These results suggest that the incorporation of CPC in the electrospinning process change the cristallinity of CPC and/or confers to this drug a new amorphous state (Figure 4).

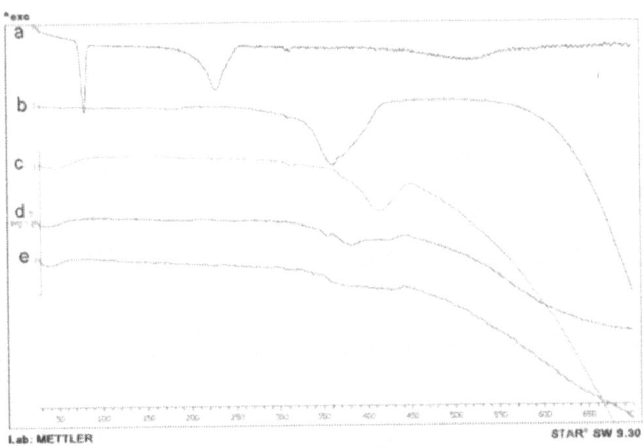

Figure 4: (a) CPC; (b) PMMA Polymer; (c) PVP Polymer;(d) PVP/PMMA nanofiber without CPC and (e) PVP/PMMA nanofiber with CPC (5% wt/wt) .

The FT-IR and DSC were techniques used in order to know the interactions among the polymers and the CPC.

IR spectra of the nanofibers without CPC presented a similar spectra to the nanofiber with CPC (Figure 5).

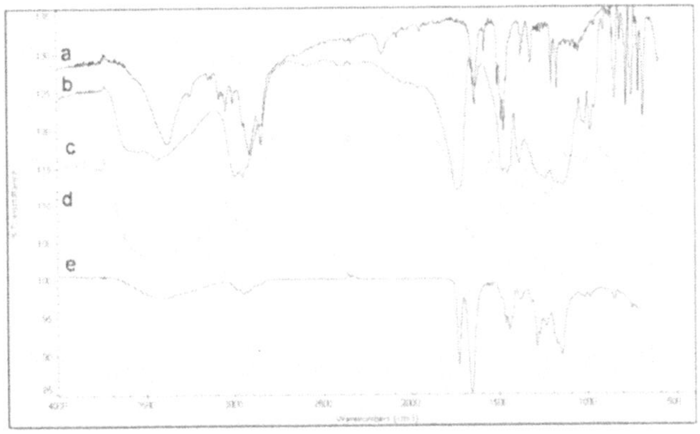

Figure 5: (a) CPC; (b) PMMA Polymer; (c) PVP Polymer;(d) PVP/PMMA nanofiber without CPC and (e) PVP/PMMA nanofiber with CPC (5% wt/wt) .

The in vitro release of CPC from the fiber in PBS buffer (pH 7.4) showed a sustained release over the experimental time of 24 h (Figure 6) and the CPC did not achieve the medium receptor, according to the permeation study.

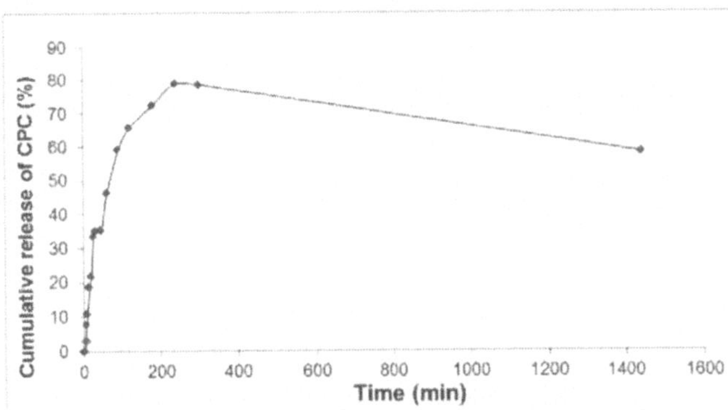

Figure 6: Release profile of CPC from nanofiber comprising PVP/PMMA with CPC (5% wt/wt).

4 CONCLUSIONS

An alternative to buccal delivery of drug, such as cetylpyridinium chloride (5% wt/wt), was developed through the electrospinning technology using the polymers polyvinylpyrrolidone (PVP) and ammonio methacrylate copolymer (PMMA). The nanofiber presented an average diameter varying from 200 nm to 500 nm. The electrospinning process seems to change the cristallinity of the CPC after its nanofiber incoporation.

The *in vitro* release of CPC from the fiber in PBS buffer (pH 7.4) showed a sustained release over the experimental time of 24 h and the drug did not permeated all the skin achieving the medium receptor. This system open a new possiblity to control the CPC release at buccal mucosa. In addition to this, this new system can improve the drug bioavailability in contact with the oral mucosa.

ACKNOWLEDGMENTS

The authors acknowledge the financial support by Conselho Nacional de Desenvolvimento Científico e Tecnológico (CNPq) to the infrastructure to do development of the nanofiber and for the Institute for Technological Research of the State of São Paulo – IPT.

REFERENCES

[1] Giuliana G, Pizzo G, Milici ME, Musotto GC, Giangreco R. In vitro antifungal properties of mouthrinses containing antimicrobial agents. J Periodontol. 68(8):729-33, 1997.

[2] Kim J, Sudbery P. Candida albicans, a major fungal pathogen. J Microbiol., vol. 49, 171-177, 2011.

[3] Santos, VA., VIEIRA, P V A., Oliveira, AM., Zanin, MHA., Borsatti, MA. Antifungal effect of electrospun nanofibers containing cetylpyridinium chloride against Candida albicans. Brazilian Oral research, v 28, 1-6, 2014.

[4] Consuelo, ID; Pizzolato, GP; Falson, F; Guy, RH; Jacques, Y. Evaluation of Pig Esophageal Mucosa as a Permeability Barrier Model for Buccal. Tissue journal of pharmaceutical sciences, vol. 94, 2777-2788, 2005.

[5] Benamor M., Aguersif, N., Draa, MT. Spectrophotometric determination of cetylpyridinium chloride in pharmaceutical products. Journal of Pharmaceutical and Biomedical Analysis. Vol.26, 151-154, 2001

[6] Krajisnik ,D.; Dakovic, A, Milojevic, M; Malenovic, M; Kragovic, M; Bogdanovic, DB; Dondur, V, Milic, J. Properties of diclofenac sodium sorption onto natural zeolite modified with cetylpyridinium chloride. Colloids and surfaces B: Biointerfaces, vol.83, 165-172, 2011.

HA-decorated nanoparticles as tumour-targeting nucleic acids carriers

A. Tirella[*], J. M. Rios de la Rosa[**], A. Gennari[***] and N. Tirelli[***]

[*]Manchester Pharmacy School/Institute of Inflammation and Repair
University of Manchester, M13 9PT, UK, annalisa.tirella@manchester.ac.uk
[**]Manchester Pharmacy School, University of Manchester, M13 9PT, UK,
juliomanuel.riosdelarosa@postgrad.manchester.ac.uk
[***] Institute of Inflammation and Repair, University of Manchester, M13 9PT, UK,
arianna.gennari@manchester.ac.uk
[****] Manchester Pharmacy School/Institute of Inflammation and Repair
University of Manchester, M13 9PT, UK, nicola.tirelli@manchester.ac.uk

ABSTRACT

In the context of nucleic acids delivery, an ideal carrier should not only provide the means for the transport of such payloads, but also protect their activity by limiting their exposure to harmful agents leading to either physical or enzymatic degradation. We here report on the development of nanoparticles designed to target CD44; which is a membrane protein overexpressed in many tumours, while also being the major receptor of hyaluronic acid (HA). The herein described particles present a chitosan (CS) core for nucleic acids complexation and a surface rich in HA.

Nanoparticles CD44-HA mediated endocytic process, and the intracellular localisation of nucleic acids and their effects were observed using fluorescently labeled nanoparticle components. Sections and 3D reconstructions of the target (HT-29, colorectal cancer cells) and off-target (THP-1 macrophages) CD44-expressing cells were registered at different time points.

Keywords: hyaluronic acid, tumour-targeting carriers, gene therapy, siRNA

1 INTRODUCTION

In designing a nucleic acid carrier for therapeutic purposes, it is necessary to identify a strategy for both stabilisation and protection of the payload, and selective targeting [1]. Viral carriers provide both aspects, but are often considered with suspicion due to the potential dangers associated (i.e. mutagenic and immunogenic responses) [2]. In turn, the use of synthetic vectors, typically cationic lipids and polymers, is typically marred by their high toxicity [3] and low efficiency in overcoming biological barriers [4]. The decoration of cationic complexes with targeting ligands is a strategy that helps overcome the above mentioned problems, as long as they also mask the typical membrane-disrupting and ultimately toxic action of positively charged materials.

We design nanoparticles that achieve targeting through the interactions between hyaluronic acid (HA) and CD44; this approach has been explored for cancer targeting, due to the overexpression of CD44 in tumours [5, 6, 7], and also employing the CD44 behaviour as an endocytic ligand [8].

Confocal microscopy is a valuable tool for analyzing the delivery of nucleic acids to cultured cells at the resolution of a single cell. In recent years, the design of labeled nanoparticle was used to study the endocytotic mechanism and their intracellular localisation in living cells [9, 10], quantification of uptaken is also performed but no observation of delivered cargos are measured [11]. The interest to observe the delivery mechanism of nucleic acids carriers intracellularly over time, from the membrane, through endocytosis, to the entry and delivery into the cytosol, and finally determining the induced effect still proves challenging. To our knowledge, there are no studies reporting the tracking of both nanoparticles and the biological effect of the cargo.While a number of reports describe the reported effect.

Chitosan is first complexed with nucleic acids, adding then HA to produce nanoparticles with an HA-displaying surface; depending on the molecular weight of chitosan, HA can be complexed with a different strength, allowing for a different surface presentation of HA and thus also for a modulation of HA/CD44 interactions [12, 13].

In this study we have evaluated the loading capacities of the nanoparticles and then focused on the internalization of CS//HA nanoparticles in two CD44-positive cellular models: a human colorectal-cancer cell line (HT-29) and a human macrophage cell model (THP-1-M, as a model for off-target interactions). Fluorescently labeled CS//HA nanoparticles were then used to track the nanoparticle-fate from the internalisation to the intracellular delivery, and finally to monitor the transfection efficiency.

2 MATERIALS AND METHODS

2.1 Nanoparticle preparation and characterisation

Chitosan (CS) obtained from crab shells, viscosimetric average molecular weight of 656 kDa (#51009219, Lot: WE44069811; Sigma Aldrich, Basingstoke, UK) was degraded via nitrite-mediated degradation obtaining also a low average molecular weight chitosan (MW 35 kDa). Hyaluronic acid (HA) with average molecular weight of 305 kDa (GPC with triple detection) was purchased from Novozymes Biopharma A/S (Bagsvaerd, Denmark). Salmon sperm DNA (≤ 2000 bp, 10 mg/mL) was purchased from Sigma-Aldrich (UK). Silencing RNA (siRNA, 23 nt, 5 mg/mL, Standard A4 purification) was purchased from GE Healthcare (Dharmacon, UK). Please note that all the solutions were prepared with RNase-free reagents and sterilised using appropriate filters.

Nanoparticles were prepared as described by Almalik et al. [12]. Briefly, a 0.069 wt% CS solution was prepared dissolving the polymer (low and high MW) in 4.6 mM HCl RNA-free solution. After complete dissolution, the pH was adjusted to 5 by the addition of appropriate volumes of NaOH 0.1 M. A 3 mg/mL HA solution was prepared in water and adjusted to pH 5 by addition of 0.1 M HCl solution. For the nanoparticle preparation, a first complexation was obtained by mixing the CS solution with a nucleic acid solution (at different concentrations, accordingly to the target payload). A volume of the HA solution was placed under vigorous magnetic stirring; the same volume of CS/nuclei acid complex was pipetted in the vial and left stirring for 30 min at room temperature. All the procedures were performed in sterile and RNA-free conditions.

The Z-average size and the ζ-potential of nanoparticles were both measured at room temperature with a Zetasizer Nano ZS (Model ZEN3600; Malvern Instruments Ltd., UK) equipped with a solid state HeNe laser (λ=633 nm) at a scattering angle of 173°. Nanoparticles were suspended in deionised water at a concentration of 1 mg/mL for all the measurement performed. The encapsulation efficiency was determined by measuring the concentration of non-encapsulated nucleic acids with the PicoGreen or RiboGreen assay (Life Technologies, UK) for DNA and RNA, respectively. DNase I was used to investigate the protection of encapsulated nucleic acids. Recovered supernatant was assayed usign the BioAnalyzer (DNA capillary electrophoresis, Agilent Technologies).

2.2 *In vitro* models of CD44-mediated cell uptake

THP-1 monocytes were differentiated towards macrophages by addition of 50 ng/mL PMA in cell culture medium (THP-1-M). HT-29 cells were cultured in complete medium. The total CD44 content was quantified through western blotting for both the cell lines (data not shown). For equal loading and normalisation of protein levels, ACTB (β-actin) was chosen as a housekeeping gene for it has been described as one of the most stable genes during THP-1 differentiation. Membrane-bound CD44 was also observed staining cells with Alexa594-conjugated anti-human CD44 (BioLegend, UK). Cells were stained on ice and then fixed with 4% PFA.

Nanoparticles were prepared with fluoresceinamine-labeled HA or CS, in case of uptake or live imaging for nanoparticle tracking respectively. siGLO Cyclophilin B Control siRNA and siGLO Green Transfection Indicator (Dharmacon, UK) were used to track nucleic acid fate and transfection efficiency. Cell nuclei were stained with Hoechst, cell membrane was stained with Alexa633-ConcanavalinA or CellMask Deep Red, while Alexa488-Phalloidin was used to stain actin on fixed cells (Invitrogen, UK). Confocal acquisitions were performed using a Leica TCS SP5 AOBS inverted confocal using a 63x/0.60-1.40/HCX PL Apo objective. The confocal settings were as follows, section thickness 0.571 μm, scan speed 200 Hz, pixel size 125 nm, optimal Z section number determined by the confocal software. To eliminate any possible cross-talk between channels, images were collected with a sequential scan, usign the following laser lines and mirror settings: 405(30%) 460-475nm; 488(30%) 495-580nm; 594(50%) 615-665nm; 633(40%) 650-700nm.

3 RESULTS

We have monitored the Z-average size and ζ-potential of the particles (Figure 1) as a function of the amount of nucleic acids, in order to highlight any sign of decrease of stability due to neutralisation of surface charge and resulting agglomeration at high loads. Independently on their size, nucleic acids (siRNA, plasmids, genomic DNA) were successfully encapsulated in the nanoparticles up to a load of 25% in relation to the chitosan content, with an encapsulation efficiency >95% and negligible effects on the nanoparticle physical properties.

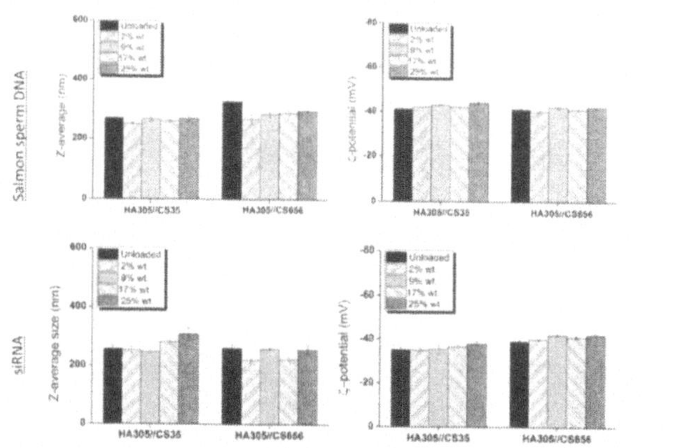

Figure 1: Z-average size (left handside) and ζ-potential (right handside) of low and high MW CS nanoparticles decorated with HA. Nucleic acids were loaded with different weigh ratios against CS (from 2 to 25%, grey shades) and un-loaded nanoparticles (black) were used as a reference value. No influence of nucleic acid size was observed, as shown in CS//HA loaded with salmon sperm (top row) and siRNA (bottom row).

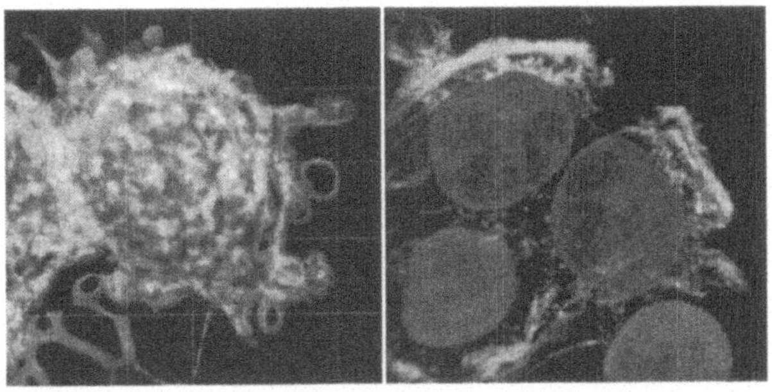

Figure 2: 3D rendering of membrane-bound CD44 on THP-1-M (left) and HT-29 (right) cells. Nuclei (blue), actin (green) and CD44 (red) are highlighted; macrophages (left hand side) were also stained with CellMask^{TM} to detect cell membrane (white). Volumes of about $(30\times30\times15)\mu m^3$ are rendered.

The nanoparticles were also prepared by separately labelling HA, CS, and nucleic acids with different fluorescent tags, which allowed to separately follow: i) the CD44-mediated internalisation, ii) the intracellular trafficking and iii) their individual components fate in a colorectal cancer (HT-29) and a macrophage (THP-1M) model. Sections showing the CD44 expression in colorectal cancer cells and macrophages are shown in Figure 2. The volumetric reconstruction of the living cells via confocal microscopy allows to follow in real time the cell-nanoparticle interactions. An example of HA/CD44 interaction in HT-29 after 2 hours treatment with nanoparticles is shown in Figure 3. After treatment cells were fixed and then sections were acquired using confocal microscope. As shown, both low and high MW CS nanoparticles are uptaken after 2 hours, about 0.1 µg NPs/ µg protein and 0.2 µg NPs/ µg protein respectively.

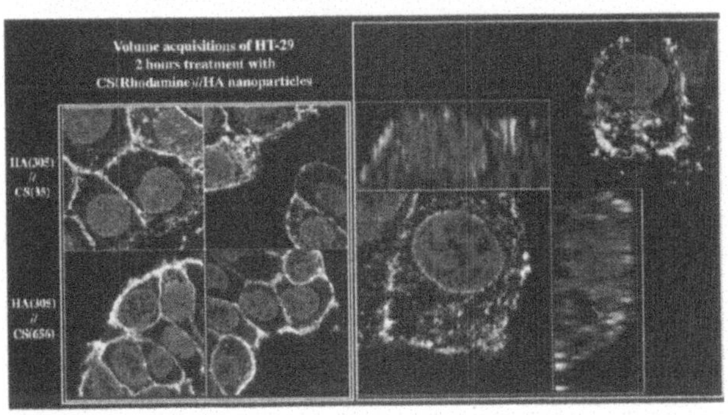

Figure 3: HT-29 treated for 2 hours with fluorescently labeled and unloaded CS//HA nanoparticles (red). Maximum projections of $(40\times40)\mu m^2$ and 1 µm thich of low and high MW CS (lef hand side); orthogonal projection and 3D rendering of a single cell exposed to low MW CS nanoparticles for 2 hours. Cells were fixed after the treatment and nuclei (blue), actin (green) and glycosaminoglycans (white) were stained to locate nanoparticles within the cells.

4 CONCLUSION

In summary, the study demonstrated that CS//HA nanoparticle are a suitable cargo for nucleic acid in the CD44-mediated endocytic process. It is believed that nanoparticles enter cells through the endosomal pathway, eventually escaping into the cytoplasm, allowing for the release of the loaded content. Moreover, nucleic acid with different size can be successfully encapsulated and then delivered to CD44 expressing cells. We observed that the uptake of CS//HA nanoparticles by THP-1 macrophages is about 10% lower compare to that of HT-29. Human macrophages were also evaluated as a potential off-target; however an increased uptake in HT-29 was observed.

5 ACKNOLEDGEMENTS

Financial support from the Technology Strategy Board (TSB) under the grant agreement 33362-239223. The Bioimaging Facility microscopes used in this study were purchased with grants from BBSRC, Wellcome and the University of Manchester Strategic Fund.

REFERENCES

[1] Ding Y, Jiang ZW, Saha K, Kim CS, Kim ST, Landis RF, et al. Gold Nanoparticles for Nucleic Acid Delivery. Mol Ther 2014;22(6):1075-83.

[2] Glover DJ, Lipps HJ, Jans DA. Towards safe, non-viral therapeutic gene expression in humans. Nat Rev Genet 2005;6(4):299-U29.

[3] Lv HT, Zhang SB, Wang B, Cui SH, Yan J. Toxicity of cationic lipids and cationic polymers in gene delivery. J Control Release 2006 10;114(1):100-9.

[4] Juliano R, Bauman J, Kang H, Ming X. Biological Barriers to Therapy with Antisense and siRNA Oligonucleotides. Mol Pharmaceut 2009;6(3):686-95.

[5] Misra S, Heldin P, Hascall VC, Karamanos NK, Skandalis SS, Markwald RR, et al. HA/CD44 interactions as potential targets for cancer therapy. The FEBS Journal 2011, 278(9):1429–1443

[6] Sy MS, Guo YJ, Stamenkovic I. Inhibition of tumor growth in vivo with a soluble CD44-immunoglobulin fusion protein. The Journal of Experimental Medicine 1992, 176(2):623–627.

[7] Platt VM, Szoka FC. Anticancer Therapeutics: Targeting Macromolecules and Nanocarriers to Hyaluronan or CD44, a Hyaluronan Receptor. Molecular Pharmaceutics 2008, 5 (4):474–486.

[8] Ouasti S, Kingham PJ, Terenghi G, Tirelli N. The CD44/integrings interplay and the significance of receptor binding and re-presentation in the uptake of RGD-functionalized hyaluronic acid. Biomat 2012;33:1120-1134.

[9] Mickler FM, Möckl L, Ruthardt N, Ogris M, Wagner E, Braüchl C. Tuning Nanoparticle Uptake: Live-Cell Imaging Reveals Two Distinct Endocytosis Mechanisms Mediated by Natural and Artificial EGFR Targeting Ligand. Nano Lett. 2012;12:3417-3423.
[10] Zhang Y, Hensel M. Evaluation of nanoparticles as endocytic tracers in cellular microbiology. Nanoscale 2013;5:9296-9309.
[11] Portis AM, Carballo G, Baker JL, Chan C, Walton SP. Confocal microscopy for the analysis of siRNA delivery by polymeric nanoparticles. Microsc Res Tech. 2010;73(9):878–885.
[12] Almalik A, Day PJ, Tirelli N. HA-coated chitosan nanoparticles for CD44-mediated nucleic acid delivery. Macromol Biosci 2013;13:1671-1680.
[13] Almalik A, Karimi S, Ouasti S, Donno R, Wandrey C, Day PJ, Tirelli N. Hyaluronic acid (HA) presentation as a tool to modulate and control the receptor-mediated uptake of HA-coated nanoparticles. Biomat 2013;34(21):5369-5380.

Efficient drug delivery for cancer treatment using SiO_2-LDH nanocomposites

L. Li[*1], Z. Gu[*1], W.Y. Gu[*], J. Liu[**] and Z. P. Xu[*]

[*]Australian Institute for Bioengineering and Nanotechnology, The University of Queensland, Brisbane, QLD 4072, Australia,

l.li2@uq.edu.au; z.gu@uq.edu.au; w.gu@uq.edu.au; gordonxu@uq.edu.au

[**]Department of Chemical Engineering, Curtin University, Perth, WA 6845 Australia, jian.liu@curtin.edu.au;

ABSTRACT

Well-dispersed SiO_2 nanodot-coated LDH nanohybrids have been prepared via a nanodot-coating strategy. SiO_2@LDH nanohybrids show good dispersion in aqueous solution and cell culture medium. The anticancer delivery test demonstrates significant inhibition of cancer cell proliferation using SiO_2@LDH nanohybrids to deliver methotrexate (MTX).

Keywords: layered double hydroxide (LDH), functionalization, drug delivery, self-assembly

1. INTRODUCTION

Target-specific delivery with sustained release of anticancer agents has attracted considerable research interest in cancer chemotherapy.[1, 2] For this purpose, layered double hydroxides (LDHs), a family of anionic clay materials, have been considered as a promising candidate for drug and gene delivery due to their unique properties, such as high anion exchange capacity, low cytotoxicity, pH-controlled release, good biocompatibility, tunable particle size, and protection of drugs and genes in the interlayer.[3-8] Many studies have shown that various biofunctional molecules, including DNA, siRNA, drugs, and vitamins, have been successfully stabilized and preserved by incorporating into LDHs, showing high delivery efficiency and bioactivity in in-vitro tests.[3, 6, 9-14] However, LDH nanoparticles easily form a random agglomeration in phosphate buffer solution (PBS) and cell culture medium. This issue severely affects the bioavailability, circulation, and target delivery of LDHs in animal tests.

To circumvent this issue, surface functionalization of LDH is highly desirable. Recently, Oh et al[15] functionalized LDH nanoparticles through condensation of aminopropyltriethoxy-silane (APTES) on the LDH nanoparticle surface and then conjugated with the cancer-cell-specific ligand, folic acid (FA) to amine-functionalized SiO_2. They observed that coupling FA to MTX/LDH hybrids enhanced cellular uptake and inhibited cancer-cell proliferation in FR overexpressed cells compared to MTX/LDH. However, the ionic nature of the LDH material makes surface functionalization time-consuming and unstable, which limits the successful application in drug and gene delivery. Therefore, cost-effective rational design of mutilfunctional LDH nanocarriers with high effiency, good bioavailability, low cytotoxicity and long circulation time is necessary.

As is well known, silica nanoparticles (SiO_2 NPs) with defined size, morphology and surface properties can be easily functionalized with various functional groups and cancer-cell targeting moieites. Thus, coating LDH nanoparticles with SiO_2 nanodots can circumvent the complicate and time-comsuming modification process,[15] which would render the possibility of designing a new generation of drug/gene delivery system.

In this work, uniform SiO_2 nanodots with a negative zeta potential (~10 nm, zeta potential of -38.8 mV, **Fig. 1**) were coated onto the surface of LDH nanoparticles (80-150 nm, zeta potential of +42.5 mV, **Fig.** 1) to form SiO_2@LDH nanohybrids. These nanohybrids are monodispersed in PBS and cell culture medium, in sharp contract with LDH nanoparticles that are readily aggregated. These nanohybrids can be further conjugated with target moieties for targeting delivery. To study the drug delivery efficiency of the nanohybrid systems, methotrexate (MTX), a widely used anti-cancer agent, was used in this work as a model drug, and thus MTX-LDH suspension and nearly monodispersed SiO_2@MTX-LDH nanohybrid suspension were prepared in the similar way and further testes in cancer cell treatment.

2. EXPERIMENTAL

2.1 Synthesis of LDH and MTX-LDH suspension

Layered double hydroxide (LDH) and MTX-LDH suspensions were synthesised via coprecipitation-hydrothermal method, a procedure published elsewhere [1]. Briefly, $Mg(NO_3)_2$ (3.0 mmol) and $Al(NO_3)_3$ (1.0 mmol) were dissolved in 10 mL distilled water and quickly added to 40 mL NaOH solution (6.0 mmol) under vigorous stirring, following by 10 min stirring. The LDH slurry was collected via centrifuge separation. After washing with distilled water twice via centrifugation, LDH slurry was resuspended in 40 mL distilled water. The resulting inhomogeneous suspension was transferred to an autoclave

[1] Address correspondence to Dr Li Li and Dr Zi Gu , E-mail: l.li2@uq.edu.au; z.gu@uq.edu.au

(stainless steel with a Teflon lining) and heated to 100 °C for 16 h. After hydrothermal treatment, a transparent, homogenous suspension Mg_2Al-LDH was obtained.

Preparation of MTX-LDH suspension was similar. After collecting the LDH slurry via centrifuge separation, the slurry was exchanged with 0.1 mmol MTX in 40 mL solution (neutralized with dilute NaOH solution, with pH of 8-9) for 1 h. After washing and separation via centrifugation, the MTX-LDH slurry was dispersed in 40.0 mL distilled water and transferred to an autoclave for hydrothermal treatment at 100 °C for 16 h.

2.2 Synthesis of SiO_2@LDH and SiO_2@MTX-LDH colloidal suspension

To prepare SiO_2@LDH and SiO_2@MTX-LDH suspension, SiO_2 nanodots were pre-prepared via a microemulsion method. Briefly, 0.73 g octane and 0.014 g L-arginine were dissolved in 14 mL distilled water at 60 °C and 0.2 g TEOS was added to the solution under vigorous stirring at 60 °C for 4 h. Then, 14 ml of as-synthesised LDH and MTX-LDH was added to the above solution at 60 °C under stirring for 20 h, respectively. After washing with distilled water for 4 times, the as-obtained SiO_2@LDH and SiO_2@MTX-LDH nanohybrids were resuspended in distilled water, respectively.

2.3 Characterization

Photon correlation spectroscopy (PCS, Nanosizer Nano ZS, MALVERN Instruments) was used to analyse the particle size distribution of LDH suspensions. The same instrument was also used to measure the Zeta potential of LDH nanoparticles in as-prepared suspensions. Powder X-ray diffraction (XRD) patterns were recorded on a Rigaku Miniflex X-ray diffractometer with variable slit width at a scanning rate of 2°/min with 2θ ranging from 2.5° to 80° using Co Kα radiation (λ=0.17902 nm). Mg and Al concentrations in all samples were determined by inductive coupled plasma atomic emission spectroscopy (ICP-AES) on a Varian Vista Pro instrument. Transmission electron microscopy (TEM) images were obtained on a JEOL 1010A transmission electron microscope at an acceleration voltage of 100 kV. Element analysis of C, H and N was performed by Flash EA 1112 CHNS-O analyser (Thermo Electron Corp., US)

3. RESULTS AND DISCUSSION

SiO_2 nanodots coated LDH nanohybrids were synthezised via self-assemmbly by electrostatic interactions. As shown in **Fig. 2A and 2B**, LDH and MTX-LDH nanoparticles have hexagonal morphology with the particle size in the range of 80-150 nm **(Fig. 1A)**. After coated with SiO_2 nanodots, SiO_2@LDH and SiO_2@MTX-LDH nanohybrids **(Fig. 2C** and **2D)** retained the hexagonal morphology. The average size of silica nanodots was around 10-12 nm **(Fig. 1A, 2E** and **2F)**. Note that uniform SiO_2 nanodots were evenly attached on the surface of LDH nanosheets **(Fig. 2C and 2D)**. XRD patterns of LDH, SiO_2@LDH, MTX-LDH and SiO_2@MTX-LDH **(Fig. 3)** exhibit the characteristics of LDH structure with a rhombohedral symmetry, as reflected by (003), and (006) peaks. After being coated with SiO_2 nanodots, the (003) and (006) peaks of the LDH phase became weaker and broader, suggesting the crystallinity of LDH nanoparticles was lowered during SiO_2 dot-coating. An additional peak at 5.9° corresponding to an interlayer spacing of approx. 1.75 nm was obseved in the MTX-LDH and SiO_2@MTX-LDH samples **(Fig. 3B** and **3D)**, suggesting MTX was intercalated into the LDH layers.[14, 16, 17]

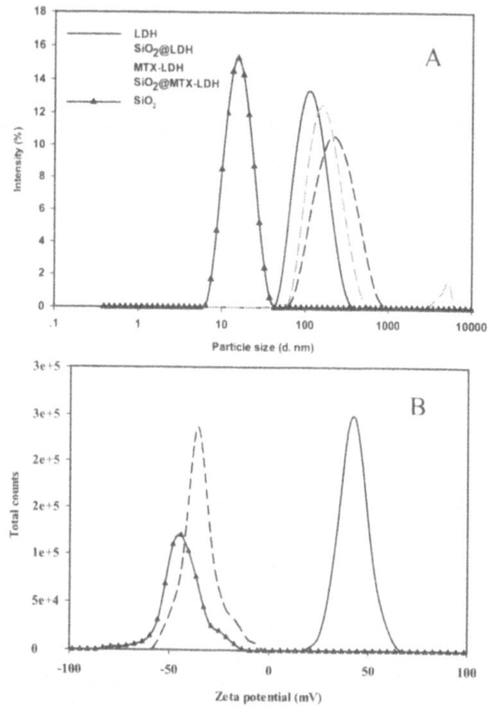

Fig. 1 (A) Particle size distribution of LDH (line), SiO_2 (line with triangle), SiO_2@LDH (dash), MTX-LDH (dot) and SiO_2@MTX-LDH (grey dash), (B) zeta potential of LDH (line), SiO_2 (line with triangle) and SiO_2@LDH (dash)

Element analysis using CHN analyzer and ICP-MS measurement revealed the composition of as-obtained samples, as summarized in Table 1. Note that the Mg/Al atomic ratio was 2.1, close to the nominal ratio (2.0). The MTX loading amount was 9.7 wt%, close to the designed MTX loading, i.e. 10% of the total anion capacity in LDH. The molar ratio of Al/Si was about 0.28, approx. equivalent to a mass ratio of 1:1 between the two constituents (LDH:SiO_2).

As shown in **Fig. 2B**, LDH nanoparticles in suspension possess a positive zeta potential of +42.5 mV and SiO_2 nanodots in suspension a negative zeta potential of -43.5 mV.Therefore, when mixing LDH nanoparticle suspension with silica nanodot suspension, LDH acts as a substrate to attract the negatively charged silica nanodots via electrostatic attractions. Meanwhile, the electrostatic repulsion between silica nanodots makes themsleves to repel one another so as to evenly distribute on the LDH surface, as seen in **Fig. 2C** and **2D**. These SiO_2@LDH nanohybrids exhibit a negative zeta potential of -31.1 and -32.5 mV (**Fig. 2A and Table 1**), respectively. We noted that SiO_2@LDH and SiO_2@MTX-LDH nanohybrids were able to form stable and well-dispersed suspension in water, PBS and culture medium after ultrasonication.

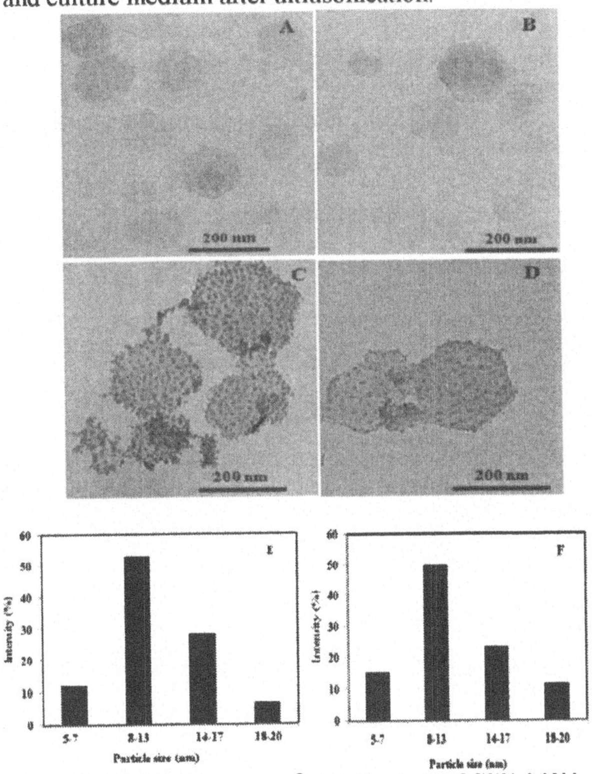

Fig. 2 TEM images of LDH (A), MTX-LDH (B), SiO_2@LDH (C) and SiO_2@MTX-LDH (D); and particle size distribution of SiO_2 nanodots (E and F) on the LDH and MTX-LDH surface.

In summary, SiO_2 dot-coated LDH nanohybrids were prepared following our proposed strategy, which can be nearly monodispersed in PBS and culture medium. It should be mentioned that using this method, the size of silica nanodots and their coverage on the LDH surface can be easily tuned by varying TEOS amount. Moreover, by virtue of diverse surface properties of SiO_2 nanodots, SiO_2@LDH nanohybrids can be further functionalised with various biofunctional molecules, which would allow the fabrication for advanced multifuntional drug and gene delivery system.

To test if the nanohybrids can be effectively taken up by cancer cells, we use human cervical cancer cell line HeLa as a model. The LDH and SiO_2@LDH nanoparticles were loaded with 21-bp dsDNA tagged with Cy3 (red flourescence) to signal the uptake.[12, 19] As shown in **Fig. 4**, both LDH and SiO_2@LDH nanoparticles can be effectively taken up by HeLa cells at the LDH:dsDNA mass ratio of 40:1, demonstrating that short gene segments can be efficiently delivered to cells by both systems. Note that the cellular uptake of LDH nanoparticles has been reflected by bright red spots in the cells (**Fig. 4A**), suggesting the internalized LDH particles were aggregated to some degree in the cytoplsma. In contrast, only faintly reddish scattered dots were observed within the cells in the case of SiO_2@LDH nanohybrids (**Fig. 4B**), indicating less aggregation of SiO_2@LDH particles in the cytoplasma. This clearly shows that SiO_2@LDH nanohybrids exhibit better dispersion within the cells. In addition, comapred to LDH alone, SiO_2@LDH nanoparticles show the weaker red signal, which may be caused by the slower cellular uptake due to the negative charges of SiO_2@LDH nanoparticles.

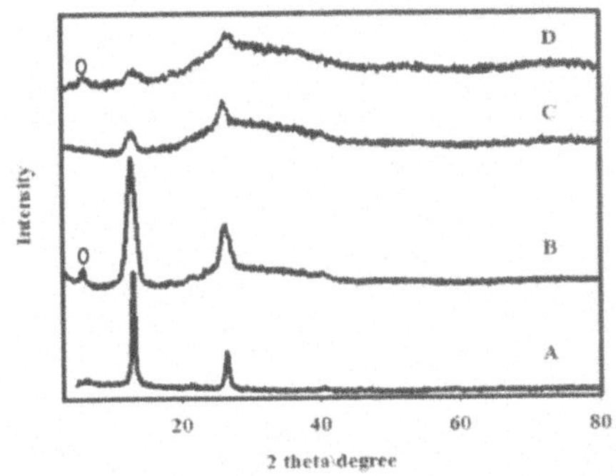

Fig. 3 XRD patterns of LDH (A), MTX-LDH (B), SiO_2@LDH (C) and SiO_2@MTX-LDH (D) nanohybrids. '0' indicates the new basal diffraction peak of the MTX-intercalated LDH phase.

Table 1 The element analysis results of LDH and SiO_2@LDH nanoparticles

samples	Mg/Al ratio	N (wt%)	Particle Size (nm)	Zeta potential (mV)
LDH-Cl	2.1	-	109.4	+42.5
SiO_2	-	-	15.0	-43.5
SiO_2@LDH	2.1	-	207.7	-31.1
MTX-LDH	2.1	1.91	102.2	-
SiO_2@MTX-LDH	2.1	1.08	174.3	-32.5

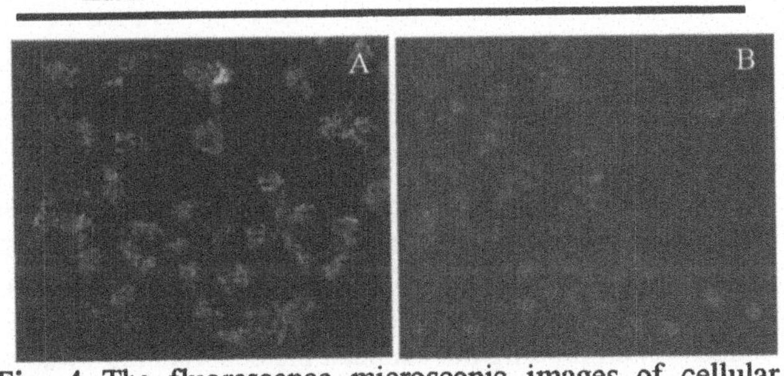

Fig. 4 The fluorescence microscopic images of cellular uptake of LDH (A) and SiO_2@LDH (B) in HeLa cells. using oligo DNA-Cy3 as the marker with the mass ratio of LDH:DNA = 40:1.

We further studied the MTX delivery efficiency in the chemotherapeuic treatment of human osteosarcoma cell line 143B PMLBK TK (ATCC Cat No CRL-8304) using the nanohybrids as the delivery vehicle. As shown in **Fig. 5**, both MTX-LDH and SiO_2@MTX-LDH nanohybrids exhibited the effective suppression of cancer cell growth in a does-dependent manner, which can be attributed to quick uptake of LDH and SiO_2@LDH nanoparticles. In comparison, MTX-LDH hybrids with or without SiO_2 modification showed a similar tumor suppression efficiency. Relatively, MTX-LDH caused more cancer cell death (90-93%) than SiO_2@MTX-LDH (85-90%) at the MTX dose of 0.064-1.02 μg/ml, and the supression efficiency in both cases was higher than MTX itself. This observation reveals that SiO_2 modification on LDH has slightly slowed down the internalisation of LDH nanohybrids due to the negative charges of SiO_2@LDH nanohybrids. As SiO_2 nanodots in SiO_2@LDH nanohybrids are further functionalized with PEG and targeting moieties, the surface negative charges of SiO_2@LDH nanohybrids would be much less and thus quick cellular uptake and targeting would be both achieved. Therefore, SiO_2@LDH nanohybrids can be the promising efficient delivery vehicles.

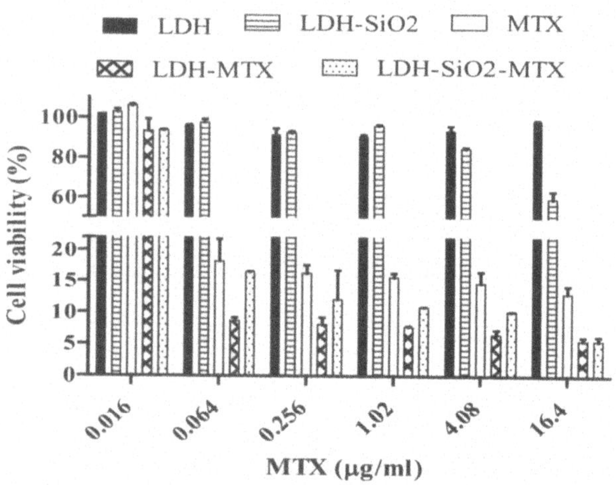

Fig. 5 The cell viability of osteosarcoma cells exposed to different amounts of MTX associated with or without LDH/ SiO_2@LDH. Cells without any treatment were used as the positive control (100% viability). The concentration of the particle control group (LDH or SiO_2@LDH) was the same as that in LDH-MTX or SiO_2@LDH-MTX hybrids, approx. 10 times as the MTX concentration.

In addition, both LDH and SiO_2@LDH nanoparticles showed a low cytotoxicity (cell viability >85%) when the concentration was lower than 40 μg/ml (10 times as the MTX amount), but SiO_2@LDH nanohybrids at higher concentration (200 μg/ml) reduced the cell viability to 60% (**Fig. 5**), which could be attributed to the trace amount of surfactants associated with SiO_2.

4. CONCULSIONS

In summary, SiO_2@LDH nanohybrids were successfully synthesized through a facile, versatile and reproducible route. The negatively charged SiO_2 nanodots were uniformly distributed on the positively charged LDH nanoparticles. As-synthesised SiO_2@LDH nanohybrids retained the LDH structure and formed stable suspension in aqueous solution, PBS and culture medium. The cellular uptake tests showed that the SiO_2@LDH nanohybrids exhibited better dispersion within the cells after internalisation, with effective inhibition of cancer cell prolification even less uptake amount of SiO_2@LDH-MTX nanohybrids compared with LDH-MTX. Therefore, monodispersed SiO_2@LDH nanohybrids can be an excellent nonviral delivery carrier for drug and gene deivery.

REFERENCES

[1] M. E. Davis, Z. Chen and D. M. Shin, Nat. Rev. Drug Discov, 7, 771, 2008.
[2] M. De, P. S. Ghosh and V. M. Rotello, Adv. Mater., 20, 4225, 2008.
[3] K. Ladewig, Z. P. Xu and G. Q. Lu, Expert Opin. Drug Del., 6, 907, 2009.
[4] P. S. Brateman and Z. P. Xu, "Handbook of layered materials", S.M. Auerbach, K.A. Carrado and P.K. Dutta, 2004, 373.
[5] J. H. Choy, "Science and Technology of Hybrid Materials", C. J. Sun, J. Ding, M. Gupta, G. M. Chow, L. Kurihara and L. Kabacoff, 2006, 1.
[6] Z. P. Xu, T. L. Walker, K. L. Liu, H. M. Cooper, G. Q. M. Lu and P. F. Bartlett, Int. J. Nanomed., 2, 163, 2007.
[7] Z. P. Xu, G. Stevenson, C. Q. Lu, G. Q. Lu, P. F. Bartlett and P. Gray, J. Am. Chem. Soc., 128, 36, 2006.
[8] Z. P. Xu and G. Q. Lu, Pure Appl. Chem., 78, 1771, 2006.
[9] Z. P. Xu, Q. H. Zeng, G. Q. Lu and A. B. Yu, Chem. Eng. Sci., 61, 1027, 2006.
[10] K. Ladewig, M. Niebert, Z. P. Xu, P. P. Gray and G. Q. Lu, Appl. Clay Sci., 48, 280, 2009.
[11] K. Ladewig, M. Niebert, Z. P. Xu, P. P. Gray and G. Q. M. Lu, Biomaterials, 31, 1821, 2010.
[12] Y. Y. Wong, K. Markham, Z. P. Xu, M. Chen, G. Q. Lu, P. F. Bartlett and H. M. Cooper, Biomaterials, 31, 8770, 2010.
[13] S. J. Choi, J. M. Oh and J. H. Choy, J. Nanosci. Nanotechno., 10, 2913, 2010.
[14] J. H. Choy, J. S. Jung, J. M. Oh, M. Park, J. Jeong, Y. K. Kang and O. J. Han, Biomaterials, 25, 3059, 2004.
[15] J. M. Oh, S. J. Choi, G. E. Lee, S. H. Han and J. H. Choy, Adv. Funct. Mater., 19, 1617, 2009.
[16] J. M. Oh, M. Park, S. T. Kim, J. Y. Jung, Y. G. Kang and J. H. Choy, J. Phys. Chem. Solids, 67, 1024, 2006.
[17] M. Chakraborty, S. Dasgupta, S. Sengupta, J. Chakraborty, S. Ghosh, J. Ghosh, M. K. Mitra, A. Mishra, T. K. Mandal and D. Basu, Ceram. Int., 38, 941, 2012.
[18] A. S. D. Zeta potential of colloids in water and waste water, American Society for testing and materials, 4187, 1985.
[19] Y. Y. Wong, H. M. Cooper, K. Zhang, M. Chen, P. Bartlett and Z. P. Xu, J. Colloid Interface Sci., 369, 453, 2012.

Thermally Responsive Lipid Coated Superparamagnetic Nanoparticles for Frequency Triggered Drug Delivery

Yogita Patil-Sen[1,2] *, Maneea Eizadi Sharifabad[1,2], Hussein Bux[1,2], Tim Mercer[3], Gary Bond[1] and Tapas Sen[1,2]

[1]Centre for Materials Science, [2]Nanobiomaterials Research Group, [3]Jeremiah Horrocks Institute for Mathematics, Physics & Astronomy

University of Central Lancashire, Preston, PR1 2HE, United Kingdom

Email: YPatil-sen@uclan.ac.uk, Tel: +44(0)1772894371, Fax: +44(0)1772894981

ABSTRACT

Herein we report a fabrication method of core-shell type silica coated superparamagnetic iron oxide nanoparticles (SPIONs) with thermally responsive lipids for controlled and targeted drug delivery using Alternative Current (AC) magnetic field for potential applications in cancer therapy. Core-shell SPIONs, loaded with anticancer drug, Doxorubicin (DOX) were coated with lipids such as dipalmitoylphosphatidylcholine (DPPC) and cholesterol so as to cap the mesopores. The particle size of the bare and coated SPIONs was measured using the dynamic light scattering technique, scanning and transmission electron microscopy (SEM and TEM). The change in surface morphology of bare and coated SPIONs was evidenced from SEM. The *in vitro* drug loading and release studies were carried in phosphate buffered saline (PBS) under the AC magnetic field as well as under thermal incubation condition at 37°C. The concentration of DOX in solution was determined by measuring the absorbance at 484 nm by UV-Visible spectrophotometer. The preliminary results indicate that the drug release behaviour under the AC magnetic field is relatively controlled compared to that of normal thermal incubation condition at 37°C.

Keywords: superparamagnetic iron oxide nanoparticles, lipid nanostructures, magnetoliposomes, drug delivery, hyperthermia cancer therapy

1. INTRODUCTION

Superparamagnetic iron oxide nanoparticles (SPIONs) have gathered a lot of attention in recent years because of their various potential biomedical applications, including contrast agent for magnetic resonance imaging, separation of biomolecules and drug delivery [1-5]. Due to their superparamagnetic properties, SPIONs are becoming increasingly popular in hyperthermia cancer therapy [6]. The nanoparticles can be targeted specifically to the cancer cell by applying an external magnetic field which can then release the drug to the target cell by inducing local hyperthermia. The size, morphology, surface charge and stability in suspension are the most important parameters which need to be cleverly controlled while considering the drug delivery application of nanoparticles. In order to have a good magnetic response the size of SPIONs need to be extremely small (< 10 nm). The small size of nanoparticles, however, lead to agglomeration of the particles which in turn causes poor stability and limited drug loading. Another major problem that can occur while using SPIONs as drug delivery vehicle is the leakage of drug molecules from the surface of the nanoparticles during the transport process in the blood stream which can reduce the drug efficacy. In order to overcome these problems, various materials such as silica [1, 2, 7-9], lipids [10] or commercial dispersing agents [11] have been used to modify the surface properties of SPIONs.

Amphiphilic molecules such as lipids possess a unique property of self-assembling in aqueous medium to give rise to different types of nanostructures [12]. Although bilayered lamellar is the most common nanostructure formed by lipid molecules in the presence of water, other structures such as hexagonal and cubic are also commonly seen. The type of nanostructure formed depends upon the lipid's structure and shape, its composition in water as well as on the physico-chemical conditions employed such as temperature and pressure. These self-assembled nanostructures have been extensively studied as model membrane systems in order to understand the various biological processes such as signal transduction, cell migration, cell proliferation [13]. Moreover, lipid based nanostructures have been employed in numerous disciplines including food, cosmetic and the pharmaceutical industry as carriers of nutrients, flavours, perfumes, drugs and other bioactive molecules [14, 15]. The nanoscale size, excellent biocompatibility, non-toxicity and their ability to solubilise both hydrophobic and hydrophilic molecules make lipid nanostructures a great component for drug delivery systems. These nanostructures not only protect the functional molecules from chemical and enzymatic degradation but also release bioactive molecules to the specific target and in a controlled manner [16]. However, in order to improve the applicability, the non-fluid viscous lipid phases, such as hexagonal and cubic, are dispersed into excess water to form oil-in-water type emulsions which are kinetically stabilized using stabilizers such as carbon nanotubes [17] or silica nanoparticles [18].

Multifunctional hybrid nanoparticle such as magnetoliposomes which consist of lipids and SPIONs (Figure 1), are emerging as a new class of nanomaterials for cancer therapy. These hybrid assemblies exhibit superior properties compared to the individual systems. Moreover, they provide an alternative method to the traditional chemo and radiation therapy to treat cancer as the latter two methods lack specificity and hence reportedly have harmful side effects [19].

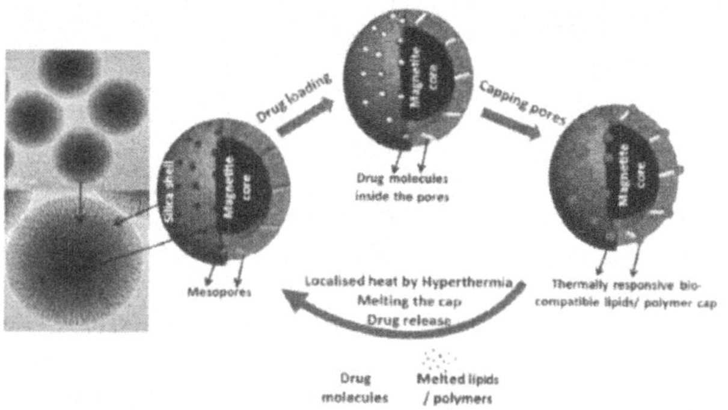

Figure 1 Schematic representation of multi-functional hybrid nanoparticles containing lipids and SPIONs for drug delivery applications.

2. EXPERIMENTAL

2.1 Synthesis of SPIONs

Bare SPIONs (Fe_3O_4) were prepared following a previously reported method [11].

2.2. Synthesis of mesoporous silica coated SPIONs

The core-shell type mesoporous silica coated SPIONs ($Fe_3O_4@nSiO_2$) were synthesized using a modified Stober system. 20 mg of Fe_3O_4 nanoparticles were dispersed into the solution containing 25 ml de-ionized water, 15 ml ethanol, 75 mg cetyl trimethyl ammonium bromide (CTAB) and 0.275 ml concentrated ammonia solution (28%). 0.125 ml tetraethoxy silane (TEOS) was added after vigorous stirring for 0.5 h and the reaction continued for 16 h. The product was separated and collected with a magnet, followed by washing with deionized water and ethanol for 3 times, respectively.

2.3 Preparation of lipid coated Doxorubicin loaded core-shell SPIONs

20 mg of a pure lipid dipalmitoylphosphatidylcholine, DPPC or a 3:2 (w/w) mixture of DPPC to cholesterol was dissolved in 2 ml chloroform in a 50 ml round bottom flask. The flask containing lipid solution was attached to a rotary evaporator and immersed in a water bath at 45 ºC and the solvent was evaporated under reduced pressure (Buchi Rotavapor R-114, Bauchi, Switzerland) to form a thin lipid film.

100 µg/ml Doxorubicin (DOX) solution was prepared by dissolving 1 mg DOX in 10 ml dimethyl sulphoxide (DMSO)/water (1:1).

30 mg of silica coated SPIONs were taken in 50 ml centrifuge tube and separated and washed with ethanol (x3, 5ml) followed by deionized water (x3, 5ml). 7.5 ml DOX solution was then added to the tube containing 30 mg of silica coated SPIONs and the mixture was vortexed for 5 minutes.

The solution of core-shell SPIONs with DOX was added to the flask containing the lipid film in which another 2.5 ml deionized water was added. The mixture was vigorously shaken manually for 5 minutes and then sonicated for 10 minutes in a continuous pulse mode at a maximum power of 40% using Vibra Cell Sonicator (Sonics and Materials Inc., USA) with a 13-mm diameter titanium probe.

2.4 Drug loading and release

The lipid coated DOX loaded core-shell SPIONs were centrifuged at a speed of 10,000 rpm for 10 min to separate the debris formed after sonication and non-entrapped DOX and core-shell SPIONs. The concentration of DOX in the supernatant was determined by measuring the UV absorbance at λ= 484 nm and comparing it with the pre-established calibration curve of known DOX concentrations in DMSO/water. The amount of DOX incorporated into the lipid coated core-shell SPIONs was obtained by subtracting the amount of free drug in the supernatant from the total amount of drug initially added.

Release of DOX from lipid coated nanomaterials was studied in PBS buffer under the AC magnetic field as well as under thermal incubation condition at 37°C. 4 mg of washed nanoparticles were treated with 1 mL of PBS buffer (pH = 7.1) at 37 °C under end-over-end rotation for up to 48 h. The nanoparticles were separated at different time intervals from the solution by magnetic separation and the absorbance was measured at 484 nm. The concentrations of DOX in the solution were determined by comparing the absorption value with a pre-established standard curve of known DOX concentrations in PBS buffer. Similarly, 4 mg of DOX loaded lipid coated nanoparticles were treated with 1 mL of PBS buffer and were placed under alternative AC magnetic field.

2.5 Characterization of SPIONs, core-shell SPIONs and lipid coated core-shell SPIONs

The particle size of SPIONs, core-shell SPIONs and lipid coated core-shell SPIONs was measured using the dynamic light scattering techniques (Zetasizer Nano, Malvern Instruments, UK and Mastersizer 2000, Malvern Instruments, UK), scanning electron microscopy (SEM) (FEI Quanta 200ESEM) and transmission electron microscopy (TEM) (JEOL JEM-2000 EX electron microscope Operated at 200 kV, images processed using Gatan Digital Microscope software). The surface morphology of the nanoparticles was observed using SEM. The heating ability of bare SPIONs and coated SPIONs was evaluated by DM100Series Nanoheating

Instrument (nanoScale Biomagnetics, Spain) using the supplied software "Maniac".

3. RESULTS AND DISCUSSION

The average particle size for the DPPC coated core-shell SPIONs with DOX reduced significantly after sonication. The particle diameter was found to be ~100 μm before the ultra-sonication and ~4 μm for sonicated particles (Figure 2). This suggested that the lipid coated core-shell SPIONs are highly aggregated even after sonication.

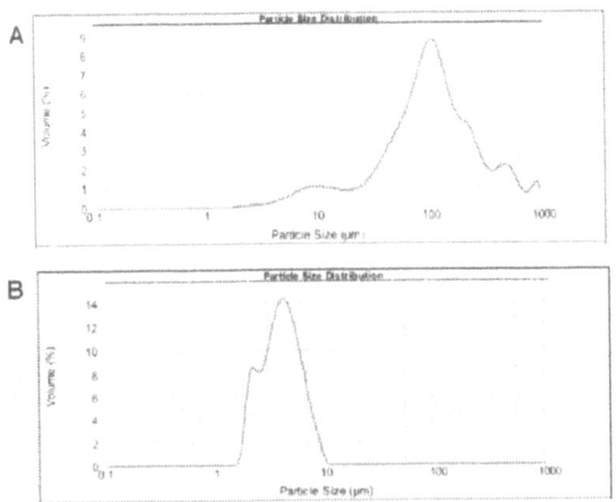

Figure 2 Particle size distribution of DPPC coated core-shell SPIONs with DOX (A) before sonication, (B) after ultra-sonication for 10 minutes using titanium probe.

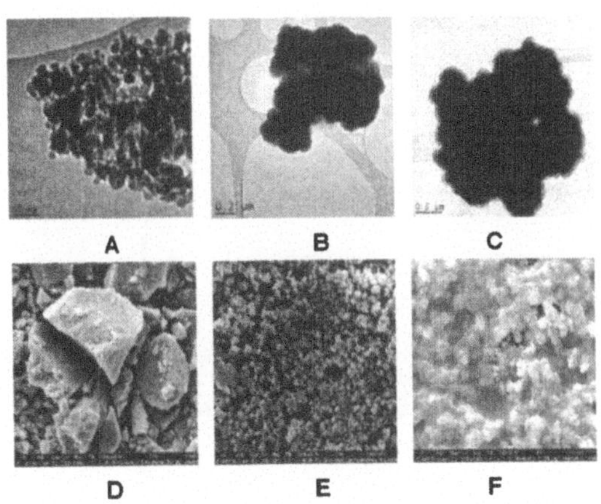

Figure 3 TEM images (A) bare SPIONs, (B) core-shell SPIONs, (C) DPPC coated core-shell SPIONs with DOX sonicated. (D), (E) and (F) represent their respective SEM images.

The observation was supported by TEM and SEM as shown in Figure 3. TEM shows that the bare SPIONs have particle size ~10 nm, which increased to 20 nm after coating them with silica. A thin film around the aggregated SPIONs was clearly seen which confirmed the lipid coating (Figure 3C). The change in surface morphology of the bare SPIONs after coating with silica and DPPC was also clearly observed by SEM images as shown in Figure 3 D, E and F.

The heating ability of SPIONs and core-shell SPIONs was tested for hyperthermia which indicated good heating response (Figure 4) and hence their suitability for drug delivery applications in cancer therapy.

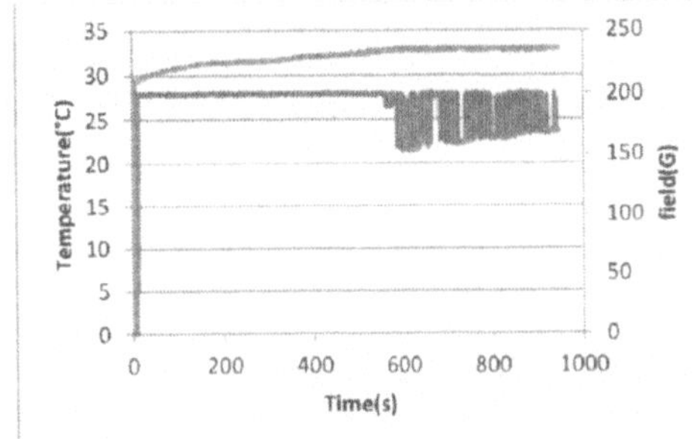

Figure 4 The magnetic heating profile of bare SPIONs (blue), silica coated SPIONs (red) at field frequency of 406 KHz.

Preliminary data suggest that drug release behaviour was different under hyperthermia compared to that for thermal heating using incubation (data not shown).

REFERENCES

1. Mahmoudi, M., et al., Advanced Drug Delivery Reviews, 2011. **63**(1-2): p. 24-46.
2. Sen, T., A. Sebastianelli, and I.J. Bruce, Journal of the American Chemical Society, 2006. **128**(22): p. 7130-7131.
3. Pankhurst, Q.A., et al., Journal of physics D: Applied physics, 2003. **36**(13): p. R167.
4. Langer, R. and D.A. Tirrell, Nature, 2004. **428**(6982): p. 487-492.
5. Perez, J.M., Nature nanotechnology, 2007. **2**(9): p. 535-536.
6. Hergt, R., et al., Journal of Physics-Condensed Matter, 2006. **18**(38): p. S2919-S2934.
7. Philipse, A.P., M.P. Van Bruggen, and C. Pathmamanoharan, Langmuir, 1994. **10**(1): p. 92-99.
8. Bruce, I.J., et al., Journal of Magnetism and Magnetic Materials, 2004. **284**: p. 145-160.
9. Sen, T. and I.J. Bruce, Microporous and Mesoporous Materials, 2009. **120**(3): p. 246-251.
10. Sen, T., et al., Rsc Advances, 2012. **2**(12): p. 5221-5228.
11. Sen, T., et al., Micro & Nano Letters, 2006. **1**(1): p. 39-42.
12. Mashaghi, S., et al., International Journal of Molecular Sciences, 2013. **14**(2): p. 4242-4282.
13. Patil-Sen, Y., et al., Physical Chemistry Chemical Physics, 2004. **6**(7): p. 1562-1565.
14. Mezzenga, R., et al., Nature materials, 2005. **4**(10): p. 729-740.

15. Pardeike, J., A. Hommoss, and R.H. Müller, International Journal of Pharmaceutics, 2009. **366**(1): p. 170-184.
16. Kulkarni, C.V., et al., International Journal of Pharmaceutics, 2015. **479**(2): p. 416-21.
17. Gaunt, N.P., et al., Nanoscale, 2015. **7**(3): p. 1090-1095.
18. Salonen, A., F.o. Muller, and O. Glatter, Langmuir, 2010. **26**(11): p. 7981-7987.
19. Harris, F., et al., Medicinal Research Reviews, 2013. **33**(1): p. 190-234.

Nanomedicine in Neurodegenerative Disorders: Understanding the journey

Giovanni Tosi[1], Barbara Ruozi[1], Antonietta Vilella[2], Andreas M. Grabrucker[3,4], Daniela Belletti[1], Maria Angela Vandelli[1], Tobias M. Boeckers[1], Flavio Forni[1], Michele Zoli[2], [5]Aruna Sharma, [6]Dafin F Muresanu, [5]**Hari S Sharma***

[1] Pharmaceutical Technology, Te.Far.T.I. group, Department of Life Sciences, University of Modena and Reggio Emilia, Italy
[2] Department of Biomedical, Metabolic and Neural Sciences, University of Modena and Reggio Emilia, Italy
[3] WG Molecular Analysis of Synaptopathies, Neurology Dept., Neurocenter of Ulm University, Ulm, Germany
[4] Institute for Anatomy and Cell Biology, Ulm University, Ulm, Germany
[5] Laboratory of Cerebrovascular Research, Dept. Surgical Sciences, Anesthesiology & Intensive Care Medicine, University Hospital, Uppsala University, SE-75185 Uppsala, Sweden
[6]Dept. of Clinical neuroscience, University of Medicine & Pharmacy, Cluj-Napoca, Romania

Abstract

Nanocarriers can be useful tools for delivering drugs to the central nervous system (CNS). Their distribution within the brain and their interaction with CNS cells must be assessed accurately before they can be proposed for therapeutic use. We investigated these issues by employing poly-lactide-co-glycolide nanoparticles (NPs) specifically engineered with a glycopeptide (g7) conferring to NPs the ability to cross the blood brain barrier (BBB) at a concentration of up to 10% of the injected dose. g7-NPs display increased *in vitro* uptake in neurons and glial cells. *in vivo* administration of g7-NPs leads to a region- and cell type-specific enrichment of NPs within the brain. Moreover, g7-NPs are endocytosed in a clathrin-dependent manner and transported into a specific subset of early endosomes positive for Rab5 *in vitro* and *in vivo*. Moreover, in order to understand the journey of NPs, we demonstrated that g7-NPs can be transported intra- and intercellularly inside vesicles. Cell-to-cell transport is mediated by tunneling-nanotube (TNT)-like structures in cell lines and most interestingly in glial as well as neuronal cells in vitro. These in vitro findings were in part confirmed by in vivo evidence after i.p. administration in mice. We also tested Ab-modified g7-NPs both in vitro and in vivo to investigate the possibility of a specific targeting.

1. INTRODUCTION

The development of nanocarriers, such as liposomes and nanoparticles (NPs), able to reach brain cells after having crossed the blood brain barrier (BBB), is an important topic of the pharmaceutical research [1]. The crossing of the BBB has been widely investigated and approached by means of surface engineering of nanocarriers in healthy and diseased states allowing also to propose the pathway involved in BBB crossing (i.e. endocytosis/transcytosis) [2]. In this paper we would like to investigate in detail the fate of glycopeptide-enginereed poly(D,L-lactide-co-glycolide (PLGA) NPs (g7-NPs), that are able to cross the BBB after in vivo administration in rodents [3], by assessing their tropism for specific brain regions and cell types. In addition, we studied whether g7-NP uptake in neurons occurs through the same mechanisms in vivo and in vitro. Another topic, required for a more complete overview of the fate of NPs inside the CNS is their intra/inter-cellular trafficking and the possibility of governing and driving this trafficking.

2. MATERIALS and METHODS

2.1 Preparation of nanoparticles and characterization

Nanoparticles were prepared as described in literature with some modifications in the preparation procedure described before [3]. All the batches of NPs were characterized regarding their surface, chemico-physical, morphological properties, by photon correlation spectroscopy, scanning electron microscopy, atomic force microscopy and X-ray photoelectron spectroscopy.

2.2Animal procedures, immuno-histochemistry and in vivo endocytosis inhibition

g7-NPs or C-NPs suspended in saline solution were injected i.p. in C57Bl6 mice. After 10 min, 30 min, 1 hr, 2 hr, 6 hr, or 5 days, animals were sacrificed. For histological and histochemistry processing, we refer to established procedure [4,5]. A total of six C57Bl6 mice hippocampi/group were infused unilaterally with 1 uL of chlorpromazine hydrochloride (Sigma-Aldrich) (30 μM) or staurosporine (Sigma-Aldrich) (60 μM) to inhibit clathrin-dependent and caveolin-dependent endocytosis, respectively.

2.4 Primary cell cultures, cell measurements and statistical analysis

Hippocampal neuronal cultures were prepared using a modified Banker culture protocol as reported in previous papers [4,5]. Immunofluorescence and immune-histochemistry were applied. Images were taken with a spinning disk confocal microscope from Zeiss and a fluorescence microscope. Quantification of fluorescence data was performed using ImageJ 1.47d for Macintosh and Axiovision.

3. RESULTS and DISCUSSION

3.1 Nanoparticle characterization

All the batches of NPs produced were analyzed for the surface properties (size and surface charge) and morphology. The results obtained by PCS analysis showed that both g7-NPs and C-NPs were sizing around 170 nm (±18 nm), with a polydispersity index (PDI) with values close to 0.02 (± 0.01) indicating a good homogeneity in dimensions. The surface charge, expressed as zeta potential (z-p), showed slightly negative charge values (-18 mV ± 5 mV). The SEM and AFM images demonstrated that g7-NPs are featured by spherical shapes with regular surfaces, also confirming the size distribution assessed by PCS analysis.

3.2 Nanoparticle accumulation in brain areas and cell type: in vivo evidence

In order to clarify the possible accumulation of g7-NPs in specific brain areas or cell populations, we injected g7-NPs and analyzed samples of different brain areas by confocal microscopy. Six hours after the injection, g7-NPs were widely distributed across brain regions, with higher concentrations in grey matter areas than in white matter areas (**Fig.1a,b** cerebral cortex vs. corpus callosum). Next, we performed detailed analysis within the hippocampal region, using different cell-type specific markers (NeuN, NPY for neuronal populations, GFAP for astroglia and Iba-1 for microglia). By far the largest population of cells with clear g7-NP accumulation were NeuN-ir neurons (**Fig.1i**). A clear accumulation of g7-NPs was also observed in some Iba-1-ir microglia (**Fig.1k**), whereas g7-NPs could only rarely be attributed to GFAP-ir astroglia (**Fig.1l**). g7-NP neuronal accumulation was heterogeneous, with preferential accumulation in interneurons of the stratum oriens and lacunosum-moleculare of Cornu Ammonis (CA) subfields as well as of the hilus of the dentate gyrus (DG), especially NPY interneurons (**Fig.1j**). Regarding the other major classes of hippocampal neurons, the g7-NP labeling was clearly present in CA3 pyramidal neurons, though at a level lower than in NPY interneurons, but labeling was almost undetectable in granule cells of the DG or CA1 pyramidal neurons.

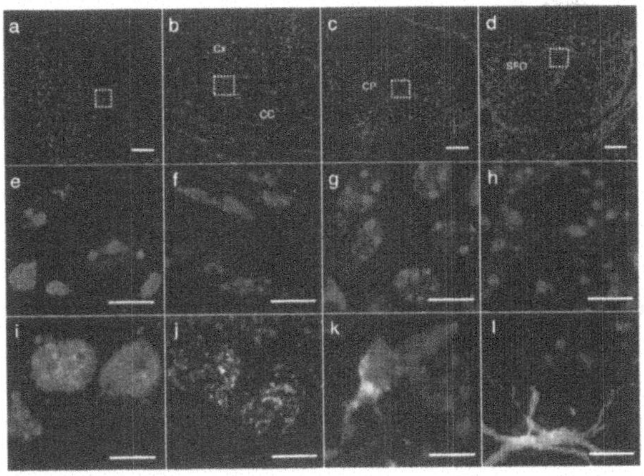

Figure 1

3.3 Nanoparticle uptake mechanisms

Recent studies reported that layered double hydroxide NPs internalization in neurons is mediated by clathrin-dependent endocytosis [6] and that NPs are targeted mainly to the clathrin endocytotic machinery [6,7]. Accordingly, g7-NP-containing vesicles showed characteristics of early endosomes derived from clathrin coated vesicles in primary hippocampal cultures and were negative for the caveola-related early endosome marker EEA1 (**Fig.2a-e**), but strongly positive for the clathrin-related early endosome marker Rab5 (**Fig.2f-j**).

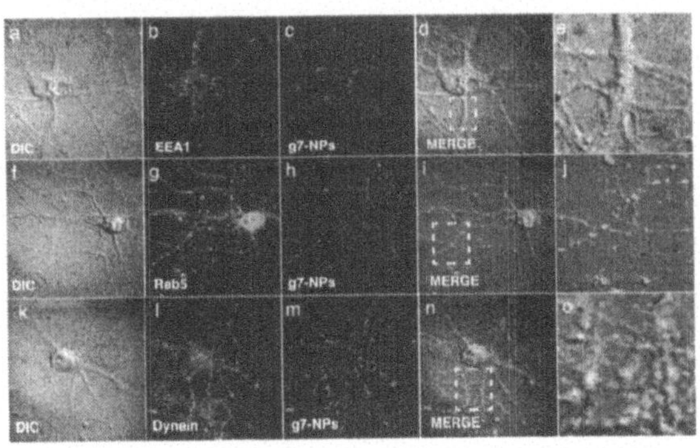

Figure 2

Analysis of the staining for clathrin revealed that a subset of g7-NPs in closer proximity to thc plasma mcmbranc colocalizcs with clathrin positive signals (**Fig.3a-d**). In line with previous studies showing that inhibition of clathrin-coated pit endocytosis decreases intracellular uptake, whereas inhibition of caveola-coated pit endocytosis does not affect NP uptake [4,5], chlorpromazine hydrochloride significantly reduced the amount of endocytosed g7-NPs, while nystatin had no significant effect (**Fig.3e-m**). However, some g7-NPs were still found intracellular, suggesting that additional mechanisms, such as membrane permeabilization or fluid-phase endocytosis [7] may provide additional mechanisms for g7-N cell entry.

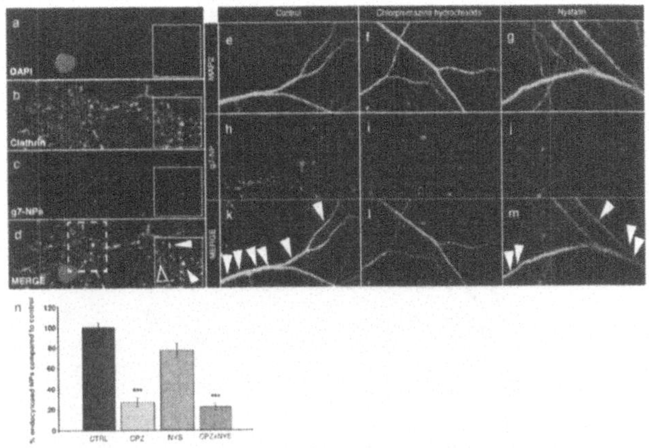

Figure 3

3.4 In vivo evidence

Based on these *in vitro* data, we sought to verify whether these uptake mechanisms of g7-NPs are active *in vivo* in the mouse brain upon peripheral g7-NP administration. First, we studied the possible co-localization of g7-NPs with the early endosome markers Rab5 and EAA1 (**Fig.4a-d**). In accordance with the *in vitro* data, there was an impressive co-localization of g7-NP and Rab5 signals in hippocampal interneurons (**Fig.4a,b**), but only minimal co-localization with EAA1 signals in the same cells (**Fig.4c,d**). Next, we tested whether neuronal uptake of g7-NPs *in vivo* is clathrin-dependent and/or caveolin-dependent. Thus, we infused chlorpromazine hydrochloride or staurosporine (an inhibitor of caveolin-dependent endocytosis) unilaterally within mouse hippocampus and 15 min later, injected g7-NPs (300 ag/300 μL /mouse) peripherally. Infusion of the respective vehicles was performed into the contra-lateral hippocampus. In accordance with the *in vitro* data, we observed a marked decrease of g7-NP signal accumulation within hippocampal neurons in chlorpromazine-treated hippocampus (**Fig.4e,f**) but not in staurosporine-treated hippocampus (**Fig.4g,h**).

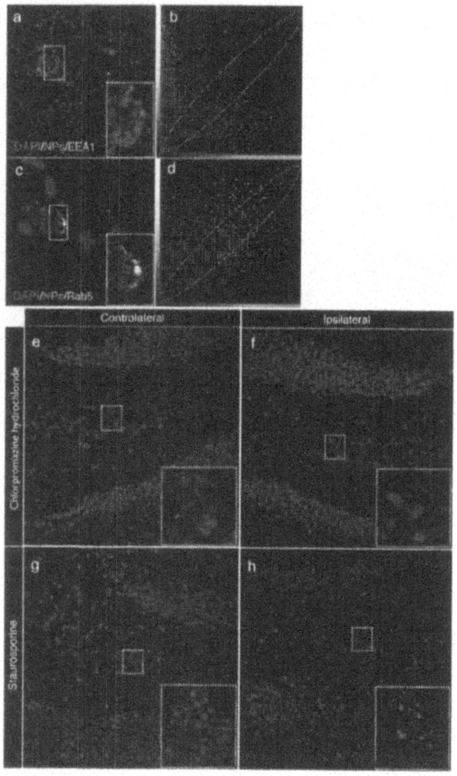

Figure 4

4. Conclusion

Drug delivery and targeting to the CNS represent a major challenge for pharmaceutical nanotechnologists. The fate of nanocarriers, once they have crossed the BBB, remains an open question, due to the lack of *in vivo* evidence and attempts to correlate *in vitro* and *in vivo* findings. Several insights were given in this research, regarding the dose/time-dependent accumulation of NPs in the brain, their regional distribution and cell tropisms, uptake mechanisms, inter- and intra-cellular transport, to finally show existing in vivo/in vitro correlations.

References

1. Tosi G. et al. Polymeric nanoparticles for the drug delivery to the Central Nervous System. Exp. Opin. Drug Deliv. 5 (2008) 155-174.

2.Sahay DY et al. Endocytosis of nanomedicines. J. Control. Rel. 145 (2010) 182–95.

3.Tosi G. et al. Investigation on the mechanisms for Blood-Brain Barrier crossing of brain targeted glycopeptides nanoparticles. Nanomedicine 6 (2011) 423-436.

4.Vilella A et al. Insight on the fate of CNS-targeted nanoparticles. Part I: Rab5-dependent cell-specific uptake and distribution. J Control Release 174 (2014) 195-201.

5.Tosi G et al. Insight on the fate of CNS-targeted nanoparticles. Part II: Intercellular neuronal cell-to-cell transport. J Control Release 177 (2014) 96-107.

6. Harush-Frenkel O et al. Surface charge of nanoparticles determines their endocytic and transcytotic pathway in polarized MDCK cells. Biomacromolecules 9 (2008) 435-443

7. Miaczynska M et al. APPL proteins link Rab5 to nuclear signal transduction via an endosomal compartment. Cell 116 (2004) 445-456

Acknowledgements. This investigation is partially supported by grants from the Air Force Office of Scientific Research (EOARD, London, UK), and Air Force Material Command, USAF, under grant number FA8655-05-1-3065; Swedish Medical Research Council (Nr 2710-HSS). The U.S. Government is authorized to reproduce and distribute reprints for Government purpose notwithstanding any copyright notation thereon. The views and conclusions contained herein are those of the authors and should not be interpreted as necessarily representing the official policies or endorsements, either expressed or implied, of the Air Force Office of Scientific Research or the U.S. Government or any of the granting organizations or collaborating entities mentioned above.

---

[5]*Hari S Sharma, Dr. Med. Sci. (UU), Director Int. Expt. CNS Injury & Repair (IECNSIR), University Hospital, Uppsala University, Prof. Neurobiology (MRC); Docent Neuroanatomy (UU); Frödingsgatan 12:28, SE-75421 Uppsala, Sweden, Phone & Fax: +46 18 243899, Cell Phone: +46 70 2011 801; Email: Sharma@surgsci.uu.se

Nanodelivery of cerebrolysin induces profound neuroprotection in heat stroke following chronic hypertension in combination with SiO2 and carbon nanoparticles induced exacerbation of brain damage

Dafin F Muresanu[1], Aruna Sharma[2], Ranjana Patnaik[3], Ala Nozar[4], Herbert Mössler[5], José Vicente Lafuente[6], Z Ryan Tian[7], Asya Ozikzilcik[7], **Hari S Sharma*[2]**

[1]Dept. of Clinical Neurosciences, University of Medicine & Pharmacy, Cluj-Napoca, Romania
*[2]Laboratory of Cerebrovascular Research, Dept. of Surgical Sciences, Anesthesiology & Intensive Care Medicine, University Hospital, Uppsala University, SE-75185 Uppsala, Sweden,
Email: Sharma@surgsci.uu.se, Aruna.sharma@surgsci.uu.se
[3]School of Biomedical Engineering, Indian Institute of Technology, Banaras Hindu University, Varanasi, India,
[4]Anesthesiology, Massachusetts General Hospital, Harvard University, Boston MA, USA,
[5]Ever NeuroPharma, Oberburgau, Austria,
[6]Dept of Neurosciences, University of Basque Country, Bilbao, Spain.
[7]Dept. Chemistry & Biochemistry, University of Arkansas Fayetteville, AR, USA

ABSTRACT

Military personals are often exposed to combat stress induced hypertension in an environment rich in various nanoparticles (NPs) emanating from different sources e.g., gun powder explosion or blast injuries. Furthermore, combat or peacekeeping activity in the Middle East region also exposes them to high environmental heat and silica dust. Thus, it is quite likely that all these combinations of hypertension, silica dust (SiO2 NPs) and carbon NPs following missile explosion or blast injuries in desert high heat lead to exacerbation of brain damage following heat stroke. This suggests that hypertension and/or NPs exposure in high environmental heat is dangerous for our soldiers as a combination of these factors may induce brain dysfunction and damage along with functional disturbances. In this innovation we have shown that a combination of hypertension, SiO2 NPs and carbon nanoparticles (CNPs) at hot environment aggravates brain damage. In this complex situations TiO2-nanowired delivery of multimodal drugs e.g., cerebrolysin is needed to contain neuronal damage and restoring the functional disturbances effectively.

Keywords: TiO2 nanowired Cerebrolsyin, Heat Stroke, SiO2 NPs, Carbon NPs, hypertension, brain pathology, blood-brain barrier, brain edema, neuroprotection

1 INTRODUCTION

Our military personals are often exposed to combat stress induced hypertension in an environment rich in various nanoparticles emanating from different sources e.g., gun powder explosion or blast injuries [1,2]. Furthermore, combat or peacekeeping activity in the Middle East region also exposes them to high environmental heat and silica dust [3,4]. Thus, it is quite likely that all these combinations of hypertension, silica dust (SiO2 NPs) and carbon NPs following missile explosion or blast injuries in desert high heat lead to exacerbation of brain damage following heat stroke.

Previously, we have shown that hypertensive animals when exposed to heat stress exhibit massive brain damage and deterioration of their sensory motor dysfunctions [4,5]. In addition, rats exposed to engineered metal nanoparticles (NPs) i.e., Ag, Cu or Al also exacerbated brain pathology following heat stroke. This suggests that hypertension and/or NPs exposure in high environmental heat is dangerous for our soldiers as a combination of these factors may induce brain dysfunction and damage along with functional disturbances.

This hypothesis was tested in our rat model of heat stroke where renal hypertensive rats were intoxicated with SiO2 and single walled carbon nanotubes (SWCNT) before heat exposure. Furthermore, to find out novel therapeutic measures in these complex situations with several co-morbidity factors, we used a multimodal drug Cerebrolysin (Ever Neuro Pharma, Austria). Cerebrolysin is a well-balanced composition of multiple neurotrophic factors and active peptide fragments needed to enhance neurorepair and neuroregeneration [1-3]. Thus, nanodelivery of cerebrolysin could be useful in inducing neuroprotection following heat stroke associated with hypertension and NPS exposure more effectively [1-4].

2 MATERIALS & METHODS

Experiments were carried out on Male Wistar Rats (200-300 g) housed at controlled ambient temperature (21±1°C) with 12 h light and 12 h dark schedule. Food and water were provided ad libitum before experiment. All the experiments were carried out according to the Guidelines & Care for laboratory animals as described by national Institute of Health and approved by local Institutional Ethics Committee.

2.1 Exposure to heat

Rats were subjected to heat exposure in Biological Oxygen Demand Incubator (BOD) maintained at 34°C for 4 h per day for 8 days. The relative humimiduty 45-47 % and wind velocity 20-25 cm/Sec were kept constant. Control rats were either placed in identical incubator at 21°C for 4 h for 8 days or kept at room temperature (21±1°C).

On the 9th day rats from various grous were exposed ti heat stroke at 38°C in the BOD incubator for 4 h [6].

2.2 Chronic hypertension

Chronic hypertension was produced by using a silver clip to constrict one renal artery leaving both the kidneys intact (2K1C). After 4 to 6 weeks, rats develop hypertension similar to those seen in clinical situations (Mean Arterial Blood Pressure, MABP 180±8 torr, Controls 110±6) [4].

2.3 Exposure to nanoprticles

Intoxication of SiO2 NPs and SWCNT (50-60 nm) was done by administering 50 mg/kg, i.p. dose of each NPs once daily for one week either at room temperature (21°C) or at 34° C (wind velocity 20-25 cm/sec, relative Humidity 45-47 %) for 8 days [3].

2.3 Treatment with Cerebrolysin

Cerebrolysin (Ever NeuroPharma, Austria) 5 ml/kg or 10 ml/kg was used once daily for 3 days from day 5,6 and 7th of heat exposure either at 34°C or in control rats after NPs intoxication or saline administration in normal or hypertensive rats. On 9th day brain pathology was determined in these treated groups [1-3].

In separate group of rats TiO2-nanowired cerebrolysin was also administered in a dose of 5 ml/kg under identical conditions. Cerebrolysin was tagged with TiO2- nanowires (Fig. 1) using standard procedures as described earlier [2].

In control groups, TiO2 nanowires alone was administered to examine any neurotoxicity of nanowires alone. Our observations show that TiO2 nanowires do not induce any brain pathology (results not shown).

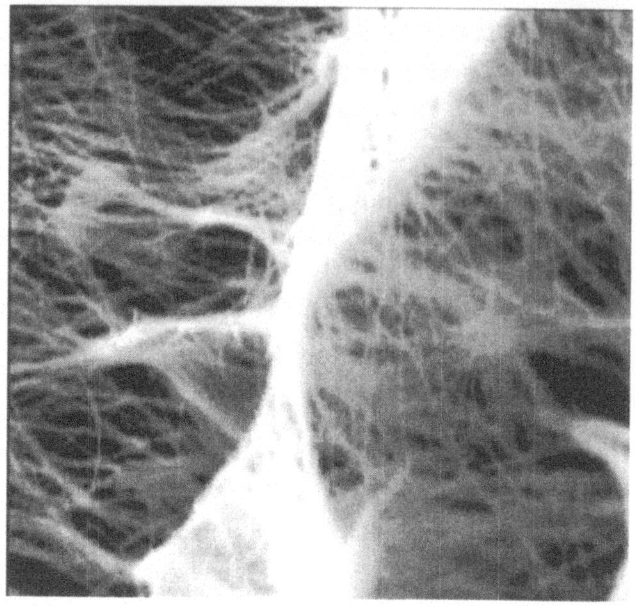

Fig. 2. TiO2 nanowired for Nanodelivery of Cerebrolysin

2.3 Parameters measured

The following parameters were measured in normal or hypertensive rats exposed to hot environments intoxicated with SiO2 NPs and SWCNT.

2.3.1 Blood-brain barrier

The blood-brain barrier (BBB) leakage was measured using Evans blue albumin (EBA, 2% of 0.3 ml/100g i.v.) and radioiodine ($^{[131]}$-Iodine], 10 μCi/100g i.v.) extravasation in the brain [2,3]. After washing out of intravascular tracer with 0.9 % saline through heart the brain were dissected out and examined for blue staining. After that tissue pieces from selected brain areas were then dissected and radioactivity determined in a Gamma Counter (Packard, USA). Leakage of these tracers was expressed as percentage increase in the brain over blood concentration [2,3].

2.3.2 Brain Edema formation

The brain edema was measured using brain water content. Desired tissue pieces from brain were dissected out and weighed to determine their wet weight. After that these tissue pieces were kept in an oven maintained at 90° C for 72 h to obtain their dry weight. The percentage water content was calculated from the differences between wet and dry weight of the samples [2,3].

2.3.3 Neuronal injury

Neuronal injury was evaluated using Nissl or Haematoxylin & Eosin (HE) staining using standard histopathological techniques [3]. For this purpose, animals were perfused in situ with 4 % buffered paraformaldehyde preceded with a brief saline rinse though cardiac puncture. Coronals sections of the brain were the cut passing through the hippocampus and embedded in paraffin. About 3 µm thick sections were cut and stained with HE or Nissl using commercial protocol [1-3]. The number of damaged or distorted neurons in specific anatomical brain areas were counted manually.

2.4 Statistical analyses

ANOVA followed by Dunnett's test for multiple group comparison with one control group was used to analyze statistical significance of the data obtained. A p-value less than 0.05 was considered significant.

3 RESULTS

3.1 Brain pathology in heat exposure

Hypertensive rats exposed to SiO2 NPs and SWCNT in combination at high environmental temperature exhibited 3- to 5-fold higher increase in BBB permeability in 10 brain regions to EBA and radioiodine aftre heat challenge (38°C for 4 h) as compared to those rats placed at room temperature.

The brain edema in 8 regions showed a 6- to 8-fold increase in NPs treated hypertensive rats at 34°C and neuronal and glial damages were enhanced by 4- to 6-fold that their counterparts placed at 21° C (results not shown). The heat exposed hypertensive rats following NPs exposure also showed severe functional disturbances in behavioral tests as compared to those placed at 21°C [Teble 1].

Interestingly, a combination of SiO2 NPs and SWCNT in hypertensive animals exhibited about 2-fold increase in BBB disruption, edema formation and brain pathology as compared to normotensive rats under identical conditions. [Table 1].

3.2 Cerebrolysin treatment

Cerebrolysin treatment (5 ml or 10 ml/kg, i.v.) did not reduce brain pathology or behavioral disturbances in hypertensive rats with NPs exposure following heat stress. However, mild to moderate reduction in behavioral functions and brain pathology was seen in this group of rats at 21°C.

On the other hand TiO2 nanowired cerebrolysin (5 ml/kg) was able to reduce brain pathology and behavioral dysfunction in heat exposed hypertensive rats treated with SiO2 and SWCNT. [See Fig. 2, Table 1].

Neural injury, neuronal loss, distorted neurons; perineuronal edema and sponginess of the neuropil were also markedly reduced by TiO2-nanowired cerebrolysin in heat-exposed animals after NPs intoxication [See Fig. 3].

In these animals treated with TiO2 nanowired cerebrolysin, glial and myelin reactions ware also reduced (Result not shown).

Table 1. Nanoparticles intoxication in hypertension exacerbates hyperthermia induced brain pathology.

Expt. Type	BBB [131]-Iodine %	Brain Edema Water Content %	Neural Injury Nr./section	Astrocytic Reaction GFAP+ cells Nr
Control	0.23±0.02	74.34±0.12	2±2	6±2
4 h Heat Stress 38°C	2.34±0.32**	80.56±0.12**	180±24**	204±28**
Hypertension+HS	2.98±0.12*#	81.78±0.11*#	240±14*#	267±21*#
HS+SiO2	2.89±0.12*#	81.45±0.23*#	220±18*#	256±12*#
HS+CNT	2.96±0.23*#	81.98±0.34*#	234±19*#	289±19*#
HY+HS+SiO2	3.34±0.12*#	82.34±0.56*#	298±12*#	304±21*#
HY+HS+CNT	3.67±0.22*#	82.98±0.32*#	356±32*#	402±29*#

*** P < 0.01. ** P <0.01 from control, # P <0.05 from Hypertensive (HY) group. Values±SD (5 to 8 rats at each point)**

4 DISCUSSION

Our observations are the first to show that nanowired cerebrolysin is the most powerful agent in reducing brain pathology in hypertensive rats reared at hot environments following their exposure to a combination of factors e.g., SiO2 and SWCNT after heat stroke. This indicates that military personnel that are exposed to various kinds of external (NPs) or internal insults (hypertension) at battlefield when gets additional heat stress they could develop greater brain pathology and abnormal behavioral dysfunction. In such situation, normal drugs are unable to treat them effectively [1-3]. Thus, the multimodal drug Cerebrolsyin that is a balanced composition of several neurotrophic factors and active peptide fragments could be of immense help [see 1-3]. In addition, our observation further show that nanodelivery of cerebrolysin could offer even much superior neuroprotection in multiple combinations of insults during heat stroke.

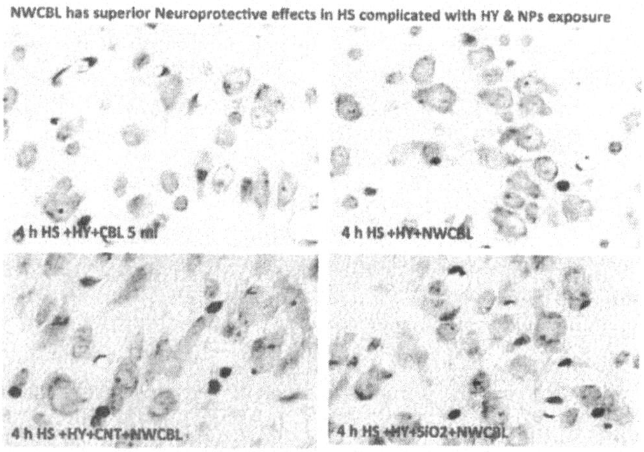

Fig. 2. Shows marked neuroprotection by TiO2-nanowired cerebrolysin (NWCBL) in hypertensive rats after NPs intoxication at hot environment. Nissl Stain x 40

The mechanisms by which a superior neuroproetction even in lower dose by nanodelivered cerebrolysin is seen in heat stroke is not well understood. However, it appears that TiO2 nanowired could reach the deeper brain tissues faster due to breakdown of the BBB [2]. Obviously, a greater BBB breakdown in hypertensive rats exposed to NPs following heat stress is in line with this idea [2,3,6]. Another possibility for superior effects of TiO2-nanodelivery of cerebrolsyin could be due to the fact that the drug could resist biological degradation because of its tight binding with TiO2 nanowires. This would results in gradual release of the drug from nanowires over a long period of time [2,3].

However, it remains to be seen whether TiO2 nanodelivery of cerebrolsyin following heat stroke in hypertensive group intoxicated with several NPs could also reduce brain pathology if given after loger therapeutic window, a feature that is currently being investigated in our laboratory.

5 CONCLUSION

In conclusion, our observations point out a new role of TiO2 nanowired cerebrolsyin in terateing complicated heat stroke cases in clinical situations. Since TiO2-nanodelivery of cerebrolsyin appears to be superior in nature, new studies are needed to evaluate safety and relaiability of TiO2 nanowires over longer time periods for its clinical usage.

6 ACKOWLEDGEMENTS

This investigation is supported by grants from the Air Force Office of Scientific Research (EOARD, London, UK), and Air Force Material Command, USAF, under grant number FA8655-05-1-3065; Swedish Medical Research Council (Nr 2710-HSS), Göran Gustafsson Foundation, Stockholm, Sweden (HSS), Astra Zeneca, Mölndal, Sweden (HSS/AS), Ministry of Science & Technology, People Republic of China (LF/HSS); The University Grants Commission, New Delhi, India (HSS/AS), Ministry of Science & Technology, Govt. of India (HSS/AS), Indian Medical Research Council, New Delhi, India (HSS/AS) and India-EU Co-operation Program (RP/AS/HSS) and IT 794/13 (JVL), Government of Basque Country and UFI 11/32 (JVL) University of Basque Country, Spain. The U.S. Government is authorized to reproduce and distribute reprints for Government purposes. The views and conclusions are those of the authors and should not reflect the official policies of granting organizations.

7 REFERENCES

[1] Sharma HS, Sharma A, Mössler H, Muresanu DF. Neuroprotective effects of cerebrolysin, a combination of different active fragments of neurotrophic factors and peptides on the whole body hyperthermia-induced neurotoxicity: modulatory roles of co-morbidity factors and nanoparticle intoxication. Int Rev Neurobiol. 2012;102:249-76.

[2] Sharma A, Muresanu DF, Mössler H, Sharma HS. Superior neuroprotective effects of cerebrolysin in nanoparticle-induced exacerbation of hyperthermia-induced brain pathology. CNS Neurol Disord Drug Targets. 2012 Feb;11(1):7-25. Review.

[3] Sharma HS, Muresanu DF, Patnaik R, Stan AD, Vacaras V, Perju-Dumbrav L, Alexandru B, Buzoianu A, Opincariu I, Menon PK, Sharma A.Superior neuroprotective effects of cerebrolysin in heat stroke following chronic intoxication of Cu or Ag engineered nanoparticles. A comparative study with other neuroprotective agents using biochemical and morphological approaches in the rat. J Nanosci Nanotechnol. 2011 Sep;11(9):7549-69.

[4] Muresanu DF, Zimmermann-Meinzingen S, Sharma HS. Chronic hypertension aggravates heat stress-induced brain damage: possible neuroprotection by cerebrolysin. Acta Neurochir Suppl. 2010;106:327-33.

[5] Sharma HS, Zimmermann-Meinzingen S, Johanson CE. Cerebrolysin reduces blood-cerebrospinal fluid barrier permeability change, brain pathology, and functional deficits following traumatic brain injury in the rat. Ann N Y Acad Sci. 2010 Jun;1199:125-37.

[6] Sharma HS, Hoopes PJ. Hyperthermia induced pathophysiology of the central nervous system. Int J Hyperthermia. 2003 May-Jun;19(3):325-54. Review.

[2]*Hari S Sharma, Dr. Med. Sci. (UU), Director Int. Expt. CNS Injury & Repair (IECNSIR), Professor of Neurobiology (MRC), Docent in Neuroanatomy (UU), University Hospital, Uppsala University, Frödingsgatan 12:28, SE-75421 Uppsala, Sweden, Phone & Fax: +46 18 243899, Cell Phone: +46 70 2011 801; Email: Sharma@surgsci.uu.se

TiO2 nanowired delivery of a Chinese traditional medicine DL-3-n-butylphthalide (DL-NBP) induces profound neuroprotection in concussive head injury

Lianyuan Feng[1], Aruna Sharma[2], Huang Yin[3], José Vicente Lafuente[4], Dafin F Muresanu[5], Z Ryan Tian[6], Asya Ozikzilcik[6], **Hari S Sharma*[2]**

[1]Department of Neurology, Bethune International Peace Hospital, Zhongshan Road (West), Shijiazhuang, Hebei Province, China;

*[2]Laboratory of Cerebrovascular Research, Dept. of Surgical Sciences, Anesthesiology & Intensive Care Medicine, University Hospital, Uppsala University, SE-75185 Uppsala, Sweden, Email: Sharma@surgsci.uu.se, Aruna.sharma@surgsci.uu.se

[3]CSPC NBP Pharmaceutical Medicine, Zhongshan Road (West), Shijiazhuang, Hebei Province, China

[4]Dept of Neurosciences, University of Basque Country, Bilbao, Spain

[5]Dept. of Clinical Neurosciences, University of Medicine & Pharmacy, Cluj-Napoca, Romania;

[6]Dept. Chemistry & Biochemistry, University of Arkansas Fayetteville, AR, USA

ABSTRACT

Our military personnel are often engaged in combat or peacekeeping operation across the World. These soldiers are prone to traumatic head or spine injury in the battlefield. Blunt head trauma is one of the most devastating injuries leading to instant death in more than 80% of the victims. Those who survive show lifetime disability. Thus, novel therapeutic strategies to treat blunt head trauma is the need of the hour. In this innovation we have explored the role of a Chinese traditional medicine DL-3-n-butylphthalide (DL-NBP) in a rat model of concussive head injury (CHI). We delivered this compound either alone or using Titanium nanowires delivery to explore the benefit of nanodrug delivery in CHI. Since CHI induces much more pronounced death and disability and so far therapeutic approaches are still elusive we showed that this traditional Chinese medicine has some values in CHI in reducing brain pathology. These neuroprotective effects of NBP are far superior when the drug was delivered using nanobiotechnology in CHI.

Keywords: DL-3-n-butylphthalide (DL-NBP), concussive head injury, brain patholoy, TiO2 nanowired drug delivery, blood-brain barrier, brain edema, neuroprotection,

1 INTRODUCTION

DL-3-n-butylphthalide (DL-NBP) a yellowish oil-like liquid (99.6% purified) is a Traditional Chinese Medicine used for various neurological ailments since ages. The basic mechanisms behind such a neuroprotective effects of DL-NBP are related to its possible strong antioxidative properties [1]. Since CHI cuased by motor vehicle accident or blunt trauma on the head during combat operations induces severe brain pathology and oxidative stress, use of NBP in such circumstances is needed [1-3]. In CHI cases rehabilitation and supportive strategies costs several millions of USD with only a mild improvement in the victims over time. Thus, we investigated DL-NBP in CHI to reduce brain pathology and behavioral dysfunction using nanotechnologies.

The DL-NBP is an extract from Chinese celery and is used in stroke patients showing marked improvement in their cognitive and mental health [1]. Since concussive head injury (CHI) is often fatal in more than 90% of cases involving our soldiers in the battlefield and the remaining 10 % of cases show lifetime disability there is an urgent need to find out better therapeutic use of novel medicine or use of available medicine in a more effective way. Keeping these views in mind we wanted to explore the role of DL-NBP in CHI using nanodrug delivery employing TiO2 nanowires in our rat model of CHI that induced profound brain pathology similar to that of clinical cases.

2 MATERIALS & METHODS

Experiments were carried out on Male Wistar Rats (200-300 g) housed at controlled ambient temperature (21±1°C) with 12 h light and 12 h dark schedule. Food and water were provided ad libitum before experiment. All the experiments were carried out according to the Guidelines & Care for laboratory animals as described by national Institute of Health and approved by local Institutional Ethics Committee.

2.1 Concussive head injury (CHI)

The CHI was induced by dropping a weight of 114.6 g on the right parietal skull bone over a distance of 20 cm in anesthetized rats resulting an impact of 0.224 N on the skull surface [Fig. 1]. This impact induces severe brain pathology over 4 h to 24 h. [3]. The animals were allowed to survive either 8 h or 24 h aftre trauma in this study.

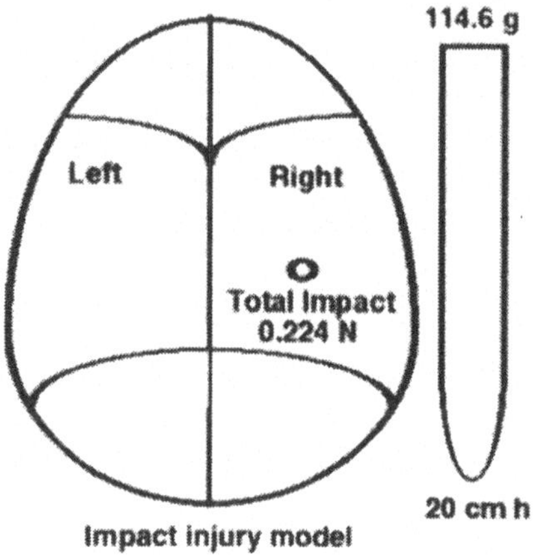

Fig. 1. Concussive head injury (CHI) model in rats

2.2 Treatment with DL-3-n-butylphthalide

DL-NBP was administered in a dose of 40 mg or 60 mg/kg, i.p. following 2 h and 4 h after injury in 8 h survival group and 8 h and 12 h after trauma in 24 h survival group. In separate group if animals DL-NBP was tagged with TiO2 nanowires using standard procedures [Ref 2; See Fig. 2] and TiO2-nanowired NBP was given in identical conditions in CHI in a dose range of 20ng or 40 mg/kg, i.p. [See Table 1].

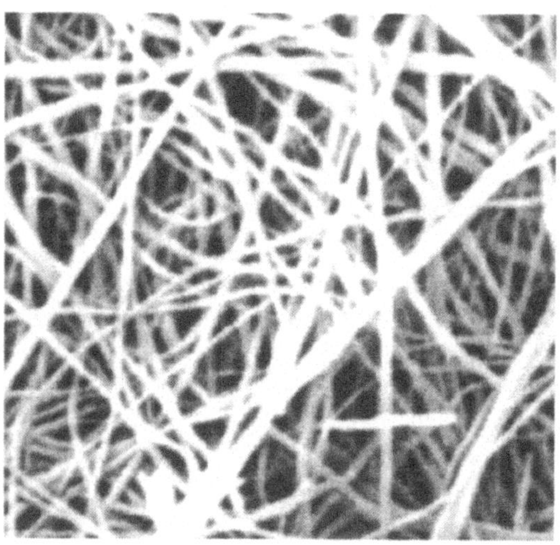

Fig. 2. TiO2 nanowired for tagging DL-3-n-butylphthalide

2.3 Parameters measured

The following parameters were measured in untreated or NBP terated traumatized rats and the data were compared with saline treated control groups under identical conditions.

2.3.1 Blood-brain barrier

The blood-brain barrier (BBB) leakage was measured using Evans blue albumin (EBA, 2% of 0.3 ml/100g i.v.) and radioiodine ($^{[131]}$-Iodine], 10 μCi/100g i.v.) extravasation in the brain [2,3]. After washing out of intravascular tracer with 0.9 % saline through heart the brain were dissected out and examined for blue staining. After that tissue pieces from selected brain areas were then dissected and radioactivity determined in a Gamma Counter (Packard, USA). Leakage of these tracers was expressed as percentage increase in the brain over blood concentration [2,3].

2.3.2 Brain Edema formation

The brain edema was measured using brain water content. Desired tissue pieces from brain were dissected out and weighed to determine their wet weight. After that these tissue pieces were kept in an oven maintained at 90° C for 72 h to obtain their dry weight. The percentage water content was calculated from the differences between wet and dry weight of the samples [2,3].

2.3.3 Neuronal injury

Neuronal injury was evaluated using Nissl or Haematoxylin & Eosin (HE) staining using standard histopathological techniques [3]. For this purpose, animals were perfused in situ with 4 % buffered paraformaldehyde preceded with a brief saline rinse though cardiac puncture. Coronals sections of the brain were the cut passing through the hippocampus and embedded in paraffin. About 3 μm thick sections were cut and stained with HE or Nissl using commercial protocol [3]. The number of damaged or distorted neurons in specific anatomical brain areas were counted manually.

2.4 Statistical analyses

ANOVA followed by Dunnett's test for multiple group comparison with one control group was used to analyze statistical significance of the data obtained. A p-value less than 0.05 was considered significant.

3 RESULTS

3.1 CHI induced Neurotoxicity

We observed a progressive increase in the BBB breakdown from 4 h after CHI that continued to enhance by 200 % at 8 h, 400 % after 12 h and 600 % at 24 h after injury. The neuronal damages were also showed massive increase by 8 h after CHI (+4 to 6 fold) followed by +7 to +8 fold at 12 h and +9 to +10 fold at 24 h in the cortical and subcortical areas. There was a voluminous increase in brain swelling by 1 % at 4 h, 2 % at 8 h, 4% at 12 h and 6 % at 24 h as compared to intact controls.Leakage of Evans blue and radioioidine across the BBB was also increased by 4 to 8 fold in CHI group as compared to saline treated controls [Table 1].

3.2 NBP induced Neuroprotection

In separate group of rats with CHI, we administered NBP (40 or 60 mg/kg, i.p.) 2h and 4 h after injury in 8 h survival group and 8 h and 12 h after trauma in 24 h survival group. We found that NBP in 40 mg was able to reduce BBB breakdown, brain edema formation and brain pathology after 8 h CHI, whereas, 60 mg dose of NBP was needed for identical neuroprotection in 24 h CHI group. Interestingly, TiO2 Nanowired NBP requires only 20 mg/kg in 8 h group and 40 mg/kg NBP in 24 h group for effective neuroprotection indicating a potential enhanced effectiveness of DL-NBP following TiO2 nanowired delivery in CHI [See Fig. 3, Table 1].

Neural injury, neuronal loss, distorted neurons; perineuronal edema and sponginess of the neuropil were also markedly reduced by TiO2-nanowired NBP at 12 and 24 h with lower doses of the compound (20 and 40 mg respectively) as compared to normal NBP drug [See Fig. 3].

The left uninjured half exhibited better neuroprotection by TiO2-Nanowired NBP as compared to right injured side.

Also reduction in leakage of Evans blue, radioiodine and volume swelling of the brain correlated well with decreased neuronal injury in TiO2-nanowired NBP treated CHI group (Result not shown).

Table 1. Neuroprotective effects of TiO2-nanowired delivery of DL-3-n-butylphthalide (DL-NBP) in concussive head injury (CHI)

Expt. Type	Control	CHI 12 h	CHI 24 h	CHI 24 h+ (DL-NBP) 60 mg	CHI 24 h+ TiO2-(DL-NBP) 40 mg
EBA mg%	0.24±0.06	1.86±0.11**	2.76±0.18**#	1.04±0.12*a	0.74±0.11*#b
[131]-Iodine %	0.30±0.08	1.94±0.13**	2.96±0.14**#	1.15±0.10*a	0.90±0.09*#b
Brain water %	74.76±0.18	77.64±0.24**	81.33±0.16**#	76.19±0.27a	75.06±0.13#b
Brain Swelling % *f*	0	+12**	+28**#	+8*a	+2*#b
Neuron Injury Nr	2±3	189±25**	350±36**#	98±14*a	36±14*#b

Values are Mean±SD of 5 to 6 rats, CHI = Concussive head injury; DL-NBP = DL-3-n-butylphthalide;
** P <0.05, ** P <0.01 from control, # P <0.05 from CHI 12 h, a P <0.05 from CHI 24 h, b P <0.05 from CHI+NBP. For details see text.*

4 DISCUSSION

Our observations are the first to show a new role of DL-NBP in treating CHI with regard to pathophysiology of brain injury. In addition, the nanowired delivery of NBP appears to be far superior to NBP alone in treating CHI induced brain pathology, not reported earlier.

The probable mchansims behind NBP induced neuroprotection in CHI is not well understood. However, available evidences suggests that NBP could play a key role in reducing oxidative stress, production of free radicals and generation of nitric oxide in the brain [1]. Obviosuly, a reducion in these oxidative stress parameters in CHI could be largely responsible for the neurprotective effects of NBP in this investigation [1-3]. However, to furtejr confirm this point, measurement of oxidative parameters in CHI is needed.

Another important point came out from this investigation is a better neuroprotection of the contralateral half of the brain aftre CHI [Fig. 3] aftre NBP treatment. This suggests that NBP could prevent oxidative stress in the uninjured brain much better than the directly injured side. This property of NBP could be of great clinical importantce as the drug may be able to reduce brain pathology clinical cases of CHI where uninjured side showes greater damage.

The mechanisms by which a superior neuroproetction even in lower dose by nanodelivered NBP is seen in CHI is not well understood from this investigation. However, it appears that NBP when given with TiO2 nanowired material, the drug could reach the deeper brain areas faster due to breakdown of the BBB [2]. Obciously, a greater BBB breakdown in CHI is seen in the uninjured half. Another possibility for superior effects of TiO2-nanodelivery of NBP is due to the fact that the drug could resisi biological degradation due to its tight binding with TiO2 nanowires. Thus, the drug may be acting for long time once enetered into the brain [2,3].

However, it remains to be seen whether TiO2 nanodelivery of NBP aftre CHI could also reduce behavioral and cognitive functions, a feature that is currently being investigated in our laboratory.

5 CONCLUSION

In conclusion, our observations point out a new role of NBP in terateing CHI cases in clinical situations. Furtehrmore TiO2-nanodelivery of NBP appears to be superior in nature. Thus, new studies are needed to evalute safety and relaiability of nanodelivery of NBP in clinical situations.

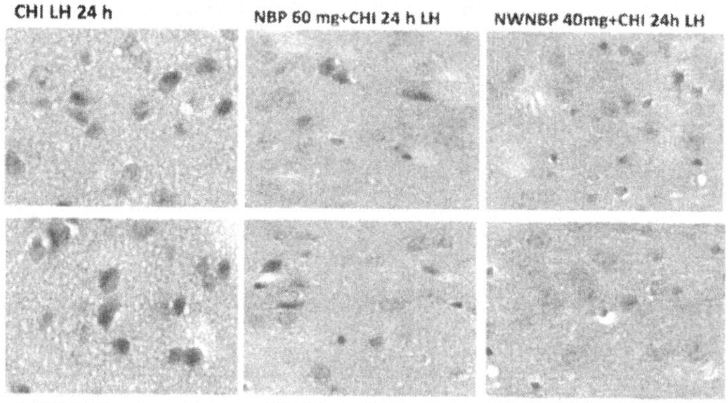

Fig. 3. Shows superior neuroprotection by TiO2-nanowired NBP at 24 h after CHI in lower doses as compared to normal NBP at identical period with higher doses. H&E stain on paraffin sections on parietal cerebral cortex. x 40.

6 ACKNOWLEDGEMENTS

This investigation is supported by grants from the Air Force Office of Scientific Research (EOARD, London, UK), and Air Force Material Command, USAF, under grant number FA8655-05-1-3065; Swedish Medical Research Council (Nr 2710-HSS), Göran Gustafsson Foundation, Stockholm, Sweden (HSS), Astra Zeneca, Mölndal, Sweden (HSS/AS), Ministry of Science & Technology, People Republic of China (LF/HSS); The University Grants Commission, New Delhi, India (HSS/AS), Ministry of Science & Technology, Govt. of India (HSS/AS), Indian Medical Research Council, New Delhi, India (HSS/AS) and India-EU Co-operation Program (RP/AS/HSS) and IT 794/13 (JVL), Government of Basque Country and UFI 11/32 (JVL) University of Basque Country, Spain. The U.S. Government is authorized to reproduce and distribute reprints for Government purposes. The views and conclusions are those of the authors and should not reflect the official policies of granting organizations.

7 REFERENCES

[1] Li L, Zhang B, Tao Y, Wang Y, Wei H, Zhao J, Huang R, Pei Z. DL-3-n-butylphthalide protects endothelial cells against oxidative/nitrosative stress, mitochondrial damage and subsequent cell death after oxygen glucose deprivation in vitro. Brain Res. 2009 Sep 22;1290:91-101.

[2] Sharma HS, Patnaik R, Patnaik S, Mohanty S, Sharma A, Vannemreddy P. Antibodies to serotonin attenuate closed head injury induced blood brain barrier disruption and brain pathology. Ann N Y Acad Sci. 2007 Dec;1122:295-312.

[3] Sharma HS, Sharma A. Nanowired drug delivery for neuroprotection in central nervous system injuries: modulation by environmental temperature, intoxication of nanoparticles, and comorbidity factors. Wiley Interdiscip Rev Nanomed Nanobiotechnol. 2012 Mar-Apr;4(2):184-203. doi: 10.1002/wnan.172. Epub 2011 Dec 8. Review.

---

[2]*Hari S Sharma, Dr. Med. Sci. (UU), Director Int. Expt. CNS Injury & Repair (IECNSIR), Professor of Neurobiology (MRC), Docent in Neuroanatomy (UU), University Hospital, Uppsala University, Frödingsgatan 12:28, SE-75421 Uppsala, Sweden, Phone & Fax: +46 18 243899, Cell Phone: +46 70 2011 801; Email: Sharma@surgsci.uu.se

Mn nanoparticles exacerbates concussive head injury induced blood-brain barrier disruption, brain pathology, and sensory motor dysfunctions. Neuroproetction by Nanowired delivery of 5-HT2 receptor antagonists Ketanserin and Ritanserin

[1]**Aruna Sharma**, [2]Lianyuan Feng, [3]Dafin F Muresanu, [4]Jose Vicente Lafuente, [5]Z Ryan Tian, [5]Asya, Ozkizilcik, [6]Mark A Smith, [1]Hari S Sharma*

[1*]Cerebrovascular Research Laboratory, Department of Surgical Sciences, Anesthesiology & Intensive Care Medicine, Uppsala University Hospital, SE-75185, Uppsala, Sweden; Email. Aruna.Sharma@surgsci.uu.se; Sharma@surgsci.uu.se

[2]Department of Neurology, Bethune International Peace Hospital, Zhongshan Road (West), Shijiazhuang, Hebei Province, China;

[3]Dept. Clinical Neuroscience, University of Medicine & Pharmacy, Cluj-Napoca, Romania

[4]Lab Neurociencias Clínicas y Experimentales (LaNCE)), Dpto. de Neurociencias, Universidad del País Vasco - EuskalHerriko Unibertsitatea, Apdo. 699, 48080-Bilbao, Spain;

[5]Department of Chemistry and Biochemistry, J. William Fulbright College of Arts and Sciences, University of Arkansas Fayetteville in AR, USA

[6]Dept of Pathology, Case Western Reserve University School of Medicine, Cleveland, OH, USA

ABSTRACT

This innovation deals with adverse effects of Mn nanoparticles (NPs) on brain pathology in concussive head injury. Also Mn nanoparticles treatment alters therapeutic modalities. The novel data presented in the abstract show that Mn exposure to our military personal from various industries, welding or deep mine work could make them more vulnerable to any additional head injury. This work also for the first time showed that neuroprotective agents in Mn treated traumatized group require in high concentration to treat effectively the brain pathologies. These data are strategically very important for further development of industrial or military medicine for health benefits. We have shown that nanowired delivery of 5-HT2 receptor antagonists e.g., ketanserin and ritanserin remarkably attenuated exacerbation of brain pathology after CHI in MN NPs treated group, not reported earlier.

Keywords: Mn NPs, brain pathology, blood-brain barrier, military, ketanserin, ritanserin, neuroprotection

1 INTRODUCTION

Manganese (Mn) is an essential component for human brain function. However, excessive exposure of Mn may induce neurotoxicity [1-3]. Our military personnel are often exposed to Mn nanoparticles (NPs) from industrial sources, welding machine, battery assembly, hazardous wastes and/or deep mines resulting in physical disability and/or mental sickness that correlates well with their blood level of Mn [2,3]. However, whether additional injury or head trauma in these Mn exposed personals will induce exacerbation of brain pathology or alter treatment modalities are still unclear [3,4]. In present investigation effects of Mn NPs on Blood-brain barrier (BBB) breakdown, edema formation and cell injury was examined in a rat model of closed head injury (CHI). In addition, disturbances in sensory motor functions are also evaluated in Mn NPs treated normal and traumatized rats.

2 MATERIALS & METHODS

Experiments were carried out on Male Wistar rats (200-250 g body weight) housed at controlled room temperature (21±1°C) with 12 h light and Dark Schedule. Food and water were supplied ad libitum before experiment. All the experiments were carried out according to the Guidelines & Care for laboratory animals as described by national Institute of Health and approved by local Institutional Ethics Committee.

2.1 Exposure to MnNPs

Mn NPs (30-40 nm size, Product Nr # 45508, Alfa Aesar, Karlsruhe, Germany) were administered in a dose of 10 or 20 mg/kg, i.p. once daily for 7 days. Control rats received saline instead of MnNPs. On the 8th day, one group of Mn NPs or saline treated treated rats were subjected to CHI.

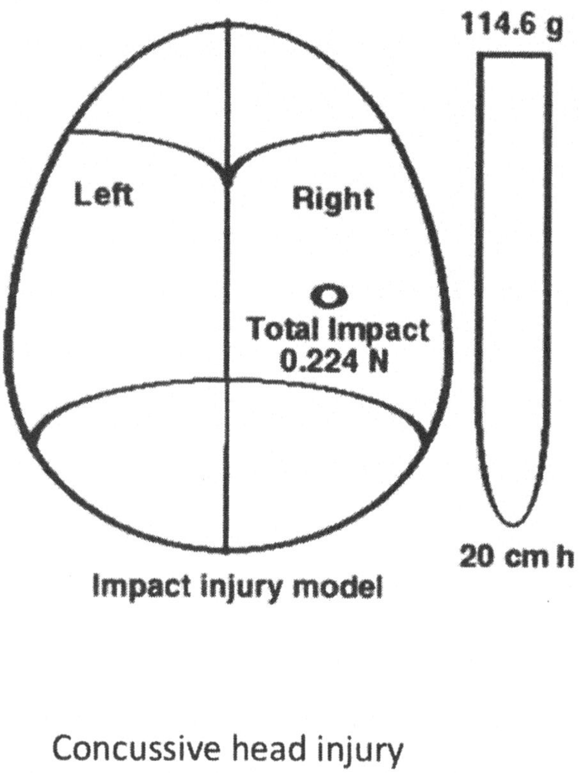

Fig. 1. Rat model of Concussive head injury (CHI). A metal weight is dropped over the right parietal skull inducing an impact force of 0.224 N resulting in more severe damage on the left side due to concussion [For details see Text].

2.2 Concussive head injury

CHI was produced by dropping a weight of 114.6 g from a 20 cm height through a guided tube over the right parietal bone under Equithesin anesthesia (3 ml/kg, i.p.) [Fig. 1]. This combination delivers an impact of 0.224 N on the skull surface resulting in very similar to clinical concussion as evident with severe damage of contralateral side (left half) aftre injury [4,5].

2.3 Nanodelivery of drugs

We used normal (2 or 4 mg/kg, i.p.) or TiO2 nanowired Ketanserin or Ritanserin (1 or 2 mg/kg, i.p.) in normal or brain injured rats following MnNPs exposure. The drugs were given in separate groups of animals once daily starting from 3 days after MnNPs treatment and continued until 7th days [5,6]. In these treated and untreated traumatized rats neuropathological and behavioral parameters were evaluated (see below).

2.3.1 Blood-brain barrier

The BBB leakage was measured using Evans blue albumin (EBA, 2 % 3 ml/kg i.v.) and radioiodine ([131]-Iodine] 100 μCi/kg, i.v.) extravasation in the brain given 10 min before termination of the experiment [5]. After intravascular washout of the tracers with 0.9 % saline in situ, the brains were dissected out and examined for blue staining and the radioactivity determined in a Gamma Counter (Packard, USA). Leakage of these tracers in brain was expressed as percentage increase over blood concentration [3,4,6].

2.3.2 Brain Edema formation

The brain edema formation was measured using water content in the brain by obtaining wet and dry weights of samples. According to standard protocol [5,6].

2.3.3 Neuronal injury

For this purpose, animals were perfused in situ with 4 % buffered paraformaldehyde preceded with a brief saline rinse though cardiac puncture and coronals sections of the brain were cut and embedded in paraffin. Using Nissl or Haematoxylin & Eosin (HE) staining on 3-μm thick paraffin sections obtained from brains of different groups neuronal damages were analyzed. [5,6].

2.3.4 Rota Rod performance

In all these animals, behavioral dysfunctionswas examined on a Rota Rod apparatus (Harvard Apparatus) manually using a speed of 16 rpm as described earlier [6].

2.3.5. Inclined plane angle test

Animals from different groups wer placed on an inclined plane angle apparatus (at 60°) and exmimed their stay on the plane for 5 sec. Those animals that could not stand on this steep plane the angle of the plane was lowered down to record their saty and noted. [6]

2.3.5. Walking on a mesh grid

Animal were allowed to walk on a steel mesh grid inclined at 45° and faulty placement of forepaw was recorded over a period of 120 sec. Normal animals could walk easily without any problems [5,6].

2.4 Statistical analyses

ANOVA followed by Dunnett's test for multiple group comparison with one control group was used to analyze statistical significance of the data obtained. A p-value less than 0.05 was considered significant.

Table 1. Neuroprotective effects of TiO2-nanodelivered 5-HT2 receptor antagonists Ketanserin and Ritanserin on Mn NPs induced brain pathology in CHI.

Expt. Type	Control	CHI 5 h LH	CHI+MnNPs LH	CHI+MnNPs +KTN LH	TiO2-KTN+ +CHI+MnNPs LH
		A. Neuropathological Parameters			
EBA mg%	0.24±0.08	2.34±0.10**	3.67±0.14**#	1.04±0.09*a	0.54±0.16*#b
[131]-Iodine %	0.34±0.08	2.83±0.16**	3.98±0.12**#	1.68±0.11*a	0.64±0.13*#b
Brain Water %	75.36±0.12	81.54±0.24**	83.28±0.34**#	77.34±0.26*a	75.88±0.22*#b
Neuron Injury Nr	1±2	284±65**	485±72**#	122±43*a	24±12*#b
		B. Behavioral Parameters			
Rota-Rod Performances Sec	120±3	74±6**	56±8**#	94±8*a	112±8*#b
Inclined Plane Angle test °	60±2	34±3**	23±3**#	48±5*a	54±3*#b
Placement Error Nr	1±1	15±2**	24±4**#	8±32*a	3±2#b

*Values are Mean±SD of 5 to 6 rat; .LH = Left Half, non –traumatized; KTN = Ketanserin, * P <0.05, ** P <0.01 from control, # P <0.05 from CHI =Concussive head injury, a P <0.05 from CHI+MnNPs, b P <0.05 from CHI+MnNPs+TiO2-KTN = nanodelivered Ketanserin. For details see text.*

3 RESULTS

3.1 MnNPs induce neurotoxicity

Normal animals treated with MnNPs exhibited cognitive and motor dysfunction on the 8th day [Table 1]. At this time, BBB disruption, brain pathology and brain edema formation were most marked in the sensory-motor cortex (+134 %), hippocampus (+180 %), caudate putamen (+87%), cerebellum (+167%) and thalamus (+98%), hypothalamus, (+56%) pons (+52%), medulla (+48%) and spinal cord (+156%). Neuronal injuries were present in all these brain regions [Table 1].

3.2 MnNPs exacerbate CHI pathology

MnNPs intoxicates rats aftre CHI exhibited about 2- to 4-fold exacerbation of BBB breakdown, edema formation and neuronal injuries. In these animals, the cognitive and sensory motor functions were further worsened than CHI in normal animals [see Table 1].

These adverse effects of MnNPs in CHI are dose dependent [results not shown].

3.3 Effect of Drug Treatment

Treatment with potent neuroprotective compounds ketanserin or ritanserin (selective 5-HT2 receptor antagonists) induced marked neuroporetction in normal animals exposed to MnNPs. This protective effects of 5-HT2 receptro antagonists was dose dependent [Fig. 2]. However, when these drugs were administered in MnNPs intoxicated rats aftre CHI, high doses of the drugs (4 mg) is needed to induce appreciable neuroproetction. High dose of the drugs (4 mg instead of 2 mg) was also needed to improve behavioral responses after CHI in MnNPs exposed rats [Table 1].

On the other hand, when TiO2-nanowired ketanserin or ritanserin were adminsitered marked neuroportection was seen with lower doses of the drugs (1 or 2 mg/kg, i.p.) in CHI rats aftre MnNPs intoxication. These animals also showed remarkable improvement in behavioral functions [Table 1]. It appears that ritanserin has more superior effects than ketanserin at idenitical doses in CHI induced barin pathology in normal or MNNPs intoxicated rats [results not shown].

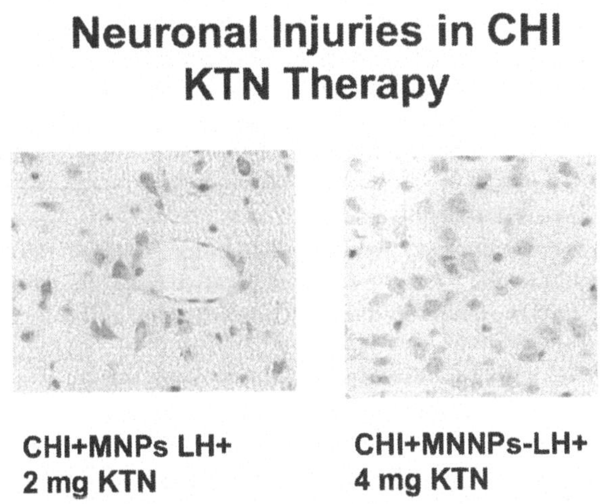

Fig. 2. Shows Higher dose of Ketanserin (4 mg) is able to reudce neuronal injuries in the contralateral half (Left half, LH) aftre concussive head injuruy (CHI). Nissle staining on Paraffin section, x 40.

4 DISCUSSION

Our novel findings clearly show that MnNPs not only induce neurotoxicity in normal animals but also exacerbate brain pathology and behavioral functions after brain injury. This suggests that exposure to MnNPs to healthy populations or military personnel during work related activities could be more dangerous in terms of their mental health following any kinds of head injury or insults to the central nervous system (CNS). Our results are this first to point out that the pathological processes following CHI are exacerbated by prior MnNPs exposure.

The probable mechanisms underlying MnNPs exposure and exacerbation of CHI induced brain pathology are still unclear. However, available evidences show that exposure of MnNPs induces oxidative stress in the brain and alters glutamine (Gln)/glutamate (Glu)-γ-aminobutyric acid (GABA) cycle (GGC) between astrocytes and neurons leading to neurotoxicity [3,4]. Furthermore, an increased production of nitric oxide synthase (NOS) occurs following MnNPs exposure hat may also contribute to damage of neurovascular units [2-4]. A breakdown of the BBB to serum proteins in this investigation is in line with this idea. Consequently brain edema develops and cell injury occurs due to exposure of neurons, astrocytes to unwanted materials from the blood [5,6].

Our observations further show that two potent serotonin 5-HT2 receptor antagonists ketanserin and ritanserin when are able to thwart MnNPs induced neurotoxicity in normal animals. This suggests that serotonin is somehow involved in Mn neurotoxicity. Blockade of 5-HT2 receptors are known to reduce BBB breakdown, edema formation and cell injury [5,6]. This suggests that apart from oxidative stress and altered amino acid metabolism, serotonin could play key roles. A massive increase in plasma and brain serotonin in CHI is associated with brain pathology and behavioral dysfunctions further support this hypothesis [5].

However, when CHI was induced in MnNPs exposed rats, these 5-HT2 receptor blockers were not enough to contain brain pathology. In this group TiO2 nanowired delivery of ketanserin or ritanserin is needed to induce desired neuroprotection. This suggests that nanodelivery of compounds may have superior neuroprotective effects than the parent compounds [see 6].

An improved behavioral function in MnNPs treated traumatized rats by TiO2-nanowoired ketanserin or ritanserin could be due to neuroprotection offered by these agents. Obviously, damage to sensory motor pathways due to edema formation and BBB breakdown resulting in neural injuries contributes to behavioral dysfunction.

5 CONCLUSION

In conclusion, our investigations for the first time point out a role of 5-HT2 receptor blockers in MnNPs neurotoxicity and further show that exacerbation of brain pathology following CHI in MnNPs exposed subjects may beconta8ned by TiO2 nanowired delivery of these drugs achieve better neuroprotection that the parent compounds. The findings have immense clinical significance in military medicine as well as in those populations who are exposed to MnNPs in their work-related environments.

6 ACKNOWLEDGEMENTS

This investigation is partially supported by grants from the Air Force Office of Scientific Research (EOARD, London, UK), and Air Force Material Command, USAF, under grant number FA8655-05-1-3065; Swedish Medical Research Council (Nr 2710-HSS), IT 794/13 (JVL), Govt. of Basque Country and UFI 11/32 (JVL) University of Basque Country, Spain. Technical assistance of Mari-Anne Carlsson and Ingmarie Olsson of Uppsala University is highly appreciated. The U.S. Government is authorized to reproduce and distribute reprints for Government purpose notwithstanding any copyright notation thereon. The views and conclusions are exclusively those of the authors and should not reflect the official policies or endorsements of the Air Force Office of Scientific Research or the U.S. Government or any of the granting organizations or collaborating entities mentioned above.

7 REFERENCES

[1] Bowman AB, Kwakye GF, Herrero Hernández E, Aschner M. Role of manganese in neurodegenerative diseases. J Trace Elem Med Biol. 2011 Dec;25(4):191-203. Review.

[2] Rivera-Mancía S, Ríos C, Montes S. Manganese accumulation in the CNS and associated pathologies. Biometals. 2011 Oct;24(5):811-25. Review.

[3] Racette BA, Aschner M, Guilarte TR, Dydak U, Criswell SR, Zheng W. Pathophysiology of manganese-associated neurotoxicity. Neurotoxicology. 2012 Aug;33(4):881-6.

[4] Sidoryk-Wegrzynowicz M, Aschner M. Manganese toxicity in the central nervous system: the glutamine/glutamate-γ-aminobutyric acid cycle. J Intern Med. 2013 May;273(5):466-77. Review.

[5] Sharma HS, Patnaik R, Patnaik S, Mohanty S, Sharma A, Vannemreddy P. Antibodies to serotonin attenuate closed head injury induced blood brain barrier disruption and brain pathology. Ann N Y Acad Sci. 2007 Dec;1122:295-312.

[6] Sharma HS, Sharma A. Nanowired drug delivery for neuroprotection in central nervous system injuries: modulation by environmental temperature, intoxication of nanoparticles, and comorbidity factors. Wiley Interdiscip Rev Nanomed Nanobiotechnol. 2012 Mar-Apr;4(2):184-203. Review.

---

[1] *Hari S Sharma, Dr. Med. Sci. (UU), Director Int. Expt. CNS Injury & Repair (IECNSIR), University Hospital, Uppsala University, Prof. Neurobiology (MRC); Docent Neuroanatomy (UU); Frödingsgatan 12:28, SE-75421 Uppsala, Sweden, Phone & Fax: +46 18 243899, Cell Phone: +46 70 2011 801; Email: Sharma@surgsci.uu.se

TiO2 Nanowired Cerebrolysin enhances neuroprotective effects of mesenchymal stem cells following concussive head injury at hot environments

Hari S Sharma*[1], Dafin F Muresanu[2], Lianyuan Feng[3], José Vicente Lafuente[4], Ranjana Patnaik[5], Z Ryan Tian[6], Asya Ozikzilcik[6], Herbert Mössler[7], Aruna Sharma[1]

*[1]Laboratory of Cerebrovascular Research, Dept. of Surgical Sciences, Anesthesiology & Intensive Care Medicine, University Hospital, Uppsala University, SE-75185 Uppsala, Sweden,
Email: Sharma@surgsci.uu.se, Aruna.sharma@surgsci.uu.se
[2]Dept. of Clinical Neurosciences, University of Medicine & Pharmacy, Cluj-Napoca, Romania;
[3]Bethune International Peace Hospital, Shijiazhuang, Hebei, China
[4]Dept of Neurosciences, University of Basque Country, Bilbao, Spain
[5]Dept. of Biomaterials, School of Biomedical Engineering, Indian Institute of Technology, Banaras Hindu University, Varanasi-221005, India
[6]Dept. Chemistry & Biochemistry, University of Arkansas Fayetteville, AR, USA
[7]Ever NeuroPharma, Oberburgau, Austria

ABSTRACT

Military personnel are often working in hot environment where concussive head injury (CHI) during combat operations is quite common. In this innovation we for the first time show that traumatic brain injury occurring in hot environment is exacerbated and this could be due to lack of neurotrophic factors. Thus, nanowired delivery of Cerebrolsyin, a balanced composition of several neurotrophic factors and active peptide fragments could be useful for reducing the brain pathology either alone or together with mesenchymal stem cells (MSCs) in CHI during hot environments. Our observations show that TiO2 nanowired delivered Cerebrolsyin potentiates MSCs induced neurorepair and neuroregeration. This research has potential clinical significance for our armed forces that are working hard in hot environment during combat operations.

Keywords: cerebrolysin, concussive head injury, brain patholoy, TiO2 nanowired drug delivery, blood-brain barrier, neuroprotection, mesenchymal stem cells

1 INTRODUCTION

Concussive head injury (CHI) could lead to either instant death or lifetime disabilities [1]. Our soldiers are highly vulnerable to CHI during combat operations [2]. Thus, to treat these soldiers effectively novel strategies using combination therapy is needed. Since stem cell therapy enhances neurorepair in brain or spinal cord injuries [3], this is quite likely that this may be effective in CHI as well. However, this is still unclear whether CHI occurring at high environmental temperature could adversely affect the outcome with regard to brain pathology or sensory motor disturbances. Since our soldiers are often engaged in combat operations in desert environments under high heat conditions, this is likely that their outcomes following CHI require special treatment using combination therapy.

In this investigation we used nanowired delivery of mesenchymal stem cells (MSCs) intravenously following CHI [3] and also added a known neuroprotective multimodal drug Cerebrolsyin [1,2] with or without nanowired delivery to see whether a synergistic better effects of this combination can result in good neurorepair following CHI at high environmental heat conditions.

2 MATERIALS & METHODS

Experiments were carried out on Male Wistar Rats (200-300 g) housed at controlled ambient temperature (21±1°C) with 12 h light and 12 h dark schedule. Food and water were provided ad libitum before experiment. All the experiments were carried out according to the Guidelines & Care for laboratory animals as described by national Institute of Health and approved by local Institutional Ethics Committee.

2.1 Concussive head injury (CHI)

CHI was inflicted in rat model using a weight drop of 114.6 g on the parietal skull bone under Equithesin anesthesia from a 20 cm height using a guide tube. This arrangement induces an impact of 0.224 N on the right parietal skull surface causing serious brain edema and volume swelling particularly in the left hemisphere due to a "counter-coup" phenomenon [1,4,5]. The animals were allowed to survive 48 h aftre the insult.

2.2 Exposure to Hot environment

Separate group of rats were exposed to hot environment in a Biological Oxygen Demand (BOD) incubator maintained at 38° C for 1 h daily for 2 weeks. The relative humidity (45-47%) and wind velocity (20-25 cm/sec) was kept constant. After 2 weeks these heat-treated rats were subjected to CHI.

2.3 MSCs Treatment

Commercially available MSCs (1 million cells, Sigma Chemical Co., St. Louis, USA) were delivered intravebously in a group of rats with CHI either at room temperature or in heat treated animals immediately after CHI. In separate group of rats TiO2 nanowired MSCs were also administered in identical conditions [3].

2.4 Cerebrolsyin Treatment

Cerebrolysin (Ever NeuroPharma, Austria) was co-administered in a doe of 2.5 ml or 5 ml/kg, intravenously together MSCs. In separate group of CHI TiO2-nanowred cerebrolysin was also co-administered with TiO2-MSCs [1-3].

2.3 Parameters Measured

The following parameters were measures in animals that received TiO2-nanowired MSCs either alone or together with TiO2-nanowred cerebrolysin following CHI.

2.3.1 Blood-brain barrier

The blood-brain barrier (BBB) leakage was measured using Evans blue albumin (EBA, 2% of 0.3 ml/100g i.v.) and radioiodine ($^{[131]}$-Iodine], 10 µCi/100g i.v.) extravasation in the brain. After washing out of intravascular tracer with 0.9 % saline through heart the brain were dissected out and examined for blue staining. After that tissue pieces from selected brain areas were then dissected and radioactivity determined in a Gamma Counter (Packard, USA). Leakage of these tracers was expressed as percentage increase in the brain over blood concentration [4,5].

2.3.2 Brain Edema formation

The brain edema was measured using brain water content. Desired tissue pieces from brain were dissected out and weighed to determine their wet weight. After that these tissue pieces were kept in an oven maintained at 90° C for 72 h to obtain their dry weight. The percentage water content was calculated from the differences between wet and dry weight of the samples [4,5].

2.3.3 Neuronal injury

Neuronal injury was evaluated using Nissl or Haematoxylin & Eosin (HE) staining using standard histopathological techniques [3-5]. For this purpose, animals were perfused in situ with 4 % buffered paraformaldehyde preceded with a brief saline rinse though cardiac puncture. Coronals sections of the brain were the cut passing through the hippocampus and embedded in paraffin. About 3 µm thick sections were cut and stained with HE or Nissl using commercial protocol [3]. The number of damaged or distorted neurons in specific anatomical brain areas were counted manually.

2.4 Statistical analyses

ANOVA followed by Dunnett's test for multiple group comparison with one control group was used to analyze statistical significance of the data obtained. A p-value less than 0.05 was considered significant.

3 RESULTS

3.1 CHI induces Neurotoxicity

When rats were exposed at 38°C for 1 h daily until 2 weeks and then the identical CHI was delivered, the magnitude and intensity of brain edema development and volume swelling was exacerbated by 2 to 3 fold as compared to the CHI delivered in rats kept at room temperature (21±1°C) [see Table 1].

Leakage of Evans blue and radioioidine across the BBB was laso much 2- to 4-fold higher in CHI group exposed to hot environments than identical injury at room temperature [Table 1].

Neural injury, neuronal loss, distorted neurons, perineuronal edema and sponginess of the neuropil were also exacerbated in animals with CHI performed at 38°C as compared to CHI inflicted on rats at 21°C [Table 1].

A general expansion of the neuropil supporting edema formation is also clearly evident in CHI at 21°C that was also exacerbated at 38°C. The uninjured left half became almost fluid in CHI group at hot environment, a feature that was seen only in a very mild form in rats subjected to CHI at room temperature treated animals.

These pathophysiological changes such as breakdown of the BBB, edema formation and neuronal injuries after CHI at normal or hot environments were progressive with time, from 12, 24 and 48 h periods examined (Result not shown).

Table 1. CHI 48 hinduced neurotoxicity and neuroproetction in contralateral left hemesphere (LH) by co-administration of TiO2 nanowired Cerebrolsyin and MSCs 12 h after injury

Expt. Type	Control	CHI 21°C 48 h	CHI 38°C 48 h	CHI 48 h+ TiO2-MSCs 38° C	CHI 48 h+ TiO2 CBL+MSCs 38°C
EBA mg%	0.24±0.06	2.38±0.14**	4.26±0.28**#	1.34±0.18*a	0.84±0.10*#b
[131]-Iodine %	0.30±0.08	2.96±0.19**	4.91±0.17**#	1.55±0.12*a	0.94±0.12*#b
Brain water %	74.76±0.18	79.84±0.14**	84.53±0.26**#	76.29±0.24a	75.16±0.23#b
Brain Swelling % *f*	0	+20**	+44**#	+8*a	+2*#b
Neuron Injury Nr	2±3	389±35**	659±46**#	108±34*a	86±24*#b

*Values are Mean±SD of 5 to 6 rats, CHI = Concussive head injury; MSCs = mesenchymal stem cells; CBL = cerebrolysin; TiO2-MSCs = TiO2-nanowied; * P <0.05, ** P <0.01 from control, # P <0.05 from CHI 21°C, a P <0.05 from CHI 38°C, b P <0.05 from CHI+MSCs. For details see text.*

3.2 MSCs induced Neuroprotection

Commercially available MSCs (1 million cells) were delivered in a group of rats with CHI either at room temperature or in heat treated animals resulted in a mild but significant reduction in volume swelling and brain edema formation [Results not shown].

On the other hand, when TiO2 nanowired MSCs are given under identical conditions CHI did not results in massive brain edema formation or volume swelling at room temperature. However, heat stressed rats did not show sufficient reduction in brain pathology [Table 1]. Thus, TiO2 nanowired MSCs alone did not reuduce neuronal injuires and edematous expansion of the neuropil when CHI was perfomed at hot environments.

3.3 Cerebrolysin induced Neuroprotection

Co-administration of Cerebrolysin 2.5 ml/kg, i.v. together with MSCs attenuated neurotoxicity in CHI at room temperature very effectively [Results not shown]. However, Cerebrolysin alone (5 ml/kg) or with MSCs did not reduce brain pathology in CHI rats at hot environments.

Interestingly when TiO2 nanowired Cerebrolsyin (2.5 ml/kg) was co-administered with nanowired MSCs either 8 or 12 h after CHI significant reduction in brain pathology and brain edema formation was seen in heat treated rats at 48 h [Fig. 1, Table 1]. Thua, thesc rats exhibited profound healthy neurons in the cortex (Fig. 1) in both the right and left halves. A significant reduction in EBA and radioiodine leakage as well as reduction in brain edema formation was also seen in this group (Table 1).

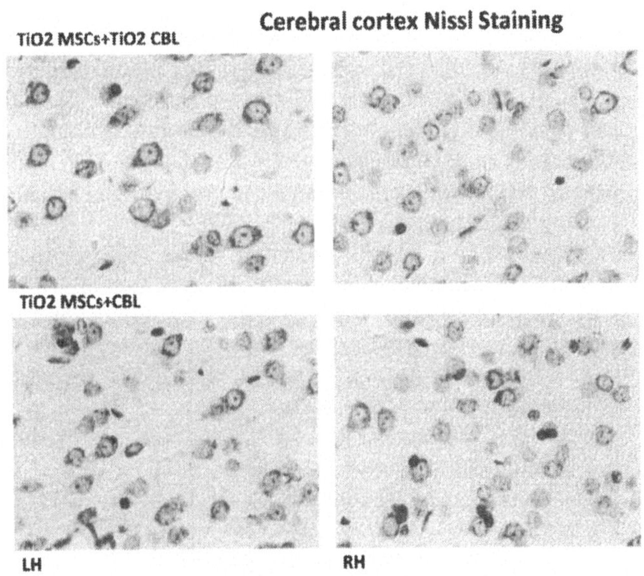

Fig. 1. Shows neuroprotection by co-administration of TiO2-nanowired mesenchymal stem cells (MSCs) and TiO2-nanowired cerebrolysin (2.5 ml/kg) 12 h after concussive head injury (CHI) in rats at hot environment (38°C, Upper panel). Most marked neuroprotection in terms of healthy neurons are seen in the left half (LH) after CHI. Whereas, TiO2 MSCs with cerebrolysin alone (lower panel) reduced neuronal damages in LH but some neurons in RH still show damage. Nissl staining on 3-µm paraffin section x 40.

4 DISCUSSION

The salient novel finding in this innovation suggests that Cerebrolysin if co-administered with MSCs 8 to 12 h after CHI is able to induce profound neuroprotection than any one of the drugs given alone. However, when the same combination was given in rats in which CHI was inflicted at hot environment, the magnitude of neuroprotection was much less evident. This suggests that efficacy of any drug also depends on the magnitude and intensity of CHI, not reported earlier. Since hot environment exacerbated brain pathology caused by CHI this is necessary to effectively reduce these additional burden on neurotrauma by effective and prolonged administration of drugs in question. To that end when TiO2 nanowired delivery of both MSCs and cerebrolysin in combination was done, most marked neuroprotection was seen in CHI at hot environment. This opens up new possibilities to deliver drugs using nanowired technologies in clinics for better healthcare.

The possible mechanisms behind a combination of nanowired MSCs and cerebrolysin is unclear from this investigation. However, available evidences point out that nanodelivery of drugs could be more effective than parent compounds [1,2]. This is due to the fact that binding with nanoparticles makes these drugs rapidly penetrable into the CNS and also make them less vulnerable to biodegradation |1-3]. As a result MSCs could survive for longer duration and prolonged continuous exposure of cerebrolysin that is a balanced composition of neurotrophic factors and active peptide fragments is able to induce neurorepair and enhance neuroregeneration [see 1-3].

Obviously, a combination of these agents using nanodelivery could synergistically enhance neurorepair in CHI. This would result in much belter neuroprotection than any drug given alone. The biological mechanisms of improved neuroprotection by the combined administration of cerebrolysin and MSCs are not well known. However, it appears that the combination of these agents may reduce exacerbation of oxidative stress that could be caused by heat treatment and CHI combination more effectively [2-4].

5 CONCLUSION

In conclusion, our reuslts show that CHI at hot environment exacerbated brain pathology indicating that military personnel when injured at hot environment their treatment regiment should be modifed than standard therapy to achieve good results. Furtehrmore, our observations are the first to point out that a combination of MSCs with a multimodal drug, Cerebrolysin if given using nanotechnology is far more effective in CHI at hot environment than normla compounds. This suggests that nanodelivery of therapeutic agents in combination with stem cells will enhance therapeutic efficacy in CHI at hot environments, not reproted earlier.

6 ACKNOWLEDGEMENTS

This investigation is supported by grants from the Air Force Office of Scientific Research (EOARD, London, UK), and Air Force Material Command, USAF, under grant number FA8655-05-1-3065; Swedish Medical Research Council (Nr 2710-HSS), Göran Gustafsson Foundation, Stockholm, Sweden (HSS), Astra Zeneca, Mölndal, Sweden (HSS/AS), Ministry of Science & Technology, People Republic of China (LF/HSS); The University Grants Commission, New Delhi, India (HSS/AS), Ministry of Science & Technology, Govt. of India (HSS/AS), Indian Medical Research Council, New Delhi, India (HSS/AS) and India-EU Co-operation Program (RP/AS/HSS) and IT 794/13 (JVL), Government of Basque Country and UFI 11/32 (JVL) University of Basque Country, Spain. The U.S. Government is authorized to reproduce and distribute reprints for Government purposes. The views and conclusions are those of the authors and should not reflect the official policies of granting organizations.

7 REFERENCES

[1] Ruozi B, Belletti D, Forni F, Sharma A, Muresanu D, Mossler H, Vandelli MA, Tosi G, Sharma HS. Poly (D,L-Lactide-co-Glycolide) Nanoparticles Loaded with Cerebrolysin Display Neuroprotective Activity in a Rat Model of Concussive Head Injury. CNS Neurol Disord Drug Targets. 2014;13(8):1475-82.

[2] Sharma HS, Menon PK, Lafuente JV, Aguilar ZP, Wang YA, Muresanu DF, Mössler H, Patnaik R, Sharma A. The role of functionalized magnetic iron oxide nanoparticles in the central nervous system injury and repair: new potentials for neuroprotection with Cerebrolysin therapy. J Nanosci Nanotechnol. 2014 Jan;14(1):577-95. Review.

[3] Sharma HS, Feng L, Lafuente JV, Muresanu DF, Tian ZR, Patnaik R, Sharma A. TiO2-nanowired delivery of mesenchymal stem cells thwarts diabetes-induced exacerbation of brain pathology in heat stroke. CNS Neurol Disord Drug Targets. 2015 Feb 25.PMID: 25714976.

[4] Sharma HS, Patnaik R, Patnaik S, Sharma A, Mohanty S, Vannemreddy P. Antibodies to dynorphin a (1-17) attenuate closed head injury induced blood-brain barrier disruption, brain edema formation and brain pathology in the rat. Acta Neurochir Suppl. 2010;106:301-6. doi: 10.1007/978-3-211-98811-4_56.

[5] Sharma HS, Patnaik R, Patnaik S, Mohanty S, Sharma A, Vannemreddy P. Antibodies to serotonin attenuate closed head injury induced blood brain barrier disruption and brain pathology. Ann N Y Acad Sci. 2007 Dec;1122:295-312

[1] *Hari S Sharma, Dr. Med. Sci. (UU), Director Int. Expt. CNS Injury & Repair (IECNSIR), Professor of Neurobiology (MRC), Docent in Neuroanatomy (UU), University Hospital, Uppsala University, Frödingsgatan 12:28, SE-75421 Uppsala, Sweden, Phone & Fax: +46 18 243899, Cell Phone: +46 70 2011 801; Email: Sharma@surgsci.uu.se

Efficiency of different chitosan and its derivative nanoparticles for delivery of recombinant plasmid of Tibetan pig interleukin-7 gene and regulation effect in vivo on immune responses of mice

Xiaoping Wan[a*], Xiao Yang[a*], Jianling Chen[a*], Ge Liang[b], Yihui Chen[a], Jiangling Li[b], Yiren Gu[b], Rui Liu[b**], Xuebin Lu[b**], Rong Gao[a**]

a. Key Laboratory for Bioresource and Eco-Environment of the Education Ministry, Key Laboratory for Animal Disease Prevention and Food Safety, College of Life Science, Sichuan University, Chengdu, 610064, China
b. Sichuan Academy of Animal Science, Chengdu, Sichuan, 610066, China

*:The authors contributed equally to the work. **: Corresponding authors: gaorong96@163.com; lake96@qq.com

ABSTRACT

Interleukin-7 is a key cytokine to regulate survival and growth of both B and T lymphopoiesis. For sake of developing safe and effective technique to deliver and express animal immune gene *in vivo* to enhance the comprehensive immunity of animal against infectious disease, the interleukin-7 gene of Tibetan pig was firstly cloned and reconstructed into eukaryotic expression plasmid,VR1020. The recombinant plasmid, VRTPIL-7, was expressed in HEK293 cells *in vitro* to confirm its bioactivity on lymphocytes of pig. Then VRTPIL-7 was respectively packed by polyethylenimine (PEI), chitosan(CS) and its two derivatives, methoxy poly(ethylene glycol) (PEG) and PEI modified CS(CS-PEG-PEI) and PEG-modified galactosylated chitosan (CS-PEG-GAL) via by ionotropic gelation method. Their nanoparticle sizes, dispersivity and zeta potentials were analyzed with a transmission electronic microscope and a Zetasizer3000 HS/IHPL instrument (Malvern Instruments Ltd., Malvern, UK). Subsequently, nanoparticles of VRTPIL-7 wrapped with CS, CS-PEG-PEI and CS-PEG-GAL were utilized to intramuscularly inoculate Kunming female mice at the age of 21 days, with a dose at 100μg recombinant plasmid per mice for studying the regulation effect on immunity *in vivo*. Their blood were respectively collected before and after inoculation on 1, 2, 3, 4 and 5 weeks to test the changes of immune cells, IgG□IgG1, IgG2a and the expression level of immune genes by ELISA and quantitative RT-PCR in mice. The results were found that the immune cells numbers in peripheral blood, immunoglobulin (IgG, IgG1, IgG2a) content was significantly increased in the serum of treated mice compared with the control group($P<0.05$). Flow cytometry and q-PCR analysis results showed that the $CD4^+$ and $CD8^+$ T cell numbers and mRNA level of TLR1, TLR4 TLR6, TLR9, TGF-β, IL-1, IL-4, IL-6 andIL-23 were significantly higher in the treated group than those of the control($P<0.05$). These results also indicated that the VTPIL-7 wrapped with CS-PEG-PEI and CS-PEG-GAL induced better innate and adaptive specific humoral and cellular immunity than chitosan, and improved the immune protection and disease-resistant ability of mice against challenge, which implying CS-PEG-PEI and CS-PEG-GAL could be promising practicable package molecules to deliver and express immune gene for prevention and control of animal infectious diseases.

Keywords: chitosan and its derivatives, nanoparticle package and delivery, gene expression, pig interleukin-7, immunity

1. Introduction

Interleukin-7 (IL-7) plays an important role in the immune system which regulates the survival and development of myeloid and lymphoid cells. It was firstly discovered in 1980s as a factor that promotes the growth of pre-B cells in bone marrow culture (Namen AE et al,1988; Kittipatarin C et al. 2007).

Evidences show that Tibetan pigs have greater resistance than domestic pigs against infectious diseases, like CSFV, PRRSV, PCV2, Haemophilus parasuis, etc. Tibetan pig was originally and exclusively found on the Tibetan plateau, a harsh and natural environment with an average altitude approximate of 4 kilometers above sea level. Tibetan pig shows distinct evolutionary scenarios compared with other pigs throughout natural selection. To further explore and clarify the immune characteristics of Tibetan pig is of great importance to understand its unique genetic characteristics for pig breeding, vaccination and prevention of diseases.

So we are interested in how IL-7 exerts unique bioactivity in animal, and try to figure out the expression effect of Tibetan IL-7 gene *in vivo* on immunity of animal for sake of development of safe and novel immunoadjuvant to promtoe the control of infectious diseases in animal in the future.

2. Materials and methods

2.1 Sample and animals.

Lymphocytes were purified from the blood of healthy Tibetan pig provided by Sichuan Academy of Animal Science according to routine method and cultured in 1640 medium with 5 ug/ml Con A for 48 hours, and the RNA was extracted from the activated leukocytes following the instruction of RNA extraction kit manual of Invitrogen.

2.2 TPIL-7 gene cloning and sequencing.

RT-PCR is used for TPIL-7 gene cloning into pMD®19-T vector (Takara), and then transformed into DH5α competent cells. Primers were designed by Primer 5.0 with *BamH I* and *Bgl II* sites according to conserved ORF sequence of IL-7 gene of Duroc□Landrace and other mammalians from NCBI/GenBank. TPIL-7-F: 5'-*ACCACGCCCGCCTCCCGCAGACCATGTTC*-3', TPIL-7-R: 5'-*GAATATTGAGAGATACGGAGTGGCAA*-3'. Plasmids were assessed for conformity by PCR and digested by *BamH I* and *Bgl II.* Subsequently, the pMD®19-T/IL-7 plasmids were sent to BGI biological company for sequencing. TPIL-7 sequence was aligned with domestic pigs and other mammalians by NCBI/Blast.

Plasmid VR1020 is a eukaryotic expression vector (Vical Company of America). We cloned the TPIL-7 into VR1020 to constructed VTPIL-7 to get secreted protein for *in vitro* and *in vivo* bioactivity test using the same fragment and methods as above mentioned.

2.3 Preparation of recombinant VTPIL-7 and copolymer/DNA nanoparticles

A recombinant DH5a *E. coli*, containing VTPIL-7 or VR1020 plasmid, was inoculated in LB broth with kanamycin (100 mg/ml) at 37□, 200rpm overnight. Bacterial cells were pelleted by centrifugation and plasmid was extracted following the spermine precipitation method described as Jason. Then the plasmid was resuspended in sterile water and stored at -20 □ until use.

Copolymers as CS, CS-PEG–PEI (Yu YY, et al. 2010), CS-PEG-LAC (Song B, et al.2009) and PEI (MW: 25000)/DNA complexes were prepared following the method of by ionotropic gelation method. Briefly, CS, CS-PEG–PEI and CS-PEG-LAC were diluted separately in CH3COOH/CH3COONa (pH 5.5) containing appropriate amount of triphosphate and heated at 65□ for 10 min with mild magnetic stirring. After that, an appropriate amount of polymer solution was added to the solution of plasmid DNA and the solution was mixed then left for 5 min. The zeta potential and average diameter of the polymeric micelles were characterized by Zetasizer 3000 HS/IHPL (Malvern Instruments Ltd., Malvern, UK).

2.4 Transfection and proliferation activity.

HEK293 cells were seeded in 12-well plates (1.0×10^5 cells/well), respectively. The cells were incubated in Dulbecco's modified Eagle medium (DMEM, Invitrogen Corporation.). The complexes of nanoparticles enwrapped with CS, CS-PEG–PEI, CS-PEG-LAC and PEI each containing 3 µg of DNA were added into the wells, respectively, then the cells were incubated with the complexes at 37□in a 5% carbon dioxide humidified atmosphere.

The bioactivity of the TPIL-7 protein was measured by its ability to provoke the proliferation of pig lymphoblast stimulated with Concanavalin A (Con A) through CCK8 colorimetry. Lymphocyte Separation Medium was used to separate of pig peripheral blood mononuclear immune (PBMI) cells for *in vitro* bioactivity test.

2.5 Animal vaccination

Forty 4-week-old female Kunming mice (purchased from the Animal Center of West China Center of Medical Sciences, Sichuan University) were randomly assigned to four groups (A1, A2, A3, C), 10 mice in each group. All the mice were muscularly injected with 6 pmol of 100µg plasmid DNA encoding TPIL-7 per mouse in the left and right quadriceps on day 0. Mice in group A1 were injected with CS/VTPIL-7, A2 were injected with CS-PEI-PEG/VTPIL-7, A3 were injected with CS-PEI-LAC/VTPIL-7 and group C were injected with CS-VR1020 as a negative control. Peripheral blood samples were collected from the tail vein of mice on days 0, 7, 14, 21, 28 and 35 after immunization.

2.6 Bioactivity assay of TPIL-7 *in vivo.*

2.6.1 Changes of cells in peripheral blood.

Peripheral blood samples were used EDTA for anticoagulation, double diluted with normal saline, then analyzed by blood testing instrument: MIND-RAY BC-3000 intelligent automatic blood cell analyzer. Blood test include red blood cells, white blood cells, hemoglobin and platelet count and so on.

2.6.2 Assay of CD3, CD4 and CD8 positive T cells by FCM.

Monoclonal antibody for FCM including Anti-Mouse CD3e FITC (0.5ug/Test), Anti-Mouse CD8a PE (0.25ug/Test), and Anti-Mouse CD4 PerCP-Cy5.5 (0.25ug/Test) for each Test contain 50µl peripheral blood (monoclonal antibody were purchased from eBioscience Company, San Diego, USA.,). FACSAria BD and FACSDiVa BD were used for FCM analysis(BD Biosciences, USA). The method is the same as manual.

2.6.3 Measurement of IgG, IgG1, IgG2a.

Mouse IgG, IgG1, IgG2a measurement ELISA kits were purchased from Bethyl Laboratory, Inc., America. The sandwich ELISA was conducted according to the manufacturer's protocols.

2.6.4 Analysis of TLRs and Cytokines.

According to reports in GenBank, 10 pairs of specific primers were designed and synthesized for mouse β-actin, TRL1, TLR4, TLR6, TLR9, TGF, IL-1, IL-4, IL-6 and IL-23 cDNA sequence.

Quantitative real-time PCR was performed to analyze the gene expression of TRL1, TLR4, TLR6, TLR9, TGF-β, IL-1, IL-4, IL-6 and IL-23 and β-actin as reference gene. The relative expression of cytokine mRNA was calculated relative to the mean expression of the cytokines/TLRs mRNA in the control pigs using the geometric means method and the formula as : relative level = $2^{-\Delta\Delta Ct}$. The ΔΔCt was calculated by the formula: $\Delta Ct_{\text{immunized group}} - \Delta Ct_{\text{control group}} = (Ct_{\text{cytokine/TLR}} - Ct_{\text{reference gene}})_{\text{immunized group}} - (Ct_{\text{cytokine/TLR}} - Ct_{\text{reference gene}})_{\text{control group}}$, where $CT_{\text{cytokine/TLR}}$ is the mean Ct value of triplicates and each $CT_{\text{reference gene}}$ is the mean Ct value of triplicates from the reference gene. Significant differences between the experimental groups were calculated for each cytokine using a Mann–Whitney rank sum test on the relative cytokines/TLRs levels.

2.7 Statistical analysis.

Statistical significance was analyzed using Student's t-test. Differences between experimental groups were considered significant when P-value was less than 0.05.

3 Results

3.1 TPIL-7 gene cloning and sequencing.

TPIL-7 gene cDNA is 571 bp of open reading frame (ORF), encoding 528 bp, and 176 amino acids. TPIL-7 cDNA, alignment with the NCBI pig IL-7 cDNA, have more than 92% homology, and the cDNA sequence of TPIL-7 was submit to GenBank, Sequence ID: gb|KF246514.1|.

3.2 Sizes and zeta potential of nanoparticles.

The cationic polymers as CS, modified-CS and PEI have been considered as excellent candidates for preparation of nanoparticles. Nanoparticles with specific size and positive charge on surface are necessary for endocytosis. The diameter and zeta potentials of the VPIL-7 nanoparticels are revealed in table 1.

Table 1. The diameter and zeta potential of package nanoparticles

Sample	Size (d.nm) X	SD(±)	Zeta Potential (mV) X	SD(±)
CS/DNA	522.2	5.15	+40.5	1.6
PEI/DNA	341	5.79	+45.5	2.04
CS-PEG-PEI/DNA	169.7	4.07	+43.9	1.11
CS-PEG-LAC/DNA	247.1	18.78	+40.2	0.78

3.3 Proliferation of pig lymphoblasts *in vitro.*

The results showed the significant proliferation of pig lymphoblasts were achieved by addition of the culture supernatant of TPIL-7 gene transfected cells in comparison with control group (Figure 1). Supernatants of CS-PEG–PEI and CS-PEG-LAC groups manifested the strongest proliferative effect on pig lymphoblasts, indicating that the TPIL-7 recombinant palsmid packed with modified chitosans can promote the expression TPIL-7 protein to stimulate better lymphocyte proliferation of pigs ($P<0.05$).

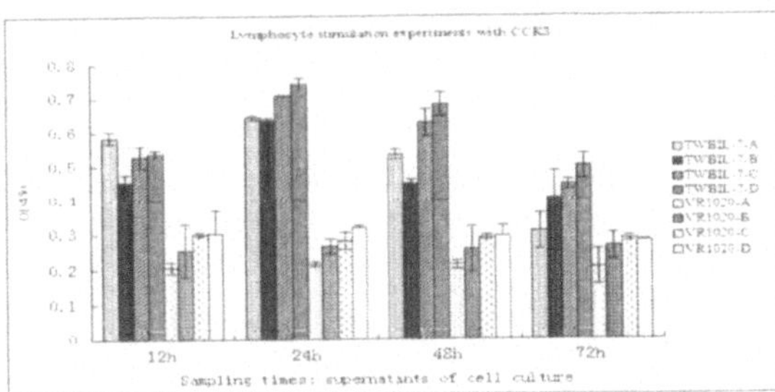

Figure 1. The proliferation results of pig lymphoblasts.
A: CS B: PEI C: CS-PEG-Lac D: CS-PEI-PEG

3.4 Promotion of immunity of mice *in vivo.*

3.4.1 Blood immunological change of mice.

During 0-35 days postinoculation, the growth rate of body weight of the treated groups was significantly higher than control group ($P<0.05$), but after 21 days of it gradually slowed. There were no significant differences among the groups treated with VTPIL-7 nanoparticles ($P>0.05$) (Figure 2), and the leukocytes of the treated mice significantly increased compared to the control group($P<0.05$).

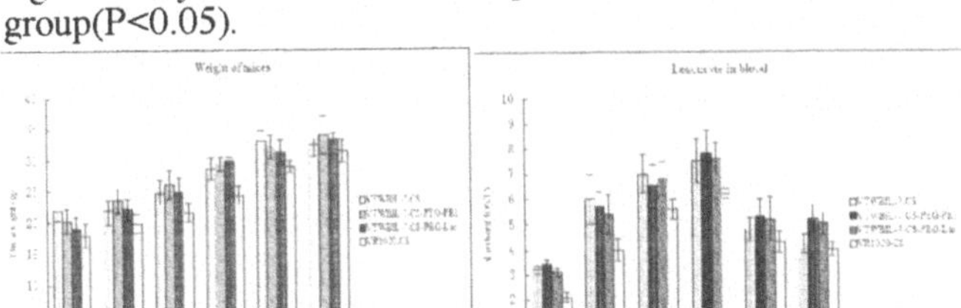

Figure 2. Changes of body weight and immune cells in the blood of the experimental mice.

3.4.2 CD4 and CD8 positive T cells.

The $CD4^+$ and $CD8^+$ T lymphocyte were both significantly increased ($P<0.05$) in peripheral blood of the treated mice, but the VTPIL-7 packed with modified chitosans resulted in higher increase of the T cells ($P<0.05$) than the control group from 21 to 35 days post inoculation (Figure 3).

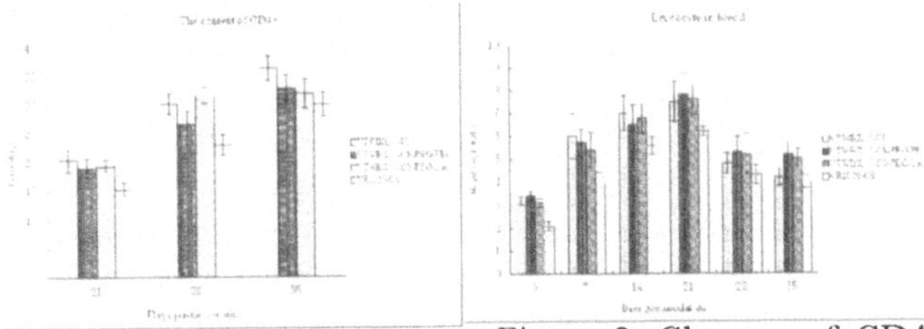

Figure 3. Changes of CD4, CD8 positive T cells in the blood of the experimental mice.

3.4.3 Increase of IgG, IgG1 and IgG2a.

Figure 4 showed that the IgG, IgG1 and IgG2a significantly increased in the serum of treated groups than the control ($P<0.05$). The ratio of IgG1/IgG2a maintained at about 1.2. This shows that the mice are more likely to induce humoral immune response.

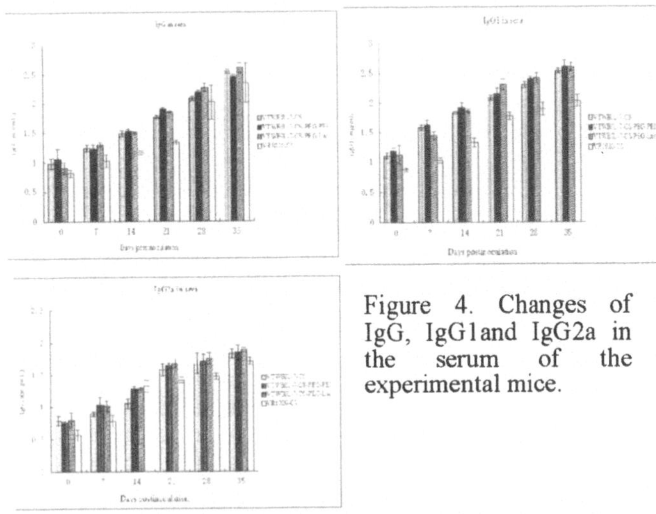

Figure 4. Changes of IgG, IgG1and IgG2a in the serum of the experimental mice.

3.4.4 Enhancement of TLRs gene expression.

The results showed that TLRs gene expression were increased significantly during 0-35 days as showed in figure 5 ($P<0.05$). TLR4 and TLR9 increased significantly than other TLRs in the treated mice. But TLR6 showed more cascade effect which could be six times higher the data before inoculation, and CS-PEG-PEI group demonstrated obviously better effect than other treated groups (Figure 5).

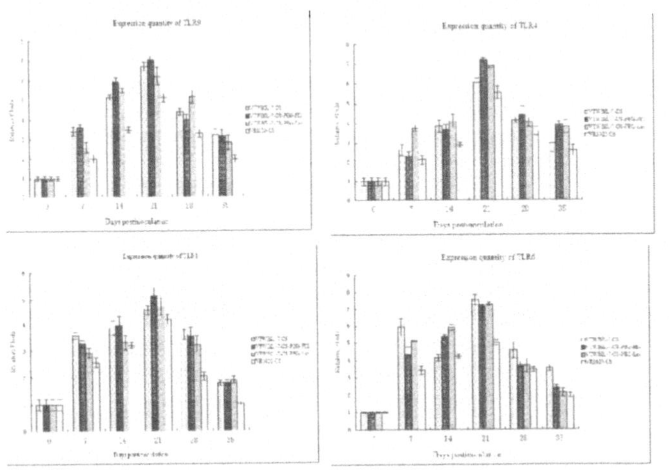

Figure 5. Change of the expression of TLR1, TLR4, TLR6 and TLR9 gene of the experimental mice

3.4.5 Elevation of cytokines gene expression.

Transcription levels of TGF-β, IL-1, IL-4, IL-6 and IL-23 genes were significantly increased in the three treated mice in comparison with those of the control during 35 days period ($P<0.05$), and CS-PEG-PEI group showed higher increases of these genes expression than the other treated mice ($P<0.05$) (Figure 6).

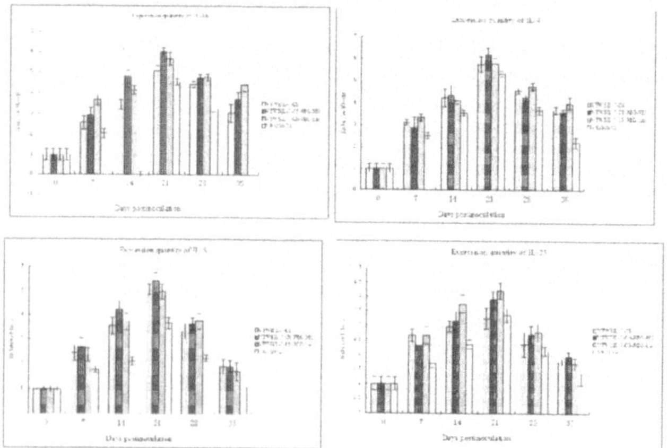

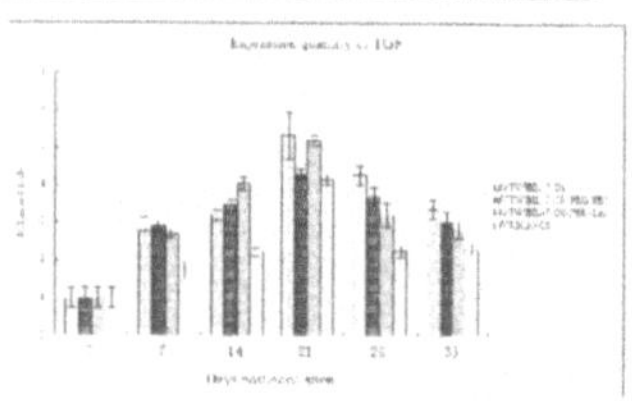

Figure 6. Change of the expression of IL-1b, IL-4, IL-6, IL-23and TGF-β gene in the leukocytes of blood from the experimental mice.

4. Discussion and Conclusion

Nanoparticles with positive charge can help the recombinant plasmid penetrate the cell membrane of transfected cells. And the average particle size ranged from 30nm to 400nm, which were qualified for efficient transfection. The supernatant of HEK293 cells was collected to stimulate lymphoblast of pigs to confirm the biological activities of TPIL-7. These results showed that the TPIL-7 gene was successfully expressed in the transfected HEK293 cells, and the supernatant had dramatically biological activity to promote the proliferation of pig lymphoblast *in vitro*.

The results were found that the immune cells numbers in peripheral blood, immunoglobulin (IgG, IgG1, IgG2a) content was significantly increased in the serum of treated mice compared with the control group ($P<0.05$). Flow cytometry and q-PCR analysis results show that the CD4 $^+$ and CD8 $^+$ T cell numbers and gene expression of TLR1, TLR4 TLR6, TLR9, IL-1, IL-4, IL-6, IL-23 and TGF-β were significantly higher in the treated group than those of the control group($P<0.05$). These results indicate that the VTPIL-7 wrapped with chitosan and chitosan derivative can significantly improve the innate immunity, humoral and cellular immunity, which could lead to stronger immune protection and disease-resistant ability of animals.

The cationic polymers as CS, modified-CS and PEI have been considered as excellent candidates for preparation of nanoparticles. Nanoparticles with specific size and positive charge on surface are necessary for endocytosis. Researches showed that many of proteins, drugs and DNA complexes were sensitive to bonding with cationic biomaterials delivery system; therefore, multi-cationic PEI was introduced onto CS to enhance the positive charge of the nanoparticles. The average diameter of CS-PEG-PEI was smaller than packaged with CS which probably due to the increase of DNA binding/condensation capability. The zeta potential of CS-PEG-PEI was in the positive range and it was higher than that of CS due to the introduction of polymeric PEI. As expected, the chitosan derivatives resulted in better expressions of the TRL1, TLR4, TLR6, TLR9, TGF, IL-1, IL-4, IL- 6 and IL-23 genes of the treated mice, indicating that the expression of TPIL-7 gene was promoted by the modified nanoparticle molecular package.

RT-qPCR results showed that the expression of TRLs and cytokine genes were significantly increased in the inoculated mice with modified chitosan VTPIL-7 ($P<0.05$) . These suggested that the TPIL-7 gene wrapped with chitosan derivatives could inspire further development of safe and effective immune adjuvant of animal for the control of infectious diseases in the future.

Reference

[1] Namen AE, Lupton S, Hjerrild K, Wignall J, Mochizuki DY, Schmierer A, et al.Stimulation of B-cell progenitors by cloned murine interleukin-7. Nature ;333:571–3,1988.

[2] Kittipatarin, C., & Khaled, A. R. Interlinking interleukin-7. Cytokine, 39(1), 75–83, 2007.

[3] Ceredig, R., and Rolink, A.G. The key role of IL-7 in lymphopoiesis. Seminars in Immunology 24, 159-164. 2012.

[4] Pellegrini, M., Calzascia, T., Toe, J.G., Preston, S.P., Lin, A.E., Elford, A.R., Shahinian, A., Lang, P.A., Lang, K.S., Morre, M., et al. IL-7 Engages Multiple Mechanisms to Overcome Chronic Viral Infection and Limit Organ Pathology. Cell 144, 601-613, 2011.

[5] Surh, C.D., and Sprent, J. Homeostasis of Naive and Memory T Cells. Immunity 29, 848-862, 2008.

[6] Yu, Y.Y., Wang, Z., Cai, L., Wang, G., Yang, X., Wan, X.P., Xu, X.H., Li, Y., and Gao, R. Synthesis and characterization of methoxy poly(ethylene glycol)-O-chitosan-polyethylenimine for gene delivery. Carbohydrate Polymers 81, 269-274, 2010.

The use of Werner complexes as remarkably versatile bioconjugation reagents in the synthesis of redox repsonsive albumin nanoparticles containing camptothecin derivatives

D.T. Nguyen,[‡] R.J. Cavazos,[‡] C.S. Morrison, J.B. Lampe, A.N. Harris, B.K. McFarlin, R.A. Petros[*]

[*]University of North Texas, Denton, TX, USA, petros@unt.edu
[‡]contributed equally to this work.

ABSTRACT

We have begun applying classic coordination chemistry to the bioconjugation of amine-containing (bio)molecules in the context of targeted drug delivery. In addition to being an efficient conjugation strategy, the chemistry is completely reversible allowing for the synthesis of redox responsive materials. Co(3+) is inert to ligand exchange; however, upon reduction to Co(2+) labile ligand exchange occurs, which provides an opportunity to capitalize on the reducing nature of cytosol and hypoxic tumor microenvironments as stimuli for degradation of the delivery vector and concomitant release of therapeutic payload. We have applied this methodology to the synthesis of albumin nanoparticles (10-500 nm) and examined particle uptake in gastric carcinoma cells (SNU-5) by image-based flow cytometry. The particles were further loaded with camptothecin derivatives via either passive encapsulation (SN-38), or cobalt-mediated bioconjugation to albumin (topotecan). Details of drug loading and release will be discussed.

Keywords: nanoparticle therapeutic, cancer, cobalt, albumin

1 INTRODUCTION

SN-38 (7-ethyl-10-hydroxycampothecin) and irinotecan are campothecin derivatives that target topoisomerase I, an enzyme that plays a vital role in DNA replication and transcription.[1] Irinotecan has been approved for the treatment of metastatic colorectal cancer and is commercially available as Camptosar (Pfizer Inc, New York, New York).[2] SN-38 is the active metabolite of irinotecan, which exhibits a much stronger therapeutic effect (100-1000x compared to irinotecan).[3-4] However, despite its promising antitumor efficacy toward a wide range of experimental tumor models, SN-38 has limited application due to its poor aqueous solubility. The development of innovative nanoparticulate systems that allow for high drug loading and improved stability have remained a challenge.[5]

Nanoparticle therapeutics derived from albumin have emerged as promising tool in oncology because of its biocompatibility, lack of immunogenicity, and biodegradability.[6] Albumin is highly soluble in water, exhibits excellent stability, and is easily purified. Albumin posseses a variety of binding sites for small molecules, and has therefore been implicated in the binding and transport of a variety therapeutics.[7-8] Recent studies have also shown that some tumor cells consume large amounts of albumin as nutrient source to support for their rapid growth.[9] Albumin is especially attractive for targeting cancer cells because the cell surface receptor responsible for binding albumin, the 60kDa glycoprotein (gp60) (albondin), is overexpressed in a host of cancer cell types.[10]

2 RESULTS

2.1 Synthesis and Characterization of SN-38 loaded Co^{3+}-Alb NPs

Bovine serum albumin was prepared in ultrapure water as a 10mg/mL solution. Cobalt(II) chloride hexahydrate (100mM) and 25mM of NaOH solution was also prepared in ultrapure water. In a small glass vial, 200uL of bovine serum albumin (10 mg/mL aqueous) was mixed with 50uL of NaOH (25mM) and the solution was colorless. Cobalt(II) chloride hexahydrate (50 L, 100 mM) was added and the solution immediately turned blue and became turbid. The solution was sonicated for 5s and left undisturbed for 15 minutes at room temperature. Hydrogen peroxide (5 L, 30% aqueous) was added to the solution followed by 1 mL of ultrapure water. The solution immediately turned yellow without any changes in turbidity. The solution was centrifuged (Eppendorf, model #5810R) for 1 minute at 21,000 x g and the supernatant was discarded. The pellet was then resuspended into 1 mL of water, centrifuged at 21,000xg and the supernatant was discarded. The pellet was suspended in water and dynamic light scattering (DLS, Microtrac Nanotrac Ultra) was used to measure the size of the nanoparticle. The nanoparticles were 140-200 nm in diameter (Figure 1).

SN-38 (7-ethyl-10hydroxycampothecin, TCI Development Co., Shanghai, China) was dissolved in dichloromethane/ethanol (1:1) at 10mg/mL. An emulsification/solvent evaporation method was used to incorporate SN-38 into Co-Alb NPs, which was accomplished by mixing 5 mL of a Co-alb NP solution with 2.5 mL of the SN-38 solution. Sonication was applied for 2-3 minutes followed by stirring at RT for three hours to allow all the organic phase to evaporate. The solution was then centrifuged at 21,000 x g for 5 minutes. The supernatant, which contained unbound SN-38 was removed and analyzed to determine drug loading. The pellet was then washed three times with ultrapure water. The particles were re-suspended in water and analyzed by DLS (Figure 1).

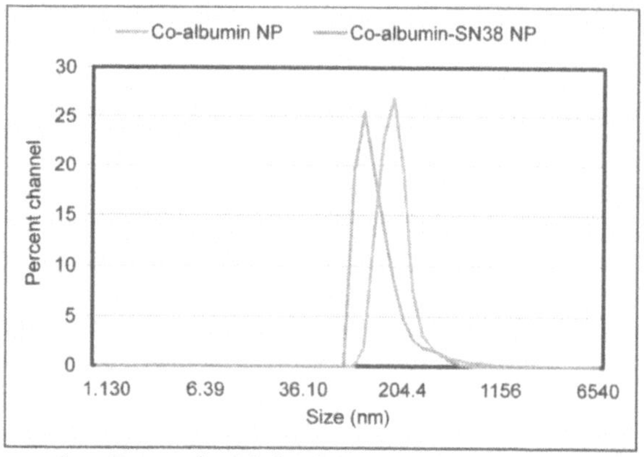

Figure 1. Dynamic Light Scattering results for Co-Alb NPs.

To determine SN-38 loading, the supernatants collected after centrifugation were analyzed by UV/Vis (Synergy 2, Biotek USA) based on a calibration curve of SN-38 (380 nm). Drug loading capacity and encapsulation efficiency was calculated according to Equations 1 and 2 below. Encapsulation efficiency was observed to be at 94% and loading capacity was approximately 31%. SN-38 loaded Co-Alb NPs exhibited minimal changes in size upon incubating in PBS; however when incubated in PBS plus reduced glutathione (GSH 10mM), the nanoparticles degraded rapidly (Figure 2). Representative SEM images of the particles are shown in Figure 3.

Encapsulation Efficiency =

$$\frac{\text{Total drug - residual drug}}{\text{Total drug}} \times 100 \qquad 1$$

Drug loading capacity =

$$\frac{\text{Loaded drug}}{\text{Total weight of NPs}} \times 100 \qquad 2$$

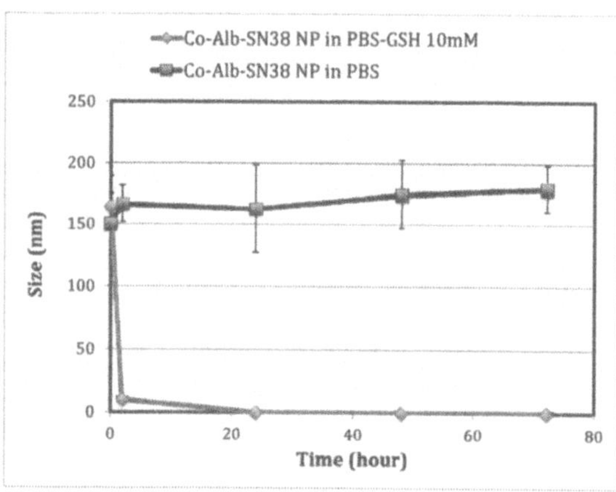

Figure 2. Stability of SN-38 loaded Co-Alb NPs

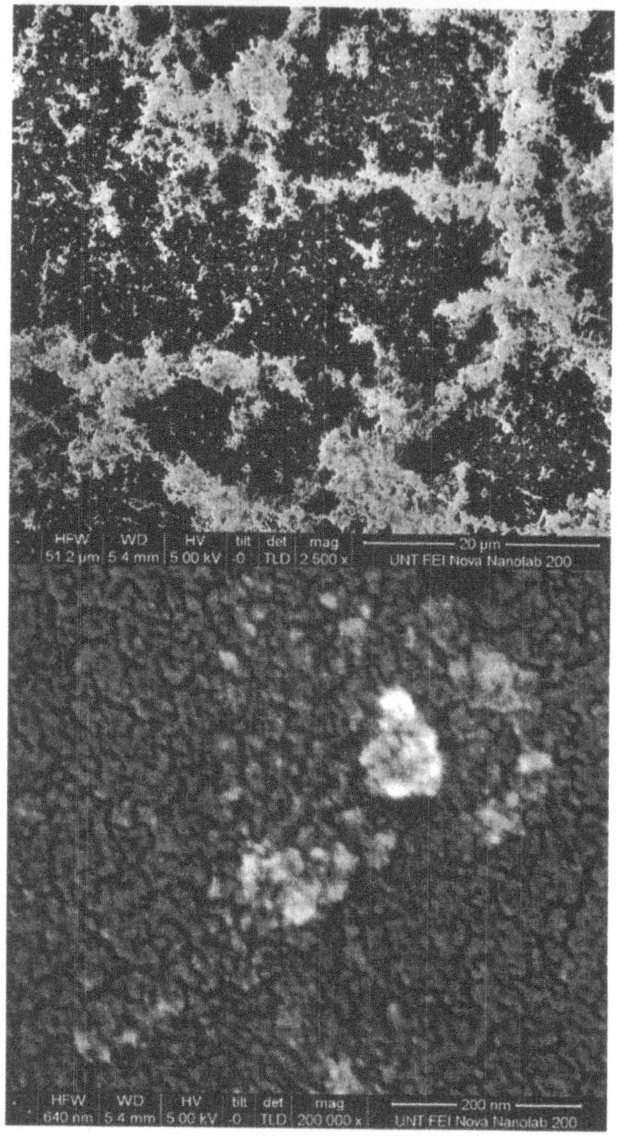

Figure 3. Representative SEM images of Co-Alb NPs.

3 CONCLUSIONS

SN-38 was successfully encapsulated in Co-Alb NPs. Drug loading could be achieved at very high levels (>30 wt%) in a highly efficient manner (>90%). The particles exhibited excellent stability under physiological conditions, but degraded rapidly under reducing conditions making this a promising formulation for further preclinical testing.

REFERENCES

[1] Roger, E.; Lagarce, F.; Benoit, J.-P., Development and characterization of a novel lipid nanocapsule formulation of Sn38 for oral administration. European Journal of Pharmaceutics and Biopharmaceutics 79 (1), 181-188, 2011.
[2] Ebrahimnejad, P.; Dinarvand, R.; Sajadi, A.; Jaafari, M. R.; Nomani, A. R.; Azizi, E.; Rad-Malekshahi,

M.; Atyabi, F., Preparation and in vitro evaluation of actively targetable nanoparticles for SN-38 delivery against HT-29 cell lines. Nanomedicine: Nanotechnology, Biology and Medicine 6 (3), 478-485, 2010.

[3] Atyabi, F.; Farkhondehfai, A.; Esmaeili, F.; Dinarvand, R., Preparation of pegylated nano-liposomal formulation containing SN-38: in vitro characterization and in vivo biodistribution in mice. Acta Pharm, 59 (2), 133-144, 2009.

[4] Kunii, R.; Onishi, H.; Machida, Y., Preparation and antitumor characteristics of PLA/(PEG-PPG-PEG) nanoparticles loaded with camptothecin. European journal of pharmaceutics and biopharmaceutics, 67 (1), 9-17, 2007.

[5] Petros, R. A.*; DeSimone, J. M.* "Strategies in the Design of Nanoparticles for Therapeutic Applications," Nature Rev. Drug Discov., 9, 615-627, 2010.

[6] Kratz, F., A clinical update of using albumin as a drug vehicle—A commentary. Journal of Controlled Release 190, 331-336, 2014.

[7] Weber, C.; Coester, C.; Kreuter, J.; Langer, K., Desolvation process and surface characterisation of protein nanoparticles. International journal of pharmaceutics, 194 (1), 91-102, 2000.

[8] Irache, J. M.; Merodio, M.; Arnedo, A.; Camapanero, M. A.; Mirshahi, M.; Espuelas, S., Albumin nanoparticles for the intravitreal delivery of anticytomegaloviral drugs. Mini reviews in medicinal chemistry, 5 (3), 293-305, 2005.

[9] Commisso C, Davidson SM, Soydaner-Azeloglu RG, Parker SJ, Kamphorst JJ, Hackett S, Grabocka E, Nofal M, Drebin JA, Thompson CB, et al: Macropinocytosis of protein is an amino acid supply route in Ras-transformed cells. Nature, 497:633-637, 2013.

[10] Müller, A.; Homey, B.; Soto, H.; Ge, N.; Catron, D.; Buchanan, M. E.; McClanahan, T.; Murphy, E.; Yuan, W.; Wagner, S. N., Involvement of chemokine receptors in breast cancer metastasis. nature, 410 (6824), 50-56, 2001.

Peptide RF482 Encapsulated Chitosan – PLGA Nanoparticles Inhibit Respiratory Syncytial Virus

S Bawage, P Tiwari, V Dennis, S Singh

Center for NanoBiotechnology Research, Alabama State University
1627, Harris way, Life science building, Montgomery
Alabama 36104, USA, ssingh@alasu.edu

ABSTRACT

Worldwide, human respiratory syncytial virus (RSV) is one of the leading causes of lower respiratory tract infections in children and older adults. Currently, an anti-viral drug ribavirin and RSV neutralizing monoclonal antibody- palivizumab are the only relief options. However, efficacy, cost and adverse effects are the problems associated with them. Nanotechnology has significantly contributed in the medicine, especially, drug delivery. Nano-encapsulation of drugs provides the advantage of longer retention and controlled release. RSV infects the host cells by F protein mediated cell membrane fusion. Chemical or protein fusion inhibitors that block this vital step of RSV infection are currently explored. In this study, we synthesized the biodegradable nanoparticles by double emulsion method, firstly we encapsulated RSV F protein derived RF482 peptide (a fusion inhibitor) with poly [lactic acid-glycolic acid] (PLGA) and then by chitosan. The size and shape of nanoparticles were determined by electron microscopy and differential light scattering. Charges were analyzed by measuring the zeta potential. Cytotoxic concentration of these nanoparticles for HEp-2 cell lines was determined by MTT assay. Fluorescence microscopy showed the attachment of nanoparticles on the HEp-2 cells. Our study showed that, the mucoadhesive chitosan-PLGA nanoparticles (without RF482 peptide) contribute in RSV inhibition. However, the RF482 peptide encapsulated chitosan-PLGA nanoparticles treatment showed higher RSV inhibition than the unloaded nanoparticles.

Keywords: RSV, Chitosan, PLGA, fusion inhibitors, nanoparticles

1 INTRODUCTION

RSV is one of the leading causes of lower respiratory tract infections in children and old adults. Currently there is no vaccine, except for the passive vaccine Palivizumab. RSV infects the cells with the help of F protein that is involved in attachment and entry of RSV into the host cell. RSV F forms a trimeric coiled-coil for fusing viral and cell membrane. This process has attracted avenues to develop fusion inhibitors (chemical and peptide) that can inhibit RSV entery into the cells [1]. There is a wide scope for the development of drugs against RSV since the drugs face the issues regarding toxicity, pharmacokinetic and pharmacodynamic stability. Various means are explored to have a reasonable mode of delivery and attain the desired results [2]. To have an efficient delivery system and protect the drugs from cellular degradation, various encapsulation methods are employed. Genereally, polymeric nanoparticles of chitosan, PLGA, polylactic acid (PLA), etc, are prefered choice for such stated task. These materials are muchoadesive, genereally biodergardable, less toxic, have controlled release properties and can act as adjuvant [3]. The present work exploits these properties and studies the effect of RF-482, a peptide fusion inhibitor encapsulated by chitosan – poly (lactic acid glycolic acid) nanoparticles (CPRF482). The nanoparticles were characterized by various biophysicsal methods and RSV inhibition was determined.

2 MATERIALS AND METHODOLOGY

2.1 Materials

Human epithelial cell lines (HEp-2) (ATCC CCL-23) were cultured in minimal essential medium supplemented with 10% fetal bovine berum (FBS), 2 mM L-Glutamine and penicillin, streptomycin and kanamycin (Sigma, USA). Respiratory Syncytial Virus (RSV) strain Long was propagated as per the instructions from ATCC. Peptide RF-482 was synthesized with the N terminal modification with H-AEEAc linker from Bacham Americas Inc.

2.2 Synthesis of Chitosan-PLGA nanoparticle

Two ml of PLGA (1%) and 500 µl (500 µg) RF-482 peptide were mixed and sonicated on ice. The resulting emulsion was added dropwise to 2% poly vinyl alcohol (PVA) and stirred for 2 h. Then 500 µl of 1% chitosan was added dropwise to stirring PVA and stired overnight and harvested at 40000 x g for 10 min. The pellet was washed twice and lyophilize. Nanoparticle stock of 1 mg/ml was prepared in MEM (2% FBS) and sonicated for 5-10 min to get uniform suspension. Control chitosan and PLGA nanoparticles were prepared similarly.

2.3 Dynamic light scattering

Nanoparticle suspension was prepared in distilled water. Hydrodynamic radius and zeta potential were measured using standard procedure for Malvern Zetasizer Nano ZS instrument.

2.4 FT-IR spectroscopy

FT-IR characteristic peaks corresponding to the nanoparticles were acquired in the spectral range of 400-4000 cm^{-1} using attenuated total reflectance (ATR) sampling accessory of Thermo Fisher Nicolet 380 instrument.

2.5 Electron microscopy

Electron microscopy was performed to image the nanoparticles. A drop of nanoparticle suspension was placed on copper grids samples and stained with 1% phosphotungstic acid. Nanoparticles were imaged using Zeiss EM10 transmission electron micrscope [4].

2.6 Cytotoxicity assay

HEp-2 cells (17000) were seeded into each well of 96 well plate and incubated at 37°C (5% CO_2) for 24 h. Cells were washed with HBSS and 100 µl nanoparticles (chitosan-PLGA and PLGA) of appropriate concentration suspended in MEM (10% FBS) were added to the cells and incubated for 48 h. MTT assay 3-(4,5-dimethylthiazol-2-yl)-2,5-diphenyltetrazolium bromide (CellTiter 96® Non-Radioactive Cell Proliferation Assay) was performed according to the manufacturer's instruction (Promega Madison, WI). The formazon crystals were disssolved and read at 570 nm with the reference wavelength of 650 nm. The cells treated with nanoaprticles were compared with control cells (non-treated) to determine the vialbity.

2.7 Fluorescence microscopy

HEp-2 cells (30000) were seeded in an 8 chambered slide and incubated at 37°C (5% CO_2) for 24 h. Next day, the cells were washed twice with 250 µl HBSS and nanoparticles were added to the cells and incubated for 1 h. Volume was made to 250 µl/chamber with MEM (2% FBS) and incubated for 48 h at 37°C (5% CO_2). Cells were fixed with 10% TCA for 30 min and washed with 70%, 90%, 100% ethanol successively for 5 min each. Cells were then washed with 1X PBS, stained with DAPI and cell mask dyes. Cover slip was placed over the slide and sealed. The slide was observed under Nikon Ti Eclipse microscope.

2.8 RSV Inhibition

In a12-well plate, $2x10^5$ HEp-2 cells per well were seeded and incubated for 24 h. Cells were washed with HBSS before adding RSV 100 µl (100 PFU) and incubated for 1 h. Respective concentration of nanoparticles were added to the cells and total volume was made with MEM (2% FBS) to 1 ml and incubated at 37°C for 48 h. Plate was centrifuged at 3000 rpm for 5 min followed by removing the medium. Cells were trypsinized and pelleted for RNA extraction (QIAamp Viral RNA Mini Kit).

cDNA was synthesized using SuperScript® II Reverse Transcriptase (Invitrogen™) as instructed by manufacturer. Absolute quantification of RSV F gene copies was done using TaqMan chemistry, Applied Biosystem ViiA 7 real time PCR (Applied Biosystem International, Foster City, CA, USA) as previously described by Mentel et al. [5]. Each 20 µl qPCR reaction was compared with the standard curve and controls, to know the RSV F copies for the treatment groups.

3 RESULTS

The zeta size and zeta potential was recorded and it was observed that the average size of nanoparticles encapsulating RF-482 was smaller (133.30 nm) than the control PBS encapsulating nanoparticles (212.17 nm). The zeta potential for CPPBS and CPRF482 was 37.0 ± 3.3 mV and 22.8 ± 1.3 mV, respectively. FT-IR spectroscopy showed the encapsulation of the RF-482 and TEM images show the layer of chitosan on to the PLGA, size and shape of nanoparticles (Figure 1).

A

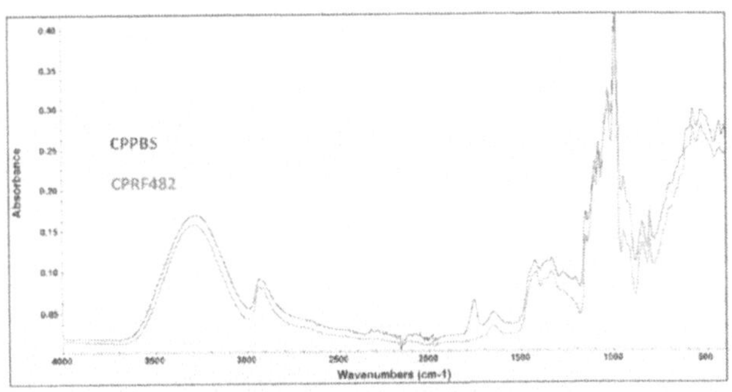

B

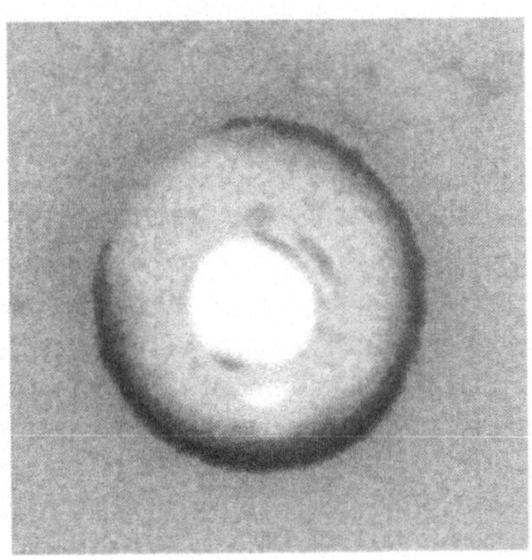

C

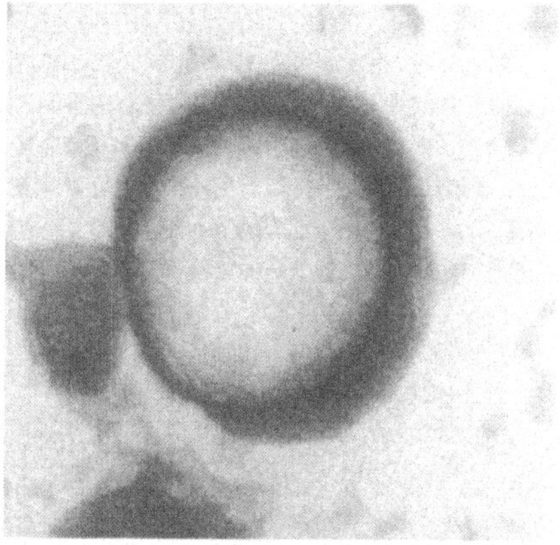

Figure 1: Synthesis and characterization of chitosan-PLGA nanoparticles to evaluate the RSV inhibition. (A) FT-IR spectrum of nanoparticles, (B) TEM image of CPPBS nanoparticle and (C) CPRF482 nanoparticle

MTT assay showed that chitosan-PLGA and PLGA nanoparticles were not toxic to HEp-2 cells up to 100 μg/ml after 48 h of exposure (Figure 2). The viability was 80% for 200 and 400 μg/ml of nanoparticles. Therefore, the nanoparticle was tested for the concentration of 25, 50 and 100 μg/ml against RSV.

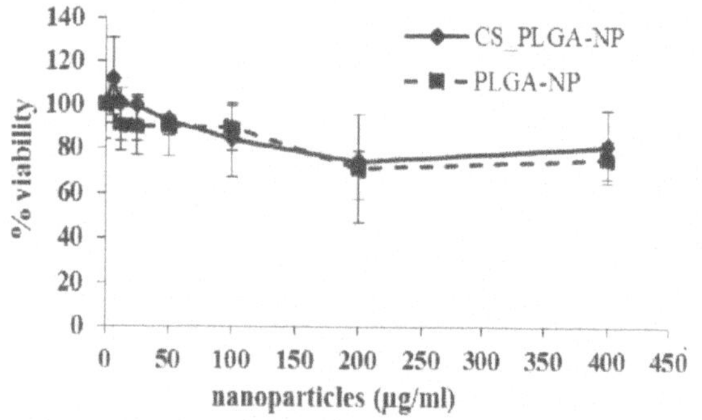

Figure 2: MTT assay for different concentration of PLGA and Chitosan-PLGA nanoparticles at 48 h

The RF-482 is 39 amino acid long and is UV invisible, hence it was FITC labeled. To know the interaction of nanoparticles with the cells fluroscence micrscopy was performed. Nanoparticles were observed adhering as aggregates over the surface of the cells (Figure 3A). Also, the uptake of CPRF482 nanoparticles makes the cell appear green (FITC labelled), as the peptide RF-482 seems distributed in the cell (Figure 3B). The nucleus appears blue due to DAPI staining and the cytoplasm appears red.

A

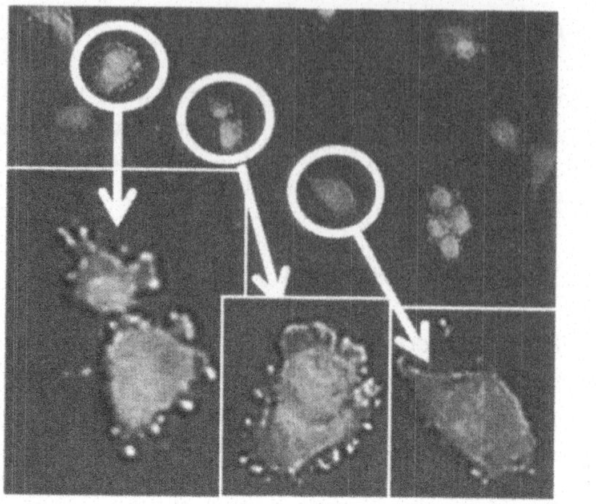

B

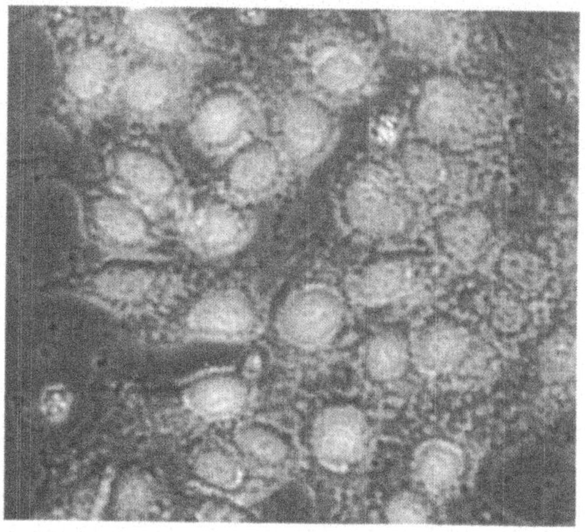

Figure 3: Uptake of chitosan-PLGA nanoparticles by HEp-2 cells after 48 h under fluorescence microscope. (A) Cells with CPPBS; inset showing nanoparticles attached to the cells; (B) cells appearing green (FTIC labeled RF-482) due to uptake of CPRF482 nanoparticles. (Blue-DAPI, Red-Cell membrane mask, Green-FTIC)

RSV was inhibited by CPRF482, i.e. chitosan-PLGA nanoparticles itself and also by RF-482 encapsulated chitosan-PLGA nanoparticles (Figure 4). The concentration of 50 and 100 μg/ml of nanoparticles (CPPBS or CPRF482) inhibited the RSV. However, at 25 μg/ml, CPRF482 nanoparticle were effective than CPPBS nanoparticles inhibiting RSV. The inhibition occurs mainly due to the fusion inhibitor RF-482, that binds the RSV F protein therby RSV can not attach to the cells. Our results also, indicates that RSV inihibition was not only due to peptide but also by the chitosan-PLGA. This inhibition can be due to the physical hinderence created by chitosan-PLGA, that does not allow RSV to interact with cell membrane. This study shows that chitosan-PLGA can be used for efficent delivery of the peptide and by itself act as an RSV inhibitor.

TechConnect Briefs 2015, TechConnect.org, ISBN 978-1-4987-4729-5

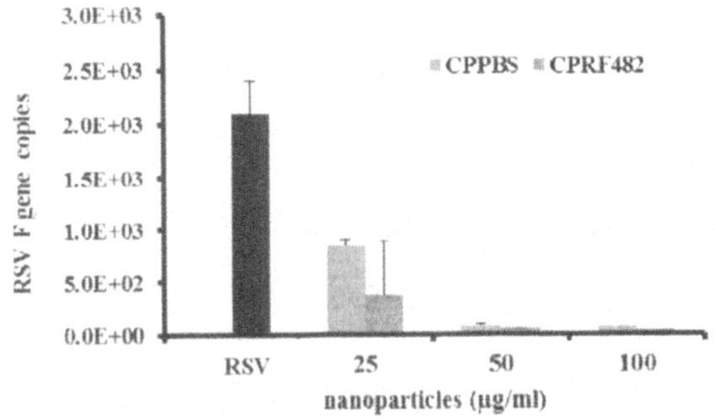

Figure 4: RSV inhibition after treatment with nanoparticles (CPPBS and CPRF482) at 25, 50 and 100 µg/ml concentration

4 CONCLUSION

Chitosan and PLGA are FDA (Food and Drug Administration) approved materials. This attracts fabrication of nanoparticles for delivery of drugs. The lower toxicity, muco-adhesive properties and controlled drug release has allured pharmaceutical research. Our study concludes that RF-482 interacts with RSV F six helix bundle and may inhibit RSV F mediated cell infection. Chitosan PLGA nanoparticles encapsulating RF-482 were characterized and found effective in inhibiting RSV at concentration of 100 µg/ml. It was surprising that chitosan-PLGA-PBS nanoparticles were also able to inhibit RSV, although the inhibition was less compared to RF-482 encapsulated nanoparticles.

REFERENCES

1. Ni L, Zhao L, Qian Y, Zhu J, Jin Z, Chen YW et al. Design and characterization of human respiratory syncytial virus entry inhibitors. Antiviral therapy. 2005;10(7):833-40.

2. Bawage SS, Tiwari PM, Pillai S, Dennis V, Singh SR. Recent advances in diagnosis, prevention, and treatment of human respiratory syncytial virus. Adv Virol. 2013;2013:595768. doi:10.1155/2013/595768.

3. Kumari A, Yadav SK, Yadav SC. Biodegradable polymeric nanoparticles based drug delivery systems. Colloids and surfaces B, Biointerfaces. 2010;75(1):1-18. doi:10.1016/j.colsurfb.2009.09.001.

4. Eroglu E, Tiwari PM, Waffo AB, Miller ME, Vig K, Dennis VA et al. A nonviral pHEMA+chitosan nanosphere-mediated high-efficiency gene delivery system. Int J Nanomedicine. 2013;8:1403-15. doi:10.2147/IJN.S43168.

5. Mentel R, Wegner U, Bruns R, Gurtler L. Real-time PCR to improve the diagnosis of respiratory syncytial virus infection. J Med Microbiol. 2003;52(Pt 10):893-6.

PLGA and PLGA-PEG nanoparticles loaded with curcumin and their effect on aflatoxin B1 *in vitro*.

A. Rivero-Gonzalez, *R. Diaz-Torres and P. Ramirez-Noguera.

Universidad Nacional Autónoma de México, Facultad de Estudios Superiores Cuautitlán, Unidad de Investigación Multidisciplinaria, Laboratorio 9: Toxicología Celular. México.

ABSTRACT

Aflatoxin B1 (AfB1) is a potent hepatotoxic and hepatocarcinogenic mycotoxin. Lipid peroxidation and oxidative DNA damage are the main manifestations of toxicity induced by aflatoxin B1 that could be counteracted by antioxidants. The aim of this study was to evaluate the effect of nanoparticles prepared with polylactic-co-glycolic acid (PLGA) and PLGA with polyethylene glycol (PLGA-PEG) loaded with curcumin in human hepatocytes exposed to aflatoxin B1 in vitro.

Some biomarkers associated to oxidative stress were estimated. The results showed an increase in GSH concentration in cells exposed to nanoparticles with curcumin. PLGA-PEG nanoparticles showed more significant difference compared to control. These results suggest that curcumin loaded nanoparticles can modulate cellular effects associated with oxidative stress in vitro.

Keywords: curcumin, nanoparticles, plga, plga-peg, aflatoxin b1.

1 INTRODUCTION

Aflatoxins are toxic metabolites produced for food during growth of the fungal groups Aspergillus flavus and Aspergillus parasiticus [1,2].

Aflatoxin B1 is a potent hepatotoxic and hepatocarcinogenic mycotoxin. Lipid peroxidation and oxidative DNA damage are the main manifestations of toxicity. Many parts of plants, such as curcumin, have been studied as protectors against liver damage associated with lipid peroxidation as in the toxicity induced by aflatoxin B1 [2]. Unfortunately, it has been shown that the low aqueous solubility and limited oral bioavailability are major obstacles in its development as a therapeutic agent [3].

1.1 PLGA and PLGA-PEG Nanoparticles

The new drug delivery systems have had a huge impact on medical technology [4]. For efficient drug release, the nanoscale range of the particles is critical, because this can facilitate the increase in drug cellular uptake, directed at target cells, which in turn, enhances the viability of therapeutic compounds [5].

PLGA is part of a family of biodegradable polymers approved by the FDA which are highly biocompatible and have been extensively studied as delivery vehicles for drugs, proteins and other macromolecules such as DNA, RNA and peptides. [1,3,6].

PLGA is a co-polymer of polylactic acid (PLA) and polyglycolic acid (PGA). Is best defined biomaterial available for drug administration with respect to design and performance [7]. However, the need for better delivery formulations that incorporate a variety of drugs and methods of administration has resulted in the development of various types of block copolymers with polyethylene glycol (PEG). This PEG layer serves as a barrier and reduces interactions with foreign molecules by steric repulsion, leading to greater storage stability.

2 MATERIALS AND METHODS

2.1 Preparation of PLGA and PLGA-PEG Nanoparticles loaded with Curcumin.

The preparation of both PLGA and PLGA-PEG nanoparticles loaded with curcumin was using the technique called nanoprecipitation. 75 mg of PLGA or PLGA-PEG and 3.8 mg of curcumin were dissolved in 10 mL of acetone, the solution was added dropwise to a solution of 1% of PVA (poly Vynil Alcohol) under agitation, the stirring was maintained for 10 min. Subsequently the organic solvent (acetone) is evaporated in a rotary evaporator. The resulting suspension was centrifuged, preparing a system containing 5 mL of glycerol, 20 mL of the nanoparticles suspension and 10 mL of deionized water, the system is centrifuged at 15000 rpm for 1 h, the supernatant was removed, collecting 5mL near to the interface and stored at 4°C protected from light.

2.2 Determination of encapsulation efficiency.

The determination of curcumin encapsulated was performed by spectrophotometry, previously performing a calibration curve. To determine the encapsulation efficiency, 2 mL of nanoparticle suspension (both PLGA and PLGA-PEG curcumin loaded) were placed in vials to constant weight and placed in an oven at 60 ° C for 24 h to allow water evaporation, subsequently the resulting solid is weighed, and it was suspended in 2 mL of acetonitrile. The system was gentle stirred for 4 h to allow curcumin diffuse into the solvent, finally it was centrifuged at 13000 rpm for 20 min, the supernatant was removed and diluted 1:5 with ethanol (80%) and the estimation was performed by spectrophotometry .

The encapsulation efficiency was calculated by the following equation:

$$Encapsulation\ efficiency\ (\%) = \frac{(\text{Total drug content in nanoparticles})}{(\text{Total drug amount})} x100$$

2.3 Characterization of Nanoparticles

Characterization of particle size and zeta potential were performed by dynamic light scattering and laser doppler electrophoresis respectively in Malvern Instruments Zetasizer device. Regarding morphology was performed by scanning electron microscopy (SEM).

2.4 Experimental Model

To determine the hepatoprotective effect of Curcumin nanoparticles against AfB1 induced intoxication, was used as a model cell line HepG2 [8,9].

2.5 Cell culture and exposure scheme.

Cells were maintained in culture medium Dulbecco's Modified Eagle Medium (DMEM) supplemented with 10% fetal bovine serum and 2% antibiotic at 37°C and 5% of CO_2. Curcumin was dissolved in dimethylsulfoxide (DMSO) and then diluted with culture medium to adjust the desired concentrations, the final concentration of DMSO in culture was not greater than 0.1%, likewise, the controls with 0.1% DMSO were included in all experiments.

Cells were seeded in 24-well plates until confluence. Initially exposure to AfB1 (25 µM) for 4 h was performed, then a second exposure with free or encapsulated curcumin at different concentrations for 4 h was done.

2.6 Cell viability (MTT)

Two hours before exposure time completion described above, 5 mL of MTT reactant were added in a concentration of 5 mg/mL. After the exposure time has elapsed, the culture medium was removed and added 0.3 mL of a buffer nonidet in isopropanol and left under gentle stirring for 20 min. Finally the solution was removed and the viability is determined by spectrophotometry in Elisa plate reader (mrc scientific instruments).

2.7 Determination of reduced glutathione (GSH)

Once the exposure time is completed, the culture medium is removed from the wells, washed 2 times with cold PBS and cells were harvested by scraping. The cell suspension was centrifuged at 12000 rpm for 10 min at 4°C, then the supernatant is removed and the button resuspended in PBS with protease inhibitor. The suspension is sonicated and 10 µL of this suspension are taken for protein determination. To the remaining suspension is added 5 µL of sulfosalicylic acid (5%) and it is incubated at room temperature for 10 min, finally centrifuged at 12000 rpm for 10 min at 4°C and the supernatant is recovered. 50 µL of this solution was taken and placed in a well of a 96-well microplate, finally 150 µL of reaction buffer are added and the GSH in the sample is quantified in Elisa plate reader at $\lambda = 412$ nm.

2.8 Statistical analysis.

The data are presented with standard error of the mean (SEM) analyzed using ANOVA of one via. All experiments were performed at least 6 times independently.

3 RESULTS AND DISCUSSION

Table 1 shows results obtained for curcumin loaded nanoparticles of PLGA and PLGA-PEG, note that the PLGA nanoparticles, despite having a larger size, have a lower encapsulation percentage.

3.1 Curcumin Nanoparticles Characterization.

Table 1. Results of the development of Curcumin nanoparticles of PLGA and PLGA-PEG

Polymer	Encapsul. Efficiency (%)	Part. Size (nm)	PDI	Zeta Potential (mV)
PLGA	22.26	261.2	0.095	-1.46
PLGA-PEG	64.46	104	0.171	-1.625

This PLGA encapsulation efficiency can be attributed to several factors. First, both PLGA and PLGA-PEG have hydrophobic nature, for that reason, it is easy for them to catch the hydrophobic curcumin. Second, the hydrophobic nature of curcumin, which results in minimal migration of the drug to the external aqueous phase during the nanoparticles formation [10].

In the PLGA-PEG nanoparticles, PEG chains are oriented toward the external aqueous phase, enveloping the PLGA-curcumin complex. This PEG layer serves as a barrier and reduces interactions with foreign molecules by steric repulsion as well as giving a hydrophilic property, so these PEG chains provide greater stability, reducing interaction and preventing migration of curcumin outwards, giving greater load stability [7].

3.2 Nanoparticles Morphology

The morphology was performed by scanning electron microscopy (Jeol SEM JSM6010LA).

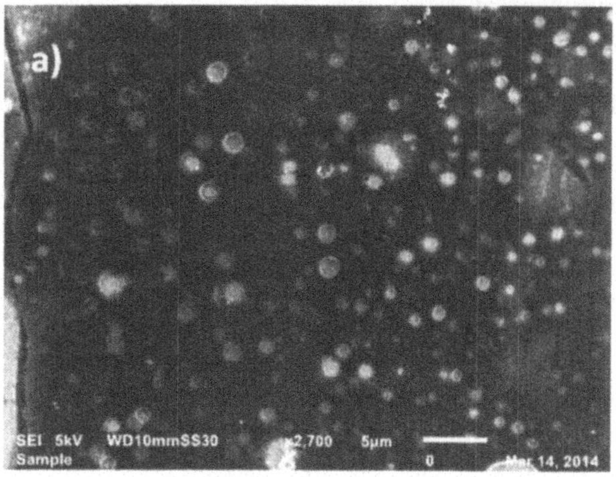

Figure 1: PLGA Curcumin Nanoparticles SEM Micrography.

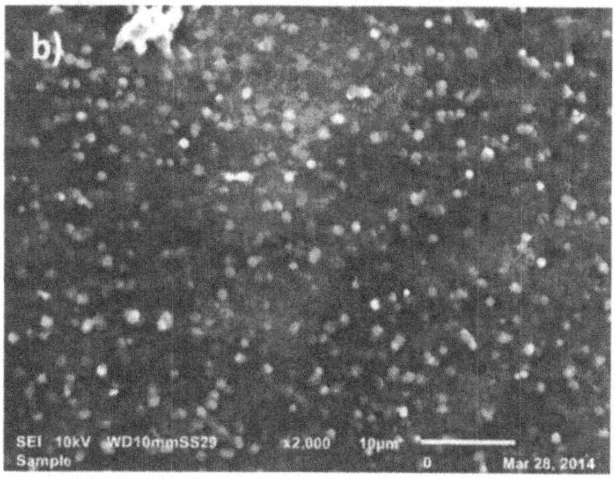

Figure 2: PLGA-PEG Curcumin Nanoparticles SEM Micrography.

In SEM micrographs for Curcumin Nanoparticles (Figure 1 and 2) we can see spherical nanoparticles with an uniform size distribution for both PLGA and PLGA-PEG nanoparticles, having a polydispersity index (PDI) of 0.095 and 0.171 respectively.

3.3 Cell viability (MTT)

Figure 3 shows the results of the exposure of HepG2 cells against a concentration of 25 µM of AfB1 for 4 h, and subsequent exposure to free curcumin and Curcumin Nanoparticles at concentrations of 0.2, 2, 5 and 10 µM. This results showed viability above of 100% in both Curcumin Nanoparticles (PLGA and PLGA-PEG).

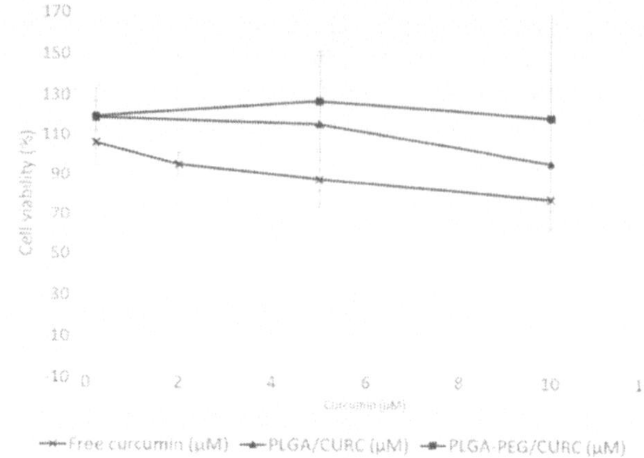

Figure 3: Viability of HepG2 cell line post exposure for 4 h to free curcumin and Curcumin Nanoparticles at different concentrations (0.2, 2, 5 and 10 µM), after an exposure with AfB1 (25 µM) for 4h.

Figure 3 shows the means of the treatments studied, which shows that there were significant differences between the values of cell viability treatments of free curcumin and treatments with Curcumin Nanoparticles for both PLGA and for PLGA-PEG, denoting those differences starting in 0.2 µM, increasing cell viability, even on the treatment of major concentration of free curcumin.

These results clearly show the hepatoprotective effect of curcumin against mycotoxin intoxication. However, due to the decrease in cell viability generated by free curcumin and AfB1 be totally different processes, as mentioned above, curcumin can change the morphology of the cell surface and lead to a pro-apoptotic process, this process is dose-dependent [11,12]; on the contrary, the process of cytotoxicity of aflatoxin, wherein the oxidative stress is the main toxic effect [1], so further studies are needed to define more accurately the hepatoprotective effect of curcumin over AfB1 intoxication.

3.4 Determination of reduced glutathione (GSH)

In Figure 4 the results for the quantification of GSH in HepG2 cells are shown. Although no statistically significant difference between treatments are seen, except between free curcumin at a concentration of 2 µM and 20 µM and Curcumin Nanoparticles of PLGA at 0.2 µM and 20 µM compared with the control, in the figure 4 we observed a tendency wherein the amount of GSH present in the sample significantly increases when concentration of free curcumin and encapsulated curcumin in PLGA and PLGA-PEG nanoparticles increases, since concentrations of 0.2, 2 and 20 µM.

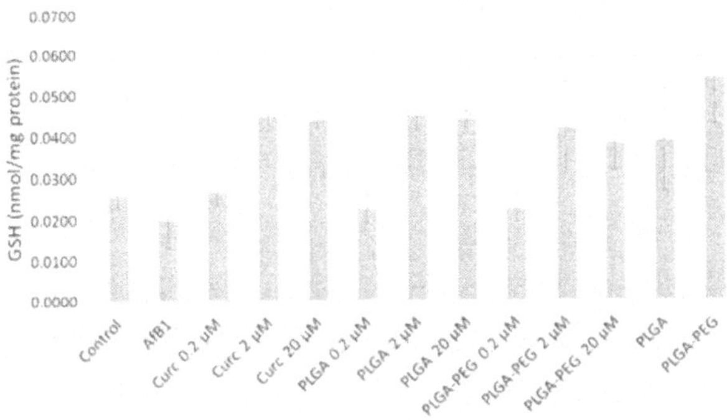

Figure 4: Quantification of GSH in HepG2 cells exposed to Curcumin Nanoparticles post an exposure of AfB1 25 μM for 4h. Error bars correspond to standard error of the means.

This is consistent with previous studies by Pou Kuan Leong et. al [13]. Where exposure of AML12 cells to curcumin was performed, finding an increase in GSH levels, coupled with a decrease in lipid peroxidation. The greatest effect of curcumin on the stimulation of GSH production may be related to their ability to directly induce dissociation of Keap 1 protein, a Nrf2 repressor, with an antioxidant modulation [13].

4 CONCLUSIONS

PLGA and PLGA-PEG nanoparticles loaded with curcumin were prepared, obtaining better results in terms of particle size and loading of curcumin with the polymer PLGA-PEG. The cell viability assay showed that the encapsulated curcumin has a better protective effect against AfB1 intoxication than free curcumin, showing an increased intracellular concentration of GSH, having a tendency in increasing GSH in cells treated with Curcumin Nanoparticles. Based on these results, we can say that the Curcumin Nanoparticles prepared with PLGA and PLGA-PEG have potential as a treatment for AfB1 intoxication.

REFERENCES

[1] El-Agamy. "Comparative effects of curcumin and resveratrol on aflatoxin B1-induced liver injury in rats. Organ toxicity and mechanisms." 2010;: p. 389–396.

[2] Fengyuan Wang. "Protective Effects of Sodium Selenite against Aflatoxin B1-Induced Oxidative Stress and Apoptosis in Broiler Spleen." International Journal of Environmental Research and Public Health. 2013;: p. 2834–2844.

[3] J. Shaikh. "Nanoparticle encapsulation improves oral bioavailability of curcumin by at least 9-fold when compared to curcumin administered with piperine as absorption enhancer." European Journal of Pharmaceutical Sciences. 2009;: p. 223–230.

[4] Indu Bala. "PLGA Nanoparticles in Drug Delivery: The State of the Art. Critical Reviews in Terapeutic Drug Carrier Systems." 2004;: p. 387-422.

[5] Boitumelo Semete. "*In vivo* evaluation of the biodistribution and safety of PLGA nanoparticles as drug delivery sistems." Nanomedicine: Nanotechnology, biology, and medicine. 2010; 6: p. 662-671.

[6] Leanne L. Bedard TEM. "Aflatoxin B1 induced DNA damage and its repair." Cancer letters. 2006;: p. 174-183.

[7] Hirenkumar K. Makadia. "Poly Lactic-co-Glycolic Acid (PLGA) as Biodegradable Controlled Drug Delivery Carrier." Polymers. 2011;: p. 1377-1397.

[8] Ruijie Liu. "Degradation of aflatoxin B1 in aqueous medium through UV irradiation." European Food Research Technology. 2011;: p. 1007–1012.

[9] Xiaohu Luo. "Effect of ozone treatment on aflatoxin B1 and safety evaluation of ozonized corn." Food control. 2014;: p. 171-176.

[10] Amalendu P Ranjan. "Scale up, optimization and stability analysis of Curcumin C3 complex-loaded nanoparticles for cancer therapy." Journal of Nanobiotechnology. 2012;: p. 10-38.

[11] Mu Wang. "Curcumin induced HepG2 cell apoptosis-associated mitochondrial membrane potential and intracellular free Ca2+ concentration." European Journal of Pharmacology. 2011;: p. 41-47.

[12] Tathagata Choudhuri. "Curcumin Selectively Induces Apoptosis in Deregulated Cyclin D1-expressed Cells at G2 Phase of Cell Cycle in a p53-dependent Manner." The Journal of Biological Chemistry. 2005 Mayo; 280(20).

[13] Pou Kuan Leong. "Prooxidant-Induced Glutathione Antioxidant Response in Vitro and in Vivo: A Comparative Study between Schisandrin B and Curcumin." Biological & Pharmaceutical Bulletin. 2012;: p. 464–472.

Schwann Cells and Dorsal Root Ganglion Neurons Are Differentially Susceptible to Oxidative Stress Induced by Silicon Dioxide Nanoparticles

James C.K. Lai[1], Ashvin R. Jaiswal[2], Vinay K. Idikuda[3], Jean Pfau[4], Alok Bhushan[5], and Solomon W. Leung[6]

[1]Department of Biomedical and Pharmaceutical Sciences, College of Pharmacy, Division of Health Sciences, Idaho State University, Pocatello, ID 83209, USA Fax: 208-282-4305; Tel: 208-282-2275; email: lai@pharmacy.isu.edu

[2]Department of Immunology, University of Texas MD Anderson Cancer Center, Houston TX 77054, USA Tel: 713-794-1179; email: ARJaiswal@mdanderson.org

[3]Department of Physiology and Biophysics, Virginia Commonwealth University, Richmond, VA 23298 USA Tel: 208-705-2081; email: idikudav@mymail.vcu.edu

[4]Department of Biological Sciences, College of Science and Engineering, Idaho State University, Pocatello, ID 83209, USA Fax: 208-282-4570; Tel: 208-282-3914; email: pfaujean@isu.edu

[5]Department of Pharmaceutical Sciences, Jefferson School of Pharmacy, Thomas Jefferson University, Philadelphia, PA 19107, USA Fax: 215-503-3082; Tel: 215-503-5039; email: Alok.Bhushan@jefferson.edu

[6]Department of Civil and Environmental Engineering, College of Science & Engineering, Idaho State University, Pocatello, ID 83209, USA Fax: 208-282-4538; Tel: 208-282-2524; email: leunsolo@isu.edu

ABSTRACT

Silicon dioxide (SiO_2) particles, including nanoparticles, have been increasing employed in diverse industrial and biomedical applications including those in drug formulations, food, and cosmetics as SiO_2 has been generally regarded as a non-toxic substance. However, the environmental safety and health impact of SiO_2 particles have not been elucidated. This study investigated the hypothesis that SiO_2 nanoparticles exert differential cytotoxic effects on DRG neurons and Schwann cells. Treatment with SiO_2 nanoparticles induced dose-related decreases in survival of DRG neurons and Schwann cells with Schwann cells being more susceptible. SiO_2 nanoparticles induced concentration-related decreases in glutathione (GSH) in Schwann cells and such decreases were related to their decreases in survival. SiO_2 nanoparticles also induced some decreases in GSH in DRG neurons. Expression of manganese superoxide dismutase (Mn-SOD) in Schwann cells showed concentration-related decreases when treated with SiO_2 nanoparticles. However, expression of Mn-SOD in DRG neurons was increased. Thus, our findings may have pathophysiological implications in the biocompatibility and health hazard of SiO_2 nanoparticles and may be critically relevant to toxicological studies prior to clinical trials of drugs formulated with and/or delivered employing agents containing such nanoparticles.

Key words: silicon dioxide nanoparticles, cytotoxicity of silicon dioxide nanoparticles, Schwann cells and dorsal root ganglion neurons, oxidative stress & glutathione, biocompatibility, nanotoxicity

1 INTRODUCTION

Silicon dioxide (SiO_2) particles, including nanoparticles, have been increasing employed in diverse industrial and biomedical applications including those in drug formulations, food, and cosmetics as SiO_2 has been generally regarded as a non-toxic substance [1-4]. Nevertheless, there have been reports that SiO_2 particles of different sizes down to nanoparticles are not as harmless as they were assumed to be [4]. Furthermore, the environmental safety and health impact of SiO_2 particles have not been elucidated [2-4].

There is evidence that the blood-brain barrier is inadequate to restrict entry of nanoparticles into the central nervous system [4]. Additionally, there are reports of nanoparticles penetrating through the skin and then entering peripheral nerve terminals and moving by retrograde transport from the nerve terminals to neuronal perikarya (or cell bodies) [4]. Consequently, it is pathophysiologically important to investigate how nanoparticles may exert their putative cytotoxicity upon entry into the nervous system and neural cells either through the blood-brain barrier or via retrograde transport upon entry through the skin [4].

For over a decade, we have been developing various neural cell models *in vitro* to facilitate the systematic investigation of the putative cytotoxicity of nanoparticles and other nanomaterials [3, 5-10]. More recently, we have developed neural cell models employing physiologically important neural cell types derived from the peripheral nervous system (PNS), namely dorsal root ganglion (DRG) neurons and Schwann cells [8-10]. We have employed these versatile models *in vitro* to elucidate the putative cytotoxicity of nanoparticles of metallic and non-metallic oxides and of other nanomaterials and the underlying cellular and molecular mechanisms [4, 8-10]. This study was therefore initiated to test the hypothesis that SiO_2 nanoparticles exert differential cytotoxic effects on DRG neurons and Schwann cells.

2 MATERIALS AND METHODS

2.1 Materials

Silicon dioxide (SiO_2) nanoparticles (STREM Chemicals, Newburyport, MA, USA; Cat. #93-1434; <12 nm particle size, 99+ % colloidal silica) were dispersed in 100 ml of sterile saline in a sealed conical flask and the suspension stirred at ambient temperature overnight before being diluted to the specified concentrations for treatment of cells. The immortalized DRG neurons were a gift from Dr. Ahmet Höke's Laboratory at Johns Hopkins University School of Medicine. The rat Schwann cell line (R3) was obtained from ATCC (Manassas, VA, USA).

2.2 Cells and Culture Conditions

DRG neurons or Schwann cells were cultured in DMEM, supplemented with 10% (v/v) fetal bovine serum, 1% (w/v) sodium pyruvate, 0.292 g/l L-glutamine, and 1.5 g/l sodium bicarbonate and were incubated at 37°C and 5% (v/v) CO_2.

2.4 MTT Assay

Cellular viability was determined using the modified MTT assay [5]. DRG neurons or Schwann cells were seeded per well in a 48-well plate in DMEM with specified concentrations of SiO_2 nanoparticles. After incubation at 37°C for 36 hours, 10% of MTT (5 mg/ml in PBS) reagent was added to each well. After incubation for another 4 hours, the medium was removed gently and the cellular reaction product was solubilized in 200 µl DMSO. Then the optical density of the contents of each well was measured at 570 nm using a plate reader [5]. The absorbance corresponds to live cells present in each well [5].

2.5 Other Assays

The assay of total cellular content of glutathione (GSH) was carried out as we had described previously [11, 12]. The expression of manganese superoxide dismutase (Mn-SOD, a key antioxidant enzyme) in DRG neurons and Schwann cells treated with or without silicon dioxide nanoparticles was determined employing Western blot analysis as described previously [6, 7, 11].

2.6 Statistical Analysis of Data

Statistical significance of experimental results was analyzed with one-way ANOVA followed by Dunnett's post-hoc test with a minimum significance level set at $p < 0.05$ using the SPSS 17 software package.

3 RESULTS AND DISCUSSION

Results of our ongoing studies indicate that, consistent with our hypothesis, treatment with SiO_2 nanoparticles induced dose-related decreases in survival of DRG neurons and Schwann cells once the nanoparticles had entered both cell types. It is noteworthy that, in comparison with DRG neurons, Schwann cells were more susceptible to the cytotoxicity of the nanoparticles, especially at the lower treatment concentrations (data not shown).

To further elucidate the molecular mechanisms underlying the cytotoxic effects of SiO_2 nanoparticles in Schwann cells and DRG neurons, we tested the hypothesis that these nanoparticles induce oxidative stress on these two neural cell types to a greater or lesser extent.

We found that SiO_2 nanoparticles induced concentration-related decreases in the cellular content of glutathione (GSH) (Figure 1), an important antioxidant, in Schwann cells and such decreases were closely related to the nanoparticles-induced decreases in their survival. By contrast, even though SiO_2 nanoparticles also induced decreases in the cellular content of GSH in DRG neurons (Figure 2), the magnitude of the decreases was less compared to those induced by the nanoparticles in Schwann cells, especially at the highest treatment concentration of nanoparticles (Figure 1).

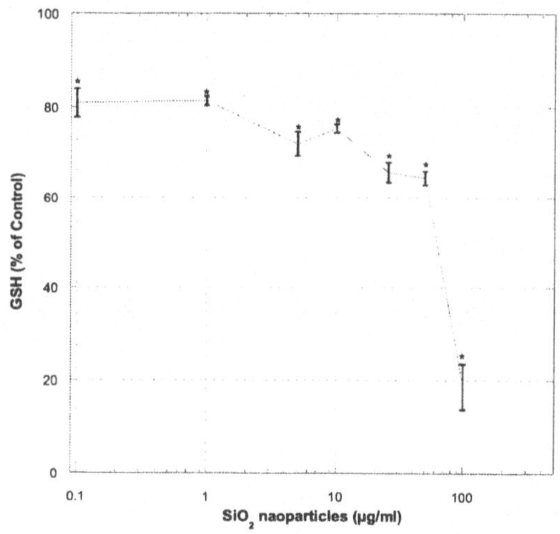

Figure 1. Effects of treatment with silicon dioxide nanoparticles on glutathione (GSH) content in Schwann cells. Schwann cells were treated with silicon dioxide nanoparticles at the specified concentrations for 36 hours and their GSH content was determined. Values are the mean ± S.E.M. of at least three separate determinations and are expressed as % of control; *p<0.05 versus control.

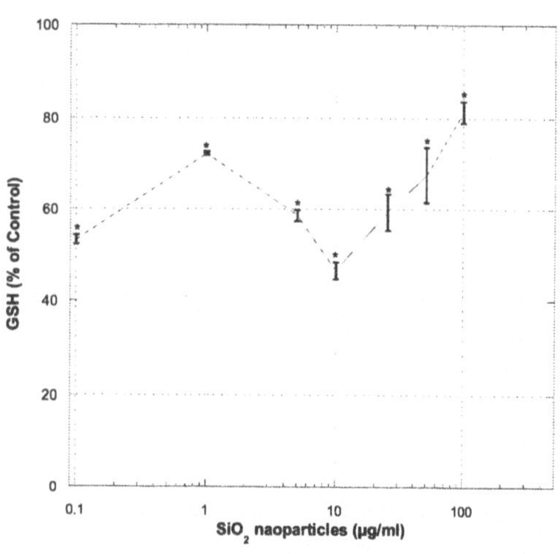

Figure 2. Effects of treatment with silicon dioxide nanoparticles on glutathione (GSH) content in DRG neurons. DRG neurons were treated with silicon dioxide nanoparticles at the specified concentrations for 36 hours and their GSH content was determined. Values are the mean ± S.E.M. of at least three separate determinations and are expressed as % of control; *p<0.05 versus control.

Consistent with the decreases in GSH in Schwann cells, their expression of manganese superoxide dismutase (Mn-SOD, a key antioxidant enzyme) also showed concentration-related decreases when treated with SiO_2 nanoparticles (data not shown). On the other hand, when DRG neurons were treated with SiO_2 nanoparticles, their expression of Mn-SOD was increased (data not shown). The latter observation suggested that the increased Mn-SOD expression in DRG neurons treated with SiO_2 nanoparticles might be a compensatory mechanism elicited by the treatment of the neurons with the nanoparticles. Indeed, compatible with this conclusion was the trend of increases in GSH content in DRG neurons treated with the nanoparticles at concentrations higher than 10 µg/ml (Figure 2). Thus, these findings prompted us to conclude that Schwann cells are more susceptible than DRG neurons to oxidative stress induced by SiO_2 nanoparticles.

4 CONCLUSIONS

Treatment with silicon dioxide (SiO_2) nanoparticles induced concentration-related decreases in survival of both Schwann cells and DRG neurons, the Schwann cells being more susceptible to the cytotoxicity of the nanoparticles. In parallel with the dose-related decreases in the survival of Schwann cells there were corresponding decreases in the content of glutathione (GSH), an important antioxidant, in Schwann cells treated with increasing concentrations of SiO_2 nanoparticles. On the other hand, even though treatment of DRG neurons with increasing concentrations of these nanoparticles also lowered their survival, their decreases in survival did not closely parallel the decreases in their GSH content induced by the nanoparticles. Furthermore, while treatment of Schwann cells with SiO_2 nanoparticles induced decreases in their expression of manganese superoxide dismutase (MnSOD), such treatments induced increases in the expression of MnSOD in DRG neurons. The latter increases in MnSOD expression in DRG neurons also coincided with the trend of increased GSH content suggesting that the concurrent increases in MnSOD expression and GSH content might be the consequence of compensatory mechanisms elicited in the DRG neurons by treatment with the higher levels of SiO_2 nanoparticles. These results strongly suggested that SiO_2 nanoparticles exerted differential cytotoxic effects on Schwann cells and DRG neurons, the two key neural cell types derived from the peripheral nervous system. Thus, our results may assume pathophysiological importance in determining how exposure to SiO_2 nanoparticles may impact on the structure and function of neural cells in the peripheral nervous system. Additionally, our findings may be critically relevant to toxicological studies prior to clinical trials of drugs formulated with and/or delivered employing agents containing such nanoparticles.

5 ACKNOWLEDGEMENTS

Our studies were supported by a DOD USAMRMC Project Grant (Contract #W81XWH-07-2-0078) and small project grants from Mountain States Tumor and Medical Research Institute (MSTMRI).

REFERENCES

[1] Lewinski N, Colvin V & Drezek R (2008) Cytotoxicity of Nanoparticles. small 4(1):26-49.

[2] Gao W, Lai JCK & Leung SW (2012) Functional Enhancement of Chitosan and Nanoparticles in Cell Culture, Tissue Engineering and Pharmaceutical Applications. Front Physiol 3, Article 321, pp. 1-13.

[3]. Lai JCK, Lai MB, Edgley KL, Bhushan A, Dukhande VV, Daniels CK & Leung SW (2007) Silicon Dioxide Nanoparticles Can Exert Cytotoxic Effects on Neural Cells. In Proceedings of 2007 Nanotechnology Conference and Trade Show, Volume 2, Chapter 8: Bio Materials and Tissues, pp. 741-743.

[4] Lai JCK, Jaiswal AR, Lai MB, Jandhyam S, Leung SW & Bhushan A (2015) Toxicity of Silicon Dioxide Nanoparticles in Mammalian Neural Cells. In: *Handbook of Clinical Nanomedicine — From Bench to Bedside* (Bawa R, Audette GF & Rubinstein I, eds.), Pan Stanford Series in Nanomedicine (Bawa, R, Series Ed.), Volume 1, Pan Stanford Publishing, Singapore (in press).

[5] Lai JCK, Jandhyam S, Lai MB, Dukhande VV, Bhushan A, Daniels CK & Leung SW (2008) Cytotoxicity of Metallic Oxide Nanoparticles: New Insights into Methodological Problems and Advances in Elucidation of Underlying Mechanisms. In Proceedings of the 12th World Multi-Conference on Systemics, Cybernetics and Informatics: WMSCI 2008, June 29th-July 2nd, 2008, Orlando, FL, USA, Volume II (Callaos N, Lesso W, Zinn CD, Baralt J, Eshraghian K, Severi S, Hashimoto S & Sahara T, eds.), pp. 10-15.

[6] Lai JCK, Lai MB, Jandhyam S, Dukhande VV, Bhushan A, Daniels CK & Leung SW (2008) Exposure to Titanium Dioxide and Other Metallic Oxide Nanoparticles Induces Cytotoxicity on Human Neural Cells and Fibroblasts. Int J Nanomed 3(4):533-545.

[7] Lai JCK, Ananthakrishnan G, Jandhyam S, Dukhande VV, Bhushan A, Gokhale M, Daniels CK & Leung SW (2010) Treatment of Human Astrocytoma U87 Cells with Silicon Dioxide Nanoparticles Lowers Their Survival and Alters Their Expression of Mitochondrial and Cell Signaling Proteins. Int J Nanomed 5:715-723.

[8] Jaiswal AR, Lu S, Pfau J, Wong YYW, Bhushan A, Leung SW, Daniels CK & Lai JCK (2011) Effects of Silicon Dioxide Nanoparticles on Peripheral Nervous System Neural Cell Models. Technical Proceedings of the 2011 NSTI Nanotechnology Conference and Expo – Nanotech 2011, Volume 3, Chapter 7: Environment, Health & Safety, pp. 541 – 544.

[9] Jain A, Jaiswal AR, Lu S, Wong YYW, Bhushan A, Leung SW, Daniels CK & Lai JCK (2011) Molecular Effects of Silicon Dioxide Nanoparticles on Cell Survival Signaling of Dorsal Root Ganglion (DRG) Neurons and Schwann Cells. Technical Proceedings of the 2011 NSTI Nanotechnology Conference and Expo – Nanotech 2011, Volume 3, Chapter 7: Environment, Health & Safety, pp. 545 – 548.

[10] Lu S, Jaiswal AR, Wong YYW, Bhushan A, Leung SW, Daniels CK & Lai JCK (2011) Differential Cytotoxic Effects of Titanium Oxide Nanoparticles on Peripheral Nervous System Neural Cells. Technical Proceedings of the 2011 NSTI Nanotechnology Conference and Expo – Nanotech 2011, Volume 3, Chapter 7: Environment, Health & Safety, pp. 533 – 536.

[11] Dukhande VV, Malthankar-Phatak GH, Hugus JJ, Daniels CK & Lai JCK (2006) Manganese Induced Neurotoxicity is Differentially Enhanced by Glutathione Depletion in Astrocytoma and Neuroblastoma Cells. Neurochem Res 31(11):1349-1357.

[12] Dukhande VV, Kawikova I, Bothwell ALM & Lai JCK (2013) Cell Death Induced by Depletion of Mitochondrial Glutathione. Apoptosis 18(6):702-712.

Co-Culturing Dorsal Root Ganglion Neurons with Schwann Cells Protects Them against the Cytotoxic Effects of Silver Nanoparticles

Wenjuan Gao*, Solomon W. Leung**, Alok Bhushan*** and James C.K. Lai****

*Department of Civil and Environmental Engineering, College of Science & Engineering, Idaho State University, Pocatello, ID 83209, USA Fax: 208-282-4538; Tel. 208-240-8606; email: gaowenj@isu.edu

**Department of Civil and Environmental Engineering, College of Science & Engineering, Idaho State University, Pocatello, ID 83209, USA Fax: 208-282-4538; Tel: 208-282-2524; email: leunsolo@isu.edu

***Department of Pharmaceutical Sciences, Jefferson School of Pharmacy, Thomas Jefferson University, Philadelphia, PA 19107, USA Fax:215-503-3082; Tel: 215-503-5039; email: Alok.Bhushan@jefferson.edu

****Department of Biomedical and Pharmaceutical Sciences, College of Pharmacy, Division of Health Sciences, Idaho State University, Pocatello, ID 83209, USA Fax: 208-282-4305; Tel: 208-282-2275; email: lai@pharmacy.isu.edu

ABSTRACT

Because of their presumed inertness, silver nanoparticles are increasingly employed in many consumer products. Nonetheless, their environmental and health hazard impact are not understood. Our previous studies led us to develop a co-culture model consisting of dorsal root ganglion (DRG) neurons and Schwann cells and to employ it to investigate our hypothesis that co-culturing DRG neurons with Schwann cells imparts protection on them against cytotoxicity induced by silver nanoparticles. Our results indicated that silver nanoparticles induced concentration- and time-related decreases in survival of DRG neurons or Schwann cells in monotypic cultures: both cell types withstood the cytotoxicity of silver nanoparticles and survived better when maintained in co-cultures. DRG neurons and co-cultures of DRG neurons and Schwann cells, but not Schwann cells alone, expressed synapsin. Silver nanoparticles suppressed synapsin expression in DRG neurons alone but not in co-cultures with Schwann cells. Schwann cells and co-cultures of DRG neurons and Schwann cells, but not DRG neurons alone, expressed glial fibrillary acidic protein (GFAP). However, silver nanoparticles markedly suppressed GFAP expression in Schwann cells alone but not in co-cultures with DRG neurons. Thus, our results provide support for our hypothesis and may be relevant to toxicological studies prior to clinical trials of drugs formulated with agents containing silver nanoparticles.

Keywords: dorsal root ganglion neurons, Schwann cells, co-culture, silver nanoparticles, cytotoxicity, nanotoxicity

1 INTRODUCTION

Recently many nanoparticles have been gainfully exploited in a variety of applications in diverse industries including biotechnology (e.g., as probes in cell & tissue imaging and in new drug delivery & targeting) because of their unique physico-chemical properties [1, 2]. Because of their presumed inertness, silver nanoparticles are increasing employed in many consumer products and are thus produced on an industrial and large scale [3-5]. Nonetheless, their environmental and health hazard impact are unknown [3-5].

Over the last decade, we have been developing a variety of neural and non-neural cell types as *in vitro* models for systematic investigation of putative cytotoxicity of various nanomaterials, including metallic and non-metallic nanoparticles [6-11]. More recently, we have developed two non-tumor neural cell models *in vitro* for systematic investigation of putative cytotoxicity of various nanomaterials including nanoparticles employing dorsal root ganglion (DRG) neurons and Schwann cells, which are physiologically important neural cell types of the peripheral nervous system [12-13].

Our previous findings that several metallic nanoparticles exerted cytotoxic effects on DRG neurons and Schwann cells [12-13] strongly suggest that they constitute excellent non-tumor neural models for elucidating the cellular and molecular mechanisms underlying the cytotoxicity of nanoparticles and other nanomaterials in neural cells and such mechanistic studies may have pathophysiological and/or pathogenic implications in neurodegeneration and nerve degeneration. Thus, our previous studies led us to develop a co-culture model consisting of DRG neurons and

Schwann cells and to employ it to investigate the hypothesis that co-culturing DRG neurons with Schwann cells imparts some protection on them against the cytotoxicity induced by silver nanoparticles.

2 MATERIALS AND METHODS

2.1 Materials

S16 Schwann cells were obtained from ATCC (Manassas,VA, USA). DRG (50B11) neurons were kind gifts from Dr Höke's Laboratory at Johns Hopkins School of Medicine. Thiazolyl blue tetrazolium bromide (MTT) and dimethyl sulfoxide (DMSO) were purchased from Sigma-Aldrich (St Louis, MO, USA). Fetal bovine serum (FBS) was obtained from Atlanta Biologicals (Lawrenceville, GA, USA). Silver nitrate ($AgNO_3$) and trisodium citrate ($C_6H_5Na_3O_7•2H_2O$) were purchased from Fisher Scientific (Pittsburgh, PA, USA). The monoclonal antibody against glial fibrillary acidic protein (GFAP) was obtained from Cell Signaling Technology (Beverly, MA, USA). The polyoclonal antibody against synapsin, goat anti-rabbit IgG secondary antibody, and goat anti-mouse IgG secondary antibody were purchased from Abcam Inc. (Cambridge, MA, USA). The monoclonal antibody against β-actin was purchased from Santa Cruz Biotechnology, Inc. (Dallas, TX, USA). All chemicals were of analytical grade and were usually obtained from Sigma-Aldrich.

2.2 Cell Culture

S16 Schwann cells or DRG neurons were cultured in Dulbecco's modified Eagle's medium supplemented with 10% (v/v) FBS and 25 mM glucose at 37°C and with 5% (v/v) CO_2.

2.3 Cell viability assay

Cell survival/proliferation was determined by using the modified MTT assay [8-11]. S16 Schwann cells were seeded (3000 cells/well) into a 24-well plate and allowed to attach and grow for 1 hour. Then the same number of DRG neurons were seeded onto the substratum layer of Schwann cells and cultured as described above. After 1 hour, cells were treated with or without specified concentrations of silver nanoparticles which were prepared as described previously [11]. Monotypic cultures (i.e., Schwann cells or DRG neurons alone) were also set up similarly. The plates so prepared were incubated for 1, 2, 3, 4, or 5 days at 37°C. At the end of the incubation period, 100 µL of 5 mg/mL MTT dye in PBS was added into each well and the plates were incubated for an additional 4 hours at 37°C. The purple-colored insoluble formazan crystals in viable cells were dissolved using DMSO and the subsequent absorbance (designated as X) of the content of each well was measured at 570 nm using a Bio-Tek Synergy HT Plate Reader (Winooski, VT, USA) [14].

The medium and silver nanoparticles by themselves had absorbance: thus, their absorbance (i.e., the control sets of wells) had to be subtracted from the absorbance of live cells with or without different concentrations of silver nanoparticles treatments. The control sets of wells were set up alongside those sets of wells in the plates as detailed in the preceding paragraph except that the control sets of wells did not contain any seeded cells. At the end of the specified culture period, 100 µL of 5 mg/mL MTT dye in PBS was added into each well and the plates were incubated for an additional 4 hours at 37°C. The subsequent absorbance (designated as Y) of the content of each well was measured at 570 nm as described above. (X-Y) was taken as the absorbance attributed to viable cells in each well. The absorbance of co-cultures of DRG neurons and Schwann cells were compared with the sum of absorbance of monotypic cultures (i.e., Schwann cells or DRG neurons alone).

2.4 Western blot analysis

Expression of synapsin and GFAP was determined by Western blot analysis [10, 15]. Cells treated with or without silver nanoparticles (125 µL) were collected and homogenized. Protein content of the homogenates dissolved by 10 M NaOH for 3 days was then determined using the bicinchoninic acid technique as described previously [10, 15]. Equal amounts of protein from the samples were loaded onto the lanes of the gels, and the proteins were separated by polyacrylamide gel electrophoresis and transferred to a polyvinylidene fluoride membrane. Monoclonal or polyclonal antibodies against the respective proteins were then used to probe the proteins of interest. The polyvinylidene fluoride membrane containing the target protein was then developed using the chemiluminescence technique on an X-ray film to assess the extent of expression of respective proteins [10, 15].

2.5 Statistical analysis of data

Experiments were performed at least three times with a minimum of 6 replicates for each set, and all data were recorded as mean ± standard deviation.

3 RESULTS AND DISCUSSION

Employing the modified MTT assay, we studied the effects of different concentrations of silver nanoparticles on survival, proliferation, and/or growth of the Schwann cells alone, DRG neurons alone, and DRG neurons co-cultured with Schwann cells. As shown in Figures 1, 2 and 3, the survival/proliferation of cells (both DRG neurons and Schwann cells) in co-cultures was better than that of cells in monotypic cultures after 4 days' culture (compare the dark line with the green line in Figures 1, 2, and 3). Silver nanoparticles were cytotoxic to DRG neurons and Schwann cells. At treatment concentrations of 25 μL and higher, silver nanoparticles induced concentration- and time-related decreases in survival of DRG neurons or Schwann cells in monotypic cultures. However, both cell types survived better in co-cultures when they were exposed to higher concentrations (75 μL and 125 μL) of silver nanoparticles compare to when they were exposed to the silver nanoparticles in monotypic cultures (compare the orange line with the brown line in Figures 2 and 3), especially at the concentration of 125 μL (compare Figures 2 and 3). Thus, we employed the concentration of 125 μL to do all subsequent experiments. All these results (Figure 1-3) indicated that co-culturing DRG neurons with Schwann cells imparted some protection on them against the cytotoxicity induced by silver nanoparticles.

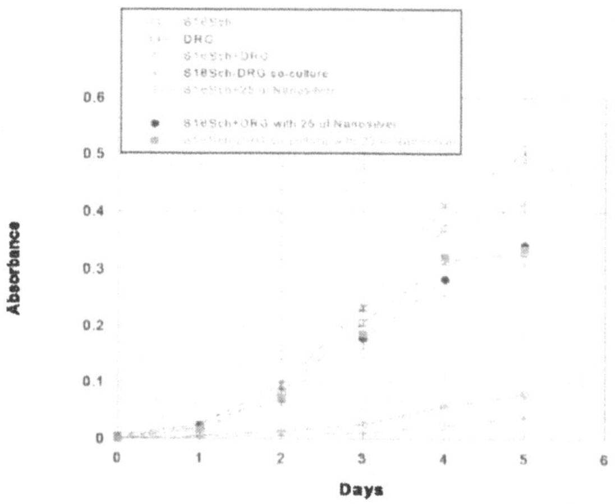

Figure 1: Effects of treatment with 25 μL silver nanoparticles for 5 days on the survival/proliferation of S16 Schwann cells alone, DRG neurons alone, or S16 Schwann cells co-cultured with DRG neurons.

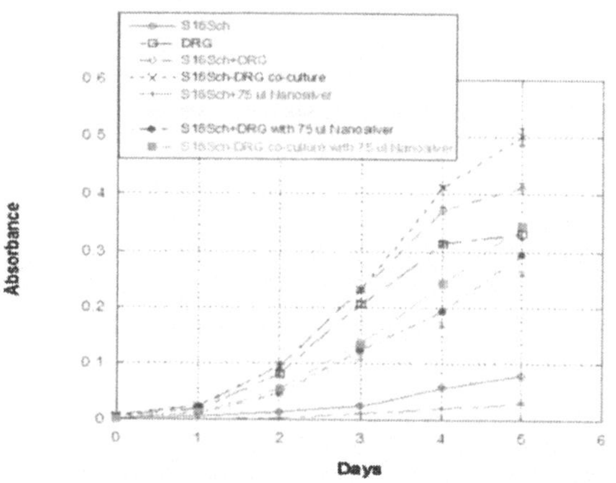

Figure 2: Effects of treatment with 75 μL silver nanoparticles for 5 days on the survival/proliferation of S16 Schwann cells alone, DRG neurons alone, or S16 Schwann cells co-cultured with DRG neurons.

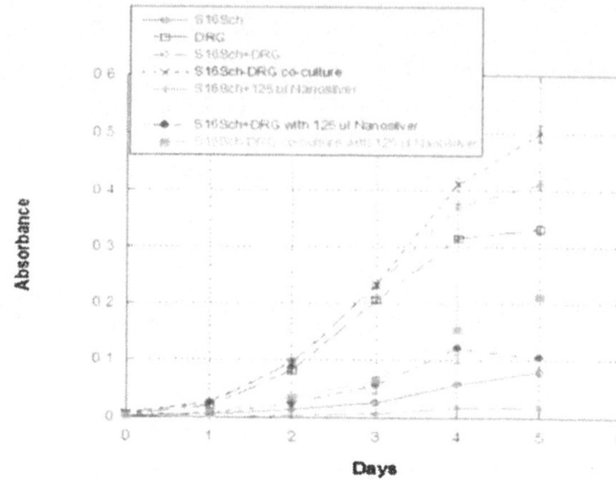

Figure 3: Effects of treatment with 125 μL silver nanoparticles for 5 days on the survival/proliferation of S16 Schwann cells alone, DRG neurons alone, or S16 Schwann cells co-cultured with DRG neurons.

To determine some of the cellular and molecular mechanisms underlying this phenomenon of "protection" in co-cultures, we examined the expression of two cellular biomarkers, namely synapsin (a marker of neuronal and synaptic function) and GFAP (a glial marker), in DRG neurons and Schwann cells either singly in monotypic cultures or in co-cultures in the presence or absence of silver nanoparticles at the concentration of 125 μL by Western blot analysis (data not shown). As expected of a neuronal biomarker, DRG neurons, but not Schwann cells, expressed synapsin. Similarly, co-cultures of DRG neurons and Schwann cells also expressed synapsin. By contrast, synapsin expression was increased in DRG neurons when they were co-cultured with Schwann cells and treated with or without silver nanoparticles, suggesting that co-culturing with Schwann cells exerted beneficial effects on DRG neurons and protect them from the cytotoxicity of silver nanoparticles. As expected of a glial marker, Schwann cells, but not DRG neurons, expressed GFAP. Similarly co-cultures of DRG neurons and Schwann cells also expressed GFAP. However, the presence of silver nanoparticles markedly suppressed GFAP expression in Schwann cells in monotypic culture but not in co-cultures of DRG neurons with Schwann cells, indicating that co-culturing with DRG neurons appeared to render the Schwann cells less susceptible to the cytotoxicity of silver nanoparticles. These results are consistent with our hypothesis that co-culturing DRG neurons with Schwann cells imparts some protection on them against the cytotoxicity induced by silver nanoparticles.

4 CONCLUSIONS

Results of our ongoing studies revealed that, while silver nanoparticles induced concentration- and time-related decreases in survival of DRG neurons or Schwann cells in monotypic cultures, both cell types appeared to withstand the cytotoxicity of silver nanoparticles and survive better when maintained in co-cultures. Synapsin expression was increased in DRG neurons when they were co-cultured with Schwann cells and treated with or without silver nanoparticles. GFAP expression was increased in Schwann cells when they were co-cultured with DRG neurons and treated with silver nanoparticles. Taken together, our results are consistent with our hypothesis that co-culturing DRG neurons with Schwann cells imparts some protection on them against the cytotoxicity induced by silver nanoparticles and may have pathophysiological implications in the biocompatibility and health hazard of silver nanoparticles. Moreover, our findings may be of relevance to toxicological studies of drugs formulated with agents containing silver nanoparticles before such drugs can be considered for human clinical trials.

5 ACKNOWLEDGMENTS

We thank Dr. Ahmed Höke (Johns Hopkins University School of Medicine) for his generous gift of DRG neurons. Our study was supported, in part, by an USAMRMC Project Grant (Contract #W81XWH-07-2-0078), University Research Committee Grant from Idaho State University, and small project grants from MSTMRI.

REFERENCES

[1] Gao WJ, Leung SW, Lai JCK. (2012) Functional enhancement of chitosan and nanoparticles in cell culture, tissue engineering, and pharmaceutical applications. Frontiers in Physiology. 3: 321-333.

[2] Lai JCK, Jaiswal AR, Lai MB, et al. (2015) Toxicity of silicon dioxide nanoparticles in mammalian neural cells. In Handbook of Clinical Nanomedicine — From Bench to Bedside (Bawa R, Audette GF & Rubinstein I, eds.), Pan Stanford Series in Nanomedicine (Bawa R, Series Ed.), Volume 1, Pan Stanford Publishing, Singapore (in press).

[3] Mueller NC, Nowack B. (2008) Exposure modeling of engineered nanoparticles in the environment. Environmental Science & Technology. 42: 4447-4453.

[4] Stensberg MC, Wei Q, McLamore ES, et al. (2011) Toxicological studies on silver nanoparticles: challenges and opportunities in assessment, monitoring and imaging. Nanomedicine (Lond). 6(5): 879-898.

[5] Blinova I, Niskanen J, Kajankari P, et al. (2013) Toxicity of two types of silver nanoparticles to aquatic crustaceans Daphnia magna and Thamnocephalus platyurus. Environmental Science and Pollution Research. 20: 3456-3463.

[6] Lai JCK, Lai MB, Edgley KL, et al. (2007) Silicon dioxide nanoparticles can exert cytotoxic effects on neural cells. In Proceedings of 2007 Nanotechnology Conference and Trade Show, Volume 2, Chapter 8: Bio Materials and Tissues, pp. 741-743.

[7] Lai JCK, Lai MB, Jandhyam S, et al. (2008) Exposure to titanium dioxide and other metallic oxide nanoparticles induces cytotoxicity on human neural cells and fibroblasts. International Journal of Nanomedicine. 3(4): 533-545.

[8] Lai JCK, Jandhyam S, Lai MB, et al. (2008) Cytotoxicity of metallic oxide nanoparticles: new insights into methodological problems and advances in elucidation of underlying mechanisms. In Proceedings of the 12th World Multi-Conference on Systemics, Cybernetics and Informatics, Volume II, pp. 10-15.

[9] Lai MB, Jandhyam S, Dukhande VV, et al. (2009) Cytotoxicity of metallic oxide nanoparticles in human neural and non-neural cells. In Technical Proceedings of the 2009 Nanotechnology Conference and Trade Show, Volume 2, Chapter 3: Nano Medicine, pp. 135-138.

[10] Lai JCK, Ananthakrishnan G, Jandhyam S, et al. (2010) Treatment of human astrocytoma U87 cells with silicon dioxide nanoparticles lowers their survival and alters their expression of mitochondrial and cell signaling proteins. International Journal of Nanomedicine. 5: 715-23.

[11] Gao WJ, Leung SW, Bhushan A, et al. (2014) Cytotoxic effects of silver and gold nanoparticles in human glioblastoma U87 Cells. In Technical Proceedings of the 2014 NSTI Nanotechnology Conference & Expo, Nanotech 2014 Vol. 3, section 2. Sustainable Nanotechnolog: Environmental Apps. & EHS Implications, pp. 134-137.

[12] Jaiswal AR, Lu S, Pfau J, et al. (2011) Effects of silicon dioxide nanoparticles on peripheral nervous system neural cell models. Technical Proceedings of the 2011 NSTI Nanotechnology Conference and Expo – Nanotech 2011, Volume 3, Chapter 7: Environment, Health & Safety, pp. 541 – 544.

[13] Idikuda VK, Jaiswal AR, Wong YYW, et al. (2012) Cytotoxicity of magnesium oxide nanoparticles in Schwann cells. In Technical Proceedings of the 2012 NSTI Nanotechnology Conference & Expo – Nanotech 2012, Volume 3, Chapter 5: Environmental Health & Safety, pp. 342-345.

[14] Dukhande VV, Malthankar-Phatak GH, Hugus JJ, et al. (2006) Manganese induced neurotoxicity is differentially enhanced by glutathione depletion in astrocytoma and neuroblastoma cells. Neurochemical Research. 31(11):1349-1357.

[15] Puli S, Lai JCK, Edgley KL, et al. (2006) Signaling pathways mediating manganese-induced neurotoxicity in human glioblastoma cells (U87). Neurochemical Research. 31(10): 1211-1218.

A New and Versatile Method for Preparing Drug Nanoparticle Formulations for Enhancing Bioperformance

John D. Higgins and Dennis Leung

Merck Research Labs

770 Sumneytown Pike, Box 4, WP14-2

West Point, PA 19380-0004

1. INTRODUCTION

It is well known that a significant percentage of contemporary drugs suffer from poor aqueous solubility, which often results in poor bioperformance (i.e., poor oral absorption and high exposure variability). With this in mind, a wide variety of drug delivery methods have evolved over the years in an effort to address this ubiquitous challenge. Towards this end, the preparation and application of drug nanoparticle formulations was first reported over 20 years ago [1]. In this approach, a drug's particle size is reduced to a median particle size in the nanometer range. This reduction in particle size results in a large increase in surface area (figure 1) and dissolution rate, which can lead to a subsequent improvement in oral absorption. Generally, the drug is maintained in its crystalline state. The success of the crystalline nanoparticle approach in enhancing bioperformance has been born-out by the appearance of several nanoparticle drug products on the market [2] for use in a variety of administration routes.

Figure 1. Decreasing particle size and associated increased surface area

One of the most effective methods of producing nanoparticles is *via* high-energy wet media milling using a high-speed impeller mill, which produces an aqueous suspension of nanoparticles. In this method, mixing shear forces and the forces generated during the drug particles' impact with the milling media [3] can reduce the drug particle size to the range of 200 nm in a few hours. It is important to note that a stabilizer (i.e., polymers such as poloxamer) must be added to the resulting aqueous suspension of nanoparticles in order to prevent agglomeration/aggregation. Although it has proven to be effective, typical impeller media-milling processes have some significant limitations. Some of these limitations are tabulated below.

Limitations of impeller media milling methods:

- Relatively large amounts of drug (>20 mg) are generally required to make even the smallest possible nanoparticle formulations.
- Since heat is generated, the milling equipment requires cooling capability.
- Although the mixing rate of a typical impeller media mill is a high-energy process, it often takes hours to reduce the particle size of an average compound into the nanometer size range.
- Impeller mills operate via a batch process, which is not amenable to screening multiple surfactants and milling conditions for optimal performance.

TechConnect Briefs 2015, TechConnect.org, ISBN 978-1-4987-4729-5

These limitations make it particularly difficult to investigate nanoparticles early in drug discovery space, where several drug candidates often need to be evaluated at once and drug quantities are very limited.

2. A NEW METHOD FOR PRODUCING CRYSTALLINE NANOPARTICLES

Here we report a useful new method [4] for generating nanoparticles, using a LabRam® [5] high-frequency resonant mixer (figure 2). The LabRam mixer utilizes Resonant Acoustic® technology that applies low frequency, high-intensity acoustic energy to create a uniform shear field throughout the mixing vessel. The result is rapid fluidization and dispersion of material inside the vessel. Resonant Acoustic® mixing (RAM) is a non-invasive mixing technology, where no impellers or wands are used.

The versatility and advantages of using the LabRam for nanoparticle preparation method are several fold. The advantages are tabulated below.

Advantages of the LabRam preparation method:

- Nanoparticle suspensions can be produced in considerably shorter timeframe (often <10 minutes) as compared to impeller media milling. This allows for far more experiments to be run in parallel (i.e., high throughput in multi-well plates).
- Resonant mixing conditions are less stressful (i.e., less heat) as compared to impeller milling, which results in better performing nanosuspensions (i.e., less particle agglomeration or drug degradation)
- A wide range of vessels and preparative scales are accessible – from a few mgs divided into a multi-well plate, through 10s of mgs in vials to multi-gram quantities. Hence, the multi-well plate capability allows for high throughput screening of stabilizer systems, while the multi-gram scale capabilities enable the preparation of nanoformulation supplies to support *in vivo* studies such as a long-term toxicology study.

Hence, using the LabRam®, we now have the ability prepare and evaluate nanoparticle formulations during the drug discovery lead optimization stages all the way to the preparation of long-term toxicology supplies and even the clinic.

Figure 2. The Resodyne LabRam Resonant Mixer

2.1 Nanoparticle Preparation and Stabilizer Screen Example Case Study

An example of the utility of the LabRam preparation method is shown with a nanoparticle stabilizer screen we executed with a highly insoluble compound. In this study, we compared the LabRam method to a process using a typical high-speed impeller mill. Firstly, using the LabRam, we were able to quickly screen 5 different stabilizers in parallel in less than an hour, as compared to an all-day sequential batch process using the impeller mill. Table 1 shows the stabilizers evaluated by the 2 milling methods along with the resulting particle sizes generated.

Stabilizer[1]	Impeller Mill D50 / D90 (nm)[2]	LabRam D50 / D90 (nm)[2]
HPMC/SDS	311 / 585	256 / 356
HPC-SL/SDS	111 / 1330	192 / 247
PVP K28-32/SDS	157 / 2510	170 / 276
Pluronic F127	1126 / 1580	391 / 599
Tween80	269 / 1031	186 / 360

Table 1. Nanoparticle stabilizer screen and resulting particle size. [1]HPMC: Hydroxypropyl methylcellulose; SDS: Sodium dodecyl sulfate; HPC: Hydroxypropyl cellulose; PVP: Polyvinyl pyrrolidone; Pluronic: Poloxamer. [2]Particle size measurement by dynamic light scattering.

From table 1, it is easily seen that some stabilizers performed better than others in preventing agglomeration (larger particle sizes generated).

Again, using the LabRam, these critical screening data were obtained in less than an hour using only a few mgs of drug, vs. an entire day using an impeller mill (with significantly more drug consumption).

Figure 3 shows light microscope images of representative nanoparticles prepared by an impeller mill and the LabRam®, using the same stabilizer. From the images, it is clear that the nanoparticles prepared by the LabRam were significantly less aggregated, even though the same stabilizer was employed.

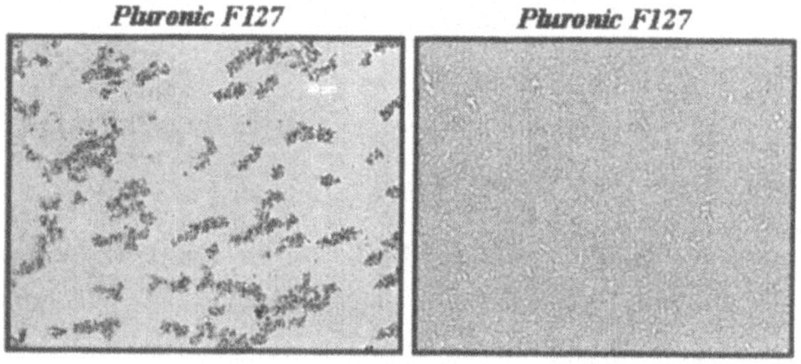

Figure 3. Nanoparticles prepared by an impeller mill and the LabRam, using the same stabilizer (Pluronic F127)

3. CONCLUSIONS

Application of a LabRam has proven to be a versatile and advantageous method for producing nanoparticle formulations. This method is compatible with a wide range of drug structural types and can efficiently produce aqueous nanoparticle formulations that are amenable to various routes of administration. With this new method in hand, the evaluation of nanoparticle formulations is now routine in early drug discovery space, using minimal drug supplies and time resource.

4. REFERENCES

1. See for example, Liversidge, E., et.al., Eur. J. Pharm. Sci., *18*, 113-120, **2003**
2. For example, oral Aprepitant (Emend®)
3. An effective milling media is, for example, Yttrium stabilized Zirconium oxide 0.5 mm beads; 95% ZrO_2 + 5% Y_2O_3.
4. D. Leung et al., Int. J. Pharm., *473*, 10-19, **2014**
5. http://www.resodynmixers.com

Dielectrophoretic (DEP) Isolation and Detection of Cancer Related Cell Free DNA, RNA and Exosome Biomarkers from Blood and Plasma (Abstract # 600)

Michael J. Heller, Jennifer Marciniak, Stuart Ibsen, Elaine Skowronski and Avery Sonnenberg

University of California San Diego, Departments of Nanoengineering and Bioengineering, La Jolla, CA, USA 92093-0448

ABSTRACT

The analysis of cell free (cf) DNA and cf-RNA nanoparticulate biomarkers continues to become more widely used for cancer diagnostics and management. We have now demonstrated a dielectrophoretic (DEP) based approach that allows cf-DNA biomarkers from chronic lymphocytic leukemia (CLL) patients to be isolated in about 10 minutes directly from a small volume (25ul) of patient whole blood and plasma. Cf-DNA isolated from fifteen CLL patient blood samples and eleven CLL patient plasma samples were PCR amplified to identify the VHL genotype and then sequenced. The DNA sequencing results for cf-DNA isolated by DEP were compared to two gold standard methods for CLL analysis. In further work, cf-RNA for ROR1 has also been isolated from CLL samples, and RNA containing exosomes excreted by glioblastoma cells have been isolated by DEP.

Keywords: DEP, Cancer, Cell Free DNA, Cell Free RNA, Biomarkers

1. Introduction

Cell-free (cf) DNA and cf-RNA are important biomarkers for early detection of cancer, residual disease, monitoring chemotherapy and other aspects of cancer management. The isolation of cancer-related cf-DNA from plasma may allow "liquid biopsies" to replace more invasive tissue biopsies for detecting and analyzing cancer mutations. However, the present methods for isolating cf-DNA from plasma are complex, time-consuming and relatively expensive processes that rule out use for point-of-care (POC) diagnostic applications. Other limitations of conventional sample preparation processes include: requirement of at least one or more milliliters of plasma; processing of blood to plasma; a large number of manipulations, which increases the chance for technician errors; decrease of recovery efficiency with decrease in sample size and concentration; degradation of cf-DNA by mechanical sheering during the processing steps; and limiting PCR analysis to shorter target DNA sequences due to the degradation of cf-DNA. Finally, other potentially important cancer-related biomarkers such as cf-RNA, exosomes and microvesicles also require relatively long and involved processes for their isolation from plasma. With regard to hematological cancers such as chronic lymphocytic leukemia (CLL) and lymphomas, DNA for PCR and sequencing can be obtained from transformed cells, as well as from cf-DNA isolated from plasma. In the case of CLL, B cells from patients can be segregated into one of at least two major subsets on the basis of whether or not the immunoglobulin (IG) variable region has somatic mutations. Patients with CLL cells that express unmutated IG heavy chain variable region genes (IGHV genes) tend to have an aggressive clinical course relative to that of patients who have CLL cells that express IGHV with somatic mutations. For CLL diagnostics and management, genomic DNA is isolated from the peripheral blood mononuclear cells (PBMCs). The PBMCs are usually purified from the CLL patient blood samples by density centrifugation using Ficoll-Hypaque 1077. This is a long and labor-intensive process, which adds considerable cost to patient management and precludes any POC applications. To assess the unique patient-specific IGHV expressed by the CLL B cells, PCR and DNA sequencing are performed on the isolated genomic DNA to determine the mutation status for the expressed IGHV gene.

Electrokinetic technologies, like AC dielectrophoresis (DEP) have long been known to provide effective separation of cells, nanoparticles, DNA and other biomolecules. However, until recently, DEP techniques remained impractical for use with high-conductance solutions (5-15 mS/cm), as well as with whole blood, plasma and serum. In earlier work, sample dilution to low-conductance conditions (<1 mS/cm) was required before effective DEP separations could be carried out. While some progress was made using DEP under high-conductance conditions, these efforts have been limited to separations of cells and micron-sized entities by negative DEP forces using hybrid electrokinetic devices. Such devices still could not be used with whole blood samples, and, more importantly, they did not provide efficient isolation of DNA from the sample. More recently, we have been able to develop electrokinetic techniques that allow nanoscale entities, including high molecular weight (hmw) DNA and nanoparticles, to be isolated from high-conductance (>10 mS/cm) buffer solutions and whole blood samples. We were also able demonstrate isolation of virus from blood and fluorescent detection of cell free (cf) DNA from CLL patient blood samples. Most recent, we were able demonstrate PCR and Sanger DNA sequencing results for cf-DNA biomarkers isolated by DEP using only 25 ul samples of unprocessed CLL patient blood . The PCR and Sanger sequencing results for the DEP process were equivalent to results obtained using conventional sample preparation of cf-DNA from 1 ml of CLL patient plasma, and to the "gold standard" DNA sequencing results obtained using an established method for isolating DNA from the leukemic cells of CLL patients that requires 15-20 ml of blood. In addition to comparing the DEP isolation of cf-DNA from CLL patient blood with CLL patient plasma, we are also showing some initial results for the DEP isolation RNA containing exosomes from Glioblastoma cell cultures.

2. Materials and Methods

New dielectrophoretic (DEP) microarray devices specifically designed for the isolation of cf-DNA and cf-RNA from high conductance buffer, blood and plasma samples were obtained from Biological Dynamics. DEP experiments were carried out by adding about 25-50 ul of blood/plasma sample to the DEP microarray device. The DEP field was applied at 10 kHz and 20 Vp-p for 5-10 minutes. The microarray was then washed with 0.5x PBS with the DEP field on and examined by epifluorescent microscopy. Cf-DNA was then removed from device and PCR amplified using VHL patient specific primers.

3. Results and Discussion

Figure 1 below shows the fluorescent image results for cf-DNA isolated by DEP from one normal blood sample (Normal-1e B) and five representative CLL blood samples (CLL-1e B, CLL-2e B, CLL-3e B, CLL4e B, CLL-5e B) on the left side, compared with one normal plasma sample (Normal-1e P) and five representative CLL plasma samples (CLL-1e P, CLL-2e P, CLL-3e P, CLL4e P, CLL-5e P) on the right side The 3D fluorescence intensity images created by MATLAB, which provide better visualization of the relative amounts of cf-DNA are also shown. Overall, the fluorescent DNA levels were higher in most of the CLL patient samples when compared to the fluorescent DNA levels obtained from the normal blood and plasma samples. Using the electrokinetic DEP microarray device, cancer related cf- DNA from only 25 ul of unprocessed patient blood or plasma could be detected within five minutes. Elution of the cf-DNA from the DEP device for quantification and PCR analysis required less than ten minutes. The PCR and DNA sequencing results for cf-DNA isolated from 15 CLL patient blood samples by the DEP process were exactly comparable to results obtained using conventional methods, which take several hours to complete, are more laborious (many steps) and required much larger plasma/blood samples i.e., one ml plasma for Qiagen cf-DNA extraction method, and 15-20ml blood for isolation of genomic DNA from CLL patient leukemic B cells. The results for eight of eleven CLL plasma samples compared well with the gold standards.

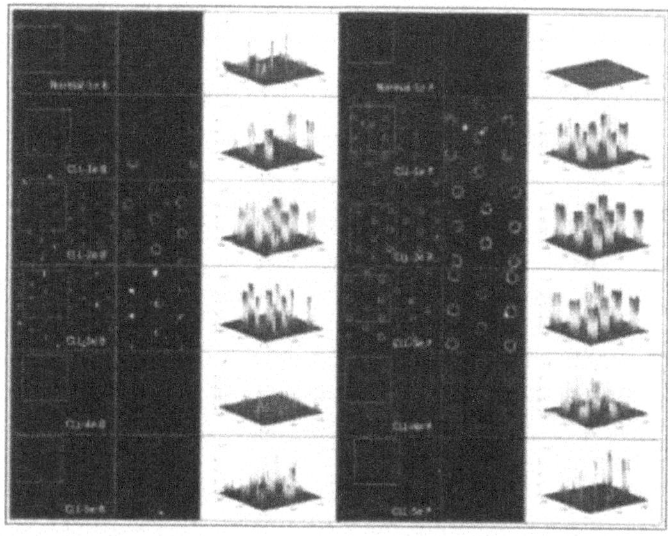

Figure 1 *- Fluorescence Detection and analysis of cf-DNA from one normal blood sample and five CLL patient blood samples (left side), compared with one normal plasma sample and five CLL patient plasma samples.*

In further work, DEP was used to isolate cf-RNA containing ROR1 mRNA from CLL patient blood samples (Figure 2), and RNA containing exosomes excreted by glioblastoma cells (Figure 3).

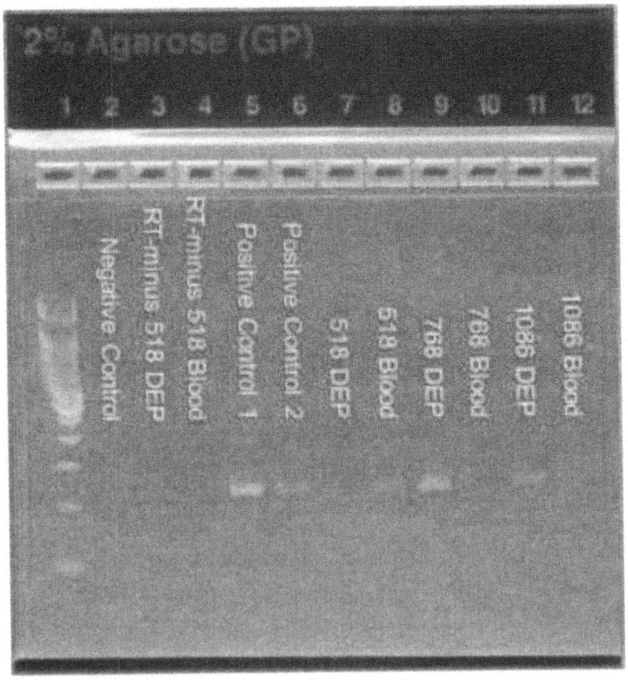

Figure 2 – PCR results showing cf-RNA isolated from CLL patient blood contains ROR1 mRNA.

***Figure 3** –DEP isolation of extracellular cf-RNA containing exosome from glioblastoma brain cell culture.*

4. Conclusions

Generally, cf-DNA and cf-RNA for "liquid biopsies" is isolated form plasma using relatively long and involved processes which are not suitable for point of care (POC) diagnostics. New DEP devices are being used to collect cf-DNA and cf-RNA directly from Chronic Lymphocytic Leukemia (CLL), including PCR identification of ROR1 mRNA from cf-RNA. Blood sample to PCR is now achieved in less than ten minutes. Further work has now also demonstrated that DEP can be used to isolate RNA containing exosomes released by glioblastoma cells. Thus, the new DEP technology and devices set the stage for "seamless sample to answer" diagnostic systems which will allow a variety of important cancer and other disease biomarkers to be rapidly isolated and analyzed directly from whole blood and other clinical samples.

4. Acknowledgements

The authors wish to thank Dr. Raj Krishan, Dr. David Charlot and Mr. Gene Tu at Biological Dynamics (San Diego, CA) for their help and cooperation, and for allowing early access to the ACE instrument and microarrays used in this study. The authors also wish to thank Adam Wright for his help with image processing. The DEP technology used in the study was the result of original research carried out by the M.J.H. lab at the UCSD Moores Cancer Center under NIH NCI NanoTumor Center Grant (U54-CA119335). The work was also funded in part by NIH grant PO1-CA081534 for the CLL Research Consortium (TJ Kipps PI).

5. References

1. Sonnenberg A, Marciniak JY, Rassenti L, Ghia EM, Skowronski EA, Manouchehri S, McCanna JP, Widhopf II GF, Kipps TJ and Heller MJ, *"Dielectrophoretic Isolation and Detection of Cancer Related Circulating Cell Free DNA Biomarkers from Blood and Plasma"*, Electrophoresis, V35, 12-13, pp. 1828-36, July 2014.
2. Sonnenberg A, Marciniak JY, Rassenti L, Ghia EM, Skowronski EA, Manouchehri S, McCanna JP, Widhopf II GF, Kipps TJ and Heller MJ, "*Rapid Electrokinetic Isolation of Cancer-Related Circulating Cell Free DNA Directly from Blood"*, Clinical Chemistry, 60:3, pp.500-509, 2014
3. McCanna JP, Sonnenberg A, and Heller MJ, Low level epifluorescent detection of nanoparticles and DNA on dielectrophoretic microarrays, J. BioPhotonics , *1–11 (2013) / DOI 10.1002/jbio.201300046*
4. Sonnenberg, A, Marciniak JY, McCanna J, Krishnan R, Rassenti L, Kipps TJ and Heller MJ "Dielectrophoretic Isolation and Detection of cfc-

DNA Nanoparticulate Biomarkers and Virus form Blood", Electrophoresis, V34, pp.1076-1084, 2013
5. Sonnenberg, A, Marciniak, JY, Krishnan, R, Heller MJ, "Dielectrophoretic Isolation of DNA and Nanoparticles form Blood", Electrophoresis V33, 2482-2490, 2012
6. Heller MJ, Krishnan R and Sonnenberg A, Detection of Cancer Related DNA Nanoparticulate Biomarkers and Nanoparticles in Whole Blood, TechConnect World 2010 Proceedings, Nanotechnology 2010, Vol 3, Chap. 6, Cancer Nanotechnology pp 372-375
7. Krishnan R, Dehlinger DA, Gemmen GJ, Mifflin RL, Esener S and Heller MJ, "Interaction of nanoparticles at the DEP microelectrode interface under high conductance conditions", Electrochemical Communications, V11, #8, 1661-1666, 2009
8. Krishnan R and Heller MJ, "An AC electrokinetic method for the enhanced detection of DNA nanoparticles", J. Biophotonics, V2, #4, pp. 253-261, 2009
9. Krishnan R, SullivanBD, Mifflin RL, Esener SC, and Heller MJ, Alternating current electrokinetic separation and detection of DNA nanoparticles in high-conductance solutions, Electrophoresis, v. 29, #9, pp. 1765-1774, May 2008.

Targeted Delivery of Anti-Cancer Drug Sorafenib through Magnetic Solid Lipid Nanoparticles

A. Grillone*, E. Redolfi Riva*, S. Moscato**, R. Sacco***, V. Mattoli* and G. Ciofani*

*Italian Institute of Technology, CMBR@SSSA, Pontedera, Italy, gianni.ciofani@iit.it
**University of Pisa, Department of Clinical and Experimental Medicine, Pisa, Italy
***Pisa University Hospital, Department of Gastroenterology, Pisa, Italy

ABSTRACT

Sorafenib is an anticancer drug approved by the Food and Drug Administration for the treatment of hepatocellular carcinoma and advanced renal carcinoma. The clinical application of sorafenib is promising, yet limited by its insolubility and severe toxic side-effects. The aim of this study is to develop and characterize sorafenib-loaded magnetic nanovectors to deliver and accumulate drug just to the disease site with the help of a remote magnetic field. Sorafenib and superparamagnetic iron oxide nanoparticles were encapsulated in solid lipid nanoparticles (SLNs) by a hot homogenization technique, by using cetyl palmitate as lipid matrix. Biological effects were evaluated *in vitro* on human hepatocarcinoma HepG2. Our results confirm the possibility to prepare stable SLNs able to kill cancer cells through sorafenib cytotoxic effect, and to enhance/localize this effect in a desired area thanks to the magnetically-driven accumulation of the drug.

Keywords: solid lipid nanoparticles, magnetic nanoparticles, sorafenib, HepG2

1 INTRODUCTION

The multi-kinase inhibitor (MKI) sorafenib (tradename Nexavar®, Bayer) has been recently approved by the FDA for the treatment of non-resectable hepatocarcinoma and advanced renal carcinoma (HCC) [1]. Preclinical studies show that sorafenib acts through several mechanisms to inhibit tumor angiogenesis and to induce tumor cell apoptosis [2]. Despite its proven survival benefit, sorafenib can lead to important side effects including hand and foot syndrome, diarrhea, and hypertension [3]. The aim of this study is the development of a magnetic nanovector able to efficiently and selectively deliver sorafenib toward cancer lesions thanks to a physical guidance mediated by magnetic nanoparticles.

The proposed system can selectively deliver sorafenib, by concentrating the drug in correspondence of a target site. Moreover, its use can increase the efficacy of the therapy avoiding the before mentioned side effects, such as drug aspecific biodistribution, which could expose healthy tissues to the drug action. The overcoming of this latter drawback could be one of the most important improvement brought by our system, since lack of specificity in cell recognition represents the bottleneck of several traditional chemotherapy approaches. Solid lipid nanoparticles (SLNs) have been chosen to selectively target sorafenib. This nanovector is composed by a solid lipid matrix dispersed in aqueous solution and stabilized by surfactant, and incorporates the lipophilic drug inside its core. SLNs combine several benefits compared to other types of carriers. First of all, they ensure a controlled and sustained release of the drug by protecting the active molecule from degradation and metabolic inactivation thanks to the hard lipid matrix. Moreover, they allow a high loading of the drug, jointly to a low cellular and systemic toxicity [4].

Cetyl palmitate, an ester manufactured from raw materials of natural origin and used in several cosmetic formulations, has been chosen as the main component of SLNs core, because of its high biocompatibility [5].

To localize sorafenib action, magnetic targeting is ensured by encapsulation of superparamagnetic iron oxide nanoparticles (SPIONs) [6], that assure magnetic momentum and drivability to the lipid carriers.

2 MATERIALS AND METHODS

2.1 SLN preparation and characterization

Solid lipid nanoparticles were prepared by using an oil-in-water homogenization process at high temperature. In details, 5 mg of sorafenib tosylate (kindly provided by Bayer International) were dissolved in methanol at 65°C. 210 mg of cetyl palmitate (kindly provided by Gattefossé Italy) and 15 mg of SPIONs (average diameter of 10 nm; EMG1300 from Ferrotec) were dissolved in chloroform and thereafter mixed with the drug solution and gently stirred. This chloroform/methanol (2:1) mixture was then added into a vial containing Tween 80 (Sigma Aldrich) water solution (6%), and sonicated for 5 min with a probe-tip ultrasonicator (SONOPLUS mini20, Bandelin). The temperature of the emulsion was kept constant at 75°C to allow organic solvents evaporation. At the end of the procedure, the obtained dispersion was maintained at 4°C for 15 min to allow the formation of the SLNs. The final product (Sor-Mag-SLNs) was then purified by gel chromatography by using Sephadex G-25 pre-packed columns (GE Healthcare Life Sciences).

Plain magnetic SLNs (Mag-SLNs), *i.e.*, not loaded with sorafenib, were prepared following the same procedure, but omitting the drug, and were used as controls in biological experiments.

Several features of the obtained nanovectors were evaluated, including size distribution, zeta potential, and morphology through atomic force microscopy (AFM).

2.2 Biological experiments

Cell viability was assessed using the crystal violet nuclear staining assay. HepG2 cells were seeded in 96-well culture plates at a density of 5000 cells *per* well for 24 h, and then treated with Mag-SLNs (0-100 µg/ml) or Sor-Mag-SLNs (0-100 µg/ml) for 72 h. At the end of the treatment, the media were removed, and the cells were washed with PBS and stained for 10 min with 0.5% crystal violet (Sigma) dissolved in methanol/water. Subsequently, the dye was solubilized with 33% acetic acid and the absorbance was read at 570 nm on a microplate reader. Triplicates for each concentration were analyzed.

In vitro testing to assess targeting and biological effects has been performed on the hepatocellular carcinoma cell line HepG2 thanks to a dynamic culturing system based on a microfluidic set-up (Figure 1).

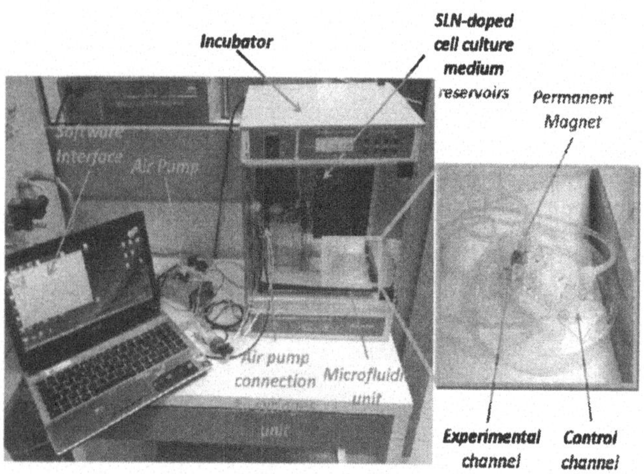

Figure 1: Photo of the set-up exploited for dynamic magnetic targeting experiments.

Accumulation of SLNs was obtained by means of an external permanent magnet (B_r = 1.32 T), and targeted internalization by cells was evaluated through confocal laser scanning microscopy (C2s, Nikon) by using fluorescent-labeled nanovectors.

Selective cytotoxic effects of Sor-Mag-SLNs (200 µg/ml) was investigated at 72 h after the targeted accumulation in the dynamic system with the Live/Dead® viability/cytotoxicity Kit (Molecular Probes), that allows live cells (stained in green by calcein) from death cells (stained in red by ethidium homodimer-1, EthD-1) to be discriminated.

3 RESULTS AND DISCUSSION

Results demonstrated the possibility to successfully fabricate stable magnetic SLNs encapsulating sorafenib. These nanovectors showed an average size of about 300 nm and a surface charge around -20 mV. AFM images of obtained drug-loaded SLNs are reported in Figure 2.

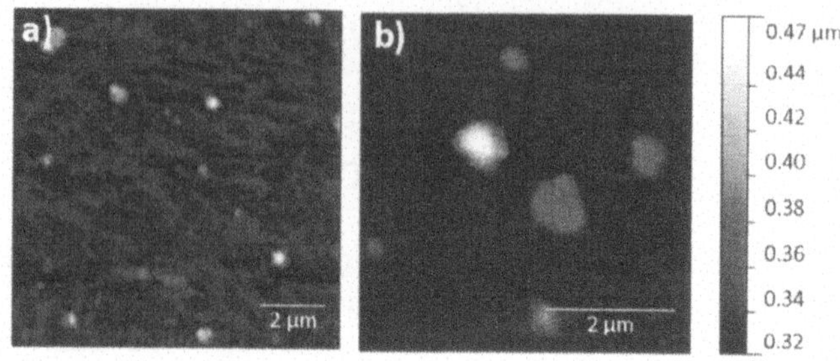

Figure 2: AFM topographical scans at low (a) and high (b) magnification of SLNs.

Antitumor activity of Sor-Mag-SLNs was investigated against HepG2 cells with the crystal violet proliferation assay and compared to that of the plain Mag-SLNs. As shown in Figure 3, Mag-SLNs did not significantly inhibit cell viability after 72 h of incubation (at all the investigated concentrations $p > 0.05$ with respect to the controls, ANOVA followed by Bonferroni *post-hoc* test). On the other hand, Sor-Mag-SLNs showed a strong cytotoxic effect at concentrations of 10, 25, 50 and 100 µg/ml, with a significant reduction of viability ($p < 0.05$), that reaches about 80% of the control for the 100 µg/ml Sor-Mag-SLN concentration (corresponding to about 3.20 µM of drug).

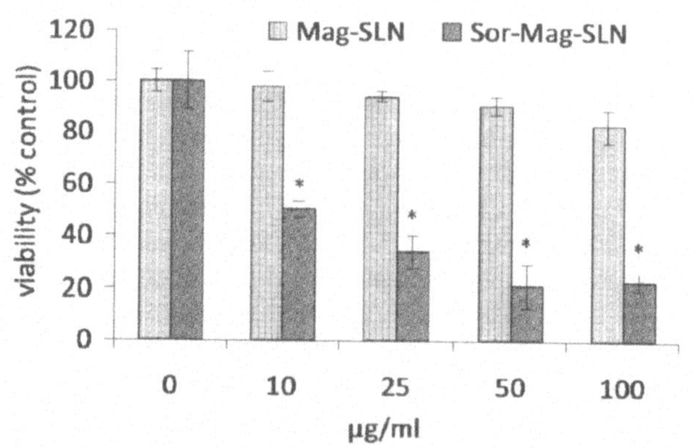

Figure 3: Crystal violet viability assay for increasing conentrations of Mag-SLNs and Sor-Mag-SLNs.

Magnetic targeting experiments were performed in dynamic conditions, and obtained results showed an enhanced Sor-Mag-SLN up-take by cells cultured in correspondence of the region of the microfluidic channel where the external permanent magnet was placed (Figure 4). Conversely, no significant nanoparticle internalization could be appreciated in cells cultured in the control channel, thus demonstrating the possibility to actively targeting the nanoparticle up-take and, consequently to selectively localize the effect of the drug.

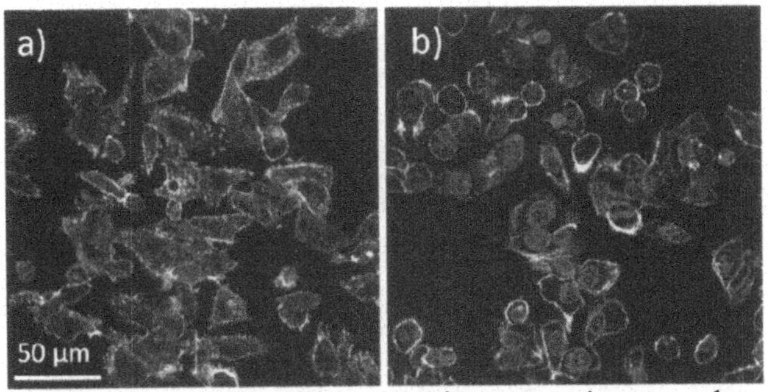

Figure 4: Magnetic targeting experiments show enhanced SLN internalization in correspondence of the region of the microfluidic channel where the external magnet was placed (a) with respect to the control channel (b). SLNs in red; f-actin in green; nuclei in blue.

The possibility of magnetically targeting the effects was demonstrated by the testing in dynamic conditions of the drug-loaded SLN efficiency, evaluated through the Live/Dead® viability/cytotoxicity assay after 72 h of incubation after the treatment. Quantitative evaluation of cell mortality, evaluated as percentage of cells positive for EthD-1, resulted of about 95% in the channel close to the magnet, while about 5 % in the control compartment.

Summarizing, in this work we have reported on a novel approach to actively target sorafenib own to its encapsulation in magnetic solid lipidic nanoparticles, prepared through a hot oil-in-water homogenization process. In particular, we focused on the problems related to the clinical use of sorafenib, related to the difficulty of localizing the drug effect, its poor solubility in aqueous environment, and its high toxicity for healthy tissues and organs. In this scenario, encapsulation of the drug into targetable nano-carriers could represent an advantageous approach to reduce undesired side-effects and to improve drug pharmacokinetics [7-9].

4 CONCLUSION

Our results confirm the possibility to prepare stable SLNs able to destroy HepG2 cancer cells through sorafenib cytotoxic effect, and to enhance this effect in a desired area thanks to the magnetically-driven accumulation of the drug.

Sorafenib-loaded magnetic solid lipid nanoparticles represent a good example of nanodevice in the field of oncology. Their potential dual function, as both contrast agent (data not shown) and therapeutic agent, make them a potential theranostic device for the treatment of cancer, and future experiments will be performed to extend analysis of the proposed nanoplatform *in vivo*.

ACKNOWLEDGEMENTS

Authors gratefully thank Bayer International and Gattefossé Italy, for providing sorafenib and cetyl palmitate, respectively.

REFERENCES

[1] Llovet JM *et al.*, N. Engl. J. Med. 2008, 359:378-90.

[2] Wilhelm SM *et al.*, Mol. Cancer Ther. 2008, 7(10):3129-40.

[3] Keating *et al.*, Drugs 2009, 69:223-40.

[4] Mehnert W, *et al.*, *Adv. Drug Deliv. Rev.* 2001, 47:165-96.

[5] Blasi P, *et al.*, *Int. J. Pharmaceut.* 2011, 419:287-95.

[6] Lu AH, *et al.*, Angew. Chem. Int. Ed. 2007, 46:1222-44.

[7] Torchilin VP, Adv Drug Deliver Rev 2012, 64:302-15.

[8] Allen TM, *et al.*, Adv. Drug Deliver. Rev. 2013, 65:36-48.

[9] Cho K, *et al.*, Clin. Cancer Res. 2008, 14:1310-6.

Cellular Uptake Behavior of Glycoconjugated and Micellar-Encapsulated Quantum Dots on HeLa Cervical Cancer Cells

J.-P. Merkl,[†] C. Schmidtke,[†] M. Safi,[#] M. Muroski,[#] D. Alpers,[‡] H. Tran,[†] J. Ostermann,[†] A.-M. Kreuziger,[†] S. Ehrich,[%] G. Strouse,[#] G.Wolters-Eisfeld,[%] H. Mattoussi[#] and H. Weller.[†,&]

† Institute of Physical Chemistry, Grindelallee 117, 20146 Hamburg, and the Hamburg Center for Ultrafast Imaging, University of Hamburg, Luruper Chaussee 149, 22761 Hamburg, Germany, email presenting author: merkl@chemie.uni-hamburg.de

‡ Institute of Organic Chemistry, Martin-Luther-King-Platz 6, 20146 Hamburg, Germany,

Department of Chemistry and Biochemistry, Florida State University, 95 Chieftan Way, Tallahassee, Florida 32306, United States of America,

% General, Visceral and Thoracic Surgery Department and Clinic, University Medical Center Hamburg-Eppendorf, Martinistrasse 52, 20246 Hamburg, Germany

& Department of Chemistry, Faculty of Science, King Abdulaziz University, P.O BOX 80203 Jeddah 21589 (Saudi Arabia).

ABSTRACT

Glyconanomaterials are of great interest for bio-imaging to understand biological processes, for instance metabolism, cell-cell or virus-cell interactions, gluconeogenesis, and cancer. Herein, we present a strategy for the biofunctionalization of fluorescent quantum dots (QDs) from continuous-flow reactor with carbohydrates. This flow reactor enables the reproducible synthesis of a large amount of QDs, with controlled surface functionalization. These QDs act as fluorescent biomarkers and as structural scaffolds for the presentation of glycoclusters to lectins, receptors and cells. Before the phase transfer into water takes place, the carbohydrates are covalently attached to an amphiphilic poly(isoprene)-*b*-poly(ethylene glycol) diblock copolymer (PI-*b*-PEG) using click-chemistry as previously described by us. These functionalized polymers are subsequently used for the encapsulation of the QDs under preservation of their unique optical properties in a continuous flow system. Binding of glycoconjugated QDs to the human cervical cancer cell line HeLa was characterized using confocal microscopy. Depending on the terminal group of the polymer (namely: D-maltose, D-glucose, carboxyl, and amine), the uptake of the functionalized QDs can be controlled and directed. Functionalization with maltose yields very high uptake in low incubation times and low concentrations. Although serum is known to inhibit the cellular response of artificial nanostructures, we observe reduced but significant cellular uptake of the maltose functionalized nanocontainers in serum containing media. This encapsulated materials have already been tested to be suitable for *in vivo* tumor targeting, due to their lack of toxicity as well as extraordinary stability. Though this method relies on highly reproducible continuous flow systems, which yield high amounts of well defined, functional, non-toxic and highly stable nanoparticles this method has extraordinary industrial, biological and medical relevance.

Keywords: cellular uptake, carbohydrates, diblock copolymer, continuous flow reactor, glycopolymer

INTRODUCTION

In the human organism carbohydrates play an essential role in many biological processes, for instance as structural device in the DNA and RNA, in the energy metabolism, as posttranslational modifications of proteins influencing tumor genesis, bacterial/ viral infection, inflammation, immune response, and cell-cell communication.[1-2] Many processes are mediated *via* interactions between sugar-binding proteins, so called glycoreceptors, and carbohydrates.[3] These interactions are difficult to analyze and to monitor with a single monosaccharide moiety due to the weak affinity.[4] This binding strength can be increased about several orders of magnitude by the combination of nanomaterials with carbohydrates. These gyconanomaterials present many sugars to the glycoreceptor. The strong binding of glycoclusters is a result of polyvalent interactions.[5] By combining the carbohydrate with a fluorescent quantum dot (QD), the localisation of the composite in the tissue can be monitored and the influence of carbohydrates on such composites may be studied. QDs are used in biomedicine for imaging, because of their excellent optical properties.[6] QDs benefit from high photostability and of their adjustable emission wavelength correlating with the particle size in contrast to organic dyes.[7-8]

Nonetheless, only a few articles in scientific literature can be found on glycoconjugated QDs.[9-11] Recently, we published a micellular encapsulation technique which is applicable for different NPs (QDs, iron oxides and gold NPs) using an amphiphilic poly(isoprene)-*b*-poly(ethylene glycol) diblock copolymer (PI-*b*-PEG).[12] The polymers can be cross-linked which leads to a high stability, preserving the physical properties of the particle.[13-14]. Due to the synthesis of the diblock copolymer by living anionic polymerization (LAP) the chain end can be tailored by different strategies.[12] More recently, we presented the attachment of azido-modified carbohydrates to PI-*b*-PEG with an alkyne end group using copper catalyzed Huisgen's 1,3-dipolar cycloaddition (*click*-chemistry).[15] The glycosylated polymers were successfully used for the encapsulation of QDs and used for lectin-binding assays.

Here, we performed uptake studies of the glycoconjugated QDs on HeLa cells *in vitro*. Their binding to the cervical cancer cell line was characterized using confocal microscopy. Depending on the terminal group of the polymer, (D-maltose, D-glucose, carboxy, and amine) the uptake of the functionalized QDs can be controlled. This may have a high potential for specific targeting of cervical cancer cells and further tumor entities *in vivo*.

EXPERIMENTAL SECTION

Synthesis. CdSe/CdS/ZnS core/shell/shell nanoparticles were synthesized in a flow reactor as described previously.[16-17] The block copolymers PI and PI-*b*-PEG (**1**) were synthesized by living anionic polymerization (LAP) as previously described.[12] PI-DETA was synthesized by CDI-activation and coupling to diethylenetriamine (DETA).[18]

Functionalization of PI-*b*-PEG. PI-*b*-PEG-NH_2 (**2**), PI-*b*-PEG-COOH (**3**), and PI-*b*-PEG-C≡CH (**4**) the glycosylation *via* click-chemistry were realized according to Schmidtke *et al.*[12, 19].

Phase Transfer into Water. The QDs were encapsulated with PI-*b*-PEG diblock copolymer as reported previously.[12]

Confocal Microscope Imaging. HeLa cells were cultured to 30% confluency in Lab-Tek thin bottom chambers in Dulbecco/Vogt modified Eagle's minimal essential medium (DMEM, c(glucose) = 4.5 g/L) medium containing 10% FCS at a temperature of 37 °C in a humidified atmosphere containing 5% CO_2. Medium was replaced with serum-containing or serum free DMEM. Subsequently, the cells were exposed to 0.1 µM of QDs for one hour and to the endosomal marker Cy5-transferrin. The cells were fixed using repeated rinsing with 4% paraformaldehyde and mountened with *Prolong Antifade Reagent with DAPI*.

Instrumentation: Confocal microscopy images were collected on an inverted Nikon TE2000-E2 Eclipse C1si (Nikon Instruments Inc., U.S.) equipped with a Nikon CFI Plan Apochromat TIRF 60x oil immersion objective (NA 1.49, 0.12 mm WD) and imaged with a Cool SNAP HQ2 monochrome camera (Photometric). DAPI, Cy5 and a QD filters (excitation 400-440 nm, emission 540-580 nm) were used.

RESULTS AND DISCUSSION

Phase Transfer

Since most high-quality NPs with a high crystallinity and monodispersity are produced in organic solvents at high temperatures *via* bottom-up procedure, a phase transfer into aqueous media is necessary.[20] In our case, we synthesized CdSe/CdS/ZnS core/shell/shell QDs in a flow-reactor.[16] These QDs can be transferred into water using an amphiphilic, cross-linkable poly(isoprene)-*b*-poly(ethylene glycol) diblock copolymer (PI-*b*-PEG) as previously reported.[18] We have already shown that the functionalization of the polymer can be done prior to the phase transfer (pre-assembly). This strategy is useful to incorporate functional groups (amine (**1**), alkyne (**2**), or carboxyl (**3**)) or small affinity molecules (e.g. biotin or carbohydrates).[12] We adopted this method for the glycosylation of the diblock copolymer with D-glucose (**5**) and D-maltose (**6**) using Huisgen's 1,3-dipolar cycloaddition[21] under copper catalyzed conditions[22],[23] between a glyco-azide (2,3,4,6-tetra-*O*-acetyl-*β*-D-glucopyranosyl azide and 2,3,6,2′,3′,4′,6′-hepta-*O*-acetyl-*α*-D-glucopyranosyl-1→4-*β*-D-glucopyranosyl azide) and a polymeric alkyne (PI-*b*-PEG-C≡CH (**4**)).[15] The reaction is not affected by other biological relevant end groups and the resulting triazole ring is very stable against hydrolysis/enzymatic degradation. Sharpless *et al.* defined this kind of reactions as click-chemistry which are simple, selective, avoiding byproducts and running under mild conditions in high yields.[24] Taking the highly selective structure/effect relationships between glycoreceptors and sugars into account, we synthesized functional polymers with the same connectivity to the anomeric carbon atom of the carbohydrate.

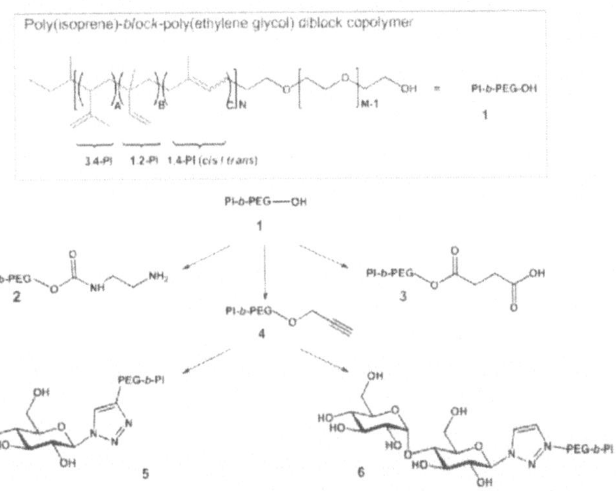

Scheme 1: Schematic representation of the functionalization / glycoconjugation strategy and chemical structures of the resulting compounds.

These functionalized polymers were successfully used for the micellar encapsulation of QDs in water with excellent fluorescence quantum yields.[13] No spectral shifts were obtained by UV-vis spectroscopy, neither in the absorption nor in the emission maximum. Furthermore, the good fluorescence properties of the glyco-QDs showing that the toxic copper catalyst is well separated during the purification of the polymers, because even smallest quantities of copper ions quench the QDs immediately.[14, 25-27]

Cellular Uptake of encapsulated QDs on human cervical carcinoma cell line HeLa

Cervical cancer rates fourth in cancer related death in women. Therefore, the human cervical carcinoma cell line

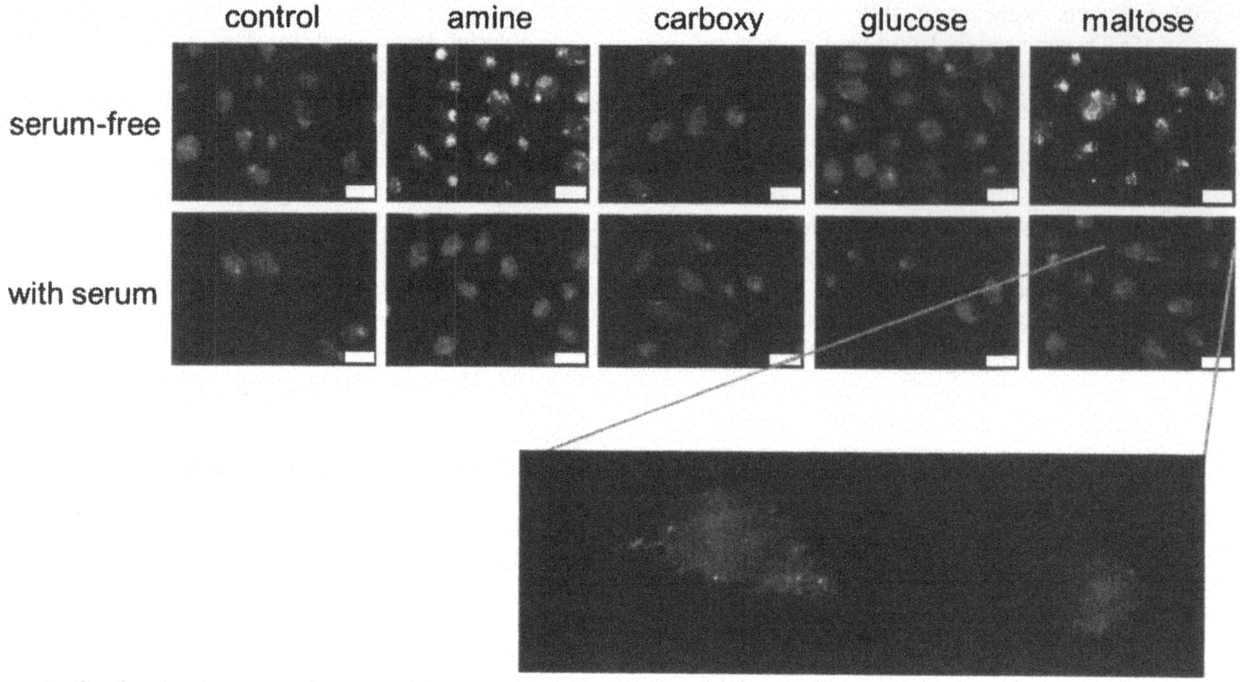

Figure 1: Confocal microscopy images of HeLa cells after exposure to different functionalized QDs. Top row: serum-free media, middle row: serum containing media, bottom: enlargement of HeLa cells incubated with maltose functionalized QD in serum-containing media. The color code represents blue: DAPI, red: Cy5, yellow: QD, the scale bar represents 20 μm.

HeLa is highly attractive for *in vitro* binding and uptake studies to achive cancer specific targeting. Many important biological processes like inflammation, cancer and cancer metastasis, are frequently mediated by glycoreceptor−glycan interactions.

The influence of different functional groups and carbohydrates on the surface of the QDs, namely: amine, carboxyl, D-glucose, and D-maltose were studied.

We performed the uptake-studies on HeLa cells, where the functional QDs were incubated on the cells for one hour in serum-containing and serum-free medium. The latter prevents a possible protein corona formation on the NP or an impaired uptake mechanism.[28] Using confocal microscopy, the cell-nanoparticle interactions were investigated and the fluorescent quantum dots were located intracellularly (see Figure 1). QDs were identified by both spectral position and by wavelength scans, if necessary. Surprisingly, we observed cellular uptake of maltose-functionalized QDs in serum-free as well as serum-containg media, even though cellular uptake in serum-containing media was significantly reduced. Furthermore, we observed a strong uptake in serum-free media for the QDs with amine, whereas the carboxylated and glucose functionalized NPs showed none. These results are in accordance to our previous published results.[29] Colocalization with the endosomal marker Cy5-transferrin indicates endosomal uptake in both cases. We interpret the cellular uptake of maltose-QDs medium as specific, due to comparable low incubation times and concentrations compared to our recent reports.[29] The reduced, but significant uptake of maltose-QDs in serum containing media supports this assumption. This is also supported by a recent report of Song *et al.*, where enhanced cellular uptake of maltose-modified liposomes in HeLa cells in serum-free medium is reported.[30] In a more recent report, the influence of the number of glucose units in a maltooligosaccharide-QD construct on the rate of nuclear import was studied. There microinjection was used to transfer QDs into the cells and an enhanced rate of nuclear import of maltooligo-QDs with more than three glucose units was found in comparison to (mono)-glucose functionalized QDs.[31] These reports underline the importance of polysaccharides in the cellular response of glyco-QD constructs. Interestingly, eukaryotic cells do not express endogenous maltose receptors since maltose is broken down to glucose by maltase (alpha-glucosidase) in the small intestine.[32] It can be hypothesed that maltose is transported via its terminal glucose moiety by the faciliative glucose receptors (GLUTs) that are frequently overexpressed in cancer cells to overcome the elevated need for energy and that the glucose functionalized QDs are not identified due to the triazole.[33] Further investigations will elucidate the effects of carbohydrate presentation on the NP's surface and their receptor interactions in detail.

CONCLUSION

In conclusion, we could show a simple method to couple different carbohydrates (D-maltose and D-glucose) to the PI-*b*-PEG diblock copolymer *via* click-chemistry. The glycosylated polymers were successfully used for the phase transfer of fluorescent QDs preserving their optical properties. Both QD synthesis as well as phase transfer were carried out in automated flow systems, allowing highly reproducible synthesis of a big amount of nanomaterials. The resulting glyco-QDs have proven to be non-toxic assessed by several toxicity assays (WST8 and LDH) at concentrations as high as 1 μM. Uptake studies presented in this article, show that these glyconanomaterials can be used to target HeLa cells. These results are promising and underline the potential of glyconanomaterials for further *in vivo* studies.

ACKNOWLEDGEMENT

This work was supported by the EU within the FP 7 program (Vibrant, EU 228933). J.-P.M. acknowledges the support of the Chemical Industry Fund, VCI: German Chemical Industry Association, the German-American Fulbright Program and furthermore travel fellowships of the Freunde- und Förderverein Chemie der Universität Hamburg e.V. and of the University of Hamburg. We would like to thank Matthias Wulf from the Institute of Organic Chemistry (University of Hamburg) for providing us the maltose-derivate.

REFERENCES

[1] T. K. Lindhorst, *Chem. Unserer Zeit* **2000**, *34*, 38-52.
[2] A. Varki, *Glycobiology* **1993**, *3*, 97-130.
[3] Y. C. Lee, R. T. Lee, *Acc. Chem. Res.* **1995**, *28*, 321-327.
[4] X. Wang, O. Ramström, M. Yan, *Adv. Mater.* **2010**, *22*, 1946-1953.
[5] M. Mammen, S.-K. Choi, G. M. Whitesides, *Angew. Chem. Int. Ed.* **1998**, *37*, 2754-2794.
[6] P. Alivisatos, *Nat. Biotech.* **2004**, *22*, 47-52.
[7] W. Chan, S. Nie, *Science* **1998**, *281*, 2016-2018.
[8] U. Resch-Genger, M. Grabolle, S. Cavaliere-Jaricot, R. Nitschke, T. Nann, *Nat. Meth.* **2008**, *5*, 763-775.
[9] P. Babu, S. Sinha, A. Surolia, *Bioconjugate Chem.* **2006**, *18*, 146-151.
[10] R. Kikkeri, B. Lepenies, A. Adibekian, P. Laurino, P. H. Seeberger, *J. Am. Chem. Soc.* **2009**, *131*, 2110-2112.
[11] H. Bavireddi, R. Kikkeri, *Analyst* **2012**, *137*, 5123-5127.
[12] C. Schmidtke, E. Poselt, J. Ostermann, A. Pietsch, H. Kloust, H. Tran, T. Schotten, N. G. Bastus, R. Eggers, H. Weller, *Nanoscale* **2013**, *5*, 7433-7444.
[13] C. Schmidtke, H. Lange, H. Tran, J. Ostermann, H. Kloust, N. G. Bastús, J.-P. Merkl, C. Thomsen, H. Weller, *J. Phys. Chem. C* **2013**, *117*, 8570-8578.
[14] J. Ostermann, J.-P. Merkl, S. Flessau, C. Wolter, A. Kornowksi, C. Schmidtke, A. Pietsch, H. Kloust, A. Feld, H. Weller, *ACS Nano* **2013**.
[15] C. Schmidtke, A.-M. Kreuziger, D. Alpers, A. Jacobsen, Y. Leshch, R. Eggers, H. Kloust, H. Tran, J. Ostermann, T. Schotten, J. Thiem, J. Thimm, H. Weller, *Langmuir* **2013**, *29*, 12593-12600.
[16] H. Tran; S. Niehaus; H. Weller; D. Ness, *Vol. WO2014033213 (A2)*, **2014**.
[17] H. Tran, University of Hamburg, PhD thesis (Munich), **2012**.
[18] E. Pöselt, C. Schmidtke, S. Fischer, K. Peldschus, J. Salamon, H. Kloust, H. Tran, A. Pietsch, M. Heine, G. Adam, U. Schumacher, C. Wagener, S. Förster, H. Weller, *ACS Nano* **2012**, *6*, 3346-3355.
[19] C. Schmidtke, A.-M. Kreuziger, D. Alpers, A. Jacobsen, Y. Leshch, R. Eggers, H. Kloust, H. Tran, J. Ostermann, T. Schotten, J. Thiem, J. Thimm, H. Weller, *Langmuir* **2013**.
[20] G. Palui, F. Aldeek, W. T. Wang, H. Mattoussi, *Chem Soc Rev* **2015**, *44*, 193-227.
[21] R. Huisgen, G. Szeimies, L. Möbius, *Chem. Ber.* **1967**, *100*, 2494-2507.
[22] C. W. Tornøe, C. Christensen, M. Meldal, *J. Org. Chem.* **2002**, *67*, 3057-3064.
[23] V. V. Rostovtsev, L. G. Green, V. V. Fokin, K. B. Sharpless, *Angew. Chem. Int. Ed.* **2002**, *41*, 2596-2599.
[24] H. C. Kolb, M. G. Finn, K. B. Sharpless, *Angew. Chem.* **2001**, *113*, 2056-2075.
[25] C. Schieber, A. Bestetti, J. P. Lim, A. D. Ryan, T.-L. Nguyen, R. Eldridge, A. R. White, P. A. Gleeson, P. S. Donnelly, S. J. Williams, P. Mulvaney, *Angew. Chem. Int. Ed.* **2012**, *51*, 10523-10527.
[26] H. Kloust, C. Schmidtke, J. P. Merkl, A. Feld, T. Schotten, U. E. A. Fittschen, M. Gehring, J. Ostermann, E. Poselt, H. Weller, *J. Phys. Chem. C* **2013**, *117*, 23244-23250.
[27] J. P. Merkl, J. Ostermann, C. Schmidtke, H. Kloust, R. Eggers, A. Feld, C. Wolter, A. M. Kreuziger, S. Flessau, H. Mattoussi, H. Weller, *Proc. Spie* **2014**, *8955*.
[28] M. Moros, B. Hernáez, E. Garet, J. T. Dias, B. Sáez, V. Grazú, Á. González-Fernández, C. Alonso, J. M. de la Fuente, *ACS Nano* **2012**, *6*, 1565-1577.
[29] J. Ostermann, C. Schmidtke, C. Wolter, J. P. Merkl, H. Kloust, H. Weller, *Beilstein J. Nanotechnol.* **2015**, *6*, 232-242.
[30] C. K. Song, S. H. Jung, D.-D. Kim, K.-S. Jeong, B. C. Shin, H. Seong, *Int. J. Pharm.* **2009**, *380*, 161-169.
[31] S. Sekiguchi, K. Niikura, Y. Matsuo, S. H. Yoshimura, K. Ijiro, *RSC Adv.* **2012**, *2*, 1656-1662.
[32] S. H. Sorensen, O. Noren, H. Sjostrom, E. M. Danielsen, *Eur J Biochem* **1982**, *126*, 559-568.
[33] M. L. Macheda, S. Rogers, J. D. Best, *J. Cell Physiol.* **2005**, *202*, 654-662.

Oxidized Graphene Nanoribbons as a Delivery System for the Bioactive Sphingolipid Ceramide

Cassandra Suhrland*, Jean-Philip Truman**, and Balaji Sitharaman***

*Department of Biomedical Engineering, Stony Brook University, NY, USA
cassandra.suhrland@stonybrook.edu
**Health Science Center, Stony Brook University, NY, USA Jean-Philip.Truman@stonybrookmedicine.edu
***Department of Biomedical Engineering, Stony Brook University, NY, USA
balaji.sitharaman@stonybrook.edu

ABSTRACT

Ceramides are a group of sphingolipids that have garnered a growing interest as possible anticancer agents due to their chain-length-dependent, potent pro-apoptotic effects. However, as lipids, they have virtually no solubility in aqueous solution. Consequently, there is currently no clinically approved delivery method for *in vivo* use of ceramide as a therapeutic agent. To this end, we have developed a novel method for loading the short-chain C_6 ceramide onto oxidized graphene nanoribbons (O-GNRs). Mass spectrometry revealed a loading efficiency of 51% C_6 onto O-GNRs. Using a Prestoblue viability assay, we determined that that 100 µg/mL of C_6-loaded O-GNRs reduced HeLa cell viability by 97% compared to controls. We confirmed that C_6-loaded O-GNRs were actually entering cells using transmission electron microscopy. Taken together, these data show that O-GNRs are a promising delivery agent for ceramide.

Keywords: graphene, cancer, drug delivery, ceramide, sphingolipid

1 INTRODUCTION

Ceramide, a type of sphingolipid, has been increasingly recognized for the important roles it plays in cell metabolism and certain diseases. For example, aberrant ceramide metabolism has been shown to play key roles in cancer survival and cell proliferation[1, 2]. As such, these metabolic pathways are promising targets for inducing apoptosis in cancer cells via externally-delivered ceramide. One candidate for targeting these pathways is the synthetic, short-chain C_6 ceramide, which has been shown to inhibit cell growth and induce apoptosis when delivered to cancer cells[1]. However, as it extremely hydrophobic, it requires a delivery system to have sufficient solubility in aqueous solution. Current methods for increasing ceramide solubility for delivery into cells, such as solubilization in DMSO and liposomes, have issues such as low efficacy, complex and error-prone loading processes, or would not be suitable for therapeutic use. Here we present a facile method for loading ceramide onto oxidized graphene nanoribbons (O-GNRs) derived from carbon nanotubes, and we demonstrate highly-reduced HeLa cell viability after exposure to ceramide-loaded O-GNRs. We also demonstrate high levels of uptake of our ceramide-loaded O-GNRs (referred to as Cer-O-GNRs) into HeLa cells via TEM.

2 METHODS

2.1 Materials

Ceramide C_6 was purchased from Avanti Polar Lipids (Alabaster, AL, USA). All other materials and reagents were purchased from Sigma Aldrich (St. Louis, MO, USA) unless noted otherwise.

2.2 O-GNR Synthesis and C6 Loading

O-GNRs were synthesized from carbon nanotubes using the longitudinal unzipping method[3]. O-GNRs (1 mg/mL) and C_6 ceramide (200 µg/mL) were dispersed in 100% ethanol using bath sonication. 500 µL of each of these stock solutions were added to a 20mL glass scintillation vial and the mixture was sonicated for 15 minutes. The mixture was then bubbled vigorously with nitrogen gas and covered with parafilm in order to prevent oxidation of ceramide. In order to force hydrophobic interaction between the O-GNRs and ceramide, 9 times the volume (9mL) of double-distilled H_20 was added to the solution containing O-GNRs and ceramide using New Era NE-300 syringe pump (New Era Pump Systems Inc. Farmingdale, NY, USA). Water was added in two phases: an initial slow phase for 1 hour at 2 mL/hr, and then a faster phase at 7 mL/hr until completion, for a total addition time of 2 hours. After completion of loading, the mixture was centrifuged for 1hr at 13000RPM and the supernatant with unloaded ceramide removed, leaving

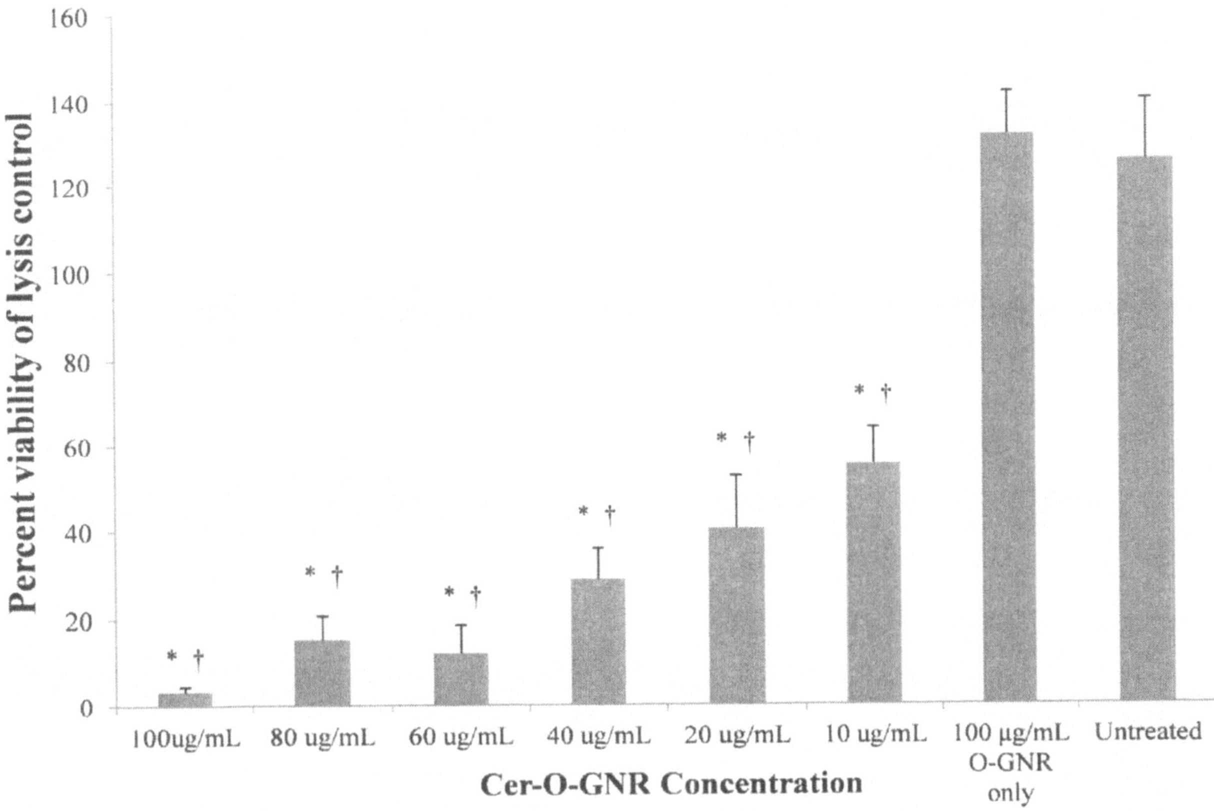

Figure 1: HeLa cell viability measured using Prestoblue after incubation with Cer-O-GNRs for 24 hours. Ceramide loading onto O-GNRs was 255 ng per microgram of O-GNRs. * indicates significant difference from "100 µg/mL O-GNRs only." † indicates significant difference from "Untreated."

ceramide-loaded O-GNRs (Cer-O-GNRs). In order to calculate loading efficiency, Cer-O-GNRs were resuspended in 100% ethanol and sonicated for 1 hour, to separate ceramide from the O-GNRs. The ceramide was isolated using centrifugation, and the concentration of ceramide in the supernatant was determined using mass spectroscopy.

2.3 Prestoblue Assay for Cell Viability

In order to determine the efficacy of Cer-O-GNRs as an anticancer agent, we incubated HeLa cells with Cer-O-GNRs and measured cell viability using Prestoblue, a resazurin-based viability dye. HeLa cells were grown at 37 °C, 5% CO_2 in DMEM supplemented with 10% fetal bovine serum and 1% penicillin streptomycin. Cells were seeded in 96 well plates at a density of 5000 cells per well and allowed to adhere overnight. The following day, HeLa cells were incubated with Cer-O-GNRs at concentrations ranging from 10-100 µg/mL. In order to ensure that any observed reduction in viability was not due to the loading process or O-GNRs themselves, O-GNRs were sent through the loading and purification processes without any ceramide, and incubated with cells at the highest concentration tested (100 µg/mL). After 24 hours the media was removed from each well and washed twice with Phosphate Buffered Saline. 100 µL of the viability reagent mixture was added to each well and placed back in the incubator. After 2 hours, the fluorescence intensity of each well was measured with an excitation wavelength of 530nm and emission wavelength of 580nm using a Molecular Devices SpectraMax M2e (Sunnyvale, CA, USA). The cell viability is expressed as a percent of the lysis control with the formula$\left((I_{test} - I_{blank})/(I_{control} - I_{blank}) \times 100\%\right) - 100\%$, where I_{test} is the fluorescence intensity of cells exposed to Cer-O-GNRs or untreated/O-GNR only controls, $I_{control}$ is the fluorescence intensity of lysed cells, and I_{blank} is the fluorescence intensity from empty wells. Statistical analysis was performed using one-way ANOVA with Tukey's post-hoc analysis.

2.4 Cellular Uptake of O-GNRs via TEM

HeLa cells were seeded on ACLAR® film (Electron Microscopy Sciences, Hatfield, PA) at a density of approximately 20000 cells per well in 6 well plates and allowed to adhere overnight. Cells were then exposed to 100µg/mL Cer-O-GNRs for one hour. After this incubation time, cells were fixed with 2.5% glutaraldehyde (Electron Microscopy Sciences, Hatfield, PA) in 0.1 M PBS for 15 minutes at room temperature. The films were then placed in 2% osmium tetroxide in 0.1 M PBS and dehydrated through graded ethanol washes. The films were then embedded using durcupan resin. The embedded specimens were examined for areas with high cell densities, and these areas were cut into

80 nm sections using an Ultracut E microtome (Reichert-Jung, Cambridge, UK), and placed on formvar-coated copper grids. Sections were imaged using a Tecnai Bio Twin G transmission electron microscope (FEI, Hillsboro, OR), at 80 kV. Images were obtained using an XR-60 CCD digital camera system (AMT, Woburn, MA).

3 RESULTS AND DISCUSSION

3.1 Loading of C_6 onto O-GNRs

Using mass spectroscopy, we calculated the loading efficiency to be approximately 51% of our initial concentration, indicating a high loading efficiency despite the concentrations of ceramide and O-GNRs being diluted 20-fold by the loading process. This is a significant improvement in efficiency over the maximum possible incorporation of C_6 into liposomes[4].

3.2 Viability of HeLa Cells Exposed to Cer-O-GNRs

After exposure to 100 µg/mL of Cer-O-GNRs for 24 hours, HeLa cells had their viability reduced their viability by approximately 97% compared to the untreated or O-GNRs control (Figure 1). Even at the lowest concentration used (10 µg/mL), HeLa cell viability was reduced by 56% compared to the untreated controls. In order to ensure that this toxicity could not be attributed to either the loading process or the concentrations of O-GNRs used we put O-GNRs through the loading process without ceramide, and incubated them with HeLa cells at the highest concentration used (100 µg/mL). This control was significantly different from all Cer-O-GNR conditions tested while not being significantly different from the untreated control, meaning that the reduction in viability observed could not be attributed to the loading process or the O-GNRs themselves.

3.3 TEM of Cer-O-GNR Uptake in HeLa Cells

We have previously demonstrated that O-GNRs get uptaken into HeLa cells in great quantities via a macropinocytotic-like, EGFR-dependent process[5, 6]. However, since we were able to load significant quantities of C_6 onto O-GNRs, it was not clear whether this loading would interfere with the uptake of O-GNRs into HeLa cells. Figure 2 illustrates that loading O-GNRs with C_6 does not appear to interfere with this uptake process. There are many O-GNRs of different sizes and shapes present in vacuoles, as well as uptake of more particles in progress. Furthermore, these regions of uptake along the outer edge of the cell demonstrate the characteristic "blebbing" of the cell membrane previously observed in the uptake of O-GNRs into HeLa cells[6]. This suggests that loading ceramide onto O-GNRs does not affect their ability to enter cells, and that they can actually deliver their cargo inside the cell itself rather than merely releasing it in close proximity.

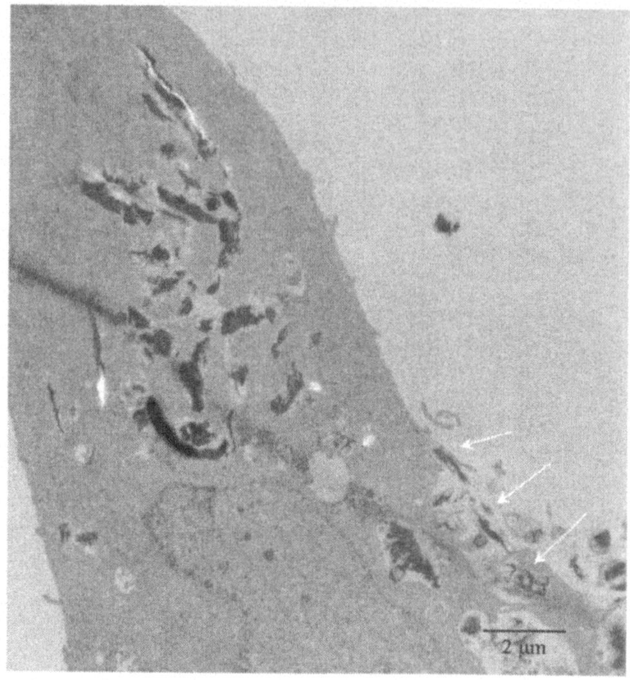

Figure 2: Transmission electron microscopy images of HeLa cells incubated with Cer-O-GNRs after one hour. There are many nanoparticles contained within vacuoles, as well as edge regions with "blebbing" of the membrane (white arrows) characteristic of the macropinocytotic-like process.

4 CONCLUSIONS

O-GNRs are a promising delivery system for the short-chain ceramide C_6. Using a fast, simple method for loading, we were able to induce high levels of toxicity and uptake in HeLa cells at low concentrations using our Cer-O-GNRs. Future studies will look at ceramides with longer variable chains such as C_{16} ceramide.

REFERENCES

[1] Barth, B.M., M.C. Cabot, and M. Kester, *Ceramide-based therapeutics for the treatment of cancer.* Anticancer Agents Med Chem, 2011. **11**(9): p. 911-9.

[2] Stiban, J., R. Tidhar, and A.H. Futerman, *Ceramide synthases: roles in cell physiology and signaling.* Adv Exp Med Biol, 2010. **688**: p. 60-71.

[3] Kosynkin, D.V., et al., *Longitudinal unzipping of carbon nanotubes to form graphene nanoribbons.* Nature, 2009. **458**(7240): p. 872-6.

[4] Khazanov, E., et al., *Physicochemical and biological characterization of ceramide-containing liposomes: paving the way to ceramide therapeutic application.* Langmuir, 2008. **24**(13): p. 6965-80.

[5] Mullick Chowdhury, S., et al., *Cell specific cytotoxicity and uptake of graphene nanoribbons.* Biomaterials, 2013. **34**(1): p. 283-93.

[6] Mullick Chowdhury, S., P. Manepalli, and B. Sitharaman, *Graphene nanoribbons elicit cell specific uptake and delivery via activation of epidermal growth factor receptor enhanced by human papillomavirus E5 protein.* Acta Biomater, 2014. **10**(10): p. 4494-504.

Characterisation of THP-1 macrophages as an *in vitro* model for CD44-targeted therapies

J. M. Rios de la Rosa*, A. Tirella** and N. Tirelli***

*Manchester Pharmacy School, University of Manchester, M13 9PT, UK, juliomanuel.riosdelarosa@postgrad.manchester.ac.uk
**Manchester Pharmacy School/Institute of Inflammation and Repair University of Manchester, M13 9PT, UK, annalisa.tirella@manchester.ac.uk
*** Manchester Pharmacy School/Institute of Inflammation and Repair University of Manchester, M13 9PT, UK, nicola.tirelli@manchester.ac.uk

ABSTRACT

CD44 is the major cell surface receptor of hyaluronic acid (HA) and mediates the response of cells to their microenvironment. Whereas the standard isoform (CD44s) is ubiquitously expressed throughout the body, variant isoforms (CD44v) are aberrantly overexpressed in certain tumours, thus CD44 has become a potential target for cancer chemotherapy. However, CD44 is also one of the most abundantly expressed macrophage receptors, rendering this cellular type a potential competitor for the uptake of targeted nanocarriers *in vivo*.

An *in vitro* human macrophage model is described herein. This model would allow to evaluate CD44 off-targeting. In particular, we have developed a procedure for the differentiation and activation of THP-1 cells. The CD44 expression of THP-1 macrophages was compared to that of the tumoural cell line HT-29 in terms of amount and isoform composition. The internalisation kinetics of HA nanoparticles was also evaluated.

Keywords: CD44s, CD44v, hyaluronic acid, CD44-targeted therapies, THP-1 macrophages

1 INTRODUCTION

CD44 is the major cell surface receptor of HA [1] and is increasingly recognised as one of the 'hot' receptors in targeted tumour therapy. The success of targeting therapies is hindered mainly by the complexity of the receptor. CD44 is exquisitely regulated at the post-transcriptional level resulting in a combination of standard (CD44s) and variant (CD44v) isoforms. Second, the different isoforms are susceptible to numerous post-translational modifications. Indeed it is well known that CD44s/v-HA interactions have a key role in tumorigenesis, metastasis, and putatively in the identity itself of cancer stem cells [1, 2]. However, CD44 is also widely expressed on other cells, such as macrophages resident in the organs of the reticuloendothelial system. Importantly, it has been reported that monocyte-derived macrophages express an active (HA-binding) state of the receptor whereas freshly isolated peripheral blood monocytes express an inactive form [3]. Therefore, in view of the development of CD44-targeted therapies and foreseeing the intravenous injection of appropriate carriers, it is chief to assess possible off-targeting effects.

The main constraint when considering experimental work with human macrophages is their difficult isolation, the invasiveness of the isolation procedures, and their inability to proliferate in culture. For these reasons a number of studies focused on CD44-targeting make use of a well established murine cell model, RAW 264.7 macrophages [4] e.g. as a positive control for uptake experiments [5, 6]. Yet, the use of a murine cell line presents obvious limitations.

The human monocytic leukaemia THP-1 cell line shows commitment towards macrophage differentiation [7-9]. THP-1 macrophages resemble native monocyte-derived macrophages in terms of morphology, expression of membrane antigens and receptors, transient induction of several proto-oncogenes, and secretory products [7]. To the best of our knowledge, it is currently not known whether macrophages exhibit a specific CD44s/CD44v expression pattern, whether this depends on the degree and mode of activation, and if this might influence the affinity towards HA. The aim of the present study was to develop an *in vitro* human macrophage model to study the role of these sentinel cells in the uptake of CD44-targeting carriers.

2 MATERIALS AND METHODS

2.1 General Cell Culture

The human monocyte cell line THP-1 (TIB-202™) and the colorectal carcinoma cell line HT-29 (HTB-28™) were purchased from American Type Cell Collection (ATCC) (Manassas, VA, USA). Both cell types were maintained at 37°C in a humidified 5% (v/v) CO_2 air atmosphere. The medium was changed every 2-3 days as recommended by the supplier.

THP-1 premonocytes were cultured in RPMI 1640 medium (Gibco®/Invitrogen, UK) supplemented with 10% (v/v) foetal bovine serum (FBS), 2 mM L-Glutamine, 1%

(v/v) antibiotic-antimycotic solution and 50 μM 2-mercaptoethanol (Sigma-Aldrich, UK). Differentiation into THP-1 macrophages was triggered by incubation with 50 ng/mL phorbol 12-myristate 13-acetate (PMA) (Sigma-Aldrich, UK) for 24 hours. Classical and alternative activation were achieved by treatment with LPS/IFN-γ or IL-4/IL-13, respectively.

The human colorectal adenocarcinoma HT-29 cell line was cultured in Modified McCoy's 5A medium supplemented with 10% (v/v) FBS and 1% (v/v) antibiotic-antimycotic solution (Sigma-Aldrich, UK).

Phase contrast images of live THP-1 macrophages were recorded with a Leica DM6000 B inverted microscope (Leica Microsystems, UK) coupled with a 5.5 Neo sCMOS camera (Andor, UK) and the *μManager* software (v.1.46).

2.2 CD44 Expression Analysis

Western blotting: whole cell lysates (20 μg/well) were separated by 7.5% sodium dodecylsulphate-polyacrylamide gel electrophoresis (SDS-PAGE) (Run settings: 1 hour, 100V, 0.5 A) and transferred to a poly(vinylidene difluoride) membrane (PVDF) using a Criterion™ Blotter (Bio-Rad, UK) (Run settings: 30 minutes, 100 V, 0.5 A). Membranes were blocked by incubation with 5% non-fat milk (Fluka, UK) in Tris-buffered 0.1% Tween-20 (TBS-T, Bio-Rad, UK) for 1 hour at RT, cut into two sections and incubated accordingly with mouse anti-human CD44 (Clone 156-3C11, #170-5061; Cell Signaling Technology, UK) (dilution 1:1000 in TBS-T) at 4°C overnight, or rabbit anti-human β-actin (ab8227, Abcam, UK) (1:5000 dilution in TBS-T) for 1 hour at RT. After several washings, membranes were incubated with goat anti-mouse IgG-peroxidase (A0168, Sigma-Aldrich, UK) or goat anti-rabbit IgG-peroxidase (A0545, Sigma-Aldrich, UK), respectively, for 1 hour at RT. Bands were detected using Clarity Western enhanced chemiluminescence (ECL) substrate (Bio-Rad, UK) with the aid of a ChemiDoc™ MP System (Bio-Rad, UK). ImageJ software (http://rsb.info.nih.gov/ij) was used to perform a densitometric analysis of protein bands.

Flow cytometry: cells were detached using Enzyme-Free Buffer formulated in PBS (Gibco®/Invitrogen, UK). $0.5x10^6$ cells per sample tube were incubated with Fc-receptor blocking inhibitor (Affymetrix/eBioscience, UK) followed by staining with anti-human CD44 APC-conjugated (ab81424, clone MEM-263; Abcam, UK) for 30 minutes at RT. Excess unbound antibody was washed off and live analysis was performed on 10,000 individual cells with a CyAn™ ADP analyser and the Summit (v4.3) software (Beckman Coulter, Inc.).

2.3 Preparation of HA-coated Nanoparticles

Middle viscosity chitosan (CS) obtained from crab shells, viscosimetric average molecular weight of 656 kDa, was purchased from Sigma Aldrich (#51009219, Lot: WE44069811; Basingstoke, UK), whereas that of 35 kDa was prepared experimentally via nitrite-mediated degradation. Hyaluronic acid (HA) with weight average molecular weight of 305 kDa (GPC with static light scattering detector, experimental) was purchased from Novozymes Biopharma A/S (Bagsvaerd, Denmark) and fluorescently-labelled (FA-HA) with fluoresceinamine (Sigma-Aldrich, UK).

Nanoparticles were synthesised as described by Almalik et al. [5]. Briefly, a 0.069 wt% CS solution and a 3 $mg \cdot mL^{-1}$ HA solution were prepared by dissolving dry purified polymer in 4.6 mM HCl or deionised water, respectively. The pH was adjusted to 5 by addition of 0.1 M NaOH in the case of the CS solution, or 0.1 M HCl in the case of the HA solution. The solutions were then filtered using either a 0.22 μm (HA) or a 0.45 μm (CS) filter. One-stage preparation foresees the dilution 1:2 of each polymer solution in pH 5 deionised water, and complexation by addition of a volume of the CS solution into an equal volume of the HA solution under magnetic stirring (~1000 rpm) for 30 minutes at RT.

2.4 Quantification of Cellular Uptake (Fluorimetry)

Cells were incubated with different concentrations of FA-HA nanoparticles using either low (HA305//CS35) or high (HA305//CS656) molecular weight chitosan for the formulation. After specified incubation times (0, 1, 2 and 4 h), cells were washed three times with PBS and lysed with RIPA buffer. The amount of membrane-bound and internalised nanoparticles was estimated on the basis of the fluorescence of the cell lysates (λ_{ex} = 485 nm; λ_{em} = 528 nm; Sensitivity: 75; Optical position: Top 50%; Light source: Xenon flash). Nanoparticles were dispersed in cell lysates to provide a calibration for the fluorescence data. The amount of membrane-bound and internalised nanoparticles was normalised by the total protein content, which was quantified using the BCA assay kit (Sigma-Aldrich, UK).

3 RESULTS

We have established protocols for the differentiation of THP-1 premonocytes into non-polarised (M0) macrophages, as well as for their activation via the classical (M1) or alternative (M2a) pathways (Figure 1). The efficiency of the protocols was assessed using CD14, CD11c and CD44 as differentiation markers, and CD206 and the production of TNF-α and IL-1β as M2a and M1 activation markers, respectively (data not showed).

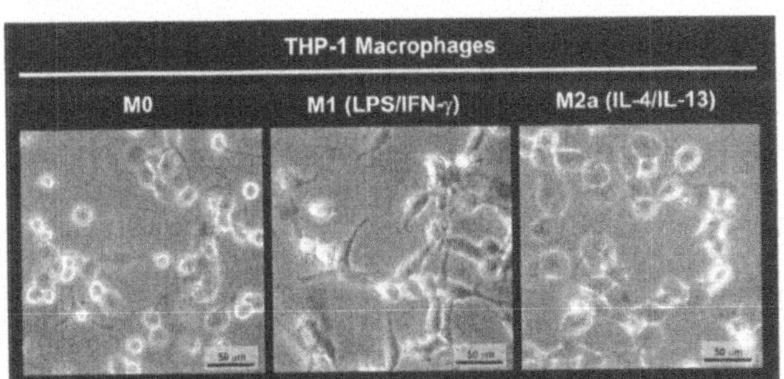

Figure 1. Morphology of THP-1 macrophages. THP-1 premonocytes were incubated with PMA to induce differentiation into non-polarised macrophages (M0). These

macrophages undergo classical (M1) or alternative (M2a) activation upon treatment with LPS/IFN-γ or IL-4/IL-13, respectively. Representative Phase Contrast images, 40X magnification.

The expression of CD44 increased after differentiation of THP-1 monocytes into M0 macrophages; flow cytometry indicated a 1.34-fold increase, Western blotting a 0.75-fold increase. Classical activation (M1) showed further upregulation of the receptor, whereas alternative activation (M2a) appeared to have a negligible effect. In all cases we were able to detect only the standard isoform, whereas the colorectal cancer cell line HT-29 mainly expresses high molecular weight isoforms (Figure 2).

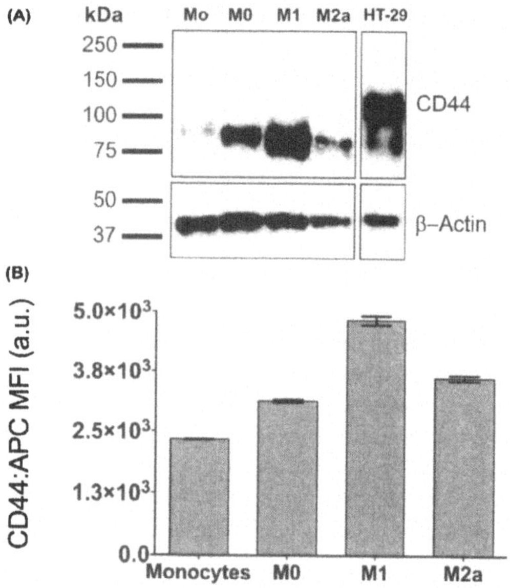

Figure 2. CD44 expression analysis. (A) Western blot showing differences in terms of expression and isoforms between THP-1 premonocytes (Mo), non-polarised THP-1 macrophages (M0), or THP-1 macrophages activated via the classical (M1) or alternative (M2a) pathways, and the HT-29 cell line. β-actin was used as loading control. (B) Flow cytometry analysis showing the CD44:APC median fluorescence intensity (MFI) for THP-1 premonocytes and THP-1 macrophages (M0, M1 and M2a).

Finally, we present a comparison of the kinetics of internalisation of HA-coated CS nanoparticles (Figure 3). As showed in previous publications from our group, the molecular weight of chitosan proves essential for the conformation of the HA on the surface of the nanoparticles, therefore for its presentation to CD44 on cell membranes. High molecular weight chitosan nanoparticles (HA305//CS656) were uptaken in larger amounts than their low molecular weight counterparts (HA305//CS35) for both cell models. The amount of nanoparticles internalised by THP-1 M0 macrophages was determined to be about 10% of that observed in the target HT-29 cancer cell model. This may be ascribed to a different affinity of the standard and the variant isoforms towards HA or to a different density of CD44 on cell membranes, or both.

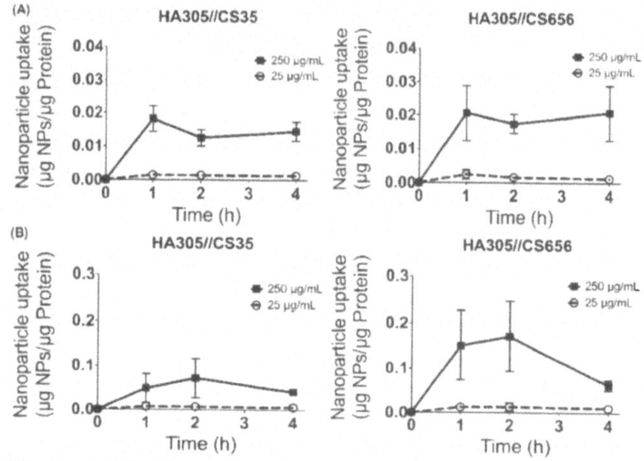

Figure 3. Uptake of HA//CS nanoparticles by (A) THP-1 M0 macrophages (B) HT-29 cells as a function of time, chitosan molecular weight and concentration of nanoparticles. The uptake was measured as the fluorescence of cell lysates (using a calibration of nanoparticles in cell lysates) and normalised against the amount of total protein content.

4 CONCLUSION

The THP-1 cell line provides a valuable human macrophage model to evaluate the interaction of HA-based structures with CD44. Our results suggest that the uptake of HA//CS NPs by THP-1 M0 macrophages is roughly 10% of that seen in the cancer cell line HT-29. The complete validation of THP-1 cells as an off-target model, however, would still require a comparison to primary and/or *in vivo* models.

5 ACKNOWLEDGEMENTS

J. M. Rios de la Rosa is fully funded by EPSRC, NoWNano DTC. We would like to thank A Spadea (*Manchester Pharmacy School, University of Manchester, UK*) for her support on protein analysis.

REFERENCES

1. Zoller, M., *CD44: can a cancer-initiating cell profit from an abundantly expressed molecule?* Nature Reviews Cancer, 2011. **11**(4): p. 254-267.
2. Williams, K., et al., *CD44 integrates signaling in normal stem cell, cancer stem cell and (pre)metastatic niches*. Experimental Biology and Medicine, 2013. **238**(3): p. 324-338.
3. Vivers, S., I. Dransfield, and S.P. Hart, *Role of macrophage CD44 in the disposal of inflammatory*

cell corpses. Clinical Science, 2002. **103**(5): p. 441-449.

4. Vachon, E., et al., *CD44 is a phagocytic receptor.* Blood, 2006. **107**(10): p. 4149-4158.
5. Almalik, A., P.J. Day, and N. Tirelli, *HA-Coated Chitosan Nanoparticles for CD44-Mediated Nucleic Acid Delivery.* Macromolecular Bioscience, 2013. **13**(12): p. 1671-1680.
6. Almalik, A., et al., *Hyaluronic acid (HA) presentation as a tool to modulate and control the receptor-mediated uptake of HA-coated nanoparticles.* Biomaterials, 2013. **34**(21): p. 5369-5380.
7. Auwerx, J., *The Human Leukemia-Cell Line, Thp-1 - a Multifaceted Model for the Study of Monocyte-Macrophage Differentiation.* Experientia, 1991. **47**(1): p. 22-31.
8. Tsuchiya, S., et al., *Induction of Maturation in Cultured Human Monocytic Leukemia-Cells by a Phorbol Diester.* Cancer Research, 1982. **42**(4): p. 1530-1536.
9. Daigneault, M., et al., *The Identification of Markers of Macrophage Differentiation in PMA-Stimulated THP-1 Cells and Monocyte-Derived Macrophages.* Plos One, 2010. **5**(1).

Inhalable Stimuli-Responsive Theranostic Nanoparticles for Targeting Cancer in Deep Lung Tissue

YH Abbas [1], HME Azzazy [1,*], S Tammam[1,2], A Lamprecht [2,3], ME Ali [2], AM Schmidt [4], S Sollazzo [4], S Mathur [5]

[1] Department of Chemistry, The American University in Cairo, Egypt

[2] Laboratory of Pharmaceutical Technology and Biopharmaceutics, University of Bonn, Germany

[3] Laboratory of Pharmaceutical Engineering, University of Franche-Comté, Besançon, France

[4] Department of Chemistry, Institute of Physical Chemistry, University of Cologne, Cologne, Germany

[5] Department of Chemistry, Institute of Inorganic Chemistry, University of Cologne, Cologne, Germany

[*] Corresponding author at: Department of Chemistry, The American University in Cairo, Egypt.

Tel: +2 02 2615 2559 (GMT+2 hr)
Cell: +2 01000 565 727 (GMT+2 hr)
Fax: +2 02 2795 7565
E-mail addresses: hazzazy@aucegypt.edu

ABSTRACT

In 2012 lung cancer accounted for 13% (1.83 million) of cancer cases and caused 19% (1.56 million) of cancer deaths worldwide according to the IARC. Despite advances in surgery and drug discovery, lung cancer remains difficult to treat as a result of unavoidable exposure to carcinogens, poor diagnosis and the lack of targeted drug delivery platforms. The aim of this study was to develop a non-invasive, patient convenient platform for the targeted delivery of chemotherapeutics to cancer in deeper lung tissue. This platform is intended to improve the therapy of deep lung tissue cancer through the combinatorial effect of tissue targeting, controlled drug release, and the possible imaging and prognostic values as a result of the USPIONs content.

Keywords: theranostic nanoparticles, stimuli-responsive nanoparticles, USPIONs, deep lung tissue cancer, chitosan

Abbreviations: CS, Chitosan; TPP, Sodium Tripolyphosphate; AA, L-ascorbic acid; FL, Fluorescein; EDAC, N-(3-Dimethylaminopropyl)-N'-ethylcarbodiimide hydrochloride; FBS, Fetal Bovine Serum; PBS, Phosphate Buffered Saline; MTT, 3-(4, 5-dimethylthiazolyl-2)-2, 5-diphenyltetrazolium bromide; DMSO, Dimethyl sulfoxide; EGFR-Antibody or Ab, Monoclonal Anti-Epidermal Growth Factor Receptor Antibody; WGA-AF, Wheat Germ Agglutinin, Alexa Fluor 488 Conjugate; MD, maltodextrin; PVP, polyvinylpyrrolidone.

1. INTRODUCTION

Lung cancer is the deadliest solid tumour [1]. As stated by the International Agency for Research on Cancer (IARC) in 2012, lung cancer accounted for 13% (1.83 million) of cancer cases and caused 19% (1.56 million) of cancer deaths worldwide [2]. Moreover, the World Health Organization estimates that lung cancer will account for 19.35% of all cancer deaths by 2020 [3]. There are two types of lung cancer: small cell lung cancer (SCLC) and non-small cell lung cancer (NSCLC). NSCLC represents 85% of the lung cancer cases [4].

It is believed that inhalation is the optimal route of administration in the case of NSCLC chemotherapy as it offers a non-invasive route for chemotherapeutic delivery lacking the undesirable attributes of I.V. based approaches in addition to the specific tumour targeting for enhanced patient conveniency. Since most of NSCLCs seem to allocate in the airways [4-5], the development of nanoparticle (NP)/microparticle (MP) carrier system capable of delivering chemotherapeutics to airway tumours is a valid approach.

The aim of this study was to develop a non-invasive, patient convenient platform for the targeted delivery of chemotherapeutics to NSCLC. Tumour cell targeting and localized chemotherapeutic delivery with the provision of a desirable -stimuli responsive- drug release profile will be allowed by the delivery platform.

2. MATERIALS AND METHODS

2.1. Materials

Medium molecular weight CS was purchased from Bio Basic Inc. (Ontario, Canada). TPP was bought from Mistral (Northern Ireland, UK). $FeCl_3.6H_2O$, $NaHCO_3$, AA, FL, EGFR-Antibody, MTT, DMSO, MD and PVP were bought from Sigma-Aldrich (Munich, Germany). Spectra/Por Float-A-Lyzer G2 (MWCO 100 kDa) was supplied from Spectrum Laboratories Inc. (Schwerte, Germany). RPMI 1640 with L-Glutamine, Pen-Strep and L-Glutamine were purchased from Lonza (Basel, Switzerland). A549 and L929 cell lines were obtained from Vacsera Holding Company (Cairo, Egypt) and the American Type Culture Collection (Manassas, VA, USA); respectively. FBS, Hoechst 33258, Pentahydrate (bis-Benzimide) and WGA-AF were supplied by Life Technologies (Carlsbad, CA, USA).

2.2. Methods

2.2.1. NPs formulation:

2.2.1.1. CS NPs:

NPs were prepared by the ionotropic gelation of chitosan (CS) and tripolyphosphate (TPP). CS (0.1 % w/v) was added to TPP (0.1 % w/v) for NPs synthesis.

2.2.1.2. USPIONs:

For synthesis of USPIONs, 0.45 M $NaHCO_3$ was added to 0.3 M $FeCl_3.6H_2O$. Subsequently, 0.3 M Vit C were added. The mixture was then transferred into a steel-lined Teflon tube to be autoclaved.

2.2.2. Encapsulation Efficiency % (EE %) of NPs:

FL was chosen as a model drug. Similarly as CS NPs, drug-loaded NPs were formulated using ionotropic gelation. FL EE % was determined fluorometrically at λ_{ex}= 490 nm and λ_{em}= 519 nm.

2.2.3. Drug Release:

2.2.3.1. Determination of drug release profile from drug-loaded NPs:

FL release rate from NPs was determined using spectra/por float-a-lyzers (MWCO=100 KDa). At assigned time intervals (0.5, 1, 2, 4, 8, 12, 24, 48, 72, 96, 120, 144 and 168 hrs) aliquots were withdrawn from the release compartment and analyzed for the released drug.

2.2.3.2. Determination of USPIONs effect on drug release upon the application of high frequency magnetic field (HFMF):

USPIONs and FL-loaded CS NPs were used to determine the effect of magnetic field application on drug release profile in comparison to FL-loaded CS NPs. The release profiles of the samples were compared under a constant temperature of 32 °C (thermostat) and under the periodic influence of a HFMF.

2.2.4. NPs-cell interaction:

2.2.4.1. MTT assay:

MTT toxicity assay was performed on A549 and L929 cells with different concentrations of CS NPs, CS-USPIONs and USPIONs. The absorbance was measured using microplate fluorometer at λ= 570 nm.

2.2.4.2. Effect of NPs concentration and modification on the cells' uptake magnitude:

Different concentrations of tagged and non-tagged CS NPs and CS-USPIONs loaded with FL were used to study NP cellular uptake in A549 cells. Amount of uptaken NPs were determined by fluorometery (λ_{ex}= 495 nm and λ_{em}= 519 nm).

2.2.4.3. NPs Uptake confirmation:

For cellular uptake, purified FL-loaded CS NPs were used. Cells were labelled with Hoechst and WGA-AF.

2.2.5. Spray freeze drying (SFD) and Lung deposition evaluation by next generation impactor (NGI):

2.2.5.1. Incorporation of NPs in PVP: MD MPs:

CS NPs and CS-USPIONs were formulated and suspensions were SFD in the presence of cryopreservatives (5% (w/v) of PVP: MD mixture).

2.2.5.2. Evaluation of aerodynamic properties of MPs using NGI:

MPs were discharged from the dry powder inhaler (DPI) into the NGI passing by seven stages with aerodynamic diameters cut-offs 9.1, 5.2, 3.3, 1.9, 1.1, 0.6, and 0.4 μm; respectively at rate of 45 L/min. The contents of different stages were determined spectrophotometrically (λ_{es}: 485 nm and λ_{em}: 535 nm) from which the aerodynamic diameter of MPs were calculated.

2.2.6. Statistical analysis:

Statistical analysis of data was performed with one way analysis of variance (ANOVA). *P*-values less than 0.05 were considered significant.

3. RESULTS

3.1. NPs formulation:

3.1.1. CS NPs:

NPs obtained were less than in 150 nm in hydrodynamic diameter and positively charged.

3.1.2. USPIONs:

USPIONs synthesized were of size and surface potential 5 nm and -26.6 mV; respectively. According to the JCPDS-ICDD Card No. 24-0081 the crystal structure was related to cubic γ-Fe_2O_3. They possessed magnetic saturation (Ms), coercivity (Hc) and retentivity (Mr) of 48.4 Am^2/Kg, 9.9×10^{-4} T and 0.5 Am^2/Kg; respectively.

3.2. EE % of NPs:

NPs entrapped 23.9 μg/ml of FL with the absence of NP final size alteration. Incorporation of USPIONs into CS NPs reduced concentration of FL entrapped to 13.0 μg/ml.

3.3. Drug Release:

3.3.1. Determination of drug release profile from drug-loaded NPs:

Only 5 and 25 % of FL was released from CS NPs and CS-MNPs; respectively after 168 hrs compared to the control that provided 100 % release during the same period.

3.3.2. Determination of MNPs effect on drug release upon the application of HFMF:

Upon the application of alternating magnetic field (AMF) the rate of release of FL-loaded CS-USPIONs significantly increased. However, for FL-loaded CS NPs the release rate was not affected by AMF application.

3.4. NPs-cell interaction:

3.4.1. MTT assay:

The cytotoxicity of CS NPs, CS-USPIONs and USPIONs were evaluated for A549 and L929 cell lines. At high concentrations for CS NPs and CS-USPIONs (2000 μg/ml) preferential cytotoxicity towards cancer cells is observed. Whereas, at high concentration of USPIONs (200 μg/ml) cytotoxicity was visualized on both cell lines.

3.4.2. Effect of NPs concentration and modification on the cell's uptake magnitude:

Cellular uptake increased with increased NPs concentration while modification of NPs with Abs has lowered the cellular uptake significantly.

3.4.3. NPs Uptake confirmation:

Green fluorescence attributed to FL loaded NPs appeared to be inside the cell in the perinulclear area. The visualized overlap of red (WGA-AF) and green (FL) fluorescence indicated that the green fluorescence seen perinuclearly is in fact owed to NPs rather than noise or released FL.

3.5. SFD and evaluation of Lung deposition by NGI:

3.5.1. Incorporation of NPs in PVP: MD MPs:

MPs obtained are spherical in shape and nearly of a size ~ 3 μm.

3.5.2. Evaluation of aerodynamic properties of MPs using NGI:

Inhaled percentage of MPs (passed the throat) is ~37 % for both NPs. However, ~50 % were deposited past the trachea. Both types of particles have also possessed fine particle fraction (FPF≤ 5.2 μm) of ~40 % indicating successful delivery to deep lung tissue. Processing the NGI data also allowed the determination of the MMAD which corresponded to 6.1 μm for CS-USPIONs and 5.1 μm for CS NPs.

4. CONCLUSION

In conclusion, lung cancer is the deadliest solid tumour among all types of cancer. The number of cases and deaths caused cancer is increasing by time. The optimum route of chemotherapeutic drug delivery was thought to be inhalation since it lacks systemic toxicity and invasiveness experienced by I.V. route. The combination of initial spike dose of chemotherapy followed by sustained supply of the drug was thought to provide an efficient therapeutic platform for tumour eradication and recurrence prevention. Thus, based on these thoughts a non-invasive, patient convenient platform for the targeted delivery of chemotherapeutic drugs to cancer in deeper lung tissue was developed and studied. MPs containing either CS NPs or CS-USPIONs were developed. USPIONs incorporation has enhanced drug release by 1.7 fold in response to external magnetic field while CS NPs provided a slow sustained release. The prepared MPs had fine particle fraction (FPF≤5.2 μm) of ~40 % w/w and MMAD of 5-6 μm as determined by the NGI. The targeted delivery to the lung cancer using the developed formulation seems to be a promising approach.

REFERENCES

1. Pope-Harman, A. and M. Ferrari, *Medical Nanotechnology and Pulmonary Pathology BioMEMS and Biomedical Nanotechnology*, M. Ferrari, Desai, T., and Bhatia, S., Editors. 2007, Springer US. p. 193-212.
2. *World cancer factsheet*. 2014, International Agency for Research on Cancer and Cancer Research UK: London.
3. Mathers, C.D. and D. Loncar, *Updated projections of global mortality and burden of disease, 2002-2030: data sources, methods and results.* Geneva: World Health Organization, 2005.
4. Celikoglu, F., S.I. Celikoglu, and E.P. Goldberg, *Bronchoscopic intratumoral chemotherapy of lung cancer.* Lung Cancer, 2008. **61**(1): p. 1-12.
5. Lam, B., S.Y. Lam, M.P. Wong, C.G.C. Ooi, D.Y.T. Fong, D.C.L. Lam, A.Y.K. Lai, C.-m. Tam, C.B.Y. Pang, M.S.M. Ip, and W.-k. Lam, *Sputum cytology examination followed by autofluorescence bronchoscopy: A practical way of identifying early stage lung cancer in central airway.* Lung Cancer, 2009. **64**(3): p. 289-294.

Influence of conjugating folic acid and methotrexate on physicochemical properties of P(CL)-mPEG nanoparticles

O. Issarachot[*], J. Suksiriworapong[*,**], K. Sripha[***], R. Yumoto [****], M. Takano[****] and **V.B. Junyaprasert**[*,**]

[*] Department of Pharmacy, Faculty of Pharmacy, Mahidol University, 447 Sri-Ayutthaya Road, Rajathavee, Bangkok 10400, Thailand

[**] Center of Excellence in Innovative Drug Delivery and Nanomedicine, Faculty of Pharmacy, Mahidol University, 447 Sri-Ayutthaya Road, Rajathavee, Bangkok 10400, Thailand

[***] Department of Pharmaceutical Chemistry, Faculty of Pharmacy, Mahidol University, Rajathavee, Bangkok 10400, Thailand

[****] Department of Pharmaceutics and Therapeutics, Graduate School of Biomedical and Health Sciences, Hiroshima University, 1-2-3 Kasumi, Minami-ku, Hiroshima 734-8553, Japan

Corresponding author: varaporn.jun@mahidol.ac.th

ABSTRACT

The objective of this study was to investigate an effect of conjugating folic acid (FOL) as a targeting ligand and conjugating methotrexate (MTX) as an anticancer drug on the formation of nanoparticles (NPs). MTX-conjugated nanoparticles were prepared using 10, 20 and 30 % MTX conjugated poly(ε-caprolactone)-co-methoxy poly(ethylene glycol) (P(MTXCLCL)-mPEG) copolymers by dialysis method. Additionally, the targeted nanoparticles were obtained by the physical adsorption of FOL-conjugated (P(FOLCL)-mPEG) copolymer on the surface of P(MTXCLCL)-mPEG nanoparticles. The particle size of P(MTXCLCL)-mPEG nanoparticles increased with an amount of conjugating MTX. After the adsorption of P(FOLCL)-mPEG, the nanoparticles became larger and the zeta potential turned to more negative by approximately two times. All nanoparticles were spherical in shape and uniform size distribution. The % yield of nanoparticles ranged between 52-58%. The release profile showed higher release of MTX in acidic pH medium.

Keywords: methotrexate, folic acid, targeted nanoparticles, poly(ε -caprolactone)-co-methoxy poly(ethylene glycol)

1 INTRODUCTION

Generally, various nanoparticulate carriers have been exploited in drug delivery research including micelles, dendrimers, liposomes, polymeric nanoparticles and solid lipid nanoparticles. Among these carriers, the polymeric nanoparticles have gained much attention as a promising delivery system because they possess several essential advantages for drug delivery system such as sub-cellular size, biocompatibility with tissues and cells, and sustained- and controlled-release properties [1-2]. Nowadays, the polymeric nanoparticles are preferably developed from biodegradable materials which are non-toxic and removable from the body. In addition, the surface properties of polymeric nanoparticles can be modified with targeting molecules for chemotherapy. These nanoparticles exhibit the versatility for targeting delivery of small drug molecules as well as macromolecular therapeutic agents to the tissues of interest.

In general, the drugs are encapsulated in the polymeric nanoparticles by physical interaction; however, the initial burst release and uncertain amount of drug payload are frequently encountered. Hence, the polymer-drug conjugates is introduced to overcome these problems. The nanoparticles made of polymer-drug conjugates have received much attention in drug delivery due to their advantages such as increasing drug stability and improving pharmacokinetics of drugs, leading to diminish toxicity and multi-drug resistance [3-4].As approved by U.S. FDA, P(CL) have been increasingly investigated as biomaterials for pharmaceutical and biomedical applications. To increase the use of P(CL) in biomedical application, poly(ethylene glycol) (PEG) has been employed to copolymerize with P(CL) resulting in amphiphilic copolymer or PEGylated P(CL). In the previous report, small bioactive molecules could be successfully conjugated onto the P(CL) backbone at various amounts by click reaction. Therefore, it is of our interest to further investigate the conjugation of cytotoxic drugs and targeting ligands on the PEGylated P(CL) copolymers for chemotherapy.

In our previous study, P(CL)-methoxy-PEG or P(CL)-mPEG were synthesized and conjugated with methotrexate (MTX) (P(MTXCLCL)-mPEG) and folic acid (FOL) (P(FOLCL)-mPEG) as anticancer drug and targeing moiety, respectively [5]. MTX was conjugated at 10, 20 and 30 % mole fraction whereas FOL was fixed at 5 mole per polymer chain. Therefore in this study, the targetd and non-targeted NPs using these conjugated copolymer are prepared. The characteristics of NPs were charcterized in term of particle size, size distribution, surface charge, % yield, morphology and release characteristics.

2 MATERIALS AND METHOD

2.1 Materials

MTX- and FOL-conjugated P(CL)-mPEG copolymers defined as P(MTXCLCL)-mPEG and P(FOLCL)-mPEG, respectively, were synthesized according to our previous report [5]. MTX was obtained from Suzhou Rovathin, Pharmatech Co., Ltd., China. FOL was purchased from Fluka (Steinhheim, Germany). All solvents of analytical grade. Acetoniltrile (ACN), dimethylformamide (DMF), and dimethyl sulfoxide (DMSO) were purchased from RCI Labscan Ltd., Bangkok, Thailand.

2.2 Preparation of MTX-conjugated NPs

Briefly, MTX-conjugated copolymers were dissolved in DMF. Then, the polymeric solution was dropped into deionized water as an aqueous phase. The mixture was equilibrated by magnetically stirring and then transferred into a dialysis bag (MWCO 6000-8000 Da, Boron GmbH, Ludwigshafen, Germany). The colloidal dispersion was dialyzed against deionized water to remove an organic solvent. The dialysis medium was changed every 3 h. The NPs were kept as a dispersion form for further analysis.

2.3 Preparation of FOL-decorated MTX-conjugated NPs

To prepare the targeted NPs, P(MTXCLCL)-mPEG NPs were decorated with P(FOLCL)-mPEG copolymer. The colloidal suspension of P(MTXCLCL)-PEG NPs was initially formed by slowly dropping the polymeric solution into an aqueous phase. Subsequently, a solution of P(FOLCL)-mPEG in DMSO was added to the dispersion and magnetically stirred for 2 h to allow the adsorption of P(FOLCL)-mPEG onto the surface of nanoparticles. Then, the rest procedure was similarly performed as mentioned in section 2.2.

2.4 Characterization of NPs

2.4.1 Particle size and size distribution

The particle size (z-ave) and polydispersity index (PI) were determined by ZetasizerNanoZS (Malvern Instrument, Malvern, UK) withHe-Ne laser at a wavelength of 633 nm, 25 °C. The intensity of the scattered light was detected with a photomultiplier at a scattering angle of 173°. Each experimental was performed in triplicate.

2.4.2 Surface charge

The surface charge of NPs was determined by electrophoretic mobility of particles and calculated by the Helmholtz-Smoluchowsky equation. The measurements were conducted by ZetasizerNanoZS. The sample was diluted with deionized water to yield a conductivity of 50 µS/cm. Each experimental was repeated five times at 25 °C with a field strength of 20 V/cm.

2.4.3 Particle morphology

The morphological examination of NPs was conducted using scanning electron microscope equipped with field emission cathode using 15 kV upper detector(Hitachi S4500, Hitachi, Tokyo, Japan). The sample was dried and sputtered with gold before observation.

2.4.4 Yield of NPs

The yield of NPs was calculated according to Eq. (1). Briefly, the collected NPs after dialysis were transferred into 5 mL vial and then pre-frozen overnight. The frozen NPs were sequentially lyophilized for at least 24 h. Then, an actual weight of lyophilized NPs was recorded.

$$\%\ \text{Yield} = \frac{\text{Actual weight of lyophilized NPs} \times 100}{\text{Initial weight of polymer}} \quad (1)$$

2.4.5 *In vitro* drug release study

The *in vitro* release measurement of MTX was performed in a 0.01 M phosphate buffer solution. The release of MTX from NPs was investigated by dialysis method. The 2 mL of NPs dispersion was put into a dialysis bag (MWCO 6000-8000 Da) and then introduced into 30 mL phosphate buffer solution pH 4.5 or 7.4. The release chamber was incubated at 37 °C with gentle shaking. At determined intervals, 2 mL of release medium was withdrawn and replaced with an equal volume of fresh release medium. The content of MTX was analyzed by HPLC method as described in the following section. The experiments were carried out in triplicate.

2.4.6 High performance liquid chromatography (HPLC) analysis

The amount of MTX in samples was quantified by HPLC method. The analysis was performed at 25 °C on a reverse phase Hypersil ODS column, 5 µm particle size, 250×4.6 mm (Thermo Fisher Scientific Inc., Massachusetts, USA) using a mixture of ACN and 0.05 M phosphate buffer pH 6 (10:90 v/v) as a mobile phase pumped through the column at a flow rate of 1.2 mL/min. A chromatogram was recorded by UV detector at a wavelength of 304 nm. The retention time of the drug was 15 min. The linearity and precision were also determined. The standard curve of MTX was linear with r^2 of > 0.9995 over the concentration range of 0.05-20 µg/mL. The intra- and inter-day precisions were less than 2 %. The recovery was 98 ± 1 % which is accepted for accuracy criteria (90-110%). Before analysis, all samples were filtered through 0.22 µm syringe filter.

2.4.7 Statistical analysis

The data were analyzed for statistical significance by one way ANOVA. The level of significance was at p-value < 0.05. All data are expressed as mean ± SD from at least three measurements. For the *in vitro* release results, the different and similarity factors (f_1 and f_2, respectively) were calculated and used for statistical comparison [6].

Typically, f_1 and f_2 values between 0-15 and 50-100, respectively, indicate define the similarity of two release profiles.

3 RESULTS AND DISCUSSION

3.1 Formation of MTX-conjugated NPs

Due to the di-block pattern of P(MTXCLCL)-mPEG, the feature of P(MTXCLCL)-mPEG nanoparticles could be formed as core-shell NPs by an aggregation of P(CL) segment inside the core of NPs whereas PEG segment presented as hydrophilic shell on the surface of NPs as demonstrated in Figure 1(a). The characteristics of all nanoparticles are summarized in Table 1.

The particle size of P(MTXCLCL)-mPEG NPs increased depending on the % conjugating MTX. 10 % P(MTXCLCL)-mPEG NPs had the smallest particle size 101±1 nm whereas 30 % P(MTXCLCL)-mPEG NPs had the largest size of 141±6 nm. It has been reported that the amount of drug molecules conjugated along the polymer chain had an effect on the formation of NPs [7]. The increasing % conjugating MTX led to the increased bulkiness of MTX molecules inside the core of NPs. However, these NPs were still smaller than 200 nm. This particle size range has been reported that the NPs can be passively targeted to the tumor tissue [8]. In addition, the particle size distribution of 10 % and 20 % P(MTXCLCL)-mPEG NPs was narrow with the PI value of 0.117-0.157. However, 30 % P(MTXCLCL)-mPEG NPs showed large PI due to more hydrophobicity of 30 % P(MTXCLCL)-mPEG copolymer leading to aggregation during dialysis. The surface charge of 10 and 20 % P(MTXCLCL)-mPEG NPs was negative and closed to zero because the PEG chain surrounded the surface of NPs. The ZP value of 30 % P(MTXCLCL)-mPEG NPs was slightly more negative indicating some of MTX molecules available on the surface of NPs. All formulations showed similar % yield ranging from 52.17-58.12 %. The NPs demonstrated spherical in shape and uniform size distribution as seen by SEM micrograph in Figures 2(a).

3.2 FOL-decorated MTX-conjugated NPs

FOL-decorated MTX-conjugated NPs (FOL-P(MTXCLCL)-mPEG NPs) were also prepared in order to develop the targeted nanocarriers by physical adsorption between PEG chains of both P(MTXCLCL)-mPEG and P(FOLCL)-mPEG. The anticipated architecture of FOL-P(MTXCLCL)-mPEG NPs is illustrated in Figure 1(b). The characteristics of FOL-P(MTXCLCL)-mPEG NPs are summarized in Table 1. All NPs after decoration had the significantly larger particles due to the adsorption of P(FOLCL)-mPEG on the surface of NPs. The mPEG chain of P(FOLCL)-mPEG inserted in the PEG shell of NPs tethering FOL molecules outside the NPs and thus enlarging the particles of FOL-decorated NPs. The ZP of FOL-P(MTXCLCL)-mPEG NPs became more negative being attributable to the presence of FOL molecules of P(FOLCL)-mPEG copolymer on the surface of P(MTXCLCL)-mPEG NPs. The % yield of FOL-P(MTXCLCL)-mPEG NPs were 53.89-56.98 %. These results showed that P(FOLCL)-mPEG had no impact on the yield of NPs. The morphology of FOL-P(MTXCLCL)-mPEG NPs also exhibited spherical in shape and uniform size distribution as shown in Figures 2(b).

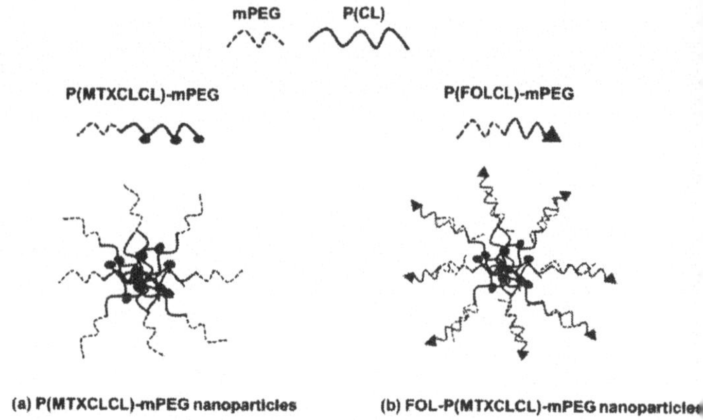

Figure 1 : The postulated architectures of P(MTXCLCL)-mPEG NPs (a) and FOL-P(MTXCLCL)-mPEG NPs (b).

Table 1 : Particle size (z-ave), polydispersity index (PI), zeta potential (ZP), and % yield of P(MTXCLCL)-mPEG NPs and FOL-P(MTXCLCL)-mPEG NPs

Formulation	z-ave (nm)	PI	ZP (mV)	% yield
10 % P(MTXCLCL)-mPEG	101±1	0.117±0.001	-3.7±0.8	58.12±2.35
20 % P(MTXCLCL)-mPEG	128±3	0.157±0.051	-4.7±1.1	53.23±3.21
30 % P(MTXCLCL)-mPEG	141±6	0.508±0.056	-10.7±1.1	52.17±2.24
FOL-10 % P(MTXCLCL)-mPEG	122±1	0.180±0.008	-12.8±0.1	56.98±1.23
FOL-20 % P(MTXCLCL)-mPEG	161±1	0.114±0.022	-8.1±0.1	55.16±0.58
FOL-30 % P(MTXCLCL)-mPEG	172±1	0.303±0.031	-18.0±0.9	53.89±1.08

3.3 *In vitro* release study

The release of MTX from P(MTXCLCL)-mPEG NPs was performed in buffers pH 4.5 and 7.4 representing the endosomal environment and in the blood stream, respectively [9]. From Figure 3, MTX showed the sustained release over 20 days by less than 50 % release of total drug.

At pH 4.5, the total MTX released from NPs increased with % conjugating MTX. The slow release of MTX at pH 7.4 suggested that the conjugating MTX may prevent the leakage of drug in the blood circulation before reaching the target site.

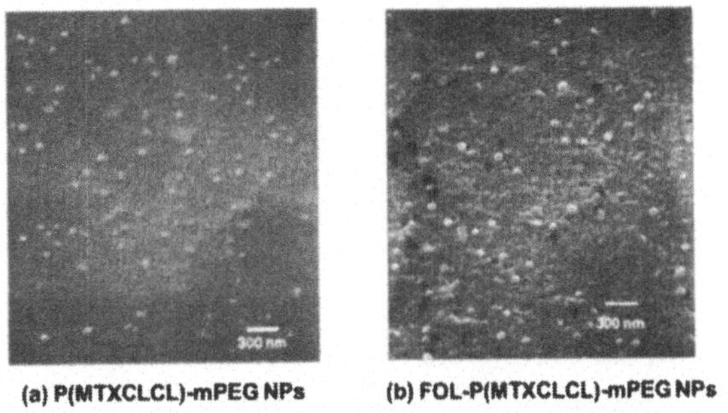

Figure 2 : SEM micrographs of P(MTXCLCL)-mPEG NPs (a) and FOL-P(MTXCLCL)-mPEG NPs (b).

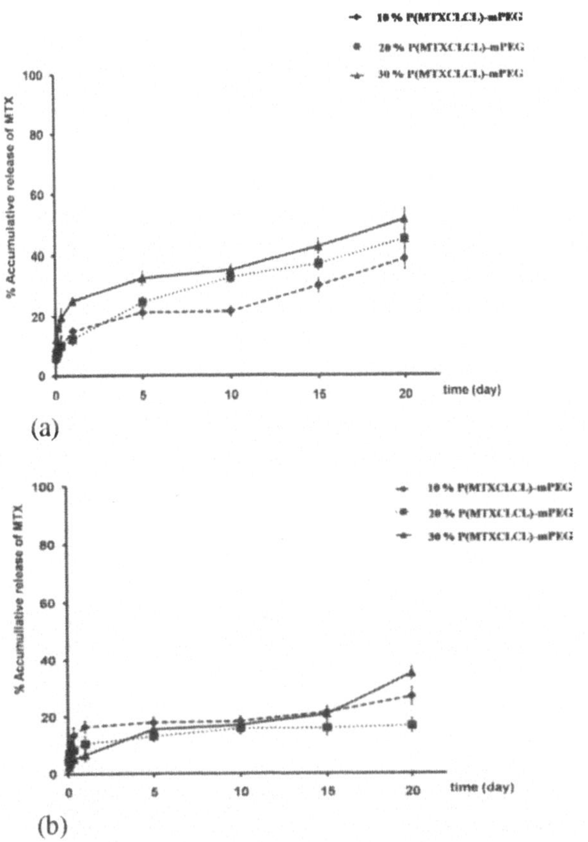

Figure 3: The release profiles of MTX from P(MTXCLCL)-mPEG NPs in phosphate buffer solutions pH 4.5 (a) and 7.4 (b).

4 CONCLUSIONS

The preparation of P(MTXCLCL)-mPEG NPs and FOL-P(MTXCLCL)-mPEG NPs were achieved by dialysis method. The particle size of P(MTXCLCL)-mPEG NPs relatively increased with the content of conjugating MTX. The targeted FOL- P(MTXCLCL)-mPEG NPs were successfully prepared by means of physically adsorptive interaction of PEG chains. The adsorption of P(FOLCL)-mPEG resulted in larger particle size and more negative surface charge whereas it did not affect PI and % yield. The release of conjugating MTX was pH-dependent with higher extent in acidic medium and could be sustained over 20 days. All of results led to the conclusion that P(MTXCLCL)-mPEG NPs and FOL-P(MTXCLCL)-mPEG NPs can potentially be used as targeted anticancer drug delivery system.

Acknowledgments

This study is financial supported from the Thailand Research Fund (TRF) through the Royal Golden Jubilee Ph.D. Program (Grant No.PHD/0238/2548) and the Office of the Higher Education Commission and Mahidol University under the National Research Universities Initiative.

REFERENCES

[1] K.S. Soppimatha, T.M. Aminabhavia, A.R. KulkarniaandW.E. Rudzinski. Biodegradable polymeric nanoparticles as drug delivery devices. J Control Release 70(1-2), 1-20, 2001.

[2] J.K. Vasir and V. Labhasetwar. Biodegradable nanoparticles for cytosolic delivery of Therapeutics. Adv Drug Deliv Rev 59(8), 718-728, 2007.

[3] G. Pasut and F.M. Veronese. Polymer-drug conjugation, recent achievements and general strategies. Prog Polym Sci 32(8-9), 933-961, 2007.

[4] J.Khandare and T. Minko. Polymer-drug conjugates: progress in polymeric prodrugs. Prog Polym Sci 31(4), 359-397, 2006.

[5] O.Issarachot,J. Suksiriworapong,K. Sripha and V.B. Junyaprasert.Modification of Tricomponent and DicomponentPoly(e-caprolactone)-co-Poly(ethylene glycol) with Methotrexate and Folic Acid. J. APPL. POLYM. SCI. 129(2), 721-734, 2013.

[6] United States Food and Drug Administration (1997) Guidance for industry: dissolution testing of immediate release solid oral dosage forms. U.S. Department of Health and Human Services Food and Drug Administration Center for Drug Evaluation and Research (CDER).http://www.fda.gov/cder/guidance/1713bp1.pdf. Accessed 23 Oct 2012.

[7] Y. Wang, X. Yang, J. Yang,Y. Wang, R. Chen, J. Wu, Y. Liu and N. Zhang. Self-assembled nanoparticlesof methotrexate conjugated O-carboxymethyl chitosan: preparation, characterization and drug releasebehavior in vitro. CarbohydrPolym 86, 1665-1670, 2011.

[8] A. Albanese, P.S. Tang, W.C. Chan. The effect of nanoparticle size, shape, and surface chemistry on biological systems. Annu Rev Biomed Eng 14, 1–16, 2012.

[9] K. Ulbrich and V. Ŝubr.Polymeric anticancer drugs with pHcontrolledactivation. Adv Drug Deliv Rev 56, 1023-1050, 2004.

Mitochondria-targeting Gold Nanoparticles for Improved Therapeutic Efficacy of Photodynamic Therapy of Breast Cancer

Yamin Yang*, Ning Gao**, Yue Hu**, Chao Jia*, Henry Du**, and Hongjun Wang*

*Department of Chemistry, Chemical Biology, and Biomedical Engineering, Stevens Institute of Technology, Hoboken, NJ 07030, USA, hongjun.wang@stevens.edu
**Department of Chemical Engineering and Materials Science, Stevens Institute of Technology, Hoboken, NJ 07030, USA

ABSTRACT

In recognition of the unique contribution of gold nanoparticles (Au NPs) and critical role of mitochondria in photodynamic reaction, mitochondria-targeting Au NPs are prepared and employed to photodynamic therapy (PDT) of breast cancer. Au NPs with positive surface charge are synthesized and modified with functional groups (triphenylphosphonium, TPP). The integration of positive surface charge with mitochondrion-targeting domain (*i.e.*, TPP) onto Au NPs allow their selective accumulation in the mitochondria of breast cancer cells, leading to elevated reactive oxygen species (singlet oxygen $^1O_2$, in particular) formation and significant improved destruction of breast cancer cells by 5-aminolevulinic acid (5-ALA)-assisted PDT under light irradaition. These findings provide guidance for the improved design of intracellular organelle-targeted therapeutic nanoparticles for PDT.

Keywords: Gold nanoparticles, photodynamic therapy, mitochondria, triphenylphosphonium, reactive oxygen species

1 INTRODUCTION

Photodynamic therapy (PDT), which combines a photosensitive drug (*i.e.* photosensitizer, PS) with light irradiation, is an emerging minimally invasive therapeutic modality for a variety of oncologic diseases. Despite its benefits, PDT has not yet become a mainstream cancer therapeutic technique partially due to limited efficacy and poor selectivity of commercial PSs. There is still a great need for further enhancing the therapeutic effectiveness and specificity of current PDT. In recognition of their superior optical and surface plasmonic properties, as well as good biocompatibility and robust versatility in surface functionalization, gold nanoparticles (Au NPs) can be utilized as the enhancers for PDT by promoting the generation of reactive oxygen species (ROS), mainly composed of singlet oxygen ($^1O_2$), under light irradiation. Our recent findings have shown that Au NPs can enhance 5-aminolevulinic acid (5-ALA)-induced ROS formation and the enhancement is size-dependent, which suggest that plasmonic Au NPs may amplify photonic energy absorbed and then transfer the energy to neighboring PSs for enhancing ROS formation. Compared to conventional PDT, the use of Au NPs as an enhancer and mediator to elevate ROS formation represents a promising PDT strategy [1-4].

In the meantime, given that mitochondria are the primary site for preferential accumulation of PS and $^1O_2$ production, together with their crucial roles in regulating cell apoptosis and cancer metabolism, they are of particular interest as potential target for improved PDT efficiency. In fact, some PSs such as protoporphyrin IX (PpIX) and phthalocyanine preferentially accumulate in mitochondria [5]. In this regard, targeted accumulation of Au NPs into mitochondria is expected to further enhance the generation of $^1O_2$. Various approaches have been explored to target mitochondria using targeting moieties, among which lipophilic cationic triphenylphosphonium (TPP) has been widely applied in drug design or surface modification for targeting mitochondria. Combinations of phosphonioalkanethiol and triphenylphosphonium have shown a remarkable ability to pass through mitochondrial membranes and accumulate inside mitochondria [6]. In the present study, mitochondria-targeting Au NPs were fabricated by conjugating TPP functional groups onto Au NPs with positive surface charge. It is hypothesized that the combination of positive surface charge with mitochondria-targeting domain (*i.e.*, TPP) onto Au NPs can selectively accumulate into mitochondria of breast cancer cells. Consequently, ROS formation will be elevated under light irradiation, efficient and selective destruction of breast cancer cells by 5-ALA enabled PDT will be better achieved.

2 MATERIALS AND METHODS

2.1 Synthesis of Au NPs, surface modification with TPP and characterization

Positively charged Au NPs (Au NPs (+)) were prepared by reducing chloroauric acid ($HAuCl_4$) with branched polyethyleneimine (BPEI) under UV irradiation as described in our previous study [1]. Briefly, a mixture of BPEI with a molecular weight of 10,000 g.mol^{-1} and $HAuCl_4$ solutions was stirred for 5 min in an ice bath and reacted for 1 hr under UV irradiation. Mitochondria-targeting TPP molecules with thiolated ends were synthesized and conjugated onto Au NPs surface to form the TPP-Au NPs *via* the high affinity of thiol groups (SH-) to Au [6]. Physicochemistry properties of prepared Au NPs were characterized by UV-visible spectrophotometer, dynamic light scattering and laser Doppler electrophoresis, accordingly, and TPP conjugation onto Au NPs was confirmed by Surface-enhanced Raman spectroscopy (SERS) and Fourier transform infrared spectroscopy (FTIR).

2.2 Cell culture and cellular PDT

Human breast cancer cell line MDA-MB-231 (ATCC) was cultured in L-15 medium (Leibovitz) (Sigma-Aldrich, Saint Louis, MO) with 10% fetal bovine serum (FBS) and 1% penicillin/streptomycin (pen/strep) at 37°C without CO_2. Human mammary epithelial cell line MCF-10A (ATCC) was maintained in DMEM/F-12 media supplemented with 5% horse serum, 1% pen/strep, EGF (20 ng/mL), hydrocortisone (0.5 mg/mL), cholera toxin (100 ng/mL) and insulin (10 μg/mL) in a humidified 5% CO_2 atmosphere at 37°C.

Cellular PDT treatment was performed according to the procedures described in our previous study [1]. Briefly, cells were seeded (8,000 cells/well) and cultured in 96-well plates for 24 hrs and then the cells were incubated with serum-free cell culture medium containing 1 mM 5-ALA with/without Au NPs for 4 hrs. Then the cells were carefully washed with HBSS and refreshed with 100 μL HBSS for each well. The cells were irradiated with a broadband light source using a halogen lamp (100 mW/cm^2 at the location of the electrode, Dolan-Jenner Fiber-Lite MI-150, Dolan-Jenner Industries, MA) for 1 min. After irradiation, the cells were further cultured with the complete medium for 24 hrs before cell viability evaluation.

2.3 Singlet oxygen measurement

To measure the intracellular ROS level and specifically $^1O_2$ in the PDT treatment, Singlet Oxygen Sensor Green reagent (SOSG, Molecular Probes), was used. Briefly, SOSG reagent was prepared at a final concentration of 1.33 μM in HBSS. MDA-MB-231 and MCF-10A cells treated with 5-ALA with/without different Au NPs were incubated with SOSG reagent at working concentration for one hour before light irradiation. Then the cells were washed with PBS to remove excess SOSG reagent and 100 μL of HBSS was added into each well for irradiation. The fluorescence intensity was measured using the Synergy™ fluorescence microplate reader with an excitation BP filter at 485/20 nm and emission BP filter at 528/20 nm. Since SOSG detected the singlet oxygen free radicals from both PDT treatment and intrinsic baseline, the fluorescence intensity was subtracted with that of the sham control (treated in the same way but no 5-ALA and Au NPs).

2.4 Cell viability evaluation after various PDT

After various PDT treatments, cell viability was determined by thiazolyl blue tetrazolium bromide (MTT) assay and Live/Dead Viability/Cytotoxicity staining. For MTT assay, cultures after different treatments were incubated with cell culture medium containing MTT (0.5 mg/mL) for 2 hrs and then the reaction product of formazan crystal was extracted with dimethyl sulfoxide (DMSO). Absorbance of the extract was measured at 570 nm with the Synergy™ multi-mode microplate reader. Cultures without 5-ALA and Au NPs under the same light irradiation condition were used as control.

For Live/Dead staining, MDA-MB-231 cells cultured on glass cover slips after PDT treatment were gently washed with HBSS and then incubated with the Live/Dead Viability/Cytotoxcity solution for 30 min. Viable cells were stained green by calcein acetoxymethyl (Calcein AM, 0.05%), while the nuclei of dead cells were stained red by ethidium homodimer-1 (EthD-1, 0.2%).

The PDT results were also evaluated based on a co-culture system with the presence of both breast cancer cells and normal cells. MDA-MB-231 cells were first labeled with Cell Tracker™ Green CMFDA for 30 min. Both MDA-MB-231 cells and MCF-10A cells of the same amount were suspended, pooled and co-cultured in the medium composed of 1:1 original cell culture medium for each cell type. The co-cultured cells were conducted for PDT with various Au NPs (40 μM) similarly to that described above. The fluorescence images merged with bright field images of cells were captured before PDT and 24 hrs after PDT using a Nikon Eclipse 80i epifluorescent microscope.

2.5 Mitochondrial localization of TPP-Au NPs

To determine the intracellular localization of different Au NPs after uptake, MDA-MB-231 cells incubated with various Au NPs as described above were rinsed with HBSS to remove any free particles, trypsinized, washed and centrifuged into cell pellet. Cell pellets were fixed with 4% paraformaldehyde (EM Sciences, Hatfield, PA) for 2 hrs at room temperature, post-fixed with 1% osmium tetroxide for 1 hr, dehydrated in a graded series of ethanol solutions and then embedded in epoxy resin. Ultrathin sections (~ 70 nm) were obtained and examined with a transmission electron microscope (TEM).

2.6 Statistical analysis

All quantitative results were obtained at least in triplicate. Data were expressed as the mean ± SD. Unpaired student's t-test was used for statistical significance analysis. A value of $p<0.05$ was considered statistically significant.

3 RESULTS AND DISCUSSION

After conjugation of TPP molecules onto Au NPs, positively charged TPP-Au NPs (45 ± 10 mV) with diameter of 50 ± 16 nm showed negligible cytotoxicity to both breast cancer and normal cells. TPP-Au NPs were preferably taken up by breast cancer cells, as much as about two times of that ingested by normal cells. It is most likely due to a lower membrane potential with cancerous cells that provides an intrinsic selective uptake of positively charged TPP-Au NPs in cancer cells.

In PDT treatment, selective destruction of breast cancer cells is a synergy between preferred uptake of TPP-Au NPs and subsequently elevated ROS ($^1O_2$, in particular) formation by TPP-Au NPs with the presence of PS under light irradiation. As shown in Figure 1A, the highest $^1O_2$ formation occurred in MDA-MB-231 cells treated with TPP-Au NPs. With the presence of Au NPs, a significantly higher level of $^1O_2$ was always measured in MDA-MB-231 cells compared to MCF-10A cells.

With the increase of intracellular $^1O_2$ level in PDT, higher cell destruction was expected. Indeed, the presence of Au NPs caused a significant increase in cell killing by PDT. The highest cell destruction was observed with TPP-Au NPs, which is consistent with the $^1O_2$ measurement. Compared to 5-ALA-PDT only, more MDA-MB-231 cells were killed with TPP-Au NPs (Figure 1B).

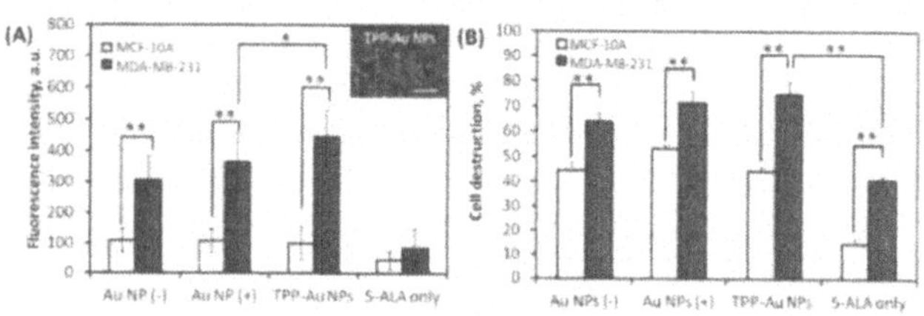

Figure 1 (A) Elevated generation of singlet oxygen species ($^1O_2$) in MCF-10A and MDA-MB-231 cells treated by various Au NPs in combination with 5-ALA or 5-ALA only after PDT treatments. Inset: Representative fluorescence images of MDA-MB-231 cells after PDT treatments with TPP-Au NPs. Singlet Oxygen Sensor Green (SOSG) probe inside of cells with green fluorescence in the presence of intracellular $^1O_2$. Scale bar: 100 μm. (B) Cell destruction of MCF-10A and MDA-MB-231 cells after PDT treatments with various Au NPs. * statistically significant, $p < 0.05$; ** statistically significant, $p < 0.001$.

To further confirm the MTT result, MDA-MB-231 cells after various PDT treatments were also fluorescently stained for viability (live/dead staining). The staining revealed that the presence of TPP-Au NPs caused the most cell death (Figure 2A), consistent with MTT results.

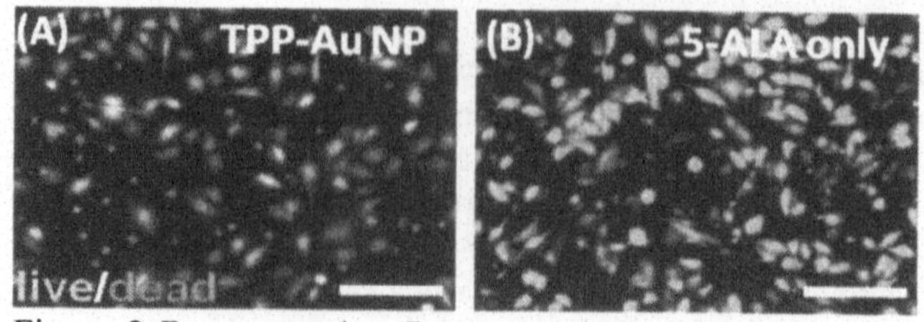

Figure 2 Representative fluorescent images of MDA-MB-231 cells stained with a Live/Dead kit after various PDT treatments. Live cells were stained green with Calcein-AM and dead cells stained red with EthD-1. Cells were treated with (A) 5-ALA and TPP-Au NPs and (B) 5-ALA only after light irradiation and further cultured for 24 hrs. Scale bar: 200 μm.

In the study on the selectivity of PDT, a co-culture system was used to emulate the *in vivo* circumstances where cancer cells coexist with normal cells. As shown in Figure 3, examination of the co-culture by fluorescence microscopy showed that a significant decrease of MDA-MB-231 cells (green-labeled) after PDT treatment in the TPP-Au NP-assisted groups. To quantify the difference, fluorescently labeled cells from 5 randomly selected regions were counted before and after PDT treatment. It was found that about 70% MDA-MB-231 cells were killed in TPP-Au NPs group compared to control group. A significant decrease of breast cancer cells was observed after TPP-Au NP-enabled PDT, while normal cells remaining their normal spindle shape and became confluent after culture for 24 hrs (Figure 3A). Clearly, under the same treatment, more breast cancer cells were destroyed than normal cells, implying the potential selectivity of TPP-Au NP-enabled PDT.

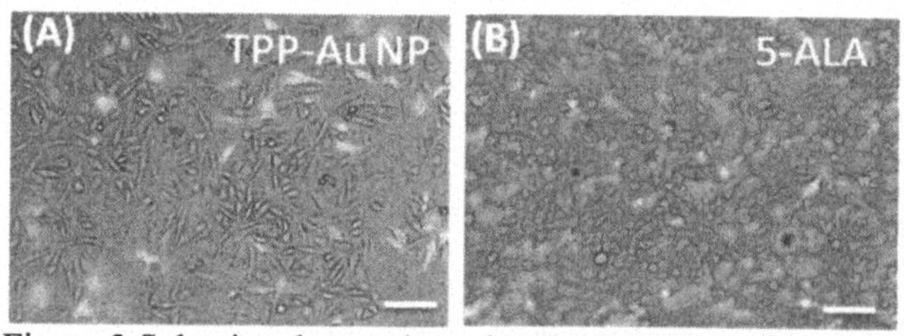

Figure 3 Selective destruction of MDA-MB-231 cells in the co-culture experiment. Merged bright field and fluorescence microscopic images of co-cultured MCF-10A (non-labeled, bright field) and MDA-MB-231 (green, fluorescence). Cells were treated with (A) 5-ALA and TPP-Au NPs and (B) 5-ALA only after light irradiation and further cultured for 24 hrs. Scale bar: 200 μm.

To determine the intracellular destination of ingested Au NPs, MDA-MB-231 cells loaded with Au NPs were sectioned and examined by TEM. As shown in Figure 4, after incubation for 4 hrs, TPP-Au NPs mainly accumulated in mitochondria, more specifically in the mitochondrial intermembrane space.

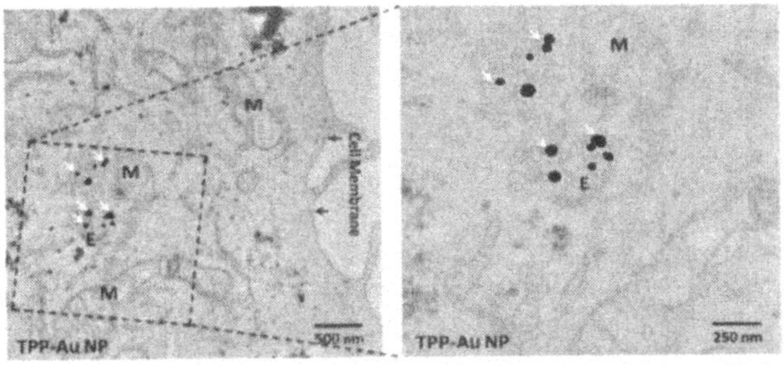

Figure 4. Representative TEM micrographs of intracellular localization of TPP-Au NPs in MDA-MB-231 cells. Au NPs were indicated by white arrows. E: endosome; M: mitochondrion.

Clearly, TPP-Au NPs yielded the most efficient killing of breast cancer cells with less damage to normal cells. This enhancement might be attributed to the specific delivery of TPP-Au NPs into mitochondria. Conjugation of TPP lipophilic cation, a commonly used mitochondrotropic molecule, onto Au NPs is expected to guide the Au NPs into mitochondria. Recognizing that mitochondria are the primary sites for converting 5-ALA into PpIX and forming $^1O_2$ during photodynamic reaction, over 90% of the total body oxygen is consumed by mitochondria, and localization of PSs in the mitochondria is more efficient to kill cells, thus, accumulation of Au NPs in mitochondria would be beneficial for enhancing the formation of $^1O_2$. Indeed, taking advantage of their mitochondria targeting feature, TPP-Au NPs showed the elevated generation of ROS and $^1O_2$ and resulted in much higher killing of breast cancer cells (74.8% cell destruction) than the cells treated with 5-ALA alone under the same PDT condition. Taken together, the enhanced PDT efficiency by TPP-Au NPs is a combined effect of 1) superior optical and plasmonic properties of Au NPs and 2) the localization of Au NPs in mitochondria for enhanced ROS formation.

4 CONCLUSION

Conjugation of Au NPs with mitochondria-targeting TPP leads to targeted delivery of Au NPs into mitochondria of breast cancer cells for further enhanced PDT efficacy but with minimal damage to normal cells. TPP-Au NPs, combining its mitochondria-targeting domain with positive surface charge, represent a promising strategy to significantly enhance therapeutic efficacy in 5-ALA-assisted PDT and and selective destruction of breast cancer cells. Targeted delivery of Au NPs into mitochondria provides an important guidance and critical evidence on the effectiveness of intracellular targeting for further optimization of drug delivery and improvement of therapeutic efficacy. Understanding the cellular uptake and intracellular destination of Au NPs in relation to their physicochemical properties is of particular importance for their potential applications as a functional therapeutic agent.

5 REFERENCES

[1] M. K. K. Oo, X. Yang, H. Du, H. Wang, Nanomedicine, 3, 777, 2008.

[2] M. K. K. Oo, Y. Yang, Y. Hu, M. Gomez, H. Du, H. Wang, ACS Nano, 6, 1939, 2012.

[3] Y. Yang, Y. Hu, H. Du, H. Wang, Chem Commun. 50(55), 7287, 2014.

[4] Y. Yang, N. Gao, Y. Hu, C. Jia, T. Chou, H. Du, H. Wang, Therapeutic Delivery, 6(3), 307, 2015.

[5] M. Wachowska, A. Muchowicz, M. Firczuk, M. Gabrysiak, M. Winiarska, M. Wańczyk, K. Bojarczuk, J. Golab. Molecules. 16(5), 4140, 2011.

[6] Y. Ju-Nam, N. Bricklebank, D. W. Allen, P. H. E. Gardiner, M. E. Light, M. B. Hursthouse. Organic & Biomolecular Chemistry, 4(23), 4345, 2006.

Synthesis of a pH Responsive Polymeric Drug Delivery – tailoring functionalization for effective delivery

X. Li[*] and C. Malardier-Jugroot[**]

[*] Royal Military College of Canada, Kigston, ON, Canada, xia.li@rmc.ca
[**] Royal Military College of Canada, Kigston, ON, Canada, cecile.malardier-jugroot@rmc.ca

ABSTRACT

The molecular modelling characterization and synthesis of the folate functionalized biomaterial poly(styrene-alt-maleic anhydride), PSMA predicted the folic acid functionalized SMA as a good candidate for targeted drug delivery. Folate was attached to the pH responsive SMA polymer via a biodegradable linker 2,4-diaminobutyric acid, DABA. The computational results predicted DABA as the optimal linker and showed that the functionazlied PSMA is linear at neutral condition and its linearity remains pH responsive. However, the earlier synthesis route displayed difficulties in characterization and purification due to the complexity of polymeric products. In this paper, a different synthesis route is shown along with NMR and IR characterization. Finally, we show some preliminary results of applying biomaterials in cell lines. Curcumin, a natural ingredient was selected as a fluorecent marker to study polymer-cell binding efficiency as well as cellular uptake.

Keywords: folate receptor, drug delivery, pH-responsive, amphiphilic co-polymers, curcumin

1 INTRODUCTION

Drug delivery systems capable of unloading drugs in response to pH changes have received much attention in recent years. Poly(styrene-alt-maleic anhydride), PSMA, is an amphiphilic alternating copolymer that self-assembles into nanotubes in aqueous solution, physiological-pH environments[1]. Its self-assembled structure with a 2nm hydrophobic inner diameter is stable in neutral pH solution, but collapses when pH is increased or decreased. These unique features of PSMA make it a great candidate for controlled release drug delivery vehicle for cancer cells for which the pH is significantly lowered compared to primary cells. PSMA has been considered as drug delivery vesicle as early as 1980s by Maeda and coworkers[2]. Its been shown that PSMA has low toxicity in cells and binds to albumin and therefore has an increased circulation time in blood stream[3-5].

To increase the specificity of the drug delivery vessel, folic acid is attached. The folic acid has extermely high affinity (K_D~10M) towards its receptors which are over-expressed in certain types of cancers[i]. Indeed, the folate-conjugated nanotubes could potentially transport small, highly toxic, hydrophobic chemotherapy agents safely through the body to be released selectively inside target cells. Once the drug is unloaded, the biocompatible SMA, folate, and linker groups would clear through the excretory system.

The synthesis was conducted in our lab using DCC/NHS coupling reagents linking carbolylic acids with amines. Amide bonds were therefore made among SMA, linker and folic acid. Previously we reported a synthesis route where polymer SMA was the starting material, linker and folic acid were added subsequently. However, due to the complexity of characterizing polymeric compounds, the second method starts with folic acid and linker and SMA polymer were attached in order. All structures were purified in each step and characterized using IR and NMR techniques.

The polymers were labelled using curcumin, which is the active ingredient in tumeric. Curcumin is an effective and safe anticancer agent, however its hydrophobicity inhibits its clinical application[6,7]. Nanotechnology provides an effective method to improve the water solubility of hydrophobic drug. In this research, curcumin is incorporated to the hydrophobic interior of polymeric biomaterial by diffusion as a drug mimic and its fluorecent property serves as a biomarker. Human pancreatic cancer cell lines PANC-1 was culcutred in the lab and the polymer-cell binding efficiency and cellular uptake activities were analysed by fluorecent microscopy

2 METHODS

2.1 Materials

13% poly(styrene-alt-maleic anhydride) (Mw=350,000) was purchased from Sigma Aldrich and freeze dried into white powder. Boc-2,4-diaminobutyric acid (DABA), N-hydroxy-succimide (NHS), dicyclohexylcarbodiimide (DCC), and folic acids were purchased from Sigma Aldrich and were used directly without further processing.

2.2 Synthesis of folic-conjugate

First, folic acid was dissolved in DMSO and allowed to react with DCC/NHS at room temperature overnight. The product was filtered using 0.2 μl filter. The resulting

TechConnect Briefs 2015, TechConnect.org, ISBN 978-1-4987-4729-5

activated folic-NHS was added dropwise to a solution of boc-protected linker boc-2,4-diaminobutyric acid (DABA). DCC/NHS were added as coupling reagents along with boc-2,4-diaminobutyric acid at a 1:10 ratio. The reaction was carried out at room temperature over night. The crude product was dialysed against water to remove DMSO solvent and the product freeze dried.

The boc protecting group was removed by trifluoroacetic acid and dichloromathane TFA/DCM at 30 degrees for 6 hours. TFA was then evaporated, and the remaining product was taken up by dichloromethane (DCM) and recrystalized in ether. The product was then filtered, air dried and characterized by ^{1}H NMR.

PSMA as added to the deprotected FA-daba oligamer and the reaction was carried out over night with DCC/NHS coupling reagent in DMSO. The product was dialysed against water and the final product was obtained after freeze drying. The dry product was characterized using ^{1}H NMR in DMSO and IR by making KBr pellets.

2.3 Cell lines

Human pancreatic cancer PANC-1 cells were maintained in monolayer culture at 37 degrees in DMEM containing 10% heat-inactivated newborn calf serum and 3 μg/ml plasmacin. The cells were seeded to wells one day prior to treatment.

2.3 SMA-curcumin nanoparticle

Curcumin was added to various SMA concentraions in media to saturation. The excedss curcumin was removed by centrifuge at 3000 rpm speed in 7 mins. The supernant was obtained and added to the wells containing cells lines.

2.4 Fluorecent microscopy

The SMA with saturated tumeric was dissolved in DMEM media of concentrations of 0.3 uM, 1uM, 3uM and 10 uM. Spectra were recorded with a fluorescence spectrophotometer

(FP-6600; JASCO) with excitation at 480 nm and emissions between 500 and 700 nm In vitro cytotoxicity assay.

3 RESULTS

Polymers are complicated to analyze, especially when only a few of the bonds are being traced. Therefore starting the synthesis with folic acid allows us to fully characterize FA-daba prior to PSMA attachment. Figure 1 shows the NMR spectrum after folic acid was attached to DAB(boc)-OH (after deprotection). The existence of folate is confirmed by peaks at 6.61 and 7.64 ppm which correspond to its para-aminobezoic acid group and the signal at 8.63 ppm, attributable to the pteridine moity proton from the folate. The peaks at 4.31and 4.47 ppm confirmed the formation of amide bond between the two. The clearance around 1.6 ppm showed that the boc protecting group was sucessfully removed.

PSMA was added to the system by coupling reagents DCC/NHS and the resulting product was characterized by NMR (Figure 2) and IR (Figure 3). The presence of folate is confirmed by peaks at 6.61 and 7.64 ppm, which correspond to its para-aminobezoic acid group, and the signal at 8.63 ppm, corresponding to its pteridine moity proton.

In Figure 3, the loss of peaks 1850 cm^{-1} and 1780 cm^{-1} indicates the conversion of maleic anhydride groups; the strong peak at 1731 cm^{-1} is produced by the remaining carboxylic acid. Furthermore, bands at 1626 and 1530 cm^{-1} are attributed to the formation of amide C=O and N–H functionalities respectively, demonstrating the link between SMA and folic acid [7].

The biomaterial PSMA and functionalized PSMA-daba-FA were briefly tested in PANC-1 cell lines with curcumin serving as a biomarker. As figure 4 shows, 10 μM SMA with curcumin were added to the cell lines, after 12 hours the cells seem to internalize some curcumin and cells appear relatively healthy. However when functionalized FA-daba-SMA were used in the treatment, cells are quickly loosing their morphology and the tumeric uptake is a lot higher. It shows that SMA-daba-FA has great potential as a drug delivery vesicle.

4 CONCLUSION AND FUTURE WORK

In this study, we presented an alternative synthesis route of SMA-daba-FA. The synthesis was characterizaed by NMR and IR and the results confirmed the structure of the biomaterials. The polymers were tested in human pancreatic cancer PANC-1 and monitored by fluorescent microscopy. The preliminary results presented here showed that FA-daba-SMA has great drug delivery potential. The study is still ongoing and cell viability assays and drug release will be tested to provide quantitave data.

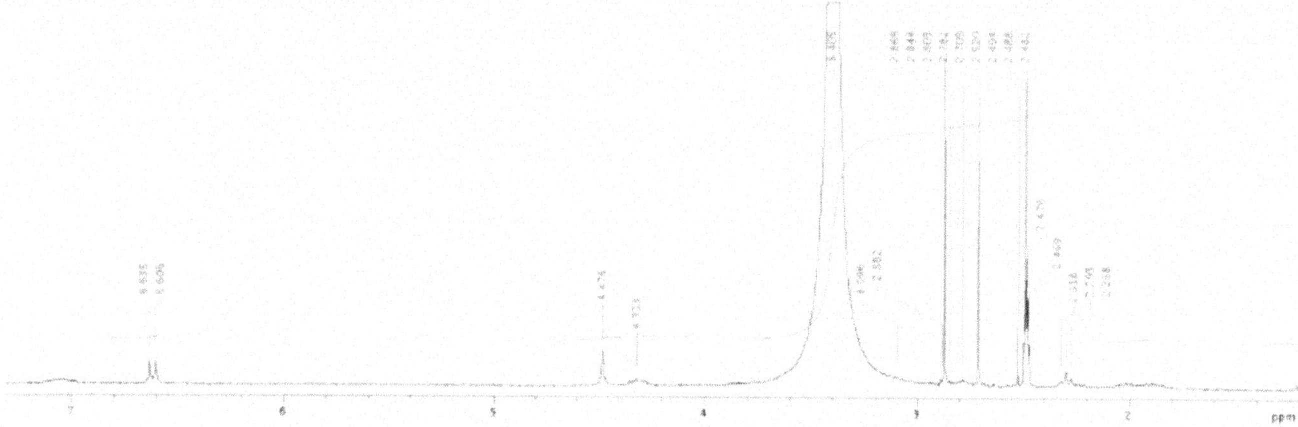

Figure 1: the NMR spectrum after folic acid was attached to DAB(boc)-OH, after deprotection

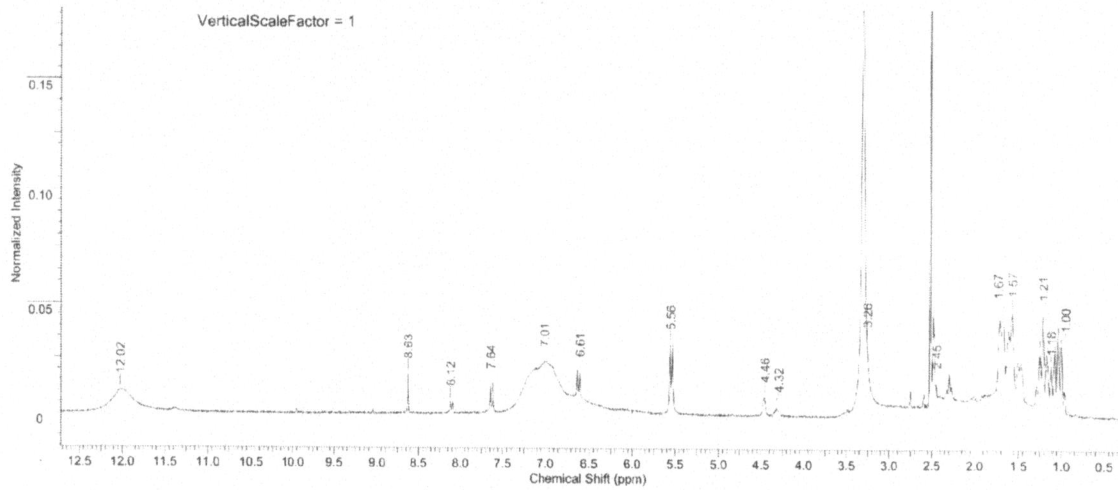

Figure 2: ^{1}H NMR spectra of folic acid-DABA-SMA conjugates in d-DMSO

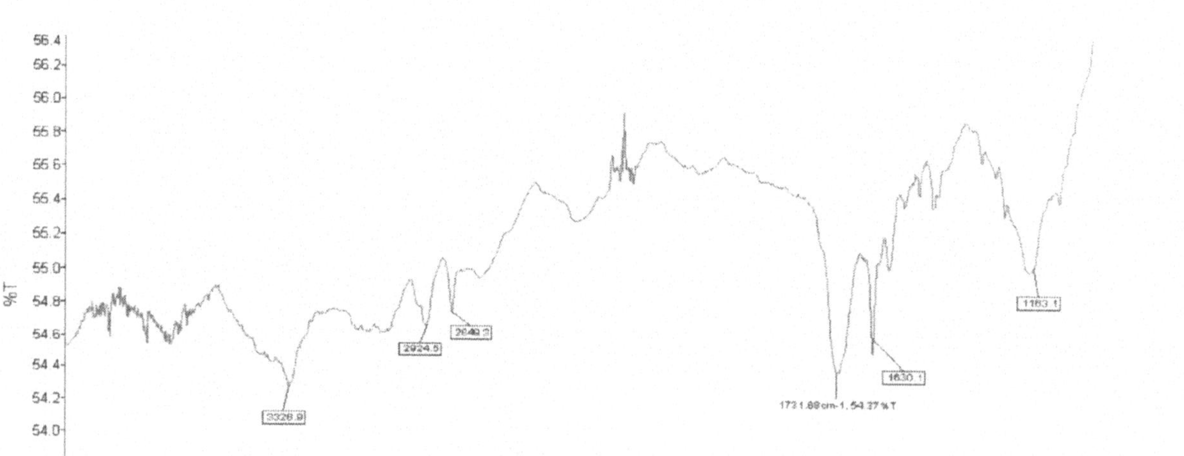

Figure 3: IR spectra of folate-DABA-SMA conjugate

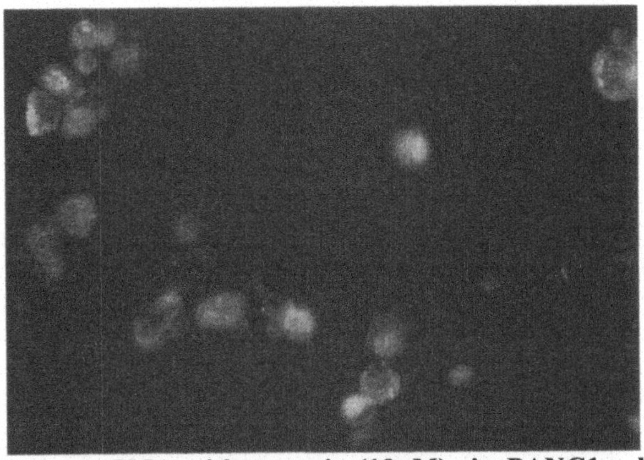

Figure 4: SMA with tumeric (10µM) in PANC1 cells after 12 hours

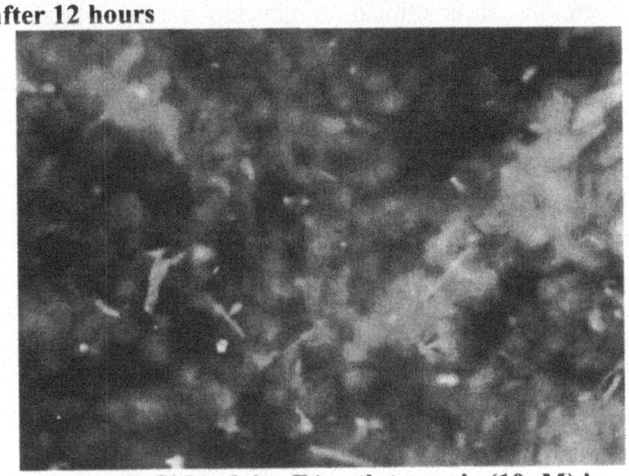

Figure 5: SMA-daba-FA wth tumeric (10µM) in PANC-1 cells after 12 hours

Acknowledgment. The authors would like to thank Dr. Myron Szewczuk for his generous guidance and support during cell culture and fluorescent microscopy studies.

REFERENCES

1. Malardier-Jugroot, C., van de Ven, T. G. M., Cosgrove, T., Richardson, R. M. & Whitehead, M. A. Novel Self-Assembly of Amphiphilic Copolymers into Nanotubes: Characterization by Small-Angle Neutron Scattering†. *Langmuir* **21,** 10179–10187 (2005).
2. Maeda, H., Ueda, M., Morinaga, T. & Matsumoto, T. Conjugation of poly (styrene-co-maleic acid) derivatives to the antitumor protein neocarzinostatin: pronounced improvements in pharmacological properties. *J. Med. Chem.* **28,** 455–461 (1985).
3. Iyer, A. K., Greish, K., Fang, J., Murakami, R. & Maeda, H. High-loading nanosized micelles of copoly(styrene–maleic acid)–zinc protoporphyrin for targeted delivery of a potent heme oxygenase inhibitor. *Biomaterials* **28,** 1871–1881 (2007).
4. Daruwalla, J. *et al.* In vitro and in vivo evaluation of tumor targeting styrene-maleic acid copolymer-pirarubicin micelles: Survival improvement and inhibition of liver metastases. *Cancer Sci.* **101,** 1866–1874 (2010).
5. Henry, S. M., El-Sayed, M. E. H., Pirie, C. M., Hoffman, A. S. & Stayton, P. S. pH-Responsive Poly(styrene- *alt* -maleic anhydride) Alkylamide Copolymers for Intracellular Drug Delivery. *Biomacromolecules* **7,** 2407–2414 (2006).
6. Bisht, S. *et al.* Polymeric nanoparticle-encapsulated curcumin ('nanocurcumin'): a novel strategy for human cancer therapy. *J. Nanobiotechnology* **5,** 3 (2007).
7. Chen, M. Curcumin-loaded nanoparticles induce apoptotic cell death through regulation of the function of MDR1 and reactive oxygen species in cisplatin-resistant CAR human oral cancer cells. *Int. J. Oncol.* (2013). doi:10.3892/ijo.2013.2050

A Multifunctional Theranostic Nanoplatform for Image-Guided Surgery and Intraoperative Therapy

O. R. Taratula, C. Schumann, T. Duong, K. L. Taylor, and O. Taratula

Department of Pharmaceutical Sciences, College of Pharmacy, Oregon State University, Portland, Oregon 97225, United States, Oleh.Taratula@oregonstate.edu

ABSTRACT

This work is focused on the development of a multifunctional agent that, when injected, can accumulate at the cancer site and help surgeons see non-visible tumor cells during surgery. Through the addition of light irradiation, the same agent can be used immediately after surgery to kill off any hidden cancer cells that may have been left behind. A single naphthalocyaninebased theranostic nanoplatform, therefore, was developed for concurrent NIR fluorescence imaging and combinatorial phototherapy with dual photodynamic (PDT) and photothermal (PTT) therapeutic properties . The conversion of a hydrophobic SiNc into a biocompatible nanoplatform (SiNc-NP) was realized by SiNc encapsulation into the hydrophobic interior of a polypropylenimine (PPI) dendrimer following surface modification with polyethylene glycol (PEG). The efficiency of SiNc-NP as an NIR imaging and phototherapeutic agent was confirmed in vitro and in vivo.

Keywords: naphthalocyanine, photothermal therapy, photodynamic therapy, near-infrared imaging, theranostic

1 INTRODUCTION

Each year there are about 22,000 new cases of ovarian cancer in the USA, and more than 14,000 women die because of it [1]. The high mortality rate is attributed to the fact that 60% of cases are diagnosed at an advanced stage, after the cancer has spread to the abdominal cavity and adjacent organs. Currently, optimal surgical resection of all disease sites is the only way to improve survival of these patients. Even with the best microsurgical techniques, however, surgeons, using visual inspection for malignant tissue identification, leave behind microscopic tumors that lead to cancer relapse. If these tumors could be detected, resected and treated more efficiently during surgery, survival rates would significantly increase. It is important, therefore, to improve tumor delineation during surgery enhancing surgical resection efficiency and provide an effective intraoperative treatment, to further eliminate any unresectable tumors. A recent clinical trial proved that the number of ovarian cancer tumors resected by surgeons under the guidance of a visible fluorophore (FITC) increased by 5.3-fold [2]. However, FITC is not optimal for surgery because high absorption and scattering of visible light by body tissue allow detection of only the surface of the tumors; and, importantly, high autofluorescence from healthy tissue reduces contrast. Since biological tissue demonstrates minimal autofluorescence and light absorption in the NIR spectral region (750 – 900 nm), NIR fluorophores can overcome these challenges [3]. The available NIR fluorophores, like indocyanine green (ICG), however, are not cancer specific and undergo irreversible photobleaching under exposure to light, which would give false negative results to the surgeon [4].

Therefore, it is critical to develop a cancer-targeted nanomedicine platform that enables maximal surgical resection by optically imaging tumor margins and can be simultaneously employed intraoperatively as a cancer therapeutic agent. The main building block of our delivery system is a photosensitizer (naphthalocyanine, SiNc) that upon exposure to NIR light generates strong NIR fluorescence signal for imaged guided surgery and destroys cancer cells by producing intercellular reactive oxygen species (ROS) and heat via PDT and PTT therapeutical mechanisms respectively. In vitro and in vivo studies of the developed SiNc-NP are discussed.

2 RESULTS AND DISCUSSION

Here, we employ silicon naphthalocyanine (SiNc) as the main building block to prepare the single agent-based nanomedicine platform (Fig. 1).

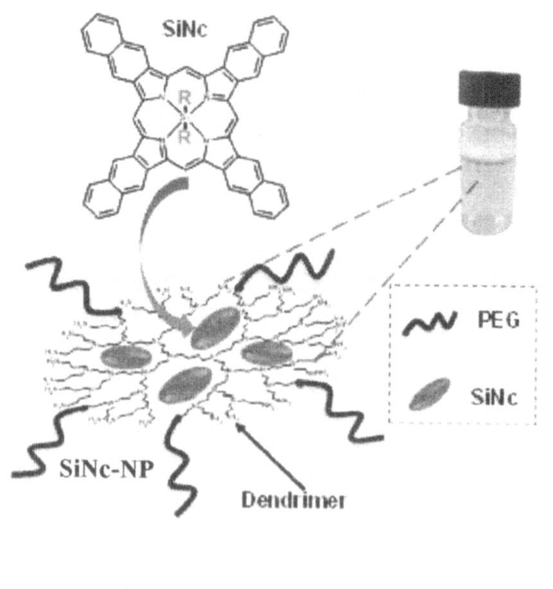

Figure 1: Nanomedicine platform (NP) based on SiNc-loaded dendrimer. The inset: SiNc-NP solution in water.

2.1 Nanometer-Sized Complex

Owing to its unique intrinsic properties, including enhanced absorption in the spectral range 750 - 800 nm, SiNc can be efficiently activated with NIR light to produce fluorescence, ROS and heat for imaging and combinatorial treatment of tumors [5]. However, potential clinical application of SiNc is substantially limited by poor water solubility, severe aggregation and lack of cancer specificity [5]. Particularly, an intrinsic tendency of SiNc molecules to aggregate in aqueous environment, through π-π stacking and hydrophobic interactions, results in the self-quenching effect of the excited states [6]. Consequently, after exposure to NIR light, aggregated SiNc molecules can only dissipate the absorbed energy through nonradiative decay (heating), while fluorescence (imaging) and ROS production (PDT) are diminished [7]. To attain water solubility and preserve fluorescence, ROS and heat generation properties, we encapsulated SiNc molecules into the interior of the water-soluble polypropylenimine (PPI) dendrimers (Fig. 1). The specific structure of PPI dendrimers, having a combination of interior hydrophobic pockets and peripheral hydrophilic primary amines, offers the possibility to encapsulate and separate the SiNc molecules thereby both decreasing aggregation and enhancing water solubility. Further, to enhance stability and biocompatibility of the dendrimer surface of the SiNc-PPI was modified with a hydrophilic polymer (polyethylene glycol, PEG). The PEG layer enhances water solubility, minimizes aggregation, prevents the immune system recognition , decreases toxicity, and leads to improved tumor accumulation via EPR effect [8].

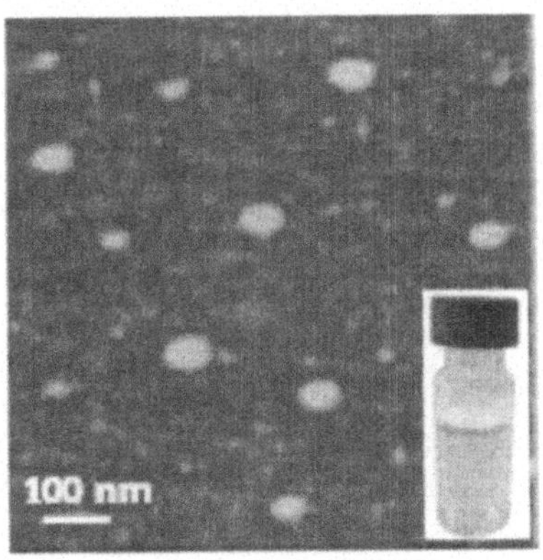

Figure 2: AFM image of the SiNc-NP.

The encapsulation procedure is based on a simple mixing of SiNc dissolved in THF with methanol solution of PPI dendrimers. After evaporation of the organic solvents, resultant complexes can easily be dissolved in water, providing SiNc aqueous solubility of at least 1 mg/mL. We also discovered that modification of SiNc with hydrophobic organic linkers (R) improves solubility in organic solvents, thus enhancing loading efficiency into the hydrophobic interior of the dendrimer up to 20 %. According to AFM images (Fig. 2), the dendrimer-based nanoplatform loaded with SiNc has a diameter of 64.6 ± 9.1 nm, which is within the desired range of 10 − 100 nm to prevent elimination by the kidneys (>10 nm), recognition by macrophage cells (<100 nm), and to enhance tumor-targeted delivery via the EPR effect (<200 nm) [9].

2.2 NIR Fluorescence Imaging

The SiNc(R) encapsulated in the dendrimer-based nanocarrier exhibited a strong NIR absorption in water with an absorbance peak at 782 nm (Fig. 3A, black curve).

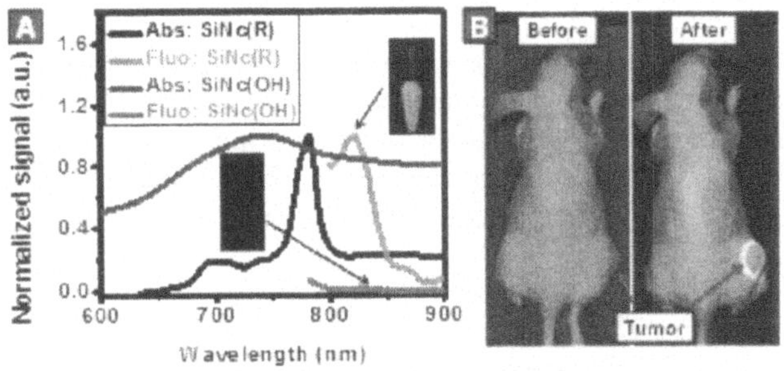

Figure 3: (A) Normalized absorbance (Abs) and fluorescence (Fluo) spectra of substituted SiNc(R) and non-substituted SiNc(OH) encapsulated in PPI dendrimer. Insets indicate fluorescence images of SiNc(R) and SiNc(OH) complexes in water recorded with Li-COR Pearl Animal Imaging system. (B) Images of mouse with ovarian tumor before (left) and after (right) injection with SiNc-NP.

The detected absorption fits within the NIR optical window (750-900 nm) required for an efficient fluorescence imaging and phototherapy [3]. Importantly, upon excitation with NIR light, the developed complex showed distinctive fluorescence emission at 820 nm in aqueous media (Fig. 3A, red curve and inset). In contrast, no detectable fluorescence signal was observed for the complexes based on unsubstituted SiNc(OH) (Fig. 3A, green curve and inset), which indicates the formation of non-fluorescence aggregates during the encapsulation process. Efficiency of SiNc(R) complexes as an NIR-imaging agent was confirmed in vivo by recording the strong fluorescence signal in the tumor area (Fig. 3B, right image). No signal was detected in the mouse before complex injection (Fig. 3B, left image), revealing that the recorded signal is not related to tissue autofluorescence. Our data revealed that encapsulation of substituted SiNc(R) within the PPI dendrimer interior decreases aggregation and preserves fluorescence intensity after dispersion of the nanocarriers in water.

2.3 Heat and ROS Generation

We have revealed that only a nanoplatform based on substituted SiNc(R) can simultaneously activate both PTT (heat) and PDT (ROS) mechanisms in ovarian cancer cells exposed to 785 nm laser light of 1.3 W/cm^2 power density. The SiNc(R)-treated cancer cells exhibit rapid heating to 52 °C upon exposure to NIR light (Fig. 4A, black curve). In contrast, the temperature change for non-transfected cells was less than 1 °C under the same experimental conditions (Fig. 4A, red curve), indicating that 785 nm laser (1.3 W/cm^2) alone is not capable of heating. We have also confirmed efficient intracellular ROS generation by the substituted SiNc(R)-based nanoplatform upon NIR light activation. The cancer cells incubated with the SiNc(R) nanoplatform, then subsequently irradiated with the NIR laser light (785 nm, 1.3 W/cm^2), showed a 6-fold elevation in intracellular ROS levels, compared to the controls (Fig. 4B). In contrast, the unsubstituted SiNc(OH) encapsulated in the dendrimer was not able to increase intracellular ROS levels significantly under the same experimental conditions, which is related to aggregation of SiNc(OH) during the encapsulation process and the quenching of the excited states.

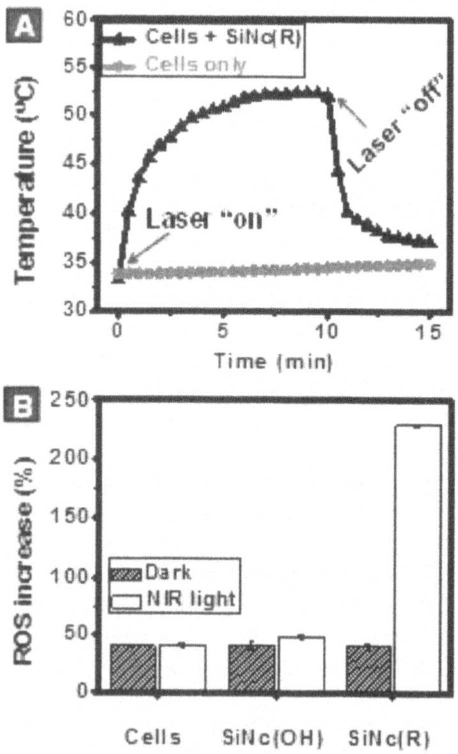

Figure 4: (**A**) Temperature profile of A2780/AD ovarian cancer cells transfected with SiNc(R)-NP (25 µg/mL, black curve) exposed to the laser diode (785 nm, 1.3 W/cm^2). Non-transfected cells exposed to the laser diode were used as the control (red curve). (**B**) Relative intracellular ROS level in A2780/AD cells incubated with SiNc(R) and SiNc(OH) complexes (25 µg/mL) and irradiated for 10 min using the 785 nm laser diode (1.3 W/cm^2).

2.4 Combinatorial Phototherapy

We revealed the superior efficacy of the combinatorial phototherapy mediated by SiNc(R)-based nanoplatform on ovarian carcinoma cells (A2780/AD) with more than a 99% cancer cell death (Fig. 5A, yellow bar), in comparison to PDT (50% cells death, cyan bar) and PTT (87% cell death, magenta bar) alone. Notably, the toxicity of the nanoplatform was insignificant under dark conditions (no laser irradiation), indicating safety and specificity of the combinatorial phototherapy (Fig. 5A, red and green bars). We have also confirmed anticancer efficacy of the combinatorial phototherapy in nude mice bearing a human ovarian cancer xenograft.

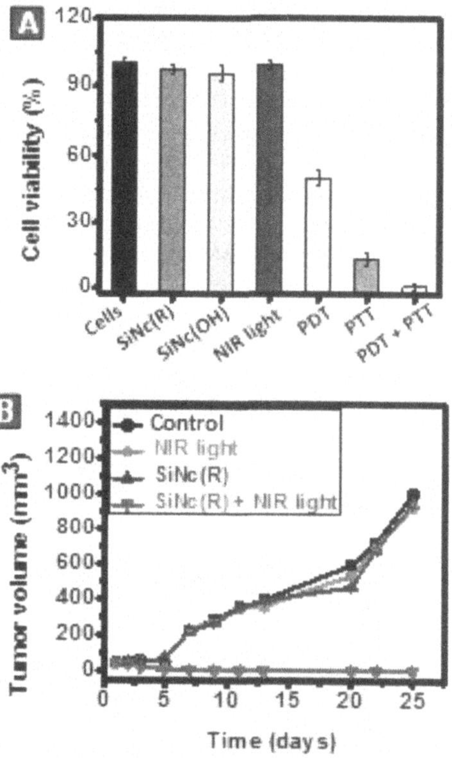

Figure 5: (**A**) Viability of DOX-resistant A2780/AD ovarian cancer cells after exposure to the following treatments: (1) no treatment (black bar); (2) SiNc(R)- and SiNc(OH)-based nanoplatforms under dark conditions (red and green bars); (3) NIR light (785 nm, 1.3 W/cm^2, 10 min, blue bar); (4) PDT: SiNc(R) platform + 785 nm NIR light (0.3 W/cm^2, 10 min, cyan bar); (5) PTT: SiNc(OH) platform + 785 nm NIR light (1.3 W/cm^2, 10 min, cyan bar); and (6) combinatorial phototherapy- SiNc(R) platform + 785 nm NIR light (1.3 W/cm^2, 10 min, cyan bar). (**B**) Growth of A2780/AD tumors in mice after treatment with: (1) saline (black); (2) 785 nm laser diode (1.3 W/cm^2, 10 min, red); (3) SiNc(R)-based nanoplatform (blue curve); and (4) combinatorial phototherapy: SiNc(R)-based platform combined with laser irradiation (1.3 W/cm^2, 10 min, green curve).

It is remarkable that solid tumors treated with a single dose of SiNc(R)-based nanoplatform combined with 10 min of NIR light irradiation (785 nm, 1.3 W/cm^2) shrank gradually and were completely eradicated from the mice with no evidence of cancer recurrence in the 25 day follow up period (Fig. 5B, green curve). Finally, not even slight skin burn was noticed after irradiating the non-treated tumors for 15 min, suggesting that 785 nm laser light at a power density 1.3 W/cm^2 is safe for phototherapy, making the application for the SiNc nanoplatform even more attractive. In vitro and in vivo studies revealed that combinatorial phototherapy mediated by SiNc(R)-based nanoplatform can efficiently destroy chemotherapy-resistant ovarian cancer cells and can potentially maximize the treatment effect of surgery by killing unresected cancer cells.

2.5 No need for Drug Release

Thus, the SiNc(R) nanoplatform can provide an efficient anticancer effect without releasing photoactive agent. The release of anticancer agents is one of the most important and limiting steps for drug delivery systems to exert efficient antitumor activity. Even if enhanced tumor accumulation of a drug carrier is accomplished, high antitumor efficacy cannot be achieved without reasonable drug release. The main advantage of the developed combinatorial phototherapy is that its anticancer efficacy is not limited by SiNc released from the dendrimer-based nanoplatform. To prove this point, the SiNc(R) release profile has been evaluated in PBS buffer at pH 5.5 containing 10 mM of reduced glutathione (GSH), mimicking intracellular (acidic and reduced) conditions.

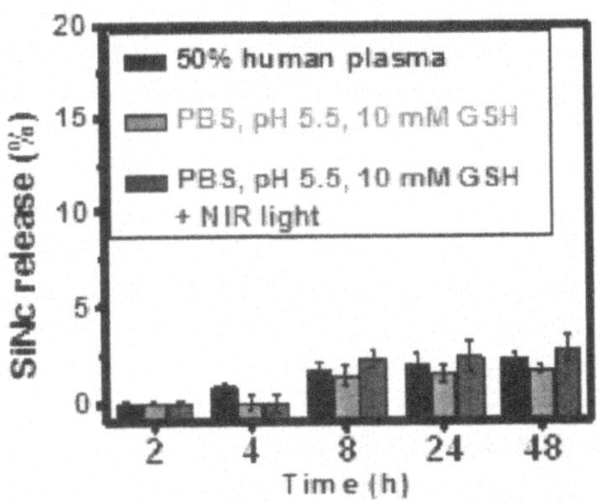

Figure 6: The release profile of SiNc(R) from dendrimer-based nanoplatform at 37 °C in (1) 50% human serum; (2) PBS buffer at pH 5.5 with 10 mM GSH; and (3) PBS (pH 5.5, 10 mM GSH) + 785 nm light (1.3 W/cm^2, 20 min).

Our results, represented in Fig. 6 revealed that SiNc release was also minimal (less than 2%) under studied conditions. Taking into consideration the therapeutic efficacy studies (Fig. 5) and the SiNc release profiles (Fig. 6), one can conclude that the dendrimer encapsulated SiNc(R) molecules do not have to be released in the cancer cells to cause the PDT and PTT effects. Thus, due to the strong hydrophobic interactions between SiNc and the dendrimer interior, the release of SiNc is limited in aqueous setting, which prevents leaching of photoactive agents from the carriers in systemic circulation and, thus, minimizes side effects and maximizes delivery to the targeted cancer [10].

3 CONCLUSIONS

We reported a simple and effective approach for constructing a single SiNc-based theranostic agent for cancer imaging and treatment. The developed encapsulation strategy offers the possibility of separating the SiNc molecules thus decreasing their aggregation, preserving their NIR fluorescence signal, stabilizing their PDT and PTT properties and enhancing their water solubility. Furthermore, SiNc-NP exhibits minimal dark cytotoxicity and remarkable combinatorial phototherapeutic effects on ovarian cancer in vitro and in vivo, through its cytotoxic effects that cause a high production of ROS and heat without releasing the photoactive agent from the nanoplatform. Finally, the potential of SiNc-NP as an NIR imaging agent was confirmed by recording the strong fluorescence signal in the tumor area. We anticipate that the developed theranostic nanoplatform can be potentially applied for NIR fluorescence image-guided surgery and intraoperative cancer treatment.

4 ACKNOWLEDGEMENTS

Medical Research Foundation of Oregon, PhRMA Foundation and the College of Pharmacy at Oregon State University.

REFERENCES

[1] Cancer facts and figures, American Cancer Society.;http://www.cancer.org/research/cancerfacts statistics/cancerfactsfigures2014/, 2014.
[2] van Dam GM, Themelis G, Crane LM, Harlaar NJ, Pleijhuis RG, Kelder W, et al. Nature Med. 17, 1315-9, 2011.
[3] Gibbs SL. Quant. Imag. Med. Surg. 2, 177-87, 2012.
[4] Schaafsma BE, Mieog JS, Hutteman M, van der Vorst JR, Kuppen PJ, Lowik CW, et al. J. Surg. Oncol. 104, 323-32, 2011.
[5] Josefsen LB, Boyle RW. Theranostics. 2, 916-66, 2012.
[6] Josefsen LB, Boyle RW. Metal-Based Drugs. 2008, 276109, 2008.
[7] Lim CK, Shin J, Lee YD, Kim J, Oh KS, Yuk SH, et al. Theranostics. 2, 871-9, 2012.
[8] Allen TM, Cullis PR. Science. 303, 1818-22, 2004.
[9] Longmire M, Choyke PL, Kobayashi H. Nanomedicine-UK. 3, 703-17, 2008.
[10] Taratula O, Schumann C, Naleway MA, Pang AJ, Chon KJ, Taratula O. Mol Pharm. 10, 3946-58, 2013.

Microarray-based Functional Nanoproteomics for Cancer

C. Nicolini*, NL. Bragazzi** and E. Pechkova**

*Fondazione EL.B.A. Nicolini (FEN), Pradalunga, Largo Redaelli 7, Bergamo 24020, Italy, president@fondazioneelba-nicolini.org
**Laboratories of Biophysics and Nanobiotechnology (LBN), Department of Experimental Medicine (DIMES), Genoa University, Via Antonio Pastore 3, Genoa 16132, Italy, eugenia.pechkova@gmail.com robertobragazzi@gmail.com

Abstract

Using the New England BioLabs (NEBL) SNAP-based Genes Expression in conjunction with our "sub-micron arrays" (Anodic Porous Allumina and/or Kapton based Nanopores), we exploit our proprietary microarrays scanner (DNASER, DNA analyzer) and Label Free Nanotechnologies to carry out the following tasks:

1) Construction of SNAP-based Genes Nanoarrays, using gold surface coated for 10 minutes with 2% solution of 3-Aminopropyltriethoxysilane (APTES) in acetone, rinsed in acetone and dried with filtered air. Full length complementary DNAs (cDNAs) for oncosuppressor 53 (p53), Cyclin-dependent kinase 2 (CDK2), SH2 (Src Homology 2) domain of the proto-oncogene tyrosine-protein kinase (Src) and tyrosine-protein phosphatase non-receptor type 11 (PTPN11) were amplified and cloned. Printing mix was prepared with 0.66 μg/μl DNA capture reagent BG-PEG-NH2 for the one-step synthesis of SNAP-tag substrates from esters on labels or surfaces;

2) Determination of Protein-Protein Interaction for the chosen cancer following the identification of leader genes (or hub genes, investigated with theoretical *ab initio* bioinformatics analysis using in-house software and algorithms, and then experimentally confirmed via DNASER). These genes are expressed by Pure (Protein synthesis Using Recombinant Elements) Express in spots less than 1 micron size piezo-microdispensed and then characterized via Label Free proprietary Autoflex Mass Spectrometry (MS) integrated with *ad hoc* software, namely the Spectrum Analyzer and Data Set manager (SpADS) and a proprietary Quartz Crystal Micro-balance with Dissipation factor monitoring (QCM_D) Nanoconductimetry, enabling to describe properties such as changes in frequency and conductance, viscoelasticity and dissipation factor. Solutions without DNA were prepared (called Master Mix, MM), as negative controls, in printing mix. Negative controls were prepared with a varying concentration range of SNAP capture reagent. As a positive control (for fluorescence analysis) mouse IgG or rabbit IgG (Pierce, IL, USA) were added in a printing mix instead of DNA.

Keywords: Quartz Crystal Microbalance with Dissipation Factor monitoring (QCM_D) and nanoconductimetry, SNAP microarrays, mass spectrometry (MS)

1 INTRODUCTION

On the light of the sixteen recently published experimental papers, referenced in the last section [1-16], we have successfully completed the feasibility study on two distinct but highly correlated lines of research in progress since 2012. On the basis of these obtained findings, we have subsequently prepared several proposals to be submitted as grant applications to the Small Business Innovation Research (SBIR) program, along with a small company based in the United States of America (USA), being presently identified.

We have identified the technologies we intend to use: namely, the New England BioLab (NEBL) proprietary "SNAP based Genes Expression" in conjunction with our proprietary "Sub-micron Arrays" (either our Anodic Porous Allumina – APA – Nanopores and/or Kapton based Nanopores, in cooperation with Arts Engineering), our proprietary microarrays scanner DNASER (DNA Analyzer) and Label Free Nanotechnologies resulting from a collaboration between the Laboratories of Biophysics and Nanobiotechnology (LBN) at Genoa University, Department of Experimental Medicine (DIMES), Genoa, Italy and Fondazione EL.B.A. Nicolini (FEN), Pradalunga (Bergamo), Italy.

Namely, we intend to carry out:

1) Construction of SNAP Genes Nanoarrays, using gold surface coated for 10 minutes with 2% solution of 3-Aminopropyltriethoxysilane – APTES – (Pierce, Rockford, IL, USA) in acetone, rinsed in acetone and dried with filtered air. Full length complementary DNAs (cDNAs) for oncosuppressor protein 53 (p53) and Cyclin-dependent kinase 2 (CDK2) were both purchased form DNASU Central Plasmid Repository located in the Biodesign Institute, Arizona State University (ASU), USA (https://dnasu.org/DNASU/Home.do). Full length cDNAs for the SH2 domain of the proto-oncogene tyrosine-protein kinase (Src) and tyrosine-protein phosphatase non-receptor type 11 (PTPN11) were purchased from Open Biosystem, Thermo Scientific. cDNAs were amplified and cloned into NdeI and XhoI sites in pCOATexp SNAPf vector, a

derivative of pCOATexp and pSNAPf (further details can be found at https://www.neb.com/products/n9183-psnapf-vector#tabselect0). Printing mix was prepared with 0.66 μg/μl DNA [bovine serum albumin, BSA, bound], capture reagent: BG-PEG-NH2 (an amine-terminated building block for the one-step synthesis of SNAP-tag substrates from NHS esters or other activated carboxyl esters on labels or surfaces, https://www.neb.com/products/s9150-bg-peg-nh2), ranging from 80 to 800 ng/μl [sBS3 bound];

2) Determination of Protein-Protein Interaction for the chosen cancer following the identification of genes leaders (hub genes, being highly interconnected and investigated via bioinformatics analysis and then experimentally confirmed via DNASER) expression by PureExpress in spots sized less than 1 micron size piezo-microdispensed (as well proprietary) and characterized via Label Free proprietary dedicated Autoflex Mass Spectrometry (MS) integrated with *ad hoc* software such as the proprietary Spectrum Analyzer and Data Set manager (SpADS) and a proprietary Quartz Crystal Micro-balance with Dissipation factor monitoring (QCM_D) Nanoconductimeter for performing nanoconductimetry. Solutions without DNA (master mix, or MM) were prepared, negative controls, in printing mix. Negative controls were prepared with a concentration range from 80 to 800 ng/μl of SNAP capture reagent. As a positive control (for fluorescence analysis) mouse IgG or rabbit IgG (Pierce, IL, USA) were added in a printing mix instead of DNA.

We present results obtained applying our innovative conductometer, realized by combining Nucleic Acid Programmable Protein Arrays (NAPPA) technology with QCM_D, to the characterization of protein-protein and protein-sterol interactions in a multiparametric way, taking advantage of the multiple information provided by the analysis of the conductance curves (i.e. conductance, viscoelasticity and adsorbed mass, which can be estimated using the Sauerbrey equation).

Moreover, through our nanoconductometer we acquired information on the kinetic constant of enzymatic interaction. The protocol for properly performing nanoconductimetric assay is shown in Figure 1. Piezoelectric inkjet printing offers the possibility of spotting on complex surfaces and using clinical samples of limited volume and amount, while QCM_D discriminates several genes per spot contrary to the single one with fluorescence (Figure 2). We investigated well known pairs of interacting molecules like Cholesterol side-chain cleavage enzyme 11A1 (CYP11A1) and Cholesterol, Jun and Activating Transcription Factor 2 (ATF2), p53 and MDM2. An interesting implication for potential clinical applications concerned furthermore the possibility to drastically reduce the time of protein expression and capture under our experimental conditions.

The QCM_D instrument we used allowed us to monitor *in real time* the trend of Dissipation factor (D factor) and frequency (f) during the interaction between CYP11A1 and Cholesterol, both in solution and in blood. Assuming a Michaelis-Menten like behavior and fitting these experimental data, we were able to compute a constant (K) of about 100 μM, a value which is in good agreement with the values found in the extant literature. Finally, in order to verify the possibility to analyze simultaneously more interactions in a single NAPPA-Quartz Crystal (QC), we immobilized on the same QC up to three cDNAs. In our hands, we were able to successfully identify all of them and, subsequently, we analyzed the response to multi-protein interactions. Jun&CDK2 and Jun&CDK2&p53 co-expressed in the same QCs were indeed tested for ATF2 interaction, both in flow and statically. Taken all together, we demonstrated the versatility of the NAPPA-QC biosensors for the detection of protein-protein interactions and protein-sterol interaction in cancer control as well as for controlling other diseases.Moreover, we measured also interaction between genes and their products and drugs (such as Temozolomide, commercially known as Temodar, an antiblastic used for brain cancer), showing that our proprietary instrument is able to perform both genomics and proteomics/pharmacogenomics and pharmacoproteomics (as can be seen in Figure 3). QCM_D is multi-purpose and, furthermore, is a very quick and effective tool for characterizing a biochemical system.

The analysis of self-assembling protein microarray NAPPA expressed with the SNAP tag in *Escherichia coli* coupled self free expression system prove capable to develop a highly standardized procedure in order to analyze the protein-protein interaction occurred on the array combining label free MS. We employ in the process PURE (Protein synthesis Using Recombinant Elements) system.

For the first time an improved version of NAPPA, that allows for functional proteins to be synthesized *in situ* – with a SNAP tag – directly from printed cDNAs prepared just in time for the assay, has been expressed with a novel cell-free transcription/translation system reconstituted from the purified components necessary for *Escherichia coli* translation – the PURE system – and analyzed in a label free manner by three different mass spectrometry techniques, namely two Matrix Assisted Laser Desorption Ionization Time-of-Flight (MALDI-TOF), a Voyager and a Bruker Ultraflex, and a Liquid Chromatography-Electrospray Ionization Mass Spectrometry (LC-ESIMS/MS). Autoflex Mass Spectrometry shows discrimination of NAPPA versus SNAP arrays (Figure 4) and of different genes as p53 versus CdK2 genes expressed using PURE E,Coli system (Figure5). Due to the high complexity of the system, very difficult results were obtained and, therefore, in order to reduce this complexity and make sense of the findings, there was the need to develop and design an *ad hoc* bioinformatics tool for the analysis. (Figure 6) Furthermore, a contemporary fluorescence analysis of NAPPA, expressed by means of PURE system, has been performed to fully characterize this new NAPPA SNAP system and correlated to pure MS results..

2 results

Here we show what we experimentally obtained.

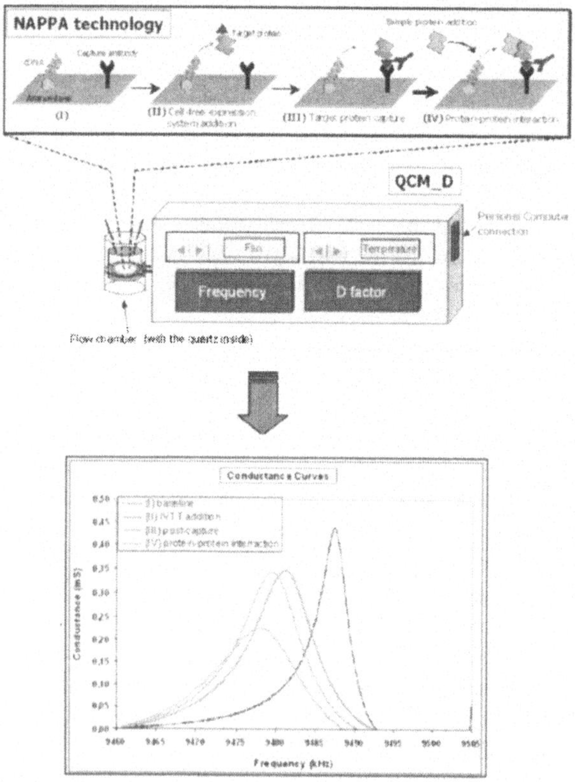

Figure 1: Our proprietary nanoconductomer for performing nanoconductimetry, developed by successfully coupling Quartz Crystal Micro-balance with Dissipation factor monitoring (QCM_D) with Nucleic Acid Programmable Protein Arrays (NAPPA) technology.

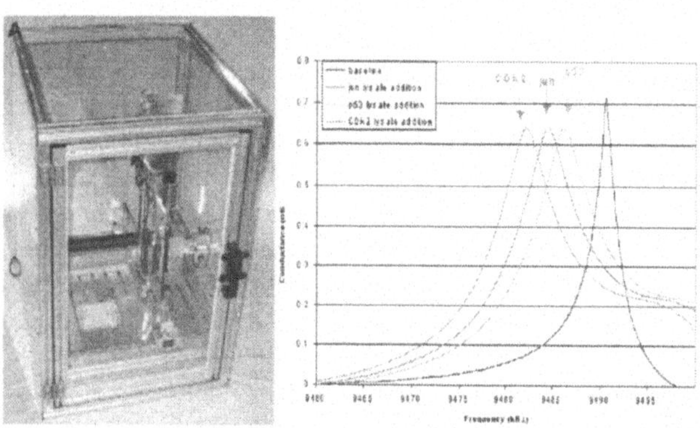

Figure 2. Piezoelectric liquid dispensing (left) for printing and probing of NanoProbeArrays of June, CDK2 and p53

Figure 3. QCM_D conductance curves for MM_MLH1 and POLB interacting with Temozolomide at increasing concentrations

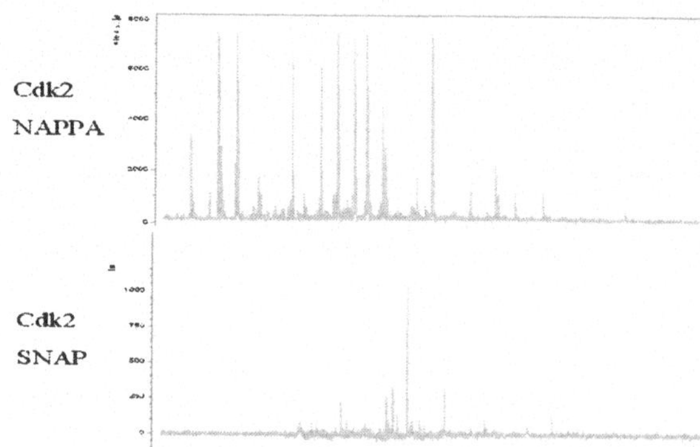

Figure 4. Mass Spectrometry NAPPA versus SNAP arrays

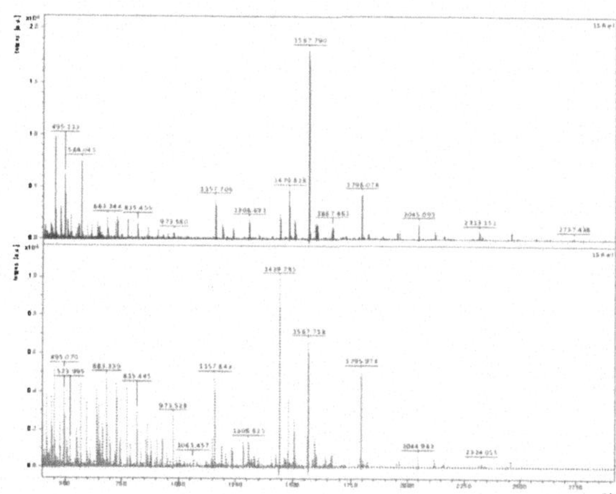

Figure 5. Mass Spectrometry of Snap-Nappa Arrays of p53 versus CdK2 genes expressed using PURE E,Coli system.

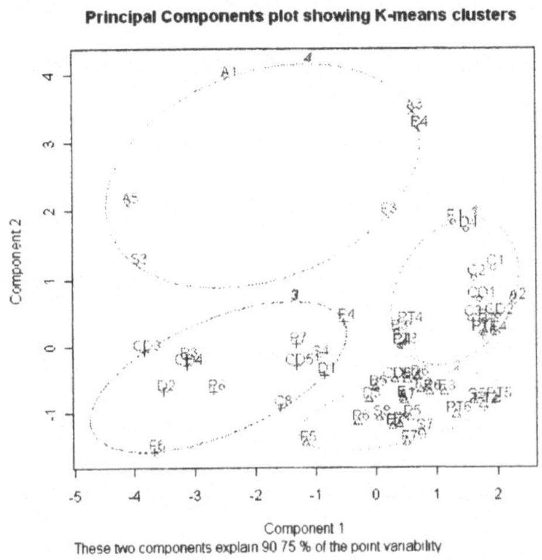

Figure 6. Cluster assignment for each known protein sample on a specimen of 56 spectra. Statistics are based on the SpADS results coupled with K Means clustering given in figure below. In bold assignment of clusters by human interpretation of cluster results. Highlited with * striking recognition.

3 CONCLUSIONS

The two combined Label Free technologies, MS integrated by proprietary software and technology, new *ad hoc* proprietary bioinformatics tools, and QCM_D Nanoconductometry integrated by proprietary hardware and software, as applied to optimized microarray with micron sized genes spots using SNAP and bacterial lysate technologies further optimized in a future application for SBIR program, may be constructed in such further optimized way to overcome multiple gene expression and interaction in a fashion unmatchable by current fluorescence technology using single gene, oversized spots.

references

[1] Bragazzi NL, Nicolini C. Nanogenomics for Personalized Nanomedicine: An Application to Kidney Transplantation. Cell Mol Biol. 2014;60:115.

[2] Nicolini C, Singh M, Spera R, Felli L. Analysis of gene expression on anodic porous alumina microarrays. Bioengineered. 2013 Sep-Oct;4(5):332-7.

[3] Nicolini C, Bragazzi NL, Pechkova E. Determination of Protein-Protein Interaction for Cancer Control via Mass Spectrometry and Nanoconductimetry of NAPPA SNAP Arrays: An Overview. NanoWorld J. 2015;1(1): 9-17.

[4] Spera R, Festa F, Bragazzi NL, Pechkova E, LaBaer J, Nicolini C. Conductometric monitoring of protein-protein interactions. J Proteome Res. 2013 Dec 6;12(12):5535-47.

[5] Nicolini C, Adami M, Sartore M, Bragazzi NL, Bavastrello V, Spera R, Pechkova E. Prototypes of newly conceived inorganic and biological sensors for health and environmental applications. Sensors (Basel). 2012 Dec 12;12(12):17112-27.

[6] Nicolini C, Correia TB, Stura E, Larosa C, Spera R, Pechkova E. Atomic force microscopy and anodic porous allumina of nucleic acid programmable protein arrays. Recent Pat Biotechnol. 2013 Aug;7(2):112-21.

[7] Nicolini C, Bragazzi N, Pechkova E. Nanoproteomics enabling personalized nanomedicine. Adv Drug Deliv Rev. 2012 Oct;64(13):1522-31.

[8] Belmonte L, Spera R, Nicolini C. SpADS: An R Script for Mass Spectrometry Data Preprocessing before Data Mining. J Comput Sci Syst Biol. 2013;6:298-304.

[9] Spera R, Labaer J, Nicolini C. MALDI-TOF characterization of NAPPA-generated proteins. J Mass Spectrom. 2011 Sep;46(9):960-5.

[10] Bragazzi NL, Nicolini C. A Leader Genes Approach-based Tool for Molecular Genomics: From Gene-ranking to Gene network Systems Biology and Biotargets Predictions. J Comput Sci Syst Biol. 2013;6:165-176.

[11] Pechkova E, P. Wiktor, Bragazzi N, Fernanda F, Nicolini C. Nanoprobe NAPPA Arrays for the Nanoconductimetric Analysis of Ultra-Low-Volume Protein Samples Using Piezoelectric Liquid Dispensing Technology. NanoWorld J. 2015;1(1): 26-31.

[12] Bragazzi NL, Spera R, Pechkova E, Nicolini C. NAPPA-Based Nanobiosensors for the Detection of Proteins and of Protein-Protein Interactions Relevant to Cancer. J Carcinog & Mutagen. 2014;5:166.

[13] Bragazzi NL, Pechkova E, Nicolini C. Proteomics and proteogenomics approaches for oral diseases. Adv Protein Chem Struct Biol. 2014;95:125-62.

[14] Nicolini C, Spera R, Festa F, Belmonte L, Chong S, Pechkova E, LaBaer J. Mass Spectrometry and Florescence Analysis of SNAP-NAPPA Arrays Expressed Using E. coli Cell_Free Expression System. J Nanomed Nanotechnol. 2013;4(5):181-95.

[15] Nicolini C, Spera R, Pechkova E. SpADS and SNAP-NAPPA Microarrays towards Biomarkers Identification in Humans: Background Subtraction in Mass Spectrometry with E.coli Cell Free Expression System. J Mol Biomark Diagn. 2015;6:214.

[16] Nicolini C, Spera R, Bragazzi NL, Pechkova E. Drug-Protein Interactions for Clinical Research by Nucleic Acid Programmable Protein Arrays-Quartz Crystal Microbalance with Dissipation Factor Monitoring Nanoconductometric Assay. American Journal of Biochemistry and Biotechnology. 2014;10(3):189-201.

From Nanogenomics and Structural-Functional Proteomics to Personalized Medicine for Oral Lichen Planus (OLP) as a Translational Model of Oral Cancer Useful to Introduce an Industrial Technical Innovation

C. Nicolini*, E. Pechkova** and NL. Bragazzi***

*Fondazione EL.B.A. Nicolini (FEN), Pradalunga, Largo Redaelli 7, Bergamo 24020, Italy, president@fondazioneelba-nicolini.org

**Fondazione EL.B.A. Nicolini (FEN), Pradalunga, Largo Redaelli 7, Bergamo 24020, Italy and Laboratories of Biophysics and Nanobiotechnology (LBN), Department of Experimental Medicine (DIMES), Genoa University, Via Antonio Pastore 3, Genoa 16132, Italy, eugenia.pechkova@gmail.com

***Laboratories of Biophysics and Nanobiotechnology (LBN), Department of Experimental Medicine (DIMES), Genoa University, Via Antonio Pastore 3, Genoa 16132, Italy, 2896313@studenti.unige.it

ABSTRACT

Oral Lichen Planus (OLP) is a common chronic inflammatory disease, which involves the mucous membranes of the oral cavity, with an overall age-standardized prevalence of 1.27% (0.96% in men and 1.57% in women) as reported in the literature by McCartan and Healy. In our previous work, we exploited a bioinformatics approach, namely the Leader Gene Algorithm (LGA), enabling to underpin the main hub genes (termed as Leader Genes) involved in biological processes. In the case of OLP, we found a complex network made up of 132 genes and, in particular, we found five Leader Genes (namely, JUN, EGFR, FOS, IL2, and ITGB4). Using a subsequent bioinformatics algorithm, we managed to find the 48.39% of the already established OLP-related microRNAs (miRNAs), suggesting that at least half of the OLP-related microRNAome (miRNAome) finely tunes few, highly interconnected hub genes. Now, we would use real clinical samples in order to validate our predicted biomarkers, using molecular biology techniques, mass-spectrometry (MS) and *ad hoc* in-house developed instruments, such as nanoconductimetry via Quartz Crystal Microbalance with Dissipation factor monitoring (QCM_D). A unique combination of genomics and proteomics approaches can indeed represent a promising innovation for a personalized treatment of OLP and oral cancer.

Keywords: nanogenomics and the Leader Gene Algorithm (LGA), Oral Lichen Planus (OLP) as translational model of oral cancer, Quartz Crystal Microbalance with Dissipation Factor monitoring (QCM_D) and nanoconductimetry, SNAP microarrays, mass spectrometry (MS)

1 INTRODUCTION

Oral Lichen Planus (OLP) is a common chronic inflammatory disease [1], which involves the mucous membranes of the oral cavity, with an overall age-standardized prevalence of 1.27% (0.96% in men and 1.57% in women), as reported in the literature by McCartan and Healy [2].

The histopathological features of OLP are hydropic degeneration of the basal cell layer, hyperkeratosis, acanthosis, irregular ridges and a dense band-like infiltration of T lymphocytes mainly in the *lamina propria*. Although the aetiology of OLP is still unknown, it has been widely accepted that immunological impairments are very critical among the multiple aetiological factors. Previous studies have suggested that it may represent a cell-mediated immunologic response to an induced antigenic change in the oral mucosa [3-4].

In our previous work [5], we exploited a bioinformatics approach, namely the Leader Gene Algorithm (LGA), enabling to underpin the main hub genes (termed as Leader Genes) involved in biological processes. Only human genes were considered. In this way, it was possible to identify a list of candidate genes potentially involved in OLP pathogenesis, by mining different genomics databases and datasets.

The preliminary set of genes was then expanded using the web-available software STRING (version 9.1, freely available at http://string-db.org/) [6], considering only direct interactions (i.e.: physical contact between encoded proteins, gene expression microarray data, or direct linkage in the same pathway), with a high degree of confidence (above 0.9 – confidence value in STRING ranges between 0 and 0.99, with 0.99 being the highest confidence). In this way, it is possible to identify new genes directly linked to those with an already established role in OLP, and therefore potentially involved in this disease. In order to discard false positives, results were then filtered using a further search in literature and gene databases. The process was repeated until no new gene potentially involved in OLP was identified. Then, an interaction map among the identified genes was calculated using STRING. This software can give a combined association score to each interaction, representing the degree of confidence for each interaction. For every gene identified, we summed the different combined association scores with the other genes. The sum of all these scores is defined as the weighted number of links (WNL).

Genes were then clustered, using hierarchical or K-means algorithms, according to their WNL. The genes belonging to the highest rank are defined as Leader Genes; these genes have a significant higher WNL if compared with the other ones. The other ranks are termed class B, class C, class D genes and so on, according to their WNL scores. Genes with no identified interactions (i.e. WNL=0) are defined as orphan genes. Differences among various classes in terms of WNL were statistically evaluated using an Analysis of Variance (ANOVA) test, with a Tukey-Kramer test as *post-hoc* test. Statistical significance was set at a p-value < 0.001, in order to ensure a high level of data reliability.

LGA was already successfully applied for investigating hub genes in different pathologies, such as periodontitis [7], dental diseases and their risk of malignant transformation [8-9], and kidney transplantation [10-11], among others.

Moreover, interacting genes were classified as up-regulated, down-regulated or neutral in respect to OLP pathogenesis. For neutral genes, we mean that they do not exhibit fold expression changes in the disease *versus* health control condition or genes for which there is not a universal consensus in the literature and in the databases.

In the case of OLP, we found a complex network made up of 132 genes and, in particular, we found five Leader Genes (namely, JUN, EGFR, FOS, IL2, and ITGB4). The interactions in the obtained network showed power law behavior, in good agreement with the scale-free topology theory of the biological graphs.

Interestingly, all of them but EGFR were up-regulated. Evidence concerning EGFR regulation is indeed controversial and not conclusive. Moreover, the Leader Genes were widely distributed in the network (in term of topological parameters, such as stress, eccentricity and radiality) and showed higher topological coefficients than the other genes.

2 BIOINFORMATICS SECTION

In the current work, in order to predict the potential microRNAs (miRNAs) network related to OLP, we used our previously identified "Leader Genes", namely both the Class A genes (JUN, EGFR, FOS, IL2, and ITGB4) and the Class B genes (CASP3, CD247, IL2RA, IFNG, MMP2, LAMC2).

Genetic and genomics research is rapidly increasing our understanding of the molecular basis of some diseases and may also suggest new diagnostic and treatment strategies, paving the way for technological and industrial innovations. Many oral diseases are complex, multifactorial disorders and, as such, have a genetic basis. Studies of these pathological conditions suggest that multiple gene interactions are important determinants of susceptibility.

miRNAs are a family of small and short (usually 19-25 nucleotides long), single-stranded, endogenous, non-coding RNA molecules. They play a key role both in physiology and in pathology, and their role in pathogenesis of oral diseases is emerging. However, despite the importance of incorporating the oral microRNAome (miRNAome) in the study of oral disorders, so far only few miRNAs related to them have been discovered and described in the literature. For this reason, we predicted OLP-related miRNAs.

This could accelerate research in the field, stimulating researchers to test and verify our hypothesis and comparing wet-lab findings with computational predictions.

Then, we mined the miRGen database using the "Targets" option (freely available at http://www.microrna.gr/mirgen). This software relies on a relational database technology and enables a comprehensive research, integrating both the most widely used databases and repositories with the list of experimentally validated miRNAs and the bioinformatics tools for the miRNAs prediction (namely, DIANA-microT, miRanda, miRBase, PicTar, and TargetScanS) [12, and references therein]. We checked the biological significance of our obtained miRNAs networks mining the extant literature and using *ad hoc* bioinformatics tools (such as the miR2Disease Base, freely accessible at http://watson.compbio.iupui.edu:8080/miR2Disease/searchDiseasePre.jsp) [12, and references therein].

In order to verify the statistical significance of the enrichment of our miRNA-related list, we randomly generated a list of 11 genes (5 for Class 1 and 6 for Class 2). We used the RSA tool for this purpose (freely accessible at http://rsat.ulb.ac.be/rsat/random-genes_form.cgi), selecting "*Homo sapiens*" as source organism.

Using the above-mentioned tools we generated a network of miRNAs associated to the obtained list of random genes. The two miRNAomes were compared using the statistical test for comparing two proportions/percentages. This computation was done with the commercial software MedCalc (version 15, down-loadable at http://www.medcalc.org/) and using the R environment (freely available at http://www.r-project.org/).

Topological properties of the obtained graphs portraying the OLP-related miRNAomes have been also studied. We investigated the clustering coefficient (a measure of degree to which nodes in a graph tend to cluster together), the network diameter and radius, the network centralization, the number of shortest paths (in percentage), the characteristic path length, the average number of neighbors, and the network density (the proportion of all the possible ties that are actually present). Network density reflects the speed at which signaling information diffuse among the nodes.

All the parameters were studied using Cytoscape software (freely down-loadable at http://sourceforge.net/projects/graph-medusa/) [12, and references therein]. Networks are visualized using Medusa software [12, and references therein].

3 RESULTS

miRNAs that have been already experimentally validated represent the 35.48% (11/31) of the entire list. Class A leader genes-related miRNAs predict the 54.17% (104/192) of the OLP + oral diseases-associated miRNAs validated in the literature. Compared to the related randomly generated miRNAome, the numbers of experimentally validated miRNAs did not statistically differ (9/31 *versus* 11/31), but the numbers of oral disease-associated miRNAs yielded a statistical significance (p-value <0.05).

miRNAs that have been already found to be associated with OLP are the 32.26% (12/31) of the entire list, against 9/31 for the randomly generated miRNAs (not statistically significant). Class B leader genes-related miRNAs predict the 46.03% (58/126) of the OLP + oral diseases-associated miRNAs validated in the literature. Taken together the Class A and Class B leader genes-related miRNAs, they can predict the 48.39% (15/31) of the OLP and the of OLP + oral diseases-associated miRNAs validated in the literature.

In table 1, the topological properties of the OLP-related miRNAome are reported. The two sub-sets related to Class A and Class B Leader Genes are shown in Figure 1.

Topological parameters	**Entire OLP-related miRNAome**	**Class A Leader Genes-related miRNAome**	**Class B Leader Genes-related miRNAome**
Clustering coefficient	0.19	0.96	0.73
Network radius	3	2	2
Network diameter	2	1	1
Network centralization	0.94	0.99	0.99
Shortest paths (%)	99	99	99
Characteristic path length	1.99	1.99	1.99
Average number off neighbours	41.49	5.78	7.70
Network density	0.04	0.01	0.01

Table 1: Comparison of the topological properties among the entire OLP-related miRNAome, and the ones associated to Class A and B leader genes.

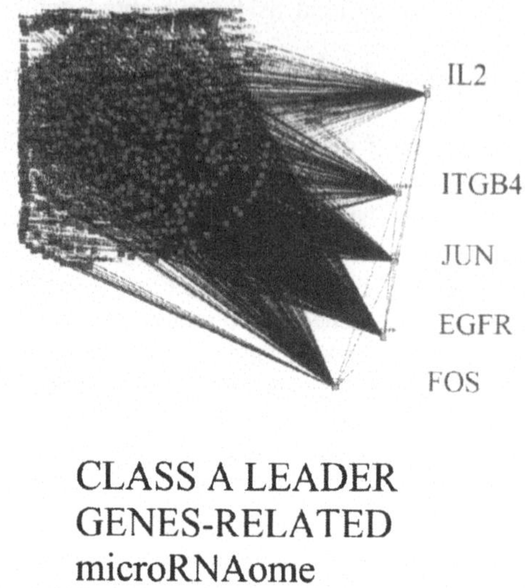

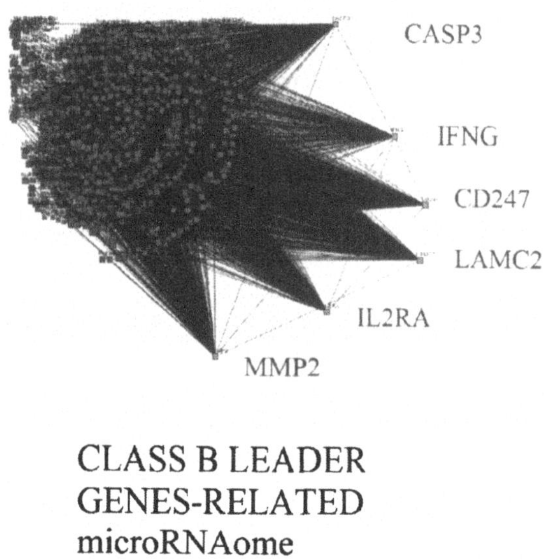

Figure 1: Class A leader genes-related miRNAome (upper), class B leader genes-related miRNAome (below). In red the OLP-related genes, in blue the OLP-related miRNAs.

4 DISCUSSION

Here, in the current contribution, we introduced a novel biomolecular strategy for predicting diseases-related miRNAs, based upon our previous published LGA. We managed to find most of the already established OLP-related miRNAs. Moreover, we found also other miRNAs that have not been directly linked with OLP yet, but have a role in other oral diseases. For the remaining miRNAs their role is still unknown.

This biomolecular strategy can foster further research in the field of the biology of oral diseases, suggesting to the researchers and molecular biologists some targets to focus on and to explore with *ad hoc* targeted experiments. Moreover, it may lead to new bioinformatics software for both commercial and academic applications.

5 CONCLUSIONS

These data could further confirm that an approach based on bioinformatics and data-mining of existing databases could be a starting point to improve our knowledge about cellular processes and molecular mechanism of diseases and to plan targeted experimentation. In particular, the detailed analysis of gene interaction maps and the ranking of genes according to their number and confidence of interactions might have great value in the identification of new targets for a focused experimental analysis, which may confirm each hypothesis and suggest potential risk factors and therapy targets. Noteworthy, a proper combination of experimental and theoretical results is necessary to draw a significant picture of a complex phenomenon, such as gene expression in a particular biologic system. In this study, some genes with a potential major role in OLP were identified and are preliminarily divided into three different groups according to their function. Their miRNAs are predicted and studied. These results might suggest targeted DNA or protein micro-arrays such as SNAP micro-arrays, as well as RT-PCR and mass spectrometry (MS) experiments and nanoconductimetry via Quartz Crystal Microbalance with Dissipation factor monitoring (QCM_D) [13-15], focused on significant genes and simpler to be analyzed than mass scale molecular genomics.

There are many relevant technological and industrial implications of our study: from the possibility of creating panels of salivary or serum biomarkers [16], or of using stem cell therapy delivering particular genes for healing oral mucosal lesions [17]. A unique combination of genomics and proteomics approaches can indeed represent a promising innovation for a personalized treatment of OLP and oral cancer [18].

REFERENCES

[1] Anuradha Ch, Reddy BV, Nandan SR, Kumar SR. Oral lichen planus. A review. N Y State Dent J. 2008 Jun-Jul;74(4):66-8.

[2] McCartan BE, Healy CM. The reported prevalence of oral lichen planus: a review and critique. J Oral Pathol Med. 2008 Sep;37(8):447-53.

[3] Sugerman PB, Savage NW. Oral lichen planus: causes, diagnosis and management. Aust Dent J. 2002 Dec;47(4):290-7.

[4] Sugerman PB, Savage NW, Walsh LJ, Zhao ZZ, Zhou XJ, Khan A, Seymour GJ, Bigby M. The pathogenesis of oral lichen planus. Crit Rev Oral Biol Med. 2002;13(4):350-65.

[5] Orlando B, Bragazzi N, Nicolini C. Bioinformatics and systems biology analysis of genes network involved in OLP (Oral Lichen Planus) pathogenesis. Arch Oral Biol. 2013 Jun;58(6):664-73.

[6] Franceschini A, Szklarczyk D, Frankild S, Kuhn M, Simonovic M, Roth A, Lin J, Minguez P, Bork P, von Mering C, Jensen LJ. STRING v9.1: protein-protein interaction networks, with increased coverage and integration. Nucleic Acids Res. 2013 Jan;41(Database issue):D808-15.

[7] Covani U, Marconcini S, Giacomelli L, Sivozhelevov V, Barone A, Nicolini C. Bioinformatic prediction of leader genes in human periodontitis. J Periodontol. 2008 Oct;79(10):1974-83.

[8] Giacomelli L, Oluwadara O, Chiappe G, Barone A, Chiappelli F, Covani U. Relationship between human oral lichen planus and oral squamous cell carcinoma at a genomic level: a datamining study. Bioinformation. 2009 Dec 31;4(6):258-62.

[9] Giacomelli L, Covani U. Bioinformatics and data mining studies in oral genomics and proteomics: new trends and challenges. Open Dent J. 2010 Jul 16;4:67-71.

[10] Bragazzi NL, Nicolini C. A Leader Genes Approach-based Tool for Molecular Genomics: From Gene-ranking to Gene-network Systems Biology and Biotargets Predictions. J Comput Sci Syst Biol. 2013;6:165-176.

[11] Racapé M, Bragazzi N, Sivozhelezov V, Danger R, Pechkova E, Duong Van Huyen JP, Soulillou JP, Brouard S, Nicolini C. SMILE silencing and PMA activation gene networks in HeLa cells: comparison with kidney transplantation gene networks. J Cell Biochem. 2012 Jun;113(6):1820-32.

[12] Bragazzi NL, Pechkova E, Nicolini C. Proteomics and proteogenomics approaches for oral diseases. Adv Protein Chem Struct Biol. 2014;95:125-62.

[13] Nicolini C, Bragazzi NL, Pechkova E. Determination of Protein-Protein Interaction for Cancer Control via Mass Spectrometry and Nanoconductimetry of NAPPA SNAP Arrays: An Overview. NanoWorld J. 2015;1(1): 9-17.

[14] Spera R, Festa F, Bragazzi NL, Pechkova E, LaBaer J, Nicolini C. Conductometric monitoring of protein-protein interactions. J Proteome Res. 2013 Dec 6;12(12):5535-47.

[15] Nicolini C, Adami M, Sartore M, Bragazzi NL, Bavastrello V, Spera R, Pechkova E. Prototypes of newly conceived inorganic and biological sensors for health and environmental applications. Sensors (Basel). 2012 Dec 12;12(12):17112-27.

[16] Totan A, Miricescu D, Parlatescu I, Mohora M, Greabu M. Possible salivary and serum biomarkers for oral lichen planus. Biotech Histochem. 2015 Apr 3:1-7.

[17] Suma GN, Arora MP, Lakhanpal M. Stem cell therapy: A novel treatment approach for oral mucosal lesions. J Pharm Bioallied Sci. 2015 Jan-Mar;7(1):2-8.

[18] Nicolini C, Bragazzi N, Pechkova E. Nanoproteomics enabling personalized nanomedicine. Adv Drug Deliv Rev. 2012 Oct;64(13):1522-31.

Nanoparticle-based sensing voltage change across a lipid bilayer

YungChia Chen*†, Thomas J. O'Shaughnessy#, Alexander L. Efros#, Alan Huston&, Kimihiro Susumu&, Michael H. Stewart, Igor L. Medintz' and James. B. Delehanty*

* Center for Bio/Molecular Science and Engineering, Code 6900,
#Materials Science Division, Code 6300
&Optical Sciences Division, Code 5600
*† American Society for Engineering Education Washington, D.C. 20036, U.S. Naval Research Laboratory, 4555 Overlook Ave, S.W. Washington, DC 20375

ABSTRACT

Developing nanoscale probes capable of real-time sensing and imaging of neuronal action potentials would be useful in understanding neuronal communication. However, currently available tools for imaging action potentials lack either the spatial or temporal sensitivity to understand neuronal networks with single cell resolution. Two methods were used here to determine how voltage-induced changes in nanoparticles' emission spectra could be employed to image membrane action potentials. First. voltage was applied to a polymeric matrix loaded with quantum dots (QDs) was used to assess luminescence responses to voltage. Second, QDs were loaded into cultured cells coupled with the controlled induction of action potentials.

Keywords: quantum dots, voltage sensing, action potential

1. INTRODUCTION

Understanding neuronal network activity with single cell resolution is critical to understanding how the brain works. Taking advantage of the inherent photophysical and electronic properties of various nanoparticle (NP) materials (e.g., semiconductor quantum dots (QDs), gold NPs) and NP-peptide hybrid constructs that allow for specific localization to cell plasma membrane [1]–[4], voltage changes due to action potentials can be directly detected. QDs can be engineered such that their spectral properties can be discretely tied to changes in a surrounding electric field and do not require genetic manipulation of the cell for function. This latter attribute makes them attractive alternatives to fluorescent methods such as genetically-encoded calcium indicators (GECI) which uses calcium signaling as a proxy for action potentials [5].

Neurons communicate through action potentials – the modulation of the electrical potential across the 5-7 nm expanse of the neuronal cell plasma membrane. Action potentials that traverse down the axon result in neurotransmitter release at the synapse which results in activation of nearby neurons resulting in action potentials in these cells.

Much is know about how an action potential occurs. But to understand whole brain activity and behavior it is necessary to understand neuronal network behavior – the activity of hundreds to thousands of neurons at single cell resolution. This goal is currently hindered by several critical technological limitations. There are a number of methods available to understand brain functionality from the synapse level to the whole brain level, but the limitations of each make it difficult draw concrete conclusions.

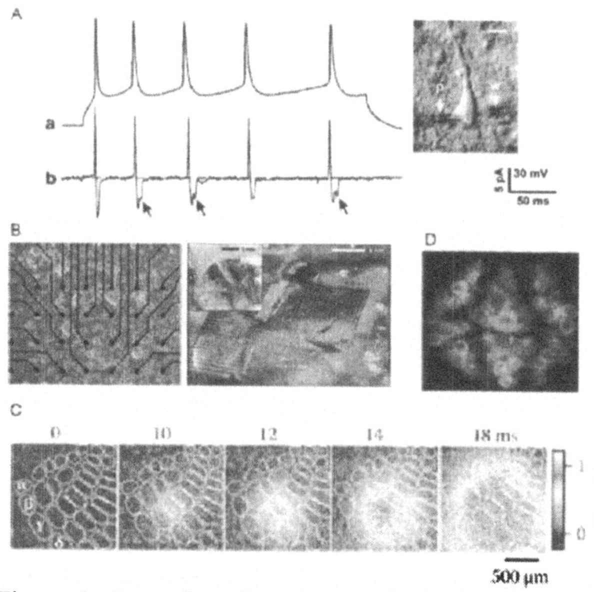

Figure 1. Examples of current methods for detecting neuronal activity. (A) Patch clamp detects single cells activity (image from [6]). (B) Typical microelectrode array for an *in vitro* preparation is along with a new generation of flexible electrodes for *in vivo* implantation (image from [7]). (C) Example of neuronal activity in the barrel cortex of an anesthetized mouse as detected with commercial VSD (image from [8]). (D) Labeling with a new type of VSD that has better spatial labeling *in vitro* (image from [9]).

Currently neuronal electrical activity can be measured through electrophysiology, voltage sensing dyes (VSD) or electrode arrays. Electrophysiology, especially patch clamp can fully delineate the electrical activity of a single neuron but is hampered by low throughput, high degree of difficulty, and invasiveness. Furthermore, if conducting these experiments *in vivo*, animals need to be anesthetized, which has shown to alter brain activity [10].

VSDs can measure the activity of whole neuronal populations but currently suffer from spatial resolution and cause significant toxicity. Other optical imaging techniques typically use calcium indicators in lieu of reading electrical activity but the dynamics of calcium imaging is much slower than that of action potentials [11]. Furthermore, disease states may cause uncoupling of second messenger systems from action potentials.

Electrode arrays combat the low throughput of electrophysiology by employing electrodes that can measure the fast spiking dynamics of neurons but the nature of the electrode means that one cannot differentiate the origin of the signal to distinguish between type of neuron and thus lack single cell resolution [11].

2. MATERIALS AND METHODS

2.1 QD synthesis

CdSe–ZnS core–shell QDs with emission maxima centered at 530, or 550 nm were synthesized and made with the hydrophobic ligands in Table 1.

Table 1. Hydrophobic ligand and QD combinations

QD	Ligand
550nm	Hexylphosphonic acid (HPA), Trioctyl phosphine oxide (TOPO), Trioctyl phosphine (TOP)
530nm	zwitterionic d-penicillamine (DPA), Trioctyl phosphine oxide (TOPO), Trioctyl phosphine (TOP), Oleylamine
530nm	Hexane Thiol

2.2 Cell culture

Cell culture was performed using PC12 cells maintained in RPMI-1640 (ATCC) supplemented with 10% HS (ATCC), 5% FBS (ATCC) and 1% (v/v) antibiotic/antimycotic (Sigma). Cultures were grown in T25 flasks in 5%CO2 at 37C and passaged every seven days. PC12 cells were plated on collagen coated MatTek dishes with 2/3 media replacement every 2-3 days.

3. RESULTS

3.1 QD fluorescence to changes in voltage

QDs have been theorized and shown experimentally to be sensitive to changes in voltage/electric field strengths [12], [13]. Preliminary data collected in our laboratory show that QDs' emission intensity decreases in a voltage-dependent fashion (Figure 2). The higher the voltage the lower the number of photons emitted per second with an almost linear relationship. Based on signal detection theory calculations, QDs have been shown to be capable of detecting neuronal membrane voltage dynamics with millisecond precision (Figure 3) and with greater sensitivity than with traditional methods such as VSDs and genetically encoded calcium indicators [13]. The QDs display reproducible photoluminescence changes with respect to electrical fields that do not suffer from hysteresis effects.

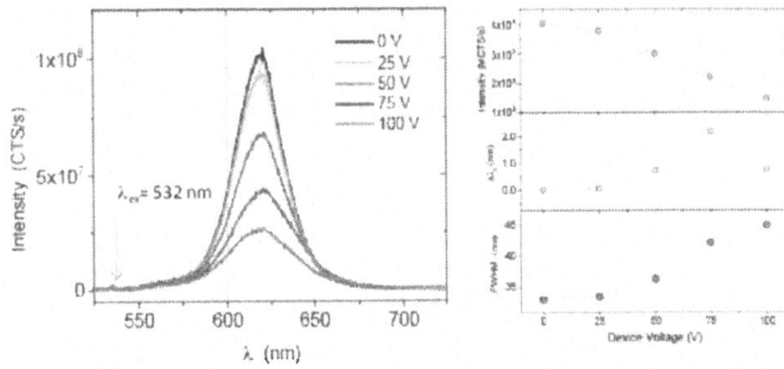

Figure 2. Preliminary data showing the voltage-dependent fluorescence quenching of quantum dots as measured in photon counts/second. The quantum dots yielded specific intensities that decreased linearly with increasing voltage. The full width half maximum (FWHM) also changed with voltage.

4. CONCLUSION

Potentiometric nanoparticles can be a useful tool in the field of neuroscience for measuring neuronal activity. Their fast response time, small size, and resistance to photobleaching make them ideal for delivery to and measuring voltage changes across a plasma membrane (Figure 3).

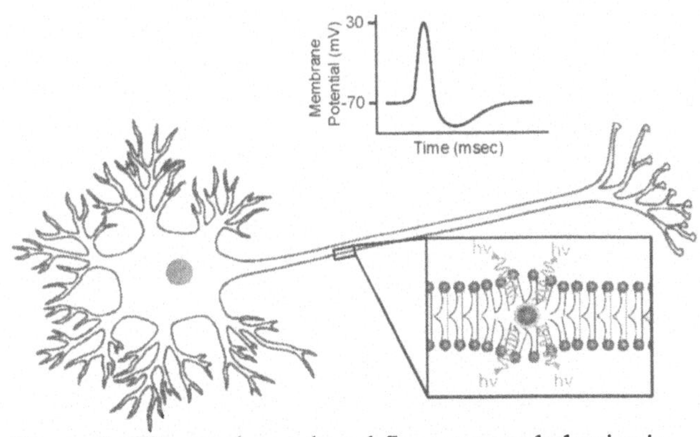

Figure 3. QDs can have altered fluorescence behavior in the presence of an electric field and this effect can be used to measure the change in electric field in the plasma membrane during an action potential.

REFERENCES

[1] J. B. Delehanty, J. B. Blanco-Canosa, C. E. Bradburne, K. Susumu, M. H. Stewart, D. E. Prasuhn, P. E. Dawson, and I. L. Medintz, "Site-specific cellular delivery of quantum dots with chemoselectively-assembled modular peptides,"

Chem. Commun., vol. 49, no. 72, pp. 7878–7880, Aug. 2013.

[2] G. Gopalakrishnan, C. Danelon, P. Izewska, M. Prummer, P.-Y. Bolinger, I. Geissbühler, D. Demurtas, J. Dubochet, and H. Vogel, "Multifunctional Lipid/Quantum Dot Hybrid Nanocontainers for Controlled Targeting of Live Cells," *Angew. Chem. Int. Ed.*, vol. 45, no. 33, pp. 5478–5483, Aug. 2006.

[3] J. B. Delehanty, K. Boeneman, C. E. Bradburne, K. Robertson, J. E. Bongard, and I. L. Medintz, "Peptides for specific intracellular delivery and targeting of nanoparticles: implications for developing nanoparticle-mediated drug delivery," *Ther. Deliv.*, vol. 1, no. 3, pp. 411–433, Sep. 2010.

[4] J. B. Delehanty, I. L. Medintz, T. Pons, F. M. Brunel, P. E. Dawson, and H. Mattoussi, "Self-Assembled Quantum Dot–Peptide Bioconjugates for Selective Intracellular Delivery," *Bioconjug. Chem.*, vol. 17, no. 4, pp. 920–927, Jul. 2006.

[5] C. Grienberger and A. Konnerth, "Imaging Calcium in Neurons," *Neuron*, vol. 73, no. 5, pp. 862–885, Mar. 2012.

[6] J. Kang, J. R. Huguenard, and D. A. Prince, "Voltage-gated potassium channels activated during action potentials in layer V neocortical pyramidal neurons," *J. Neurophysiol.*, vol. 83, no. 1, pp. 70–80, 2000.

[7] J. Viventi, D.-H. Kim, L. Vigeland, E. S. Frechette, J. A. Blanco, Y.-S. Kim, A. E. Avrin, V. R. Tiruvadi, S.-W. Hwang, A. C. Vanleer, D. F. Wulsin, K. Davis, C. E. Gelber, L. Palmer, J. Van der Spiegel, J. Wu, J. Xiao, Y. Huang, D. Contreras, J. A. Rogers, and B. Litt, "Flexible, foldable, actively multiplexed, high-density electrode array for mapping brain activity in vivo," *Nat. Neurosci.*, vol. 14, no. 12, pp. 1599–1605, Dec. 2011.

[8] I. Ferezou, F. Matyas, and C. C. H. Petersen, "Imaging the Brain in Action: Real-Time Voltage-Sensitive Dye Imaging of Sensorimotor Cortex of Awake Behaving Mice," in *In vivo optical imaging of brain function.*, 2nd ed., 2009.

[9] E. W. Miller, J. Y. Lin, E. P. Frady, P. A. Steinbach, W. B. Kristan, and R. Y. Tsien, "Optically monitoring voltage in neurons by photo-induced electron transfer through molecular wires," *Proc. Natl. Acad. Sci.*, vol. 109, no. 6, pp. 2114–2119, Feb. 2012.

[10] D. S. Greenberg, A. R. Houweling, and J. N. D. Kerr, "Population imaging of ongoing neuronal activity in the visual cortex of awake rats," *Nat. Neurosci.*, vol. 11, no. 7, pp. 749–751, Jul. 2008.

[11] A. P. Alivisatos, A. M. Andrews, E. S. Boyden, M. Chun, G. M. Church, K. Deisseroth, J. P. Donoghue, S. E. Fraser, J. Lippincott-Schwartz, L. L. Looger, S. Masmanidis, P. L. McEuen, A. V. Nurmikko, H. Park, D. S. Peterka, C. Reid, M. L. Roukes, A. Scherer, M. Schnitzer, T. J. Sejnowski, K. L. Shepard, D. Tsao, G. Turrigiano, P. S. Weiss, C. Xu, R. Yuste, and X. Zhuang, "Nanotools for Neuroscience and Brain Activity Mapping," *ACS Nano*, vol. 7, no. 3, pp. 1850–1866, 2013.

[12] J. D. Marshall and M. J. Schnitzer, "Optical Strategies for Sensing Neuronal Voltage Using Quantum Dots and Other Semiconductor Nanocrystals," *ACS Nano*, vol. 7, no. 5, pp. 4601–4609, 2013.

[13] K. Park, Z. Deutsch, J. J. Li, D. Oron, and S. Weiss, "Single Molecule Quantum-Confined Stark Effect Measurements of Semiconductor Nanoparticles at Room Temperature," *ACS Nano*, vol. 6, no. 11, pp. 10013–10023, Nov. 2012.

Cellular Delivery Modalities for Semiconductor Quantum Dots: A Comparative Analysis

Lauren D. Field*†, James B. Delehanty*, and Igor L. Medintz*

* Center for Bio/Molecular Science and Engineering, Code 6900,
U.S. Naval Research Laboratory, 4555 Overlook Ave, S.W. Washington, DC 20375
†Department of Bioengineering, University of Maryland, College Park, MD 20741

ABSTRACT

Semiconductor nanocrystals or quantum dots (QDs) have been demonstrated to be superior alternatives to traditional fluorophores (dyes and fluorescent proteins) due to their size-tunable photoluminescence, nanoscale size which produces a high surface area to volume ratio, broad excitation profiles coupled with narrow emission spectra and resistance to photobleaching. The development of delivery modalities to introduce these nanoscale probes to discrete cellular locations with control continues to be an active area of research. These methods can range from passive uptake to facilitated delivery (using bioactive ligands such as peptides) to active manipulation of the cell (microinjection). Each of these methods comes with inherent advantages and liabilities. Here, we have conducted a comparative study utilizing these various cellular delivery strategies to illustrate how a particular labeling/imaging application can be enabled and/or enhanced by choosing the optimal delivery strategy.

Keywords: quantum dots, microinjection, cell penetrating peptides

1. INTRODUCTION

Intracellular delivery of theranostic nanoparticles is a critical area of drug research, the study of which can increase understanding of intracellular process and can increase the efficacy and specificity of designed therapeutics. Current methods of examining these processes include appending fluorescent molecules onto the desired proteins for visualization. The effectiveness of these imaging moieties is hampered by the inherent optical properties of the fluorescent dyes used. Mainly, many fluorescent dyes have a high degree of photobleaching and long maturation times, which limits the time-scales that intracellular processes can be visualized over effectively [1]. An alternative method of labeling utilizes semiconductor nanocrystals, or quantum dots (QDs) which have unique photophysical properties. These include high quantum yields (QYs), broad excitation profiles and narrow emission spectra at wavelengths correlated to the size of the synthesized QDs [2]. Additionally, QDs are resistant to chemical degradation and due to the nanometer size, provide a large surface area-to-volume ratio that is ideal for attachment and display of localization peptides or drug cargo. Due to these ideal properties, use of QDs as delivery modalities is a growing field, with many diverse opportunities for study. Delivery can vary between modification with biological molecules that allow for cellular uptake or active administration via microinjection for more specified delivery.

1.1 Cellular-Penetrating Peptides

Cellular-penetrating peptides are a burgeoning field that have allowed for internalization and specified subcellular delivery of nanoparticles. The ease of synthesis, small size and vast variation in sequence makes CPPs ideal for appending to carriers [3]. QDs allow for visualization of the final location of the cargo that is facilitated by the designed CPP. One of the current issues with CPPs is endosomal sequestration, which occurs due to the internalization methods inherent in cellular function [4]. Release from the endo-lysosomal pathway is critical for cytosolic delivery, and is a critical area of research to determine new methods of escape. Some peptides have been developed that harness the physiological changes that occur as the particles move from early to late endosomes, mainly the lowering of the vesicular pH from 7 to around 5.5 [5]. A peptide sequence rich in glutamate–alanine–leucine repeats dubbed GALA was designed to mimic viral fusion proteins and has been shown to facilitate endosomal release with various carriers [6]. Future work in this area will be facilitated by attempting various peptide changes and determining the most effective sequences for cytosolic delivery.

1.1 Microinjection

In addition to their abilities as carrier molecules, QDs can be used as FRET donors due to their tunable optical properties and their ability to attach to target molecules intracellularly after delivery [2]. We have previously shown that the addition of a His6 tag allows for metal-affinity coordination between tagged proteins and the QD surface and can produce FRET-based sensors (Figure1A) [7]. Addition of a Caspase3 cleavage site between the His6 tag and the mCherry acceptor molecule provided a mechanism for elimination in FRET after protein administration. Due to the FRET potential between 550 nm emitting QDs and mCherry, it can be extrapolated that such a system can be used for intracellular sensors [8]. As mentioned previously, cytosolic delivery of QDs is a difficult process. This can be mediated through the use of microinjection, which allows for specified delivery to desired cells in an individual manner (Figure1B). Using such a localized and precise delivery method can increase understanding of intracellular functions.

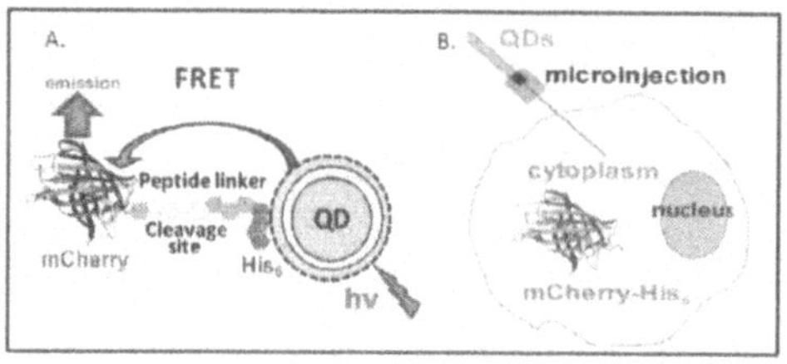

Figure 1A. Design for Caspase 3 specific sensor utilizing QDs as a donor and mCherry as an acceptor. The labeled His6 tag allows for the mCherry to be appended to the QD surface and a Caspase 3 cleavage site is localized to allow for mCherry to be released after Caspase 3 administration, and subsequently reducing FRET efficiency [8]. B. Schematic of microinjection process. Expressed mCherry containing a His6 residue is present intracellularly due to transfection and QDs are injected into the cytoplasm and allowed to complex and produce FRET. Adapted from ref. [7].

2. MATERIALS AND METHODS

2.1 QD synthesis and Peptides

QDs were synthesized as CdSe/ZnS (core/shell) with emission spectra centered around 550nm or 625nm. Various capping ligands were appended by exchanging the native hydrophobic capping shell with the desired ligand as previously described [9]. Peptides were synthesized and characterized as described previously [10]. Sequences used are described in Table 1.

Peptide	Focus	Sequence	Description	Mw / charge pH 7.4
JB577	Parent	WGDap$_{2}$VKIKKP$_{9}$GGH$_{6}$	Parent peptide	3034 / +3
JB729	Fatty acid	WGDap$_{C}$VKIKKP$_{9}$GGH$_{6}$	Palmitoyl → C_8F_{17}	3270 / +3
SI-09160	Fatty acid	WGDap$_{R}$VKIKKP$_{9}$GGH$_{6}$	Palmitoyl → cholesterol	3400 / +3
JB872	Fatty acid	WGC$_{f}$VKIKKP$_{9}$GGH$_{6}$	Palmitoyl → farnesyl	3017 / +3
JB865	Pro$_{x}$-linker	WGDap$_{p}$VKIKKP$_{12}$GGH$_{6}$	Pro$_{9}$ → Pro$_{12}$	3326 / +3
JB869	Charge	WGDap$_{p}$VRLP$_{3}$VRLP$_{3}$VRLP$_{3}$GGH$_{6}$	VKIKKP$_{9}$ → amphipathic-(VRLP$_{3}$)$_{3}$	3543 / +3
JB719	Control	KETWWETWWTEWSQPK$_{k}$RKVSGAAi-bA$_{5}$GGH$_{6}$	Chariot peptide	4297 / +4

Table1. Peptides studied for potential as CPPs. Parent peptide is JB577, with described sequence. All other peptides are modified versions with singular changes as labeled. Control peptide is commercially available Chariot peptide.

2.2 Cell Culture and QD delivery

Cell culture was performed using COS-1 cells maintained in DMEM (ATCC) supplemented with 10% FBS (ATCC) and 1% (v/v) antibiotic/antimycotic (Sigma). Cultures were grown in T25 flasks in 5%CO_2 at 37° C and passaged at 80% confluency. To perform QD deliveries, QD-peptide bioconjugates were produced by incubating the desired quantity of peptide with QDs in HEPES/DMEM and allowed to complex. Deliveries were performed on cells plated onto LabTek 8-well chambered cover glass (Nalge Nunc). Detailed description of delivery conditions is described here [10].

2.3 Microinjection and Imaging

COS-1 cells were plated and transfected using pmCherry-His6 N1 as previously described [7]. Injection was performed using an Eppendorf FemtoJet® Microinjector controlled by an InjectMan® NI 2 micromanipulator and 565nm ITK carboxyl/Ni2+ QDs. Images were taken with an Olympus IX-71 total internal reflection fluorescence microscope equipped with a 60X oil immersion lens, excitation was performed using a Xe lamp, and specific excitation and emission filters were used for all conditions.

3. RESULTS

3.1 Cellular-Penetrating Peptides

Efficient cellular uptake and endosomal release was described previously in relation to a peptide dubbed JB577. To determine the effect of certain moieties within the peptide sequence, variations on this peptide were synthesized and are described in Table 1. Of particular interest was the palmitoyl group in the parent peptide, and other changes were made within the sequence such as a lengthening of the proline chain. To determine the effectiveness against a commercially available control (Chariot), cells were administered QDs appended with the synthesized peptides and localization was viewed (Figure 2).

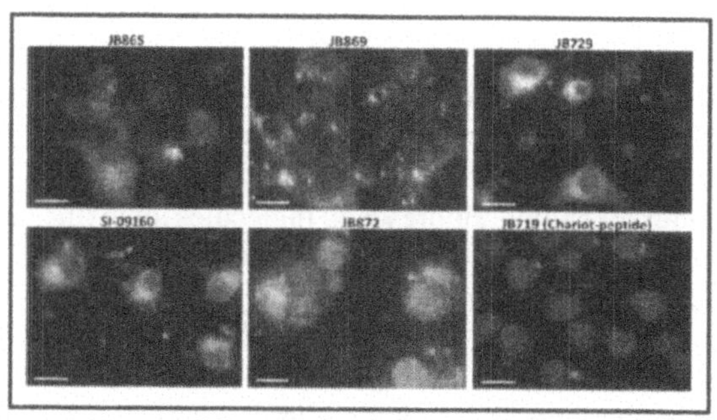

Figure 2. Modified peptides conjugated with 550 emitting QDs (Green) were visualized with DAPI as a co-stain (blue). Administration of the QD-peptide bioconjugates was performed at a QD concentration of 100nM with 75 peptides per QD. Cells were incubated with the constructs for 2 hours, after which the cells were washed and allowed to culture for 48 hours. Chariot peptide was used as a control. Image adapted from ref. [10].

All peptide variants increased the intracellular distribution of the QD-peptide bioconjugates over JB577 (not shown). Additionally, the synthesized peptide sequences are markedly better than the Chariot peptide at endosomal release, as can be seen by the disperse QD presence in all images except the bottom right, which shows punctate sequestration of the internalized QDs.

3.2 Microinjection

Cells transfected with mCherry containing a His6 segment were injected with 565 nm emitting ITK-COOH QDs. The cells were then imaged for both mCherry excitation and emission and FRET between the QD and the mCherry complex over a period of time to determine the effects of photobleaching (Figure 3). Addition of the QD allows for mCherry visualization for longer periods of time, reducing the photobleaching and producing efficient FRET over 60 seconds.

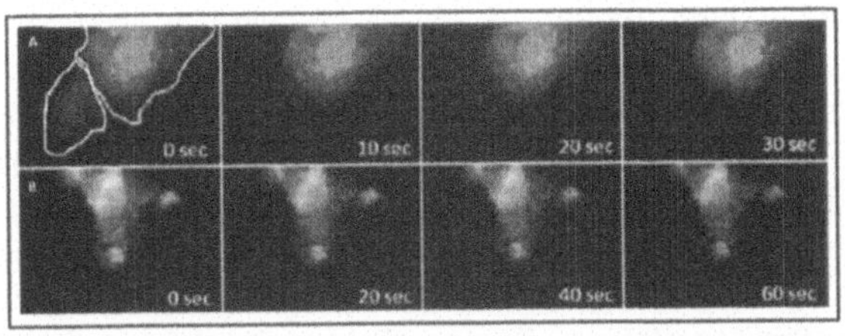

Figure 3. COS-1 cells transfected to express mCherry-His_6 (Red) and FRET between QDs and mCherry (Green). Images were taken at labeled time points during continuous excitation. Note the photobleaching of mCherry in an uninjected cell (white outline) vs. the photobleaching in an injected cell (yellow outline) [7].

4. CONCLUSION

Semiconductor QDs are a viable and effective method of visualizing administration methods, both through passive administration and internalization through conjugation with peptides and active administration using microinjection. Their photophysical properties allow for easy determination of intracellular localization after internalization and for FRET sensors to be designed and implemented effectively.

REFERENCES

1. Resch-Genger, U., et al., *Quantum dots versus organic dyes as fluorescent labels.* Nat Methods, 2008. **5**(9): p. 763-75.
2. Medintz, I.L. and H. Mattoussi, *Quantum dot-based resonance energy transfer and its growing application in biology.* Physical Chemistry Chemical Physics, 2009. **11**(1): p. 17-45.
3. Sapsford, K.E., et al., *Functionalizing Nanoparticles with Biological Molecules: Developing Chemistries that Facilitate Nanotechnology.* Chemical Reviews, 2013. **113**(3): p. 1904-2074.
4. Duncan, R. and S.C. Richardson, *Endocytosis and intracellular trafficking as gateways for nanomedicine delivery: opportunities and challenges.* Mol Pharm, 2012. **9**(9): p. 2380-402.
5. Licht, S.S., et al., *A rugged energy landscape mechanism for trapping of transmembrane receptors during endocytosis.* Biochemistry, 2003. **42**(10): p. 2916-2925.
6. Futaki, S., et al., *Unique features of a pH-sensitive fusogenic peptide that improves the transfection efficiency of cationic liposomes.* J Gene Med, 2005. **7**(11): p. 1450-8.
7. Boeneman, K., et al., *Intracellular Bioconjugation of Targeted Proteins with Semiconductor Quantum Dots.* Journal of the American Chemical Society, 2010. **132**(17): p. 5975-+.
8. Boeneman, K., et al., *Sensing Caspase 3 Activity with Quantum Dot-Fluorescent Protein Assemblies.* Journal of the American Chemical Society, 2009. **131**(11): p. 3828-+.
9. Susumu, K., et al., *Multifunctional compact zwitterionic ligands for preparing robust biocompatible semiconductor quantum dots and gold nanoparticles.* Journal of the American Chemical Society, 2011. **133**(24): p. 9480-96.
10. Boeneman, K., et al., *Selecting improved peptidyl motifs for cytosolic delivery of disparate protein and nanoparticle materials.* Acs Nano, 2013. **7**(5): p. 3778-96.

SU-8 Based Flexure-FET Biosensor to Achieve Ultra sensitive Response

Durgesh Chaurasiya[1], Srinivasan B.[2], Siva Vanjari[3] and Shiv Govind Singh[4]
[1,2,3,4] Department of Electrical Engineering, Indian Institute of Technology, Hyderabad, India 502205
E-mail: [1]ee12m1013@iith.ac.in, [2]ee12m1036@iith.ac.in, [3]svanjari@iith.ac.in, [4]sgsingh@iith.ac.in

ABSTRACT

This work presents a SU-8 based Flexure-FET biosensor to achieve ultra-sensitive response. Here, gate of the FET is replaced with the fixed-fixed beam. Beam is biased near to pull-in instability (V_{pi}) and FET channel is biased in sub-threshold regime (V_{th}) to take advantage of non-linear response of beam and FET respectively. We simulated the beam using polySi, gold and SU-8 as beam material in real environment using Coventorware to compare their Vpi. After selecting SU-8 as beam material, we optimized the dimensions of beam to achieve low V_{pi} using Coventorware and then modeled the FET whose $V_{th} \approx V_{pi}$. We simulated the electromechanical coupling using COMSOL Multiphysics. The non-linear deflection in beam due to biomolecules causes the change in potential of channel which results upto 4th order change in I_d of FET i.e. I_d at equilibrium is 1.1 pA and due to biomolecules, I_d becomes 17.3 nA. This result can be used for ultrasensitive detection of malignant molecules.

Keywords: Flexure-FET, Pull-in instatbility, Young's modulous, Label free detection, Cantilever , Biosensors

1 INTRODUCTION

Recent advancement in nano-electronics have resulted in a new class of exciting devices such as Flexure-Fet [1] (flexure sensitive field effect transistors). As Moore's law is expected to see an end, heterogeneous integration would be future of Microelectronics. Integration of MEMS, Nanosensors etc. with traditional MOSFET help in realizating Biochips and Lab-on-a-chip devices. Flexure-FET is an excellent example of heterogeneously integrated device. It has been theoritically proven to be more sensitive biosensor [1] than any other types of biosensor. Flexure-FET is potential candidate for detecting bio-molecules. It is regarded as ultrasensitive, label free detection of biomolecules.

Sensitivity of classical biosensors such as electrical [2] and mechanical [3] biosensors suffer from fundamental limitations. Sensitivity of electrical biosensors may be severely affected if the biomolecules to be detected are charge neutral. In addition, it also suffers from electrostatic screening due to the presence of charged molecules in the solution.

On the other hand, sensitivity of mechanical biosensors aren't limited by charged neutral biomolecules or electrostatic screening effect. Rather they are afflicted by the complex scheme of optical setup necessary to detect deflection of beam/cantilever due to capturing of the biomolecules. When a biomolecule attaches with the functionalized beam, it changes the mechanical properties of the beam by modulating the mass, stiffness and/or surface stress of the beam. This change in the mechanical properties is detected by change in resonance frequency (dynamic mode operation of cantilever/beam) or change in deflection of beam or change in resistance of peizoelectric material(static mode) [4]-[5]. Moreover, its response is linear [6] or logarathmic [7]-[9] with change in mass and surface stress of beam/cantilever.

So we need a methodology which can combine advantages of nanomechanical and electrical biosensors to give ultrasensitive response. Flexure-FET is one which combines advantages of both technolgies and doesn't suffer from their limitations. The sensing involves the static nano-mechanical response of the suspended gate or Flex-gate due to the adsorption of biomolecules. Nano-mechanical response can be generated either by neutral or charged biomolecules. The magnitude of nano-mechanical response could be attributed to the concentration of the target biomolecules in an analyte. The present day challenge is to detect even the lowest concentration of specific types of biomolecules which are usually so less in the early stages of diseases like Cancer. Using Flexure-FET even the smallest of the nano-mechanical response can be transduced to a significant change in an electrical signal and hence the lowest concentration of the target bio-molecules.

2 THEORY

Flexure-FET utilizes the advantages of classical electrical and mechanical biosensors to give ultrasensitive response. Here, gate of FET is replaced with the fixed-fixed beam. Beam is biased near to pull-in instability (V_{pi}) and FET channel is biased in sub threshold regime (V_{th}). This is done to operate both devices (FET and beam) in their non-linear regime simultaneously so that sensitivity is maximum. Flexure-FET is biased in sub-threshold regime below pull-in (i.e., $V_{th} \approx V_{pi}$). It is so because displacement of the fixed-fixed beam will be non-linear near to pull-in instability. In addition to this, drain current of FET is exponentially proportion to gate potential in subthreshold regime i.e. Id $\propto \exp(V_{gs}-V_{th}/ mkT)$ where Vgs is gate to source potential, m= 1 + (C_{dm}/C_{ox}), C_{dm} is depletion capcitance and C_{ox} is gate oxide capacitance. Thus overall effect will be a highly non-linear change in drain current. It is important to keep the pull-in potential of beam low, owing to the fact that low V_{th} FETs can be operated at a higher speed (as V_{pi} of beam and V_{th} of FET should be approximately equal, implying that low V_{pi} will require design of a FET having low V_{th}). Hence it necessitates the design and optimization of beam dimensions to have low stiffness, which in turn decides low value of V_{pi}. One of the key parameters to determine stiffness is the Young's modulus (E) of material. Furthermore a material having the smallest young's modulus exhibits the lowest V_{pi}.

2.1 Selecting beam material

As Young's modulous of material plays significant role in determining stiffness , which in turn decides V_{pi}. So first challenge is to select the material for fabricating beam. We chose polySi, gold (Au) and SU-8 as beam material and simulated the beam Fig1(a) in real environment using Coventorware to compare their V_{pi}. Obtained result is shown in Fig1(b) and value of V_{pi} is listed in *Table1*. Simulation results show that the SU-8 based beam exhibits the lowest V_{pi}.

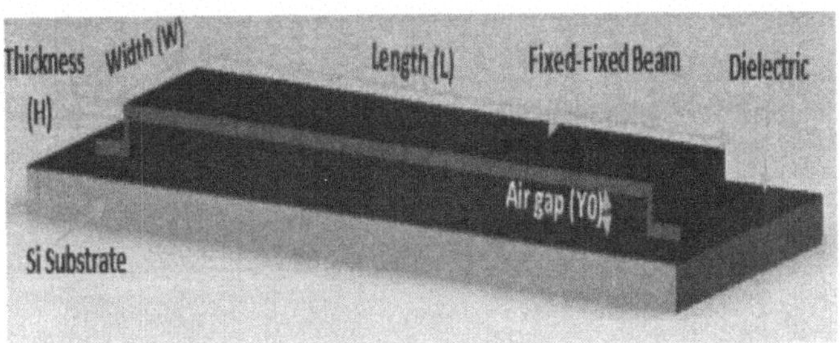

(a)

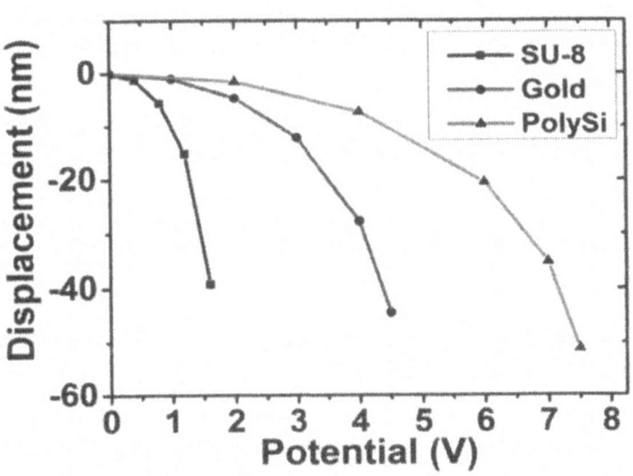

(b)

Fig.1 (a) Beam model for pull-in comparison. Parameters of beam: L= 4 µm, W= 1 µm, H= 40 nm, Y_0= 100 nm and oxide thickness T_{ox}= 10 nm. (b) Displacement of beam (Y) vs. Potential applied (V). Here we have changed the material of beam, and simulated it in Coventorware using polySi, gold and SU-8, one at a time.

Table 1 : Pull-in potential

Material	E [GPa]	V_{pi} [V]
SU-8	2~4.4	1.75
Au	33~54	4.5
PolySi	160	7.5

2.2 Optimization of Beam's dimensions

After selecting beam material (SU-8), we had to optimize the beam dimensions to keep V_{pi} of beam as low as possible. For beam , stiffness is given by k = (αEWH^3 / $12L^3$) where α = 480 for fixed-fixed beam. Fig 2(a) and (b) show the variation of stiffness with respect to beam length by keeping beam thickness (h) and beam width (w) constant respectively. Fig 2 shows that as beam length increases by keeping other parameter constant , stiffness decreases by third power of L and becomes constant after a particular value of L. Fig 2(a) shows stiffness increses linearly with the width (W) of beam whereas in Fig2(b), stiffness increses by third power of beam thickness (H).

In Flexure-FET, length and width of beam will be along the width and chanel length of MOSFET respectively. Wider MOSFET lead to strong current but it increases the overall gate capcitance. This can lead to slow response of the device so we can't keep very long length of beam. From Fig 2(a), change in stiffness becomes negligible after L = 10 µm. So we chose beam length between 6 to 10 µm. We also observe that higher value of beam width increases the stiffness. As beam's width lies across the channel length of MOSFET , so lower value of width is desirable as It will lead to stronger current in FET. So we chose beam's width w = 1µm.

Fig2(b) shows that as film thickness of beam increases its stiffness increases too. So we have to choose lower value of beam thickness. We choose beam thickness between 250-500 nm. Thus, we chose final values of beam dimensions which is listed in *Table 2*. Air gap (Y_0) also plays a vital role to decide the pull-in potential of beam. As SU-8 is an insulator so we used very thin layer of gold (thickness of gold = 10 nm) as electrode for beam contact. We simulated this fixed-fixed beam (Fig1a). During simulation we observed that higher the value of air gap , larger the V_{pi}. For dimensions , listed in *Table 2*, we obtained V_{pi} = 6.2 V.

Table 2: Optimized beam dimensions

Parameter	Value	Description
L	8 [µm]	Beam length
W	1 [µm]	Beam width
H	250 [nm]	Beam thickness
E_t	10 [nm]	Gold electrode thickness
Y_0	100 [nm]	Air gap
T_{ox}	10 [nm]	Dielectric thickness

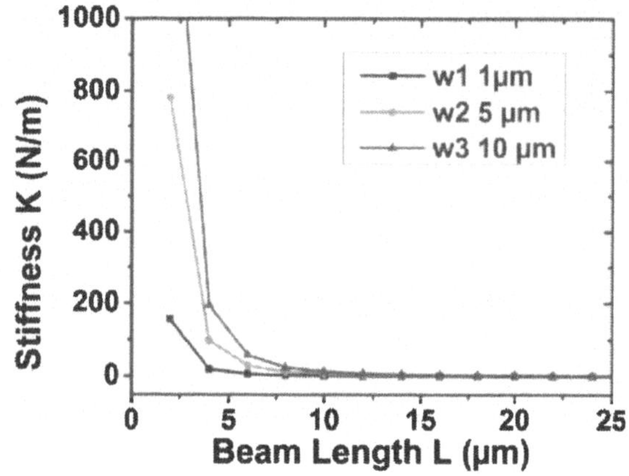

(a)

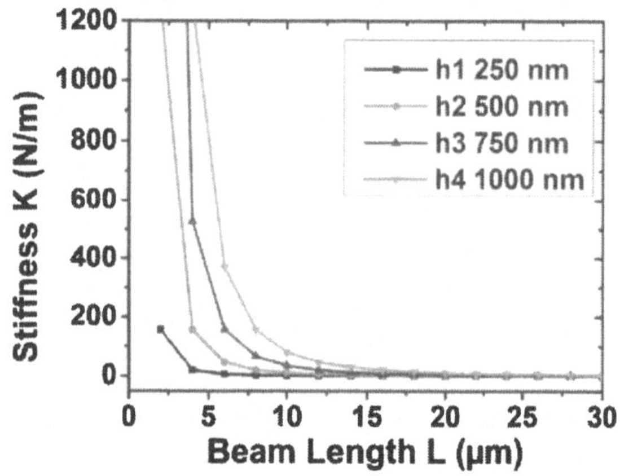

(b)

Fig.2 (a) Stiffness (k) vs. Beam Length (L) : $h = 250$ nm. (b) Stiffness (k) vs. Beam Length (L): $w = 1$ μm. Here we have chosen Young's modulus of SU-8 E=2 GPa.

2.3 Parametrs of FET

As V_{pi} of beam structure is 6.2 V. So our target is to design a FET whose V_{th} is nearly equal to V_{pi}. Different parameters of MOSFET which sets its V_{th} = 6.2 V, are listed in *table 3*. This structure Fig3(a) is simulated in COMSOL multiphysics 4.4 and result in Fig3(b) I_d vs. V_{gs} plot shows that V_{th} of FET is equal to 6.2 V.

Table 3 : MOSFET Parameter

Parameter	Value	Description
N_a	1.1e16[1/cm^3]	acceptor doping concentration
N_d	1e18[1/cm^3]	donor doping concentration of source/drain
T_{ox}	10 [nm]	gate oxide thickness
ε_{sio2}	4.5	dielectric constant of Insulator
φ_m	5.1 [eV]	metal work function

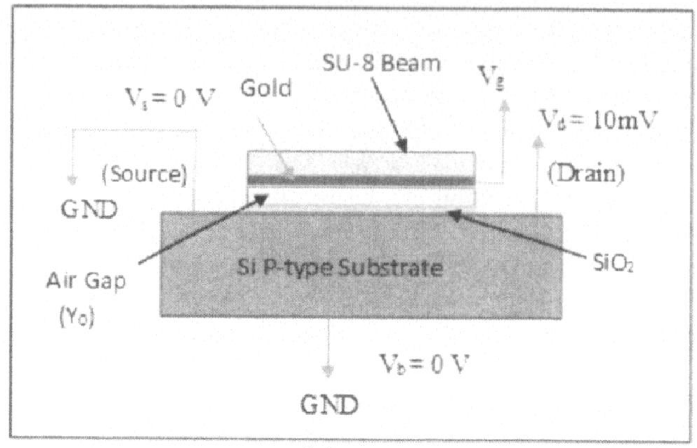

(a)

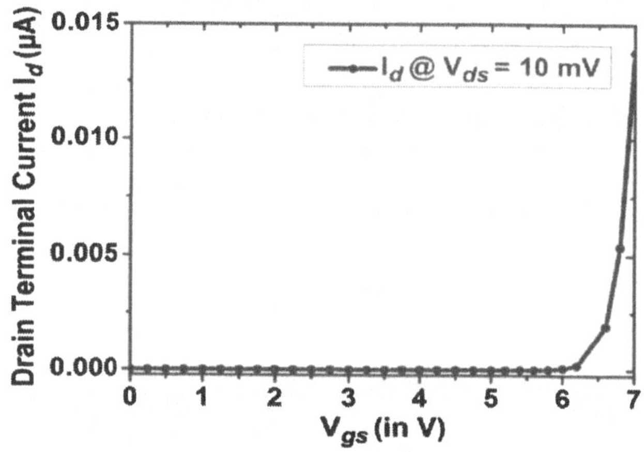

(b)

Fig.3 (a) cross sectional view of Flexure-FET. (b) I_d vs V_{gs} plot of FET.

2.4 Simulation of Electromechanical coupling in COMSOL Muliphysics 4.4

At clinically threshold concentration of biomolecules , the change in free energy density on functionalized beam due to biomolecules were reported in range of 1 to 50 mJ/m^2 [10]-[13]. This change in free energy density/surface stress (σ_s) is defined as integral of the normal stress or bulk stress (σ_m) in monolayer over its thickness:

$$\sigma s = \int_0^{tm} \sigma m \, dz \qquad (1)$$

If we substitute the surface stress 1 to 50 mJ/m2 in equation(1), we get normal stress in monolayer (thickness ~ 20 nm) in range of 50 kPa – 2.5 MPa. This bulk stress value may vary with the different type of material used for beam. We applied these bulk stress values over the top of beam and observed the change in deflection of beam.

Firstly, we simulated Fig.3(a) in COMSOL using electromechanics (emi) physics by applying bulk stress, which causes beam to deform. Then we imported the deformed beam structure into semiconductor physics and calculated the new current after deflection of the beam. This we performed over the range of bulk stress values and result

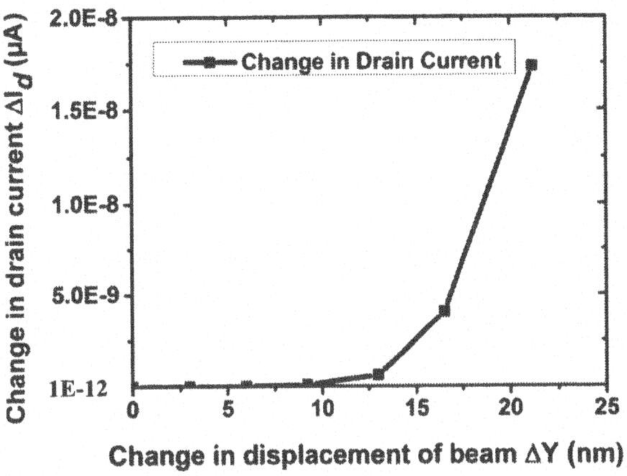

Fig.4 ΔI_d vs ΔY of the optimized beam used as gate of the FET. When SU-8 beam captures biomolecules, its stiffness changes which causes further displacement of beam from its equilibrium position at V_{pi} i.e. $\Delta Y = Y_0 - Y$.

(change in current $\Delta I_d = I_2 - I_1$, where I_2 is drain current after deflectioon of beam due to attachment of biomolecules and I_1 is drain current before attachment of biomolecules) is shown in Fig.4.

3 RESULTS

When a biomolecule attaches with the functionalized beam of SU-8, it modulates the stiffness of the beam causing a non-linear deflection of beam. The non-linear deflection is attributed to the biasing of beam closed to pull-in instability. This non-linear deflection causes the change in potential of channel which results in order of change in drain current of FET (Fig.4). Fig.4 shows upto 4th order change in drain current I_d.

4 CONCLUSIONS

Here, we demonstrated the Flexure-FET response using COMSOL Multiphysics 4.4. We observed that Flexure-FET can have a great outlook in biomedical field due to subtle change in drain current I_d of FET. We used gold as electrode for gate contact because SU-8 is insulator. If we can make SU-8 conductive then we can perform the same operation at much lower potential.

REFERENCES

[1] Ankit Jain, Pradeep R. Nair, and Mohammad Alam , "Flexure-FET biosensor to break the fundamental sensitivity limits of nanobiosensors using nonlinear electromechanical coupling", PNAS 2012 109 (24) 9304-9308

[2] P. Bergveld, "Thirty years of ISFETOLOGY: What happened in the past 30 years and what may happen in the next 30 years," Sensors and Actuators B: Chemical, Volume 88, Issue 1, 1 January 2003, Pages 1-20, ISSN 0925-4005

[3] Arlett JL, Myers EB, Roukes ML, "Comparative advantages of mechanical biosensors" Nat Nanotechnol. 2011 Apr;6(4):203-15.

[4] Wee K, et al. (2005), "Novel electrical detection of label-free disease marker proteins using peizoresisti -ve self-sensing micro-cantilevers" Biosens Bioelectron 20:1932–1938.

[5] Boisen A, Thundat T (2009), "Design & fabrication of cantilever array biosensors" Material Today 12:32 -38.

[6] Ekinci K,Roukes M (2005),"Nanoelectromechanical systems. Rev Sci Instrum" 76 (061101):1 -12.

[7] Wee K, et al. (2005), "Novel electrical detection of label-free disease marker proteins using piezoresisti-ve self-sensing micro-cantilevers" Biosens Bioelectr -on 20:1932–1938.

[8] Wu G, et al. (2001), "Bioassay of prostate-specific antigen (PSA) using microcantilevers" , Nat Biotech -nol 19:856–860.

[9] Zhang J, et al. (2006), "Rapid and label-free nano-mechanical detection of biomarker transcripts in human RNA", Nat Nanotechnol 1:214–220.

[10] J. Fritz, M. K. Baller, H. P. Lang, H. Rothuizen, P. Vettiger, E. Meyer, H. -J. Güntherodt, C. Gerber, and J. K. Gimzewski, "Translating biomolecular recognition into nanomechanics," Science, vol. 228, pp. 316–318, 2000.

[11] G. Wu, H. Ji, K. Hansen, T. Thundat, R. Datar, R. Cote, M. F. Hagan, A. K. Chakraborty, and A. Majumdar, "Origin of nanomechanical cantilever motion generated from biomolecular interactions," in Proc. Nat. Acad. Sci., vol. 98, 2001a, pp. 1560–1564.

[12] G. Wu, R. Datar, K. Hansen, T. Thundat, R. Cote, and A. Majumdar, "Bioassay of prostate specific antigen (PSA) using microcantilevers," Nature Biotechnol., vol. 19, pp. 856–860, 2001b.

[13] K. Hansen, H. Ji, G. Wu, R. Datar, R. Cote, and A. Majumdar, "Cantilever-based optical deflection assay for discrimination of DNA singlenucleotide mismatches," Analyt. Chem., vol. 73, pp. 1567–1571, 2001.

Improving the Sensitivity of Pedot:Pss Modified Gold Electrode Using Gold and Silver Nanoparticles for *Ganoderma boninense* DNA Detection

S. W. Dutse*, N. A. Yusof** and H. Ahmad***

* Department of Science Laboratory Technology, HussainiAdamu Federal Polytechnic, Kazaure, Nigeria, swdutse@yahoo.com

** Department of Chemistry, Faculty of Science, Universiti Putra Malaysia, 43400 UPM Serdang, Selangor, Malaysia, azah1973@gmail.com

*** Institute of Advanced Technology, Universiti Putra Malaysia, 43400 UPM Serdang, Selangor, Malaysia, haslina_ahmad@upm.edu.my

ABSTRACT

The sensitivity of a designed electrochemical DNA biosensor was improved using gold and or silver nanoparticles.. A gold electrode modified with a conductive nanocomposite of poly(3,4-ethylenedioxythiophen)–poly (styrenesulfonate) (Pedot-Pss) and gold or silver nano particles enhanced the conductivity of the electrode surface area. Bare and modified gold electrode surfaces were characterized using cyclic voltammetry (CV) technique in ethylenediaminetetraacetic acid (TE) supporting electrolyte. Immobilization of a 20-mer DNA probe was achieved by covalent attachment of the amine group of the capture probe to a carboxylic group of an activated3,3'-dithiodipropionic acid layer using EDC/NHSS for. Hybridization. The effect of hybridization temperature and time was optimized and the sensor demonstrated specific detection for the target concentration ranged between 1.0×10^{-15} M to 1.0×10^{-9} M with a detection limit of 9.70×10^{-19} M. Control experiments verified the specificity of the biosensor in the presence of mismatched DNA sequence.

Keywords: biosensor, pedot-Pss, DNA, nanocomposite, ruthenium complex

1 INTRODUCTION

The new ruthenium complex $[Ru(dppz)_2(qtpy)]Cl_2$ has emerged as a novel redox marker for the selective detection of DNA probe (ssDNA) in solution via electrochemical sensor. Interaction of ruthenium complex with DNA has been studied by Photophysical techniques [1]. The ruthenium complex has a high affinity to DNA intercalation due to its interacting properties that include its extended aromatic heterocyclic surface that extrude from its central core [2]. Fundamentally, conducting polymers have emerged as potential candidates for biosensors [3], and in this development Gerard et al., have reviewed the literature on applications of conducting polymers to biosensors [4]. However, this study was designed to investigate the effect of gold and or silver nanoparticles in improving signal enhancement of PEDOT-PSS modified gold electrode in DNA biosensing. The use of nano-material was considered for its advantageous properties, such as hydrophilic, excellent biocompatibility, unique characteristics in the conjugation with biological recognition elements and multiplex capacity for signal transducer [5]. Ideally, nano-materials in DNA biosensors are used as substrates for DNA attachment and as signal amplifiers for hybridization. Consequently, a considerable amount of literature that described electrode modification using nanoparticles and conducting polymers has been published.

2 METHODOLOGY

2.1 Reagents and Solvents

Synthesized new ruthenium complex of $[Ru(dppz)_2$ $(qtpy)Cl_2$; dppz = dipyrido [3,2-a: 2', 3' - c] phenzine; qpty = 2, 2', - 4, 4" . 4', 4"' - quaterpyridyl was obtained from the Department of Chemistry, Faculty of Science, Universiti Putra Malaysia. It was synthesized by Haslina Ahmad according to the literature [2] with slight experimental modification.

2.2 Reagents and Solvents

Stock solution of 25 μM ruthenium complex $[Ru(dppz)_2(qtpy)]Cl_2$ = $[Ru(dppz)]^{2+}$was prepared in a mixture of 50 mM Tris-HCl, 20 mM NaCl (volume 90%) and methanol (volume 10%). A washing buffer and supporting electrolyte of 10mM Tris-HCl containing 1 mM EDTA (TE) (pH 7.15) were prepared. Another supporting electrolyte of a 5 mM Potassium Hexacyano ferrateIII ($K_3Fe(CN)_6$) was prepared in 10mM Tris-HCl containing 1mM EDTA TE (pH 7.15). An activation solution of 5 mM N-hydroxysulfosuccinimide (NHSS) sodium salt in a mixture of 2 mM 1-(3-dimethylaminopropyl)-3-ethylcarbodiimide (EDC) was prepared in deionized water. More solution of 10mM Tris-HCl containing 1mM EDTA (TE) (pH 8.0) was prepared for the dilution of DNA oligonucleotide. Another solution of 1.0 mM $K_3Fe(CN)_6$ was prepared for the characterization of electrode active surface area and a 3 mM solution of 3, 3'-dithiopropionic acid (DPA) was formed in absolute ethanol as a linker between the modified electrode and probe DNA.

Oligomers 20-mers probe DNA modified at the 5' end with an amino (NH_2) group (5'-NH_2-CCT GCT GCG TTC TTCTTC AT-3'), 20-mers complementary target (5'-ATG AAG AAG AAC GCA GCA GG-3') and20-mers single-mismatch (5'-ATG AAG AAG TAC GCA GCA GG-3') sequences were synthesized by First Based Laboratories Sdn Bhd, Selangor, Malaysia.

2.3 Appratus and Electrodes

Voltammetric measurements were obtained using µAutolab (Ecochemie, The Netherlands) Potentiostat incorporated with a General-Purpose Electrochemical System (GPES 4.9, Eco Chemie) software. The electrochemical cell was a three electrodes system of a Metrohm gold electrode as the working electrode, platinum wire as counter electrode, and Ag/AgCl/KCl (3.0M) as a reference electrode.

2.4 Pretreatment of Gold Electrode (AuE)

The bare gold electrode (bare AuE) was polished using alumina slurry and sonicated in deionized water to remove stacked particles. It was then dipped in a concentrated sulphuric acid for 10 min and rinsed with a copious amount of deionized water. The electrode was further sonicated in absolute ethanol for 5 minutes, rinsed with deionized water and then in TE washing buffer respectively. It was subsequently dried under nitrogen gas (N-gas) flow for 10 min and kept for further drying at 25 °C for 45 min.

2.5 Gold Electrode Modification with PEDOT -PSS, PEDOT-PSS/AgNps and PEDOT-PSS/AuNps

The pretreated bare gold electrode (AuE) was drop coated on the gold electrode and exposed to oven dry at 70 °C for 15 h. This was to obtain a cured film of PEDOT-PSS on the gold electrode surface. The film surface was rinsed with deionized water and TE washing buffer to remove the unbound remnants on the film surface of the gold electrode. It was dried gently under N-gas flow for 10 min and further dried for 45 min at 25 °C. The PEDOT-PSS/AuE modified electrode was then obtained.

Some prepared PEDOT-PSS containing AgNPs solution was drop coated on pretreated bare gold electrode and the curring procedure performed as with PEDOT-PSS modification protocol. Then AgNPs/PEDOT-PSS/AuE modified electrode was obtained.

The gold electrode modification with PEDOT-PSS and gold nanoparticles was obtained by immersing PEDOT-PSS modified gold electrode in a colloidal gold nanoparticles solution and allowed for 24 h to achieve a nanocomposite film on the surface of the PEDOT-PSS modified gold electrode. The film surface was rinsed with deionized water and TE washing buffer respectively to remove the unbound gold nanoparticles from the film surface. It was dried gently under N-gas flow for 10 min and further dried for 45 min at 25 °C.

2.6 Electrochemical Characterization for the Active Surface Area of Bare and Modified Gold Electrodes

Cyclic voltammetry (CV) experiments were performed in a solution of 1.0 mM $K_3Fe(CN)_6$ prepared in double distilled water. The CV voltammograms of bare AuE, gold electrode modified with PEDOT-PSS, gold electrode modified with PEDOT-PSS/AuNPs, and gold electrode modified with PEDOT-PSS/AgNPs were carried out at scanned potential between -0.3 V to 0.7 V varying the scan rates using 50, 100, 150, 200,250, 300 and 350 mV/s. The sensitivity effect of each modification was observed based on the peak current responses from the different scan rate.

2.7 Immobilization of Probe DNA on PEDOT-PSS with Gold Nanocomposite Modified Electrode

Layer of 3,3'-dithiopropionic acid (DPA) was formed on the AuNPs/PEDOT-PSS/AuE modified electrode surface by submerging the modified gold electrode into an ethanolic solution of 3 mM DPA and allowed for 45 min. The unbound DPA was removed by washing the layer with an absolute ethanol, rinsed with Millipore water and TE washing buffer respectively. The drying process was carried out in N-gas gentle flow for 10 min.

The layered DPA modified electrode (AuNPs/PEDOT-PSS/AuE) was activated by immersing in solution containing a mixture of 2 mM EDC and 5 mM NHSS for 1 hr. This was to activate the carboxylic group of the DPA for easy bonding with the amine group of the probe DNA sequence related to *G. boninense* as shown in Fig. 1.

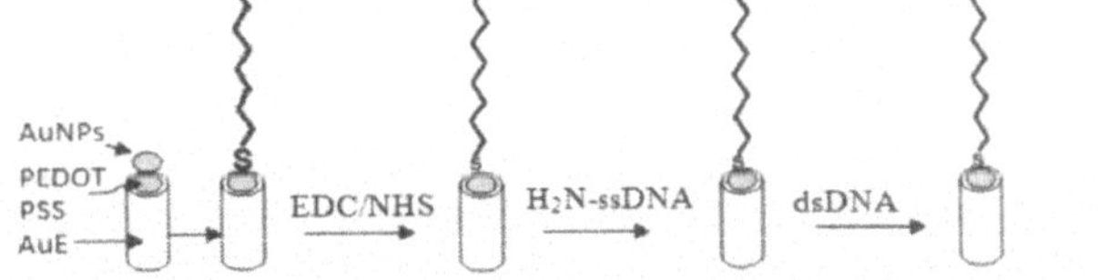

Figure 1: Schematic representation of electrode modification and covalentattachment of DNA probe for hybridization events

It was rinsed with TE washing buffer and dried via 10 minutes gentle flow of N-gas and further dried at 25°C for 45 min. The 30 μL of 100 μM probe DNA aseptically accumulated on the EDC-activated modified electrode surface for at least 12 hrs at room temperature. It was then washed with a washing TE-buffer solution to remove any unbound ssDNA. The probe-captured electrode was then denoted as ssDNA/AuNPs/PEDOT-PSS/AuE. The CV and DPV measurements were carried out utilizing the ruthenium $[Ru(dppz)]^{2+}$ metal complex, using TE containing$K_3Fe(CN)_6$supporting electrolyte.

2.8 Effect of hybridization temperature and time

Probe-modified electrode was hybridized with the complementary target DNA sequence related to *G. boninense*, by accumulating 30 μL of 100 μM target DNA containing 625μmole ruthenium complex. The target DNA was heated (denatured) in a water bath at temperature of 35 °C, 45 °C, and 55 °C each in 25 min, 45 min, and55 min respectively. The respective denatured target DNA was casted on the probe-modified electrode with addition of 625 μmole ruthenium indicator, and then cooled gradually at room temperature for annealing. The electrode surface was then rinsed with TE washing buffer to remove any unbound target DNA and dried for 5 minutes under gentle flow of N-gas and further dried at 25 °C for 45min. The protocol allowed the specific target sequence of the DNA to hybridize with the probe ssDNA and formed a double strand of the DNA (dsDNA) on the surface of the gold electrode.

The effects of the hybridization temperature and time was optimized on ssDNA/AuNPs/PEDOT-PSS/AuE sensor using CV voltammogram utilizing $K_3Fe(CN)_6$ contained in TE supporting electrolytes with the ruthenium $[Ru(dppz)]^{2+}$ complex as hybridization monitoring indicator.

2.9 Selectivity study

A 30 μL of 100 μM target DNA was collected in a vial and heated in a water bath at 45 °C for 35 min to denature. It was quickly castes on the probe-modified electrode surface, and a 625 μmole ruthenium indicator was added, and then allowed to anneal for 90 min. The electrode surface was then rinsed with TE washing buffer to remove any unbound target DNA. The hybridized DNA on the probe DNA modified electrode was dried for 5 min in a gentle flow of N-gas and further dried at 25 °C for 45 min. The protocol allowed for the specific target sequence of the DNA related to *G. boninense* to hybridize with the probe DNA. The hybridized DNA to the probe DNA was labeled dsDNA/AuNPs/PEDOT-PSS/AuE. Similar procedure was also followed to test the hybridization of mismatch target DNA sequence

The CV voltammograms are obtained for the hybridization responses. The hybridizations were monitored using the ruthenium complex $[Ru(dppz)]^{2+}$as the hybridization redox indicators in the $K_3Fe(CN)_6$ containing TE supporting electrolytes

2. 10 Effect of target DNA concentrations

The probe-modified electrode surface of ssDNA/AuNPs/PEDOT-PSS /AuE was hybridized with different prepared complementary target DNA sequences related to *G. boninense*. The target DNA concentrations were prepared in TE dilution buffer (pH 8.0) in the range 1.00 x 10^{-9} M to 1.00 x 10^{-15}M. Hybridization protocol was carried out by accumulating 30μL of the target concentration and subsequent addition of 625 μmole ruthenium[Ru(dppz)]2+ complex to the probe DNA. The hybridization was performed on probe DNA under temperature and time of45 °C and 35 minutes, and then cooled gradually at room temperature for annealing. The electrode surface was then rinsed with the TE washing buffer to remove the unbound DNA and dried for 5 min under gentle flow of N-gas and further dried at 25°C for 45 min.

The DPV measurements were performed for the hybridization responses of the prepared different concentrations in the $K_3Fe(CN)_6$ containing TE supporting electrolytes. The DNA hybridizations were monitored via the ruthenium metal complex $[Ru(dppz)]^{2+}$.

3 RESULTS AND DISCUSSION

3.1 Electrochemical characterization of bare and modified gold electrode active surface area

Cyclic voltammograms of bare, PEDOT-PSS/AuE, AuNPs/PEDOT-PSS/AuE, AgNPs/PEDOT-PSS/AuE modified gold electrodes in 1.0 mM$K_3Fe(CN)_6$ are shown in Fig. 3(ai-di). Peak current obtained is proportional to the square root of the scan rate, suggesting a sensitive diffusion-control electrochemical process [6]. Active surface area of the bare and modified electrodes can be calculated according to the Randles-Sevcik formula [5].

$$i_p = (2.19X10^5)n^{\frac{3}{2}}AD^{\frac{1}{2}}CV^{\frac{1}{2}} \quad (2)$$

where n is the number of electron participating in the redox reaction, A is the surface area of the electrode (cm^2), D is the diffusion coefficient of the molecules in the solution (cm^2s^{-1}), C is the concentration of $K_3Fe(CN)_6$ (M) and v is the scan rate ($V\ s^{-1}$). From the slope of the plot of i_{pc} versus $v^{1/2}$ (Fig. 3aii-dii), the sensitive active surface area of the electrodes can be calculated.

For bare AuE, the active surface area was calculated to be 0.052 cm^2. There was a decrease in the effective surface area to 0.047 cm^2 when bare AuE was modified with PEDOT-PSS. In further modification with gold nanoparticles, the effective surface area increased to 0.104 cm^2 whereas the modification with silver nanoparticles had increased the active surface area to 0.093 cm^2. Therefore, further investigations are carried out using AuNPs/PEDOT-PSS/AuE modified electrode.

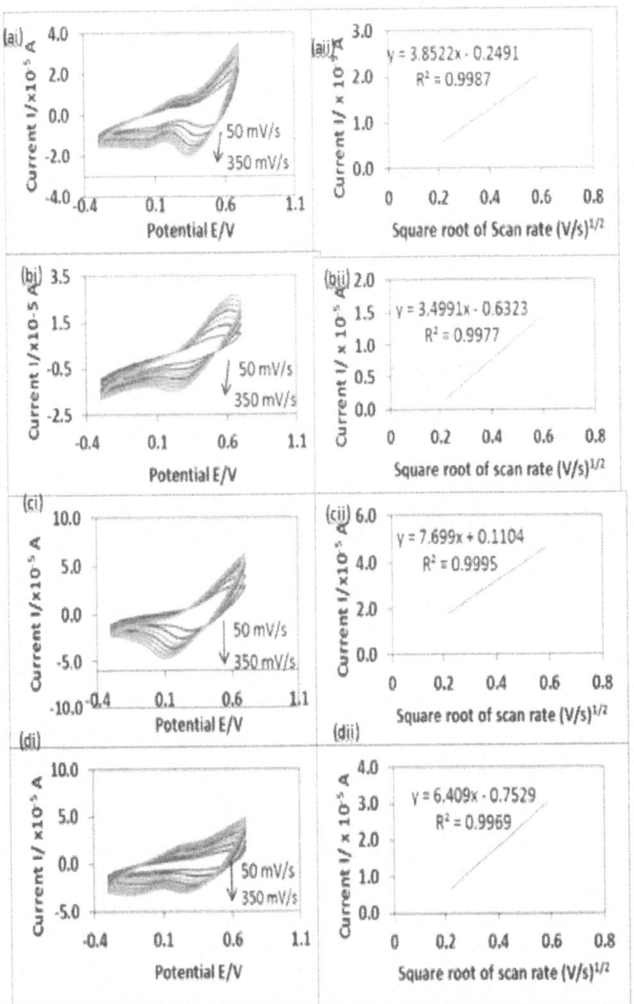

Figure 3: Cyclic voltammograms and current versus square root of scan rates (50, 100, 150, 200, 250, 300, 350 mV/s) of (a) bare gold electrode (b) PEDOT-PSS/AuE (c) AuNPs/PEDOT-PSS/AuE (d) AgNPs/PEDOT-PSS/AuE modifications in a 1.0 mM $K_3Fe(CN)_6$

3.3 Effect of hybridization temperature and time

The efficiency of hybridization response is dependent on the hybridization temperature and time, these parameters are optimized by cyclic voltammetry (CV) technique.The plots of time versus current of the CV oxidation peaks [replicates (n=3)] are shown in Fig. 5. The developed biosensor show higher hybridization current at 45 °C and 35 min.

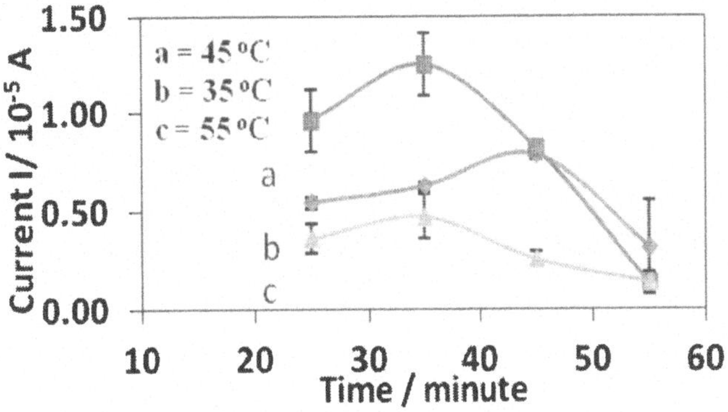

Figure 5: Reduction peaks for hybridization of dsDNA/AuNPs/PEDOT-PSS/AuE in 5 mM $K_3Fe(CN)_6$ containingTE with $[Ru(dppz)]^{2+}$ for target DNA of 100 µM concentration.

Based on the results obtained in Fig. 5., the optimum hybridization conditions are at 45 °C and 35 min, and are therefore chosen for further hybridization analysis.

3. 4 Selectivity of the developed DNA based biosensor

The selectivity of the developed DNA biosensor AuNPs/PEDOT-PSS/AuE was examined based on the electrochemical transduction of the probe ssDNA/AuNPs/PEDOT-PSS/ AuE hybridized with target DNA sequence related to *G.boninense*. The hybridization events are monitored using ruthenium[Ru(dppz)]2+ complex at optimized hybridization temperature of 45 °C and time 35min. The CV hybridization selectivity responses between capture probes, complementary and mismatch Target DNA sequences related to *G. boninense*, at 100 μM concentrations are shown in Fig. 6.

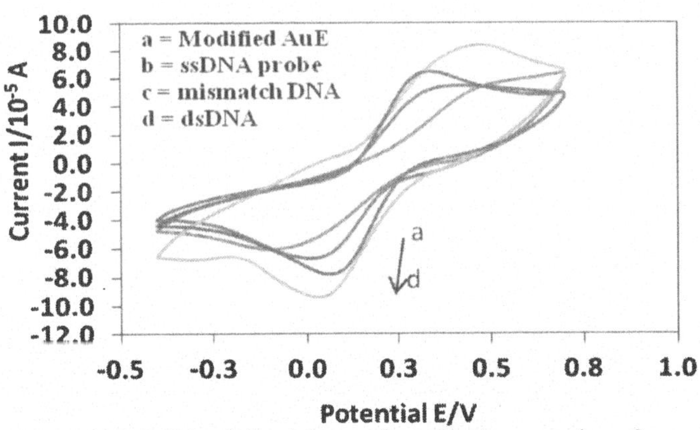

Figure 6: CV Selectivity of reduction peaks for hybridization of dsDNA/AuNPs/PEDOT-PSS/AuE in 5 mM $K_3Fe(CN)_6$ containing TE with $[Ru(dppz)]^{2+}$

3. 5 Effect of different concentrations of target DNA

Differential pulse voltammogram (DPV) technique was used to detect different concentrations of target DNA sequence to examine the sensitivity of the developed biosensors AuNPs/PEDOT-PSS/AuE. The DNA hybridization was monitored by ruthenium [Ru (dppz)]$^{2+}$ complex as indicator, results are shown in Fig. 7. The hybridization conditions applied are 45 °C and 35 min hybridization temperature and time. It is observed that the DPV peakcurrent increases with the increased target DNA concentration.

Consequently, the variations of the DPV peaks current observed before and after hybridization are linearly related with logarithmic value of target DNA concentrations in the range of 1 x 10^{-9} M to 1 x 10^{-15} M. The correlation between the DPV peak current and the target DNA concentrations are clearly explained from a plot of peak current versus log[concentration] of the target DNA shown in Fig. 8.

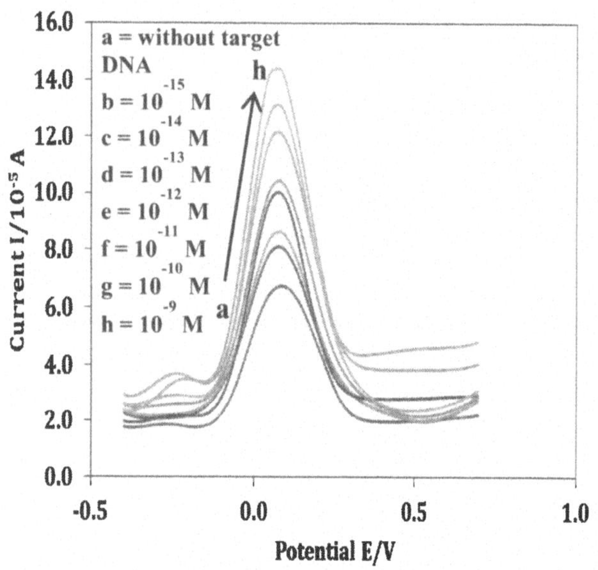

Figure 7: DPV of dsDNA/AuNPs/PEDOT-PSS/AuE in 5 mM $K_3Fe(CN)_6$containing TE with [Ru(dppz)]$^{2+}$ of different target DNA concentrations (ranged 1.0 x 10^{-15}M to 1.0 x 10^{-9}M) at scan rate 100 mV/s.

The detection limit (LoD) of the sensor was calculated as 9.70 x 10^{-19}mol L^{-1} from a well-known equation (3.3σ/S), where σ is the standard deviation of the blank measurements (n=5) and S is the slope of the linear regression equation of the plot of current versus log[target DNA].

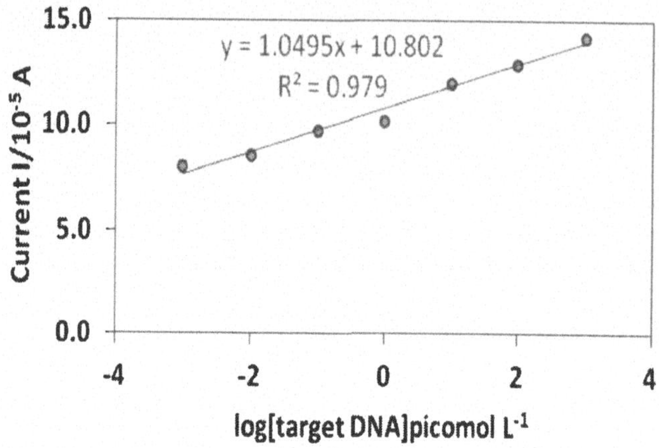

Figure 8: Linear plot of current against log[con.] of the target DNA of dsDNA/AuNPs/PEDOT- PSS/AuE in 5 mM $K_3Fe(CN)_6$ containing TE with [Ru(dppz)]$^{2+}$ of different target DNA concentrations (ranged 1.0 x 10^{-15}M to 1.0 x 10^{-9}M) at scan rate 100 mV/s.

4 CONCLUSION

The application of nanoparticles with PEDOT-PSS improves the sensitivity of modified gold electrode for *Ganoderma boninense* DNA detection. The selectivity of the developed biosensor allowed for the recognition of labeled DNA and could differentiate between ssDNA and dsDNA of fully complemented and mismatched target DNA sequences. The new ruthenium ruthenium [Ru (dppz)2 (qtpy) Cl2 ; dppz =dipyrido [3, 2-a:2', 3' - c] phenazine; dpty = 2, 2', -4, 4'' . 4', 4''' – quaterpyridyl complex was found to be a good candidate as DNA intercalating indicator.

REFERENCES

[1] Zeglis, B. M., Pierre, V. C. and Barton, J. K. Chemical Communications 44, 4565–4579, 2007.

[2] Ahmad, A., Meijer, A. J. H. M. and Thomas, J. A. Chemistry Asian Journal 6, 2339–2351, 2011.

[3] Malhotra, B. D., Chaubey, A. and Singh, S. P. AnalyticaChimicaActa 578, 59–74, 2006.

[4] Gerard, M., Chaubey, A. and Malhotra, B. D. Biosensors and Bioelectronics 17, 345–359, 2002.

[5] Liu, J., Liu, J., Yang, L., Chen, X., Zhang, M., Meng, F., Luo, T. and Li, M. Sensors 9, 7343–7364, 2009.

[6] Yu, C., Ji, W., Wang, Y., Bao, N. and Gu, H. Nanotechnology 24, 115502–115511, 2013.

Zn(II)-Mediated Gold Nanosensors for Protein Phosphatase Assay

Jin Oh Lee,* Gae Baik Kim* and Young-Pil Kim*

*Department of Life Science, Hanyang University, Seoul 133-791, Republic of Korea,
ypilkim@hanyang.ac.kr

ABSTRACT

Protein phosphatases (PPs) have served as key molecules in a myriad of biological events. Since gold nanoparticle (AuNP)-based colorimetric assays have been preferred for enzyme activity assay, a few of attempts have been designed for the simple and rapid detection of PP activity. Here we report a simple AuNP colorimetric assay of PP activity using hexahistidine-tagged phosphopeptides and divalent metal ions, which does not require antibody and rare chemical. Among metal ions including nickel (II), copper(II), cobalt(II), magnesium (II), manganese (II), and zinc(II), only zinc (II) triggered self-assembly of AuNPs in the presence of hexahistidine-tagged phosphopeptides. In contrast, PP1 caused dephosphorylation of peptides, leading to no change in the color of the AuNP solution. As a result, the PP activity was easily quantified by the extinction ratio (E520/E700) of the colloidal gold assembly, where there was a linear correlation between the PP concentration and the extinction ratio. Most importantly, this method was successfully used for analyzing protein phosphatase 2A (PP2A) activity in immunoprecipitated plant extracts. Based on the results, we anticipate that out method will find the applications to monitor the activities of various PPs and their inhibition in a rapid format.

Keywords: gold nanoparticle, protein phosphatase, metal affinity, colorimetric assay, peptides, zinc coordination

1 INTRODUCTION

Protein phosphatases (PPs) are one of the most essential enzymes underlying the regulation of many cellular functions in a coordinated manner with protein kinases [1, 2]. To this end, many methods based on radioactive, chromogenic and fluorogenic detection have been developed to analyze to PP activity [3-5]. However, in addition to labor-intensive radioactive method, non-radioactive methods still suffer from high background signal, due to interfering substances or acidic reaction conditions. This is often restricted in a rapid read-out with high detection sensitivity.

To circumvent this limitation, we demonstrate herein a gold nanoparticle (AuNP)-based colorimetric sensor for the detection of PP activity using peptide substrate and metal ion affinity. While AuNP-based colorimetric assays have attracted attention in recent years, because of the high extinction coefficient of AuNPs and detection simplicity, only a few attempts have been made to assay PP activity using AuNP-based colorimetry [6-8]. Unlike the AuNP-based assay based on simple electrostatic interactions between phosphorylated and dephosphorylated peptides, the designed AuNP nanosensor provides highly reliable detection of dephsphorylated products via a strong metal-coordinated affinity. The rationale stems from structural information of the conserved active site of common PPs, where two metal ions generally coordinates with a phosphate group of a substrate by three histidines, two aspartic acids, and one asparagine in the binding pocket [9]. Being inspired by this principle, we employed nitrilotriacetic acid (NTA)-modified AuNPs and histidine-tethered phosphopeptide to induce the metal coordination. In the present study, we investigated metal specificity and effects of metal ion concentration and pH condition on the self-assembly of AuNPs. PP activity was also investigated in plant extract to verify the usability of this nanosensor.

2 EXPERIMENTAL

2.1 Synthesis of NTA-AuNPs

To generate NTA-AuNPs, 40 mg of carboxylmethyl-polyethylene glycol-thiol (CM–PEG–SH, MW=3,400, Laysan Bio, Inc) was added to the citrate-stabilized AuNP solution (20 mL at 12 nM), followed by vigorous stirring at RT. The carboxy AuNP was further modified with Nα,Nα-bis(carboxymethyl)-L-lysinehydrate (NH_2-NTA, Sigma) via EDC. The molar concentration of NTA-AuNP was calculated by dividing the total number of gold atoms by the average number of gold atoms per AuNP after confirming the size.

2.2 AuNP-Based Colorimetric Analysis

Unless otherwise stated, NTA-AuNPs (2.5 μL at 250 nM), peptide substrate (2.5 μL at 100 μM), and divalent ion (10 μL at 10 mM) were mixed to give a final volume of 100 μL in 20 mM Tris-HCl buffer (pH 7.4) and incubated at RT for 1 h, after which extinction spectra were measured using a plate reader (Varioskan^{TM} Flash, Thermo Scientific). The peptide substrates (His_6-pPep; H_6GLRRASpLG, His_6-Pep; H_6GLRRASLG, and pPep; GLRRASpLG) were synthesized by Peptron Inc., Korea. For the divalent ion experiments, $NiCl_2$, $CoCl_2$, $CuCl_2$, $MnCl_2$, $MgCl_2$, or $ZnCl_2$ was added to the reaction mixture at a final concentration of 1 mM.

2.3 Protein Phosphatase Assay

Protein phosphatase 1 (PP1, New England Biolabs) (5 μL at different concentrations) was incubated at 30 °C for 1 h with peptide substrate (2.5 μL at 100 μM) and $MnCl_2$ (5 μL at 10 mM) in the reaction buffer (20 mM Tris-HCl, pH 7.4). For the AuNP-based assay, the reactant was transferred to 96-well plate and further mixed with NTA-AuNPs (2 μL at 250 nM) and $ZnCl_2$ (10 μL at 10 mM) at RT to give a final volume of 100 μL in 20 mM Tris-HCl buffer (pH 7.4). After completing the reaction at RT for 1 h, the extinction of the solution was determined using a plate reader.

2.4 Protein Phosphatase Assay in Plant Extracts

Protein extracts from wild-type Columbia-0 (Col-0) or transgenic *Arabidopsis thaliana* that expressed PP2A-A1 were prepared in extraction buffer (50 mM Tris-HCl, pH 7.5, 50 mM NaCl, 300 mM sucrose, 1% Triton X-100, 0.2 mM PMSF and 1× protease inhibitor cocktail). For immunoprecipitation (IP), soluble supernatant from protein extracts was incubated for 75 min with protein A agaose bead (Thermo Scientific Inc.) conjugated with polyclonal anti-PP2A1 tag (YFP) antibody (homemade, 4.6 μg μL^{-1}). For the AuNP-based assay, immunoprecipitated proteins from wild-type and PP2A-A1-expressed fractions were washed four times with extraction buffer and then incubated with His_6-pPep substrate (2.5 μL at 100 μM) in the presence of $MnCl_2$ (5 μL at 10 mM) dissolved in reaction buffer (20 mM Tris-HCl buffer, pH 7.4) at RT for 1 h with vigorous shaking. Aliquots (50 μL) of the reaction mixture were transferred to 96-well plates and were mixed with NTA-AuNP and $ZnCl_2$. After additional incubation at RT for 1 h, the extinction of the mixture was measured using a plate reader.

3 RESULTS AND DISCUSSION

As illustrated in Fig. 1, we hypothesized that NTA-AuNPs (21.3 ± 1.1 nm in core diameter) can be bound strongly to the hexahistidine and phosphate group of His_6-tagged phosphopeptide (His_6-pPep: $\underline{H_6}$GLRRA$\underline{Sp}$LG) via Zn(II) coordination because Zn(II)-chelated ligand (NTA) is known to have strong affinity for hexahistidine [10, 11] and phosphate group [12-14]. This cross-link brings about the self-assembly of NTA-AuNPs, resulting in a color change from reddish-brown to violet. In contrast, dephosphorylation of peptide substrates by PP activity could not cause the cross-link of NTA-AuNPs, due to lack of Zn(II)-phosphate coordination.

To check this possibility, different divalent ions including Ni(II), Co(II), Cu(II), Mn(II), Mg(II), and Zn(II) were tested in the presence of His_6-pPep and NTA-AuNPs (Figure 2). Among the tested divalent ions, only Zn(II) induced self-assembly of NTA-AuNP, resulting in changes in extinction and color from reddish-brown to reddish-violet (Figs 2A and 2B), whereas other divalent ions did

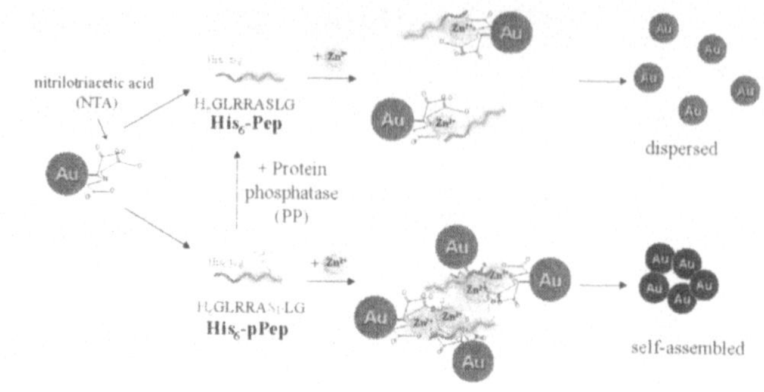

Figure 1: Schematic of colorimetric PP activity assay using NTA-modified AuNPs and His_6-tagged phosphopeptides (His_6-pPep) in the presence of Zn^{2+}. The PP activity inhibits the assembly of NTA-AuNPs by dephosphorylation of His_6-pPep (top right), whereas NTA-AuNPs are self-assembled by Zn^{2+}-coordination (bottom right).

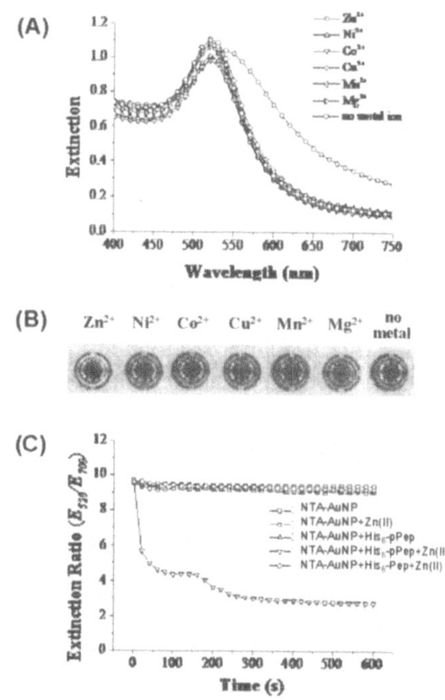

Figure 2: Effect of different divalent ions on the self-assembly of NTA-AuNPs with His_6-pPep (H_6GLRRAS(p)LG) in (A), (B), and (C) or His6-Pep (H_6GLRRASLG) in (C): (A) extinction spectra, (B) photograph of the NTA-AuNP solutions, and (C) time-lapse changes in the extinction ratio (E_{520}/E_{700}) of NTA-AuNPs. The final concentrations of NTA-AuNP, peptide, and divalent ion were 5 nM, 2.5 μM, and 1 mM, respectively.

not give rise to a significant color change at the same concentrations. Importantly, time-lapse extinction graph showed that Zn(II)-induced self-assembly rapidly occurred approximately 5 min after the addition of Zn(II) ions (Fig. 2C).

To examine whether the Zn(II)-mediated self-assembly of NTA-AuNPs is induced by His_6-pPep, different synthetic

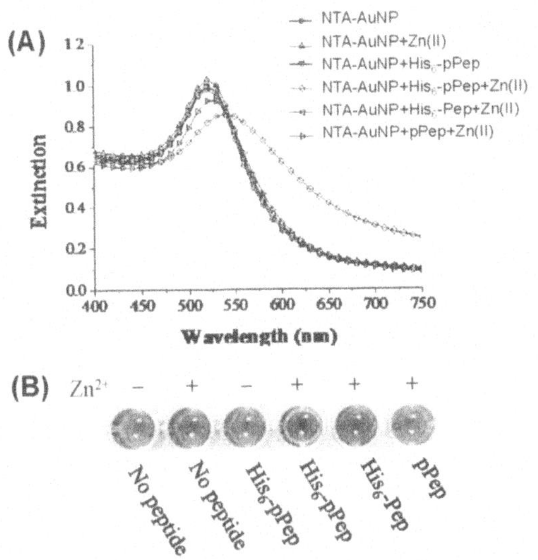

Figure 3: (A) Extinction spectra and (B) photograph of NTA-AuNPs in the presence and absence of Zn^{2+} with different peptides (His_6-pPep, His_6-Pep, and pPep (GLRRAS(p)LG).

peptides (His_6-Pep; H_6GLRRASLG and pPep; GLRRAS(p)LG) were tested. In Fig. 3, while the original color of NTA-AuNPs was not affected by addition of either Zn(II) or His_6-pPep, significant self-assembly of NTA-AuNPs was only observed in the presence of both His_6-pPep and Zn(II). There was no color change when unphosphorylated peptide (His_6-Pep) or His_6-deficient peptide (pPep) was used under the same conditions. Although the coordination structure is not clearly understood in this study, these results strongly support that Zn(II) can provide better binding affinity for both a substrate $P–O^-$ group and a His_6-tag than other metal ions. Consequently, the self-assembly of NTA-AuNPs in the presence of Zn(II) discriminated phosphorylation of his-tagged peptides.

To assay protein phosphatase activity using this AuNP-based colorimetry, extinction of NTA-AuNPs was monitored as a function of protein phosphatase 1 (PP1) concentrations in the presence of His_6-pPep and Zn(II). In Figs 4A and 4B, there was a linear correlation between the extinction ratio and PP1 concentration in the range of 0.05 U to 0.5 U. The maximal recovery yield was 83.5 % at 5 U PP1, based on the relative extinction ratio of His_6-Pep (control). Most importantly, bovine serum albumin (BSA), of which concentrations were equivalent to those of full active PP1, caused a negligible change under the same conditions (Fig. 4C). When the extinction ratio was converted to the activity level (%), a critical difference in activity level was observed between PP1 and BSA (Fig. 4D), indicating that the self-assembly of NTA-AuNPs/Zn(II) in the presence of pep1 is controlled by the enzymatic activity of PP1. The detection sensitivity for PP1 in Fig. 4A was as low as 0.01 U, which is greater than that 0.4 U of the traditional colorimetric assay using *p*-nitrophenyl phosphate [15].

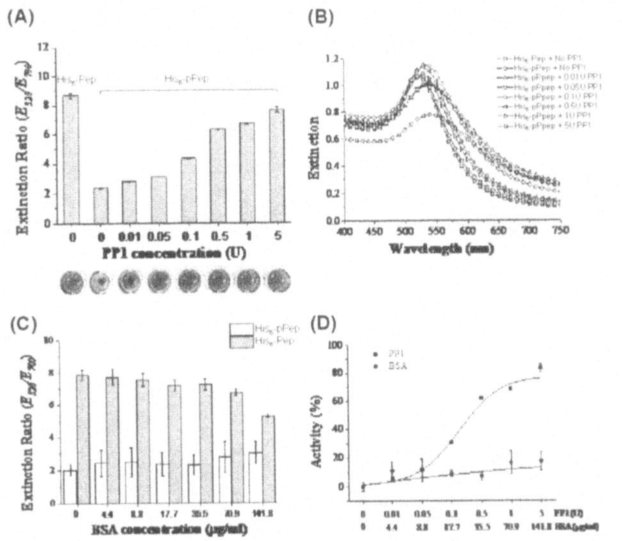

Figure 4: Colorimetric assay of protein phosphatase 1 (PP1) activity using NTA-AuNP and His_6-pPep in the presence of Zn^{2+}. Changes in (A) extinction ratio and (B) extinction spectra of the NTA-AuNP solution were displayed as a function of PP1 concentration, ranging from 0.01 to 5 U. (C) Control experiment as a function of bovine serum albumin (BSA) concentration. (D) The activities from PP1 and BSA were calculated and plotted from the extinction ratios.

To further investigate whether this method is useful for analyzing PP activity in real samples, PP activity from plant extracts was measured using this gold nanosensors (Fig. 5). Total proteins was initially extracted from transgenic *Arabidopsis thaliana* overexpressing PP2A-A1 (one of three isoforms of PP2A) fused to YFP and wild-type *A. thaliana* (Columbia-0, abbreviated Col-0), which was followed by enrichment by immunoprecipitation (IP) using anti-PP2A tag polyclonal antibody (anti-YFP pAb) attached to the protein A agarose beads (Fig. 5A). When each aliquot of immunoprecipitated beads was incubated with His_6-pPep and NTA-AuNP/Zn(II) in a microwell plate (Fig. 5B), a relatively higher recovery of the color of NTA-AuNPs, which was responsible for PP2A activity, was observed in immunoprecipitated proteins obtained from transgenic PP2A extracts than wild-type (Fig. 5C). Compared to the control (the first well in Fig. 5C), an increased level of PP2A activity was detectable in wild-type extracts, probably due to endogenous PP2A. This result indicates that IP-combined colorimetric detection would be very useful for *in vitro* analysis of PP activity in complex samples.

Despite the reversible sensing ability to phosphorylation, our colorimetric format was limited to detect protein kinase activity, because abundant adenosine triphosphates in a reaction buffer of protein kinase critically impeded the phosphopeptide-Zn(II) interaction on the NTA-AuNP (data not shown). Therefore, the assay format described in this paper is more suitable for measuring PP activity, than protein kinase activity.

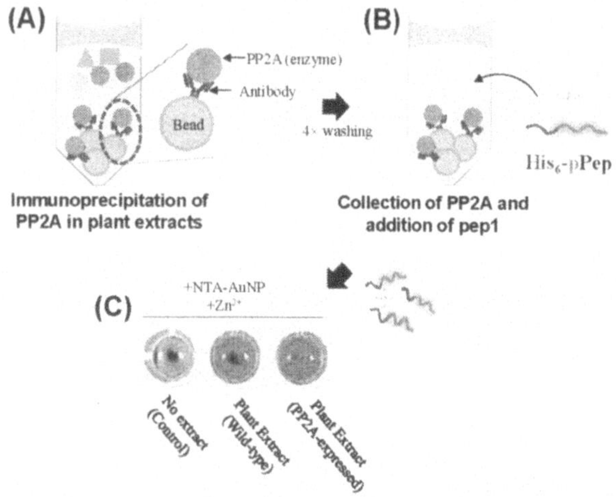

Figure 5: Immunoprecipitation (IP)-combined colorimetric assay of PP2A activity in plant extracts using NTA-AuNPs: (A) IP of PP2A using anti-PP2A-tag antibody and (B) PP2A reaction with His_6-pPep after IP. (C) Photographs of NTA-AuNPs on PP2A activity in control (free extract), extract (wild-type Col-0), and tranformant extract (PP2A-expressed).

Due to their high surface-to-volume ratio, NTA-AuNPs were self-assembled rapidly via strong binding affinity with Zn(II) ions in the presence of His_6-tagged phosphopeptides, which would not be possible using other NTA-coated beads or metal-chelating microparticles.

4 CONCLUSIONS

We reported the simple and rapid detection of PP activity using Zn(II)-mediated colorimetric gold nanosensors with His_6-tagged phosphopeptides. As a result of specific Zn(II)-coordination between the carboxy groups of NTA-AuNPs and the oligohistidine and phosphate groups of His_6-tagged phosphopeptides, the self-assembly of NTA-AuNPs was controlled by PP1 activity-driven dephosphorylation, which allowed for the quantification PP activity based on the extinction ratio (E_{520}/E_{700}) of NTA-AuNP. Furthermore, the gold nanosensor combined with IP was capable of detecting PP2A activity in plant extracts. We anticipate that our method will be useful for studying the physiological activities of PPs in a simple and rapid manner.

ACKNOWLEDGMENTS

This work was supported by Mid-career Researcher Program (No. 2013R1A2A2A03015161) and Nano·Material Technology Development Program (No. 2012M3A7B4035286) through the National Research Foundation (NRF) funded by the Ministry of Science, ICT, and Future Planning (MSIP).

REFERENCES

[1] A. L. Bauman, and J. D. Scott, Nat. Cell Biol., 4, E203-E206, 2002.

[2] A. Bononi, C. Agnoletto, E. De Marchi, S. Marchi, S. Patergnani, M. Bonora, *et al.*, Enzym. Res., 329098, 2011.

[3] T. McAvoy and A. C. Nairn, "Serine/threonine protein phosphatase assays. Current protocols in molecular biology" edited by Frederick M Ausubel [et al] Ch. 18, Unit18, 2010.

[4] U. Lorenz, "Protein tyrosine phosphatase assays. Current protocols in immunology" edited by John E Coligan [et al], Ch 11, Unit 11, 2011.

[5] M. Brune, J. L. Hunter, J. E. Corrie, and M. R. Webb, Biochemistry, 33, 8262-8271, 1994

[6] Y. Choi, N. H. Ho, and C. H. Tung, Angew. Chem. Int. Ed., 46, 707-709, 2007..

[7] T. Serizawa, Y. Hira, and M. Aizawa, Mol. BioSys., 6, 1565-1568, 2010.

[8] J. Zhou, X. Xu, X. Liu, H. Li, Z. Nie, M. Qing, et al., Biosens. Bioelectron., 53, 295-300, 2014.

[9] Y. Shi, Cell, 139, 468-484, 2009.

[10] S. Y. Park, S. M. Lee, G. B. Kim, and Y. P. Kim, Gold Bull., 45, 213-219, 2012.

[11] G. B. Kim, K. H. Kim, Y. H. Park, S. Ko, and Y. P. Kim, Biosens. Bioelectron. 41, 833-839, 2013.

[12] E. Kimura, Pure Appl. Chem., 65, 355-359, 1993.

[13] E. Bobyr, J. K. Lassila, H. I. Wiersma-Koch, T. D. Fenn, J. J. Lee, I. Nikolic-Hughes, et al., J. Mol. Biol., 415, 102-117, 2012.

[14] T. H. Evers, M. A. M. Appelhof, F. W. Meijer, and M. Merkx, Protein Eng. Des. Sel., 21, 529-536, 2008.

[15] H. Sotoud, P. Gribbon, B. Ellinger, J. Reinshagen, P. Boknik, K. Kattner, et al., J. Biomol. Screen, 18, 899-909, 2013.

Determination of the bi-substrate kinetic coefficients for the β-D-glucose- NAD-GDH enzymatic reaction

J. Berthier[1], M. Cubizolles[1], P. Laurent[1], V. Poher[2], P. Pouteau[2], M-L. Cosnier[1]
[1] University Grenoble Alpes, F-38000, Grenoble,
Department of Biotechnology, CEA, LETI, MINATEC, 38054 Grenoble, France
[2] Avalun, Grenoble, France

ABSTRACT

Colorimetric detection of glucose in sample liquids such as human plasma is made by using enzymatic reactions. In the multi reactional scheme, the first enzymatic reaction is determinant. Either glucose oxidase (GOX) or glucose dehydrogenase (GDH) can be used to convert glucose. We focused here on the study of the enzyme GDH together with the enzymatic cofactor NAD (nicotinamide adenine dinucleotide, oxidized form).

This reaction falls in the category of ternary enzymatic reactions. Such reactions depend on four parameters. A method to determine these four parameters is presented in this work, based on a comparison between a series of experiments and the theory. The best values of the parameters are indicated.

Keywords: ternary enzymatic reaction, bi-substrate kinetics, glucose.

INTRODUCTION

Many glucose tests are now currently available on the market. However, the accuracy of detection can still be improved. The most convenient detection is colorimetry. Colorimetric detection of glucose in sample liquids such as human plasma is made by using enzymatic reactions. In the multi reactional scheme, the first enzymatic reaction is determinant. Either glucose oxidase (GOX) or glucose dehydrogenase (GDH) can be used to convert glucose. We focused here on the study of the enzyme GDH together with the enzymatic cofactor NAD (nicotinamide adenine dinucleotide). The reaction can be written as

$$\begin{aligned}\beta - D - Glu\cos e + NAD + GDH \leftrightarrow \\ Glucono - lactone + NADH + GDH\end{aligned} \quad (1)$$

This reaction falls in the category of ternary enzymatic reactions (figure 1). In a first reactional step, the coenzyme NAD (nicotinamide adenine dinucleotide) binds to the GDH (glutamate dehydrogenase) and in a second step, the β-glucose binds to the enzymatic complex. The reaction then occurs, and the product (glucono-lactone) is released together with the NADH (nicotinamide adenine dinucleotide dehydrogenase, reduced form).

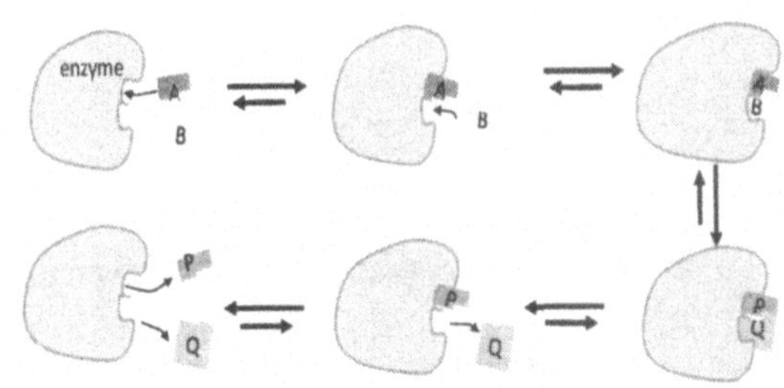

Fig.1. Principle of enzymatic ternary reaction: A=β-glucose, B=NAD, P= glucono-lactone, Q= NADH.

Ternary reaction models—also called bi-substrate models, because one enzyme and one coenzyme intervene—depend on four parameters [1-4].

In order to optimize the reaction, these parameters must be knowned. A method to determine these four parameters is presented in this work. The determination of the four reaction parameters stems from a comparison between a series of experiments and the theory. The best values of the parameters are indicated.

THEORY

The theory of ternary enzymatic reaction is well documented. It is derived from the Michaelis-Menten approach [1-4]. The reaction velocity is given by

$$V = \frac{K_0[E_0]}{1+\frac{K_{MB}}{[B]}} \frac{[A]}{[A]+\frac{K_{MA}[B]+K_{SA}K_{MB}}{[B]+K_{MB}}} \quad (2)$$

where $[E]$ is the concentration of enzyme (GDH), $[E_0]$ is the initial enzyme concentration ($[E_0]$=1.4 10^{-5} mM); $[A]$ is the β-D-glucose concentration, $[B]$ the NAD concentration, and the coefficients K's are the Michaelis-Menten constants. In our case the NAD concentration is fixed and equal to $[B_0]$ = 0.4 mM.

Relation (2) generalizes the Michaelis-Menten law: if we note

$$V_{max,app} = (K_0[E_0]) \Big/ \left(1+\frac{K_{MB}}{[B]}\right) \quad (3)$$

and

$$K_{M,app} = \frac{K_{MA}[B]+K_{SA}K_{MB}}{([B]+K_{MB})}, \quad (4)$$

relation (2) reduces to the Michaelis-Menten law.

For the concentration $[B]$ the reaction can be simply characterized by the two Michaelis-Menten constants V_{max} and K_M. But the knowledge of the four ternary reaction constant bares a larger generality.

Reaction kinetics (2) is valid if both concentrations $[A]$ and $[B]$ are such that $[A] >> [E_0]$ and $[B] >> [E_0]$, which is our case for reaction (1) and which has been reproduced in the experimental set-up.

MODEL

Relation (2) has three unknowns V, $[A]$ and $[B]$, and four parameters K_0, K_{MA}, K_{MB} and K_{SA}. By definition the reaction velocity is

$$V = \frac{d[A]}{dt} \quad (5)$$

Using the law of mass action [1,2], we can write

$$[B] = [B_0] - [A_0] + [A] \quad (6)$$

Relation (2) can then be cast in the form of a differential equation in $[A]$, with the initial values $[E_0]$, $[B_0]$

$$\frac{d[A]}{dt} = \frac{K_0[E_0]\{[A]^2+[A]([B_0]-[A_0])\}}{[A]^2+[A]([B_0]-[A_0]+K_{MA}+K_{MB})+K_{SA}K_{MB}+K_{MA}([B_0]-[A_0])} \quad (7)$$

Equation (7) is of the form:

$$\frac{d[A]}{dt} = -cste\,\frac{P_2([A])}{Q_2([A])}, \quad (8)$$

where P_2 and Q_2 are two second order polynomials.

We have programmed the integration of (8) with the software MATLAB, and used a conventional Runge-Kutta approach to solve for the NAD concentration.

Finally, the product (glucono-lactone) concentration $[Q]$ is given by

$$[Q] = [A_0] - [A] \quad (9)$$

EXPERIMENTS

A series of six kinetics using increasing concentrations of β-D-glucose has been performed (figure 2). Briefly, final concentrations of β-D-glucose, ranging from 0mM to 20mM, were incubated together with NAD and glucose dehydrogenase (*Pseudomonas sp* from Sigma Aldrich) at concentrations of 0.4 mM and 0.5U/mL respectively, using a 25mM Tris, 192mM Glycine buffer pH 8.05 at 37°C. For each β-D-glucose concentration, the reactions have been done in triplicates, Kinetics were obtained by following the OD (optical density) at 340nm, corresponding to NADH spectrum.

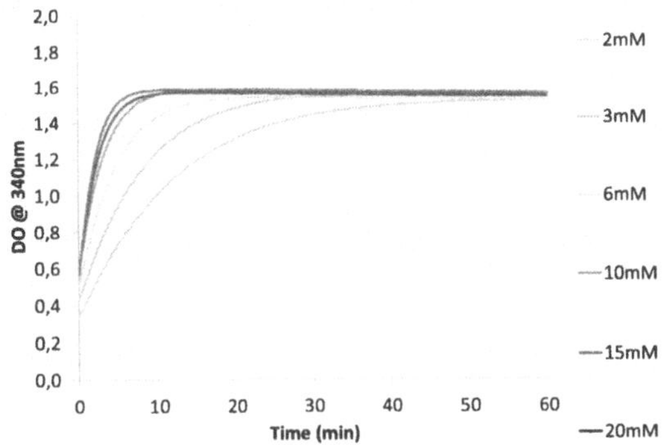

Fig.2. Reaction kinetics for different concentration of β-D-glucose {2,3,6,10,15,20 mM} in Tris/Glycine buffer.

In a Michaelis-Menten type of reaction, the reaction is characterized by the affinity of the substrate (β-D-glucose) for the enzyme (k_M) and the enzyme turn over (k_{cat}), which represent a global view of the reaction.

On the basis of the Lineweaver-Burk plot (figure 3), the affinity of the β-D-glucose for the enzyme can be determined: k_M= 9.0 mM. In addition, the enzyme turn over has the value k_{cat} = 89 s^{-1}.

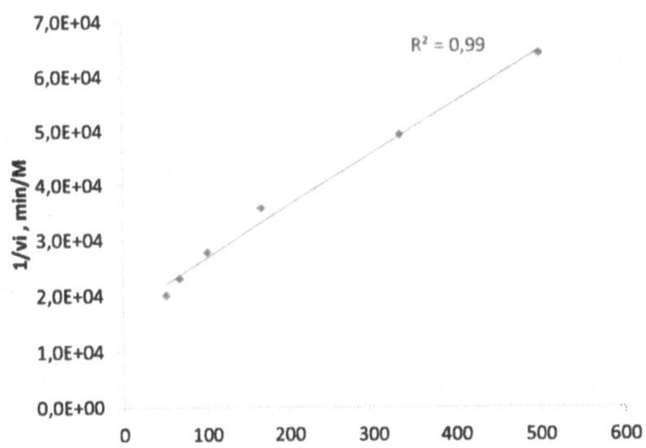

Fig.3. Lineweaver-Burke plot used for the determination of k_M and k_{cat}. The slope of the linear fit is k_M/V_{max}, and the intercept with the vertical axis is $1/V_{max}$. The notation [*S*] stands for the β-D-glucose concentration.

The ratio k_{cat} / k_M, reflecting enzyme efficiency, equals to 9.8 x 10^3 $M^{-1}.s^{-1}$. These results are in the same order of magnitude with that obtained in previous works at pH 8.0, performed on wild type and mutants of GDH isolated from *Bacillus megaterium,* known to be specific for glucose determination in body fluids [5]. In these studies, k_M values ranged from 2.7 to 55 mM and k_{cat} values are comprised between 23 and 430 s^{-1} [6-8]. Depending on the mutants, the enzyme efficiency previously reported varies from 3.6 x 10^3 to 42.8 x 10^3 $M^{-1}.s^{-1}$. The results that we obtain for the enzyme efficiency using GDH from *Pseudomonas sp* compare well with the previous ones reported on the various forms of GDH from *Bacillus megaterium*.

DETERMINATION OF THE REACTION PARAMETERS

In reality, the reaction is a Michaelis-Menten ternary reaction. In order to characterize the reaction in details, we follow a double reciprocal plot approach. The velocities *V* are calculated from the kinetic curves of figure 2. In a {*1/[A], 1/V*} coordinates plot, the experimental data points are satisfactorily aligned (figure 4).

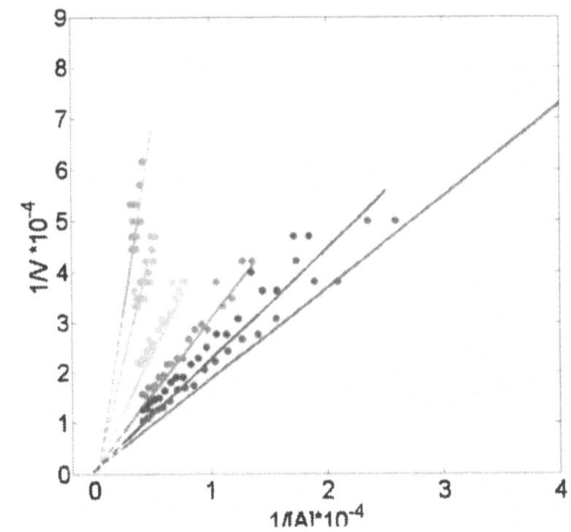

Fig.4. Representation of the experimental velocities in a double reciprocal diagram.

The four bi-substrate Michaelis-Menten parameters (K_0, K_{MA}, K_{MB} and K_{SA}) can be determined by comparison with experimental results.

A linear fit with the experimental results produces the values of the four reaction constants (figure 5). The fitted values of the four bi-substrate reaction constants are: K_0=1.1 10^5 s^{-1}, K_{MA}= 1 mM, K_{MB} = 2.9 mM, K_{SA}=12 mM. Reporting these values in (3) and (4), with

$[B]$=0.4 mM, we find $K_{M,app}$= 10.7 mM, and $V_{max,app}$= 0.003 mM/s, and $k_{cat,app}$= $V_{max,app}$ /$[E_0]$= 222 s^{-1}. The values of apparent K_M and k_{cat} deduced from the ternary model are in the same range as the corresponding experimental Michaelis-Menten kinetic constants.

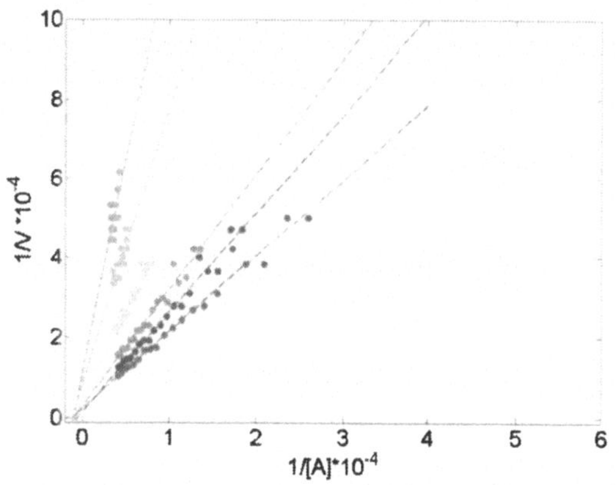

Fig.5. Fit of the theoretical model on the experimental results in the double Lineweaver-Burke diagram.

Reversing the approach, and reporting the fitted parameters in relation (5), we obtain a satisfactory comparison of the kinetics between experiments and model (figure 6), confirming the approach.

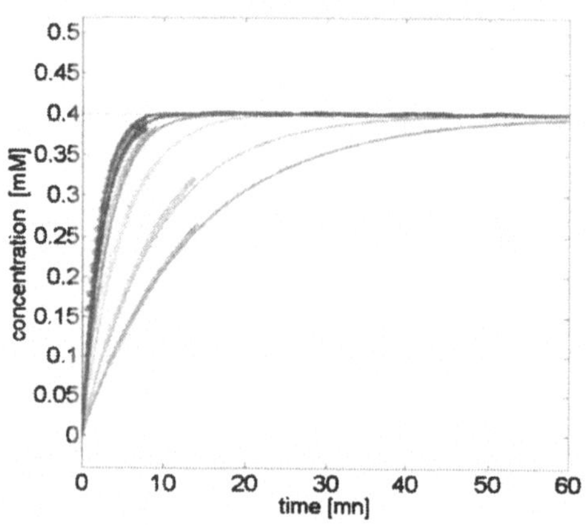

Fig.6. Kinetics comparison between model and experiments: continuous lines correspond to the experimental measures and the dots to the model.

CONCLUSION

In this work, we propose a method for the determination of the kinetic coefficients for the β-D-glucose-NAD-GDH enzymatic ternary reaction, based on the comparison between the theory of bi-substrate enzymatic reaction and experiments.

In order to determine the four constants of the reaction, we place ourselves in a double reciprocal plot approach {*1/[A], 1/V*} where the kinetic curves are linear. Such an approach facilitates the fit of the constants. A reverse reconstruction reproduces the kinetic signals.

The knowledge of the coefficients of the reaction is essential to optimize glucose detection systems. Moreover, the method can be generalized to other bi-substrate reactions.

REFERENCES

[1] I.H. Segal, Enzyme kinetics, John Wiley & Sons, New York, 1975.

[2] A. Cornish-Bowden, Fundamentals of enzyme kinetics, 3rd Edition, 2004, Portland Press

[3] P.A. Frey, A.D. Hegeman Enzymatic reaction mechanisms, Oxford University Press, New York, 2007.

[4] C.A. Janson, W.W. Cleland, The kinetic mechanism of glycerokinase, J Biol Chem. **249**(8), 2562-2566, 1974.

[5] H. Rindfrey, R. Helger and H. Lang, Kinetic Determination of Glucose Concentrations with Glucose Dehydrogenase, J. Clin. Chem. Clin. Biochem., **15**, 217-220, 1977.

[6] S.-H. Baik, T. Ide, H. Yoshida, O. Kagami, S. Harayama, Significantly enhanced stability of glucose dehydrogenase by directed evolution, Appl Microbiol Biotechnol **61**:329–335, 2003.

[7] T. Nagao, T. Mitamura, Xiao Hui Wang, S. Negoro, T. Yomo, I. Urabe, H. Okada, Cloning, Nucleotide Sequences, and Enzymatic Properties of Glucose Dehydrogenase Isozymes from Bacillus megaterium IAM1030, J. Bacteriology, **174**(15), 5013-5020, 1992.

[8] K. Yamamoto, T. Nagao, Y. Makino, I. Urabe, H. Okada, Characterization of Mutant Glucose Dehydrogenases with Increasing Stability, Ann N Y Acad Sci. **613**, 362-365, 1990.

A Simple Process for Selective Bio-functionalization of SU-8 surface for Lab-on-a-Chip applications

Srinivasan .B[1], Durgesh Chaurasia[2], Sivaramakrishna Vanjari[3] and Shivgovind Singh[4]

Department of Electrical Engineering,
Indian Institute of Technology, Hyderabad
Telangana, 502 205, INDIA
[1]ee12m1036@iith.ac.in, [2]ee12m1013@iith.ac.in, [3]svanjari@iith.ac.in, [4]sgsingh@iith.ac.in

ABSTRACT

This paper presents a simple process of bio-functionalizing the surface of UV crosslinked SU-8 surfaces. SU-8 is a popular photo-patternable epoxy based polymer used extensively in Lab-on-a-Chip and MEMS applications. In this work we realized a SU-8 surface with Hydroxyl (–OH) groups by adding a copolymer Glycidol. Biotin Hydrazide is immobilized over this surface after APTES and Glutaraldehyde treatment. FTIR spectroscopy analyses were performed on the samples at various stages of bio-functionalization process.

Keywords: su-8, bio-functionalization, glycidol, ftir, biotin Hydrazide.

1 INTRODUCTION

SU-8 is a popular epoxy based negative photoresist developed by IBM. It has been widely used as structural material in Bio-MEMS and Lab-on-a-Chip (LoC) devices due to its bio-compatible nature, ease of fabrication and inertness to most chemicals (and bio-fluids).

Various procedures were proposed earlier for bio-functionalizing SU-8 surfaces for Biosensing applications. Often SU-8 surface has to be made hydrophilic for microfluidic transport and bio-functionalization purposes. Plasma treatment, the most frequently used technique is neither selective nor it can retain the hydrophilicity over an extended period of time. Selective functionalization and permanent hydrophilicity are most desired for LoC application and increased shelf life. If this could be achieved using a procedure which involves low temperature and less corrosive chemicals, it would enable fabrication of structures on Back End of Line (BEOL) layers of Integrated Circuit by CMOS post processing. Several processes have been demonstrated to achieve selective and stable hydrophilic SU-8 surface using low temperature methods [1] [2].

Photo patterned hydrophilic surfaces can be realized using a mix of SU-8 photoresist and Glycidol, a hydrophilic copolymer [1]. The resulting hydrophilicity is due to hydroxyl groups (-OH) present in Glycidol. The SU-8 and Glycidol mix (here onwards referred to as SU-8 + Glycidol) can be patterned using conventional SU-8 Photolithography process and hence involves lower temperatures only. It was proven that the hydrophilicity remains stable for a long duration when compared to plasma treatment process by following [1] and [2].

In this work, we extend the techniques proposed by [1] and [2] and bio-functionalize the SU-8 surface in order to develop a generic LoC platform using hydrophilic SU-8 based micro-channels. As a proof of concept, immobilization of Biotin Hydrazide molecules is demonstrated. For achieving this APTES molecules were covalently attached to the –OH groups on the SU-8+Glycidol surface. Then the Biotin Hydrazide molecules were immobilized over this layer by using Glutaraldehyde molecules as crosslinkers. The immobilized Biotin can be used to detect Avidin or Streptavidin molecules.

2 EXPERIMENTAL

2.1 Chemicals and Molecules

The process was carried out on a Silicon wafer. The following were the chemicals used in this process. Microchem SU-8 2002, SU-8 developer (PGMEA), Sigma Aldrich Glycidol 96% (G5809), Biotin Hydrazide (B7639), Phospate Buffer Saline (PBS) tablets (P4417), Technic France high purity Dimethyl Sulfoxide (DMSO), Alfa APTES((3-Aminopropyl)triethoxysilane) 98% (Alfa A10668), Glutaraldehyde 25% aq. Soln (A17876), Toluene (high purity without moisture), Tween 20 and Isopropyl Alchohol (IPA).

2.2 Procedure

Initially a 4 inch Piranha cleaned Silicon wafer was diced into small pieces of about 2 cm^2 for using as substrate. These substrates were baked at 200°C on hot plate for 1 hour for completely removing moisture.

Step 1: Each sample was spin coated with SU-8 2002 and crosslinked using UV lithography. The procedure is as follows:

Spin coating rate: Step1- 500 rpm, 10 sec with acceleration of 300 rpm/sec and Step2- 3000 rpm, 30 sec with acceleration of 300 rpm/sec.
Soft bake: 65°C for 3 minutes and 95°C for 6 minutes.
Exposure: Flood exposure with UV (365nm) for 80 sec (80mJ/ cm^2) using Suss MicroTec MA6/BA6 aligner system.
Post Exposure Bake (PEB): 65°C for 3 minutes and 95°C for 6 minutes. The samples were cooled down to room temperature gradually in a time of about 1 hour to avoid internal stresses in the film.

Then the SU-8 coated samples were rinsed in IPA for dissolving uncrosslinked residues and dried using purified compressed air (CDA). This first layer of SU-8 was used as a base for SU-8+Glycidol layer which would be functionalized.

Step 2: Next a polymer mix of SU-8 2002 and Glycidol was made with a ratio of 1:0.3 (by wt. %) using micro-balance. This mixture was thoroughly probe sonicated in an ice bath for 30 minutes to avoid lumps and ensure uniformity. This mix was used to form patterned layer of SU-8+Glycidol over the pure SU-8 layer. The following steps were followed in fabricating this layer.
Spin coating rate: Step1- 1000 rpm, 40 sec with acceleration of 300 rpm/sec. This parameters helps in getting uniform film of SU-8+Glycidol polymer.
Soft bake: 65°C for 3 minutes and 75°C for 6 minutes. The flash point of Glycidol is 81°C and hence the highest temperature used was 75°C.
Exposure: Exposed using mask with square patterns (1cm^2) with UV (365nm) for 200 sec using Suss MicroTec MA6/BA6 aligner system. This dosage was chosen as the film thickness is larger than pure SU-8 2002.
Post Exposure Bake (PEB): 65°C for 3 minutes and gradually increased to 95°C in 15 minutes and holded at the same temperature for 6 minutes. The samples were cooled down to room temperature gradually in a time of about 1 hour to avoid internal stresses in the film. In PEB, 95°C was used as it is necessary for the SU-8 monomers to crosslink. However, gradual temperature raise is important, so that the SU-8 and Glycidol's epoxy structures open up and corsslinks to form a stable polymer matrix below the flash point of Glycidol.

Later the samples were developed using SU-8 developer (PGMEA), vigorously washed in IPA and dried using CDA. Contact angle measurement was done using goniometer and a lower angle (~ 65°) was observed on this layer than a pure SU-8 2002 (~78°) layer due to the presence of Hydroxyl groups on the sufrace [1].

Step 3: APTES solution was prepared using Toluene as solvent with a concentration of 2.0% (v/v). Toluene was chosen as it supports in forming covalent bonds between APTES and –OH groups on the SU-8+Glycidol surface [7]. The samples coated with SU-8+Glycidol layer and one sample with only pure SU-8 layer were incubated in this solution for 2 hours at room temperature. Then the samples were thoroughly washed in pure Toluene to detach the unbonded APTES and dried with CDA.

Step 4: Next the samples were smeared with Glutaraldehyde solution and left for 2 hours at room temperature. Later the samples were washed using copious amout of DI water and dried using CDA.

Step 5: Further step is to immobililze Biotin Hydrazide over the surface. Biotin Hydrazide powder is dissolved in DMSO(1mL)+PBS(9mL) buffer (pH ~ 7.4) solution with a concentration of 0.5mg/mL. The samples were incubated in this Biotin Hydrazide solution for 12 hours at room temperature. Then the samples were washed thoroughly in PBS+Tween 20 solution to remove excess and non-specifically adsorbed Biotin Hydrazide molecules. The procedure for preparing the PBS+Tween 20 solution is as follows, 1 PBS tablet was dissolved in 200mL DI water to obtain a solution of pH 7.4 which is used for Biotin Hydrazide solution. Tween 20 (0.5% v/v) was mixed in this PBS solution to use for washing purpose.

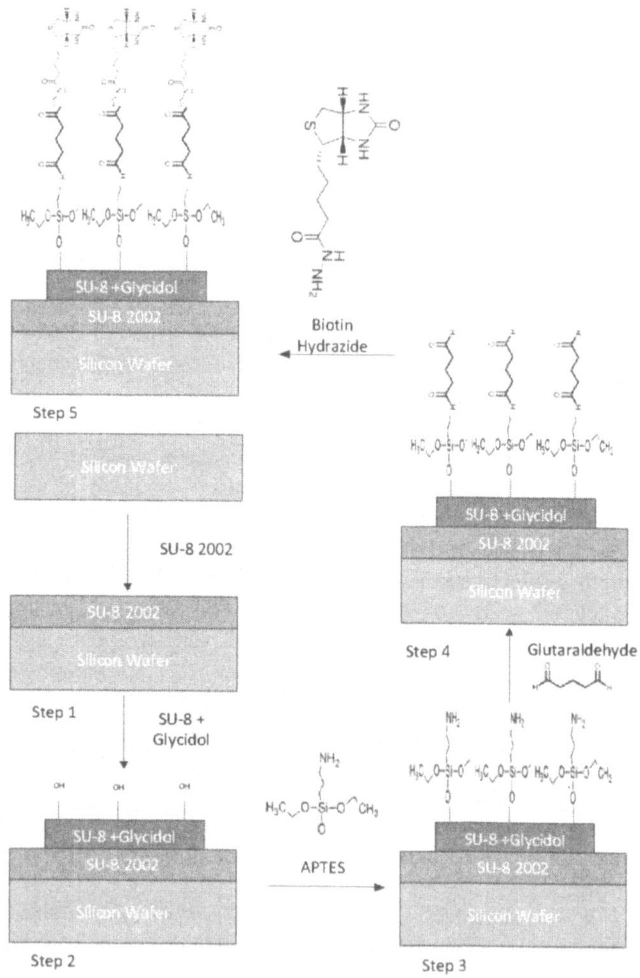

Figure 1: Schematic illustration of the SU-8 Bio- functionalization process

3 RESULTS AND DISCUSSION

A qualitative Fourier Transform Infra Red (FTIR) spectroscopy analysis to study the process was carried out using Bruker TENSOR37 system. FTIR Transmittance spectra characteristics were studied on SU-8+Glycidol film, Glutaraldehyde treated surface and Biotin Hydrazide immobilized surface.

A typical SU-8 film's FTIR transmittance spectra has epoxy-group and solvent signatures between 700 to 1800 cm^{-1} [8]. The Figure 2 is the spectra of the SU-8+Glycidol film which shows the presence of Hydroxyl groups by decreased transmittance around 3500 cm^{-1}. These Hydroxyl groups are part of Glycidol and acts as binding sites for the APTES molecules in the crosslinked layer.

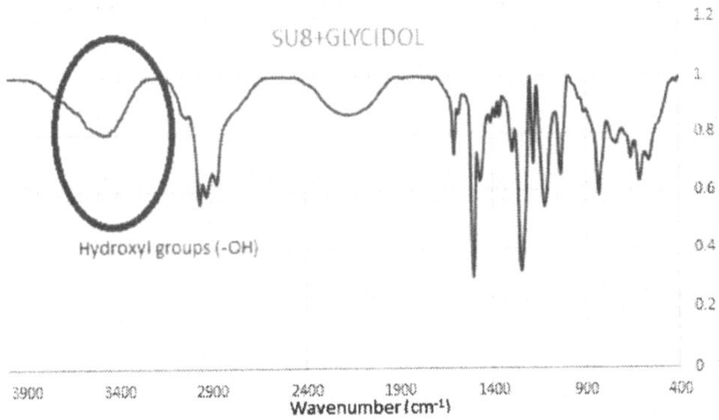

Figure 2: FTIR Transmittance spectra of SU-8+Glycidol film

Figure 3 indicates the presence of Aldehyde groups (2830–2695 cm^{-1}) after treating the SU-8+Glycidol surface with APTES and Glutaraldehyde solutions (GA curve). One of the Aldehyde groups in the Glutaraldehyde reacts with the Amine group ($-NH_2$) in APTES to form a bond. The other Aldehyde group of Glutaraldehyde is available for immobilizing molecules with $-NH_2$ at the surface.

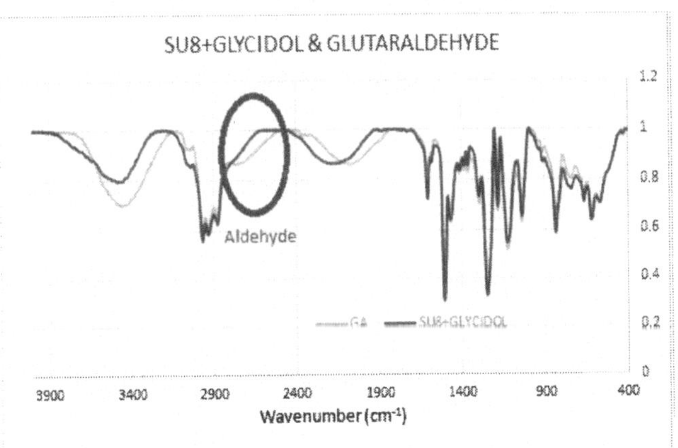

Figure 3: FTIR Transmittance spectrums of SU-8+Glycidol film and APTES+Glycidol (GA) treated films

The trasmittance at Aldehyde region of the spectrum has increased after immobilizing Biotin Hydrazide molecules. This can be observed in Figure 4.

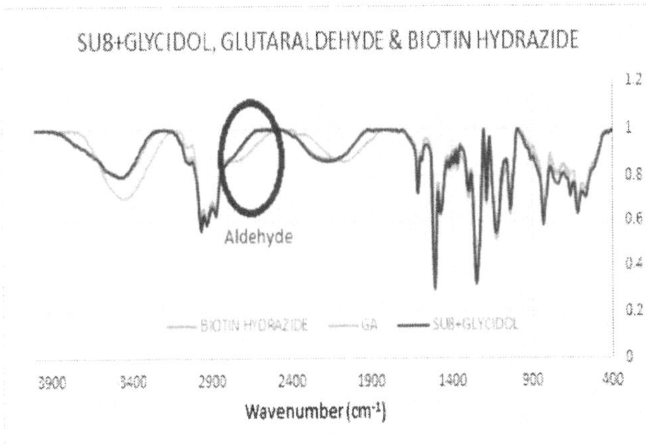

Figure 4: FTIR Transmittance spectrums of SU-8+Glycidol, GA and Biotin Hydrazide immobilized films

Figure 5 is the magnification of Aldehyde region of the FTIR spectrum. It can be observed that the Aldehyde region transmittance has increased to the level of normal SU-8+Glycidol film after Biotin Hydrazide immobilization.

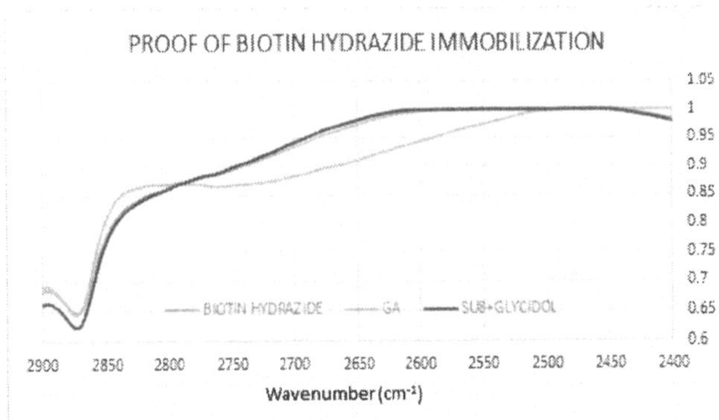

Figure 5: Magnified view of Aldehyde region of FTIR spectrum of SU-8+Glycidol, GA and Biotin Hydrazide

This is due to the well known fact that Aldehyde fuctional group in the Glutaraldehyde linker reacts with the Amine in the Hydrazide part of Biotin Hydrazide (in pH 6.5-7.5) to form Hydrazone bond [10]. This results in the reduction of Aldehyde signature in the FTIR spectrum. Hence we could achieve a stable Biotin functionalized SU-8 surface could be used to further immobilize other bio-molecules or sense Avidin and Streptavidin molecules.

It was also observed that the Aldehyde region transmittance has reduced only in SU-8+Glycidol regions of the sample and not in the background layer made up of pure SU-8. This is due to the lack of binding spots (-OH groups) for APTES molecules over the pure SU-8 surface. Hence selectivity in bio-functionlization has been achieved.

4 CONCLUSION

Selective bio-functionalization of SU-8 surface was performed successfully using a simple, low temperature chemical process. We believe that this procedure has tremendous potential in fabrication of Lab-on-a-Chip and Biosensor devices. Further research on using this process in fabricating Bio-MEMS structures and immobilizing various other molecules are on going in our group.

REFERENCES

[1] Chun-Lung Wu, Ming-Hung Chen, and Fan-Gang Tseng, "SU-8 Hydrophilic Modification by Forming Copolymer with Hydrophilic Epoxy Molecule," Proceedings of µTAS 2003, pp. 1117-1120, October 5 – 9, 2003.

[2] Maria Nordstrom, Rodolphe Marie, Montserrat Calleja and Anja Boisen, "Rendering SU-8 hydrophilic to facilitate use in micro channel fabrication," J. Micromech. Microeng. 14 (2004) 1614–1617.

[3] Vincent T. Moy, Ernst-Ludwig Florin and Hermann E. Gaub, "Intermolecular Forces and Energies Between Ligands and Receptors," Science, New Series, Vol. 266, No. 5183 (Oct. 14, 1994), pp. 257-259

[4] Manoj Joshi, Richard Pinto, V. Ramgopal Rao, Soumyo Mukherji, "Silanization and antibody immobilization on SU-8," Applied Surface Science 253 (2007) 127–3132.

[5] Z. Wang, M. Stangegaard, M. Dufva, J.P. Kutter, and A. Wolff, "A simple hydrophilic treatment of SU-8 Surfaces for cell culturing and cell patterning," Proceedings of µTAS 2005, pp. 1117-1120, October 5 – 9, 2003.

[6] Peng Xue, Jingnan Bao, Yon Jin Chuah, Nishanth V. Menon, Yilei Zhang, and Yuejun Kang, "Protein Covalently Conjugated SU−8 Surface for the Enhancement of Mesenchymal Stem Cell Adhesion and Proliferation," dx.doi.org/10.1021/la500048z | Langmuir 2014, 30, 3110−3117

[7] Joonyeong Kim," Investigation of the Formation and Structure of APTES Films on Silicon Substrates," www.piketech.com

[8] Stephan Keller, Gabriela Blagoi, Michael Lillemose, Daniel Haefliger and Anja Boisen, "Processing of thin SU-8 films," J. Micromech. Microeng. 18 (2008) 125020 (10pp)

[9] V.S. Goudar, S. Suran, M.M. Varma, "Photoresist functionalisation method for high-density protein microarrays using Photolithography," Micro & Nano Letters, 2012, Vol. 7, Iss. 6, pp. 549–553

[10] Greg T. Hermanson,"Bioconjugate Techniques, 2nd Edition," Academic Press, pp. 506-545

Comprehensive Analysis of High Performing Electrochemical Biosensors and Their Applications

David Assan[1], James C.K. Lai[2] and Solomon W. Leung[3]

[1]Department of Civil & Environmental Engineering, Idaho State University, Pocatello, ID 83209, USA, email: assadavi@isu.edu
[2]College of Pharmacy and Biomedical Research Institute, Idaho State University, Pocatello, ID 83209, USA, email: lai@pharmacy.isu.edu
[3] Corresponding author, Department of Civil & Environmental Engineering and Biomedical Research Institute, Idaho State University, Pocatello, ID 83209, USA. Fax: 208-282-4538; tel: 208-282-2524; email: leunsolo@isu.edu

ABSTRACT

We have mounted 5 different biocatalysts on a sensor platform to examine the performance of this electrochemical sensing system for the detection of different biomolecules/metabolites and environmental important molecules, with such we also compared how this sensing system fares with literature results of similar measurements. The biocatalysts are LDH (lactate dehydrogenase), GDH (glutamate dehydrogenase), human IgG (human immunoglobulin G), Hb (hemoglobin), and PSA (prostate-specific antigen). The sensor platform constitutes of a layer of biocomposite mounted on different electrodes made out of Au, Ag, Pt, and glass carbon; the biocomposite is fabricated with polymers and sol-gel Au nanoparticles with or without an extra layer of biomolecules. The targeting species for measurements include NH_4^+, NO_3^-, CN^-, H_2O_2 and the biomolecules specific to the biomolecules/antigens coated on the surface of the biocomposites. In this report, we will provide a comprehensive analysis of this sensing system, including the unique identification potentials and sensitivities. This sensing system can be a valuable tool in biomedical diagnosis and environmental forensics research.

Keywords: metabolites, diagnosis, nanoparticles, biosensor, environmental monitoring

1 INTRODUCTION

For more than two decades, research into the uses, functionality, and fabrication of immunosensors have received considerable interest from the biochemical and biomedical communities. Different methods of fabrication and combination of reagents are being employed to obtain efficient biosensors for a variety of purposes. In particular, focus has been concentrated on biosensors for the early detection of diseases in the human body such as cancers and congenital disorders. With such, the need for a highly sensitive and selective biosensor is therefore the key to unlock this avenue of research.

Biosensors are generally composed of immobilized layers of biomolecules such as proteins that are attached to support materials, these biomolecules are selectively coupled with the targeted substrates, thus define the selectivity of the sensors. Sensitivity of the biosensors often depends on the efficacy of the availability of the biomolecules that will bind with the targeted substrates, thus modification of how the biomolecules are linked to the supporting materials and their (distribution) configurations can improve sensitivity of the biosensors. For example, implanting of nanogold particles between the supporting materials and biomolecules enhances the binding surface availability for the substrates, thus increases the sensitivity of the biosensors [1, 2, 3, 4, 5].

In this study, we summarized the results of our findings of 5 different enzymes that we used for developing an electrochemical sensor platform that we can be used for varieties of biomedical applications and environmental monitoring. The resulting biosensors have superior detecting sensitivity and in general are orders of magnitude better than those found in literature reports

2. MATERIALS AND METHODS

2.1 Materials

The electrodes used in this study were highly modified and the surface of the electrodes were Gold (Au), Platinum (Pt) and Glassy carbon (GCE), they were purchased from Tianjin Aida Heng Sheng Co, Tianjin, China. The electrodes had a diameter of 0.2 cm. The platinum counter electrode had diameter of 0.1 cm and length of 0.5 cm. The biocatalysts LDH (lactate dehydrogenase), GDH (glutamate dehydrogenase), human IgG (human immunoglobulin G), Hb (hemoglobin), and PSA (prostate-specific antigen), and other chemicals such as cysteamine, melamine, bovine serum albumen (BSA), $AuCl_3HCl{\cdot}4H_2O$ (Au %> 48 %), ethanolamine (EA), and sodium citrate were purchased from Sigma-Aldrich Chemical Co, St. Louis, MO, USA. All the other chemicals used were of analytical grade. All experiments were carried out in a deoxyginated 0.1M phosphate buffer solution at pH 7.0.

2.2 Methods

The preparation of electrodes were the same as previously reported [2, 3], except that the biocomposite layers were comprised of the said enzymes as mentioned previously depending on applications. To enhance the performance of the sensors, BSA or EA was coated as a final surface layer to eliminate all the possible sites that might be competitive with the reactive bindings of the enzymes.

The electrochemical measurements were carried out on the Gamry 600 potentiostat.

3 RESULTS AND DISCUSSIONS

Table 1 summarizes the results of the 5 enzymes that we used for the analysis of different metabolites and environmental pollutants. The lowest detection limits for the sensors are in general lower than $1x10^{-18}$ M which allowed these sensors to be used for many applications that were not considered in biomedical diagnosis and environmental monitoring, as well as homeland security.

Thus far, the best anchoring material for the electrode sensor is Pt, however, GCE may be preferred in certain applications.

Table 1: Performance summary of Au, GCE, and Pt electrode and the associated enzymes.

Anchoring Material	Target species	Enzyme	Characteristic Peak (V)	Other Peaks (V)	Current Magnitude*, μA
Platinum	α-ketoglutarate /NH_4^+	GDH	0.75 (O)	0 (R)	12.6 (O)
Gold			0.75 (O)	-	7.27 (O)
Glassy Carbon			0.75 (O)	-	3.0(O)
Platinum	Lactate	LDH	0(R)	0.6 (O)	0.4 (R)
Gold			0.6 (O)	0.5 (R)	0.15 (O)
Glassy Carbon			0.5 (R)	0.6 (O)	0.37 (R)
Platinum	H_2O_2	Hemoglob in	0 (R)	0.75 (0)	4.33 (R)
Gold			0.485 (R)	1.1 (O)	2.04 (O)
Glassy Carbon			0.45 (R)	0.6 (O), 1.1 (O)	5.5 (O)
Platinum	NO_2^-	Hemoglob in	0 (R)	0.8 (O), 0.5 (R)	3.71 (O)
Gold			0.5 (R)	0.8 (O)	4.20 (O)
Glassy Carbon			0.45 (R)	0.8 (O)	2.24 (O)
Platinum	HIgG	Anti-HIgG	0.04 (R)	0.6 (O)	4.55 (R)
Gold			0.5 (R)	0.3 (O)	0.073 (R)
Glassy Carbon			1.1 (O)	0.4 (R)	1.95 (O)
Platinum	PSA	Anti-PSA	0.04 (R)	-	2.05 (R)
Gold			-0.2 (R)	-	0.032 (R)
Glassy Carbon			0.5 (R)	-	0.17 (R)

(O) = Oxidative peak
(R) = Reductive peak
*Concentration dependent of enzymes used.

4 CONCLUSIONS

Our research group has developed some of the most sensitive biosensors ever reported. These biosensors are selective and durable, that allow researchers and practitioners to explore applications in medical diagnosis and monitoring that were not possible in the past. Applications for these biosensors are yet to be developed.

5 ACKNOWLEDGMENT

This study was partially supported by a DOD USAMRMC Project Grant (Contract #W81XWH-07-2-0078).

REFERENCES

[1] S. W. Leung, D. Assan, J. C. K. Lai, "Evaluation of Anchoring Materials for Ultra Sensitive Biosensors Modified with Nanogold Particles and Enzymes", *Sensors & Transducers journal*, Vol.15, Special Issue, October 2012, pp.59-70.

[2] J.C.K. Lai, Y. Wang, W. Gao, H. Gu, and S. W. Leung, "Performance Comparisons of Nanoparticle Modified Sensor Electrodes for the Detection of Nitrite and Peroxide", in Technical Proceedings of the 2009 Nanotechnology Conference and Trade Show, Volume 2: Chapter 4: Biosensors and Diagnostics, pp. 233-235.

[3] Solomon W. Leung, James C.K. Lai and David Assan, "Performance Comparison of Immunosensors Modified with Polymers, Nanoparticles and Antibody", in proceeding of the NSTI Nanotechnology Conference and Expo – Nanotech 2013, May 12-15, Washington, D.C., Vol 3, Chapter 1, Biosensing, Diagnostics and Imaging, pp. 53-56.

[4] Solomon W. Leung, David Assan· and James C.K. Lai, "Characterization of an Ultra-High Performance Immunosensor Modified with Sol-Gel Nanogold Particles and Its Applications", in proceeding of the International Conference on Mathematics, Engineering

& Industrial Applications, May 28 to 30, 2014 (ICoMEIA 2014), Penang, Malaysia, to be

[5]. David Assan, X. Gao, James C.K. Lai, and Solomon Leung, "Biocomposite Electrode Sensor for the Detection of Oxidative Reactions", in Technical Proceedings of the 2014 NSTI Nanotechnology Conference & Expo, Vol 1, section 6, pp. 505-508.

A Fluorescent Nanosensor Based on Magnetic Silica Nanoparticles as Illicit Drug Vapor Logic Gate

Shohre Rouhani*, Soheila Haghgoo **

*Center of Excellence for Color Science and Technology (CECST), Institute for Color Science and Technology, Tehran, Iran,rouhani.ac.ir@icrc

*Department of Organic Colorants, Institute for Color Science and Technology (ICST), Tehran, Iran

**Pharmaceutical Department, Food & Drug Laboratory Research Center, Food & Drug Organization (FDO), Ministry of Health, Tehran, Iran, shaghgoo@hotmail.com

ABSTRACT

Herein we report a fluorescent nanosensor base on magnetic silica nanoparticle as illicit drug logic gate. A novel naphthlimide derivative was synthesized (**F**) , characterized and doped into the magnetic silica nanoparticles (Fe_3O_4@SiO_2/**F**). It Fluorescence intensity markedly enhanced by adding of metamphetamine.

Keywords: naphthalimide,sensor,logic gate ,emission,nano magnet, silica nanoparticle, Illicit Drug.

1 INTRODUCTION

Growing up abuse of amphetamine-type stimulants (ATS) over the 20 last years, particularly,involving amphetamines (amphetamine and methamphetamine) and substances of the "ecstasy"-group , among teen-agers has become a global problem. Methamphetamine "meth" can be found in the form of pills, capsules, powder or chunks, called crank, speed, crystal or ice. It can be smoked,snorted, injected or eaten. Meth is a powerful, highly toxic, addictive drug that is illegally "cooked"in household labs. Abuse of methamphetamine can result in a stimulant psychosis which may present with a variety of symptoms (e.g. paranoia, hallucinations, and delusions).

Different techniques for determination of metaaphetamine and related compounds are used such as gas GC/MS, HPLCLMS, FTIR, voltammetry, capillary electrophoresis and Surface ionization [1-4].These conventional methods, however, require large and expensive instruments, complicated experimental procedures. Thus, there is a need for development of a simple, rapid, and sensitive MA detection method. In recent years fluorescent sensing has been considerable attention for its high sensitivity, real time, online detection and easy operation. Our group has reported the design and synthesis of fluorescence dyes for sensor application [5-7].

In recent years nanoparticles have been of great interest because of their potential applications in cell targeting, drug delivery, bio separation, and magnetic resonance imaging. Magnetic and fluorescent nanoparticles are of particular importance due to their broad range of potential applications [8]. It is expected that the combination of magnetic and fluorescent properties in one nanocomposite would enable the engineering of unique multifunctional nanoscale devices, which could be manipulated using external magnetic fields. In this way, there are only few reports about the fluorescence sensing of metaaphetamine . Cheng and coworkers reported a highly fluorescent polyfluorene capable of producing highly sensitivity for metaaphetamine [9]. They also were developed a fluorescent chemosensor for methamphetamine vapor sensing based on fluorene–thiophene-based thin-film[10].Recently we reported the synthesized of a novel naphthalimide flourescence derivative for selectve recognition of methamphetamine[11] .

Herein we report a fluorescent nano chemosensor base on magnetic silica nanoparticles as illicit drug logic gate based on naohthalimide derivative.

2 EXPERIMENTAL

2.1 Materials

Ferric chloride hexahydrate ($FeCl_3$), ferrous chloride tetrahydrate ($FeCl_2$), dichloromethane (DCM), ammonium hydroxide (28wt%),tetraethyl orthosilicate (TEOS), and potassium carbonate, Acenaphthene and 3-amino propyl triethoxysilane (APTES) ,were obtained from Merck. All used amines obtained from Merck Chemical Co. and used as received.

2.2 Synthesis

Fe_3O_4 NPs were prepared according to reported literature [12]. $FeCl_2 \cdot 4H_2O$ (2g, 0.01mol) and $FeCl_3 \cdot 6H_2O$ (5.4 g, 0.02 mol) were dissolved in 120mL of deionized water. $NH_3 \cdot H_2O$ (60 mL, 28 wt%) was added under vigorous mechanical stirring. The color of the suspension turned black immediately. Afterward, the mixture was kept at 70°C for 30 min. The mechanical stirring and nitrogen atmosphere were used. After cooling down, the precipitated powders were collected by magnetic separation and washed with deionized water. The core-shell $Fe_3O_4@SiO_2$ NPs were synthesized by a sol-gel process though the hydrolysis and condensation of TEOS in ethanol and ammonia mixture. Briefly, 0.56 g citrate-functionalized Fe3O4 NPs and $NH_3 \cdot H_2O$ (5.0 mL,28 wt%) were stirred in a flask charged with 120mL ethanol.

Afterward, 4.0mL TEOS was added drop wise. The mixture was kept at room temperature for 8 h under N_2 with violent mechanical stirring. The products were collected with a magnet and washed with ethanol and toluene.

The intermediate, compound 1, was synthesized from ace-naphthene, according to the reported method in literature [13].The intermediate compound 2 was prepared at 76% yield bythe reaction of 0.52 mmol of compound 1 with 2.88 mmol of 3-aminopropyl-triethoxysilane, in ethanol. The solvent evaporatedby rotary evaporation, and the crude product recrystalized with ethanol to give compound 2 (Scheme 1)

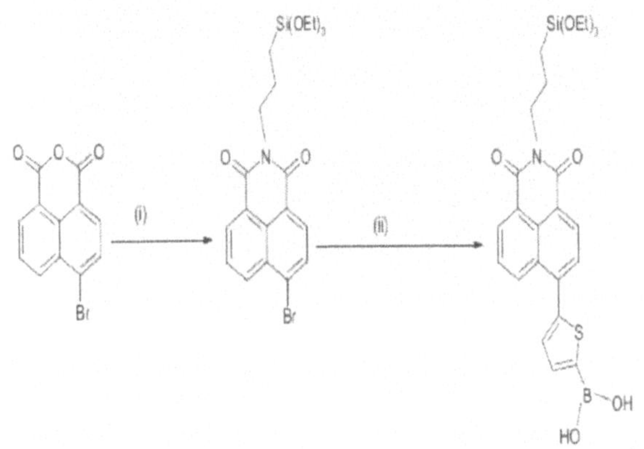

1 2 F

Scheme 1.Synthesis rute for flourescence sensor

Compound **F** was synthesized through thepalladium-catalyst Suzuki reaction [14]. The intermediate (0.5 mmol), $Pd(PPh_3)$ 4(0.011 mmol), K_3PO_4(0.78 mmol) and 2,5-thiophenediyl-bisboronicacid (0.8 mmol) in toluene were mixed, stirred and reflux for 24 h.The crude products were concentrated under vacuum, and purified by chromatography . Synthesis compounds were characterized by HNMR,CNMR,FTIR and elemental analysis.

3 RESULTS AND DISCUSSION

The TEM images of $Fe_3O_4@SiO_2$/**F** NPs were obtained. The average diameter of magnetic silica NPs was about 28 nm . The $Fe_3O_4@SiO_2$/**F** NPs have uniform spherical morphology.

The variation of functional groups on $Fe_3O_4@SiO_2$/**F** NPs was characterized by FT-IR (Figure 1). Fig. 1, shows new bands refer to bonding of compound **F** on the silica surfacfter and befor bounding on nanoparticles. The peak at 1661 is corresponds to C=O stretching.The band at 1544 $cm^{-1}$is due to the C=C stretching, which comes from the aromatic ring. The band at 1343 $cm^{-1}$is C-N stretching which is corresponded to naphthalimide ring. As expected, strong Si-O-Si stretching band in 1115 $cm^{-1}$and Si-OH stretching bandin 810 $cm^{-1}$are observed in $Fe_3O_4@SiO_2$/**F**.

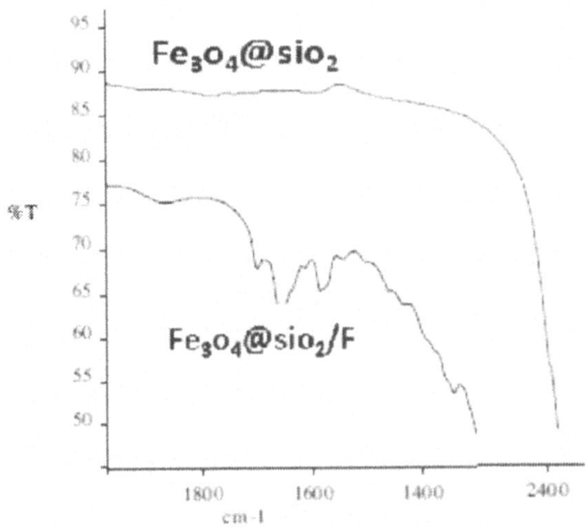

Fig 1. FTIR spectra of $Fe_3O_4@SiO_2$/F and $Fe_3O_4@SiO_2$

The UV-Vis and fluorescence property of synthesized compound in chloroform and solid state were investigated. The synthesized nanohybrid can emit blue emission against metaaphetamine vapor(scheme 2).

It exhibited selective recognition towards metaaphetamine among different amines.It fluorescence emission changes by the inputs of metaaphetamine can be viewed as a logic gate at the nanoscale level. Fluorescence intensity markedly changed as 190% by adding of metaaphetamine , the detection limits for metaaphetamine vapor were estimated to be 16 ppm.

In order to investigate the fluorescence sensing of synthesized fluorophore (**F**) against metaaphetamine , titration of **F** (1×10^{-8}mol) was performed in acetonitril solution with1×10^{-1}mol of metaaphetamine . As can beseen in Fig. 2 a) and b) fluorescence emission of nanosensor displayed enhancements upon the addition of 0–100 equivalent of metaaphetamine . Fluorescence of $Fe_3O_4@SiO_2$/**F** upon addition of metaaphetamine under 360 nm excited UV lamp gradually increases with 190% enhancement at emission maximum wavelength. We suggest that the fluorescence "turn-on" sensing mechanism could be due to the binding induced weakening of internal heavy-atom effect as previously reported by Cheng . Sulfur, as a heavy atom, may strengthens the spin-orbit coupling and increases the intersystem crossing transitions from S1 → T1, so leading to decrease the fluorescence quantum yield of the molecules. Thus, **F** emits weak fluorescence because of thiophene units. However,when amines such as metaaphetamine bind to thiophene unit, their interaction weakens the internal heavy atom effect and lead to fluorescence enhancement

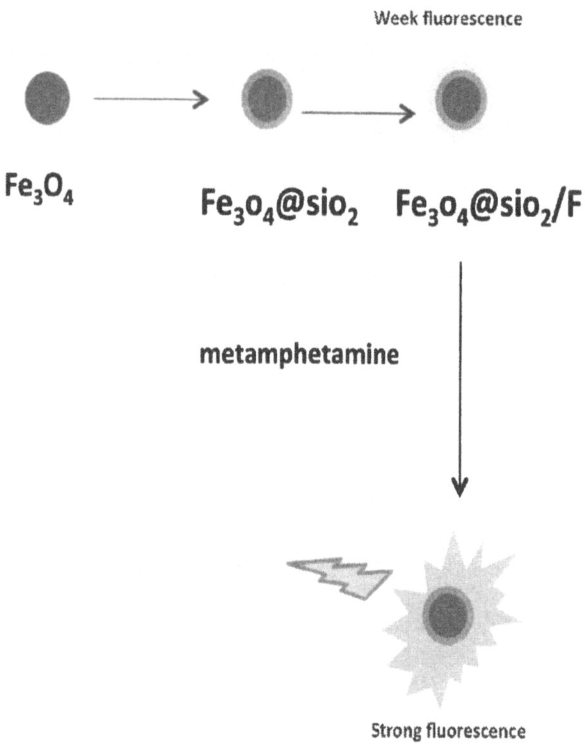

Scheme 2. Shematic response of nanosensor

To evaluate the selectivity of the nanosensor, fluorescence emission changes were obtained in the presence of different amines.The fluorescence intensity shows a little chang against several different amines . But florescence intensity considerably increased just against metaaphetamine . In order to study the interfering effects, the intensity changesof Fe_3O_4@SiO_2/**F** upon addition of different amines in the presence of metaaphetamine (MA:amine, 1:10) were investigated . As can be seen in fig. 2 the excellent selectivity of Fe_3O_4@SiO_2/**F** for metaaphetamine against different amines obtained. Therefore, it seems these amines do not interfere with the determination of metaaphetamine. Consequently, itis concluded that the new nanohybrid Fe_3O_4@SiO_2/**F** can be used as a chemosensor for the detection of different concentrations of metaaphetamine with high selectivity.

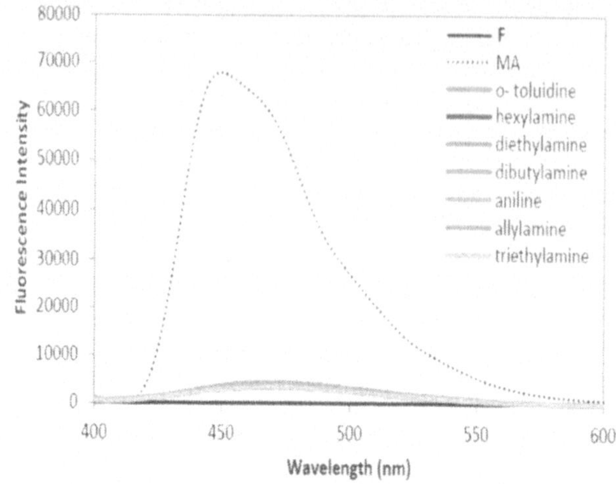

a)

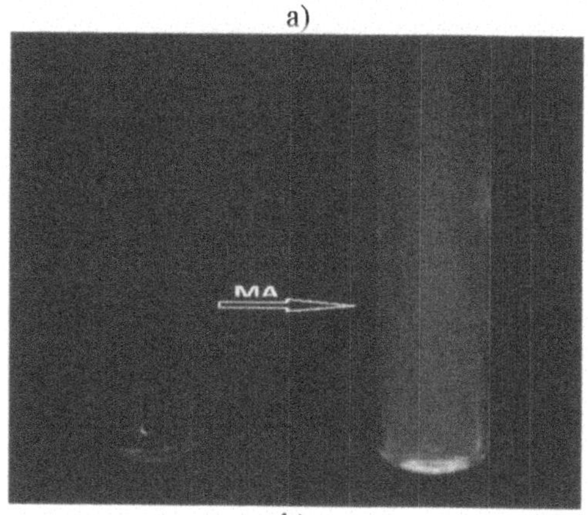

b)

Fig. 2. Fe_3O_4@SiO_2 –**F** in the absence and presence of metaaphetamine. a) emission spectra in the presence of various amine, b) visual change under uv irradiation.

In order to evaluate the applicability of the proposed nanosensor, metaaphetamine was determined in laboratory prepared samples containing different amounts of interfering amines by two method of HPLC and proposed nanosensor. Recovery of two tecnique illustrated in Table 1. As can be seen there is a satisfactory agreement between results of the the presented method and those obtained by (HPLC) . The results given in Table 1 show that the recovery of MA from samples are good (relative standard deviation (RSD) less than 4.16% (n = 6)).

Sample	Known amount	Found amount	RSD%
1	12	13±0.5	2.74
2	60	59±0.4	3.1

Table 1. Real sample analysis of metaaphetamine (μmol L^{-1}).

4 CONCLUSION

A fluorescence "turn-on" molecular logic gate system (Fe_3O_4@SiO_2/F) was developed based on a functionalized naphthalimide dye derivative linked to the magnetic silica nanoparticles surface. Reusable logic gate nanosensor shows high selectivity toward metaaphetamine vapor in ultra trace level.

REFERENCES

[1] M. Terada, J. Chromatogr. 237 (1982) 285–292.

[2] L.W. Chung, K.L. Lin, T.C.C. Yang, M.R. Lee, , J. Chromatogr. A 1216 (2009)4083–4089

[3] S. Meng Wang, H.L. Chiang, Y.S. Giang, J. Forensic Sci. 2 (2003) 27–34.

[4] V. Pavlova, S. Petrovska-Jovanovi´c, Acta Chromatogr. 18 (2007)157–167.

[5] M. Nieddu, G. Boatto, M.A. Pirisi, E. Baralla, Rapid Commun. Mass Spectrom. 23 (2009)3051–3056.

[6] M.D.R. Fernandez, S.M.R. Wille, V. di Fazio, M. Gosselin, N. Samyn, J. Chromatogr. B 878 (2010)1616–1622.

[7] B.P. Noriyuki, B.P. Susumu Fujita, H. Ohta, M.P. Makoto Fukuba, A. Toriba,K. Hayakawa, J. Forensic Sci.53 (2008) 1367–1371.

[8] H. Dai, Y.M. Wang, X.P. Wu, L. Zhang, G.N. Chen, 24 (2009)1230–1234.

[9] Z.M. Cai, Z.J. Lin, X.M. Chen, T. Jia, P. Yu, X. Chen, , Luminescence 25 (2010) 367–372.

[10] W. Matapatara, P. Thongnopnua, V. Lipipun, Asian Biomed. 1 (2007) 161–179.

[11] S. Rouhani, S. Haghgoo , Sensors and Actuators B 209 (2015) 957–965

[12] Y. Zhou, S. Wang, Y. Xie et al., Nanotechnology,vol. 19, no. 17, Article ID 175601, 2008.

[13] P.H. Grayshan, A.M. Kadhim, A.T. Peters, J. Heterocycl. Chem. 11 (1974) 33–37.

[14] L. Xiea, J. Cui, X. Qian, Y. Xu, J. Liu, R. Xu, Bioorg. Med. Chem. 19(2011) 961–967.

Environmental Health Science and SBIR/STTR Small Business Grants

D. Shaughnessy, H. Henry, K. Ahlmark, K. Mease, D. Balshaw

National Institute of Environmental Health Sciences
Division of Extramural Research and Training
Research Triangle Park, NC USA 27709, www.niehs.nih.gov/sbir

ABSTRACT

The NIEHS sets aside a percentage of extramural funds to support technology development for environmental health research through Small Business Innovation Research (SBIR) and Small Business Technology Transfer (STTR) grants. SBIR/STTR technologies supported by NIEHS reflect the Institute's broad mission, ranging from assays to detect DNA damage and mutations to high-throughput assays for large-scale toxicant screening to improved methods for assessing both external exposures and internal dose of chemicals. Small businesses are also developing innovative technologies, products, and devices that can be used at Superfund or other contaminated sites for the detection and remediation of hazardous substances. The NIEHS Worker Training Program (WTP) program focuses on the development of Advanced Training Technology (ATT) products for the health and safety training of hazardous materials (HAZMAT) workers, emergency responders, and skilled support personnel. The NIEHS supports the development of tools and materials to educate and train individuals and groups on environmental health topics. Technologies developed by NIEHS-supported small businesses are used in laboratory research, high-throughput screening efforts including ToxCast and Tox21, environmental epidemiology studies and community-based research to improve public health.

Keywords: NIEHS, small business, SBIR, environmental technologies, funding

1 NIEHS SBIR/STTR AREAS OF INNOVATION AND RESEARCH

NIEHS small business grant programs, SBIR and STTR, use a combination of research, technology transfer, and communication strategies to aid the mission of NIEHS. NIEHS SBIR/STTR aims to fund products that improve human health in the areas of:

- **Exposure Assessment** – technologies include wearable sensors, remote sensing, bio-monitoring, exposure biology, and Exposome research tools
- **Exposure-Disease Relationship** - toxicity screening tools such as high-content imaging, medium-high throughout *in vitro* assays, organotypic and computational models
- **Susceptibility** – tools which aide in the understanding of gene-exposure relationships, combined exposures, comorbidity, and life stage windows of susceptibility
- **Prevention and Intervention** – Technologies which detect, reduce, or eliminate exposures to environmental toxicants
- **Education/Training** - Tools which can assist HAZMAT, emergency responders, and skilled support personnel
- **Outreach** – educational tools for health literacy, community based participatory research, and citizen science

2 WHO WE HAVE FUNDED

Aligned with the previously mentioned areas of innovation and research at NIEHS, the following are examples of small businesses and their technologies that NIEHS has funded. These examples are only a small sampling of the technologies that NIEHS funds. Each technology described is a solution to a problem that has a sizable market and thus has significant commercial potential.

2.1 Exposure Assessment

R44ES021678
TF Health, Co
Tempe, AZ
Program Directors: Erica Forzani,
Francis Tsow (VP of R&D)
erica.forzani@breezing.co

PROBLEM it is challenging and expensive to measure environmental toxicants to study relationships between chemical exposures and diseases for more effective disease prevention

SOLUTION low cost, easy to use, pocket-sized, UL certified device, with high temporal and high spatial resolutions, and wireless communication for personal exposure monitoring

MARKET petrochemical industry, epidemiologists and exposure science professional, fire departments

2.2 Exposure Assessment

R44ES021678
TF Health, Co
Tempe, AZ
Program Directors: Erica Forzani,
Francis Tsow (VP of R&D)
erica.forzani@breezing.co

PROBLEM it is challenging and expensive to measure environmental toxicants to study relationships between chemical exposures and diseases for more effective disease prevention

SOLUTION low cost, easy to use, pocket-sized, UL certified device, with high temporal and high spatial resolutions, and wireless communication for personal exposure monitoring

MARKET petrochemical industry, epidemiologists and exposure science professional, fire departments

2.3 Exposure-Disease Relationship

R44ES018017
Litron Laboratories, LTD
Rochester, NY
Program Director: Stephen D Dertinger
sdertinger@litronlabs.com

PROBELM current in vivo gene mutation assays require standalone toxicology studies that use expensive transgenic animals and large blood volumes

SOLUTION In Vivo MutaFlow® kit is a flow cytometry Pig-a gene mutation assay kit; based on low blood volume samples and an endogenous gene reporter, it readily integrates into existing in vivo toxicology studies that use standard rodent models

MARKET commercially available kits and fee-for-service blood analysis services support pharmaceutical and chemical companies, also CROs and government laboratories

2.4 Susceptibility

R44ES020074
MatTek Corporation
Ashland, MA
Program Director: Yulia Kaluzhny
ykaluzhny@mattek.com

PROBLEM current animal-based test methods (Draize rabbit eye test) have animal welfare concerns and in vitro alternatives lack the ability to completely model in vivo responses

SOLUTION EpiOcular™ human-derived epidermal keratinocytes cultured to form a stratified squamous epithelium similar to that found in the cornea. EpiOcular™ Eye Irritation Test (EIT) will be approved as a "validated Draize replacement" in EU in 2014. OECD draft test guideline has been submitted for review.

MARKET direct sales of kits, training services, and POC studies to cosmetic, personal care, and ophthalmologic companies.

2.5 Prevention and Intervention

R43ES020634
Seacoast Science, Inc
Carlsbad, CA
Program Director: Sanjay V Patel
sanjay@seacoastscience.com

PROBLEM conventional remediation detection methods are expensive, require repeated user intervention, and need frequent replacing

SOLUTION low-cost, sensitive, unattended remote vapor intrusion monitoring using a sorbent preconcentrator to collect air samples, a gas chromatograph to separate chemical species, and a MEMS microsensor for detection of trichloroethylene, perchloroethylene, and benzene

MARKET remediation specialists and government agencies for remote tracking

2.6 Education/Training

R44ES020135
Nicolade R&D LLC
West Lebanon, NH
Program Director: Roberto Javier Nicolalde
roberto.nicolalde@gmail.com

PROBLEM process of translating existing reference materials into mobile-friendly formats is manual and labor intensive

SOLUTION mRendering™ mobile method delivers just-in-time, short, incident specific safety training, and reference materials using a back-end document processing engine that automatically parses, analyzes, mark-ups, and organizes documents

MARKET mRendering™ will be a product and service to training organizations that support first responders, skilled support personnel, and volunteer emergency personnel

2.7 Outreach

R41ES023706
Science Take-Out, LLC
Pittsford, NY
Program Director: Dina Grossman Markowitz
contact@sciencetakeout.com

PROBLEM nearly 25% of death and disease is linked to environmental factors (WHO, 2006). The general public needs a basic understanding, but this can be difficult to incorporate into curriculums

SOLUTION six fully assembled Science Take-Out® kits, which include activities, materials, and clear/concise teacher guides on environmental health science topics

MARKET existing Science Take-Out® kits sell > 17,000 units annually, ranging from $8 - $18/kit, to middle and high schools nationally and globally

3 HOW TO APPLY

For detailed application, instructions please visit www.sbir.nih.gov. Below is a brief overview of the application process.

Confirm your small business meets the eligibility requirements and contact an NIH Program Officer to ensure your technology is aligned with the mission of the institute.

- Complete the five required registrations (6-8 weeks to complete)
    - DUNS
    - SAM.gov
    - Grants.gov
    - eRA Commons
    - SBA Registry
- Choose the funding opportunity announcement for your application
    - SBIR Omnibus (small business)
    - STTR omnibus (small business + non-profit research institute)
    - Targeted SBIR/STTR Requests for Application (RFA)
- Submit your SBIR/STTR grant application to the NIH electronically
- NIH Center for Scientific Review evaluates your grant on scientific merit and commercialization potential
- Funding decisions and awards are made (Application to Award is ~ 6 months)

4 FUNDING PHASES

For detailed information about the funding phases, please visit www.sbir.nih.gov. Below is a brief overview of the funding phases.

Phase I – feasibility study for 6 months - up to $150K direct costs

- Niche assessment program – provides active Phase I awardees with market research, consumer, and competition analysis

Phase II – Full R&D for 2 years – up to $1M in direct costs

- Commercialization Assistance Program (CAP) – Trains Phase II awardees on developing tailored market entry strategies, building strategic alliances and partnerships, and developing FDA regulatory and reimbursement paths, as well as financing strategies and expertise on intellectual property matters

Fast Track – One application for Phase I and Phase II that is submitted and reviewed together

Phase IIB – Additional funding for awards or projects that require extraordinary time and effort in the R&D phase. Must have initial Phase II award to apply. Up to $1M total costs per year for up to 3 years.

Phase III – Commercialization - not funded by SBIR/STTR funds

Cyanide Detection by Highly Modified Sol-Gel Biocomposite Sensor

Seyedeh Mozneb[1], James C.K. Lai[2] and Solomon W. Leung[3]

[1]Civil & Environmental Engineering Department, Box 8060, Idaho State University, Pocatello, ID 83209, USA. Fax: 208-282-4305; tel: 208-282-2524; email: moznseye@isu.edu

[2]Biomedical and Pharmaceutical Sciences Department and Biomedical Research Institute, Idaho State University, Pocatello, ID 83209, USA Fax: 208-282-4305; tel: 208-282-2275; email: lai@pharmacy.isu.edu

[3]Corresponding author, Civil & Environmental Engineering Department and Biomedical Research Institute, Idaho State University, Pocatello, ID 83209, USA Fax: 208-282-4538; tel: 208-282-2524; email: leunsolo@isu.edu

ABSTRACT

Cyanide is an important industrial chemical but is also very toxic. Cyanide binds strongly with hemoglobin than that of oxygen which causes histoxic hypoxia, if cyanide is ingested into the bloodstream. This pairing reaction can thus be utilized for the development of a cyanide detection sensor. Our research group has been developing a sensor platform that can utilize different enzymatic couplings for the detections of metabolites and species that are of health and environmental concerns at extremely low concentrations. In this study, we report the detection of cyanide using our sensor platform with a biocomposite layer made up of polymers, nanogold sol-gel, and hemoglobin. The biocomposite is coated on the surface of anchoring materials such as Au, Pt, and glass carbon electrode. This ultra high sensitive biosensor can detect cyanide at concentration levels orders of magnitude lower than any reports found in the public domain. This report will reveal results in twofold: at high concentration levels of $1x10^{-6}$ M and low levels of $1x10^{-18}$ M, depending on the applications. We will also discuss factors that affect the sensitivity, interference, durability, and performance enhancement of this sensor. This cyanide sensor can have unique applications in homeland security, health and environmental monitoring.

Keywords: cyanide, biocomposite, performance, Au nano particles, biosensor, homeland security.

1 INTRODUCTION

Rapid development of microelectronics and nanotechnology enable fabrication and functionality of immunosensors to territories that were inconceivable decades ago. Different methods of fabrication and combination of reagents are being studied to obtain optimal and efficient biosensors for a variety of purposes. Our research group has been development a platform for electrochemical sensing that is versatile with high detection sensitivity [1, 2, 3, 4, 5], it can be used for detecting various chemical species and bioelectrolytes for environmental monitoring, as well as for health care diagnostic testings.

Cyanide molecule is made of a carbon atom triple bonded to nitrogen atom and is named the cyano group. The cyano group can react with several organic and non-organic materials, such as sodium, potassium, hydrogen, methyl and so on. Depending on how the bonding occurs and what is the material it binds to, it could show as a colorless gas or in a crystal form. Solid cyanides like potassium cyanide or sodium cyanide might or might not have a bitter almond smell when reacting with solutions. Cyanide can happen in many forms. It is found in nature, bounded to a sugar molecule in fruits like apples and peaches. It is also produced by actions of fungi and bacteria. Due to its strong affinity to hemoglobin, once it enters the human body, hemoglobin in the body will be saturated with cyanide and the victim will consequentially suffocate within minutes.

Metal cyanides have very high aqueous solubility and thus cyanide also is a widely used industrial solvent; but the acute toxicity of cyanide is deadly even in a very small amount. Recent homeland security issue heightens the need for a fast and accurate cyanide detection device. The goal of this study is to introduce a cyanide detection biosensor that is capable of detect cyanide concentrations down to $1x10^{-18}$ M, which is orders of magnitude lower than any report in the literature. Table 1 in the following summarizes the recent development in high performance cyanide detection.

2 MATERIALS AND METHODS

2.1 Materials

The electrodes used in the modification were Gold (Au), Platinum (Pt) and Glassy carbon (GCE), purchased from Tianjin Aida Heng Sheng Co, Tianjin, China. The electrodes had a diameter of 0.2 cm. The platinum counter electrode had diameter of 0.1 cm and length of 0.5 cm. The electrochemical measurements were carried out on the Gamry 600 potentiostat. cysteamine, melamine, bovine serum albumen (BSA), hemoglobin (Hb), $AuCl_3HCl\cdot 4H_2O$ (Au %> 48 %), ethanolamine (EA), and sodium citrate were purchased from Sigma-Aldrich Chemical Co, St. Louis, MO, USA. All the other chemicals were of analytical grade. All experiments were carried out in a deoxyginated 0.1M phosphate buffer solution at pH 7.0. The electrode surface was composed of Pt, Au, or glossy carbon.

Table 1: Summary of electrochemical cyanide sensing and the detection limits from recent literature.

Substances detected	Method	Detection limit	Year	Source
Cyanide compounds	Cytochrome C biosensor with electrochemical method	4.3-9.1 mu M.	2012	Fuku et al.
Cyanide groups	Hybrid Au/CNT systems with electrochemical deposition and Raman spectroscopy	Good sensitivity	2011	Orlanducci et al.
Organophosphates (cyanide ions)	Silicon-based microsensor with a teflon nanoporous membrane supported on silicon	1.5×10^{-11}M	2011	Sayyah et al.
Cyanide	miniaturized silicon.-based sensor system	μ molar	2010	Turek et al.
Cyanide	Hydroxyapatite nanowires array (HANWA) biosensor performed by template-assisted electrode-position	0.6 ng/ml	2010	Wang et al.
Cyanide	Silver doped silica nanocomposite synthesized via a sol-gel technique combined with a nanoparticle preparing method	1.4×10^{-8} M	2009	Taheri et al.
Cyanide	Fused indoline and benzooxazine fragment chemosensor	1 μ M	2008	Ren et al.
Cyanide	luminescent bacteria and PbO_2 electrochemical sensor	38.38 - 0.60 μg/mL for cyanide	2007	Liu et al.
Cyanide in blood	Amperometric test and an electrochemical sensor	4 μ M	2006	Lindsay et al.
Hydrogen cyanide gas	Simple electrochemical method	0.66 mg/m^3	1999	Rao et al.
Cyanide	Amperometric detection of the enzymatically generated products	2×10^{-6} M	1998	Wang et al.
Cyanide	Electrochemical minisensor:	4.9 nM or 0.317 μg/L	1997	Siontorou et al.
Hydrazine compounds	Cyclic voltammetry on a nafion/ruthenium(III) modified glassy carbon	1×10^{-7} M	1998	Casella et al.
Cyanide	A photoelectrochemical sensor	0.001–0.100 M	1996	Licht et al.
Cyanide ions	Simple electrochemical method	10^{-6} M	1997	Nikolelis et al.
SCN^-, $SeCN^-$	Amperometric flow sensor	25 μM - 0.8 μM (at pH 2.0) for SCN^-, and 0.1-20 μM (at pH 7.0) for $SeCN^-$	1994	Cookeas et al.

2.2 Methods

The preparation of electrodes were the same as previously reported [2, 3], except that the biocomposite layers were comprised of the Hb only. To enhance the performance of the sensors, BSA or EA was coated as a final surface layer to eliminate all the possible sites that might be competitive with the reactive bindings of Hb and cyanide.

3 RESULTS AND DISCUSSIONS

3.1 Cyanide detection in extremely low concentrations

As shown in Figure 1, a typical potentiogram of cyanide detection with the characteristic reductive peaks at -25 mV. Figure 2 is the reduced results of cyanide concentrations from 1×10^{-18} M to 1×10^{-14} M for a Pt electrode that was coated with and without BSA. BSA enhanced the performance the electrode in the extreme low concentrations, and eventually both coated and non-coated electrodes converged at about 1×10^{-6} M. A rerun of the same coated electrode after 31 days idling in neutral buffer solution at pH 7.0 at 4°C (blue curve) gave about the same readings as a freshly prepared electrode, which also demonstrates that this biosensor is durable and implies the biocomposite layer where the enzymatic reaction takes place remains active after a long period of idling.

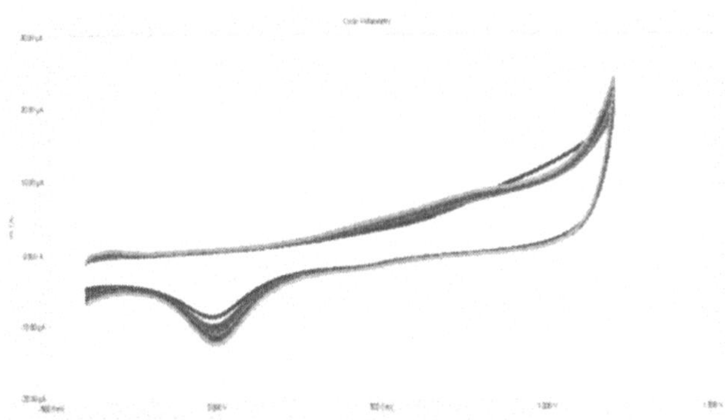

Figure 1: Potentiogram of cyanide detection in pH 7.0 phosphate buffer with a Pt electrode biosensor. Note that reduction peak increases with concentration at -25 mV.

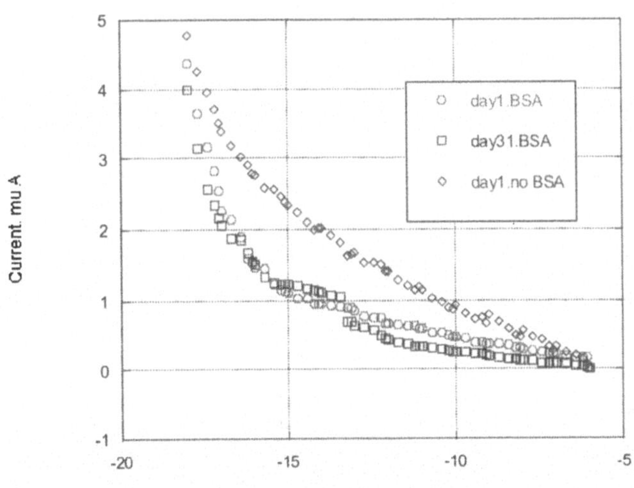

Figure 2: Cyanide detections from 1×10^{-14} M to 1×10^{-18} M.

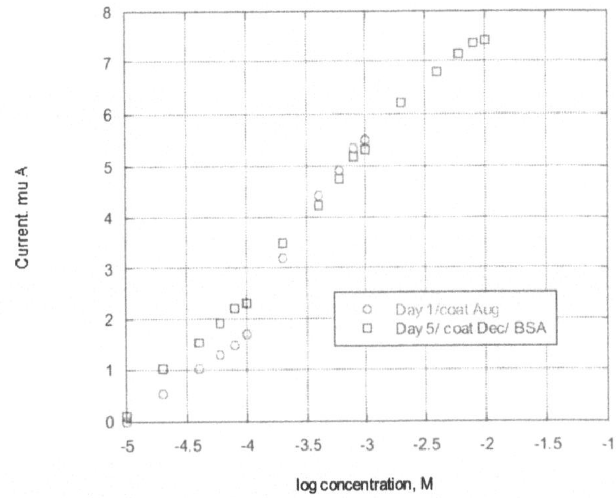

Figure 3: Cyanide detections from 1×10^{-5} M to 1×10^{-2} M

3.2 Cyanide detection in low concentrations

Figure 3 is the measurements of cyanide ion at concentrations of 1×10^{-5} M to 1.0×10^{-2} M of a Pt electrode sensor. Note that the reductive measurements became oxidative, which started at about 1x10-6 M (data not shown). The concentration range illustrated above is important because that is the measurement range that industrial applications often apply. The coating of BSA does not seem to have much effect to the performance of the sensor at this concentration range.

3.3 Performance of electrode sensor anchored with different materials.

Performance of this ultra-high sensitive electrode biosensor was compared using Pt, Au, and GCE as anchoring material. In spite of that all said anchoring materials would work in our experiments, the preferred material ranking is Pt > Au > GCE. Figure 4 is the performance of the Au electrode

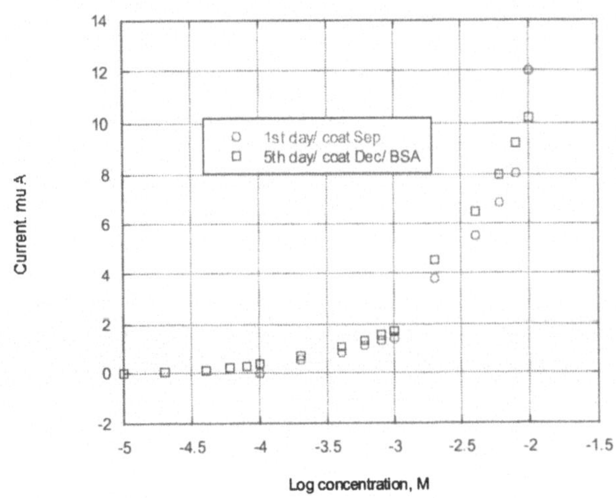

Figure 4: Measurements of cyanide ions with an Au sensor electrode.

4. CONCLUSIONS

Our results indicated that this ulta-high performance cyanide biosensor cyanide is extraly sensitive, versatile, and durable. It can be used for measuring cyanide concentrations from $1x10^{-2}$ to $1x10^{-18}$ M. As always, Pt is the best anchoring material for sensor fabrication. Because of the extreme low detection limit, many industrial and security systems can be developed uiltizing this ultra sensitive sensor.

5 ACKNOWLEDGMENT

This study was partial supported by an USAMRMC Project Grant (Contract #W81XWH-07-2-0078) and College of Science and Engineering of Idaho State University.

REFERENCES

[1] S. W. Leung, D. Assan, J. C. K. Lai, "Evaluation of Anchoring Materials for Ultra Sensitive Biosensors Modified with Nanogold Particles and Enzymes", *Sensors & Transducers journal*, Vol.15, Special Issue, October 2012, pp.59-70.

[2] J.C.K. Lai, Y. Wang, W. Gao, H. Gu, and S. W. Leung, "Performance Comparisons of Nanoparticle Modified Sensor Electrodes for the Detection of Nitrite and Peroxide", in Technical Proceedings of the 2009 Nanotechnology Conference and Trade Show, Volume 2: Chapter 4: Biosensors and Diagnostics, pp. 233-235.

[3] Solomon W. Leung, James C.K. Lai and David Assan, "Performance Comparison of Immunosensors Modified with Polymers, Nanoparticles and Antibody", in proceeding of the NSTI Nanotechnology Conference and Expo – Nanotech 2013, May 12-15, Washington, D.C., Vol 3, Chapter 1, Biosensing, Diagnostics and Imaging, pp. 53-56.

[4] Solomon W. Leung, David Assan' and James C.K. Lai, "Characterization of an Ultra-High Performance Immunosensor Modified with Sol-Gel Nanogold Particles and Its Applications", in proceeding of the International Conference on Mathematics, Engineering & Industrial Applications, May 28 to 30, 2014 (ICoMEIA 2014), Penang, Malaysia, to be

[5]. David Assan, X. Gao, James C.K. Lai, and Solomon Leung, "Biocomposite Electrode Sensor for the Detection of Oxidative Reactions", in Technical Proceedings of the 2014 NSTI Nanotechnology Conference & Expo, Vol 1, section 6, pp. 505-508.

The influence of electrode on sensitivity of pH sensors fabricated on AlGaN/GaN heterostructures

Chang Taek Lee[*,**], Sang Min Jung[*,**] and Moo Whan Shin [*,**]

[*]School of Integrated Technology, Yonsei University 162-1 Songdo-dong, Yeonsu-gu, Incheon, 406-840, Korea

[**]Yonsei Institute of Convergence and Technology School of Integrated Technology, Yonsei University 162-1, Songdo-dong, Yeonsu-gu, Incheon, 406-840, Korea

ABSTRACT

In this paper, we report on metallization stack as an ohmic contact to gateless AlGaN/GaN high electron mobility transistors (HEMTs). As compared with Ion-Selective Field Effect Transistor (ISFET) based on Silicon, AlGaN/GaN HEMT has several attractive points for various chemical and biological sensor applications. One of the unique features in AlGaN/GaN, which is a high-density Two Dimensional Electron Gas (2DEG), is known to allow highly sensitive detection of surface phenomenon. Many researchers have elucidated the mechanism and improved sensitivity with perspective of 2DEG. However, there are no researches that explain why Ti/Al/Ti/Au stack is used as an ohmic contact in pH sensors based on AlGaN/GaN HEMTs. We focused on the relationship between ohmic materials on AlGaN/GaN HEMT and sensitivity of pH sensors.

Keywords: pH sensor, Sensitivity, Surface Potential Charge, Ohmic contacts, AlGaN/GaN HEMT

1 INTRODUCTION

Measurement of pH plays an important role in many fields such as pharmacy, chemistry, and biology. Therefore, pH sensor devices are required to compatible with electronic devices and to bear harsh environments. Most of the pH sensors is consisted of a type of glass electrodes. The main issues on improvement of pH sensor are miniaturization, stability, and sensitivity so far. Among candidates for pH sensors, the Si based Ion-Selective Field Effect Transistor (ISFET) has been studied for many years. However, Si-ISFET still has a limitation on operating at high-temperature since maximum operation temperature of Si-ISFET is around 60°C [1]. In contrast, GaN, a wide bandgap semiconductor, is chemically stable, non-toxic and thermally stable. The GaN has a high electron mobility, a wide band gap energy (3.4 eV) and a large sheet carrier concentration. Especially, AlGaN/GaN HEMT has been widely used in sensing technology due to sensitive analysis resulting from characteristics of a Two Dimension Electron Gas (2DEG) structure. The density of 2DEG is modulated by the surface potential charge of the channel gate on AlGaN/GaN HEMT. Since Steinhoff published the first report of a GaN-based pH sensor with an open-gate structure [2], many researchers have reported the sensing mechanism and various way to imporve pH sensitivity. However, research on Ti/Al/Ti/Au ohmic stack consisted of drain and source in devices has not yet been done. In this work, we evaluated relationship between specific contact resistance and sensitivity of pH sensor by changing the ohmic contact metallization consisted of source and drain.

2 EXPERIMENT DETAILS

The AlGaN/GaN heterostructure in this experiment was fabricated for analysis of electrical characteristics and measurement of pH sensitivity. The AlGaN/GaN structure was a commercial wafer provided by NTT-AT and grown on a 6-inch Si(111) wafer by MOCVD. The structure consisted of a ~4-um C-doped GaN buffer layer, a C-doped i-GaN layer, a ~300-nm i-GaN layer and a ~20-nm i-$Al_{0.25}Ga_{0.75}N$ layer with a ~2-nm n-GaN layer. The drain and source electrodes were formed by deposition of ohmic multilayers. Being multilayers of Ti/Al/Ti/Au (50/200/40/40 nm), Ti/Al/Ni/Au (50/200/40/40 nm), and Ti/Au (50/40 nm). Annealing process at 850°C for 30s by Rapid Thermal Annealing (RTA) system and conventional lift-off process as shown in Figure 1.

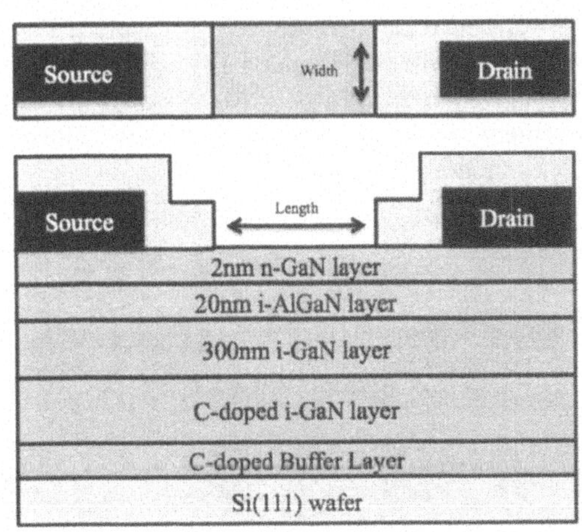

Figure 1. The cross section of a pH sensor on AlGaN/GaN HEMT structures

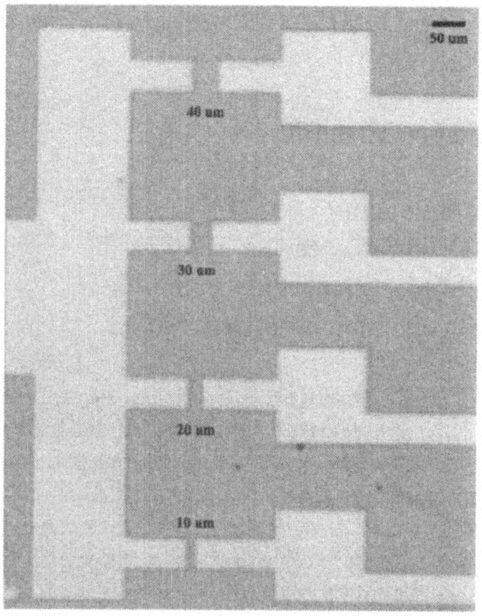

Figure 2. Sensor Device SEM images of open-gate

The device was covered with 300-nm-thick SiO_2 passivation using PECVD to prevent chemical reaction except at the open-gate region. The open-gate region was fabricated through photolithography and wet etching processes[3]. The 50×50 um^2 region of ohmic contacts were separated with gaps of 10, 20, 30, and 40 um. Each seperated regions took a role of the source and the drain part in AlGaN/GaN HEMTs. It means that the exposed open-gate area were 10×50 um^2, 20×50 um^2, 30×50 um^2, and 40×50 um^2. Figure 2. shows the scanning electron microscopy (SEM) image.

When starting the pH measuremet, we used pH 4, pH 10, and pH 7 buffer solution to calibrate the electrode with LE438 Mettler pH meters and DC suppler. All the measurements were processed at room temperature of 25 ^{0}C. Calibration process were consisted of titration using HNO_3, KOH, and de-ionized water (DI water). For the fast measurement of pH, we dropped 1 ~ 2 pH solutions on the open gate area like Figure 3. Also, Figure 4. shows the measurement system used to calculate more correct sensitivity on the pH sensor.

Figure 3. conceptual scheme to measure sensing

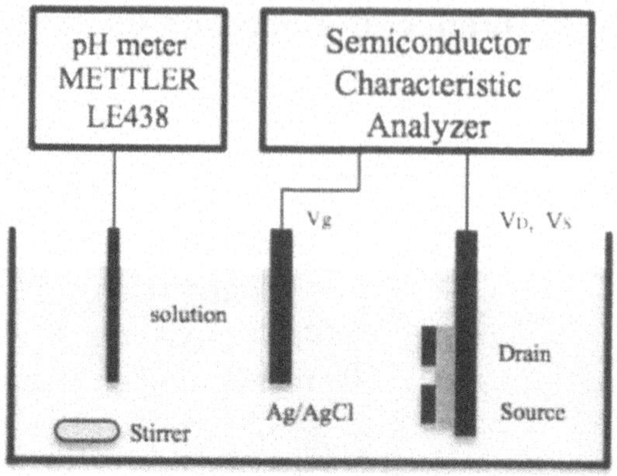

Figure 4. The measurement system of the pH sensor

We measured electrical properties, these are 2DEG and Ohmic contact characteristic, of pH sensors. At first, we assumed the open-gate HEMTs as a resistors. Through the features of 2DEG, we can find the relationship between surface potential charge and 2DEG characteristics. The absoption of OH^- affected the surface potential charge and the sheet concentration on the AlGaN/GaN HEMTs.

Next, we assumed total resistance of pH sensor means adding 2DEG and Ohmic contact resistances. The I-V characteristic was measured by four-probe method. All datas were measured at room temperature of 25 ^{0}C.

3 RESULTS AND DISCUSSION

These open-gate HEMT structure can be regarded as a resistor, linked with source, drain, and 2DEG parts. Simply, source and drain are regarded as ohmic contact, which can be calculated using I-V curve. The following equation is about 2DEG characterisitics. ρ is the specific contact resistance of 2DEG. A, section area of current, can be calculated from multiplication of the width of HEMT (W) and the depth of 2DEG (d_i) formed with AlGaN/GaN interface. L is the length between drain and source. N_S is the sheet concentration of 2DEG[4].

$$R_{2DEG} = \rho\frac{L}{A} = \frac{1}{qu_n(N_S/d_i)}\frac{L}{A} = \frac{1}{qu_n(N_S/d_i)}\frac{L}{Wd_i} \quad (1)$$

$$= \frac{1}{qu_n(N_s/d_i)}\frac{L}{Wd_i} = \frac{L}{qu_nN_SW}$$

The adsorption of OH^- on the AlGaN/GaN surface affects the surface potential charge (V) and the sheet concnetration (N_S)[6]. Figure 5. is one of the expected results with the measurement condition of 20 × 50 um^2 gate section. We measured the slope of I-V curve and calculated each component of the equation. The slope of I-V curve decreased when pH was increased.

TechConnect Briefs 2015, TechConnect.org, ISBN 978-1-4987-4729-5

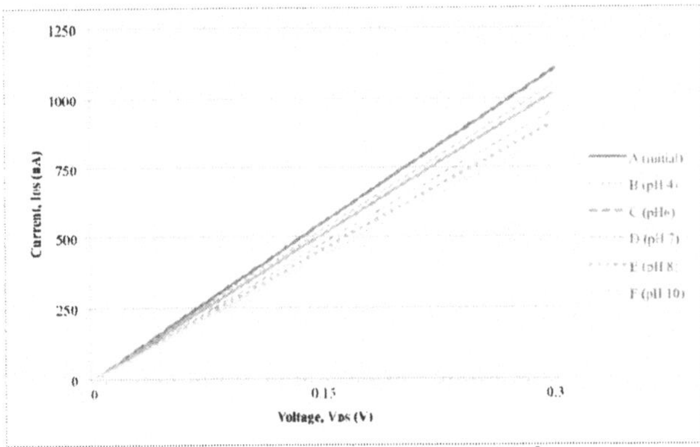

Figure 5. The slope of I-V curve for 20×50 um^2 open-gate

The change of source and drain materials in AlGaN/GaN HEMTs causes different total device resistance and sensitivity. We can improve the sensitivity of pH sensor from decreasing total resistance to induce efficiency increase.

Acknowledgement

"This research was supported by the MSIP(Ministry of Science, IT and Future Planning), Korea, under the "IT Consilience Creative Program" (NIPA-2014-H0201-14-1002) supervised by the NIPA(National IT Industry Promotion Agency)

REFERENCES

[1] Niigata, K., Narano, K., Maeda, Y., & Ao, J. P. (2014). Temperature dependence of sensing characteristics of a pH sensor fabricated on AlGaN/GaN heterostructure. Japanese Journal of Applied Physics, 53(11S), 11RD01.

[2] Steinhoff, G., Hermann, M., Schaff, W. J., Eastman, L. F., Stutzmann, M., & Eickhoff, M. (2003). pH response of GaN surfaces and its application for pH-sensitive field-effect transistors. Applied Physics Letters, 83(1), 177-179.

[3] Abidin, M. S. Z., Hashim, A. M., Sharifabad, M. E., Rahman, S. F. A., & Sadoh, T. (2011). Open-gated pH sensor fabricated on an undoped-AlGaN/GaN HEMT structure. Sensors, 11(3), 3067-3077.

[4] Huang, H. S., Lin, C. W., & Chiu, H. C. High sensitivity pH sensor using Al X Ga 1-X N/GaN HEMT heterostructure design. In *Electron Devices and Solid-State Circuits, 2008. EDSSC 2008. IEEE International Conference on* (pp. 1-4). IEEE.

Beyond Biomarkers: Array-Based Profiling for Diagnostics and High-Throughput Screening

V. Rotello*

*University of Massachusetts-Amherst, Department of Chemistry, Amherst MA 01003, rotello@chem.umass.edu

ABSTRACT

Array-based sensing is a versatile alternative to biomarker-focused strategies for biosensing. This "chemical nose" approach generates signatures that can be used for classifying complex systems for both diagnostic and screening applications. The use of nanoparticle-fluorophore sensing arrays for diagnostic and high content screening will be discussed.

Keywords: nanoparticles, diagnostics, sensors, serum

1 INTRODUCTION

The "chemical nose/tongue" approach presents a potential alternative to specific biomarker approaches. In this strategy a sensor array is generated to provide differential interaction with analytes via *selective* receptors, generating a stimulus response pattern that can be statistically analyzed and used for the identification of individual target analytes and also for profiling of complex mixtures. In our research, we have applied this methodoloy to sensing of proteins and cell surfaces, focusing on areas of biomedical importance.

2 RESULTS AND DISCUSSION

2.1 Serum Sensing

While sensing of proteins in buffer described above provides excellent proof of concept, real-world diagnostics for serum require the ability to address complex mixtures. The sensitivity of array-based sensing to subtle changes in analyte profile generates an excellent platform for sensing complex mixtures. In prior studies that provide the foundation of our proposed research, we developed a fluorophore displacement strategy for sensing of proteins in undiluted human serum using green fluorescent protein (GFP), a strategy that allowed us to identify proteins "spiked" into serum at concentrations of 500 nM in undiluted serum (serum features *~1 mM* total protein concentration) (Figure 1).[1] This sensitivity corresponds to the ability to sense 0.06-5% changes in the five most abundant proteins These studies demonstrate the ability of array-based sensing to discriminate minute changes in serum protein levels, i.e. we have a very sensitive "nose" .

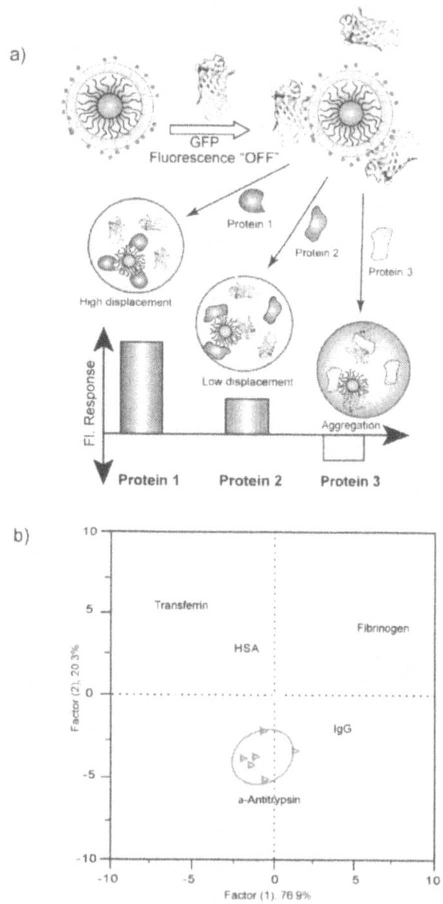

Figure 1. a) Competitive binding between protein and quenched GFP-NP complexes and protein aggregation leading to sensor response. b) Fluorescence responses for five most abundant serum proteins (500 nM) spiked into undiluted human serum c) Canonical score plot for the fluorescence patterns of the GFP-NP adducts.

2.2 Cell Surface Sensing

Our initial studies in cell surface sensing focused on differentiation of cancer cell genotypes using chemical nose sensors (Figure 7).[2] In these studies we used an array of nanoparticle-fluorescent polymer sensors. The differential interactions of the functionalized nanoparticles with the different cell genotypes was transduced through displacement of a multivalent polymer fluorophore that is quenched when bound to the particle and fluorescent after release. Using this sensing strategy we could rapidly (minutes/seconds) and effectively distinguish: 1) different cell types; 2) normal, cancerous and metastatic human breast cells; 3) *isogenic* normal, cancerous and metastatic murine epithelial cell lines. (Figure 2 c,d).

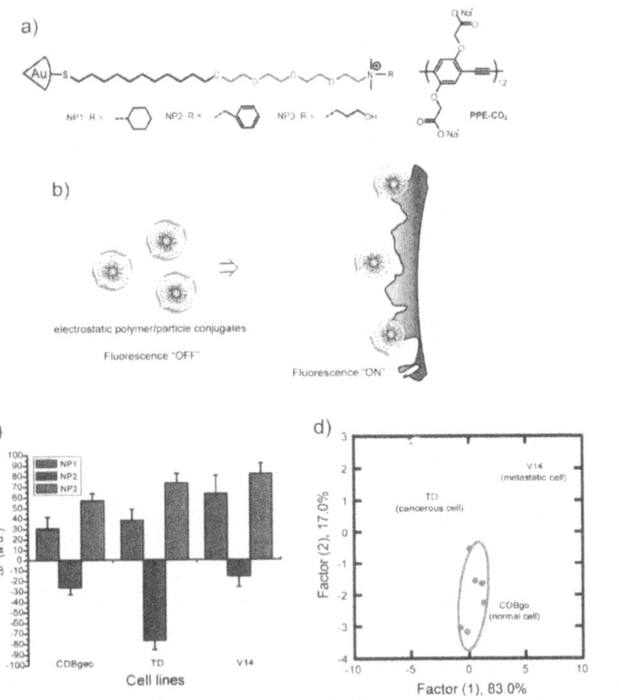

Figure 2. Array-based sensing of cells. a) polymer and particles used for sensing. b) The sensing process c) Fluorescence output of sensor array using three isogenic breast cell lines derived from BALB/c mouse. d) LDA plot for the fluorescence response patterns.

Multi-channel sensing and cell-surface phenotyping are key enabling technologies for our research. We recently introduced a simple and powerful high-throughput multi-channel sensor platform comprised of supramolecular complexes of a gold nanoparticle with three fluorescent proteins. (Figure 3a).[3]

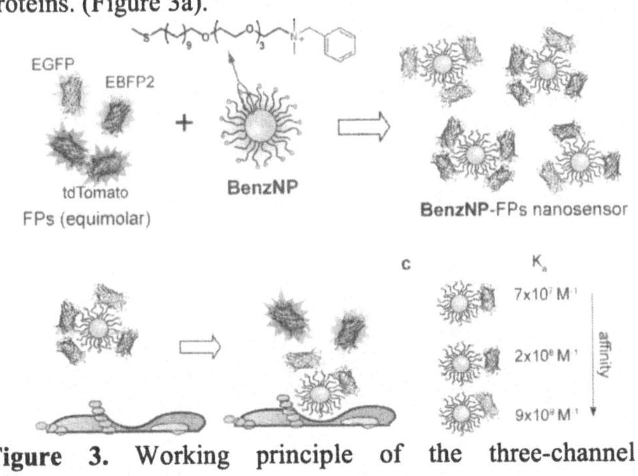

Figure 3. Working principle of the three-channel multiplexed sensor.

This sensor is engineered for selective cell surface recognition, allowing direct measurement using cell culture plates, (Figure 4a). We tested this sensor system using anti-cancer drugs. Our studies demonstrated that the sensor system responds reproducibly to physicochemical changes at the cell surface induced by these drugs, generating characteristic patterns that identify specific mechanisms of cell death (Figure 4b). This capability overcomes hurdles in developing chemotherapeutics, providing phenotypic profiling of drug mechanisms as a strategy for greatly increasing throughput in the drug discovery process. Notably, this whole-cell based high-throughput technique uses a *single well* of a microplate for drug treatment and analysis, with rapid (minutes) determination of chemotherapeutic mechanism.

In our study we used the sensor to:

1) Identify the mechanism of blinded drugs with high (90%) accuracy.
2) Differentiate between "new" mechanisms and ones in the training set using statistical methods.
3) Determine when synergistic combination therapies proceeded through the mechanism of one of the components or through a new mechanism.

The sensor platform is generalizable to any other type and does not require processing steps such as extracting biomarker or labeling cells prior to analysis.

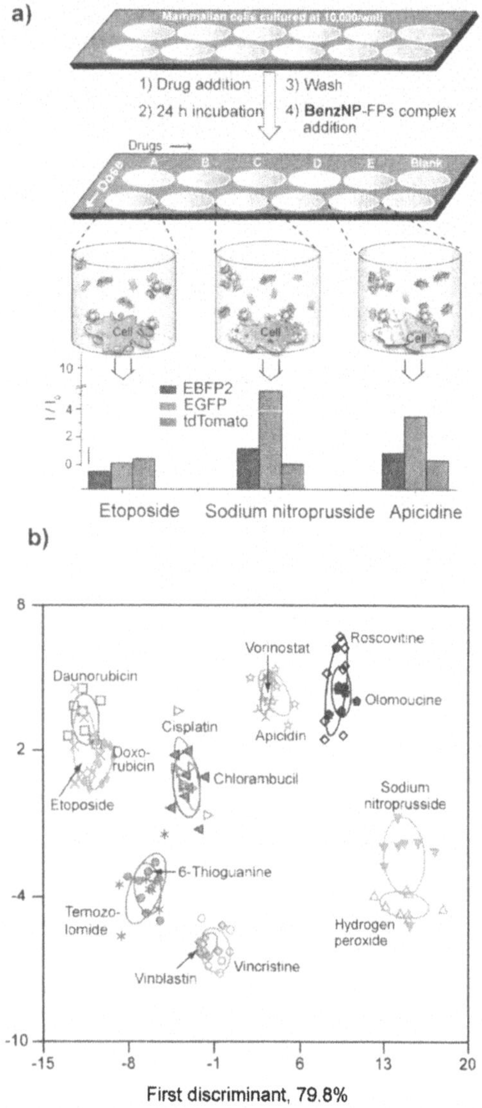

Figure 4. a) Drug screening workflow. The fluorescence outputs corresponding to the three drug candidates are actual experimental results. b) Linear discriminant analysis of the fluorescence responses plotted with 95% confidence level ellipses around the centroid of each group.

References

1. De. M.; Rana, S.; Akpinar, H.; Miranda, O.R.; Arvizo, R. R.; Bunz, U. H. F.; Rotello, V. M. "Sensing of Proteins In Human Serum using Conjugates of Nanoparticles and Green Fluorescent Protein" *Nature Chem.* 2009, 1, 461-465.

2. Bajaj, A.; Miranda, O. R.; Kim, I.-K.; Phillips, R. L.; Jerry, D. J.; Bunz, U. H. F.; Rotello, V. M. "Detection and Differentiation of Normal, Cancerous, and Metastatic Cells using Nanoparticle-Polymer Sensor Arrays" *Proc. Nat. Acad. Sci.*, 2009, *106*, 10912-10916.

3. Rana, S.; Le, N. D. B.; Mout, R.; Saha, K.; Tonga, G. Y.; Bain, R. E. S.; Miranda, O. R.; Rotello, C. M.; Rotello, V. M. "A Multichannel Nanosensor for Instantaneous Readout of Cancer Drug Mechanisms" *Nat. Nanotechnol.* **2015**, *10*, 65-69.

Enhanced Kinetics of Enzymatic Nanosensors

Joyce Breger[1,4], Scott Walper[1], Mario Ancona[3], Michael Stewart[2], Kimihiro Susumu[2], and Igor Medintz[1*]

[1]Center for Bio/Molecular Science and Engineering, Code 6900,

[2]Optical Sciences Division, Code 5600,

[3]Electronic Science and Technology Division, Code 6800, U.S. Naval Research Laboratory, Washington, DC 20375,

[4]American Society for Engineering Education, 1818 N Street NW, Suite 600 Washington, DC

*igor.medintz@nrl.navy.mil

ABSTRACT

Nanosensors employing quantum dots (QDs) and functional moieties such as enzymes are promising for diagnostics and chemical/biological threat activity. Their small size permits cell penetration and their inherent photochemical properties are well-suited for rapid, optical measurements. The effectiveness of biorecognition agents immobilized on QDs are not completely understood, hindering development of chemical/biological sensors and remediation materials. We analyze an engineered enzyme construct's effectiveness when attached to QDs illustrating an important biological threat application: neutralization of paraoxon, a simulant nerve agent. Phosphotriesterase trimer (PTE_3) was engineered to contain 3 PTE monomers attaced to a V-domain, collogen linker, and a oligohistidine (His_6) tail. Various ratios of PTE_3 were attached to two sizes of QDs, 525 or 625 nm via metal coordination. Characteristic kinetic parameters of PTE_3-QD were compared to free enzyme.

Keywords: Quantum dots, enzyme, phopsphotriesterase, paraoxon, kinetics

1 INTRODUCTION

As nature's catalysts, enzymes rapidly and efficiently transform numerous chemicals in chemical synthesis far beyond our capabilities. Harnessing enzymes' ability to catalyze chemicals of interest would circumvent creating difficult synthetic processes. Indeed, much effort has been focused in engineering enzymes with modulated activity for a variety of industrial processes.[1, 2] One phenomenon that is poorly understood but has been observed to increase catalytic activity is attaching enzymes to nanoparticles.[3] Understanding how and why enzyme activity increases at the nanoparticle surface could lead to developments in creating microfluidic devices that detect biomarkers or chemical warfare agents.

Nanosensors composed of QDs and functional moieties such as enzymes are promising for disease surveillance/diagnostics and chemical/biological threat activity. Numerous advantages of QDs over other nanoscaffolds include their inherent photochemical properties such as size tunable fluorescence, ease in attaching functional moieties, resistance to photobleaching, and much more making QDs well-suited for rapid, optical measurement applications.[4] Their small size also permits cell penetration which might be useful in theranostic application development.[5]

PTE catalyzes the detoxification of organophosphate pesticides to *p*-nitrophenol. To elucidate the mechanisms that increase catalytic activity of enzymes attached to nanoparticles, we attached various ratios of PTE_3 to two distinctly sized QDs, 525 or 625 nm and study the kinetics of hydrolyzing paraoxon.

2 METHODS

2.1 Quatum Dots

Green and red emitting CdSe/Zns (core/shell) QDs whose emission was centered at ca. 525 and 625 nm were synthesized and solubilized with dihydroplipoic acid-based zwitterionic compact ligand (DHLA-CL4) as described elsewhere.[6] DHLA-CL4 renders the QDs biocompatible with high quantum yields and long term stability in a broad pH range (pH = 5-13) while still allowing the QD surface to be functionalized with a variety of biologicals through either covalent attachment or metal affinity coordination.[7, 8] QDs functionalize with these ligands have already proven themselves in a variety of challenging environments.[9-11]

2.2 Phosphotriesterase Trimer Enzyme Preparation

The PTE_3 was engineered by combining the phosphotriesterase gene from Brevundimonas diminuta [12] with the collagen-like triple helix domain and the trimerization V domain from Streptococcus pyogenes. [13, 14] The PTE_3 cassette was excised from a maintenance plasmid and cloned to the bacterial expression vector pET28 (Stratagene) via a NcoI/NotI restriction enzyme digest. The pET28 vector allowed for controlled protein expression and integration of a C-terminal hexa-histidine which could be employed for subsequent immobilized metal affinity chromatography and QD assembly. The expression construct's sequence was confirmed prior to transformation into the Escherichia coli BL21 (DE3) expression strain. Protein was expressed and purified using shake flask cultures, typically yielding ~10 mg of protein per liter of cell culture. While spontaneous, trimer formation was not immediate, therefore, purified monomeric protein (Mw~53 kDa) was incubated at 4°C for 48 h to allow the triple helix PTE_3 tertiary structure (Mw ~160 kDa) to form. The trimer was isolated from the monomers and dimers by size exclusion chromatography. The gene sequence, description of the protein engineering and purification are described elsewhere. [15]

2.3 Phosphotriesterase Trimer-QD Bioconjugate Assemblies

QDs were diluted to a final concentration of 2.22 nM in 150 μM (pH = 8) CHES buffer and molar equivalents of PTE_3 were added at the indicated ratios while maintaining the final volume of each reaction at 1350 μL within 2 mL Eppendorf tubes. The solutions were vortexed and microcentrifuged briefly to ensure solution homogeneity. Enzyme control reactions contained no QD and only equivalent concentrations of PTE_3.

2.4 PTE Trimer Enzymatic Assays

PTE_3-QD bioconjugate activity was monitored using a Tecan Infinite M1000 dual monochromator multifunction plate reader equipped with a xenon flash lamp (Tecan, Research Triangle Park, NC). Hydrolysis of paraoxon in the presence of PTE_3-QD or PTE_3 free in solution was monitored over time by measuring the absorbance at 405 nm every 20 seconds while maintaining the temperature at 25°C. To a 384-well Corning flat bottom, non-binding plate, 10 μL of free PTE_3 or PTE_3-QD bioconjugate was pipetted into each well. A serial dilution of paraoxon solution was made in CHES buffer ranging in concentration from 5000 to 20 μM in a separate 96-well plate. Right before the start of the experiment, 10 μL of the serially diluted paraoxon was added to each well of the 384-well plate. Absorbance values were converted to concentrations values using a standard curve of *p*-nitrophenol. Initial rates were determined using a macro written in excel with an R^2 value=0.995. Kinetic parameters were determined by fitting the initial rates to the Micheales-Menten equation (1) by minimizing the error between the estimated initial rates and the actual rates.

$$\frac{dP}{dt} \cong -\frac{d[S]}{dt} = \frac{V[S]}{K_M + [S]}, k_{cat} = \frac{V_{max}}{[E]_o} \tag{1}$$

2.5 Agarose Gel Electrophoresis

PTE_3 was allowed to self-assemble onto 525 nm QDs at increasing stoichiometric ratios (0 to 16 per QD).[15] Each reaction contained 5 picomoles of QD at a final volume of 20 μL and was incubated at room temperature for 30 min to facilitate assembly. The assemblies were run on a precast 1.5% agarose low electroendosmosis (EEO) gel in 1X Tris-Borate-EDTA buffer at pH = 8.3. The gel was visualized with a BioRad Gel Image Analyzer to monitor QD migration.

2.6 Dynamic Light Scattering and ζ-potential

Dynamic light scattering (DLS) measurements were carried out using a CGS-3 goniometer system equipped with a single-photon counting avalanche photodiode for signal detection (ALV, Langen, Germany) and analyzed using Dispersion Technology Software (DTS, Malvern Instruments Ltd, Worcestershire, UK).[16, 17] The data of the pre-filtered QDs (with 0.25 μm Millipore syringe filters for typically ~ 1μM) was repeatedly collected (3-5 times) at 20°C and the number profile versus hydrodynamic size was used for data presentation. For Zeta-Potential (ζ-potential) measurement, Laser Doppler Velocimetry (LDV) measurements were performed using a ZetaSizer NanoSeries (Malvern Instruments Ltd, Worcestershire, UK) at 25°C. Three runs of the measurements were performed for each sample prepared in 0.1×PBS buffer at pH 7.4.

3 RESULTS

In this work, we attempt to elucidate the mechanisms for enhanced catalytic activity when enzymes are attached to nanoparticle surfaces as compared to free enzyme in solution. Figure 1 schematically shows the ribbon and stick model of PTE_3 composed of His_6 tail, triple helix domain, V stabilization domain, linker sequence, and PTE monomer and how the assembly would attach to the QD surface.

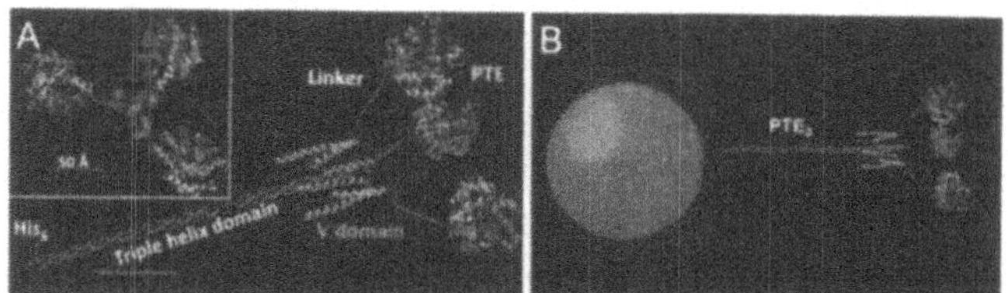

Figure 1. (A) Ribbon and stick model of PTE_3 composed of His-6 tail, triple helix domain, V stabilization domain, linker sequence, and PTE monomer. Scale bar = 5 nm (B) Schematic demonstrating PTE_3 assembly onto QD relative to scale. Adapted from Ref [15] with permission of the Royal Society of Chemistry.

To confirm PTE_3 attached to the QD surface through metal coordination, increasing discrete ratios of PTE_3 were assembled to QD and performed an agarose gel, DLS and measured the ζ-potential. The diameters of the green and red QDs were determined to be 4.3 ± 0.5 and 9.2 ± 0.8 nm, respectively corresponding to surface areas of ~58 and ~266 nm^2.[15, 18] As shown in Figure 2 A, the electrophoretic mobility shift assay (EMSA) demonstrates as the ratio of PTE_3 assembled to QD increases, the mobility of the assemblies decreases; indicating an overall increase in the

MW of the assemblies. Arrows indicate the smallest and highest ratios of assembled PTE_3/QD. DLS confirmed bioconjugates spontaneously self-assemble as shown in Figure 2 B by the increasing hydrodynamic diameter with increasing PTE_3/QD ratio. The constructs were engineered to have the $(His)_6$ tail distally located from the enzyme therefore more than two times the maximum of 12 or 24 dimeric PTE to 525 or 625 nm QDs respectively could be assembled to the QD surface. Also, the enzyme binding pocket should face outward and not be sterically hindered.

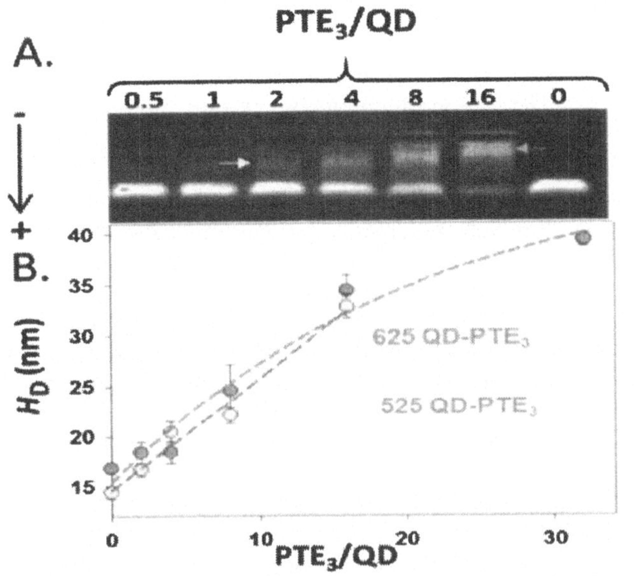

Figure 2. (A) Representative gel of increasing ratios of PTE3 assembled to 525 nm QDs. (B) Hydrodynamic diameter (H_D) determined by DLS for increasing ratios of PTE3 attached to either 525 or 625 nm QD. Adapted from Ref. [15] with permission of the Royal Society of Chemistry.

Significant enhancement in activity was seen when PTE_3 attached to either 525 nm or 625 nm QDs as compared to free enzyme at each of the ratios tested (Figure 3). Paraoxon is an insecticide and an organophosphate stimulant therefore it is commonly used as substrate for studying this family of enzymes. PTE hydrolyzes paraoxon to *p*-nitrophenol whose absorbance can be monitored overtime at 405 nm. Characteristic kinetic parameters were determined by fitting the Michaelis–Menten equation (Equation 1) to the initial rates versus paraoxon concentrations for each ratio of PTE_3 on or off the QD tested. As shown in Figure 3, assembly of PTE_3 onto the QD caused an enhancement in catalytic activity compared to free enzyme at all the ratios tested for both QD sizes.

As shown in Table 1, the K_M values should be considered apparent values as the PTE_3/QD constructs are not the same as freely diffusing enzyme. The K_M values for our engineered PTE_3 are better or at least comparable to numerous other PTE wild types and variants but the k_{cat} is decreased.[19, 20] The highest apparent activity occurs at the lowest valency of PTE_3 on the smallest QD which we have noted before with alkaline phosphatase attached to QD. [18] The k_{cat} of PTE_3 attached to QD increased by 2X compared to PTE_3 free in solution at the lowest valency and smallest QD.

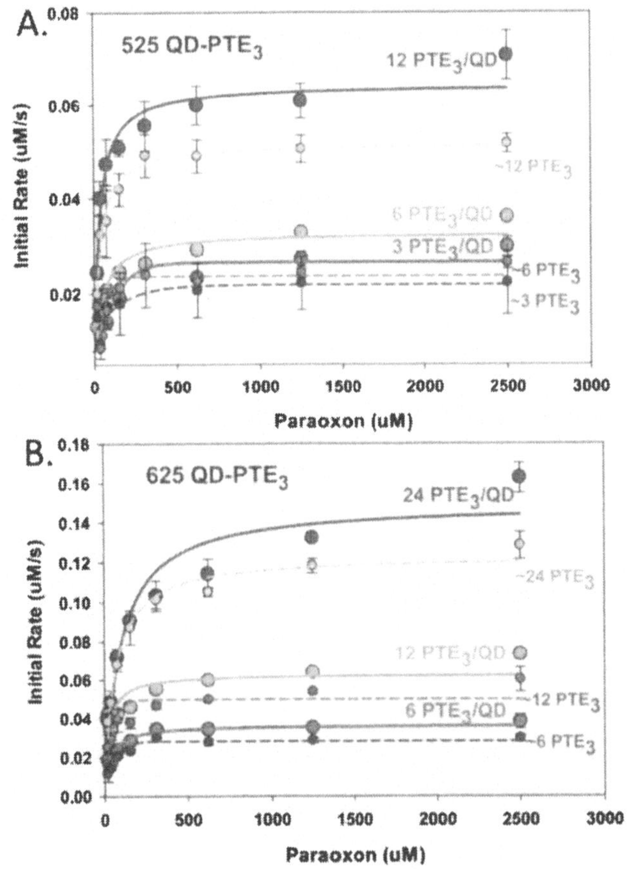

Figure 3. Trimeric PTE activity in solution and attached onto 525 nm (A) or 625 nm emitting QDs (B). Solid lines indicated activity of PTE when assembled onto QDs while dashed lines are equivalent amounts of free PTE in solution. Ratios or equivalents (~) of PTE to QD are indicated. Each data point is the average of three replicates. Note the difference in scale. Adapted from Ref. [15] with the permission of the Royal Society of Chemistry.

PTE_3/QD	k_{cat} (s^{-1})	K_M (μM)	k_{cat}/K_M ($\mu M^{-1}s^{-1}$)
525 nm QD			
3	7.9±1.0	111.9±18.9	0.07±0.01
6	4.9±0.3	33.3±7.7	0.15±0.03
12	4.8±0.4	30.0±6.9	0.16±0.03
Average:	5.9±1.8	58.4±11.2	0.13±0.05
625 nm QD			
6	5.1±0.8	28.8±1.8	0.18±0.03
12	4.7±0.1	24.6±3.3	0.19±0.02
24	5.7±0.6	167.1±39.5	0.05±0.03
Average:	5.2±0.5	73.5±20.3	0.14±0.07
PTE_3 alone			
Average:	3.9±0.8	50.1±9.8	0.10±0.03

Table 1: Summary of estimated PTE_3 kinetic parameters on or off QD. Adapted from Ref [15] with the permission of the Royal Society of Chemistry.

4 CONCULSIONS

By performing in-depth kinetic modeling we postulate an enhancement mechanism due to the substrate being captured by the bioconjugate and undergoing a translational diffusion process between closely-spaced enzymes on the QD surface. Another possibility is the solvation shell surrounding the PTE3/QD construct is influencing the kinetics. Development of these nanosensors as optical-based biosensors (e.g., within compact microfluidic devices) may greatly improve the sensitivity of conventional biological/chemical detection schemes. These enzyme/QD constructs have been engineered to have the substrate binding pocket facing outward, away from the QD surface thereby increasing favorable interactions with the substrate. Further work is being carried out to determine if these results can be replicated with other enzyme constructs and the influence of packing density as well as nanoparticle curvature in catalytic enhancement.

REFERENCES

[1] K. E. Sapsford *et al.*, "Analyzing Nanomaterial Bioconjugates: A Review of Current and Emerging Purification and Characterization Techniques," Analytical Chemistry, 83(12), 4453-4488 (2011).

[2] R. Gupta *et al.*, "Bacterial alkaline proteases: molecular approaches and industrial applications," Applied Microbiology and Biotechnology, 59(1), 15-32 (2002).

[3] B. J. Johnson, *et al.*, "Understanding enzymatic acceleration at nanoparticle interfaces: Approaches and challenges," Nano Today, 9(1), 102-131 (2014).

[4] E. Petryayeva, *et al.*, "Quantum Dots in Bioanalysis: A Review of Applications Across Various Platforms for Fluorescence Spectroscopy and Imaging," Applied Spectroscopy, 67(3), 215-252 (2013).

[5] J. Breger, *et al.*, "Continuing progress toward controlled intracellular delivery of semiconductor quantum dots," Wiley Interdisciplinary Reviews: Nanomedicine and Nanobiotechnology, (2014).

[6] K. Susumu, *et al.*, "Multifunctional Compact Zwitterionic Ligands for Preparing Robust Biocompatible Semiconductor Quantum Dots and Gold Nanoparticles," Journal of the American Chemical Society, 133(24), 9480-9496 (2011).

[7] J. B. Blanco-Canosa, *et al.*, "Recent progress in the bioconjugation of quantum dots," Coordination Chemistry Reviews, 263, 101-137 (2014).

[8] K. Susumu, *et al.*, "A New Family of Pyridine-Appended Multidentate Polymers As Hydrophilic Surface Ligands for Preparing Stable Biocompatible Quantum Dots," Chemistry of Materials, 26(18), 5327-5344 (2014).

[9] K. Boeneman, *et al.*, "Selecting Improved Peptidyl Motifs for Cytosolic Delivery of Disparate Protein and Nanoparticle Materials," Acs Nano, 7(5), 3778-3796 (2013).

[10] R. Walters, *et al.*, "Nanoparticle targeting to neurons in a rat hippocampal slice culture model," Asn Neuro, 4(6), (2012).

[11] B. K. Andrasfalvy, *et al.*, "Quantum dot-based multiphoton fluorescent pipettes for targeted neuronal electrophysiology," Nature Methods, 11(12), (2014).

[12] D. P. Dumas, *et al.*, "Purification and Properties of The Phopsphotriesterase From Pseudomonas-diminuta " Journal of Biological Chemistry, 264(33), 19659-19665 (1989).

[13] A. Yoshizumi, *et al.*, "Designed Coiled Coils Promote Folding of a Recombinant Bacterial Collagen," Journal of Biological Chemistry, 286(20), 17512-17520 (2011).

[14] A. Mohs, *et al.*, "Mechanism of stabilization of a bacterial collagen triple helix in the absence of hydroxyproline," Journal of Biological Chemistry, 282(41), 29757-29765 (2007).

[15] J. C. Breger, *et al.*, "Quantum dot display enhances activity of a phosphotriesterase trimer," Chemical Communications, (accepted), (2015).

[16] M. H. Stewart, *et al.*, "Competition between Forster Resonance Energy Transfer and Electron Transfer in Stoichiometrically Assembled Semiconductor Quantum Dot-Fullerene Conjugates," Acs Nano, 7(10), 9489-9505 (2013).

[17] T. L. Jennings, *et al.*, "Reactive Semiconductor Nanocrystals for Chemoselective Biolabeling and Multiplexed Analysis," Acs Nano, 5(7), 5579-5593 (2011).

[18] J. C. Claussen, *et al.*, "Probing the Enzymatic Activity of Alkaline Phosphatase within Quantum Dot Bioconjugates," Journal of Physical Chemistry C, 119(4), 2208-2221 (2015).

[19] C. J. Jackson, *et al.*, "Conformational sampling, catalysis, and evolution of the bacterial phosphotriesterase," Proceedings of the National Academy of Sciences of the United States of America, 106(51), 21631-21636 (2009).

[20] P. C. Tsai, *et al.*, "Enzymes for the Homeland Defense: Optimizing Phosphotriesterase for the Hydrolysis of Organophosphate Nerve Agents," Biochemistry, 51(32), 6463-6475 (2012).

Multilayer optical strip for illumination, imaging and analysis

E. Gomez-Gonzalez*, J. Marquez-Rivas**,
M.A. Perales-Esteve*** and M.J. Mayorga-Buiza**

*Department of Applied Physics III, Engineering School, University of Seville, Spain, egomez@us.es
**Service of Neurosurgery, University Hospital "Virgen del Rocio", Seville, Spain
*** Department of Electronic Engineering, Engineering School, University of Seville, Spain

ABSTRACT

This paper presents an optical device (multilayer optical strip, MOS) composed of a strip-shaped, multilayer, planar structure connected to a remote control unit. Active and passive optical elements in overlapping layers of the optical strip define different configurations for specific applications, including illumination, optical analysis, imaging and optical communication and control of other devices. The strip can be flexible, rigid or of variable shape. It can be attached to existing instruments and tools to access cavities, conduits and hard-to-reach locations with irregular or changing boundaries. It can be of small size, powered on batteries, portable and cost-effective. Main areas of application of this device span from surgery, medicine, dentistry, biology and veterinary to optical sensing of pollutants and pathogens, industrial inspection and quality control. It is also suitable for use in small spaces such as vehicles, in emergency or disaster areas and in low-resources or developing environments.

Keywords: planar optics, multilayer optics, illumination, imaging, optical analysis

1 DESCRIPTION

The presented multilayer optical strip (MOS) is composed of two main components: a so-called "optical strip" and an additional unit for power supply, thermal regulation, signal conditioning, control and, if required, data transmission and recording. The optical strip combines overlapping layers of active and passive optical elements (OEs). Several optical strips can be connected to a common control unit to define specific emitter-receiver geometries. Proper cables connect the control unit and the strip/s. Active layers of the strip may include light emitting and receiving elements, liquid crystals and linear and array image sensors, while passive layers may include color filters, diffracting grating and components, and micro and planar lenses. An additional metallic layer provides structural strenght and heat removal.

OEs in active and passive layers allow for controlling the properties (intensity, color temperature and spectral composition, polarization), spatial structure and time intervals (pulses, sequences) of the light emitted or received by the active elements. A key element in the design for intended applications is the strip (band) shape of small size. Typical dimensions of current prototypes combine widths of 5-10 mm and thicknesses of 2-5 mm with variable lengths, from 50 mm to those required by specific applications. Fast-evolving technological improvements in small, planar optical elements can be further incorporated and will allow for scaling down dimensions [1]. The control unit provides electrical power to the strip and drains heat from active OEs. This external unit also synchronizes signals and receives data, measurements and images, and allows for additional computer connections and connectivity. Depending on the structure and materials of the layers, the shape of the strip/s can be adapted so that they can be attached to instruments and tools using either mechanical fixation, magnets or adhesive layers. The strip/s can be sheathed for specific environments and sterilization or be disposable. Depending on the number and type of layers and OEs, the whole system can be portable, very cost-effective and easy to manufacture [1-2].

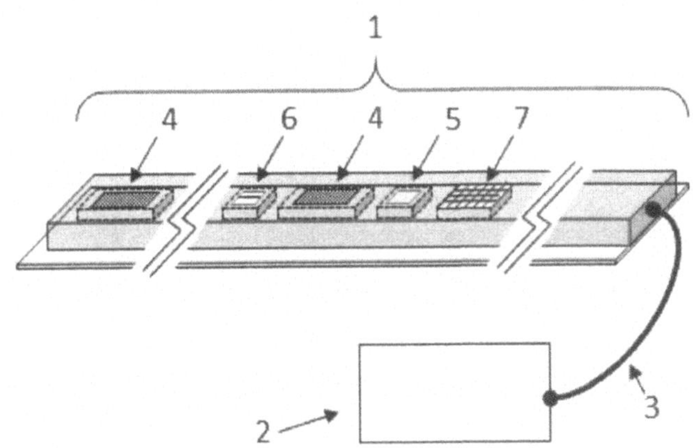

Figure 1: General overview of proposed MOS device showing optical strip (1), control unit (2) and connecting cables (3). Light-emitting elements include LEDs (4) and a laser diode (5). Receiving elements include a sensing photodiode (6) and a matrix detector array (7).

In the following sections, four cases of use are presented: deep-field surgical illumination, optical analysis of materials, inner imaging of cavities and optical control of devices. Background outlines and technology readiness level (TRL) values are given for each application. This device is currently patent pending (University of Seville, Spain, 2014) [3].

2 APPLICATION TO ILLUMINATION IN DEEP SURGICAL FIELDS

Surgical illumination devices use large-size lamps and headlights, and optical fiber and liquid light guides. They define a directional light beam projected onto the surgical field but instruments and surgeon's hands generate shadows and irregularly lit areas. Such inhomogeneities affect surgical performance, particularly in deep-field neurosurgery and in minimally invasive procedures. A linear array of high-intensity light emitting diodes (LEDs) in a MOS attached to conventional surgical instruments provides optimized illumination even at the very far end of cavities. Color temperature of emitted light is controlled by the types and modulation of selected LEDs.

A portable, battery-operated MOS system may also be used inside vehicles and in emergency situations. In such applications, current TRL = 5-6.

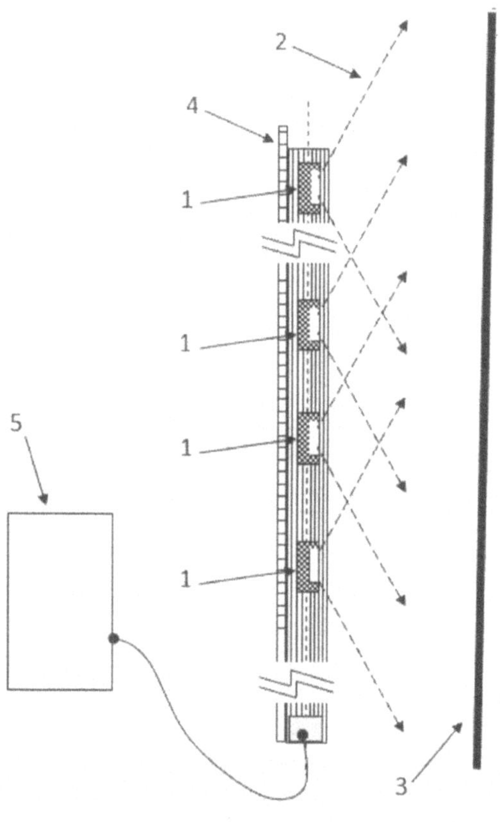

Figure 2: Application of proposed MOS as an illumination device. Each high-intensity LED (1) emitts a 120° cone of light (2). For simplicity, the illuminated surface is shown as planar (3). The supporting metal layer (4) drains heat to dissipation elements in the control unit (5).

3 APPLICATION TO OPTICAL ANALYSIS

A combined array of LEDs and receiving photosensors on one or more optical strips allows for obtaining measurements of diffuse reflectance, absorption and fluorescence –static and dynamic– spectra of the walls or the content of the cavity where the MOS device is inserted. The type and modulation of the LEDs define the emitted colors (wavelengths). Comparison with reference values or spectral signatures may identify pollutants and pathogens, especially in optically transparent media in conduits or containers. In this application, current TRL = 3.

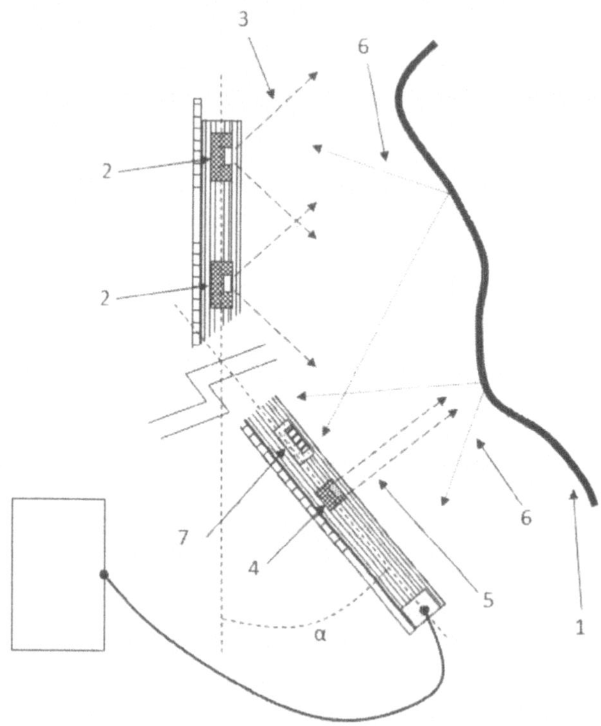

Figure 3: Application of proposed MOS for optical analysis of a surface (1). The optical strip is bended an angle α. Each LED (2) emitts a 90° cone of light (3). A laser diode (4) emits a collimated beam (5). Reflected light beams (6) are received by sensing photodiodes (7).

4 APPLICATION FOR IMAGING INSIDE CAVITIES

A MOS structure can combine image sensors with illuminating microLEDs. Active optical elements project illumination beams or patterns of structured light [4] and array sensors –with overlapping micro- and planar lenses– record images. This set-up allows for imaging inside cavities (Figure 4). Sequential scanning, stereoscopic or panoramic views and other specific reconstructions can be generated from obtained image sets. Attachment of proposed MOS to specific instruments may allow for optical imaging in locations hard to reach with current camera systems. In this application, current TRL = 2.

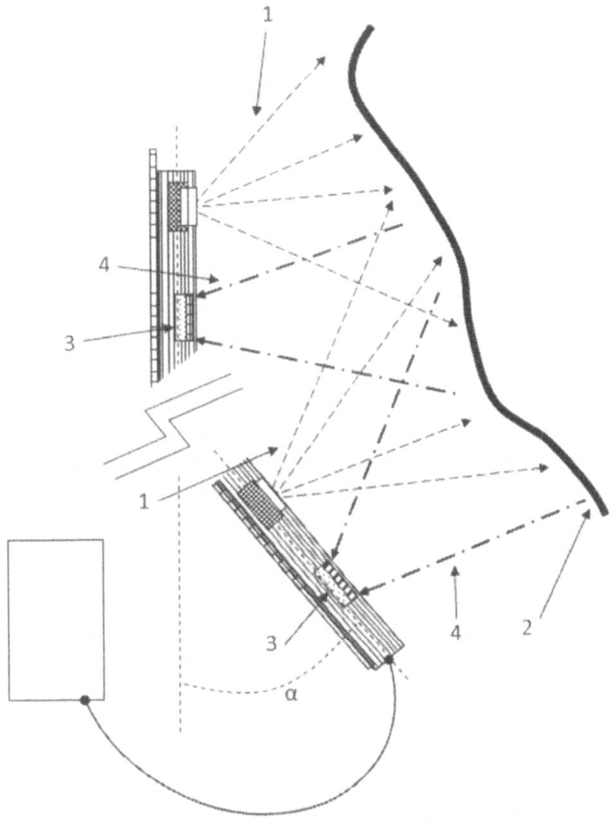

Figure 4: Application of proposed MOS for imaging inside cavities. The optical strip is bended an angle α. Illumination or structured light beams (1) are projected onto the imaged surface (2). Imaging sensors (3) receive reflected beams (4).

5 APPLICATION TO OPTICAL COMMUNICATION AND CONTROL

Optical signals emitted and received by active elements placed on the optical strip of the MOS can be used to control and communicate with other devices (sensors, switches, actuators, and even robots) in the illuminated area (Figure 5).

Light beams emitted from LEDs in the strip can be modulated to carry control signals. Data from sensors can also be sent as optical signals to be received by the optical strip and recorded or further transmitted from the control unit. In this application, current TRL = 2-3.

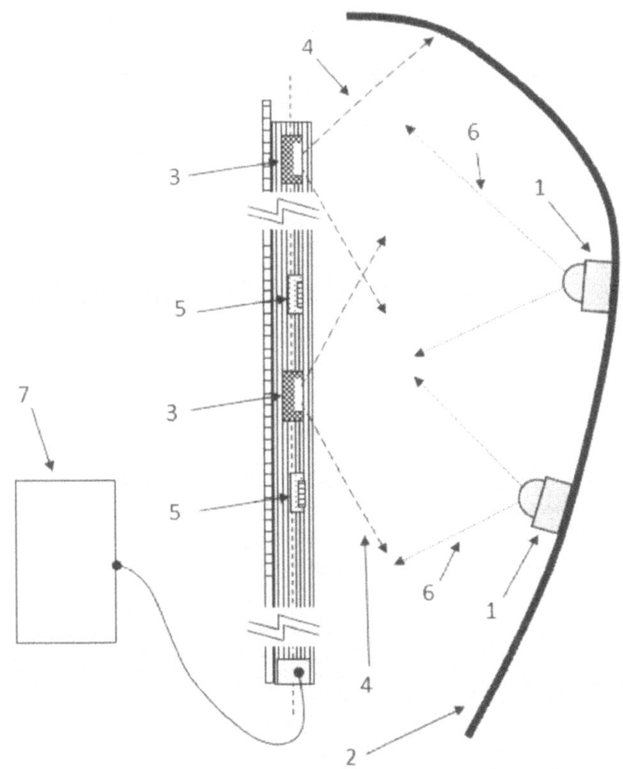

Figure 5: Application of proposed MOS for optical communication and control of devices (1) in cavities (2). LEDs (3) emit modulated light beams (4). Photodiodes (5) receive signals and data (6) from controlled devices (1) and send them to the control unit (7).

REFERENCES

[1] Milster T.D., "Miniature and micro-optics" in Bass M. et al (eds.), "Handbook of optics", Vol. 2, McGraw-Hill, 7.1-7.33, 1995.

[2] Schumann M. et al. "Hybrid 2D-3D optical devices for integrated optics by direct laser writing". Light: Science & Applications, 3, e175, 2014.

[3] Gomez-Gonzalez E., "Dispositivo Optico Planar Multicapa", Patent pending, University of Seville, 2014.

[4] Hayes T., "MicroLEDs emit light in programmable patterns", OLE, december, 15-17, 2008.

A Nanobiomimetic Neural Memcapacitor Serves as a Voltage Sensor and an Amperometry Sensor for Reagent-less Direct Detection of Sub pM Soluble Amyloid-β

S-H. Duh[1], J. Thornton[2], P.T. Kissinger[3] and E. T. Chen[4*]

[1]University of Maryland Medical System, 22 South Greene St, Baltimore, MD 21201, USA
[2] Bruker Nano, 19 Fortune Dr., Billerica, MA 01821, USA
[3]Department of Chemistry, Purdue University, West Lafayette, IN 47906, USA
[4]Advanced Biomimetic Sensors, Inc., 11140 Rockville Pike, Suite 440, Unit 422, Rockville, MD 20852, USA, *ellenchen@nanobiomimeticsensors.com*

ABSTRACT

Challenges of quantitative detection of Amyloid-beta (Aβ) in Alzheimer's brains are paramount, such as the instability of soluble $A\beta_{1\text{-}42}$ in biological specimens, the interferences among proteins and the time consuming assays. A memcapacitor device was developed with nanostructured biomimetic acetylcholinesterase (ACHE) gorge membrane on gold chips, served as an artificial normal brain motif, has detected voltage and current changes in sub pM Aβ under conditions of free from antibodies and tracers using spiked NIST SRM 965A human serum, based on the characteristics of memcapacitor/memristor. The results of Detection of Limits (DOL) are 5.0×10^{-11}M/cm^2 vs. 7.0×10^{-13}M/cm^3 over the linear range at 0.04-151 nM vs. 3.8-471 nM, for the chronoamperometry (CA) and double step chronopotentiometry (DSCPO) methods, respectively. Aβ-spiked fresh human capillary whole blood (CPWB) specimens used to verify the accuracy were traceable to NIST standards with recoveries of 100.15± 1.2% and 98.6±1.1% and the imprecision results are 3.2% vs. 6.0% (n=18, n=15) for CA and DSCPO method, respectively. The results of the signal to noise ratio for both methods are about 12.

Keywords: Nanobiomimetic Neural Memcapacitor; Voltage Sensor and Amperometric Sensor; Reagent-less.

INTRODUCTION

Amyloid-beta (Aβ) peptide accumulation and neurofibrillary tangle identified as major pathological biomarkers linked to Alzheimer's disease (AD) has been studied over decades. Besides significant progresses have been made, but lack of effective treatments and preventions addressed an urgent need for early diagnose and detection of Alzheimer's disease. It is estimated the global prevalence of dementia is about 24 million, and will double to 2040, leading to a costly burden disease to the society [1-3].

It would be more attractive to have a less invasive method to use than the CSF fluid method and to have an inexpensive method to use compared to a costly positron emission tomography with radiotracers. Therefore, plasma or human serum would be more desirable as specimens. Several road blocks have been hampered to reach the goals: the instability of Aβ in biological fluid [4-6], protein non-specific bounding caused high imprecision [4,7] and the time consuming procedures of the assay. The human biomarkers for Alzheimer's research are predominantly quantified using enzyme-linked immunosorbent assays (ELISAs) that are associated imprecision of CV% values reported in literature about 20-30% [4,7]. Calls for development of innovative tools and therapeutic approaches for better measuring preclinical and clinical biomarkers and treatment for AD is needed urgently [8-10]. Based on our experiences in development of nanostructured biomimetic sensors used for detection of toxins, blood glucose, cancers and neurotransmitters, such as acetylcholine (ACH) in biological fluid with selectivity, sensitivity and accuracy under the conditions of tracers-free, antibody-free and reagent-less [11-16], overcoming the challenges to developing nanostructured dual devices for precisely measuring preclinical Aβ in clinical useful range is encouraging to us. Providing even more useful information to clinicians and to patients is always beneficial.

Our biomimetic acetylcholinesterase (ACHE) membrane with an ACHE active site gorge deposited on a gold chip could be a best candidate to sense the presence of excess monomer Aβ, because ACHE dysregulation is well known to link to cancer, AD and other diseases [17-18]. However, direct measure Aβ, not ACH, in blood is a challenge, even though we used this "normal ACHE gorge" sensor to quantitatively detect ACHE in fM in amperometric mode [15] against a "mutated ACHE gorge" sensor, which was unable to sense ACH. Furthermore, we found the normal ACHE gorge sensor is able to serve as a memcapacitor type of memory device that clearly distinguishes conformational and neuronal circuitry change due to brain cancer cells' "bio-communicating" to the sensor [16]. Therefore, we hope the dual sensor would be able to sense the presence of Aβ in pM in order to lay a foundation for further study of the role of Aβ.

EXPERIMENTAL

Fabrication of the Nanostructure Self-Assembling Membrane (SAM) Gold Memristor Chips

The nanostructured Biomimetic "Normal Active ACHE Gorge" memcapacitor sensor with the flat bridged conformation/nanopore was freshly prepared and fabricated on gold chip. Polyethylene glycol diglycidyl ether (PEG), triacetyl-ß-cyclodextrin (T-CD), poly(4-vinylpyridine) (PVP) were purchased from Sigma. PVP was purified before use. The mono imidazolyl derivative dimethyl ß-cyclodextrin (mM-ß-DMCD) was generally synthesized according to the published procedures [19]. The procedures for mixture solution preparations of mM-ß-DMCD, T-CD, PEG, PVP and o-nitrophenyl acetate (o-NPA) and the gold chip membrane developments were followed according to the literature [20-21].

Characterization of the Membrane

The morphology of the AU/SAM was characterized using an Atomic Force Microscope (AFM) (model Multimode 8 ScanAsyst, Bruker, PA). Data Collected in PeakForce Tapping Mode. Probes used were ScanAsyst-air probes (Bruker, PA). The silicon tips on silicon nitride cantilevers have 2-5 nm radius. The nominal spring constant 0.4N/m was used. Figure 1 illustrates the 3D artificial neocortex-hippocampus memcapacitor design blocks as the neurons sensor. Figure 2 illustrates the 3D flat conformational bridge structure with "breathing nanopore" of the AFM images.

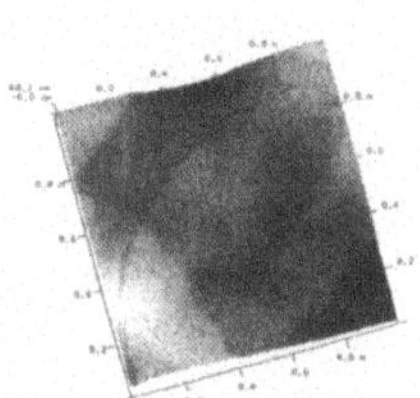

Fig. 1 (L) Illustrates the 3D memcapacitor blocks serve as the dual sensor. Greens represent conductive polymers; the oranges represent the inner "ACHE Gorge" in narrow cylinders connected through the neuronal terminals and dendrites as truncated donuts in compact flat metrics. Fig. 2 (R) 3D horizontal conformational bridge structure of the AFM images of the memcapacitor.

Overcome the Instability of Aβ

The CV method was used to identify the instability of Aβ in the aqueous media under the scan rate at 20 mV/s. After 0.1 mg/mL TCD was presented in the solution, the instability should be overcome, because β-cyclodextrin derivatives are well known stabilizers in industries [22].

Quantitation of Aβ

The CA Method. The CA method was used for quantitation of Aβ. $A\beta_{25-35}$ was purchased from Sigma. The data were acquired at room temperature under two-step fixed potentials in 8 concentration levels covering Aβ final concentrations ranging from 10^{-11}M to 10^{-7}M, with triplicates in DDH_2O with 0.1 mg/mL TCD and using an electrochemical work station (Epsilon, BASi, IN) with the companied software package. Origin 9.0 was used for all statistic data analysis and figure plotting.

The DSCPO Method. The characterization of the memcapacitor serving as a voltage sensor was conducted by the DSCPO method in ± 10 nA and 0.25 Hz in DDH_2O with 0.1 mg/mL TCD, with spiked Aβ final concentrations from 0.038 nM to 60.8 nM, and with triplicates for a calibration curve. The NIST SRM 965A human reference serum, with controlled blood glucose 70 mg/dL, spiked with Aβ having 4 levels from 3.8 to 417 nM with a single run at the same experimental conditions as in water media, and measurements without spiking Aβ were also taken for comparison.

Factors Affect on Sensor Performance

Peak duration time and Aβ concentrations were factors that affect on sensor's performance using the CA method. A healthy subject's fresh finger stick capillary whole blood (CPWB) specimens were collected, then immediately spiked Aβ in various concentrations in less than 4% water content to the blood volume without anticoagulant and without stabilizer. All measurements were in triplicates against controls at room temperature and finished in a half hour. Peak duration time had three levels: 4ms, 25 ms and 4s. The same factors were used for the voltage sensor under conditions of ± 10 nA and 0.25 Hz with data rate 1 kHz, 40, 250 Hz (both had data rate 50 kHz) for action potential peak or resting potential peak, respectively. The subject has gone through consent and was approved with the IRB.

Assessing Precision and Accuracy

Precision was evaluated by the two methods using fresh human CPWB specimens with triplicates at two levels of Aβ at 2.3 and 92 nM and three levels of frequencies at 0.25, 40 and 250 Hz in triplicates. Accuracy results were assessed by the "Point Accuracy" using signal means of the CPWB specimens against the calibration curves with the reference results that are traceable to NIST reference serum with the standard addition of the Aβ concentrations.

RESULTS AND DISCUSSIONS

Characterization of the Memristor/Memcapacitor

Literature reported electric synapse is one tenth of that of chemical synapse [23]. Fig. 3's i-V hysteresis curve was demonstrated with a switch point at the origin (0, 0). The frequency change affecting on the memristor/memcapacitor was depicted in Fig. 4 using NIST serum without Aβ. At low frequencies, the sensor has the highest *Direct Electron-relay Transfer* (DET) [10-12, 24] peak intensity than at high frequencies and all curves have hysteresis characters.

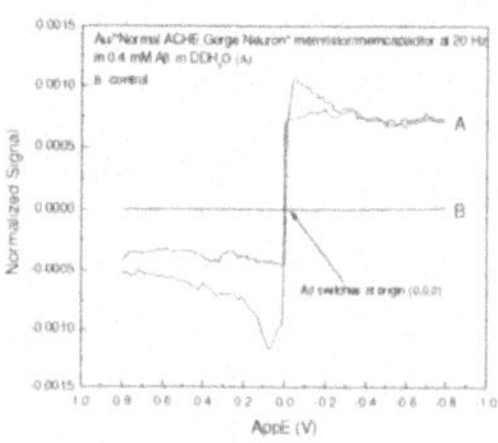

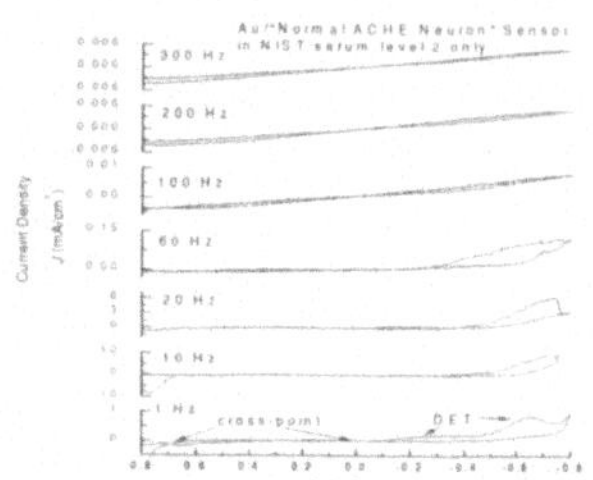

Fig 3 (L) Illustrates the hysteresis of the i-V curve of A in 400 μM Aβ in DDH_2O with a switch point at origin (0, 0, 0) against the control (B); Fig 4 (R) Depicts frequency affects on CV curves from 1 to 300 Hz in NIST serum without Aβ.

Overcome the Instability of Aβ

Instability of Aβ in various media has been reported in the literature [4-6], and our own experiments confirmed the instability in aqueous solution. The CV curves shown in Fig. 3 and Fig. 5 (L) illustrate the peaks were unstable in water. After 0.1 mg/mL TCD presences in the media, the CV curves are stabilized as shown in Fig. 6.

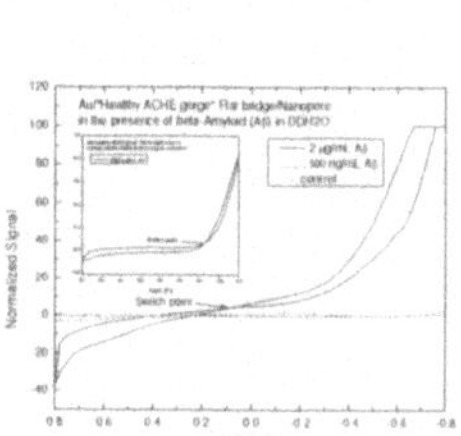

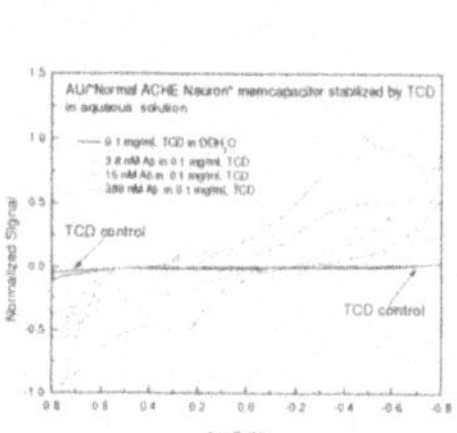

Fig. 5 (L) Illustrates the CV profiles without TCD. The control is in red; the insert is 380 nM Aβ; the solid black line is 1.52 μM Aβ. Fig. 6 depicts the i-V behaviors of the sensor in the stabilizer. Aβ at 0, 3.8, 15, 380 nM.

Quantitation of Aβ

The CA Method. Fig. 7 illustrates CA curve profiles in the presence of Aβ in aqueous solutions over the range of 0.037 to 151 nM Aβ. Fig. 8 illustrates the calibration curve with a linear regression equation Y= 0.59 +0.63X, r=0.998 (n=23), P<0.0001, Sy/x=1.96. The value of Detection of Limit (DOL) is $5.0x10^{-11}$M per cm^2 with a pooled relative standard deviation of 0.2% related to that at the mean concentration. Because this sensor is only 0.031 cm^2, hence, its DOL is 1.6 pM Aβ.

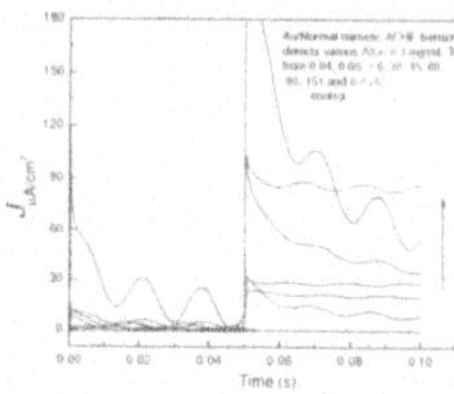

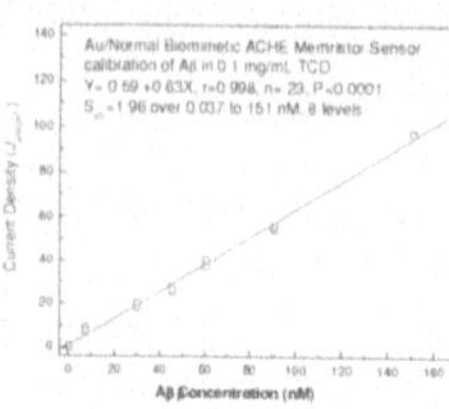

Fig. 7 (L) Illustrates CA curve profiles and Fig. 8 (R) shows the calibration plot of current density vs. Aβ concentrations.

The DSCPO Method. The magnitude of voltage change was the highest when without Aβ, as shown in the insert in Fig. 9. The calibration curve, as shown in Fig. 10, was inversely proportional to Aβ concentrations with a linear regression curve in water (in black) equation Y =1.48 - 0.018X, r = -0.992, $S_{y/x}$ =0.058, n=12, p<0.0001 over volumetric energy density from 0.41 to 1.50 μWHr/cm^3. The Detection of Limit (DOL) is $2.63x10^{-9}$M per cm^3 with a pooled relative standard deviation of 5% related to that at the mean concentration. Because this sensor is only $3.11x10^{-7}$ cm^3, hence, its DOL is $8.2x10^{-16}$M Aβ.

The Aβ calibration curve using NIST's reference human serum is shown in Fig. 10 in red and the experimental DSCPO curves profiles are shown in Fig. 11 covering clinically useful range between 3.8 to 471 nM Aβ. The linear regression equation of Y = 7.51 -0.014X, r = -0.995, Sy/x =0.36, N=4, p<0.005. The DOL value is $7.0x10^{-13}$M/cm^3.

Our results demonstrated Aβ is a strong inhibitor agent in blocking memory consolidating at Slow-Wave-Sleeping (SWS) at 0.25 Hz with the initial energy intensity decrease by 94% in 3.8 nM Aβ in serum. The energy signal without spiking Aβ is 8.5-fold higher in the serum than in the aqueous solution which indicates the serum is more suitable to the neuronal memcapacitor sensor without instability observed.

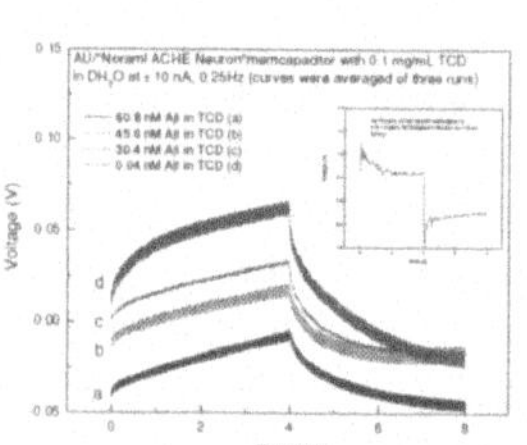

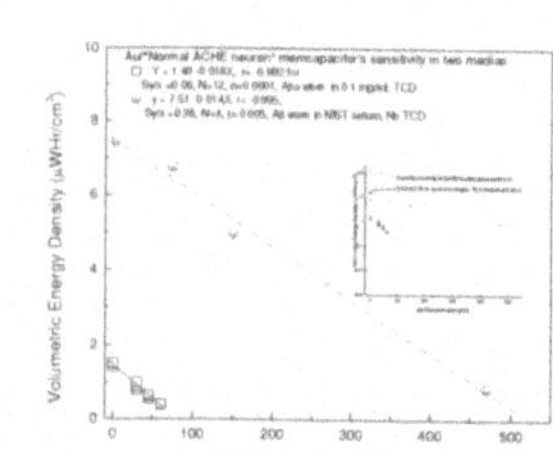

Fig. 9 (L) Depicts the DSCPO profiles in aqueous solution, and the curve with zero Aβ was in the insert. Fig. 10 (R) illustrates the calibration curves in NIST serum (Red), and in water (black).

The DSCPO profiles in the presence of Aβ in NIST serum samples were depicted in Fig. 11. The insert shows the curve without Aβ.

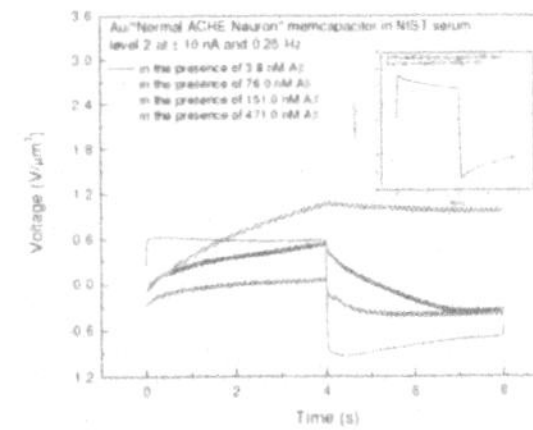

Fig. 11 Depicts DSCPO profiles of Aβ in NIST serum.

Factors Affect Sensors' Performance

Figure 12 revealed the high energy density values were associated with specific capacitance values between 1.2-2.2 F/cm^3 around zero Aβ compared with the negligible energy density at high Aβ level based on the calculations on the literature [25-26]. Fig. 13's map revealed the results obtained by a CA method using the same fresh human fresh CPWB specimens with the larger current density correlated with the highest Aβ concentration that located in the higher frequency at 250 Hz.

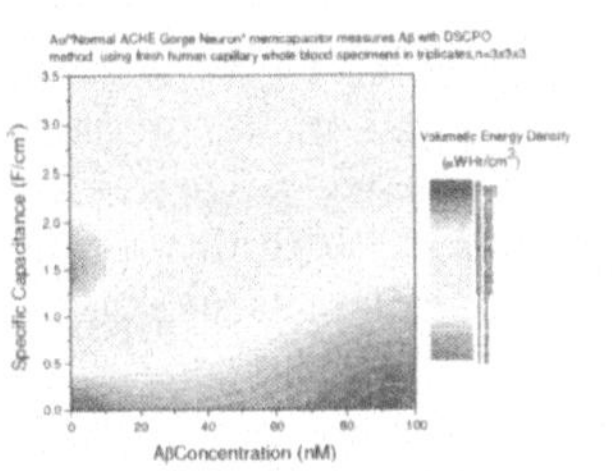

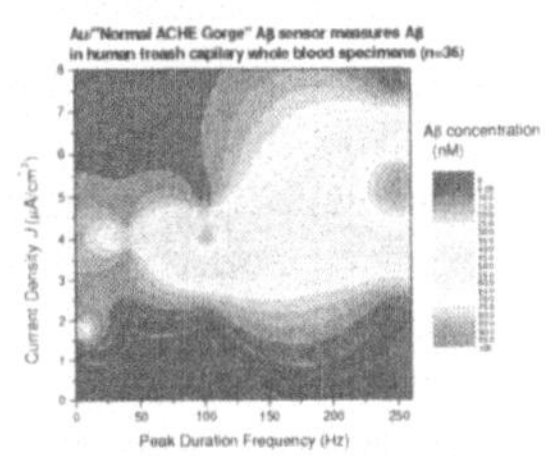

Fig. 12 (L) depicts the voltage sensor's factor map and Fig. 13 (R) depicts the CA sensor's factor map.

Assessing Precision and Accuracy

The imprecision was measured by the CPWB samples for the two sensors with the Pooled RSD of 3.2% (n=18) vs. 6.0% (n=15) over the studied ranges for the CA and the DSCPO method, respectively. The inaccuracy error was 0.1% with a recovery of 100.15± 1.2% for CA method after corrected metrics ratio. The average inaccuracy error obtained from the DSCPO method are 0.01% and 1.4% in water and in serum, respectively, that were traceable to the NIST's reference with 99.99±0.01% and 98.6±1.1% recovery.

The ratio of *Signal to Noise* (S/N) values accessed by the CA method and the DSCPO methods were calculated based on the conventional teaching [27]. The results have *S/N* values of 11.63 vs. 12.5 for the CA and DSCPO method, respectively.

CONCLUSION

The dual sensors detected sub pM Aβ with near 100% recoveries and 3-6% imprecision under antibody-free and tracer-free conditions were demonstrated. The technology may find wide applications for early monitoring of Alzheimer's disease.

REFERENCES

[1]. Reitz C, Brayne C and Mayeux R, *Epidemiology of Alzheimer disease,* Nature Reviews Neurology 7, 137-152, 2011.

[2]. John E. Morleya JE and Farr SA, Biochemical Pharmacology, 88(4), 479-485, 2014.

[3]. Jia Q, Deng Y, and Qing H, BioMed Research International Article ID 837157, 22 pages, 2014.

[4]. Toledo JB, Shaw LM and Trojanowski JQ, Alzheimer's Research & Therapy **5,** 8, 2013.

[5]. Lachno DR, Emerson JK, Vanderstichele H, et al, J Alzheimers Dis 32, 905-918, 2012.

[6]. Bibl M, Welge V, Esselmann H, Wiltfang J, Electrophoresis 33,445-450, 2012.

[7]. Esparza TJ, Zhao H, Cirrito JR, Cairns NJ et al., Ann Neurol,73(1), 104–119, 2013.

[8]. Fiandaca MS, Mapstone ME, Cheema AK, Federoff HJ, Alzheimer's & Dementia, 10, S196-S212, 2014.

[9]. Zlokovic BV, Neuron, 57(2),178-201, 2008.

[10].Chen ET, Duh SH, Ngatchou C, Thornton JT, and Kissinger PT, Nanotech (3),*101-104,* 2011.

[11].Chen ET. US Patent 6,582.583, June 24, 2003.

[12].Chen ET, US Patent 8,083,926, December 27, 2011.

[13].Chen ET, Thornton JT, Ngatchou C, Duh SH and Kissinger PT, NSTi-Nanotech (3), 115-118, 2013.

[14].Chen ET, Thornton JT, Ngatchou C, Duh SH and Kissinger PT, NSTI- Nanotech (2), 107-110, 2013.

[15]. Chen ET, Thornton J, Ngatchou C and Duh S-H. , NSTi-Nanotech, 2, 169-172, 2014.

[16].Chen ET, Thornton J and Mulchi Jr C. Sensors & Transducers, 183(12), 72-83, 2014.

[17].Nordberg A, Rinne JO, Kadir A and Långström B, Nature Reviews, 6, 78-87, 2010,

[18].Yan JJ, Cho A-Y, Kim H-S, Kim K-L, Jung J-S, British J of Pharmacology 133(1). 89–96, 2001

[19].Chen ET and Pardue HL, *Anal Chem,* 5, 2563-7, 1993.

[20].Chen ET, Thornton JT, Ngatchou C, Duh SH and Kissinger PT, NSTi- NanoTech (2), 107-110, 2013.

[21].Chen ET, Thornton JT, Ngatchou C, Duh SH, NSTi-Nanotech, 2, 200-203, 2014.

[22].Szejtli, *Cyclodextrin and Their Industrial Uses,* Editions de Sant´e, Paris, 1987.

[23].Buzsáki G et al., Nature Reviews 13, 407-420, 2012.

[24].Chen ET, *Nanobiomimetic sensing and energy storage* in the first volume of the book series of *Dekker Encyclopedia of Nanoscience and Nanotechnology* edited by Lyshevski SE, CRC Press, 2013.

[25].Hu L, Choi JW, Yang Y, Jeong S, Mantia FL, Cui L-F and Cui Y, PNAS 106 (51), 21490-21494, 2009.

[26].Thornton J, Christelle C and Chen ET, NSTi-Nanotech (2), 672-675, 2013.

[27].Skoog DA and Leary JJ, *Principles of Instrumental Analysis,* Fourth Edition, Saunders College Publishing, 1992.

In Situ Health Monitoring of Piezoelectric Sensors

Scott L. Jensen*
*NASA Stennis Space Center, John C. Stennis Space Center, Stennis, MS 39529
Electronics Engineer, Engineering & Test Directorate, Scott.l.jensen@nasa.gov

ABSTRACT

The developed technology is an in-situ measurement system for monitoring the performance of piezoelectric sensors, particularly accelerometers. The technology's dynamic response can be used to evaluate the operating range, health, and as-mounted resonance frequency of a transducer, along with the transducer's mechanical coupling with a structure. Characteristics such as sensitivity, frequency response, cable status, connectivity, bonding and linear range are determined in an instant as installed. Sensors are tested in a very wide frequency range extending to over 200 MHz, and without specialized transducers or wiring. Beyond detecting transducer failure or detachment, the technology can also identify structural changes such as deformation and microfracturing within a test article. With this monitoring system, degraded sensor performance can be quickly and economically identified with hand held unit or integrated directly into test equipment.

Keywords: in-situ, monitoring, health, resonance, as-mounted-resonance, sensitivity, frequency-response, cable-status, connectivity, linear-range, diagnostic

1 BACKGROUND

Commercial test equipment suitable for testing piezoelectric sensors is expensive and requires that the sensor be removed from the test article for evaluation. This innovative technology permits monitoring the health of piezoelectric sensors while installed on, or after thaving been removed from test articles. On several occasions with Stennis Space Center (SSC) rocket engine testing, anomalies appear in the accelerometer and dynamic pressure data recorded by the Low Speed and High Speed Data Acquisition Systems. An in-situ test system was necessarytoassess if the transducers were operating properly or if the transducers were in error. It is often unclear if anomalies in recorded signals are due to differences between the Low and High Speed Data Acquisitions Systems, differences between the transducers, a failed transducer, *or* if everything is working correctly and the systems were actually accurately recording real events. One particular data anomaly incident investigation led to conecption of this. A poor cable connection was energy to build up within an accelerometer cable due tovibrations induce from the ongoing propulsion test. Energy creation within a cable due to vibrational conditions is often known as a cable slap phenomenon. Periodically the poor connection would reconnect which dumped the cable's energy into the accelerometer. Due to the convers-piezoelectric effect, the accelerometer would vibrate in response to the stimulus; in turn, the vibrations were inducing a signal reading within other accelerometers mounted on a common tri-axial block. It was then that the idea that the converse-piezoelectric effect could be employed for introducing a stimulus energy into the structure under test while allowing the stimulus energy's response to be collected was conceived. The process possessed the potential to obtain valuable information about the sensor and the corresponding structure being tested. This is now known as the piezoelectric stimulus-response effect. The signals produced by the piezoelectric sensor under test yields an abundant amount of frequency spectrum information. The state of health of the sensor and structure can be determined by spectral analysis.

2 THE IN-SITU HEALTH MONITOR FUNCTIONALITY

The developed technology is an in-situ measurement system for monitoring the performance of piezoelectronic sensors, particularly accelerometers. With this technology, characteristics such as resonant frequency, response, cable status, connectivity, bonding and linear range, can be determined. Sensors can be tested in a very wide frequency range, extending to 200 MHz and beyond, without removing them from their mounted locations, and without requiring specially constructed transducers or special wiring. The dynamic response can be used to evaluate the operating range, health and as-mounted resonance frequency of the transducer, as well as the strength of a coupling between the transducer and a structure, and the health of the structure. The monitoring system provides the full frequency of the sensor, the linear operating range, and the resonant frequencies-both normal and as-mounted for different configuration types. The low frequency region captures the environmental noise identifying the noise levels one can expect during data collection similar to using a placebo transducer. The sensitivity indicator resonant amplitude precisely revealed the transducers sensitivity

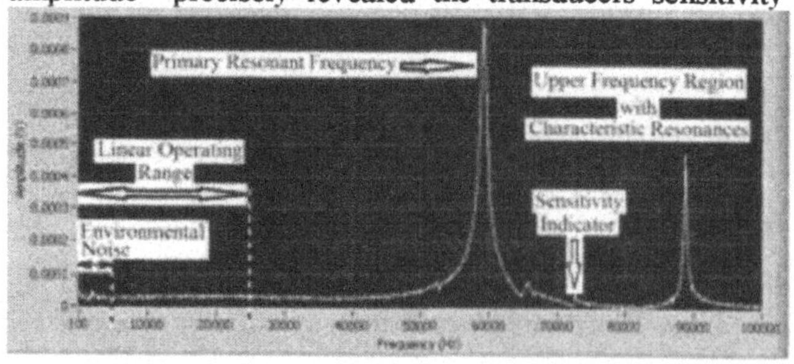

(see Figure 1). The system also provides the ability to

monitor piezoelectric transducers between propulsion tests to detect any trend indicative of transducer failure or detachment. Additionally, the system can be made portable, incorporating the unit in a battery powered sealed box, for testing in the field. Physical contact with the sensor is not necessary, therefore, monitoring can be done up to 250 feet, or even longer if certain provisions are made. With only slight modifications, this monitoring system can be used with all common transducer instrumentation. This monitoring system, can quickly and economically identify degraded sensor performance . Additionaly, installed piezoelectric sensors can be evaluated, without requiring physical contact or removing them from their mounted locations;tests are conducted through cabling. Because it is not necessary to remove the device, data that reflect the device's specific physical configuration (such as as-mounted resonant frequency) are retained, and devices that are physically inaccessible can still be tested. The testing system is not limited to identifying degraded performance in the sensor's piezoelectric elements; it can detect changes within the entire sensor, and sensor housing.

The in-situ piezoelectric test apparatus is far simpler and the system includes several enhancements and innovations over existing prior art. All of the transducer's resonant frequencies are monitored and quickly assessed, unlike other known apparatus that are only censored with the primary resonate frequency. The transducer's primary resonate frequency and sensitivity can be unknown prior to performing transducer tests. The developed test apparatus quickly performs a far more comprehensive test. It can detect changes in the transducer sensitivity, cabling, piezoelectric element, transducer housing, transducer attachment, and test article.

Some transducers have built in amplifiers that convert the generated electrical charges into voltage signals. The developed apparatus can be employed in these kinds of transducer systems. This type of transducer is widely employed and the fundamental components of previously known apparatus prevents their use in this type of transducer. This charge to voltage conversion inside of the transducer makes the signal less susceptible to changes in cable length. This is an important advantage to this class of piezoelectric devices. The developed in-situ piezoelectric test system used in conjunction with the charge converters significantly enhances the capability of piezoelectric devices.

3 THE IN-SITU HEALTH MONITOR TECHNOLOGY

Piezoelectric transducers generate electric charges in response to mechanical deformation of the materials from which they are made. Pierre and Jacques Curie discovered this phenomenon in the 1880s. The first application of the piezoelectric effect was the detection of submarines during World War 1. They also demonstrated the inverse piezoelectric effect in which an electric field applied across a piezoelectric material deforms mechanically. The research was performed at the SSC Data Acquisition and Control System Laboratory and provided a means to test piezoelectric transducers while they are installed in the field. The technology has been further refined into a portable field diagnostic tool, laboratory bench tester, and integrated system components. The hand held portable units, in a battery powered sealed box, for performing tests within the field, are known as the Piezo- Electric Transducer Testers (see Figure 2). The units have already aided in resolving questions concerning data anomalies. The patented technology has also been designed for integration within facility instrumentation systems.

Figure 2: PiezoElectric Transducer Tester Version 3.3 (PETT-3.3)

The fully functioning system consists primarily of an excitation device capable of delivering low voltage microsecond duration electrical pulses to the sensor, a high speed data acquisition device for collection data, a control circuit, and a computer with appropriate software. The in-situ health monitoring technology utilizes the primary components for executing the piezoelectric stimulus-response effect.

A simple control circuit performs a test sequence by first triggering the excitation device. The excitation device injects stimulus energy into the accelerometer under test, and the piezoelectric material inside the accelerometer physically deforms due to the converse piezoelectric effect. The piezoelectric material, commonly known as the element, inside of the accelerometer is forced to vibrate due to the deformation similar to ringing a bell (see Figure 3). The control circuit then drains residual electrical energy out of the piezoelectric element as the vibrational energy reverberates throughout the accelerometer and associated structures. The mechanical vibrations of the accelerometer correspondingly produces electrical signals due to the

Figure 3: Time domain data of vibrational ringing response within an accelerometer from injected stimulus.

piezoelectric effect. The same piezoelectric element within the accelerometer can now be used to measure the structure's mechanical reverberating response. The data acquisition device is subsequently triggered for recording the responding signals (see Figure 4). The width of

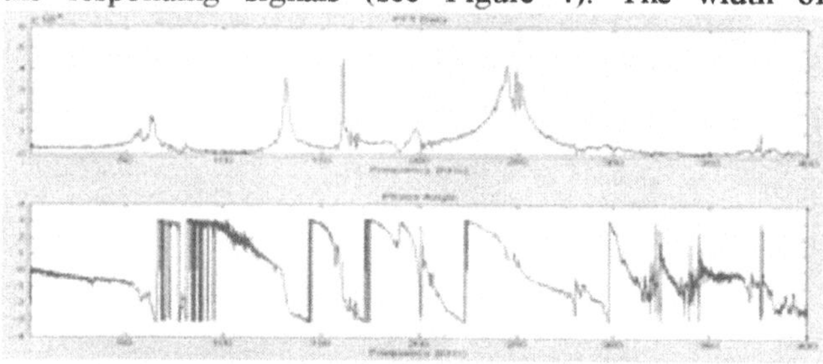

Figure 4: Frequency domain data of the vibrational ringing response within an accelerometer.

the applied pulse from the excitation device determined the frequencies contained in the pulse energy. A sharply shaped pulse contains a wide range of frequencies adequate for exciting a resonance ringing response in nearly all piezoelectric transducers even shock accelerometers and pressure transducers. One type of excitation device functions well for all kinds of transducers, and does not need exists for multiple execution configuration.

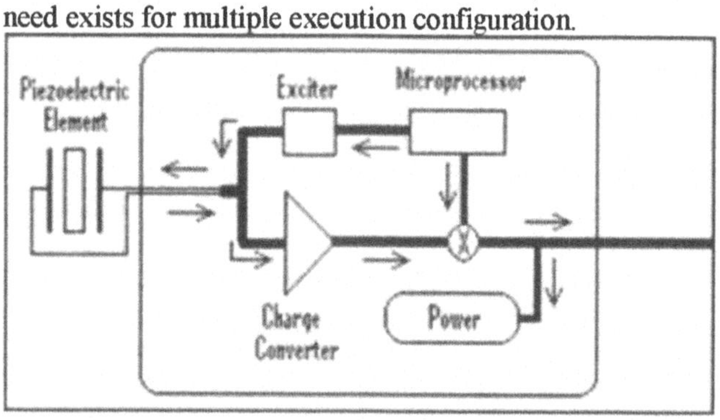

Figure 5: Generalized Test Cycle Flow Diagram

4 DATA REDUCTION AND ANALYSIS BY THE VECTOR EVALUATION

Sensor diagnoses are made by applying predefined limits to polar values that are linked to a specific structure and sensor type. The polar values are generated with a vector evaluation algorithm that transforms data sets into multidimensional vectors. The differences in the magnitudes and angles of the vectors are then computed, and results are displayed in a two dimensional polar plot with a single mean value. This process significantly reduces the massive amounts of multidimensional frequency data generated from the technology into a simplified two dimensional representation. Automated indication can also be implemented with the polar plot limits.

Multiple test cycles are consecutively perform with the sensor's response data being saved for each cycle. Each resulting data set is processed into Fast Fourier Transform (FFT) data (see Figure 6). Each value of a FFT data

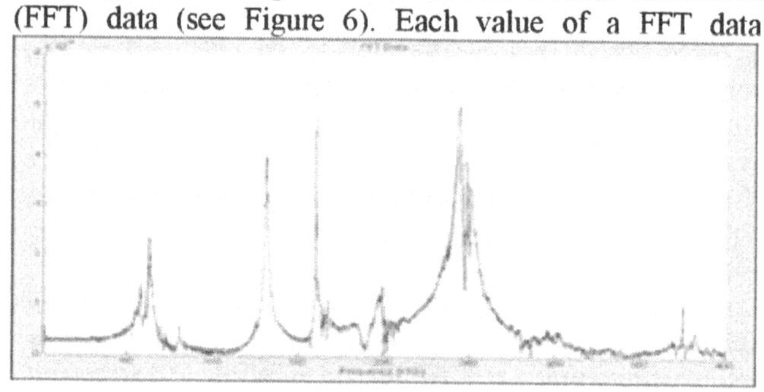

Figure 6: FFT data sets collected from multiple stimulus-response test cycles

(each frequency bin) is represented as one dimension of the set multidimensional vector. If the data sets are identical, the vector evaluation algorithm will find no difference between the data sets and produce a point in the center of a polar plot. The greater the differences between the data sets, the larger the angle and amplitude will become. The multiple test cycles will accumulate into a cluster of values (see Figure 7). The resulting cluster

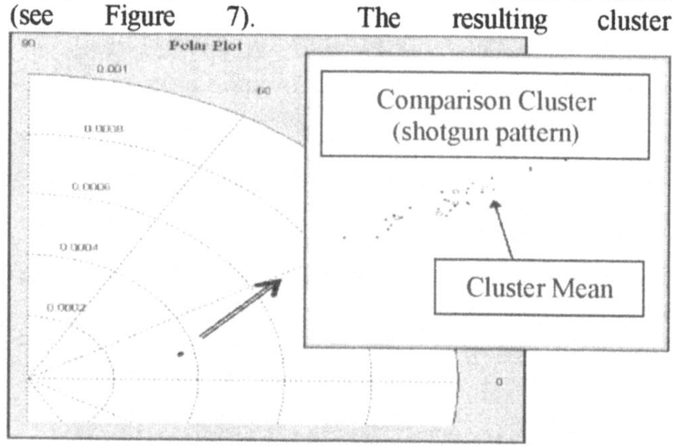

Figure 7: Vector evaluation polar plot of multipe test cycles with irregularities between data sets

should produce a linear pattern when the data sets are compared. The linear pattern indicates a good data collection was achieved, and a shotgun pattern indicates a poor data collection was performed with irregularities between data sets. This instantly validates any data reading. Fundamentally, the algorithm quickly condenses large amounts of data into a polar plot capable of quantifying small changes between the large data sets.

Diagnosing a bad piezoelectric transducer is obviously identifiable with the vector evaluation algorithm. An example of diagnosing a broken accelerometer can be seen in Figure 8. The accelerometer experienced a shock during rocket testing that slightly damaged the piezoelectric

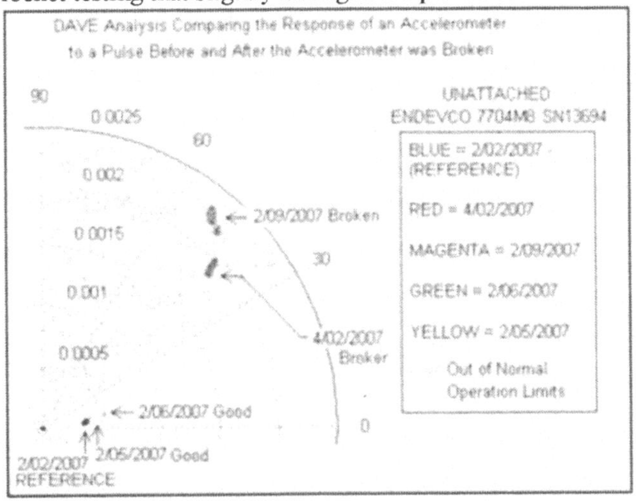

Figure 8: Example of diagnosing a broken accelerometer

element. The damage continued to progress with subsequent testing. Determining the coupling strength between the transducer and a structure is also obviously identifiable with the vector evaluation algorithm. An example of analyzing a properly mounted accelerometer can be seen in Figure 9. A significant difference is seen

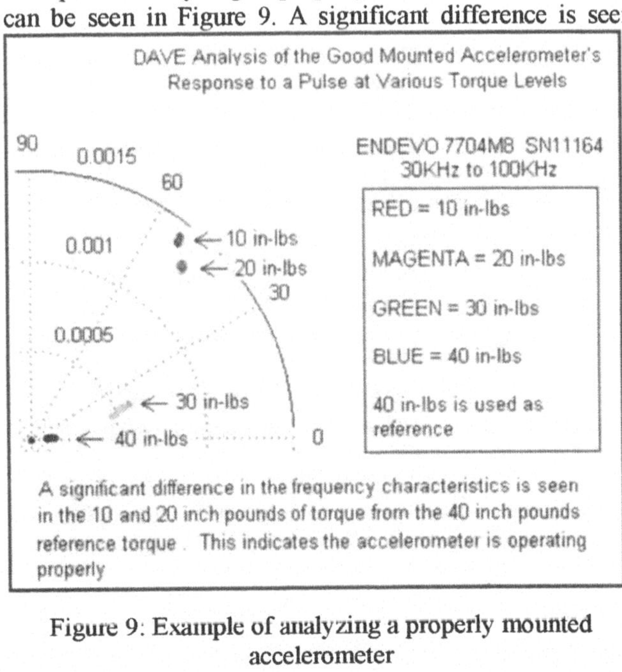

Figure 9: Example of analyzing a properly mounted accelerometer

when the structure changes the accelerometer's characteristic frequencies by mechanically becoming coupled with structure. With the vector evaluation algorithm, determining the state of a piezoelectric transducer without any prior data or information is straight forward. Diagnoses can be performed by incrementally coupling the transducer to a block while performing stimulus-response tests. a notable change will occur when the transducer mechanically becomes coupled with the block. If a notable change does not occur, then the transducer is faulty (see Figure 10). The technology

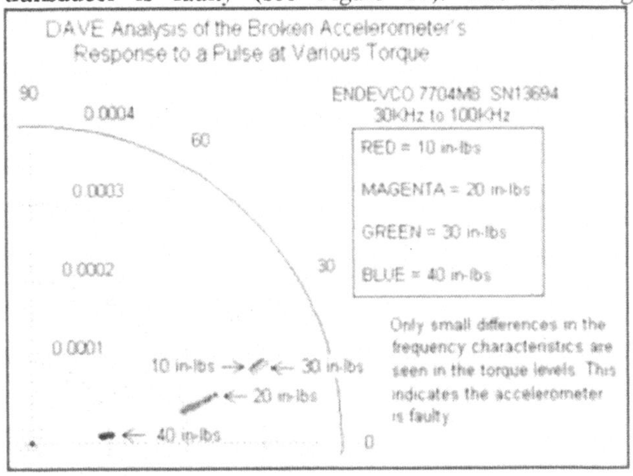

Figure 10: Example of diagnosing a broken accelerometer without any prior data or information

can also be utilized for revealing minuscule differences in a test article. The technology was successfully demonstrated for use as a smart glass monitor for solar collection mirrors. The difference between placing a dime, penny, and nickel at any location on a irregularly shaped four foot diameter mirror was reliably discernable (see Figure 11).

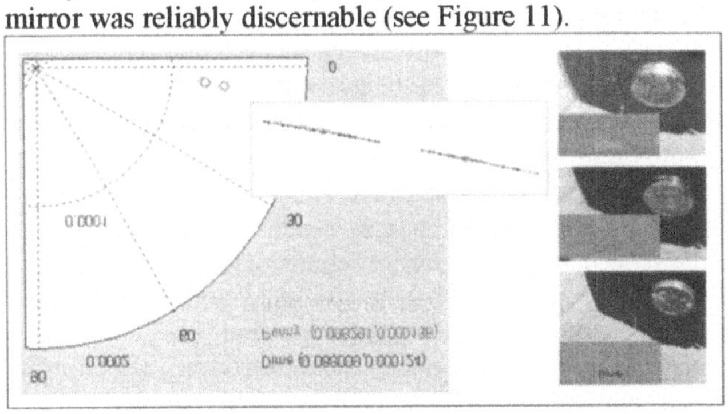

Figure 11: Example of discerning the difference between a dime, penny, and nickel on a mirror

REFERENCES

[1] Walter, P. L. (2001). The Handbook of Dynamic Force, Pressure and Acceleration Measurement. Endevco Corporation

[2] Atherton, W., Flanagan, P. (1989). A Self Diagnostic System for Piezoelectric Sensors. AIAA/ASME/SAE/ASEE 25th Joint Propulsion Conference, Monterey, CA. July 1989.

[3] Jensen, S., In-Situ Health Monitoring of Piezoelectric Sensors, US Patent 8,401,820 B2, awarded March 19, 2013.

High-sensitive, direct and label-free detection technique of nanotoxic material using QCM

Chanho Park[*], Kuewhan Jang[**], Sangmyung Lee[***], Juneseok You[****], Jinsung Park[*****] and Sungsoo Na[******]

[*]Department of Mechanical Engineering, Korea University, Seoul 136-701, Republic of Korea, chanhoo91@naver.com
[**]Department of Mechanical Engineering, Korea University, Seoul 136-701, Republic of Korea, qguy@korea.ac.kr
[***] Department of Mechanical Engineering, Korea University, Seoul 136-701, Republic of Korea, bjfbhbi@gmail.com
[****] Department of Mechanical Engineering, Korea University, Seoul 136-701, Republic of Korea, protrossko1@korea.ac.kr
[*****]Department of Control and Instrumentation Engineering, Korea University, Sejong 339-700, Republic of Korea
[******] Department of Mechanical Engineering, Korea University, Seoul 136-701, Republic of Korea, nass@korea.ac.kr

ABSTRACT

For the several decades, various nanomaterials are broadly used in industrial fields, research areas and commercialized products. Among many nanomaterials, silver ion, carbon nanotube, ZnO, and mercury ion etc are of the widely used nanomaterial in real life. However, high toxicity of those materials has been reported and the detection method is required for human health and environment. In present study, high-sensitive, label-free detection methods of nanotoxic materials are described.

We have developed a sensitive and selective quart crystal microbalance (QCM) based sensor for detection of nanotoxic materials by using frequency shift. In this study, we propose the sensitive in situ detection of nanotoxic materials by using QCM functionalized with a specific DNA for a target material. It also allows the detection of target material rapidly in real time. In contrast to the conventional methods, this method is label-free and sensitive, capable of direct rapid detection and real-time monitoring. The results suggest that QCM-based detection opens a new avenue for the development of a practical water-testing sensor.

Keywords: nanomaterial, toxicity, label-free detection, quart crystal microbalance, resonance frequency

1 INTRODUCTION

In recent years, there has been a rapid growth of nanomaterials usage from scientific interest to commercial products. Commercial products such as sun-creams, tennis rackets, solid lubricants, detergents, etc, are using nanomaterials due to unique properties of nanomaterials [1]. Among nanomaterials, single-walled carbon nanotube (SWNT), zinc oxide nanowire (ZnO NW) and silver ion (Ag^+) are the most widely used nanomaterials due to the extraordinary properties of those materials such as SWNT properties of high aspect ratio, light weight with elastic modulus of ~1.4 TPA, conductivity higher than copper, ZnO NW properties of piezoelectric and semiconductor, and Ag^+ properties of sterilization [2-4].

While SWNT, ZnO NW and Ag^+ are widely used for the commercial products, human exposure to those materials has not been concerned. Effluent from manufacturing, consumer utilization and after disposal of those nanomaterials will increase the release to the water system, and this will eventually increase human exposure to nanomaterials [5]. Moreover, recent studies have investigated the toxicity of effect of those nanomaterials [4, 6, 7]. Interstitial inflammation in the lungs was frequently observed in the high dose injection of SWNT and the decrease in cell viability of A549 and H1299 lung epithelial cell line was also observed [8]. Moreover, oxidative potential of SWNT was observed that can stimulate ROS production [3]. In the case of ZnO NW, ZnO NW started the degradation process to an ionic state of zinc ion, because of a week acid liquid state [7]. Then zinc ion leads to cytoxic phenomenon in the cell which is harmful mechanism and critical harmful concentration is about 10μg/ml of ZnO NW in weak acid solution.4 Ag^+ has greatly toxicity to living organisms by inactivating the sulfhydryl enzymes [9].

In order to protect human health from nanotoxic materials, detection of nanotoxic materials in water systems have to be concerned [1]. If we are able to monitor the concentration of nanomaterials, we can determine how harmful the concentration of nanomaterials will be acting on human health. One of approaches that can detect nanomaterial is the measurement of natural frequency of resonator sensor. There exist many resonator works for detection of biomaterials, chemical molecules, such as DNA, RNA, proteins due to the high-sensitivity and label-free method by the measurement of natural frequency [10]. Moreover, natural frequency is function of mass change and the detection is directly achieved [11].

In this paper, we have developed sensing methodology of nanotoxic materials such as SWNT, ZnO NW and Ag^+ using measurement of natural frequency of resonators. The sensing methodology is based on the shift of natural frequency upon binding of nanomaterials absorption on the DNA immobilized resonators in which the DNA sequences were specified for each case. Similarly, nanotoxic materials were also modified in order to adsorb and increase the detection performance. In case of SWNT, two different DNA sequences were used to conjugate SWNTs together for the mass amplification and the conjugate SWNTS were designed to adsorb on the DNA functionalized resonators. Similarly, ZnO NWs were wrapped with graphene and designed to adsorb on the DNA immobilized resonator. For Ag^+, Ag^+ was designed to adsorb on the poly cytosine bases DNA on the resonator through specific intercalation. By measuring the nature frequency of resonators, we were able to detect 10 ng/ml of SWNT, 100 ng/ml of ZnO NW and 1 nM of Ag^+.

2 THEORY

The natural frequency of resonator can be described by classical elastic continuum model:

$$\omega_0 = \left(\frac{\alpha}{l}\right)^2 \sqrt{\frac{EI}{\rho A}} \tag{1}$$

where α is the eigenvalue by the positive root of $1+\cos(\alpha)\cosh(\alpha)=0$, l, E ,I, ρ and A is the length, Young's modulus, cross-sectional moment of inertia, mass density and cross-sectional area of resonator. Furthermore, the natural frequency shift can be simplified by mass or stiffness change as below:

$$\Delta\omega = \frac{1}{2}\omega_0\left(\frac{\Delta k}{k} - \frac{\Delta m}{m}\right) \tag{2}$$

where ω_0 is the natural frequency, Δk, k, Δm and m is the stiffness change, stiffness, mass change and mass of resonator. From the equation, we can see that the natural frequency shift is related to stiffness and mass change. In general, the mass effect is the dominant effect and the stiffness effect can be ignored. This implies that the natural frequency shift is directly related to the mass change which caused by the adsorption of nanomaterial in our system.

Since the natural frequency of resonators were not constant, normalized natural frequency shift was used:

$$\omega_n = \frac{\omega_b - \omega_a}{\omega_b} \times 100 \tag{3}$$

where ω_b and ω_a is the natural frequency of resonator before detection and after detection.

3 RESULT

The detection was performed by the interaction of nanotoxic material with the resonator. As nanotoxic material bind on the resonator, the mass of resonator increased and the natural frequency decreased. By analyzing the natural frequency shift, we were able to detect nanotoxic material.

4. CONCLUSION

In summary, we developed detection methodology of SWNT, ZnO NW and Ag^+ by using natural frequency of resonator. The sensing mechanism is based on the measurement of the natural frequency shift arisen from the detection of the target nanotoxic materials on specified DNA immobilized resonator. The natural frequency shift was well responded to the each concentration of nanomaterial. In detail, we were able to detect SWNT, ZnO NW and silver ion with the detection limit of 10 ng/ml of SWNT, 100 ng/ml of ZnO NW and 1 nM of silver ion, respectively. Moreover, our sensing methodology is label-free and sensitive, showing the potential of natural frequency based sensing methodology as the nanotoxic material screening tool.

Acknowledgment

This study was supported by the National Research Foundation of Korea (NRF) under grant numbers of NRF-2012-0000785, NRF-2012-0006646, NRF-2014R1A2A1A11052389 and NRF-2013R1A1A1058102 funded by the Ministry of Science, ICT & Future Planning.

References

[1] Maynard, R. J. Aitken, T. Butz, V. Colvin, K. Donaldson, G. Oberdorster, M. A. Philbert, J. Ryan, A. Seaton, V. Stone, S. S. Tinkle, L. Tran, N. J. Walker and D. B. Warheit, "Safe handling of nanotechnology," Nature, vol. 444, pp. 267-9, 2006.

[2] H. Dai, "Carbon Nanotubes: Synthesis, Integration, and Properties," Accounts of Chemical Research, vol. 35, pp. 1035-44, 2002.

[3] S. George, S. Pokhrel, T. Xia, B. Gilbert, Z. Ji, M. Schowalter, A. Rosenauer, R. Damoiseaux, K. A. Bradley, L. Mädler and A. E. Nel, "Use of a Rapid Cytotoxicity Screening Approach To Engineer a Safer Zinc Oxide Nanoparticle through Iron Doping," ACS Nano, vol. 4, pp. 15-29, January 2010.

[4] H. T. Ratte, "Bioaccumulation and toxicity of silver compounds: A review," Environmental Toxicology and Chemistry, vol. 18, pp. 89-108, 1999.

[5] C. Lam, J. T. James, R. McCluskey, S. Arepalli and R. L. Hunter, "A Review of Carbon Nanotube Toxicity and Assessment of Potential Occupational and Environmental Health Risks," Critical Reviews in Toxicology, vol. 36, pp. 189-217, January 2006.

[6] G. Jia, H. Wang, L. Yan, X. Wang, R. Pei, T. Yan, Y. Zhao and X. Guo, "Cytotoxicity of Carbon Nanomaterials: Single-Wall Nanotube, Multi-Wall Nanotube, and Fullerene," Environmental Science & Technology, vol. 39, pp. 1378-83, March 2005.

[7] K. H. Müller, J. Kulkarni, M. Motskin, A. Goode, P. Winship, J. N. Skepper, M. P. Ryan and A. E. Porter, "pH-Dependent Toxicity of High Aspect Ratio ZnO Nanowires in Macrophages Due to Intracellular Dissolution," ACS Nano, vol. 4, pp. 6767-79, November 2010.

[8] S. K. Manna, S. Sarkar, J. Barr, K. Wise, E. V. Barrera, O. Jejelowo, A. C. Rice-Ficht and G. T. Ramesh, "Single-Walled Carbon Nanotube Induces Oxidative Stress and Activates Nuclear Transcription Factor-κB in Human Keratinocytes," Nano Letters, vol. 5, pp. 1676-84, September 2005.

[9] V. K. M. Poon and A. Burd, "In vitro cytotoxity of silver: implication for clinical wound care," Burns, vol. 30, pp. 140-7, 2004.

[10] B. N. Johnson and R. Mutharasan, "Biosensing using dynamic-mode cantilever sensors: A review," Biosensors and Bioelectronics, vol. 32, pp. 1-18, February 2

Substrate addition and matrix optimization increased sensitivity and selectivity of the PLGA sensor, resulting in a stronger detection signal toward bacterial contaminations

Nadira Ibrišimović[#*], Mirza Ibrišimović[#**], Aldina Kesić[*], Muhamed Mehmedinović[***]
[*]Department of Chemistry, Faculty of Science, University of Tuzla, nadira.ibrisimovic@untz.ba, aldina.kesic@untz.ba
[**]Medical School Faculty-Sarajevo School of Science and Technology, mirza.ibrisimovic@ssst.edu.ba
[***]Medical Health Center Čelić, muhamed.mehmedinovic@yahoo.com
[#] Both authors contributed equally to this work

ABSTRACT

The aim of this work was to contribute to the development of a simple and cheap biomimetic sensor, providing a higher sensitivity and selectivity toward microbiological contaminations in a test product. The biosensor is composed of two different layers: a) mirror layer and b) biomimetic poly (lactic-coglycolic acid) (PLGA) layer that can be degraded by microbial lypolitic enzymes. Changes in the thickness of the polymer layer are resulting in a colour change of the biosensor's surface. We have investigated the biology and propagation of several bacteria (*Pseudomonas*, *Enterobacter*, *Proteus* and *Salmonella*) and their metabolitic pathways. Not all bacteria produced an equal amount of lipolytic enzymes at the same stage of their reproduction, and there was no connection with a total bacterial counts. Under addition of certain substrates in the matrix of the biosensor, it was possible to stimulate the bacteria for better enzyme production, resulting higher sensitivity of the PLGA biosensor.

Keywords: lipolytic enzymes, biosensor, polymer degradation, microbial detection

1 INTRODUCTION

Most of microorganisms such as bacteria are often exposed to chemical and physical changes of their environment which directly influence their metabolism. These changes can be in pH, temperature or in environmental nutrient presentation [8][9]. Bacteria are capable of adapting their metabolic processes to the ecological conditions and to regulate differently an enzyme production in presence or absence of certain substrate components [7][8]. The metabolic characteristics of bacteria and their ability to excrete enzymes have already been used as a detecting element for the recognition of bacterial contamination by using biosensor technology [1-4]. Enzymes are proteins with a high selectivity toward substrates which can be immobilized at the surface of the transducer by covalent attachment, adsorption or being incorporated into the polymer layer of the biosensor [12]. In this work we have optimized the biosensor prototypes which setup was developed in course of our previous studies [2][4], by choosing PLGA as a polymer layer and inconnel as a mirror layer. Application of biosensors is very large and can be found in food industry, crime detection, in quality control as well as in medical diagnostic [10][11][13]. Biosensors are analytical devices which are capable of providing specific, quantitative information, based on bioreceptor that is in spatial contact with transducer. The bioreceptor is biological recognition element such as enzyme, antibody or nucleic acid [3]. A biological component in biosensor presents a sensing element that normally determines the type of the biosensor. If the sensing element is an antibody, the biosensor is called immunosensor, and if the interaction between the analyte and the biological element is influenced by some substrates followed by changes in detection signal, it can be described as a metabolism sensor [12]. The measurable biosensor signal that can be optical, magnetic or electrochemical is produced by transducer. The development of optical biosensors recently became very intensive. There are two types of optical biosensors: a) those based on the detection of visible light (or UV and Infrared light) as the consequence of certain chemical reaction, and b) optical biosensors detecting some luminescent processes [14]. The biosensors also enable rapid pathogen detection in different types of testing sample [6].

2 MATERIALS AND METHODS

2.1 Spirit Blue Agar

Spirit blue agar contains Lipase Reagent (mixture of tributyrin and polysorbate 80) or some other lipid source for detection of lipolytic microorganisms. Spirit blue is a dye without any toxic effects and that makes it suitable for many lab protocols. Peptone is also a very important component of spirit blue agar, as a source of carbon, nitrogen and vitamins.

2.2 Generation of Phosphatidylcholin-Specific Phospholipase C (PC-PLC) Line

The Amplex® Red Phosphatidylcholine-Specific Phospholipase C Assay Kit provides a sensitive method for continuous monitoring of phosphatidylcholine-specific phospholipase C (PC-PLC) activity *in vitro* using a fluorescence microplate reader. In this enzyme-coupled assay, PC-PLC activity is monitored indirectly using 10-acetyl-3, 7-dihydrophenoxazine (Amplex Red reagent), a sensitive fluorogenic probe for H_2O_2. [15]

A calibration line was generated with increasing Phosphatidylcholine-specific phospholipase C activities. For calibration line we used a stock of Phosphatidylcholine-specific phospholipase C with specific activity from 10 U/mL.

Dilutions of enzyme solutions were made with 1X working solution of Reaction Buffer.

The following protocol describes the assay of PC-PLC in a total volume of 200 µL per microplate well. The volumes recommended here are sufficient for about 100 assays:

i) Dilute the PC-PLC–containing samples in 1X Reaction Buffer. A volume of 100 µL will be used for each reaction.

ii) Prepare a positive control by diluting the 10 U/mL PC-PLC stock solution (prepared in step 1.7) into 1X Reaction Buffer to produce a 0.1 U/mL PC-PLC solution. Use 1X Reaction Buffer without PC-PLC as a negative control. A volume of 100 µL will be used for each reaction.

iii) Prepare another positive control by diluting the 20 mM H_2O_2 working solution to 10 µM in 1X Reaction Buffer.

iv) Pipette 100 µL of the diluted samples and controls into separate wells of a microplate.

v) Master mix of Amplex Red reagent was prepared containing: 200 µL of Amplex Red reagent stock solution (400 µM), 100 µL of horseradish peroxidase (2U/mL), 200 µL of alkaline phosphatase stock solution (8U/mL), 100µL of choline oxidase stock solution (0,2 U/mL and 78µL lecithin (1mM) to 9,32 mL of 1x Reaction Buffer.

Reaction was started by adding 100 µL of Master Mix to each microplate well containing the samples and controls (positive control 10 U/mL of PC-PLC and as negative control 1x Reaction Buffer). The microplate with samples was incubated for 1 hour at 37°C, protected from light. The emitted fluorescence was measured in a fluorescence microplate reader using excitation by 560 nm and emission detection at 590 nm.

2.3 Biosensor Prototypes Production

For the biosensor prototypes production, the biomimetic polymer solution of PLGA [poly(lactic-coglycolic acid)] is directly transferred onto the Inconnel layer via Gravure Printing. The polymer solution has to be applied onto the printing plate, after which the impression roller can be run over the plate resulting in a defined layer with the desired thickness over the mirror layer.

2.4 Biosensor´s Functionality and Sensitivity Test

For the functionality and sensitivity test we have prepared biosensors with different concentrations of PLGA, different additives and also different concentrations of Desmodur. All of the chosen additives were mimicking agar components in order to provide a nutrimental environment for the bacteria and to trigger the release of lipolytic enzymes which are consecutively responsible for the degradation of the polymer layer.

3 RESULTS AND DISCUSSION

3.1 Quantitative Detection of Bacterial Enzymatic Activity

The four Gram negative bacteria of *Pseudomonas*, *Enterobacter, Proteus and Salmonella* genus were examined for exogenous enzyme secretion by using Spirit Blue Agar. The bacteria were incubated at three different temperatures (4°C, 30°C and 37°C) and for three different incubation times (24h, 48h and 120h).

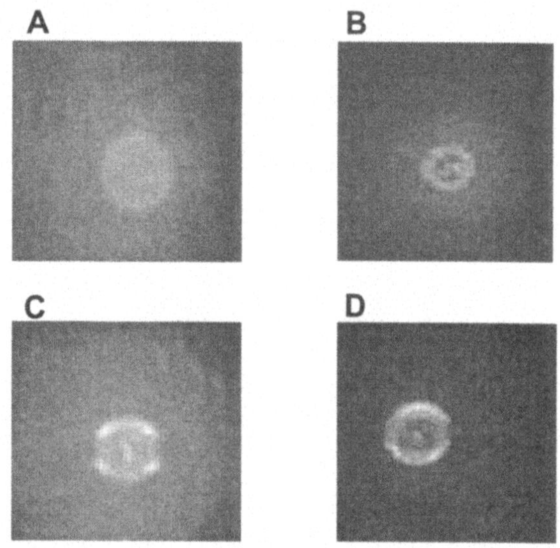

Figure 1. Quantitative evidences of enzyme activity for A: *Pseudomonas* (10^6 CFU/mL), B: *Enterobacter* (10^6 CFU/mL), C: *Proteus* (10^6 CFU/mL), and D: *Salmonella* (10^6 CFU/mL) by using Spirit Blue Agar base as a nutrient medium. Incubation at 37°C for 24 hours.

All bacterial cultures showed the strongest enzyme activity after 24 hours of incubation at 30°C and 37°C, that was observed as a white circle forming around the growing colonies (Fig. 1).

3.2 Phospholipase Enzyme Activity

The samples of two different bacteria, *Pseudomonas* and *Enterobacter*, pooled from the *log*-phase of bacterial growth were tested for the presence of PC-PLC activity. As shown in Fig. 2, the *Pseudomonas* showed the highest phospholipase activity (2.09 U/mL) after 7 hours of incubation at 37°C, at bacterial count of 1.6 x 10^7 CFU/mL. Suprisingly, the samples taken from later stages of the *log*-phase, with higher bacterial counts, did not exhibit the higher enzymatic activity as it was expected (Fig. 2).

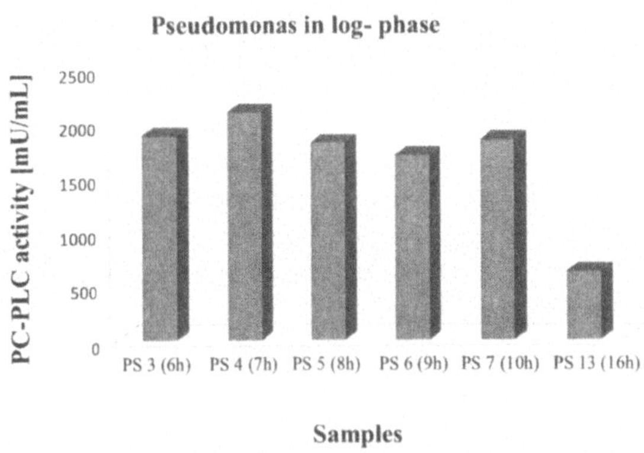

Figure 2. Phosphatidylcholin-specific Phospholipase C (PC-PLC) activity of *Pseudomonas* in *log*-phase of bacterial growth.

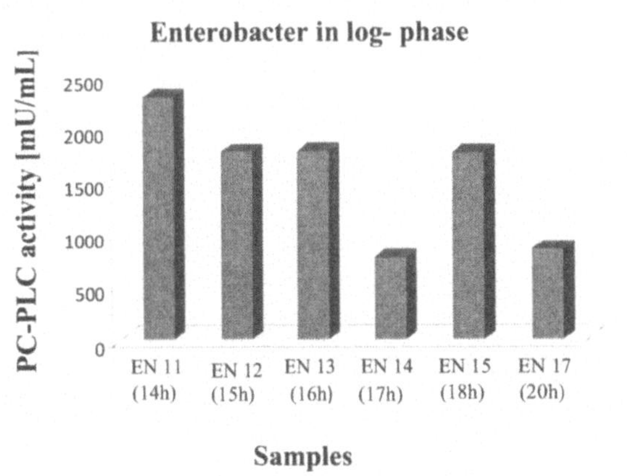

Figure 3. Phosphatidylcholin-specific Phospholipase C (PC-PLC) activity of *Enterobacter* in *log*-phase of bacterial growth.

The similar results were observed for the *Enterobacter*, displaying the highest enzymatic activity after 14 hours of incubation at 37°C, though the total bacterial count was much lower as in later points of bacterial *log*-phase, counting 1.5 x 10^5 CFU/mL (Fig. 3).

3.3 Increased Biosensor Sensitivity

The matrix of the biosensor prototypes produced according to [2][4] was changed by addition of substrates that were expected to increase the biosensor functionality and sensitivity. The choosen additives were bacterial nutritional components such as glucose and tributyrat. The substrates application stimulated bacteria for better enzyme secretion and increased the intensity of optical signal on the biosensor´s surface (Fig. 4, IIa). There was no significant changes in sensor sensitivity toward commercially available phospholipase which was used as the postive control (Fig. 4, Ib and IIb). The negative control, as expected, was not detectable on the PLGA layer of the biosensor (Fig. 4, Ic and IIc).

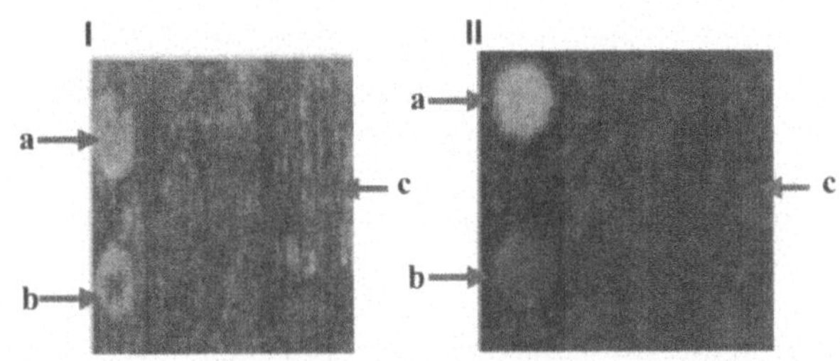

Figure 4. Substrate addition and increased biosensor sensitivity (I) in comparsion to biosensor without substrate addition (II). a: bacteria (10^8 CFU/mL), b: positive control and c: negative control.

4 CONCLUSION

An optimization of the matrix of the optical thin film biosensor-chip, which setup was explained in some of our previous work [2][4], required several parallel studies regarding bacterial growth and their enzymatic activity under different environmental conditions. The sensor is composed of two different layers: a) mirror layer (Ni-Cr composition called inconnel) and, b) biomimetic polymer [poly (lactic-coglycolic acid), PLGA]. The PLGA polymer, which is integrated in such two layer biosensor setup, can be degraded by lipolytic enzymes excreted by microorganisms. During presented study we were able to show some new findings about bacterial secretion of lytic and lipolytic enzymes. Monitoring of secretion of lipolytic enzymes from different bacteria such as *Pseudomonas*, *Enterobacter*, *Proteus* and *Salmonella*, and everything in correlation with their growth under different conditions, led to conclusion that there is no connection between a total bacterial count and amount of secreted lipolytic enzymes. A higher bacterial count did not automatically mean a better secretion of lipolytic enzymes. For example, we have seen that the highest yield of secreted aminopeptidase enzyme could be detected from bacteria in logarithmic phase in concentration of 10^7 cfu/mL (Fig. 2). Following generally known bacterial features and changing the matrix of the biosensor by addition of certain substrates, we could show that with a proper choice of substrate, its exactly defined amount and bacterial incubation conditions, was possible to induce the above mentioned bacteria for better enzyme production resulting in stronger colour change of degraded polymer layer. What was even more important, under such conditions bacteria were not forced to grow faster, in contrary they continued their normal logarithmic growth, but only the enzyme secretion and sensitivity of the sensor for secreted enzyme was higher. The substrates served as a trigger for enhanced bacterial metabolic activity.

REFERENCES

[1] N. Ibrišimović, M. Ibrišimovć, M. Barth, U. Bohrn, „Biomimetic PLGA sensor: proof of principle and application", Chemical Monthly, 141:125-130, 2010.

[2] N. Ibrišimović, M. Ibrišimović, M. Barth, U. Bohrn, F. Pittner, „Biomimetic sensor chip monitoring real-time food degradation: correlating chemical deterioration with microbiological status", Chemical Monthly, 140: 941-945, 2009.

[3] N. Ibrišimović, M. Bauer, F. Pittner, G. Bauer, R. Domnick, „Device comprising two reflecting layers and a polymer layer for analyzing the age and/or quality of natural products", U7098/DB, PCT/EP2007/063895, 2008.

[4] N. Ibrišimović, F. Pittner, M. Ibrišimović, M. Barth, U. Bohrn, "Device comprising a polymer layer and a reflecting layer" U 7145/ D, 2009.

[5] J. Woan-Fei Law, N. S. Ab Mutalib, K. G. Chan and L. H. Lee, "Rapid methods for the detection of foodborne bacterial pathogens: principles, applications, advantages and limitations", Microbiology, Volume 5, Article 770, 2015.

[6] X. Zhao, C. Lin, J. Wang, and D. H. Oh, "Advances in rapid detection methods for food pathogens", J. Microbiol. Biotechn, 24, 297-312, 2014.

[7] A. Cunha, A. Almeida, F.J.R.C. Coelho, N. C. M. Gomes, V. Oliveira, and A. L. Santos, "Bacterial extracellular enzymatic activity in globally changing aquatic ecosystems", Current Research, Technology and Education Topics in Applied Microbiology and Microbial Biotechnology, A. Mendez-Vilas (Ed.), 2010.

[8] R. J. Chróst, W. Siuda, "Ecology of microbial enzymes in lake ecosystems, In: Burns RC and Dick RP eds. Microbial Enzymes in the Environment Activity, Ecology, and Applications New York: Marcel Dekker, Inc., 35-72, 2002.

[9] D. Bisht, S. K. Yadav, P. Gautam, N. S. Darmwal, "Simultaneous production of alkaline lipase and protease by antibiotic and heavy metal tolerant Pseudomonas aeruginosa", J Basic Microbiol., 53(9): 715-22, 2013.

[10] W. M. Fakanya and I. E. Tothill,"Detection of the inflammation biomarker C-reactive protein in serum samples: Towards an optimal biosensor formula", Biosensors, 4 (4) 340-357, 2014.

[11] M. J. Abdin, Z. Altintas and I. E. Tothill, "In silico designed nanoMIP based optical sensor for endotoxins monitoring, Biosensors and Bioelectronics, 67: 177-183, 2014.

[12] J. I. Reyes De Corcuera, R. P. Cavalieri, "Encyclopedia of Agricultural, Food, and Biological Engineering", Marcel Dekker, Inc., 2003.

[13] S. S. Despande and R. M. Rocco, "Biosensors and their potential use in food quality control", Food Technol., 48 (6), 146-150, 1994.

[14] T. D. Martins, A. C. Chaves Riberio, H. Santiago de Camargo, P. Alves da Costa Filho, H. P. Mesquita Calvacante and D. Lopes Dias, "New insights on optical biosensors: techniques, construction and application, State of the Art in Biosensors-General Aspects, Dr. Toonika Rinken (Ed.), 2013.

[15] M. Zhou, Z. Diwu, N. Panchuk-Voloshina, R. P. Haugland: „A Stable Nonfluorescent Derivative of Resorufin for the Fluorometric Determination of Trace Hydrogen Peroxide: Applications in Detecting the Activity of Phagocyte NADPH Oxidase and Other Oxidases" Anal Biochem 253, 162 (1997).

Nephrotoxicity of Dextran Functionalized Graphene Nanoparticles and Their Potential as Magnetic Resonance Imaging Contrast Agents for Renal Abnormalities

S. Lee[*], S. Kanakia[*], J. Toussaint[*], P. Kulkarni[**], S.M. Chowdhury[*], S. Khan[*], K. Shroyer[***], W. Moore[****], B. Sitharaman[*]

[*]Stony Brook University, Department of Biomedical Engineering
Room 115, Bioengineering Building, Stony Brook, NY, USA, balaji.sitharaman@stonybrook.edu
[**]Northeastern University, Boston, MA, USA
[***]Stony Brook University, Department of Pathology, Stony Brook, NY, USA
[****]Stony Brook University, Department of Radiology, Stony Brook, NY, USA

ABSTRACT

Contrast-enhanced magnetic resonance angiography (MRA) employs the use of blood pool contrast agents (CA) such as gadolinium (Gd^{3+}) complexes that allow for delayed phase arterial and venous imaging and improve the sensitivity in detecting small vascular defects. However, recent association of Gd^{3+}-based CA with nephrogenic systemic fibrosis (NSF), a debilitating disease characterized by progressive and severe thickening of the skin, in patients with renal insufficiency has led to restrictions on their clinical use by the FDA. We have developed a dextran functionalized manganese intercalated graphene-based CA with relaxivity values ~20X greater than current clinically approved CAs. In this work, we report the safety and efficacy of our novel CA for renal *in vivo* MRA. The propensity for our contrast agent (at 5 and 50 mg/kg doses) to induce histopathological signs of NSF was evaluated in an experimental rat model of chronic renal failure. *In vivo* studies indicate that Mangradex formulations are safe and show no nephrotoxic effects after chronic exposure at doses ≤ 50 mg/kg. MR imaging results indicate Mangradex's excellent potential as a blood pool CA.

Keywords: MRI, graphene, nanoparticle, contrast agent, renal imaging

1 INTRODUCTION

Renal insufficiency is the loss of normal kidney function due to acute injury (via trauma, toxins, surgery, etc.), or chronic diseases such as diabetes and hypertension. Magnetic resonance angiography is an important imaging technique routinely used in the clinical diagnosis of renal failure and the detection of vascular diseases. Contrast enhanced (CE) MRA employs the use of blood pool contrast agents such as gadolinium (Gd^{3+}) complexes that allow for delayed phase arterial and venous imaging and improve the sensitivity of detecting small vascular defects. However, recent association of Gd^{3+}-based CA (GBCA) with nephrogenic systemic fibrosis, a debilitating disease characterized by progressive and severe thickening of the skin and other organs, in patients with moderate to severe renal insufficiency has fostered concerns from the Food and Drug Administration (FDA), and ultimately led to restrictions on their clinical use.[1,2] To address these safety concerns, we have developed a dextran functionalized manganese (Mn^{2+}; ~0.06 wt %) intercalated graphene-based CA (known herein as Mangradex). We previously reported these nanoparticles' excellent magnetic and relaxometric properties with relaxivity values ~ 20X greater than current clinically approved CAs.[3,4] They are disk-shaped with an average diameter of ~100 nm and thickness of 3-4 nm, hydrophilic and form stable colloidal dispersions in deionized water (up to 100 mg/mL concentration) and biological fluids that are both iso-osmol and iso-viscous to blood.[4] Our previous acute toxicity study in rodents suggested that they are hemocompatible with a maximum tolerated dose (MTD) between 50 mg/kg and 125 mg/kg, and are eliminated mainly through feces within 24 hours after intravenous (IV) administration.[5] In this work, we report the safety and efficacy of Mangradex as a novel CA for renal *in vivo* MRA.

2 MATERIALS AND METHODS

2.1 Animal Model and Treatment

The 5/6 Nephrex Wistar rat (where the left kidney has 2/3 pole regions excised and the right kidney is completely removed) has been used as an experimental animal model for chronic renal failure. Following the guidelines by the Institutional Animal Care and Use Committee (IACUC) at Stony Brook University, 5/6 Nephrex Wistar rats were injected with Mangradex at 5 and 50 mg/kg doses by tail vein IV injection after anesthetized with 5% mixture of air: isoflurane in a 1:1 air - oxygen mixture. Each animal received IV injections of Mangradex either once on the first day of the study (single injection) or five times per week for two weeks or four weeks (multiple injections). Animals were observed for vital parameters, behavior and toxicity-related adverse effects. Controls were sham and animals injected with mannitol (vehicle control) only. At the end of each study, the animals were euthanized by inhalation using 100% CO2. All major organs (heart, lungs, liver, brain, spleen) as well as the dorsal skin from each animal were harvested and prepared for histopathology analysis.

2.2 Nephrotoxicity

Nephrotoxicity was assessed by examining significant changes in skin thickness and collagen content. Measurement of dorsal skin thickness was performed by two methods: digital calipers and histomorphometry using digitized photomicrographs of H&E stained skin samples of each animal and Image J. Collagen concentration was measured and quantified per skin section obtained from 5/6 Nephrex rats using Masson's trichrome staining and Image J software.

2.3 *In Vivo* MR Imaging

In vivo T_1-weighted MRI was performed on mice using a 7 T Bruker Biospec to compare Mangradex CA (dose of 25 mg/kg, equivalent to 5 nanomoles Mn^{2+} ions/kg body weight) with FDA-approved clinical intravascular agent Ablavar® (Gadofosveset trisodium) at equivalent Gd^{3+} concentration. All images were acquired with a 7-Tesla Magnex horizontal magnet interfaced to Bruker Biopsec MR Console.

3 RESULTS AND DISCUSSION

3.1 Toxicity and Histopathology

No significant effects were observed in body weights, blood pressure and hematological parameters for all subjects (not shown). Histopathology analysis of the major organs treated with Mangradex revealed the presence of brown pigments only in lungs and liver, suggestive of graphene nanoparticles present in those organs. More brown pigmentation was observed in the rats that received multiple injections than a single injection. No diagnostic abnormalities were observed however in other major organs.

3.2 Nephrotoxicity

FDA guidelines require new pharmaceuticals under development to undergo special toxicity studies in animal models relevant to the human disease they indicated for. Since Gd^{3+}-based contrast agents present a risk to patients with renal insufficiencies, we performed single and repeat dose studies of Mangradex in 5/6 Nephrex rats, a widely used experimental animal model for chronic renal failure. Previous studies of NSF for clinically approved MRI CAs examined toxicity from 5 days to 4 weeks; thus, for short term investigation of Mangradex exposure, 2 week and 4 week time points were selected in this study. Rats administered with single and multiple IV injections of Mangradex were examined for signs of NSF that include skin thickening due to increased dermal fibrosis in the presence of unregulated collagen density.

Analysis of dorsal skin thickness and collagen concentration showed no significant differences between test and control animals. One-way ANOVA of skin thicknesses measured by histomorphometry (Figure 1) and digital calipers (not shown) revealed no statistically significant ($p<0.05$) differences among the sham control and the experimental groups.

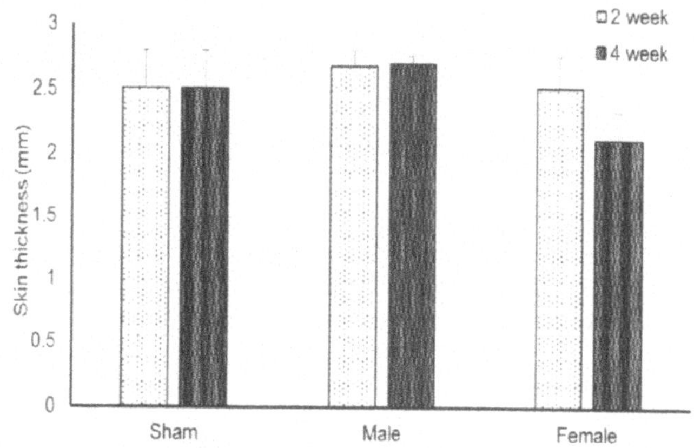

Figure 1: Dorsal skin thicknesses from 5/6 Nephrex rats given multiple doses of 50 mg/kg Mangradex for 2 or 4 weeks by histomorphometric image analysis using Image J software.

For collagen quantification, histology sections of 5/6 Nephrex rat skin were stained with Masson's Trichrome to mark collagen fibers brilliant blue. Image J software was then used to select and measure the concentration of collagen in each skin section. Analysis shows no statistically significant differences in collagen concentration in skin among test and control animals (not shown).

3.3 *In Vivo* MR Imaging

In vivo small animal MRI showed significant contrast enhancement compared to Ablavar® at potential diagnostic dosages (5 nmoles/kg) and high contrast remained in the vasculature for extended periods of time up to 85 min (not shown).

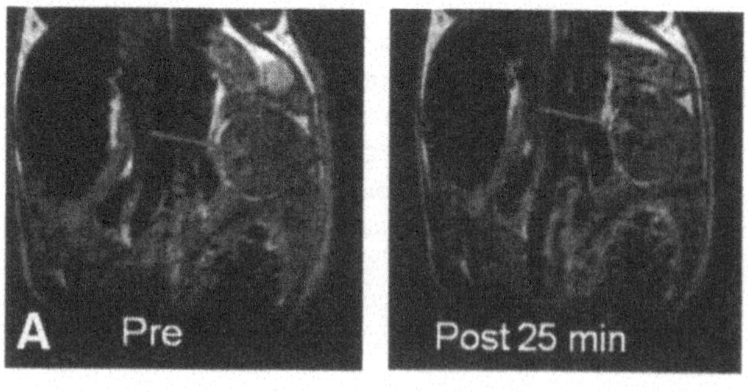

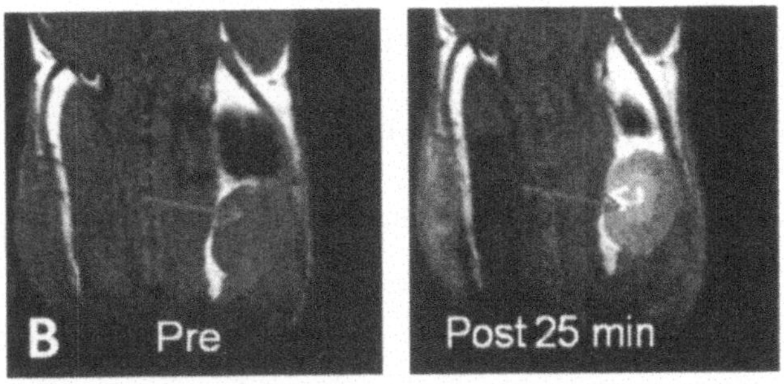

Figure 2: Representative *in vivo* MR images of pelvic region that show the nephrectomized rat kidney (red arrow) before (left) and 25 minutes after (right images) injection of Ablavar (A); before (left) and 25 minutes after (right) injection of Mangradex (B).

Compared to Ablavar®, Mangradex showed significant SNR and CNR up to 25 minutes and continued until 85 minutes post injection (not shown). The drop in intensity between the two contrast agents is likely due to the half-life of Mangradex being much longer than Ablavar®, as previously determined from our other studies. The enhanced and prolonged contrast observed in blood vessels suggests that Mangradex has great potential to be used for renal artery imaging.

4 CONCLUSIONS

In vivo small animal studies indicate that Mangradex formulations are safe and do not show any nephrotoxic effects in the animals after chronic exposure at doses ≤ 50 mg/kg. Examination of the renal vasculature at 7-Tesla indicates excellent potential of Mangradex as a blood pool MRI CA. In conclusion, these results lay the foundation for preclinical safety and efficacy studies of Mangradex in a large animals, necessary for their eventual clinical translation.

5 DECLARATION OF FINANCIAL INTEREST

Stony Brook University and the investigators have filed patents related to the technology reported in this article. If licensing or commercialization occurs, the researchers are entitled to standard royalties. S.B. has a financial interest in Theragnostic Technologies Inc., which however did not directly support this work.

6 ACKNOWLEDGEMENTS

This work was supported by Wallace H Coulter Foundation Translational Research Award. This material is also based upon work supported by the National Science Foundation Graduate Research Fellowship (under Grant No. 1315232). Any opinion, findings, and conclusions or recommendations expressed in this material are those of the author(s) and do not necessarily reflect the views of the National Science Foundation.

REFERENCES

[1] Zou, Z., Zhang, H. L., Roditi, G. H., Leiner, T., Kucharczyk, W., & Prince, M. R. (2011). Nephrogenic systemic fibrosis: review of 370 biopsy-confirmed cases. *JACC: Cardiovascular Imaging*, *4*(11), 1206-1216.

[2] Tsushima, Y., Kanal, E., & Thomsen, H. S. (2010). Nephrogenic systemic fibrosis: risk factors suggested from Japanese published cases. *The British Journal of Radiology*, *83*(991), 590–595.

[3] Paratala, B. S., Jacobson, B. D., Kanakia, S., Francis, L. D., & Sitharaman, B. (2012). Physicochemical characterization, and relaxometry studies of micro-graphite oxide, graphene nanoplatelets, and nanoribbons. *PloS one*, *7*(6), e38185.

[4] Kanakia, S., Toussaint, J. D., Chowdhury, S. M., Lalwani, G., Tembulkar, T., Button, T., ... & Sitharaman, B. (2013). Physicochemical characterization of a novel graphene-based magnetic resonance imaging contrast agent. *International journal of nanomedicine*, *8*, 2821.

[5] Kanakia, S., Toussaint, J. D., Chowdhury, S. M., Tembulkar, T., Lee, S., Jiang, Y. P., ... & Sitharaman, B. (2014). Dose ranging, expanded acute toxicity and safety pharmacology studies for intravenously administered functionalized graphene nanoparticle formulations. *Biomaterials*, *35*(25), 7022-7031.

Recent advances in capillary microflows for point-of-care devices

J. Berthier, D. Gosselin, M. Huet, M. Cubizolles, G. Delapierre

Department of Biotechnology, CEA-Leti, 17 avenue des Martyrs, 38054, Grenoble, France
Corresponding author: jean.berthier@cea.fr

ABSTRACT

Biotechnology has been recently seeing fundamental evolutions: first, a clear shift towards biological applications; second a trend towards the design and fabrication of portable devices; third the choice of inexpensive materials such as plastic or paper. The first aspect has already been reported in the literature. In this work, we deal with the two other recent—and still ongoing—evolutions. From a microfluidic stand point, the trend towards low-cost, portable systems is promoting the use of capillarity for the actuation of fluids. On the other hand, easy-to-use and inexpensive devices can be fabricated using 3D-printing of a plastic matrix or by embossing paper.

We present here the recent theoretical developments of capillary flows of biological fluids, from a static and dynamic viewpoint, in open or closed channels. On the other hand, the recent developments of stereolithography and paper-based systems are contributing to the generalization of POC and home-care systems.

Keywords: spontaneous capillary flow (SCF), Gibbs free energy, Lucas-Washburn-Rideal law, friction force, capillary force, home-care systems.

1 INTRODUCTION

Biotechnology has been recently seeing fundamental evolutions: first, a clear shift towards biological applications [1]. This is for example the case of the research on stem cell differentiation, or the mimicking of in-vivo situation in in-vitro devices. This shift has already been well documented in the literature, and we will only develop here the non-Newtonian aspects of biologic fluids.

The second trend is towards the design and fabrication of portable devices, for point-of-care and home-care applications [2-4]. The aim is towfold: first to give the patient the means to test himself easily at home or at the doctor's office; second, to improve the potentialities of portable systems to make them comparable to laboratory systems. These new devices shall combine the advantages of capillary actuation, which does not require pumps or syringes to move the fluids, with low-cost fabrication, user-friendliness, portability and telemedicine compatibility.

The third aspect of the biotechnology evolution is the use of inexpensive materials such as plastic or paper. Plastic devices are usually drilled or molded, but will soon be fabricated by using 3D-printing—in particular stereolithography. Paper substrates—as well made of paoros cellulose or embossed waterproof channels—are also solutions currently investigated

Focusing on microfluidics, we present here the recent theoretical developments of capillary flows from a static and dynamic viewpoint. On one hand, we show how the onset of spontaneous capillary flow (SCF) derives from Gibbs' thermodynamics [5,6]; on the other hand, we present a generalization of the Lucas-Washburn-Rideal law for the dynamics of capillary flow to channels of arbitrary cross-section [7,8] (figure 1). As biologic fluids—such as whole blood—are most of the time non-Newtonian [9], an analysis of the capillary behavior of these fluids is performed. Finally, we present the new developments of stereolithography for the fabrication of novel substrates.

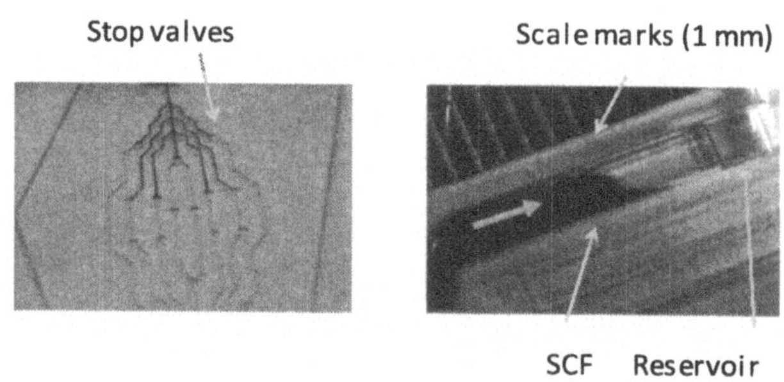

Figure 1. left: Whole blood in an open-microfluidic network; right: whole blood SCF in a narrow V-groove (Avalun®).

2 STATIC ASPECT: ONSET OF SCF

Spontaneous capillary flow occurs when the energy reduction from wetting walls outweighs the energy increase from extending the free surface. Using Gibbs thermodynamic equation, it has been shown that the general condition for SCF in wall-air systems was that the generalized Cassie angle must be less than 90° [5,6]. Let us recall that the generalized Cassie angle θ^* is the average contact angle defined in the appropriate way, i.e

$$\cos\theta^* = \sum_i \left(\cos\theta_i\ f_i\right), \tag{1}$$

where θ_k are the Young contact angles with each component i (including air) and f_i the areal fractions of each component i in a cross section of the flow (figure 2). Note that the virtual contact angle with air is 180°.

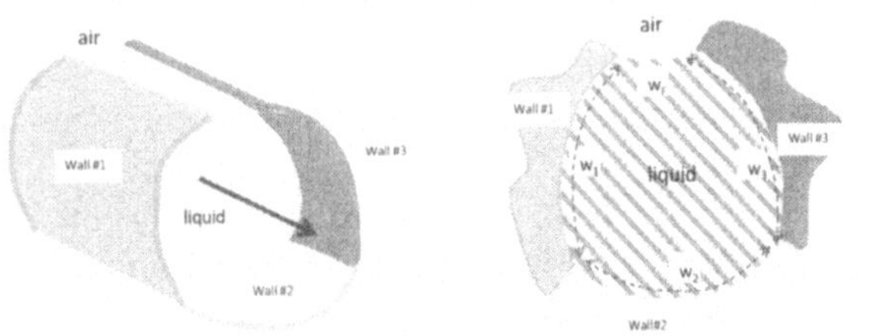

Fig.2. Sketch a cross section of the channel. The free perimeter is $p_F=w_4=w_F$; the wetted perimeter $p_W=w_1+w_2+w_3$; the areal fractions are $f_i=w_i/(p_F+p_W)$.

The condition for SCF onset is then

$$\cos\theta^* > 0 \tag{2}$$

It is straightforward to see that the relation (2) collapses to the usual criterion for confined homogeneous channels. For open-surface microfluidics, in the case where the walls are composed of a single material (with the same functionalization), relation (1) reduces to

$$p_F / p_W < \cos\theta\ , \tag{3}$$

where p_F and p_W are respectively the free—in contact with air—and wetted—in contact with wall—perimeters in a cross section of the channel.

Note that in case of contact angle hysteresis associated to the wall roughness or contamination, the contact angle θ in (3) should take into account the hysteresis. A discussion on the dynamic contact angle in capillary flows is given in another presentation [10].

3 DYNAMIC ASPECT

Once the condition for SCF fulfilled, a capillary flow advances in open or confined microchannels. The dynamics of capillary motion results from a balance between the capillary force and the wall friction force [9]. Inertial forces are negligible, as the Reynolds number for capillary flows is usually small. According to the preceding section, the capillary force can be expressed by [6,11]

$$F_{cap} = \gamma\, p\, \cos\theta^* = \gamma \sum_i p_{Wi} \cos\theta_i - \gamma\, p_F\ , \tag{4}$$

where $p = \sum_i p_{W_i} + p_F$. Relation (4) states that the capillary force on the flow is the resultant of the forward capillary force exerted on the adavancing triple line minus the backward resistance associated to the increase of the free surface.

On the other hand, the friction force is

$$F_{drag} = \int_S \tau\, ds = \left(\int_\Gamma \tau\, dl\right) z(t) = \bar{\tau}\, p_W\, z(t)\ , \tag{5}$$

where τ is the local wall friction, S the wetted surface between the origin and the front end of the liquid flow, Γ the wetted contour of the cross-section, $\bar{\tau}$ the averaged wall friction in a cross-section, and z the distance of the interface from inlet which depends on the time t. Locally, the wall fiction is

$$\tau = \mu \frac{\partial v}{\partial n} = \frac{\mu V}{\lambda} \tag{6}$$

where λ is a local friction length, V the average fluid velocity and n the coordinates perpendicular to the wall (figure 3).

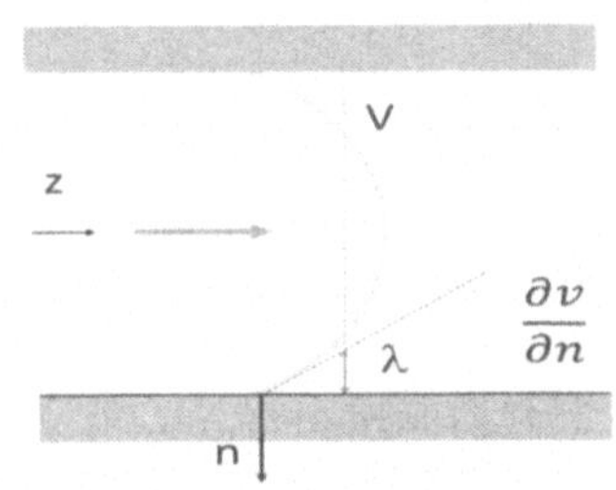

Fig.3. Definition of the friction factor $\bar{\lambda}$ from the velocity profile.

Conceptually, an averaged friction length taking into account the i different walls can be defined

$$\bar{\tau} = \sum_i \frac{1}{p_{wi}} \int_{\Gamma_i} \tau\, dl = \sum_i \frac{1}{p_{wi}} \int_{\Gamma_i} \frac{\mu V}{\lambda_i} dl = \frac{\mu V}{\bar{\lambda}}\ . \tag{7}$$

Equating (4) and (5)—after substitution of (7) —yields the general relation

$$\gamma\, p \cos\theta^* = \gamma \sum_i p_{Wi} \cos\theta_i - \gamma\, p_F = \frac{\mu V}{\bar{\lambda}} \sum_i p_{Wi}\, z(t)\ . \tag{8}$$

A first consequence is the general relation between V and z

$$V = \frac{\gamma}{\mu}\cos\theta^* \frac{p}{\sum_i p_{Wi}} \frac{\overline{\lambda}}{z} . \quad (9)$$

3.1. Confined channels

In the case of confined channels, the free perimeter is null and $p = \sum_i p_{Wi}$. In the subcase of a cylindrical tube, $\overline{\lambda} = R/4$ (where R is the radius) and (8) yields the LWR law

$$z = \sqrt{\frac{\gamma \cos\theta R t}{2\mu}} . \quad (10)$$

In the general case (8) yields

$$V = \frac{\gamma}{\mu}\cos\theta^* \frac{\overline{\lambda}}{z}, \quad and \quad z = \sqrt{\frac{\gamma}{\mu}}\sqrt{\cos\theta^*}\sqrt{2\overline{\lambda} t} . \quad (11)$$

The remaining difficulty is the determination of $\overline{\lambda}$ in function of the geometry. By analogy to forced flow pressure drop formulation, we transform (8) by introducing the Laplace pressure

$$\Delta P = \frac{\gamma p \cos\theta^*}{S_c} = \mu \frac{\sum_i p_{Wi}}{\overline{\lambda} S_c} z V = \widetilde{R}_H z V . \quad (12)$$

In the case of a single wall (i=1), $\widetilde{R}_H$ is the usual hydraulic resistance per unit length, and $\overline{\lambda}$ can be determined as a function of $\widetilde{R}_H$. It is found in tables for various geometries as a function of the Poiseuille number (*f Re/4*), and the hydraulic diameter D_H. It can be shown that $\overline{\lambda} = 8D_H/f\mathrm{Re}$, and (12) becomes

$$z = 4\sqrt{\frac{\gamma}{\mu}}\sqrt{\cos\theta^*}\sqrt{\frac{D_H}{f\mathrm{Re}} t} . \quad (13)$$

It is common to use the LWR law for any—non cylindrical—channel, replacing the radius R by the hydraulic radius. However, this replacing induces a systematic error, that may be important if the channel cross-section is very far from the circular shape [8]. The relative error is

$$\frac{z - z_{LWR}}{z_{LWR}} = \sqrt{4\frac{\overline{\lambda}}{R_H}} - 1 \quad (14)$$

For the disc $\overline{\lambda} = R/4$, but for rectangles, triangles and slit channels, the relative error can be as large as 30%.

3.1. Open channels

Open-surface microchannels are increasingly used in biotechnology, especially for point-of-care devices. For a capillary-driven flow of Newtonian liquid in open channels, assuming for simplicity that there is only a unique wall and air as boundaries, relation (8) leads to [12]

$$V = \frac{\gamma}{\mu}\frac{\overline{\lambda}}{z}\left(\cos\theta - \frac{p_F}{p_W}\right), \; and \; z = \sqrt{\frac{2\overline{\lambda}}{\mu}\frac{(p_W \gamma\cos\theta - p_F \gamma)}{p_W}}\sqrt{t} . \quad (15)$$

Different expressions have been reported in the literature for rectangular open channel [11], V-grooves [13], suspended channels [14]. Approximating the value of $\overline{\lambda}$, it has been shown that (15) reduces to the general formulation

$$V = 2\frac{\gamma}{\mu}\frac{h}{z} f , \quad (16)$$

where h is the etching depth and f a non dimensional function of the geometry and contact angle.

4 NON-NEWTONIAN BEHAVIOR

The most usual body fluid used in point-of-care devices and home-care devices is whole blood, directly obained from a finger prick. Whole blood capillary flow depends on its rheological behavior. Refering to the rheolology measurement of whole blood, there is a transition from Newtonian to non-Newtonian viscosity for a shear rate decreasing below 100 s^{-1} (figure 4) [9].

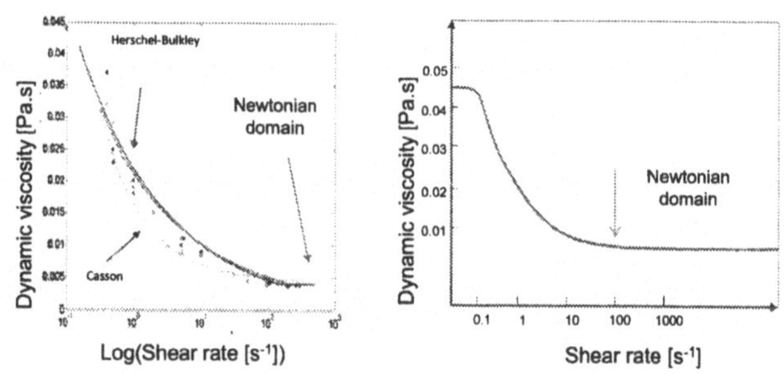

Fig.4. Viscosity of whole blood vs. shear rate

This transition is related to the formation of RBC rouleaux [15], as shown in figure 5. The reorganization of the RBCs first leaves free internal cavities filled with plasma, and later a cell-free layer at the channel walls (the Zweifach-Fung layer).

The change of rheological behavior can be seen by monitoring the advancing interface (figure 5). The penetration distance varies as the square root of time first ($z \sim t^{1/2}$) and then it shifts to a law of the form $z \sim t^{0.6}$, showing the transition to a non-Newtonian regime.

The theory of the dynamics of capillary flow of non-Newtonian fluids is presently a subject of investigation.

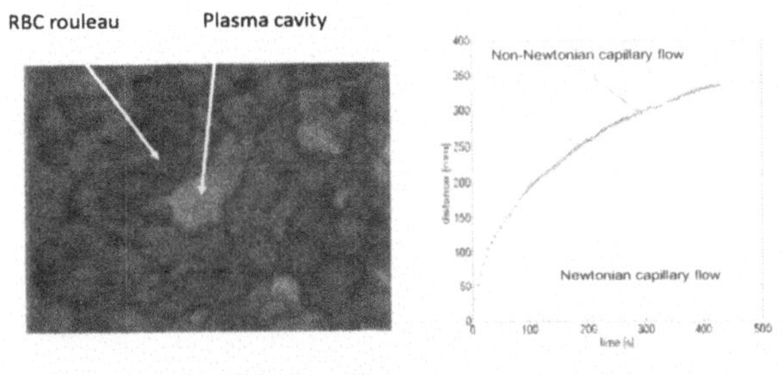

Fig.5 Left: When the shear rate decreases sufficiently, RBCs rouleaux self-assemble, often in a curved formation, leaving a cavity filled with plasma; right: experimental plot of the penetration distance vs. time showing the rheological transition.

5 MATERIALS

Portable, autonomous, and low-cost devices are ideal candidates for POC and home-care systems. Low-cost substrates compatible with capillary-driven flows are being investigated. Presently, the usual substrates are plastic (PMMA, COC or COP) etched using a milling machine. However, new substrates and fabrication methods are being developped, such as 3D printing (stereolithography) of different polymers. At the present time, limitations are the precision, the low TG (glass transition temperature), and the optical quality (figure 6). Paper substrates are also being developped as candidates for home-care systems.

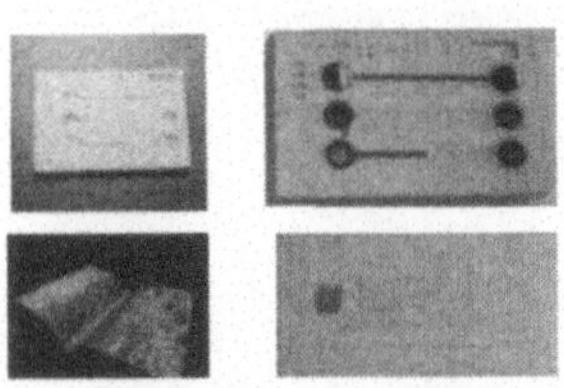

Fig.6. Top: Microchannels obtained by stereolithography, compatible with capillary flow; bottom: Embossed paper channels.

6 CONCLUSION AND PERSPECTIVES

The use of capillary flow in biotechnology is promoted by the development of point-of-care and home-care devices. The physics of spontaneous capillary flow has seen recently many new developments: geometrical conditions for SCF are now known, and the dynamics of SCF in confined and closed channels is much better mastered. In particular, the LWR law for a cylindrical tube has been generalized to any closed channel. However, SCF of non-Newtonian liquids is still a subject of investigation.

New substrates are being developped: 3D-printing—in particular stereolithography —is a promising technique, if precision, glass transition temperature and optical properties are improved. Embossed water-proof paper substrates are also thought to be a solution for the fabrication of low-cost, portable devices.

REFERENCES

[1] E.K. Sackmann, A.L. Fulton, D.J. Beebe, The present and future role of microfluidics in biomedical research, Nature, **507**:181, 2014.

[2] G.J. Kost, Principles and Practice of Point-of-Care Testing, Lippincott Williams & Wilkins, June 2002.

[3] P. Yager; T. Edwards, E. Fu, K. Helton, K. Nelson, M. R. Tam & B. H. Weig, Microfluidic diagnostic technologies for global public health, Nature, **442** (7101): 412–418, 2006.

[4] A.W. Martinez, S.T. Phillips, E. Carrilho, S.W. Thomas, H. Sindi, G.M. Whitesides Simple telemedicine for developing regions: camera phones and paper-based microfluidic devices for real-time, off-site diagnosis, Analytical Chemistry, **80**, 3699, 2008.

[5] J.W. Gibbs, A method of geometrical representation of the thermodynamic properties of substances by means of surfaces, Transactions of the Connecticut Academy of Arts and Sciences **2**, pp. 382-404, 1873.

[6] J. Berthier, K. Brakke, E. Berthier, A general condition for spontaneous capillary flow in uniform cross-section microchannels, Microfluid Nanofluid **16**:779–785, 2014.

[7] E.W. Washburn, The dynamics of capillary flow. Phys. Rev., **17**, 273-283, 1921.

[8] J. Berthier, D. Gosselin, E. Berthier, A generalization of the Lucas-Washburn-Rideal law to composite microchannels of arbitrary cross-section, Microfluid. Nanofluid., in print, 2015.

[9] E. W. Merrill, Rheology of blood, Physiological reviews, **49** (4), 863-888, 1969.

[10] J. Berthier, D. Gosselin, G. delapierre, On the dynamic contact angle in spontaneous capillary flow, Proceedings of the 2015 Nanotech Conference, 13-17 June 2015, Washington D.C.

[11] F. F. Ouali, G. McHale, H. Javed, C. Trabi, N. J. Shirtcliffe, M. I. Newton, Wetting considerations in capillary rise and imbibition in closed square tubes and open rectangular cross-section channels, Microfluid Nanofluid., **15**, 309-326, 2013.

[12] J. Berthier, D. Gosselin, N. Villard, C. Pudda, F. Boizot, G. Costa, G. Delapierre, The dynamics of spontaneous capillary flow in confined and open microchannels, Sensors & Transducers 183 (12), 123-128, 2014.

[13] R.R.Rye, F.G. Yost, Wetting Kinetics in Surface Capillary Grooves, Langmuir **12**, 4625-2627, 1996.

[14] J. Berthier, K.A. Brakke, D. Gosselin, A-G. Bourdat, G. Nonglaton, N. Villard, G. Laffite, F. Boizot, G. Costa, G. Delapierre, Suspended microflows between vertical parallel walls, Microfluid. Nanofluid., 2014, DOI 10.1007/s10404-014-1482-z.

[15] H.H. Lipowsky, Blood Rheology Aspects of the Microcirculation, Handbook of Hemorheology and Hemodynamics, O.K. Baskurt et al. (Eds.) IOS Press, p. 307, 2007.

The Automation of Protein Crystal Presentation for X Ray Diffraction Experiments using Standing Acoustic Waves in a Microfluidic Chip Environment

C. Burton* P. Docker+ M. Prince* M. Leaper* R. Morris^, M. Newton^, J. Kay+ D. Stuart+ G. Evans+

*Aston University Birmingham B4 7ET
+ Diamond light Source Harwell OX110DE
^ Notingham Trent University NG1 4BU

ABSTRACT

As the pressure continues to grow on Diamond and the world's synchrotrons for higher throughput of diffraction experiments, new and novel techniques are required for presenting micron dimension crystals to the X ray beam. Currently this task is both labour intensive and primarily a serial process. Diffraction measurements typically take milliseconds but sample preparation and presentation can reduce throughput down to 4 measurements an hour.

With beamline waiting times as long as two years it is of key importance for researchers to capitalize on available beam time, generating as much data as possible. Other approaches detailed in the literature [1] [2] [3] are very much skewed towards automating, with robotics, the actions of a human protocols. The work detailed here is the development and discussion of a bottom up approach relying on SSAW self assembly, including material selection, microfluidic integration and tuning of the acoustic cavity to order the protein crystals.

Keywords: standing acoustic waves, X-ray diffraction, protein crystallography, micro fluidics

1 INTRODUCTION

Using a Piezo actuator that has been integrated into a microfluidic chip design a Standing Surface Acoustic Wave (SSAW) can be set up laterally across a micro fluidic channel. This creates a varied acoustic field which will encourage crystals held in suspension to align at nodes or anti-nodes. Typically this is achieved using a standing half wave, providing a stable region for particulates midway across the channel. See figure 1 and figure 2 [4] for micro graphs of a channel containing 50 μm beads before and after the SSAW wave was established across it.

Figure 1: Channel with 50 micron beads
(image cropped for clarity)

Figure 2: After the SSAW was set running
(image cropped for clarity)

For the application of presenting and manipulating protien crystals for diffraction imaging, particles ordered by SSAW half waves in a controlled laminar flow along the channel the crystal may then be presented to the X-ray beam at a known and repeatable position aiding automation. The microfluidic cell designs comprise of an X-ray compatible window which will not attenuate the beam and subsequently the maximise the signal-to-noise ratio in the diffraction pattern that can be obtained from the sample.

2 CHALLENGES

2.1 Material selection

Many factors influence material selection for the project, these can be broken down into the materials used to characterize the micro fluidic architecture on chip during the design phase, and the material a device will be produced from for actual use in X-ray diffraction experiments. Characterisation of several X-ray compatible materials for such a device is planned. Hard materials such as silicon as they ensure the majority of the energy in the SSAW waves are reflected back into the sorting channel. However it is worth noting that the device that inspired the research into exploring this approach was seen at Nanotech 2013 and the chip was made from PDMS and facilitated alignment in 2D [5] (figure 3).

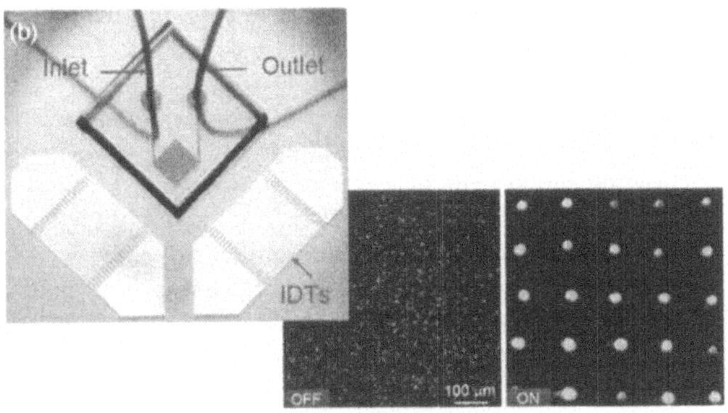

Figure 3: 2D alignment in a PDMS Device

So, there is a compromise to be evaluated in materials between diffraction capacity and manufacturability to develop an early generation design that is fit for purpose.

To achieve rapid validation in the design phase channels have been fabricated in two ways: complete design fabricated by 3D printing, and by 3D printing of a male mould and casting a device in PDMS. Both approaches have been fully characterized and initial flow properties through the devices investigated. Figures 4a and 5 give examples of chip designs that have been produced using both approaches.

Whilst the 3D printed devices offer many advantages in terms of complex geometrical possiblities, the requirement for manual polishing in order to achieve optically transparent surfaces prevent it from being a stand alone method. The 3D printed devices, despite being manufactured from a colourless and transparent UV-cured resin (Accura Si 60, 3D systems, Cal. USA) with highly polished outer surfaces, exhibit stepped inner faces on the inner faces of the channels which cannot be manually prepared. This leads to optical distortions when studied beneath a microscope and provides a barrier to optical validation of performance.

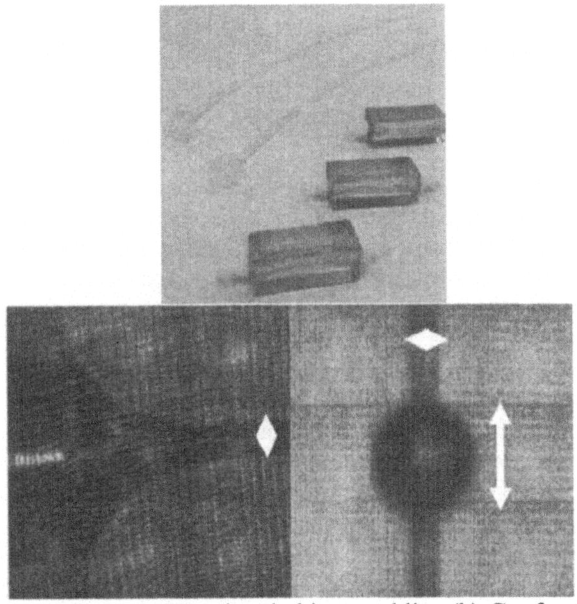

Figure 4: (a) *Top:* 3D printed chip assmblies (b) Confocal transmission scans showing dimensional accuracy of channels, scale bars show 500 micron and 2000 microns respectively

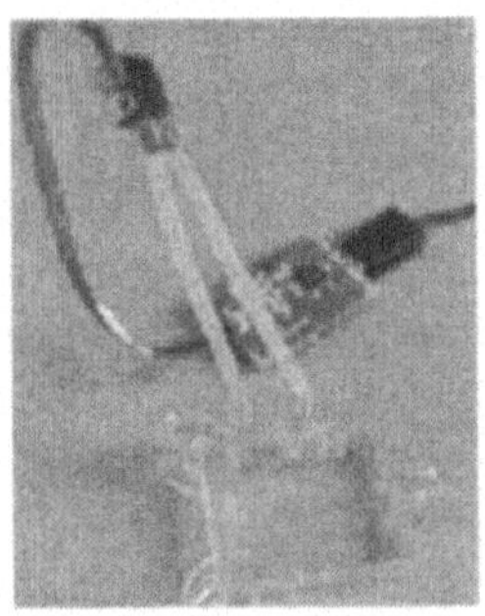

Figure 5: Cast PDMS Chip in their manifold clamp

Figure 6: KiKuchi Micropump (Scale is approximately 3:1)

The 3D printed chips in figure 4 explore Bulk Acoustic Wave (BAW) resonant cavities, each having a characteristic length and predicted acoustic mode. The devices will be explored to see if additional positional control can be achieved, and the losses that stem from the inhomogeneous structure (seen in figure 4b) can be tolerated. Figure 5 also includes with the PDMS chip a micro pump which us inder investigation with a view to integration into the final chip design. It has a dead volume of 250 nl and can run at between 1 and 350 Hz. The pump is comercially available from Kikuchi Japan. It meassures 7 mm x 7 mm (figure 6).

The small format and electronic nature of this pump is an enabling technology for the project, with rapid activation times, and full programmable control, it becomes possible to fluidically isolate desired samples, rapidly switching flow sources and velocity. Although the pumps offer a competing acoustic regime to the BAW / SSAW arrangement, the difference in operating frequency (four orders of magnitude) is expected to leave the acoustophoresis intact.

2.2 Excitation approaches

Many authors have recently explored the excitation of SSAW devices using a variety of configurations. The two methods of interest are in-plane and out-of-plane excitation. For example the half wave, BAW, excitation in figure 2 was achieved with a commercial piezo disc bonded to the bottom of the chip. Channels fabricated from high Q materials, enable far more robust BAW excitation at lower levels of acoustic power, however and as discussed by Lenshof *et al*, this need not imply a more expensive system[6]. Fabrication of Interdigitated Transducers (IDT) activated Piezo designs are currently in progress at Nottingham Trent University (UK) using standard photolithography and lift-off fabrication techniqnues; specifically 128° YX cut lithium niobate, S1318 photoresist, Ti and Au IDT deposits. Piezo material devices are also being investigated with Phisik Instrumente (MA, USA) and Ferroperm (Humlbaek, DEN) to review off the shelf BAW exciters.

A further challenge from a system and excitation perspective is the characterisation of the coupling of the SAW waves with the specific protein crystals to be acoustophoretically ordered when in service. Presented sample crystals may typically span two orders of magnitude in size and require time consuming handling to achieve cryoloop mounting. The system must also be able to autonomously prevent particle aggregation, since this may lead to multiple crystal structures being present in the diffraction data.

2.3 Excitation On/chip Off/Chip

Currently two methods are proposed for developing the desired acoustophoretic ordering effect, SSAW and BAW. The advantage of the BAW approach, externally exciting any device, is the ability to use ruggedised mount – demount proceedures, and standardised components. The core challenged posed by an otherwise ideal method is an increase in the required acoustic power to achieve the desired ordering effect, the implication of this is scheduled to be as part of the project.

Selecting an 'on-chip solution' requires that the SSAW device must be robustly connected to the fluidised channel. This connection and the fragility of the IDT microstructures may prevent re-usability, the project will investigate the effect on system reliability.

Due to the competing factors it is not decided whether excitation and fluidic manifold will be incorporated in to a single device or whether the microfluidic manifold, and exciter, may be seperated, allowing the manifold to become disposable, the BAW/SSAW exciter then being reused. The latter appears to be the most desirable due to the significantly lower power levels. In all configurations the fluidic flow section of the device will need to be disposable between experiments, allowing the processing of hazardous protiens, circumventing blockages and cleaning downtime, and most importantly providing a robust and contamination-free imaging environment.

3 PREDICTED PERFORMANCE

Resolution - From the work of Shi et al [7], 10 μm particle focusing was achieved in 2009, and more recently this was achieved in three dimensional particle streaming [8]. While theoretically far higher resolution can be achieved with SSAW techniques at higher excitation frequencies, the practical limitation of the maximum particle size approaching the wavelength of the excitation signal, means a 10 μm particle spread after ordering remains realistic.

Flow throughput – comparing the requirements of the synchrotron to a cytometer, Chen et al were able to detect flowing blood cells at a rate of 1000 events s^{-1}, analysing fluid at a rate of 15 μl min^{-1} [9]. The impact of this already achieved figure, would be the ability to process a 1 ml sample of crystals in a little more than an hour (67 minutes). Given only 10 viable samples in this volume, it becomes evident that the intended system should out-perform hand loading of samples by an order of magnitude or better.

Low force handling – The need for samples to be conveyed at power levels that are below what is needed to cause structural change is one of the confounding factors for optical tweezing. The work of Ding et al successfully demonstrated that a living organism (C. elegans) may be restained by ultrasonic acoustic forces without causing damage [10]. The added success of Zeng et al in enhancing the output of interdigitated transducers [11] through the use of Reflective Bragg Gratings (RBG) demonstrates it is possible to actuate particles at reduced levels of piezo excitation, reducing system noise, and absorbed energy by the crystal sample. Figure 7 and 8 are taken from their work and demonstrates the absorbtion loss reduction based on RBG. Further to this the work highlighted the benefits of sub millimeter wall section between the exciter and the fluid medium.

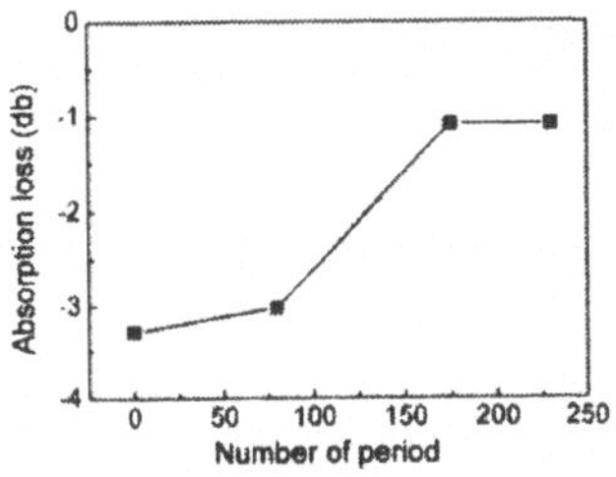

Figure 7: Absorbtion losses

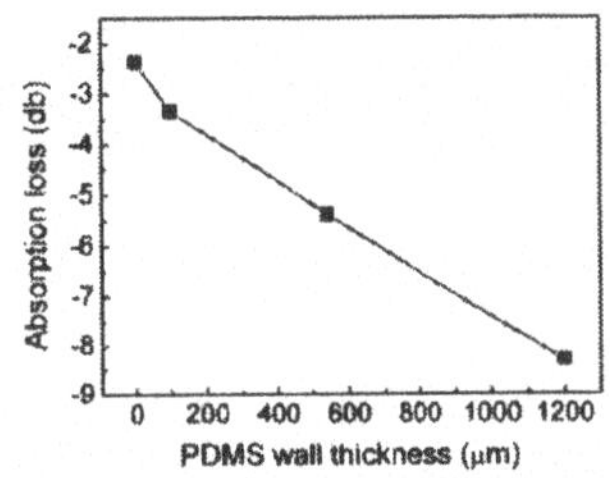

Figure 8: Wall thickness parameter

4 CONCLUSIONS

The paper covered the initial conclusions on process viability for continous mode acoustic ordered sample presentation, in-line with synchrotron beam imaging. No contra-indications were found for use, while many results demonstrate that a low complxity system will achieve desired throughput with no anticipated sample degradation.

In the feasibility assessment of the project from the literature, it is evident that the proposed system can offer useful ordering of samples for the synchrotron crystallography community. The current state of the art, developed in literature significantly since 2009, offers a robust set of tools to achieve the desired process automation and throughput improvement.

Reviewing sample handling data suggests that control of the crystals is attainable and further the construction methods are known, if customised, for each application.

Fabrication of the system has begun and the authors look forward to reporting integration and optimisation studies in the near future.

REFERENCES

[1] K. Mader, F. Marone, C. Hintermüller, G. Mikuljan, A. Isenegger, and M. Stampanoni, "High-throughput full-automatic synchrotron-based tomographic microscopy," *J. Synchrotron Radiat.*, vol. 18, pp. 117–124, 2011.

[2] P. Andrey, B. Lavault, F. Cipriani, and Y. Maurin, "Using image analysis for automated crystal positioning in a synchrotron X-ray beam for high-throughput macromolecular crystallography," *J. Appl. Crystallogr.*, vol. 37, pp. 265–269, 2004.

[3] G. David and J. Pérez, "Combined sampler robot and high-performance liquid chromatography: A fully automated system for biological small-angle X-ray scattering experiments at the Synchrotron SOLEIL SWING beamline," *J. Appl. Crystallogr.*, vol. 42, pp. 892–900, 2009.

[4] A. Neild, S. Oberti, F. Beyeler, J. Dual, and B. J. Nelson, "A micro-particle positioning technique combining an ultrasonic manipulator and a microgripper," *Journal of Micromechanics and Microengineering*, vol. 16. pp. 1562–1570, 2006.

[5] J. Shi, D. Ahmed, X. Mao, S.-C. S. Lin, A. Lawit, and T. J. Huang, "Acoustic tweezers: patterning cells and microparticles using standing surface acoustic waves (SSAW).," *Lab Chip*, vol. 9, pp. 2890–2895, 2009.

[6] a Lenshof, M. Evander, T. Laurell, and J. Nilsson, "Acoustofluidics 5: Building microfluidic acoustic resonators.," *Lab Chip*, vol. 12, no. 4, pp. 684–95, Feb. 2012.

[7] J. Shi, X. Mao, D. Ahmed, A. Colletti, and T. J. Huang, "Focusing microparticles in a microfluidic channel with standing surface acoustic waves (SSAW).," *Lab Chip*, vol. 8, pp. 221–223, 2008.

[8] J. Shi, S. Yazdi, S.-C. S. Lin, X. Ding, I.-K. Chiang, K. Sharp, and T. J. Huang, "Three-dimensional continuous particle focusing in a microfluidic channel via standing surface acoustic waves (SSAW).," *Lab Chip*, vol. 11, pp. 2319–2324, 2011.

[9] Y. Chen, A. A. Nawaz, Y. Zhao, P.-H. Huang, J. P. McCoy, S. J. Levine, L. Wang, and T. J. Huang, "Standing surface acoustic wave (SSAW)-based microfluidic cytometer.," *Lab Chip*, vol. 14, pp. 916–23, 2014.

[10] X. Ding, S.-C. S. Lin, B. Kiraly, H. Yue, S. Li, I.-K. Chiang, J. Shi, S. J. Benkovic, and T. J. Huang, "On-chip manipulation of single microparticles, cells, and organisms using surface acoustic waves," *Proceedings of the National Academy of Sciences*, vol. 109. pp. 11105–11109, 2012.

[11] Q. Zeng, H. W. L. Chan, X. Z. Zhao, and Y. Chen, "Enhanced particle focusing in microfluidic channels with standing surface acoustic waves," *Microelectron. Eng.*, vol. 87, pp. 1204–1206, 2010.

Christian Burton Aston University
burtoncg@aston.ac.uk

The application of 3D printing to study microfluidic architecture for 'on-chip' mixing systems for SRCD and UV spectroscopy

M Prince*, P Docker+, C Burton*, R Hussain+, J Kay+, G Silgardi+

+ Diamond light source Harwell OX11 0DE
*Aston University Birmingham B4 7ET

ABSTRACT

This paper details methodologies that have been explored for the fast proofing of on-chip architectures for Circular Dichroism techniques. Flow-cell devices fabricated from UV transparent Quartz are used for these experiments. The complexity of flow-cell production typically results in lead times of six months from order to delivery. Only at that point can the on-chip architecture be tested empirically and any required modifications determined ready for the next six month iteration phase. By using the proposed 3D printing and PDMS moulding techniques for fast proofing on-chip architectures the optimum design can be determined within a matter of hours prior to commitment to quartz chip production.

Keywords *3D printing, fast proofing, micro pumps, pdms chips*

1 INTRODUCTION

Circular dichroism (CD) is the differential absorption of left- and right-handed circularly polarized light. It is a form of spectroscopy used to determine the optical isomerism and secondary structure of molecules, and to study a wide variety of chiral materials in solution, particularly biologically important molecules such as proteins, nucleic acids, carbohydrates, lipids and drugs. It is an ideal technique to investigate protein/ligand binding interactions, such as those involved in signal transduction of normal and tumour cells, without the need for labelling or immobilizing any of the components. The benefit of carrying out such experiments using synchrotron radiation is that the light available is several orders of magnitude higher in intensity than that available using conventional CD instruments, thereby providing a much higher signal-to-noise ratio over a wide wavelength range (140-700 nm). The combination of high photon flux and small spot size at the Diamond Light Source (Didcot, Oxford, UK) will substantially reduce the measurement time for CD time-resolved experiments and the amount of chiral material required.

There is however an ever-growing need for different architectures to be used within the microfluidic flow cells that are used with these spectroscopy techniques which cater for a variety of analyte mixing and flow requirements. Measurements are made by transmission of light through the complete cell, thus the flow cells are typically constructed from quartz or borosilicate glass to ensure minimal light absorption from the cell imaging windows, and fluidic channels of <100 μm to minimise scattering as the light passes through the fluid medium. These exacting requirements can mean that the process of manufacturing a new cell with bespoke microfluidic architecture may take up to six months. Only at this point can the cell be tested in order to observe whether the desired mixing, a notable challenge as a result of the low Reynold's number flows in micro-scale flows [1], and other flow characteristics predicted from CFD modelling have been achieved by the new design.

A number of rapid prototyping techniques have been employed previously for the construction of microfluidic systems. Ranging from casting patterns from laser printed acetate transparencies [2] to scoring polymeric surfaces with a sharp blade [3]. The most widely employed method is the patterning of silicon and glass wafers with photoresist and casting from elastomeric resin requires costly photomasks and precision lithographic mask alignment tools [4]. This process is also time consuming and requires specialist processing equpment.

For these reasons the authors decided to explore techniques that could yield more simplistic structures that could be used to mimic a final quartz cell in order to review microfluidic flow characteristics without the protracted lead time. The chips described in this paper cover devices completely produced by 3D printing, including the integration of a truly microfluidic comercially available pump, and the use of 3D printing to produce moulds and patterns to cast elastomeric microfluidic structures.

This paper will detail the development of these proposed techniques to convey their suitability for reviewing the rapid and low-cost evaluation and iteration of microfluidic channel geometries. Ultimately permitting the flow through novel microfluidic devices to be interrogated thoroughly and the design optimized prior to making the significant financial and, most crucially, time commitment to the production of a UV compatible flow cell for presentation to the synchrotron beamline.

2 CHIP DEVELOPMENT

2.1 Direct 3D Printing of Flow Chips

The first approach developed by the authors was a process route that involved 3D printing a complete flow cell chip with 3 dimensionally patterned channels contained within. The chip was produced using a commercial SLA system (Viper Si2 from 3D Systems, Cal. USA) from colourless Accura Si60 UV-cured resin. This approach required careful orientation of the chip during construction in order to minimise the supporting structures required within the channels, and then careful and laborious clearing of uncured resin from the channels using manual and ultrasonic agitation in an isopropyl alcohol (IPA) bath. An example of early chips produced this way can be seen in figure 1.

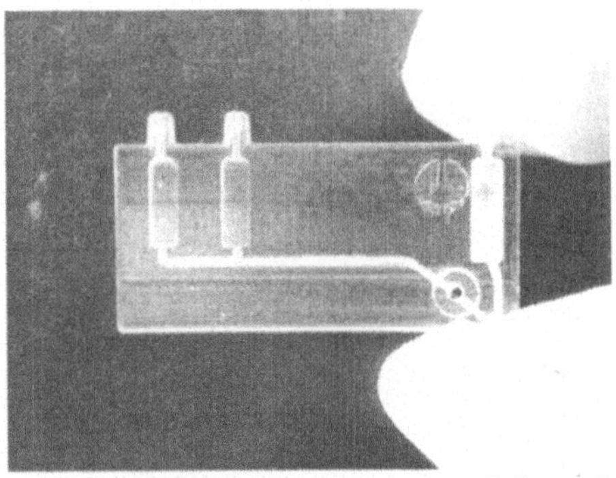

Figure 1: 3D printed chip with integrated flow channels

The channels in these chips were 300 µm diameter and contained reservoirs, ports and positioning for an UV compatible window. The small scale of the channels offers the benefit of requiring minimal support structures. The viscosity of the resin, however, poses considerable challenges for purging uncured resin from the channels.

The SLA printing process relies upon the selective curing of a photosensitive resin, which upon reaching a critical exposure energy (E_c) will solidify. The cured resin is then built up in layers to form complex 3D solid geometries. Jacob [5] demonstrated that knowledge of this and the penetration depth (D_p) of the incident light for the resin enables precise control of the solid layer depth (critical depth C_d) and subsequently the layer thickness through the expression:

$$C_d = D_p \ln(E_{max}/E_c)$$

The photopolymerisation reaction will however continue to partially occur within the resin in regions exposed to energies below E_c. These areas of partial curing produce a 'print-through' or 'overcure' effect which is typically equivalent to $3 \times C_d$, or 3 layers at z build resolution. This leads to poor control of the curing of undercut surface geometries (Figure 2) further impacting on the potential for reducing the scale of directly 3D printed chips with embedded channels.

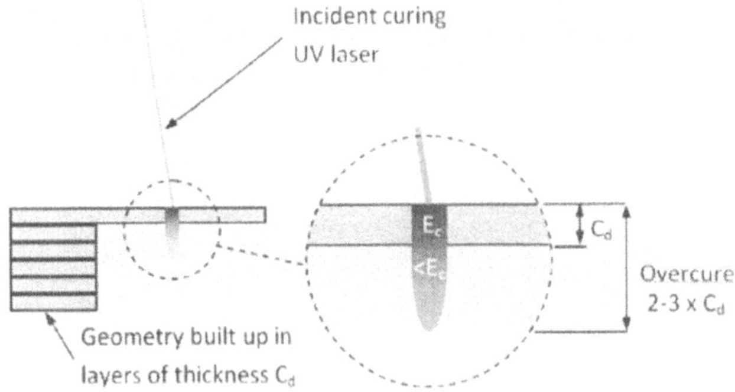

Figure 2: 'print-through' in regions of undercut surfaces leads to poor geometric control of internal channel features

The standard commercial high-resolution settings of the Viper Si2 SLA process is capable of z layer thickness of 50 µm. Engineers at Aston University have been successful in producing features with a z step resolution of 25 µm, but the overcure effect of the laser still subsequently limits the enclosed channel geometries to a minimum vertical channel height of > 70 µm.

In addition to the simple flow-channel designs, others were fabricated via the same process but with integrated pumps. A review paper [6] concluded how there were no truly microfluidic reliable on-chip pumps available. To address this the authors worked with Kikuchi (Japan) to integrate their comercially available 'stand alone' microfluidic pump (figure 3) into a 3D printed microfluidic chip. The Kikuchi M07ST-ES2-10 piezoelectric diaphragm pump has a 250 nl dead volume, can be controlled to operate between 1 and 350 HZ and the manufacturers claim they are capable of a flow rate of up to 3 ml/min [7].

Figure 3: The 7 x 7 mm micro pumps

By printing the manifold for the pump directly onto the chip there are no additional dead volumes typically produced when using macro, off-chip pumps with fluidic connectors and pipework. Figure 4 shows one of the 3D printed flow chips with its integrated micro pump as it pumps a blue-dye water solution from a reservoir.

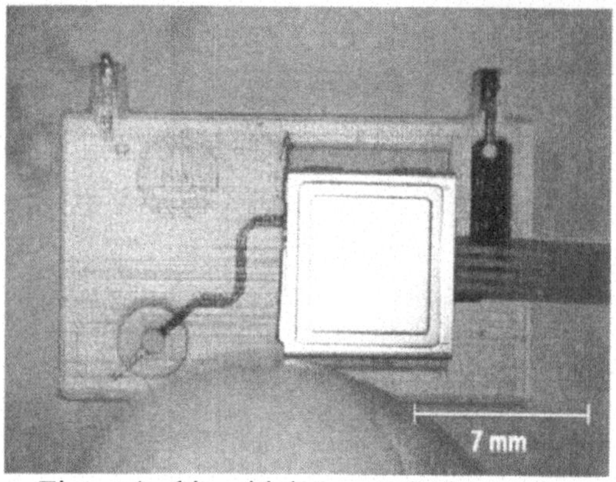

Figure 4: chip with integrated micro pump

Characterization of these pumps is ongoing along with the development of other drive technologies, including smartphone applications to control them. Additional research is also being carried out into surface coatings and treatments that can be applied to improve the wettability of the 3D printed resins in order to negate the requirement for elevated hydraulic pressures to initiate flow.

2.2 Modular 3D printed flow elements

For completeness an approach is being explored to 3D print modules with shaped cavities for 3D mixing (figure 5) and jigs to hold flat quartz windows (figure 6) through which the analyte may be imaged. Thereby offering a modular solution to prototyping the flow channel geometry whilst facilitating preliminary testing on the synchrotron beamline. It is expected that this could be an approach to additionally reduce lead times for producing working beamline compatible devices for CD development tests.

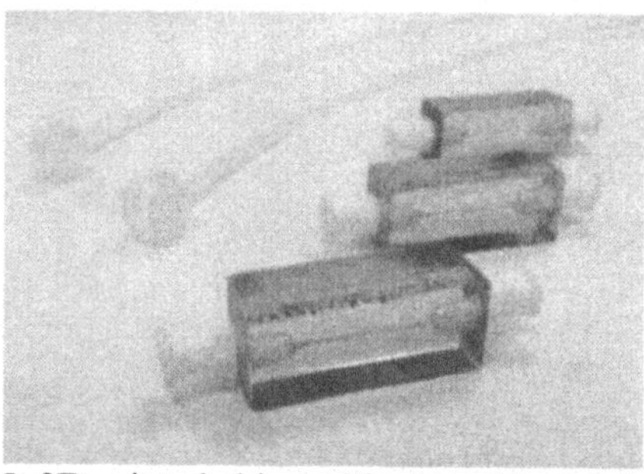

Figure 5: 3D printed chips with shaped internal cavities

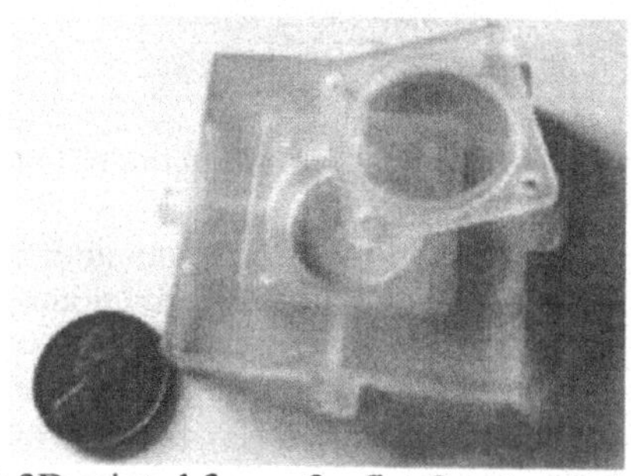

Figure 6: 3D printed frame for flat Quartz windows

2.3 Moulding from a 3D printed pattern

An approach used to great effect for microfluidic device patterning is that described by Bonyar *et al* [8,9] and Cormina *et al* [10] wherein an open mould is 3D printed with embossed fine channel and flow features (figure 7 shows a CAD image of a typical mould design). This mould is then used to cast the microfluidic channels from a conformal elastomer such as PDMS (polydimethylsiloxane) and the channels are then closed by clamping or bonding to a rigid substrate. This methodology is reported as benefitting from a cycle time of the order of 3-5 hours, including the 3D printing process [8]. This approach also takes advantage of the versatility and precision afforded by 3D printing with features as little as a single layer in height (approx. 25 μm), and enables the investigation of finer channel structures without the associated difficulties of clearing uncured material.

Beyond this a single mould can be used to produce numerous copies of a chip design with different surface treatments to compare flow behaviours, and permits the cost-effective production of single-use devices in order to minimise potential complications arising from sample contamination.

Figure 7: 3D CAD of a printed mould

Figure 8: (left) a mould housing was used to position the pattern, (middle) the pattern for the moulding and (right) the final PDMS cast flow-chip prior to trimming

Further advantages of this process lie in the material properties of PDMS. The low viscosity in its liquid state enables it to be used to replicate micro-scale geometries very precisely, it is optically clear, and it can be permanently bonded to glass through plasma or corona discharge treatment [11,12]. Finally the PDMS surface can be temporarily altered from its native hydrophobic state to a more hydrophilic nature through oxygen plasma treatment [ref 11,13,14].

The authors used this process to develop a meso-scale flow cell, with a square 500 × 500 μm profile serpentine channel. The cast chip was clamped between two machined polycarbonate jigs complete with hydraulic connection points to enable fluid to be introduced to the chip.

3 MIXING CAPABILITIES

The purpose of the chip was to demonstrate the persistence of laminar flow conditions through the gradual mixing of two concurrent fluid flows. The chip was manufactured from Sylgard 184™ (Dow Corning, USA) PDMS elastomer resin which was vacuum degassed prior to pouring and cured for 60 minutes at 100 °C prior to release. The initial flow channel was designed in a serpentine pattern with each waveform being 500 x 500 μm x 36 mm in order to mitigate the effects of the innate hydrophobicity of surface of the PDMS.

The flow chip inlets were connected using 1 mm PVC tubing to two 5 ml syringes (Becton Dickinson, USA) each loaded with a dyed distilled water solution. Pigments of yellow and blue were selected in order to provide a clear visual indication when mixing of the flow had started and completed. The outlet port was similarly connected using 1 mm PVC tubing to waste. To ensure balanced flow the two syringes were loaded with the same volume of fluid, the connective tubes were of identical length and the contents of the syringes were dispensed same rate using a dual syringe pump (KD Scientific 789210). The subsequent flow was imaged at 30 fps using a 16 MP camera (Sony NEX 5n) camera equipped with a 10 × magnification macro lens.

The initial trials using PDMS flow-chips have proven promising. The process allowed different channel designs to be investigated along with different flow rates. Figure 11 demonstrates an example where the fluids pass through the channels at a combined flow rate of (a) 1.5 ml/min without mixing and a flow rate of (b) 0.15 ml/min results yields a complete mixture of the fluids at exit.

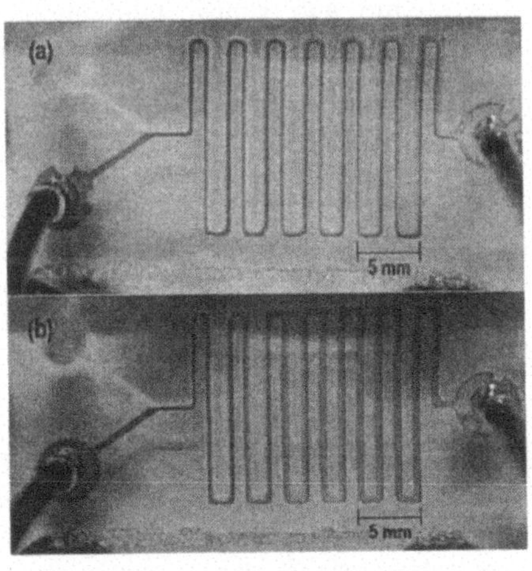

Figure 11: Mixing study within PDMS chip showing flows of (a) 1.5 ml/min and unmixed at exit and (b) 0.15 ml/min and fully mixed at exit

4 CONCLUSIONS

The work described here shows great promise in developing techniques for the fast-proofing of on-chip geometries prior to embarking on fabrication methods that are expensive in both time and financial cost. Although the directly-printed microfluidic channel methods still require much improvement the PDMS technique described offers considerable advantages. The channel geometries produced in this way are limited in comparison to direct printing as they are effectively two dimensional unless multiple layers of patterned PDMS features are employed. In this way they have much in common with those photolithographically produced from structural resists [4]. This method is currently unable to compete with photolithographic methods in terms of feature and channel resolution. However, the fabrication routes and a standardised manifold arrangement has been determined which can allow the design, fabrication, evaluation and iteration of new on-chip architecture in hours for what previously would have taken many months.

REFERENCES

[1] L. Capretto, W. Cheng, M. Hill, and X. Zhang, Top. Curr. Chem. 304, 27–68, 2011.
[2] N. Bao, Q. Zhang, J-J. Xu, and H-Y Chen, Journal of Chromatography A. 1089, 270-275, 2005.
[3] J. Do, J-Y. Zhang, and C. Klapperich, Robotics and Computer-Integrated Manufacturing, 27, 245-248, 2011.
[4] G. Whitesides, and A. Stroock, Physics Today 54(6), 42, 2001.
[5] P. Jacobs, Rapid Prototyping & Manufacturing: Fundamentals of Stereolithography, Society of Manufacturing Engineers, 1992.
[6] N. Nam-Trung, Fluids Eng 124(2), 384-392, 2002.
[7] Kikuchi pump manual supplied with devices
[8] A. Bonyár, H. Sántha, B. Ring, M. Varga, J. Kovács, and G. Harsányi, Procedia Engineering, 5, 291–294, 2010.
[9] A. Bonyar, H. Santha, M. Varga, B. Ring, A. Vitez, and G. Harsanyi, Int. J. Mater. Form. 7, 819-196, 2014.
[10] G. Cormina, A. Suska, and D. Fillipini, Lab Chip 14, 424-430, 2014.
[11] Y. Berdichevsky, J. Khandurina, A. Guttman, and Y-H Lo, Sensor Actuator B. 97, 402–408, 2004.
[12] K. Haubert, T. Drier, and D. Beebe, Lab Chip 6, 1548–1549, 2006.
[13] D. Bodhas and C. Khan-Malek, Sensors and Actuators B. 123, 368-373, 2007.

M. Prince Aston University UK
M.Prince1@aston.ac.uk

A simple method for fabrication of microarrays and microfluidic device using PDMS stamp

Hun Lee, Domin Koh, and Kwang W. Oh

SMALL (Sensors and MicroActuators Learning Lab), Department of Electrical Engineering, University at Buffalo, The State University of New York (SUNY at Buffalo), Buffalo, New York 14260, USA

ABSTRACT

We propose a simple approach to fabricate a series of microwell arrays using the conventional PDMS (Polydimethylsiloxane) stamp approach to study the cell-to-cell adhesion of Human Bone Marrow Mesenchymal Stem Cells (BM-MSCs). This approach can be applied to fabricate a microfluidic channel with glass-PDMS-glass configuration for cytologic diagnosis. To replicate the microwell arrays and the microfluidic channel, an SU-8 master mold is created, and then a PDMS stamp replica is transferred to a glass substrate. For the pattern transfer from the PDMS stamp to the glass substrate, the surface of the PDMS stamp is treated by oxygen plasma and bonded on the glass substrate. By ripping the PDMS stamp by means of applying a mechanical force, the PDMS patterns are formed on the glass substrate. For the cell-to-cell interaction studies, 4 different types of PDMS arrays are created to vary cell-to-cell contact length. For the cytologic diagnosis, the microchannel is designed, that allows for the enrichment and temporary immobilization of the cells.

Keywords: microwell array, glass-PDMS-glass fabrication, pattern transfer

1 INTRODUCTION

Patterning of microarrays on a surface has become a very attractive tool for sample detection and quantification purposes. [1] There has been a growing interest in applying this technique for various cell studies. Thus, it is important to develop efficient fabrication methods for the microarray which can offer lower costs and simpler production methods for various biological applications[1-3]. Due to the fact that PDMS can be easily fabricated by using soft-lithography techniques, it was widely used for this experiment. [4-5].

In the early stages of cell-to-cell interaction studies, Petri dishes were used to form cell-to-cell contact [5]. However, the ability of observe the cells that grew in the dish was limited due to the low desnity of each pair of cells. Moreover, under the conventional cell culture conditions, it is hard to manipulate the cell formation. To regulate the cell-to-cell contact with various shapes, densities, and contact of cells, micropatterns can be printed onto a cell culture substrate using the PDMS stamp inked by adhesive molecules [6-8]. This technology can offer a cell culture environment with well-controlled sizes, shapes, and positions on a substrate, thus providing a useful tool for cell studies. As an alternative, an agarose pattern on a glass substrate can be fabricated using the PDMS stamp method [9]. The agarose is perfused under the PDMS stamp attached on the glass. Thus, the region sealed against glass remian free of agarose. After the agarose cures, the PDMS stamp is removed, leaving behind a well defined geometry of agarose patterns. Currently, most microarrays are fabricated by the contact printing technique using a PDMS stamp for a controlled cell culture environment. Despite its huge potential, the contact printing method should be carefully controlled to avoid nonuniform patterns, as deformation can occur when handling the stamp. In the printing technology, the microarrays are typically influenced by the surface tension and viscosity of samples [10].

We describe a simple microwell array fabrication technique to isolate the effects of cell-to-cell contact by transferring PDMS patterns on a glass substrate. To overcome the technical challenges in the contact printing technique, a simple and robust fabrication method is demonstrated to study the role of intercellular adhesion forces in myogenic differentiation of stem cells and the molecular pathways governing this process. The extent of the cell-to-cell adhesion can then be examined by immunostaining various cadherin molecules which control BM-MSCs differentiation towards smooth muscle lineage. In particular, we study levels of Cadherin-11 (Cad-11) and its effect on smooth muscle genes α-SMA (Alpha- Smooth Muscle Actin), CNN-1 (Calponin) and MYH11 (Myosin-Heavy Chain). We expect that this tool can be further extended to 3 or 4 neighboring cells in micropatterns, thereby establishing Cadherin-11 as a master regulator of BM-MSCs to smooth muscle differentiation. Furthermore, we demonstrate a very useful method for a microfluidic device with glass-PDMS-glass configuration for the effectiveness of the proposed method. Thus, the proposed method can be used to generate the microarrays and

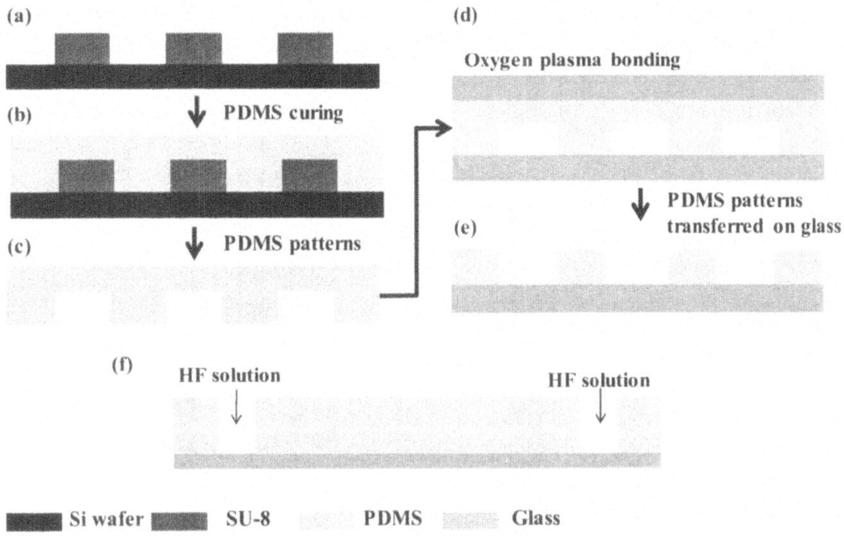

Fig. 1 Working principle for the PDMS microwell array on the glass slide. (a) Patterning of SU-8 photoresist by photolithography. (b) Replication of the PDMS stamp. (c) Peeling off the PDMS stamp from the master mold. (d) Bonding the PDMS stamp to two glass slides. (e) Transfering the patterns of the PDMS stamp to the bottom glass by applying a force to the top glass. (f) Formation of through hole by HF solution for the inlets and outlets.

microfluidic channels which provide an attractive new approach over the conventional methods.

2 METHODS/EXPERIMENT

2.1 Microarrays and Microfluidic device design

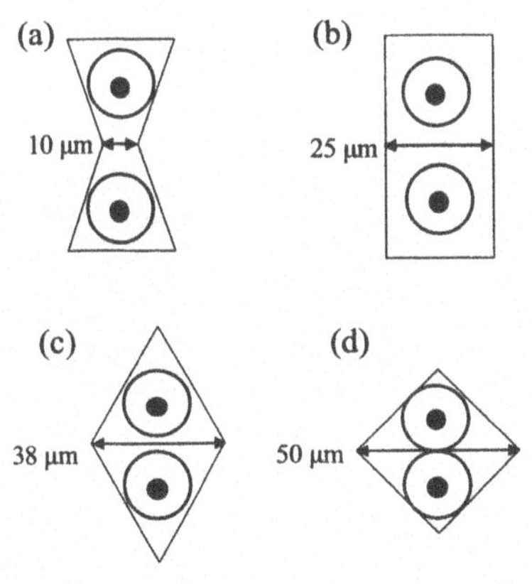

Fig. 2 4 different types of micropatterns used to trap two cells for cell-to-cell contact: (a) Bowtie (b) Rectangle (c) Rhombus and (d) Wide-rhombus.

The size of each PDMS stamp was 5 mm × 5 mm and the array of each pattern has 10 μm distance between patterns. **Fig. 1** shows the configuration of 20 μm thick PDMS pattern fabrication on the slide glass for cell-to-cell adhesion studies in the microwell array. By conventional soft-lithography processes, 4 different types of PDMS patterns (e.g., rectangle, bowtie, wide-rhombus, and rhombus) were replicated from a master mold as shown in **Fig. 2**. The density of the patterns and the contact surface area between the patterns and the glass were listed in Table 1.

To fabricate the microfluidic channel based on the proposed method, a prepolymer and curing agent were mixed at a 20:1 ratio. The microfluidic device was fabricated with the channel dimensions as follows: 6.1 cm in length, 5 mm in width and 100 μm in height.

Table 1 PDMS patterns for cell-to-cell adhesion

Shape	Rectangle	Bowtie	Wide Rhombus	Rhombus
Density of pattern (#/mm²)	937.11	797.20	641.46	946.84
Contact area of PDMS (%)	59.91	50.18	41.43	40.82

2.2 Fabrication

In order to create the SU-8 master mold, a series of flim photomasks were designed using AutoCAD commerical software (Autodesk, USA), and then printed on transparencies (CAD/Art Service Inc., Bandon, OR, USA). The PDMS patterns were fabricated by using conventional soft-lithography techniques. In order to make a master mold, negative photoresist (SU-8 2015, Micro-Chem Corp, Newton, MA, USA) was used, and then spin coated with target thickness (e.g., 20 μm) at 2000 rpm on a cleaned wafer using a spin processer (WS-650Mz NPP from Laurell Technologies, North Wales, PA, USA). Before the coating process, the silicon wafer was submerged into BHF (buffered hydrofluoric acid) at room temperature for 5 min to remove a thin oxide layer that can make a weak adhesion between the SU-8 and the surface of the wafer. Afterward, it was cleaned with acetone, followed by methanol. It was then rinsed by DI water and blown dry with filtered nitrogen gas. After the spin coating process, a soft bake process was conducted at 95 °C for 4 min. After UV exposure through the film photomask and subsequent post exposure baking process was conducted at 95 °C for 5 min, the wafer was developed in SU-8 developer (Microchem, USA) and washed with isopropyl alcohol to completely remove the developer. Using a gentle stream of nitrogen gas, the wafer with SU-8 patterns was dried. The measured thickness of SU-8 pattern was 20 ± 2 μm.

The PDMS patterns were fabricated from PDMS prepolymer and curing agent (Sylgard 184, Dow Corning Co., Midland, MI). The two materials were thoroughly mixed at a ratio of 20 : 1 (wt/wt). The mixed PDMS was degassed in a vacuum chamber to remove any air-bubbles for 20 min. The SU-8 master mold was silanized using hexamethyldisilazane (Sigma Aldrich, Saint Louis, MO, USA) in a vacuum chamber for 3 hours to easily peel off the PDMS from the SU-8 master mold. To mold the PDMS against the master mold, it was carefully poured onto the SU-8 master mold and cured at 65°C for 30 min. The PDMS replica was peeled off and bonded irreversibly to a glass by exposing it to O_2 plasma. The flat surface of PDMS was bonded to the glass, and then the surface of PDMS with the patterns was sandwiched with another glass. Finally, the glass bonded to the PDMS patterns was released from the patterns.

3 RESULT/DISCUSSION

As shown in **Fig. 3**, the patterns were successfully fabricated on the glass by causing the crack. The patterns have 20 μm thick and 10 μm gap among the patterns. The thickness was enough to trap the cells into the wells. The force was applied at the edge of the glass in order to locally initiate the crack. The crack started from a mechanically weak corner of the PDMS so that the different shapes can be transferred from the PDMS to the glass.

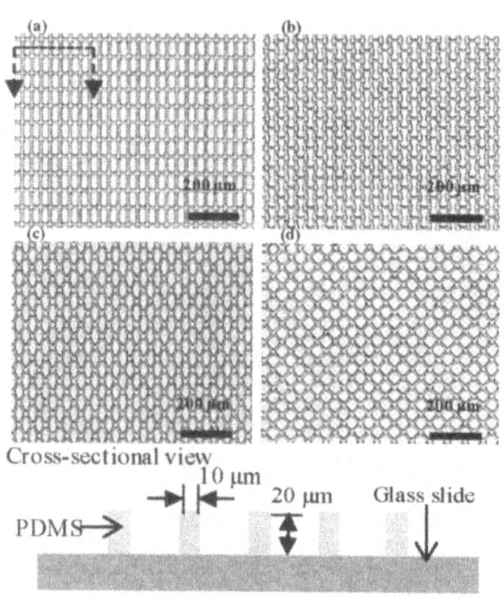

Fig. 3 Photographs of PDMS patterns on the glass substrate. (a) Rectangle. (b) Bowtie. (c) Wide Rhombus. (d) Rhombus shape.

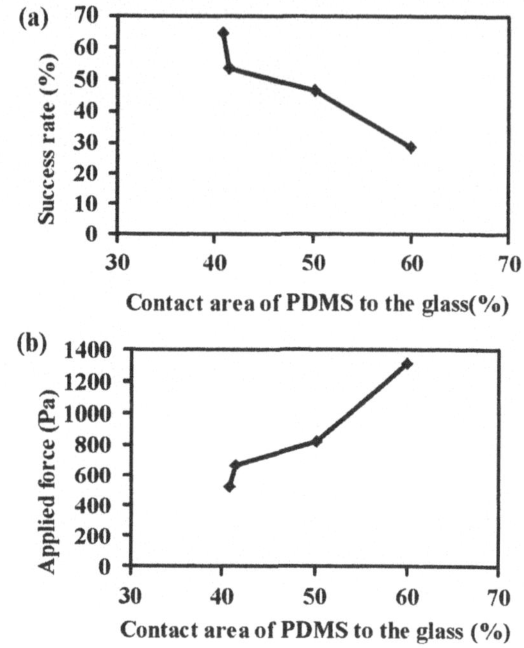

Fig. 4 (a) Success rate of patterns transferred to the glass according to the contact area of PDMS patterns. (b) Applied force to the top glass to create the crack on the patterns.

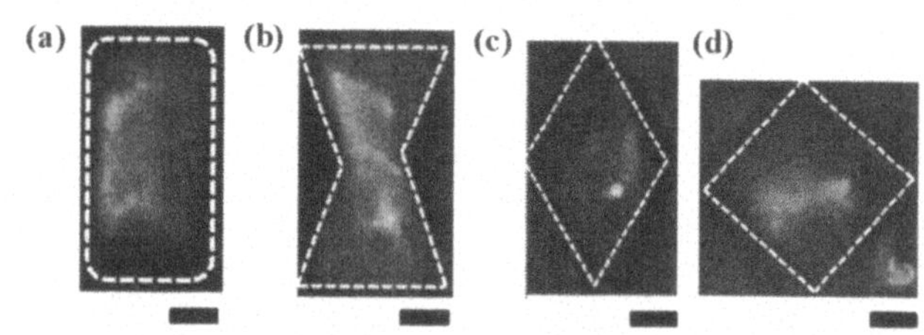

Fig. 5 Fluorescent images showing cell-to-cell adhesion in each pattern. The trapped cells in the patterns were allowed to attach for 48 hours. The samples were counterstained with DAPI for nuclei. (a) Rectangle. (b) Bowtie. (c) Wide Rhombus. (d) Rhombus shape. The scale bar is 10 μm.

After the patterns are bonded, it is simply torn from the bottom glass using the home-made equipment to release the top glass. Only patterns tear at the locations where the patterns are bonded to the bottom glass. It is best to tear the PDMS patterns, applying torque force at the position 50 mm away from the edge of the glass in order to locally initiate the crack. In the **Fig. 4 (a)**, the percentages show the success rate of patterns transferred to the glass for each condition given in Table 1. It can be seen that the patterns with smaller contact area are more easily created since the crack resistance is low so that the crack starts more uniformly along the edge of patterns at mechanically instable point. However, with increasing contact area, the success rate will gradually decrease due to the nonuniformity of the crack. We investigated the maximum force applied at the top slide glass to tear the pattern by our

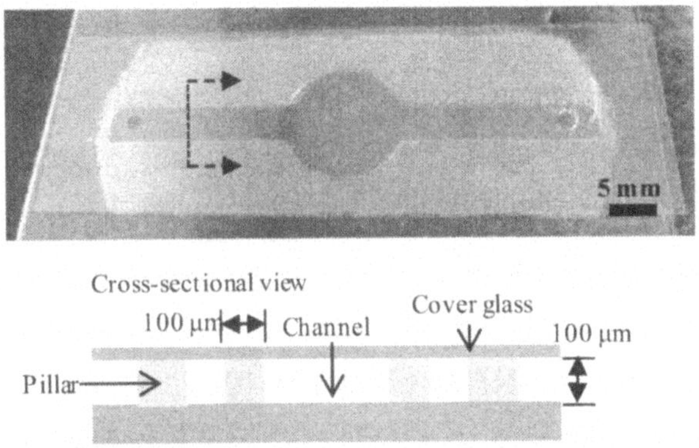

Fig. 6 Photograph of the microfluidic channel with glass-PDMS-glass configuration. The transferred channel was bonded to the cover glass with the inlet and outlet that are formed by the etching using the HF solution.

personal equipment. The mechanical force was applied at the position of glass 50 mm away from the edge of the PDMS patterns. The measured moment of force was ranged from 0.5 to 1.31 kPa as shown in **Fig. 4 (b)**.

As a demonstration of the effectiveness of this method, the cell-to-cell adhesion was performed on the each pattern. Cells were loaded into microwells so that one pair of cells was cultured within each microwell, resulting in contacting a single neighboring cell as shown in Fig. 5.

To demostrate this proposed technique for glass-PDMS-glass device, a simple microfluidic device was fabricated as shown in **Fig. 6**. This device consisted of a main channel and a circular region to allow cells to spread out. For the inlet and outlet, high concentrated HF solution of 48% was used to etch the holes on the cover glass for 20 minutes. The cover glass with the through holes was bonded onto the PDMS patterns for the microfluidic channel by an oxygen plasma treatment.

4 CONCLUSION

We have demonstrated the simple and low-cost fabrication method of microwell array on a glass slide using the conventional photolithography for cell-to-cell adhesion-mediated differentiation to smooth muscle cells. 4 different types of patterns were created on the glass substrate and the pattern transfer rate and applied force was investigated as the function of contact area of PDMS to the glass. We believe that the isolated cell pairing into the microwells and the separate analyses of the cell patterns that were cultured, allowed for a more efficient analysis from a small quantity of cells, which is highly important when using rare samples. In addition, the microfluidic channel with glass-PDMS-glass configuration was fabricated based on the proposed method, which allowed for the enrichment and temporary immobilization of cells.

REFERENCES

[1] V. N. Goral, C. F. Zhou, F. Lai and P. K. Yuen,"A continuous perfusion microplate for cell culture dagger," *Lab Chip*, 13, 1039-1043, 2013.

[2] M. Charnley, M. Textor, A. Khademhosseini and M. P. Lutolf,"Integration column: microwell arrays for mammalian cell culture," *Integr Biol*, 1, 625-634, 2009.

[3] T. Jain, R. McBride, S. Head and E. Saez,"Highly parallel introduction of nucleic acids into mammalian cells grown in microwell arrays," *Lab Chip*, 9, 3557-3566, 2009.

[4] J. H. Choi, H. Lee, H. K. Jin, J. S. Bae and G. M. Kim,"Micropatterning of neural stem cells and Purkinje neurons using a polydimethylsiloxane (PDMS) stencil," *Lab Chip*, 12, 5045-5050, 2012.

[5] C. M. Nelson, W. F. Liu and C. S. Chen,"Manipulation of Cell-Cell Adhesion Using Bowtie-Shaped Microwells," *Methods in Molecular Biology*, 370, 1-9, 2007.

[6] T. Tamura, Y. Sakai and K. Nakazawa,"Two-dimensional microarray of HepG2 spheroids using collagen/polyethylene glycol micropatterned chip," *J Mater Sci-Mater M*, 19, 2071-2077, 2008.

[7] H. Schmid, H. Wolf, R. Allenspach, H. Riel, S. Karg, B. Michel and E. Delamarche,"Preparation of metallic films on elastomeric stamps and their application for contact processing and contact printing," *Adv Funct Mater*, 13, 145-153, 2003.

[8] D. S. Gray, W. F. Liu, C. J. Shen, K. Bhadriraju, C. M. Nelson and C. S. Chen,"Engineering amount of cell-cell contact demonstrates biphasic proliferative regulation through RhoA and the actin cytoskeleton," *Exp Cell Res*, 314, 2846-2854, 2008.

[9] C. M. Nelson and C. S. Chen,"VE-cadherin simultaneously stimulates and inhibits cell proliferation by altering cytoskeletal structure and tension," *J Cell Sci*, 116, 3571-3581, 2003.

[10] A. O. Gutmann, R. Niekrawietz, R. Kuehlewein, C. P. Steinert, B. Heij, R. Zengerle and M. Daub, "Impact of medium properties on droplet release in a highly parallel nanoliter dispenser," *Sensor Actuat. A-Phys.*, 116, 187-194, 2004.

Negative Dielectrophoresis Force based High Efficiency and High Throughput Cell Sorting System Harnessing Paired Electrode Arrays and a Meso Size Channel

D. Lee*, B. Hwang* and B. Kim*

*School of Aerospace and Mechanical Engineering, Korea Aerospace University, Goyang-city, Gyeonggi-do, 412-791, Korea, bkim@kau.ac.kr

ABSTRACT

In this paper, we present a high efficiency and throughput cell sorting system employing repeated electrode arrays and a meso size channel. The paired twin electrode arrays generate a negative dielectrophoretic (n-DEP) force and deflect cells according to dielectric properties. The proposed dielectrophoresis activated cell sorter can achieve high efficiency even under strong hydrodynamic force caused by a large flow rate for high throughput because target cells are continually deflected by the n-DEP force while passing through each electrode pair. To find the optimal flow rate of the sorting system, we implemented numerical simulation under a signal-off condition. Consequently, the target cells, K562 cells, are effectively separated along an electrode pair under a voltage of 6 Vp-p with 100 kHz and a flow rate of 20 μl/min and 60 μl/min in each outlet. The proposed sorting system showed a high deflection ratio of 98.0±0.3% and high throughput over 38,000 cells/min.

Keywords: Dielectrophoresis, Negative dielectrophoretic force, Cell sorting system, Paired electrode array

1 INTRODUCTION

In the clinical pathology and diagnosis field, many cell sorting systems such as fluorescent activated cell sorters (FACS), and magnetic activated cell sorters (MACS) have been developed for the detection of cancer cells and manipulation of biological particles [1]. FACS and MACS have been major techniques for the past few decades. However, they have a limitation due to the labeling process which causes cell damage and requires high cost. Therefore, various cell sorting systems were proposed utilizing external forces, such as ultrasonic, hydrodynamic and dielectrophoretic (DEP) forces [2]. Above all, dielectrophoresis activated cell sorters (DACS), which exploit the intrinsic dielectric properties of target cells, are highlighted as an alternative cell sorting method [3-6]. Nonetheless, the DACS performing in micro channels have a limitation of throughput, even while they achieved high efficiency. In this paper, we present a high efficiency and high throughput cell sorting system which is comprised of paired electrode arrays and a meso size channel. In order to improve separation efficiency by continuous separation at each pair of eletrodes, micro electrodes were deposited sequentially on the silicon and glass plate, respectively. To enhance throughput which is representative limitation of micro channel based DACS, the channel of the system was designed with a scale of meso size. Finally, a separation experiment is carried out with a live leukemia cell line (K562 cells).

2 WORKING PRINCIPLE AND EXPERIMENTAL SYSTEM SETUP

2.1 Basic principle of cell sorting system

The direction of input cells is determined by vector summation of gravity, hydrodynamic force, and n-DEP force (Figure 1). In case of the target cells, a live leukemia cell line (K562 cells), they move toward outlet-A due to a dominant n-DEP force as shown in Figure 1-(a). In case of the non-target cells, dead K562 cells, they move toward outlet-B as shown in Figure 1-(b) since the hydrodynamic force is dominant over the n-DEP force. The hydrodynamic force is controlled by a flow regulator at each outlet and the DEP force is expressed as shown below [7].

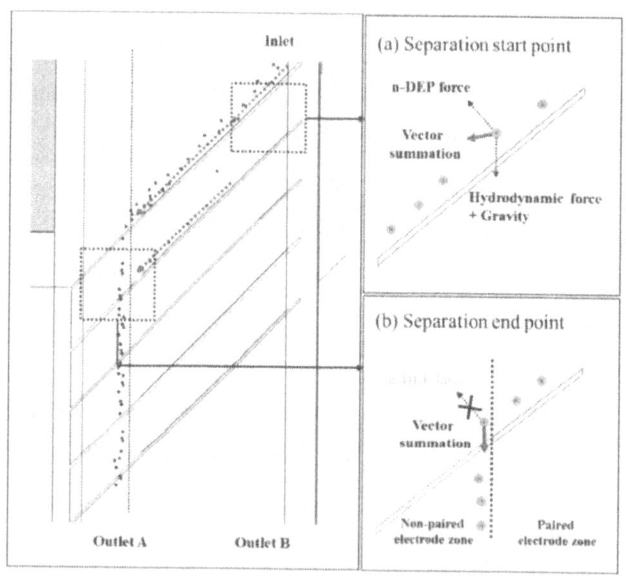

Figure 1: Working principle of the proposed sorting system: (a) the n-DEP force is dominant at the paired

electrode zone, (b) the hydrodynamic force is dominant at the non-paired electrode zone.

$$\mathbf{F}_{\mathbf{DEP}} = 2\pi r^3 \varepsilon_m \varepsilon_0 Re[f_{CM}] \nabla \mathbf{E}^2 \qquad (1)$$

Where r is the radius of the target particle, ε_m is the permittivity of the experimental buffer, ε_0 is the dielectric constant at the vacuum state, f_{CM} is the Clausius-Mossotti factor which determines the direction of the DEP force, and **E** is the electric field according to the input voltage condition.

The proposed DACS is capable of achieving high efficiency even under a strong hydrodynamic force due to a caused by large flow rate for high throughput because the target cells are continually deflected by the n-DEP force while passing through 5 electrodes pairs.

2.2 Electrode array fabrication procedure

The micro electode arrays are fabricated on two plates, which are silicon and glass as shown in Figure 2. Additionally, to minimize the change of resistance in one electrode, Au is deposited with a thickness of 6000 Å. The detailed procedure is as follows.

(a) Silicon and glass plate with chemical and mechanical planarization process (300 μm), (b) Oxide for insulation (silicon plate only) and positive photoresist (PR, AZ 7220) deposition, (c) Patterning through photolithography, (d) Developing PR, (e) Cr/Au layer (500 Å/ 6000 Å) deposition. (f) Eliminate PR through lift-off process, and completion fabrication.

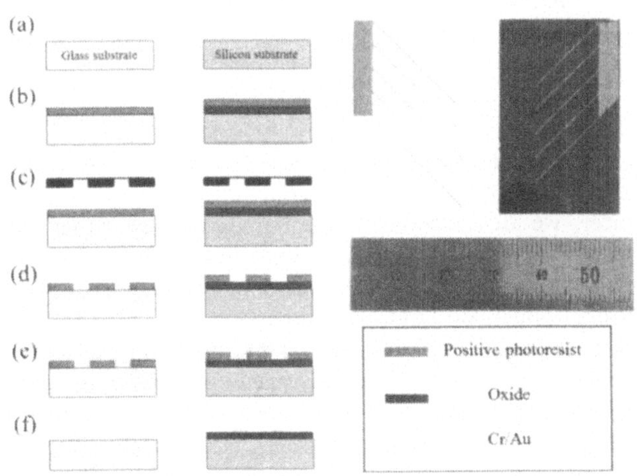

Figure 2: Micro electrode array fabrication procedure.

2.3 System configuration of cell sorting system

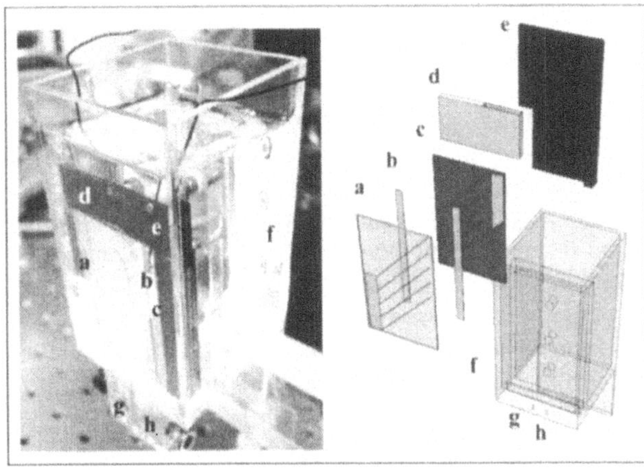

Figure 3: Configuration of the cell sorting system.

As shown in Figure 3, the proposed sorting system is comprised of a glass plate (a) and a silicon plate (c) which has a electrode array on each plate, 200 μm width PDMS side wall (b) to make a meso sized channel, Inlet (d), Supporter (e) and container (f). When assembling the two plates, electrodes on each plate are paired to generate a n-DEP barrier. The PDMS side wall enable to maintain the height of the channel of 200 μm. The inlet part for cell loading is allocated on the top of the channel and the supporter holds the assembled channel. Lastly, a reservoir keeps experimental buffer.

2.4 Experimental setup

The experimental system is set up as shown in Figure 4. It is comprised of (a) a CCD camera for monitoring cell sepration, (b) a 3-axis micro stage for fine control of focal length from the CCD camera to the position of cell separation, (c) a function generator for applying input voltage with a specific frequency, (d) an optical microscope for counting the number of cells before/after the experiment, and (e) a cell sorting system.

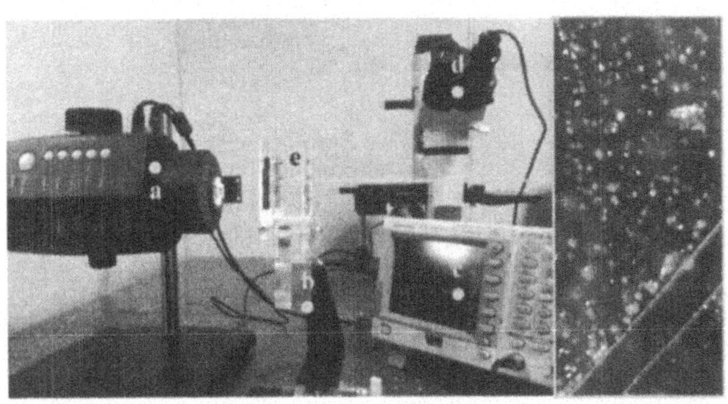

Figure 4: Experimental system setup for investgating cell separation.

3 RESULTS AND DISCUSSION

3.1 Numerical analysis

To find the optimal flow rate for the proposed sorting system, we carried out numerical simulation using COMSOL Multiphysics®. As shown in figure 5, simulation is implemented under the voltage signal-off condition. In case of the signal-off condition, all of the cells under a high flow rate must move toward outlet-B for high throughput. For the flow rate condition for each outlet, four cases are simulated as follows:

Case (a): Outlet-A: 16 µl/min, Outlet-B : 20 µl/min
Case (b): Outlet-A: 15 µl/min, Outlet-B : 20 µl/min
Case (c): Outlet-A: 12 µl/min, Outlet-B : 30 µl/min
Case (d): Outlet-A: 20 µl/min, Outlet-B : 60 µl/min

Case (a) is not a proper condition since all of the cells move toward outlet-A. Case (b), (c) and (d) meet the separation working principle which means the cells move toward outlet-B. Considering throughput, therefore, we select case (d) which has the highest flow rate of 20 µl/min and 60 µl/min in outlet-A and B, respectively, for the cell separation experiment.

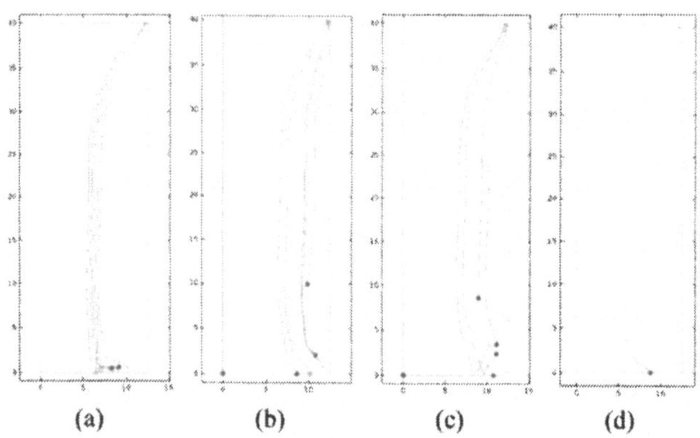

Figure 5: Particle trace under each flow rate: (a) is improper for the proposed sorting system, (b)~(d) are proper for the sorting system.

3.2 Experimental result

In order to generate an adequate n-DEP force and hydrodynamic force, we select optimal experimental input condtions which are a voltage of 6 Vp-p with 100 kHz, a flow rate of 20 µl/min and 60 µl/min in outlet-A and B respectively, and buffer conductivity of 0.2 S/m. As a target cell, K562 cell (leukemia cell line) is employed [8].

At first, when K562 cells arrived at the first electrodes pair, which consists of two electrode arrays, they are repelled by the n-DEP force, vertically against the DEP barrier. As a result, over 90% of the target cells move toward the end of the paired electrode zone (Figure 6. (a)) due to vector summation among n-DEP, hydrodynamic force and gravity. Few cells passing through the first pair are deflected at the second electrode pair or the next pair (Figure 6. (b)). Since there are 5 electrodes pairs, separation efficiency can be maximized even under a large flow rate. In the separation process, deflected target cells adhere to each other and form some groups. When groups of cells arrive at the pairing end point, they fall down toward outlet–A (Figure 6. (c)). Consequently, the proposed sorting system showed a high deflection rate of 98.0±0.3% and high throughput over 38,000 cells/min.

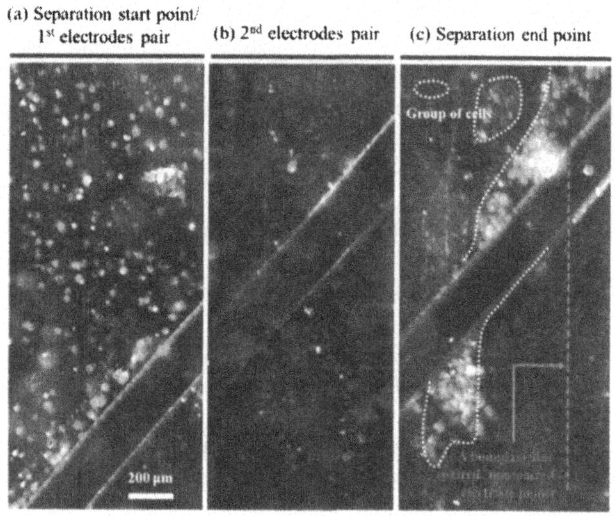

Figure 6: Movement of target cells under input condition (Voltage of 6 Vp-p with 100 kHz, flow rate of 20 µl/min and 60 µl/min in outlet-A and B, respectively): (a) 1st electrodes pair below inlet and (b) 2nd electrodes pair below inlet, (c) End of paired electrode zone.

4 CONCLUSION

In this article, we present a novel cell sorting system using the n-DEP force. Compared with a conventional cell sorter, we developed a separation technique under low voltage without a labeling process and with a simpified structure of the cell sorter. Furthermore, without a micro syringe pump, we controlled flow rate precisely using the micro flow regulator. To improve throughput and efficiency, a meso size channel as well as paired electrode arrays were employed. Conclusively, we achieve a high deflection rate of 98.0±0.3% and high throughput over 38,000 cells/min under the condition of a voltage of 6 Vp-p with 100 kHz and a flow rate of 20 µl/min and 60 µl/min in outlet-A and B, respectively. In the near future, to evaluate the performance of the proposed sorting system for various target cells, we will consistently implement the separation test utilizing different cancer cells.

ACKNOWLEDGEMENT

This research was supported by the National Research Foundation of Korea (NRF) funded by the Ministry of Science, ICT & Future Planning (No. 2005-2000206).

REFERENCES

[1] K. Bernath, M. Hai, E. Mastrobattista, A. D. Griffiths, S. Magdassi and D. S. Tawfik, "In vitro compartmentalization by double emulsions: sorting and gene enrichment by fluorescence activated cell sorting", Analytical Biochemistry, 325, 151-157, 2004.

[2] D. R. Gossett, et al, "Label-free cell separation and sorting in microfluidic systems", Analytical and bioanalytical chemistry, 397, 3249-3267, 2010.

[3] M. Alshareef, N. Metrakos, E. J. Perez, F. Azer, F. Yang, X. Yang and G. Wang, "Separation of tumor cells with dilectrophoresis-based microfluidic chip", Biomicrofluidics, 7, 011803, 2013.

[4] V. Gupta, I. Jafferji, M. Garza, V. O. Melnikova, D. K. Hasegawa, R. Pethig and D. W. Davis, "ApoStrem, a new dilectrophoretic device for antibody independent isolation and recovery of viable cancer cells from blood", Biomicrofluidics, 6, 024133, 2012.

[5] D. Lee, D. Kim, Y. Kim, K. Park, E.J. Oh, Y. Kim and B. Kim, "A Negative Dielectrophoresis and Gravity Driven Flow based High Throughput and High Efficiency Cell Sorting System", Journal of Laboratory Automation, 19, 60-74, 2014.

[6]. J. Lee, Y. Kim, D. J. Beebe and B. Kim, "Harnessing gravitational, hydrodynamic and negative dielectrophoretic forces for higher throughput cell sorting", Biochip J, 6, 229-239, 2012.

[7] H. Morgan and N. G. Green, "AC Electrokinetics: Colloids and Nanoparticles", Research Studies Press, 15-54, 2003.

[8] D. Lee and B. Kim, "A cell sorting system for the evaluation of drug sensitivity usiing negative dielectrophoresis", International Journal of Precision Engineering and Manufacturing, 16, 609-613, 2015.

The Development of an Automated Nano Sampling Handling System for Nanometre Protein Crystallography Experiments

P. Docker+, D. Axford+, M. Prince*, B. Cordovez~, J. Kay+, D. Stuart+, G. Evans+

+ Diamond light Source Harwell OX110DE
*Aston University Birmingham B4 7ET
~ Optek Systems Philadelphia, PA, 19104

ABSTRACT

As the world's synchrotrons and X-FELs endeavour to meet the need to analyse ever-smaller protein crystals, there grows a requirement for a new technique to present nano-dimensional samples to the beam for X-ray diffraction experiments.The work presented here details developmental work to reconfigure the nano tweezer technology developed by Optofluidics (PA, USA) for the trapping of nano dimensional protein crystals for X-ray crystallography experiments. The system in its standard configuration is used to trap nano particles for optical microscopy. It uses silicon nitride laser waveguides that bridge a micro fluidic channel. These waveguides contain 180 nm apertures of enabling the system to use biologically compatible 1.6 micron wavelength laser light to trap nano dimensional biological samples. Using conventional laser tweezers, the wavelength required to trap such nano dimensional samples would destroy them. The system in its optical configuration has trapped protein molecules as small as 10 nanometres.

Keywords: microfluidics, protein crystallography, X rays, optical nano traps.

1 INTRODUCTION

Typically, workers in the synchrotron community look to employ a macro, top-down approach using robots to automate human protocols [1]. This is a methodology often seen in the micro and nano engineering community at large. Currently, as the handling of nano crystals is beyond top down capabilities, X-FEL facilities are currently employing an injector approach. This requires micro litre quantities of reagant containing nanocrystals to be injected in front of the X-ray beam and the diffraction pattern is imaged by the detector. Such systems have a reported 'hit' rate of less that 10 % [2].

This paper details optical nano tweezers which are very much a bottom-up approach that facilitates self-assembly, alignment and subsequently automation. This also offers the potential to inform the user that a crystal has been trapped prior to interrogation. This techneque will also present a static array of crystals to be interrogated by the X-ray beam. This technique is not to be confused with traditional optical tweezers which can be used to manipulate a single sample. Although a very useful technique it can only be used with >1 μm sized samples as the wavelength of laser light used to manipulate the sample must be of the same order of magnitude. To achieve the trapping of nano crystals using this technique would require a wavelength of laser light with too much energy thereby destroying any sample it traps.

These nanotweezers have been developed by Optofluidics (PA, USA). By using optical wave guides 1064 nm biologically friendly laser light can be used to trap samples sizes down to 10 nm at predetermined sites micro machined into the waveguides. These can be prealigned to the X-ray source and no further alignment would be required.

2 THE TECHNOLOGY

This potential solution is offered by Optofluidics, based in the Philadelphia, USA. They have developed technology which employs a microfluidic cell that uses silicon nitride waveguides to overcome the free space limitations of traditional optical traps to capture sub-micron particles with laser light using a wavelength that is compatible with biological samples (1064 nm) whilst maintaining protein sample integrity. Briefly, light is tightly confined at the near-surface of the silicon nitride waveguides, and the technology uses the evanescent wave (light outside of the waveguides) to generate strong optical gradients which are necessary to capture such small particles. If traditional optical trapping technology were to be implemented, such technologies would either be unable to optically capture the nano crystals (because of the light diffraction limit limitations of free space traps) or they would introduce a significant heating effect tothe surrounding volume, likely destroying the analyte. The flow cell allows accurate fluidic delivery (bringing the particles to the trapping locations via pressure driven flow), and once the particles are in the vicinity of the silicon nitride traps, the light is guided through the traps which results in the particles being captured (figure 1).

Figure 1 Trapping of nano particles

The technology has further advantages as the microfluidic flow-cell can have several waveguides each with an array of trapping sites allowing the chip to form a sample grid. This would allow for a chip to be aligned to the beam and present a 'grid' of trapped protein crystals that would also remain static during interogation. In operation, a slurry of crystals are flowed through the flow cell while the 1064 nm laser light is transmitted through the waveguides. The crystals flowing over the waveguides are then captured at the trapping sites. Given that the traps are very strong, the waveguide can continuously trap crystals while the flow is maintained, but both the flow and the trapping light can be modulated on command. If the laser light is switched off the examined crystals are released and the process can then be repeated.

In addition the system will allow stationary crystals to be exposed to different types of reagents during the measurement process.

2 THE CHALLENGE

This work sets out to develop the system to meet the new challenges of X-ray protein crystallography experiments to be carried out on both Synchrotrons and the next generation of X- ray light sources, X-FEL. Often the nano crystals are constituted of up to 98 % water and suspended in an aqueous solution further complicating the selection and trapping of samples for analysis. An additional challenge is a micro engineering/fabrication one. The standard chip architecture is designed for optical microscopy where illumination and analysis measurements are carried out from one side and the system in its standard form takes place in the horizontal plane. The former requires re-engineering to the macro engineering of the chip holder and the latter the more complex issue of redesign the microfabrication route to develop chips that will cause minimal attenuation of the X-ray beam.

3 MEETING THE CHALLENGE

3.1 Trapping Experiments

The experiments required to determine whether this technique is suitable to trapping protein cryals was carried out using the standard Optofluidic chip. Figure 2 shows the microfluidic chip in its holder and micrographs of the chip architecture

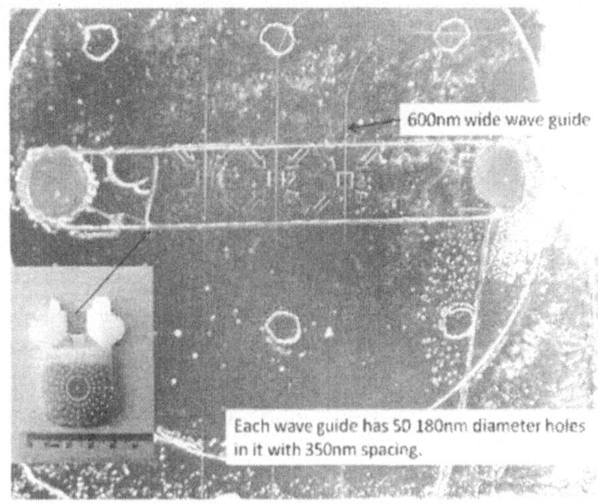

Figure 2 Standard chip architecture

Trials were conducted with 400 nm beads to clarify whether submicron specimens could be consistently trapped, and then further studies were undertaken with one μm protein crystals to clarify whether such structures could be trapped and would not perish due to the laser light being used to trap them. Figure 3 shows three 400 nm nanobeads trapped on chip. The image is taken from live footage taken using a standard upright microscope stage and a stemmer IDE UI LE Camera.

Figure 3 Trapped 400 nm beads

3.2 On Chip Modifications

Figure 4 shows how the system was configured to trap 1 μm Protein crystals. This size crystal was chosen for ease of obtaining confirmation they had been trapped without attaching a flourescent tag. Three trapped crystals can be seen in Figure 4.

Figure 4 1 μm protein crystals trapped on the wave guide.

For clarity figure 5 shows a micrograph of the waveguide when it was being processed. The dark circles are the 'holes' in the wave guide where the crystals are trapped.

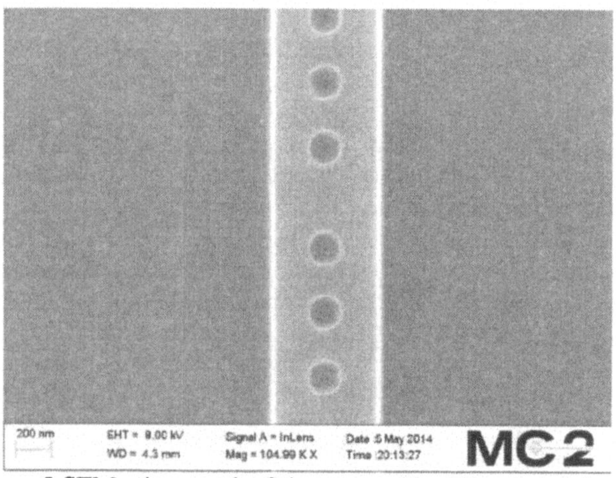

Figure 5 SEM micrograph of the waveguide. (Please note it is in the vertical orientation)

So from both a biological and size perspective the system is more than capable of meeting the demands of its new challenge. A standard chip has 320 μm of silicon below the waveguide which from standard calculations [3] would attenuate 80 % of the beam only allowing 20% to be used to obtain diffraction patterns from the sample. Predictive software was used to ensure that back-etching of this additional silicon would reduce this attenuation to an acceptable level. The new proposed chip architecture can be seen in figure 6.

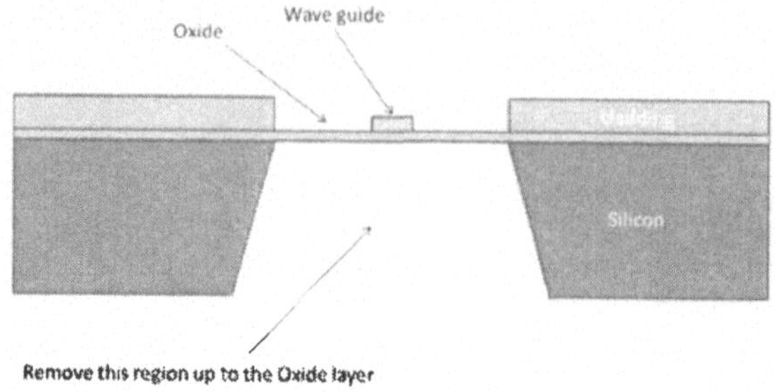

Figure 6 Proposed new chip architecture

A first run of processing the modified chips utilized wet etching as an additional stage to remove the bulk silicon below the waveguide. This approach however yielded only a small number of functional chips. The chips did however facilitate X ray tests to see if the new design had adequately reduced attenuation. The modified chip can be seen in the figure 7.

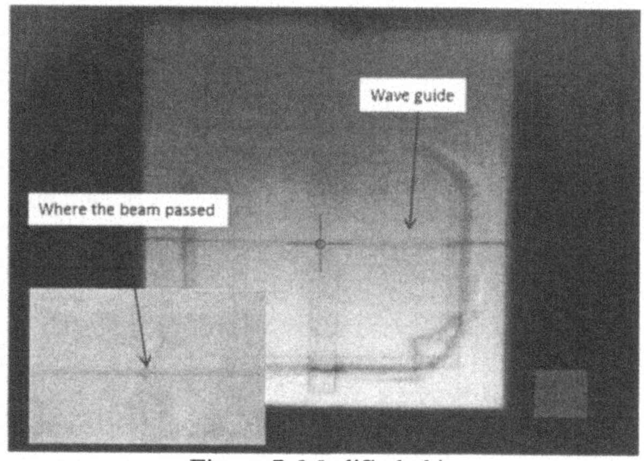

Figure 7 Modified chip

The chip was tested using 12.8 keV X-rays and proved the chip facilitated transmission of X-rays >99.9 %. It also caused negligible scatter comparable to that the air presents between the sample and the detector.

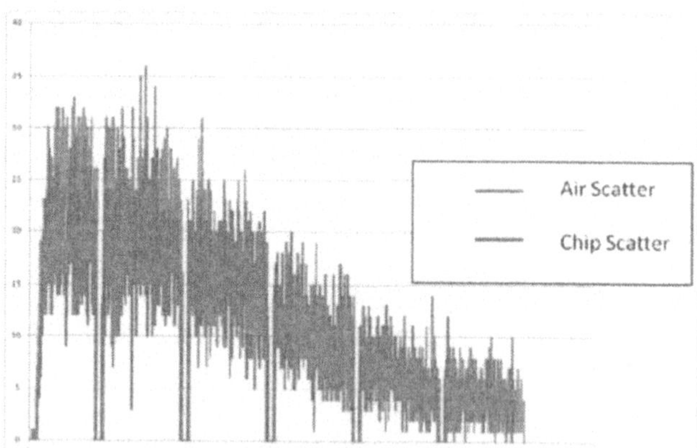

Figure 8 Scatter comparable with air between chip and detector

3.3 Hardware Modifications

The majority of the hardware is already fully compatible for X-ray diffraction experiments, the only component requiring modification is the chip holder. This component is responsible for making the fluid tight seal between the chip and the sample see figure 9.

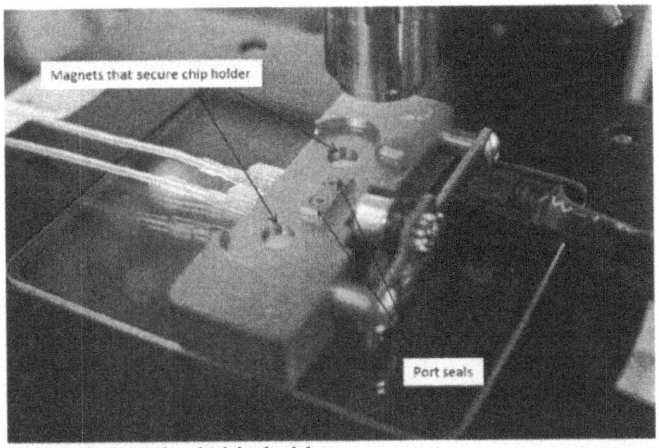

Figure 9 Standard chip holder

This chip holder requires a hole to be machined through the aluminuim between the two port seals as the X ray diffraction experiments are carried out in transmission. The beam will pass through the the hole, difract through the trapped crystals and onto the diffraction detector.

In service, laser light will pass throught the wave guides and a slurry of crystal will pass over the waveguides and will be attracted to the array of traps. The chip can then be indexed and patterns may be determined from each crystal in turn. The laser will then be turned off and the spent crystals sent to waste and the fresh analyte containing the crystals will re-enter the chip and the process repeated.

4 FUTURE CHALLENGES

The key driving challenge is to develop a completely dry etch process route in to improve yield and ensure it integrates with the modified chip holder. The last process run used wet etching to hollow the chip below the waveguides,moving forwards the next batch will employ deep reactive ion etching (DRIE) to remove the excess silicon. This approach is more expensive but more controlled. There is also no requirement to freeze dry the devices to negate issues from surface tensions caused by using wet etching techniques

A more 'pure science' challenge will investigate using manipulation of the laser light to orientate the crystals that are trapped in different or known orientations. To get the full structure of a protein the sample will need to be imaged in different orientations to obtain its full structure. By polarizing the light at the laser traps it is hypothesized that the crystals could be trapped and rotated with careful control of the laser light.

5 SUMMARY AND CONCLUSIONS

This paper details experiments that have been carried out which prove that 'typical' protein crystals can be trapped, with potential for trapping sub-micron crystals. A modified chip architecture has been developed specifically for use with a synchrotron X-ray beam. The chip is manufactured using standard semi-conductor technology which has been revised to produce a chip with a transmission of 99.98 % of the beam at energies between 6.4 - 20.0 KeV whilst maintaining the structural integrity of the flow cell part of the device. Standard chips were only capable of transmitting 20%.

This technology when fully developed for X-ray protein Crystalography experiments will truly be a game changing methodology for the nano protein crystallography community. Allowing these crystals to be held in position whilst they are imaged and allow for automation by default.

When Synchrotron and X-FEL beam time can take users two years to obtain an 8 hour slot, the ability to obtain as much high-quality data as possible is of paramount importance.

REFERENCES

[1] Cipriani, F. *et al.* CrystalDirect: a new method for automated crystal harvesting based on laser-induced photoablation of thin films. *Acta crystallographica. Section D, Biological crystallography* **68**, 1393-1399, doi:10.1107/s0907444912031459 (2012).

[2] Spence, J. C. H., Weierstall, U. & Chapman, H. N Xray lasers for structural and dynamic biology. *Reports on Progress in Physics* **75**, doi:10.1088/0034-4885/75/10/102601 (2012)

Dr Peter Docker The Diamond light Source
Peter.Docker@Diamond.ac.uk

Design, development and testing of apparatus to study gas diffusion in nanochannels

Stamatina Karakitsiou, Sabrina Daniela Eder, Bodil Holst and Alex Christian Hoffmann

Department of Physics and Technology, University of Bergen, Allégaten 55, Bergen, Norway

ABSTRACT

A device focusing on measuring flow in the molecular flow regime through nanochannels is presented. The apparatus is based on pressure driven flow, which is important for microelectromechanical (MEMS), nanomechanical -electrical systems (NEMS) and a wide range of other applications. The device is tested in the slip flow regime, but also flow in the Knudsen regime in parallel nanochannels can be measured. Knudsen numbers at the inlet between 0.08 and 0.23 are tested experimentally in this paper and is compared with theory. In addition, the functionality for Kn >10 is also discussed, based on theoretical work. The setup had been tested for a channel with $362 \pm 5\,\mu$m diameter and a length of 2 mm validating its ability to measure a wide range of flow regimes in different kind of samples. Argon gas is used for measuring the flow through the microchannel.

Keywords: Knudsen regime, slip regime, nanofluidics, microfluidics, experimental apparatus, diffusion.

1 Introduction

Flow regimes are defined on basis of the Knudsen number:

$$\mathrm{Kn} = \frac{\lambda}{D}. \qquad (1)$$

In this work the experimental Knudsen numbers are between 0.08 and 0.23, often named the "transition" flow regime, but also the molecular flow regime for Kn >10 is briefly discussed. This study was motivated from the wide range of applications where these regimes are applicable. Some of the main applications where slip and molecular flow regimes are important are mentioned below.

Chemical reactors for rarefied gases exhibit effects on catalytic reaction in micro scale of both temperature and concentration slip. The increase of the surface to volume ratio for micro-scale combustion makes for higher thermal efficiency. That is an advantage for power generation, chemical sensing and fuel conversion [1].

In solid oxide fuel cells (SOFCs) both Knudsen and bulk diffusion mechanisms are taken into account in anode diffusivity models. The characterization of anode parameters including pore size, porosity and diffusivity are critical when modelling fuel cells incorporating micro and nano-structured electrode materials, and Knudsen diffusion is important for the gas transport meaning that the total effective diffusivity in fuel cells is affected by the electrode pore size. In fact, in many micro- and nano-porous materials, the effective diffusivity is determined mainly by Knudsen diffusion [2].

Another application is shale. Pore pressure in shale gas reservoirs is crucial. Such reservoirs have a large fraction of micro and nano pores, with the majority nano sized. The transport of gas through the pores and the permeability is an important research topic in the oil and gas industry, especially during the last decade. Shale gas reservoirs have a very low permeability but they are characterised by an organic-rich deposition and they have the tendency to create clusters of mineral-filled natural fractures. Knudsen and slip regime must be considered for modelling the diffusion through shale gas reservoirs [3].

Materials like Hydrotalcites (HT) and activated carbon (AC), have interesting properties. These materials have been attractive sources of green engineering materials for solving environmental problems originating from industrial wastewater and gas pollution. HT and AC have been used as membranes for separating specific gases from gas mixtures. The effectiveness of membrane separation processes is measured by the values of permeability and separability factor. The latter is affected by a number of parameters like porosity, gas molecular characteristics surface area etc. Pressure and diffusion volumes are also part of the operational parameters, making flow regimes an important parameter in the process [4].

Carbon separation membranes for gases is also a potential application. Nanosized porous carbon membranes can be used in hydrogen recovery from raw coal gasification gas streams which can be used in coal liquefaction plants. These membranes have nano sized pores in the range from 1 to 10 nm. Gases are separated by Knudsen diffusion through the carbon membranes which have been modified with a layer of organic polymeric precursor on the microfiltration layer obtained by pyrolysis. The gas separation factors depend on several parameters including the transmembrane pressure drop [5].

Finally, In the last several years MEMS and NEMS devices have become very popular due to the low cost and high efficiency. These devices can be micro and nano sensors -actuators -reactors etc. The understanding of the flow through micro a nano channels and the prediction of the flow rates theoretically and experimentally is a challenge [6, 7].

Motivated by the above, this work focuses on an experimental method for measuring gas diffusion through micro and nano channels. Here an apparatus is introduced where the pressure-driven transport of a gas can be measured for a wide range of flow regimes, including slip and Knudsen.

2 Theoretical work for the design of the apparatus

Figure 1 shows the principle. A turbopump pumping at a volumetric rate of 66 l/s irrespective of its inlet or outlet pressure is attached to the receiver chamber. Once the molecular flow through the channel and the turbopump are equal, equilibrium is established in the receiver chamber and the pressure remains constant at p_{eq}.

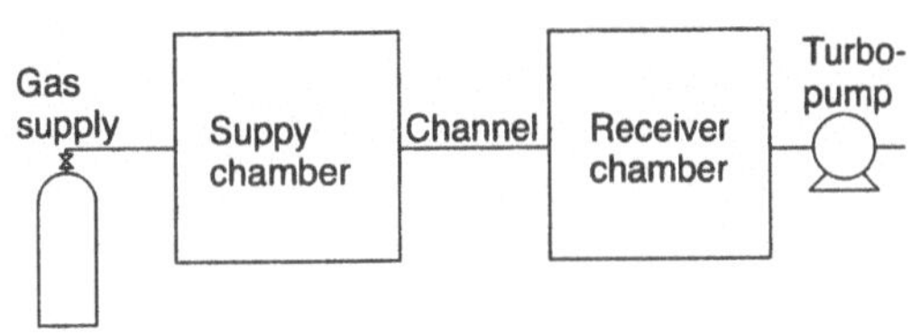

Figure 1: Principle sketch

As mentioned, the main focus here is on flow in the intermediate regime. Two different theoretical approaches to finding the flow through the channel were used for design purposes.

- The first is to apply the "Knudsen equation" [8, 9], which is, also in practice [10], applied to the entire range of Knudsen numbers from Poiseuille to molecular flow, considered by Roth [9] to be $0.009 < \mathrm{Kn} < 1$.

- The second is to apply a Maxwell slip condition at the wall and solve the resulting equation to give the flow rate (and pressure distribution, this, together with other details, will be published separately) through the channel.

Using the first method, the *conductance*, $C \equiv q_{pV}/\Delta p$ with q_{pV} the "pV flow", $p_{pV} \equiv d(pV)/dt = (dn/dt)\mathbf{R}T$ with dn/dt the molar flow and having used the ideal gas equation with the symbols having their usual meaning, was calculated from the Knudsen equation:

$$C = \frac{\pi}{128\mu}\frac{D^4}{L}\overline{p} + \frac{\frac{1}{6}\left(\frac{2\pi\mathbf{R}T}{M}\right)^{1/2}\frac{D^3}{L}\left[1+\left(\frac{M}{\mathbf{R}T}\right)^{1/2}\frac{D\overline{p}}{\mu}\right]}{\left[1+1.24\left(\frac{M}{\mathbf{R}T}\right)^{1/2}\frac{D\overline{p}}{\mu}\right]} \tag{2}$$

with M and μ the molar mass and viscosity of the gas, $\overline{p} = (p_0 + p_L)/2$ with p_0 and p_L the inlet and outlet pressures and $D = 2R$ and L the channel diameter and length.

Then the equilibrium pressure p_{eq} was computed from:

$$p_{eq} = \frac{C}{S+C}p_0 \tag{3}$$

where S is the volumetric flowrate of the turbopump and p_0 is the pressure in the supply chamber [9].

Using the second method, the theory of Maxwell slip [11], which, according to most literature, is applicable in the range $10 \times 10^{-3} < \mathrm{Kn} < 0.1$, was applied as an approximation for design purposes. The final result for the mass flow is:

$$w_s = \frac{\pi(p_0 - p_L)R^4}{8\mu L}\overline{\rho}\left(1 + \frac{4\zeta_0}{R\overline{p}}\right). \tag{4}$$

where $\overline{\rho}$ is the average gas density in the channel. $\zeta_0 = \zeta/p$, and $\zeta = \frac{2-\sigma_V}{\sigma_V}\lambda$ where σ_V is the fraction of molecules impacting on the wall that are reflected diffusively (in an arbitrary direction), found to be about 0.8 in practice [12]. λ is the mean free path in the gas, see below.

Also here the equilibrium pressure, p_{eq} in the receiver is found as the pressure at which the molar flow in through the channel equals that out through the pump (in practice allowing for leakages):

$$p_{eq} = \frac{w_s\mathbf{R}T}{MS} \tag{5}$$

The equilibrium pressure has been theoretically calculated using both the above methods for different starting pressures. The same equilibrium pressures have been determined also experimentally and the results are presented below.

Argon is used as an inactive gas for measuring the flow through the microchannel the mean free path for Argon has been found as:

$$\lambda = \frac{K_B T}{\sqrt{2}\pi d^2 p_0} \tag{6}$$

K_B is the Boltzmann constant, T is the temperature, d is the collisional diameter of the gas molecules (Ar) and p_0 is the pressure in the inlet. All experiments have been made in ambient temperature considering the flow as isothermal.

As the equations show the inlet pressure p_0 determines not only the equilibrium pressure but also the discrimination of the flow regimes (Eq. 1).

3 Introducing the experimental setup

The setup as is shown in Fig. 2 consists of two vacuum chambers of 2 liters each. Pressure gauges with different ranges are connected to each chamber. The system contains a pre vacuum line connected to both chambers and a turbopump, which pumps with a rate of 66 l/s, connected with the receiver chamber.

Before the experiment starts both chambers are pumped down to a pressure of 10^{-2} mbars. The receiver chamber is then further pumped down to a pressure of 10^{-8} mbars. Argon is allowed to slowly flow from the Argon cylinder, controlled by a pressure regulator and a manual valve. The channel is situated in between the two chambers and the equilibrium pressure is recorded by the gauge at the receiver chamber.

The channel used for this experiment is shown in Fig. 2. This is in a blank copper gasket appropriate for vacuum sealing. The channel had been made with a drill of 350 μm with a $\pm$ 20 μm divergence as shown in the photo taken in a stereo microscope.

Figure 2: Experimental apparatus for pressure-driven channel flow.

3.1 Experimental results and discussion

Table 1 shows the experimental results for the equilibrium pressure in the receiver chamber. Argon was released from the argon cylinder so as to reach different inlet pressures, p_0. The pressures in the supply chamber were recorded by a capacitance diaphragm gauge calibrated for argon[10] and in the receiver chamber by a hot cathode ionization gauge.

At the pressures used the Knudsen numbers, Eq. (1) are such that the flow from the inlet through most of the channel is in the "transition" regime. Right at the end of the channel the flow regime becomes molecular flow, but this short section does not affect the flowrate much.

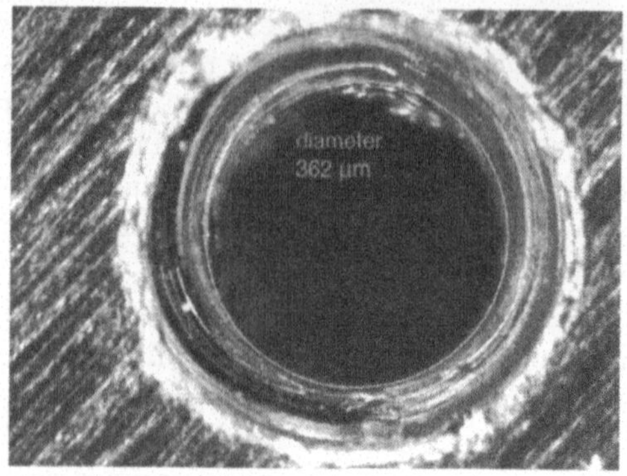

Figure 3: Blank flange with a channel of 362 μm diameter and 2 mm length (the thickness of the gasket). Sample for slip regime. Image from a stereo microscope with 7× magnification

The experimental results show very good agreement with both theories, specifically, the predicted values based on Maxwell slip, which was used as an approximation for design purposes, agree with experiment even though the flow is in the transition regime.

Figure 4 shows how the receiver pressure varies with time, as it moves toward the equilibrium pressure, and how the pressure varies along the channel, with a supply pressure of 1.5 mbar, according to Maxwell slip theory.

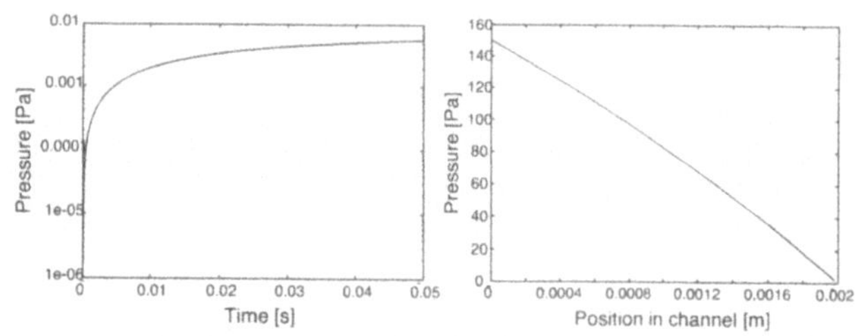

Figure 4: Left: Receiver pressure as a function of time (eqm. press.: 6.9×10^{-5} mbar). Right: pressure profile in the channel. Calculations based on Maxwell slip.

4 Molecular flow regime

Taking into consideration the theory for molecular flow, channel configurations have been designed so that the same setup can be used for measuring the Knudsen flow through nanochannels. The expression for molecular flow in terms of molecules per second through a capillary tube of radius R is:

$$G = \pi R^2 n u = -\frac{16R^3}{3m\tilde{u}}\frac{dp}{dx} \tag{7}$$

where n is the number density of gas molecules, u their velocity along the channel, m the mass of the molecules and $\tilde{u}$ the mean speed of molecules [13].

Table 1: Experimental and theoretical equilibrium pressures (in mbar) for slip regime.

Supply Press.	Experimental	Maxwell slip	Knudsen eqn.
1	2.6×10^{-5}	4.27×10^{-5}	4.04×10^{-5}
1.5	3.8×10^{-5}	6.90×10^{-5}	6.56×10^{-5}
2	5.8×10^{-5}	9.87×10^{-5}	9.41×10^{-5}
2.5	7.1×10^{-5}	1.32×10^{-4}	1.26×10^{-4}
3	1.05×10^{-4}	1.68×10^{-4}	1.61×10^{-4}

In a similar way to what was outlined in the previous sections for one microchannel, the equilibrium pressure can also be measured in this setup in several nanochannels or in a bigger channel structure.

Taking into consideration the pressure range limitations imposed by the experimental equipment and predicting the equilibrium pressure from theory, it has been calculated that, in order to achieve a measurable equilibrium pressure for the molecular-regime flow through nano-sized square channels of width 100 nm and length 20 μm, 2000 planar channels are required. For this case, the equilibrium pressure according to (7) has been predicted to be 1.103×10^{-8} mbar for a 70 mbar pressure in the supply vessel.

5 Conclusions

In this work the functionality of an apparatus for the measurement of Knudsen flow though nano and microchannels has been tested experimentally in the transition regime. Both theories used agree to a satisfactory degree with the experimental results. The differences between theory and experiment can be caused by small changes in the diameter along the channel, small changes of the ambient temperature and, of course, the error of the pressure gauges. Flow in the molecular regime has also been analysed theoretically to design a channel structure that can be analysed in the setup. One important advantage of this system is the wide range of regimes that can be measured, made possible by the fact that it can generate and measure very low pressures. Further research will include measurements of gas transport through nanochannels and various minerals in order to achieve a more clear understanding of several technologically important applications, particularly in the energy industries.

REFERENCES

[1] Bo Xu and Yiguang Ju. Concentration slip and its impact on heterogeneous combustion in a micro scale chemical reactor. Chemical Engineering Science, 60, 3561–3572, 2005.

[2] Luhan Ye, Weiqiang Lv, Yiwu Mao, Pengfei Yan, and Weidong He. An electrochemical device for the knudsen and bulk diffusivity measurement in the anodes of solid oxide fuel cells. International Journal of Hydrogen Energy, 39, 15057–15062, 2014.

[3] Chaohua Guo, Jianchun Xu, Keliu Wu, Mingzhen Wei, and Songyuan Liu. Study on gas flow through nano pores of shale gas reservoirs. Fuel, 143, 107–117, 2015.

[4] M.R. Othman. Permeability and separability of methane and carbon dioxide across meso-porous mgal hydrotalcite and activated carbon media. Chemical Engineering Science, 64, 925–929, 2009.

[5] A. S. Damle, S. K. Gangwal, and V. K. Venkataraman. Carbon membranes for gas separation: Developmental studies. Gas Separation & Purification, 8, 137–147, 1994.

[6] N. Dongari, A. Sharma, and F. Durst. Pressure-driven diffusive gas flows in micro-channels: from the knudsen to the continuum regimes. Microfluidics and Nanofluidics, 6, 679–692, 2009.

[7] I. Bargatin, E. B. Myers, J. S. Aldridge, C. Marcoux, P. Brianceau, L. Duraffourg, E. Colinet, S. Hentz, P. Andreucci, and M. L. Roukes. Large-scale integration of nanoelectromechanical systems for gas sensing applications. Nano Letters, 12, 1269–1274, 2012.

[8] M. Knudsen. Die gesetze der molekularstömung und der inneren reibungsströmung der gase durch röhen. Annalen der Physik, 333, 75–130, 1909.

[9] A. Roth. Vacuum Technology. Elsevier Science B.V., Netherlands, 1990.

[10] W. Umrath, editor. Fundamentals of Vacuum Technology. Oerlikon Leybold Vacuum, Cologne, 2007.

[11] R. B. Bird, W. E. Steward, and E. N. Lightfoot. Transport phenomena. John Wiley & Sons, New York, 2002.

[12] S. Roy, R. Raju, H. F. Chuang, B. A. Cruden, and M. Meyyappan. Modeling gas flow through microchannels and nanopores. Journal of Applied Physics, 93, 4870–4879, 2003.

[13] W. G. Pollard and R. D. Present. On gaseous self-diffusion in long capillary tubes. Phys. Rev., 73, 762–774, Apr 1948.

On the dynamic contact angle in spontaneous capillary flow

J. Berthier, D. Gosselin, G. Delapierre

University Grenoble Alpes, F-38000, Grenoble
Department of Biotechnology, CEA, LETI, Campus MINATEC, 38054, Grenoble, France
Corresponding author: jean.berthier@cea.fr

ABSTRACT

Capillary spontaneous flow is an interesting solution to move fluids either in microgravity conditions, i.e. in space, in microfluidic systems. In both cases, gravity is negligible and capillarity is the dominant force.

Spontaneous capillary flow onset as well as the dynamics of capillary flows have been recently documented in the literature for channels of different shapes, confined or open. However, the role of the dynamic contact angle is still unclear. This work aims at bringing a new light on the effect of the dynamic contact angle in the dynamics of capillary flows.

Keywords: spontaneous capillary flow (SCF), capillarics, capillary velocity, dynamic and static contact angles.

1 INTRODUCTION

Capillary actuation of fluids is an interesting solution when the gravitational forces are small: this is the case in space , or in microfluidic systems. In both cases, the ratio between gravity and capillary forces characterized by the Bond number is very small.

The onset of spontaneous capillary flow (SCF) and its dynamics have been widely investigated, first in the years 1920s [1-3], then more recently with the development of microfluidic systems for biotechnology [4-9].

Most of the time, the interpretation of the dynamics of the flow is performed using a constant contact angle. Although the study of the dynamic contact angle on the dynamics of wetting has been the subject of many investigations [10-16], the effect of the dynamic contact angle during a sponaneous capillary flow is still unclear.

In this work, we propose a correction to the capillary velocity that takes into account the dynamic contact angle, based on correlations reported in the literature. Comparison between static and dynamic velocities are presented.. It is concluded that the dynamic contact angle has an importance only on the few first millimeters of the channel. The length scale of the device is then of utmost importance. A non-dimensional number characterizing the effect of the dynamic contact angle is proposed

2 THEORETICAL APPROACH

Let us consider a closed channel of uniform cross section and arbitrary shape (figure 1).

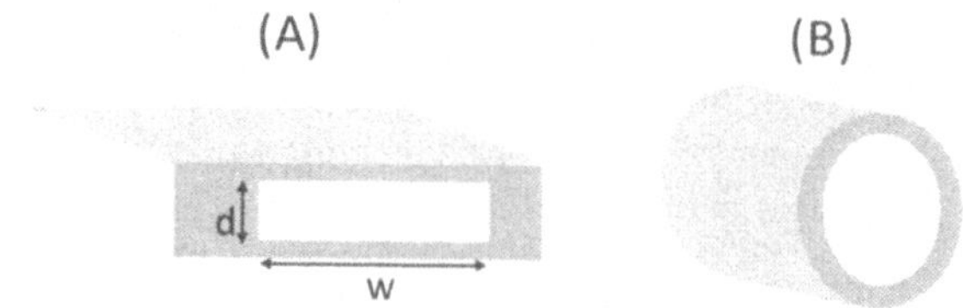

Fig.1. Sketch of a rectangular (A) and cylindrical channel (B).

It has been shown that the capillary force writes [6,7]

$$F_{cap} = \gamma \, p_W \cos\theta , \qquad (1)$$

where γ is the surface tension between the liquid and air, and p_W and θ are respectively the wetted perimeter and the dynamic contact angles in a cross-section of the channel

From a dynamic standpoint, the velocity of the capillary flow can be determined using a balance between capillary forces and friction with the wall [1-3]. The friction force is

$$F_{drag} = \int_S \tau \, ds = \left(\int_\Gamma \tau \, dl \right) z(t) = \bar{\tau} \, p_W \, z(t) \qquad (2)$$

where τ is the local wall friction, S the wetted surface between the origin and the front end of the liquid flow, Γ the wetted contour of the cross-section, $\bar{\tau}$ the averaged wall friction in a cross-section, i.e. $\bar{\tau}=(1/p_W)\int_\Gamma \tau\, dl$, and z the distance of the interface from inlet which depends on the time t.

The force balance on the fluid flow is then

$$m\frac{dV}{dt}=F_{cap}-F_{drag}\,, \qquad (3)$$

where m is the mass of the fluid in the channel and V the average velocity. The mass of fluid being proportional to the penetration distance, (3) can be written under the form

$$\rho\, z(t) S_c \frac{dV}{dt}=\gamma\, p_W \cos\theta-\bar{\tau}\, p_W\, z(t)\,, \qquad (4)$$

where S_c is the cross-section area and ρ the volumic mass of the fluid. The Reynolds number of the fluid being small, the flow is laminar. For a Newtonian fluid, the friction τ then depends on the geometry of the channel and on the average velocity V. Locally, the wall fiction is

$$\tau=\mu\frac{\partial v}{\partial n}=\frac{\mu V}{\lambda} \qquad (5)$$

where λ is a local friction length and n the coordinates perpendicular to the wall (figure 2).

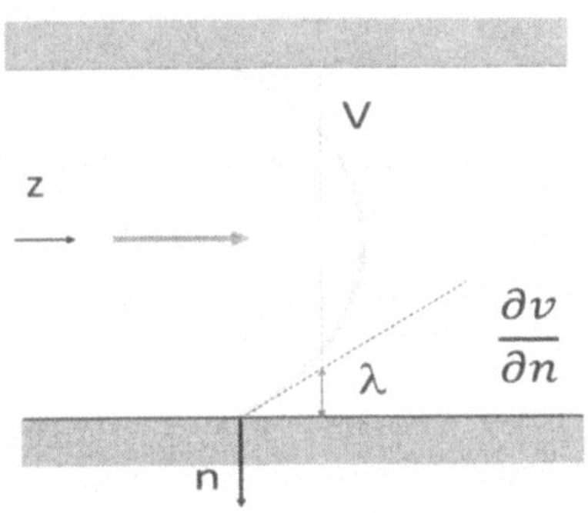

Fig 2. Definition of the friction factor from the velocity profile.

Conceptually, the friction can be averaged in a whole cross-section

$$\bar{\tau}=\frac{1}{p_W}\int_\Gamma \tau\, dl=\frac{1}{p_W}\int_\Gamma \frac{\mu V}{\lambda}dl=\frac{\mu V}{\bar{\lambda}} \quad , \qquad (6)$$

where $\frac{1}{\bar{\lambda}}=\frac{1}{p_W}\int_\Gamma \frac{1}{\lambda}dl$ is by definition the average friction length. Note that the derivation of (6) assumes a constant value of the viscosity, which is the case of Newtonian fluids. The case of non-Newtonian fluids is more complex and will be the subject of further work. Using the instantaneous relation

$$V=\frac{dz}{dt}\quad , \qquad (7)$$

Substitution of (6) and (7) in (4) yields the differential equation

$$\rho z S_c\frac{d^2z}{dt^2}=\gamma\, p_W\cos\theta-\mu\frac{p_W}{2\bar{\lambda}}\frac{dz^2}{dt}\,. \qquad (8)$$

It can be shown that, most of the time in capillary microsystems, inertia can be neglected because the Reynolds number is small, i.e.

$$\text{Re}=\frac{V w}{\nu}\leq O(1)\,, \qquad (9)$$

where w is a characteristic dimension of the channel, ν the kinematic viscosity, and O(1) means "order of 1". Instead of numerically integrating (10), a closed form expression for the travel distance z can be obtained by neglecting the inertial term on the left hand side of (8). Hence, we can write

$$\frac{dz^2}{dt}=\frac{2\gamma\bar{\lambda}}{\mu}\cos\theta\,. \qquad (10)$$

The solution of (10) is

$$z=\sqrt{\frac{\gamma}{\mu}}\sqrt{\cos\theta}\,\sqrt{2\bar{\lambda}t}\,. \qquad (11)$$

The travel distance varies as the square root of the time, in agrement with the Lucas-Washburn-Rideal (LWR) law for capillary flows inside cylinders [1-3]. The liquid velocity can be readily derived from (11)

$$V=\sqrt{\frac{\gamma}{\mu}}\sqrt{\cos\theta}\sqrt{\frac{\bar{\lambda}}{2t}}\,. \qquad (12)$$

Under this form, the fluid velocity is the product of the square root of a "physical" velocity $\sqrt{\gamma/\mu}$—related to the physical properties of the materials—by the square root of a "geometrical" velocity $\sqrt{\bar{\lambda}/2t}$, and by the cosine of the dynamic contact angle. Note that the friction length $\bar{\lambda}$ is purely a geometrical data (in the case of Newtonian fluids). The value of $\bar{\lambda}$ can found in numerous published tables [8].

On the other hand, because $dz^2/dt = 2zV$ relation (10) immediately produces a relation between the velocity and the travel distance

$$V = \frac{\gamma \bar{\lambda}}{\mu} \cos\theta \frac{1}{z}. \quad (13)$$

3 DYNAMIC CONTACT ANGLE

It has been shown that the dynamic contact angle depends on the capillary number [13-16]. In our case, the capillary number for the flow can be derived from relation (13). Let us recall that the capillary number is the ratio of the viscous forces to the surface tension forces

$$Ca = \frac{\mu V}{\gamma}. \quad (14)$$

Substitution of (14) in (13) yields

$$Ca = \cos\theta \frac{\bar{\lambda}}{z}. \quad (15)$$

Relation (15) determines the capillary number—in the case of SCF—as the product of the ratio $\bar{\lambda}/z$ by the cosine of the wetting angle. Relation (15) shows that the capillary number is proportional to the non-dimensional ratio $\bar{\lambda}/z$.

Let us now recall some empirical models for the estimation of the dynamic contact angle [12-16]. Noting θ_0 the static contact angle and θ the dynamic contact angle, Bracke and colleagues [13] proposed the formula

$$\cos\theta = \cos\theta_0 - 2(1+\cos\theta_0) Ca^{\frac{1}{2}} \quad (16)$$

Relation (16) shows that the dynamic contact angle θ is always larger than the static contact (Young) angle θ_0. The empirical correlation proposed by Seebergh and coworkers [14] is only slightly different.

If we substitute the expression (15) of the capillary number in relation (16), we find the expression of the dynamic contact angle

$$\cos\theta = \cos\theta_0 - 2(1+\cos\theta_0)\sqrt{\cos\theta}\sqrt{\frac{\bar{\lambda}}{z}}. \quad (17)$$

The cosine of the dynamic contact angle appears on both sides of (17) so that (17) is an implicit relation. Far from the channel inlet ($\bar{\lambda}/z$ <<1), we can substitute to (17) the explicit formula

$$\cos\theta = \cos\theta_0 - 2(1+\cos\theta_0)\sqrt{\cos\theta_0}\sqrt{\frac{\bar{\lambda}}{z}}. \quad (18)$$

Relations (17) and (18) are plotted in figure 3, for a flat rectangular channel (aspect ratio 7): the explicit formulation is legitimate except near the channel inlet (z=0). Note that, in the inlet region ($\bar{\lambda}/z$ >1), the expression of the contact angle is derived from (17) is

$$\cos\theta = \cos\theta_0 + 2(1+\cos\theta_0)^2 \frac{\bar{\lambda}}{z}\left[1 - \sqrt{\frac{\cos\theta_0}{(1+\cos\theta_0)^2}\frac{z}{\bar{\lambda}} + 1}\right] \quad (19)$$

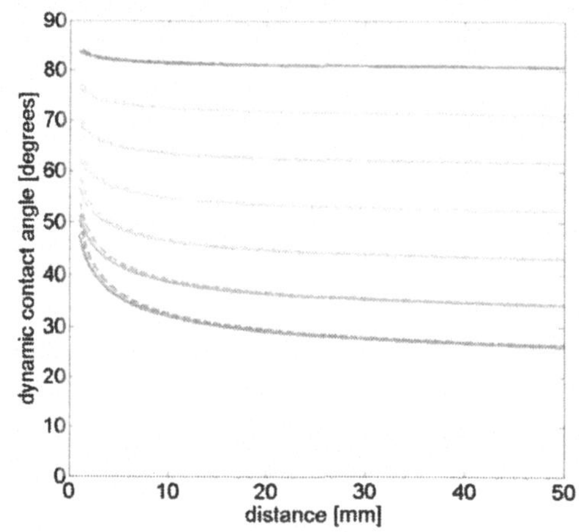

Fig.3. Variation of the dynamic contact angle as a function of the travel distance z, for seven values of the static contact angle {20, 30,40, 50, 60, 70, 80} degrees. The dotted lines correspond to the implicit solution (17) while the continuous lines correspond to the explicit relation (18).

Substituting back (18) in the velocity expression (13) we find the expression for the capillary velocity

$$V = \frac{\gamma}{\mu}\frac{\bar{\lambda}}{z}\left[\cos\theta_0 - 2(1+\cos\theta_0)\sqrt{\cos\theta_0}\sqrt{\frac{\bar{\lambda}}{z}}\right]. \quad (20)$$

Or, noting $\varepsilon = (V_{stat} - V)/V_{stat}$ the relative error due to the use of the static contact angle instead of the dynamic contact angle, where $V_{stat} = (\gamma/\mu)(\bar{\lambda}/z)\cos\theta_0$, we find

$$\varepsilon = \frac{V_{stat} - V}{V_{stat}} = 2\frac{(1+\cos\theta_0)}{\sqrt{\cos\theta_0}}\left(\frac{\bar{\lambda}}{z}\right)^{\frac{1}{2}}. \quad (21)$$

4 MODELING AND EXPERIMENTS

Relation (20) is plotted in figure 4 for $\bar{\lambda} = d / 6$—which corresponds to a flattened rectangle cross section with $w \gg d$ —and a static contact angle of 45°. The relative error is small after a few millimeters in the channel. It is however not negligible in the first few millimeters.

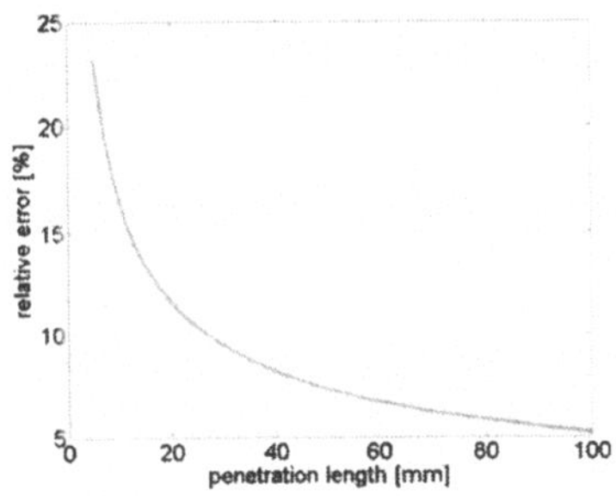

Fig.4. Relative error on the velocity ($\bar{\lambda} = d / 6$, θ=45°)

Let us now re-analyze the case of water flowing in a rectangular channel composed of COC walls and a thin plastic cover (aspect ratio 0.3), described in [8]. We find that the best fit with a static contact angle is 49°, while it is 47° using the dynamic formulation (figure 5).

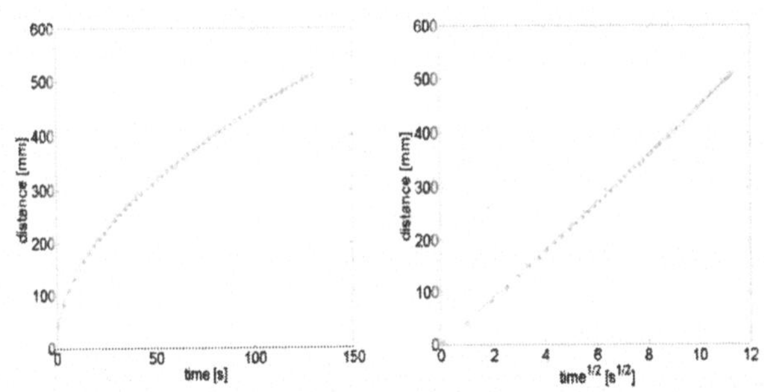

Fig.5. Comparison between experiments and model for the travel distance in a rectangular channel: left, travel distance vs. time; right, travel distance vs. square root of time. The green dots correspond to the experimental results, the yellow continuous line to the model with a static contact angle and the red crosses to the model with the dynamic contact angle.

5 CONCLUSION

The effect of dynamic contact angle is still controversial in the literature. Some papers report that the dynamic contact is not very different from the static contact angle [17,18], on the opposite some papers point out its importance [16]. In fact, all depends on the value of the capillary number.

In this work, we showed that the usual definition of the capillary number can be replaced by another one, based on the friction length, which we have shown to be $Ca = \left(\bar{\lambda}/z\right)\cos\theta$. This "channel capillary number" points out the importance of the ratio between the friction length and the channel length.

It is concluded that the effect of the dynamic contract angle strongly depends on the characteristic length of the channel.

REFERENCES

[1] R. Lucas, "Ueber das Zeitgesetz des Kapillaren Aufstiegs von Flussigkeiten, Kolloid Z., **23**, 15, 1918.

[2] E.W. Washburn, The dynamics of capillary flow. Phys. Rev., **17**, 273-283, 1921.

[3] E.K. Rideal. On the flow of liquids under capillary pressure. Philos. Mag. Ser. **6**(44):1152— 1159, 1922.

[4] R.R.Rye, F.G. Yost, Wetting Kinetics in Surface Capillary Grooves Langmuir, **12**, 4625-2627, 1996.

[5] M. Kitron-Belinkov, A. Marmur, T. Trabold, G.V. Dadheech, Groovy-drops: effect of groove curvature on spontaneous capillary flow, Langmuir, **23** (16), 8406-8410, 2007.

[6] F. F. Ouali, G. McHale, H. Javed, C. Trabi, N. J. Shirtcliffe, M. I. Newton, Wetting considerations in capillary rise and imbibition in closed square tubes and open rectangular cross-section channels, Microfluid Nanofluid., **15**, 309-326, 2013.

[7] J. Berthier, K. Brakke, E. Berthier, A general condition for spontaneous capillary flow in uniform cross-section microchannels, Microfluid Nanofluid, **16**:779–785, 2014.

[8] J. Berthier, D. Gosselin, E. Berthier, A generalization of the Lucas-Washburn-Rideal law to composite microchannels of arbitrary cross-section, Microfluid. Nanofluid., accepted, in print, 2015.

[9] M.M. Weislogel, Compound capillary rise, J. Fluid Mech., **709**, 622-64, 2012.

[10] V. Dussan, On the spreading of liquids on solid surfaces: static and dynamic contact lines, Ann. Rev. Fluid Mech., **11**, 371, 1979.

[11] Y.D. Shikhmurzaev, The moving contact line on a smooth solid surface, Int. J. Multiphase Flow, **19**, 584, 1993.

[12] T.D. Blake, Dynamics of contact angles and wetting; in J.C. Berg (Ed.), Wettability, Marcel Dekker, New-York, p. 251, 1989.

[13] M. Bracke, E De Voeght, and P. Joos, The kinetics of wetting: the dynamic contact angle, Progr Colloid Polym Sci **79**:142-149, 1989.

[14] J.E. Seebergh, J.C. Berg, Dynamic wetting in the low capillary number regime, Chem. Eng. Sc., **47** (17), 4455, 1992.

[15] T.S. Jiang, S.G. Oh, J.C. Slattery, Corrections for dynamic contact angle, J. Colloid Interface Sc., **69**, 74, 1979.

[16] S. van Mourik, A.E.P. Veldman, M.E. Dreter, Simulation of capillary flow with a dynamic contact angle, Microgravity sci. technol. XVII-3, 2005.

[17] A. Han, G. Mondin, N. G. Hegelbach, N. F. de Rooij, U. Staufer, Filling kinetics of liquids in nanochannels as narrow as 27 nm by capillary force, Journal of Colloid and Interface Science **293**, 151–157, 2006.

[18] V. D. Sobolev, N. V. Churaev, M. G. Velarde, and Z. M. Zorin, Surface tension and dynamic contact angle of water in thin quartz capillaries, Journal of Colloid and Interface Science **222**, 51–54, 2000.

Microfluidics for Multiplexing of Core Needle Biopsies

Wilfrido D. Mojica[1], Kwang Oh[2], Xiaozheng Xue[3] and Edward P. Furlani[2, 3]

1 Department of Pathology and Anatomical Sciences, School of Medicine and Biological Sciences, University at Buffalo, 100 High Street, Buffalo, NY Email: mojica@buffalo.edu
2 Dept. of Electrical Engineering, University at Buffalo SUNY, Buffalo, NY
3 Dept. of Chemical and Biological Engineering, University at Buffalo SUNY, Buffalo, NY

ABSTRACT

Minimally invasive biopsies are increasingly used to obtain tissue for diagnostic purposes. The small size of these specimens can limit mutiplexed (e.g., cytologic and molecular) diagnosis as the sample is often exhausted during a single stage of analyss, e.g., cytologic analysis, which precludes subsequent molecular testing. In this presentaion we introduce a novel mocrofluidics-on-a-slide approach that combines the concepts of cytology, microfluidics and magnetophoresis to enable multiplexing of diagnostic cells. This approach emphasizes cytology for diagnostic purposes and allows recovery of the same intact cells for downstream molecular studies. In contrast to traditional approaches that sequentially partition tissue for different studies, the proposed process allows for the interrogation of diagnostic cells by different modalities, thereby enabling extraction of a maximal amount of informative data from the cells.

Keywords: core needle biopsies, mutiplexed diagnosis of core needle biopsies, microfluidics-on-a-slide, lab-on-a-slide, cytology and molecular diagnosis of core needle biopsies.

1 INTRODUCTION

Core needle biopsies (CNB) have emerged as an attractive and viable alternative to open biopsies when tissue is needed for diagnostic purposes. In contrast to an open biopsy, a CNB is less invasive, more cost efficient and can be performed on an outpatient basis [1-4]. Using image guidance, a CNB can be used to sample almost any organ site for suspicious mass lesions. Once procured, the tissue from a CNB is typically fixed in formalin, embedded in paraffin wax, sectioned and placed on glass slides. The tissue is then stained with hematoxylin and eosin for microscopic examination. Ancillary testing can be added in the form of immuno-histochemistry for the classification and even sub-classification of tumor type, and molecular testing for the selection of targeted therapies. This is the standard order by which diagnostic material from CNB are traditionally partitioned. However, a common dilemma that has arisen with CNB is the depletion of the tissue before molecular testing can be performed. This is due to several factors, one being the diminutive size of the CNB. The tissue from a CNB typically comes from either an 18G or 20G needle, with the resultant tissue having dimensions of usually less than 1 cm in length and 1 mm in width, after tissue processing. Adding to the difficulty in making a diagnosis from these small pieces of tissue is the amount of representative tumor tissue in the CNB. The less the amount of tumor present in the CNB, the harder it is to make a diagnosis. Another issue with tissue loss is waste associated with microtome effacement. Finally, the most significant depletion of tissue occurs in ancillary testing by immunohistochemistry [5-7].

In this paper, a method is introduced that enables a multiplexed approach to the evaluation of a CNB. This involves the dissociation of cells from the tissue, then enrichment of targeted cells using immunomagnetic beads followed by temporary immobilization in a microfluidic platform that is integrated on a conventional laboratory slide (microfluidics-on-a-slide) as illustrated in **Fig. 1**. This step allows for the evaluation of the enriched cell population for cytologic diagnosis. Thereafter the same cells can be recovered and assays can be performed on them. This is in contrast to the conventional approach for CNB, where the tissue is serially partitioned for separate tests. The clinical diagnosis rendered from tissue processed in the propsed approach results in the combined assessment of the cytologic impression of the cells and their molecular findings.

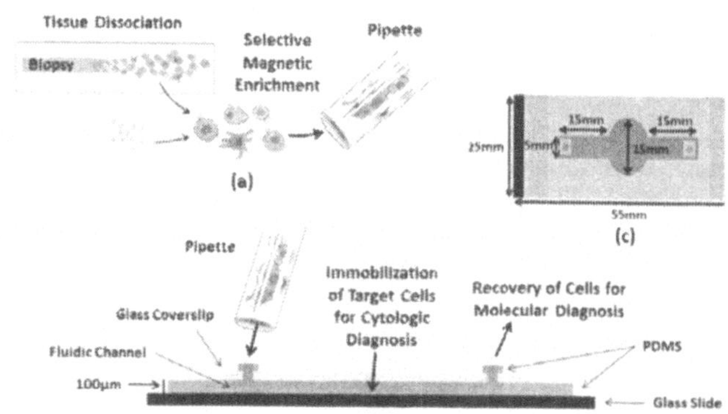

Figure 1. Diagramatic flow for processing of cytology specimens using lab-on-a-slide system: (a) processing of core needle biopsy and enrichment of target cells with immunomagnetic beads, (b) introduction of target cells via pippette into microfluidic system integrated on the top of a conventional glass slide, magnet postioned below the slide to immobilze target cells for cytologic diagnosis, (c) top viw of microfluidic system with dimensions.

2 MICROFLUIDICS-ON-A-SLIDE

The microfluidics-on-a-slide system is shown in **Fig. 2**. It was fabricated using a soft lithography technique [8-10]. To form the microchannel, a negative photoresist (MicroChem, SU-8 2050) was spin-coated on a silicon wafer substrate and patterned using a conventional photolithography technique.

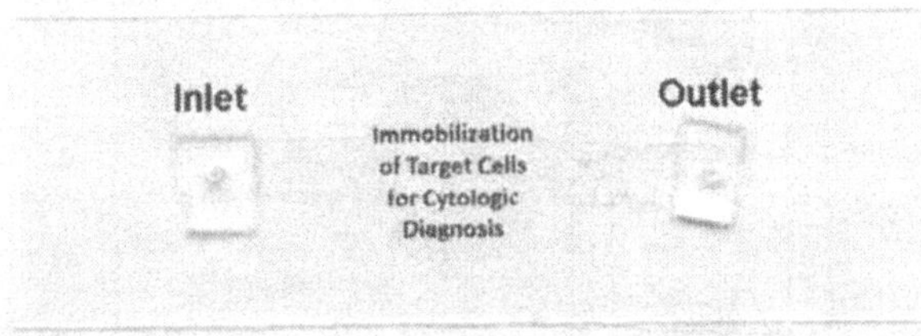

Figure 2. Fabricated microfluidc system on a glass slide.

The substrate was then silanized with hexamethyldisilazane in a vacuum chamber at room temperature for 30 minutes. This facilitated peeling the cured channel from the substrate. A base polymer of polydimethylsiloxane (PDMS) and a curing agent were thoroughly mixed at a 10:1 ratio (wt/wt). The mixture was then poured on to the substrate and a glass slide was used to push away the PDMS at a ceiling of the negative photoresist patterned with the use of a mechanical clamp [11]. After curing the PDMS was peeled off from the substrate and placed on a glass slide. The PDMS channel was then irreversibly bonded to a non-adhesive, non-charged, 25 x 75 x 1 mm glass slide (Leica Biosystems, Buffalo Grove, IL) by exposing oxygen plasma on the surface of the PDMS channel and cover glass. The channel as constructed measured 4.5 cm in length, 5 mm in width and 100 um in height. In the middle of the channel was a circular region designed for viewing that measured 1.5 cm in diameter (**Fig. 1c**). The circular design was intended to allow cells to spread out thereby preventing cells from covering and obscuring other cells. Finally, a 24 x 40 mm rectangular glass coverslip was bonded on to the top of the PDMS channel. The slightly shorter length of the glass coverslip relative to the microchannel left a small gap of 2.5 mm at both ends that could be used as inlet and outlet ports.

3 MAGNETIC ANALYSIS

The microfluidic channel was designed for placement on a small base from where a permanent magnet could be reversibly positioned just beneath the glass slide corresponding to the circular 1.5 cm viewing area (**Fig. 1 c**). A high strength (grade 42N) neodymium iron boron (NdFeB) magnet, which has a maximum remnant magnetization of B_r =1.28T, was used for the experiment. The magnet is 12.7 mm (0.5 in) in length and 6.35 mm (0.25 in) in both width and height. The magnet produces a nonuniform field distribution that induces an attractive force on magnetic particles within the microchannel. The magnet is made from grade 42N neodymium iron boron (NdFeB), which has a maximum remnant magnetization of B_r = 0.92

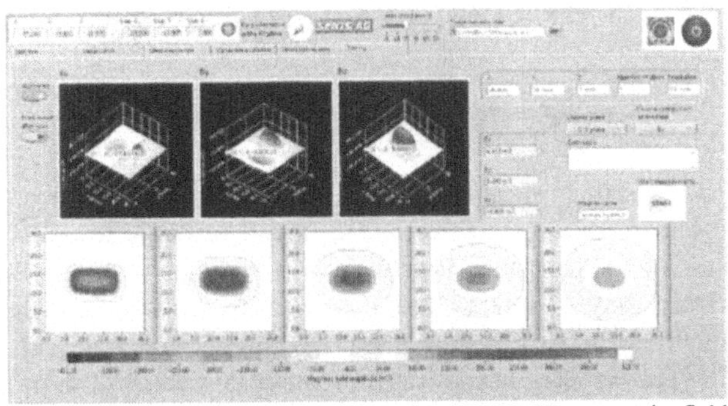

Figure 3. Measured field data from the SENIS 3D magnetic field mapping system. The three plots in the top panel are (from left to right) the spatial distribution of the field components B_x, B_y and B_z over a 26 mm × 26 mm area at a distance z = 0 mm above the magnet. The bottom panel (from left to right) contains plots of B_{total} over a 26 mm × 26 mm area at distance z = 0, 1, 2 ,3 and 4 mm, from left to right respectively.

T. The magnetic field was characterized using a 3D magnetic field mapping instrument, the MMS-1-R from SENIS GmbH (www.senis.ch) as shown in **Fig. 3**. The field distribution and force on the particles were also analysed using custom developed computational models developed using methods described by Xue and Furlani [12-15]. The force on the particles is predicted using an "effective" dipole moment approach in which a magnetized particle is replaced by an "equivalent" point dipole with a moment $\mathbf{m}_{p,eff}$. The force is given is given by

$$\mathbf{F}_m = \mu_f \left(\mathbf{m}_{p,eff} \bullet \nabla \right) \mathbf{H}_a \quad (1)$$

where μ_f is the permeability of the fluid and $\mathbf{H}_a$ is the applied magnetic field intensity at the centre of particle. The particles used in this experiment were the Dynabeads M-450 particles (4.5 µm in diameter) from CELLection™ Epithelial Enrich Dynabeads®. A surface plot of the predicted downward axial force on the particles at a distance z = 1 mm above the magnet (at the bottom of the channel) is shown in **Fig. 4**.

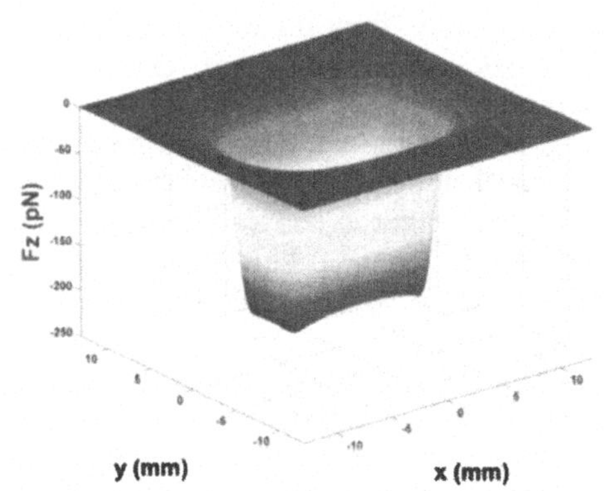

Figure 4. Surface plot of axial force F_z at z = 1 mm above the magnet.

4 MULTIPLEXED DIAGNOSIS

4.1 Sample Preparation

Thin strips of non-transformed colonic mucosa were excised from excess regions of segmental colonic resection specimens extirpated for medically indicated purposes. Colonic tissue was selected for testing based on the availability at our institution and the columnar shape of the cells. The strips of tissue were cut to approximately 2 mm in width and 1 cm in length and fixed in 25% ethanol for 2 hours. After two hours, the cells were dissociated from the tissue using the gentleMACS™ Dissociator (Miltenyl Biotec, San Diego, CA). The dispersed cell solution was filtered through Shandon™ Nylon Biopsy Bags (ThermoFischer Scientific, Waltham, MA) that allows single cells and small groups of cells to filter through, while retaining larger stromal tissue fragments. The filtered cells were then transferred to a 1.5 ml microcentrifuge tube and spun down using a table-top minifuge. The supernantant was discarded and replaced with a 1X PBS solution with 0.1% bovine serum albumin and 5 ul of resuspended CELLection™ Epithelial Enrich Dynabeads® (Life Technologies, Grand Island, NY) that had the ber-Ep4 antibody bound to them. The cells and immunomagnetic beads were incubated together at room temperature for 20 minutes with gentle mixing. The mixture was then placed in a Dynal MPC®-S magnetic particle separator for 5 minutes and the unbound cells decanted. The retained cells were stained with a 0.025% concentration of toluidine blue diluted from an originating 1% solution. The cells and stain were allowed to mix with gentle rotation. After 15 minutes, the microcentrifuge tube was spun down using the minifuge and the majority of the supernatant removed. A small amount of fluid was retained with the cells so as to be able to resupend them for eventual insertion into a microfluidic channel.

4.2 Cytologic Characteristics

Dissociation of cells from the slivers of colonic tissue resulted in single cells and small groups of cells. Immunomagnetic beads did not bind to every colonic epithelial cell. Some cells not bound to immunomagnetic beads were present in the platform immobilized with cells attached to immunomagnetic beads. Magnetophoresis was needed as a means to capture cells and immobilize them in the viewing area for cytologic examination. The small size of the immunomagnetic beads did not hinder the examination of the immobilized cells. Visual examination of the cells showed them to be almost completely columnar epithelial in nature, compatible with them being derived from the colon. The cells of interest retained their morphology as colonic epithelial cells, i.e., they remained columnar in shape with the nucleus oriented towards one pole. Dissociation from tissue, being bound to immunomagnetic beads and travelling through the microchannel did not damage the majority of cells. Staining using the supravital stain was adequate to enable evaluation of the cytologic features of the cells in the microfluidic system as shon in **Fig. 5**. There was no artifactual change in their appearance, e.g., increase in nuclear to cytoplasmic ratio or coarsening of nuclear chromatin, that would cause diagnostic confusion with malignancy.

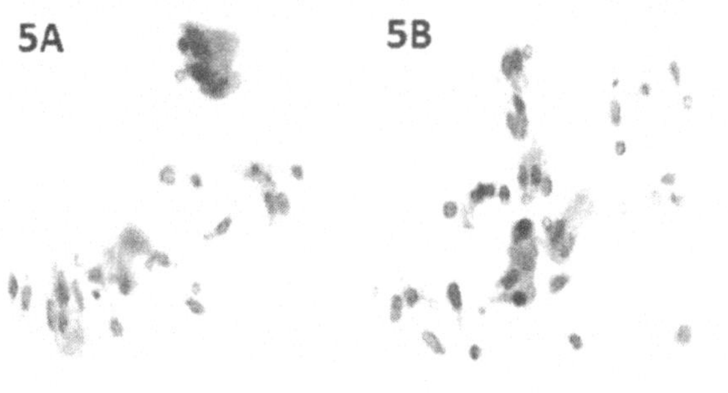

Figure 5. Two separate images of temporarily immobilized cells in the microfluidic platform's viewing stage. Cells are of colonic epithelial origin, and demonstrate preservation of columnar shape, nuclear polarity towards one pole and clear nuclear detail. Magnification 20X.

4.3 Nucleic Acid Integrity

DNA recovered from the cells was measured with A260/280 values of 1.84 and 1.87. The DNA recovered after having been fixed with ethanol and stained with the supravital dye toluidine blue could be amplified. Neither of these external factors inhibited or prevented the polymerase enzyme, which sometimes occurs with FFPE tissue. Long segments of nucleic acids measuring 1.3 and 8.5 kb could be amplified using protocols specific for long segment DNA (**Fig. 6**).

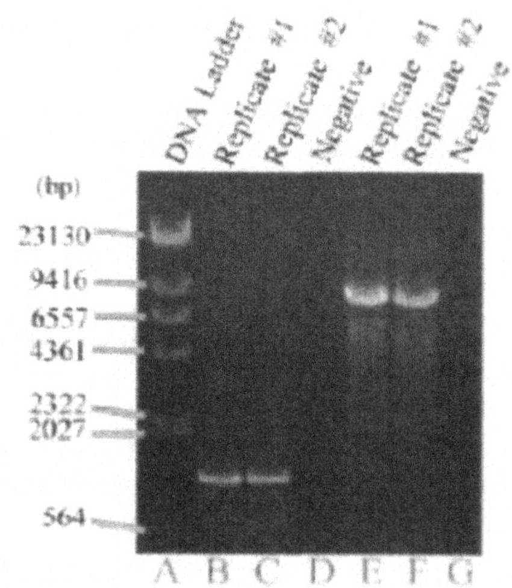

Figure 6. PCR on DNA from cells recovered from the microfluidic platform. Image demonstrates the ability to amplify long segments of DNA (1.3 kb in lanes B and C, and 8.5 kb in lanes E and F) from the recovered DNA, indicating the presence of high quality, intact, high molecular weight DNA.

5 CONCLUSION

We have introduced a mcirofluidics-on-a-slde approach for the multiplexed diagnostics of to core needle biopsies. Our results demonstrate the feasibility of using this approach for sequantial cytologic and molecular analysis. The proposed technique has several advantages relative to the conventional Formalin Fixed Paraffin Embedded (FFPE) processing approach currently practiced by virtually all pathology laboratories. Specifically, this microfluidic approach enables the multiplexing of different assays on the same cells, in contrast to the sequential partitioning of tissue sections from a CNB for different tests. Also, given a limited starting material, the ability to perform evaluations of different formats using the same diagnostic cells decreases the possibility that they are depleted before testing can be performed on them. Significantly, although we concentrated on evaluating only the elements of cytology and DNA, the recovery of intact cells would allow for the examination of other cellular components such as protein, RNA and metabolites.

ACKNOWLEDGEMENTS

The authors acknowledge financial support from the U.S. National Science Foundation, through award number CBET-1337860.

REFERENCES

[1] Yamagami, T., Ilda S., Kato, T., Tanaka, O., Kato, D., Nishimura, T. Usefulness of new automated cutting needle for tissue-core biopsy of lung nodules under CT fluoroscopic guidance. Chest. 124:147-154 (2003).

[2] Golub, R.M., Bennett, C.L., Stinson, T., Venta, L., Morrow, M. Cost minimization study of image guided core biopsy versus surgical excision biopsy for women with abnormal mammograms. J Clin Oncol. 22:2430-37 (2004).

[3] Williams, R.T., et al. Needle versus excisional biopsy for noninvasive and invasive breast cancer. Report from the national Cancer Data Base, 2003-2208. Ann Surg Oncol. 18:3802-3810 (2011).

[4] Gupta, S., et al. Quality improvement guidelines for percutaneous needle biopsy. J Vasc Interv Radiol. 21:969-975 (2010).

[5] Handorf, C.R., et al. A multicenter study directly comparing the diagnostic accuracy of expression profiling and immunohistochemistry for primary site identification in metastatic tumors. Am J Surg Pathol. 37:1067-1075 (2013).

[6] Thunnissen, E., et al. The challenge of NSCLC diagnosis and predictive analysis on small samples. Practical approach of a working group. Lung Cancer. 76:1-18 (2012)

[7] Travis, W.D., et al. Diagnosis of lung cancer in small biopsies and cytology: implications of the 2011 International Association for the Study of Lung Cancer/American Thoracic Society/European Respiratory Society classification. Arch Pathol Lab Med. 137:668-684 (2013).

[8] Lee, H., Lee, K., Ahn, B., Oh, K.W. A new fabrication process for SU-8 thick photoresist structures by simultaneously removing edge-bead and air bubbles. J Micromechanics Microengineering. 21:paper 125006 (2011).

[9] Lee, K., Kim, C., Ahn, B., Kang, J.Y., Oh, K.W. Hydrodynamically focused particle filtration using an island structure. Biochip J. 3:275-280 (2009).

[10] Ahn, B., Lee, K., Lee, H., Panchapakesan, R., Oh, K.W. parallel synchronization of two trains of droplets using a railroad-like channel network. Lab Chip. 11:3956-3962 (2011).

[11] Hsu, C.H., Chen, C., Folch, A. Microcanals for micropipette access to single cells in microfluidic environments. Lab Chip. 4:420-424 (2004).

[12] Furlani, E.P., Xue, X. Field, force and transport analysis for magnetic particle-based gene delivery. Microfluid. Nanofluid. 13(4), 589-602 (2012).

[13] Furlani, E.P., Xue, X. A model for predicting field-directed particle transport in the magnetofection process. Pharm. Res. 29(5), 1366-1379 (2012).

[14] Xue, X., Furlani, E.P. Template-assisted nano-patterning of magnetic core-shell particles in gradient fields. In Phys Chem Chem Phys. 16(26):13306-13317 (2014).

[15] Xue, X.; Furlani, E.P. Analysis of the Dynamics of Magnetic Core-shell Nanoparticles and Self-assembly of Crystalline Superstructures in Gradient Fields. The Journal of Physical Chemistry C. (2015).

Bottom-up Nanofabrication of Crystalline Structures of Magnetic Nanoparticles

Xiaozheng Xue[1] and Edward P. Furlani[1, 2]

1 Dept. of Chemical and Biological Engineering, University at Buffalo SUNY,
Email:xiaozhen@buffalo.edu
2 Dept. of Electrical Engineering, University at Buffalo SUNY,
Office: (716) 645-1194 Fax: (716) 645-3822, Email:efurlani@buffalo.edu

ABSTRACT

A bottom-up approach to the nanofabrication of three-dimensional (3D) crystalline structures of magnetic core-shell nanoparticles is presented. The approach is based on self-assembly and involves the use of soft-magnetic template elements to guide the assembly in the presence of a uniform bias field. The combination of a bias field and template-induced gradient fields provides localized regions of attractive and repulsive magnetic force that enable nanoscale control of particle placement during assembly. The method is demonstrated using a computational model that predicts particle dynamics during assembly and the final assembled structure. The analysis demonstrates that 3D crystalline superstructures can be assembled within milliseconds and that the feature resolution of the assembled structure can be tailored by controlling the process variables. This method of nanofabrication opens up opportunities for the scalable fabrication of nanostructured materials with unprecedented properties for a broad range of applications.

Keywords: Magnetic field-directed self-assembly of core-shell nanoparticle structures, template-assisted self-assembly of magnetic core-shell nanoparticles, bottom up nanofabrication, Langevin dynamics, magnetic dipole-dipole interactions.

1 INTRODUCTION

The interest in the field-directed control of magnetic particles has grown steadily in recent years due to rapid advances in particle synthesis and related enabling technologies [1-4]. Applications in this field include gene transfection [5,6], drug delivery [7,8], microfluidic-based bioseparation [9] and sorting [10]. Researchers have also explored the use of magnetic paticles' self-assembly to enable a bottom-up approach for the scalable fabrication of nanostructured materials, e.g. for photonic [11,12], magnetic, micro-optical [13] and electronic applications [14,15]. Such an approach would open up opportunities for the low-cost high-throughput production of functional nanocomposite materials.

In this presentation we demonstrate a method for controlling the self-assembly of colloidal magnetic core-shell nanoparticles into extended 3D crystalline superstructures. In this method, submicron (e.g. lithographically-patterned) template elements are used to achieve nanoscale resolution in particle placement. The templates are made from a soft-magnetic material (e.g. Permalloy, a nickel-iron alloy) and are embedded in a nonmagnetic substrate. A uniform bias field is applied to magnetize the template elements. **Fig.1** shows a portion of a self-assembly system that consists of a nonmagnetic substrate with a periodic 2D array of embedded cylindrical template elements. To achieve assembly, colloidal particles in suspension are first deposited onto the substrate in the absence of an external magnetic field. A uniform external field is then applied that saturates the soft-magnetic template elements to generate a magnetic gradient field that drives the self-assembly process. We use a computational model to study the detailed dynamics of the assembly process.

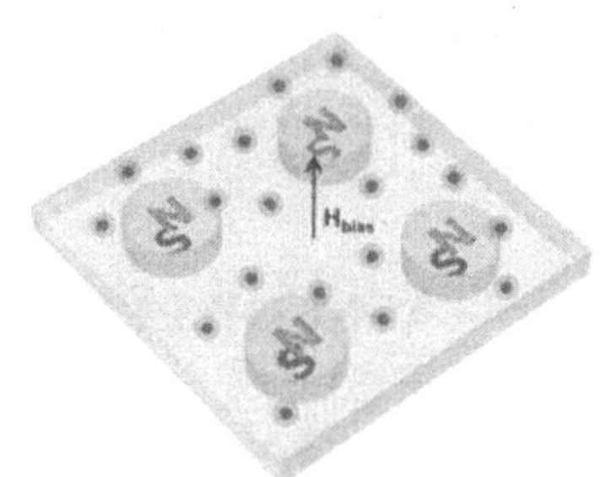

Fig.1 A template-assisted self-assembly system showing colloidal core-shell nanoparticles above a nonmagnetic substrate that contains an array of embedded soft-magnetic cylindrical template elements.

2 COMPUTATIONAL MODEL

The behaviour of colloidal magnetic particles in an applied field depends on a number of factors including hydrodynamic and magnetic forces, Brownian motion, Van der Walls force and the effects of surfactants. We predict the self-assembly of such particles using a computational model based on Langevin's equation that takes these effects into account,

$$m_i \frac{d^2 x_i}{dt^2} = F_{mag,i} + F_{vis,i} + F_{B,i}(t) + \sum_{\substack{j=1 \\ j \neq i}}^{N} \left(F_{dd,ij} + F_{vdw,ij} + F_{surf,ij} + F_{hyd,ij} \right), \quad (1)$$

Where m_i and $x_i(t)$ are the mass and position of the i'th particle. The right-hand-side of this equation represents the sum of forces on the i'th particle: $F_{mag,i}$ due to the applied magnetic field, which is a superposition of the bias field and induced gradient-fields; $F_{vis,i}$, viscous drag due to relative motion between the particles and the surrounding fluid

(Stokes drag); $F_{B,i}(t)$ a stochastic force to account for Brownian motion; $F_{dd,ij}$ the interparticle magnetic dipole-dipole force due to induced dipole moments; $F_{vdw,ij}$ Van der Waals force; $F_{surf,ij}$ a repulsive force due to surfactant contact between particles and $F_{hyd,ij}$ due to interparticle hydrodynamic interactions. We predict the particle dynamics by numerically integrating Eq.(1) using an adaptive time stepping method to accelerate and stabilize the computation. The various terms in the model are briefly described in the following sections.

2.1 Magnetic force

The magnetic force is computed using an "effective" dipole moment method in which the particle is modeled as an "equivalent" point dipole with an effective moment m_{eff}. The force on the i'th particle is given by [16]

$$F_{mag,i} = \mu_f \left(m_{i,eff} \cdot \nabla \right) H_a, \quad (2)$$

where μ_f is the permeability of the fluid and H_a is the applied magnetic field intensity at the center of particlee [16,17].

2.2 Magnetic dipole-dipole interaction

The potential energy for magnegic dipoles is given by

$$U_{dd,ij} = -\frac{\mu_f}{4\pi}\left(3\frac{(m_{i,eff}\cdot r_{ij})(m_{j,eff}\cdot r_{ij})}{r_{ij}^5} - \frac{m_{i,eff}\cdot m_{j,eff}}{r_{ij}^3} \right), \quad (3)$$

where $m_{i,eff}$ and $m_{j,eff}$ are the moments of i'th and j'th particle, respectively, and r_{ij} is the displacement vector between them. The dipole-dipole force in Eq.(1) is obtained as the gradient of the potential,

$$F_{dd,ij} = -\nabla U_{dd,ij}. \quad (4)$$

2.3 Van der Waals interaction

Van der Waals force is given by [18],

$$F_{vdw,ij} = \frac{32A}{3}\frac{(R_{p,i}\cdot R_{p,j})^3 \cdot r_{ij}}{(r_{ij}^2-(R_{p,i}+R_{p,j})^2)^2\cdot(r_{ij}^2-(R_{p,i}-R_{p,j})^2)^2}, \quad (5)$$

where A is the Hamaker constant and h_{ij} is the surface-to-surface separation distance between the i'th and j'th particle.

2.4 Surfactant force

The repulsive force caused by surfactant-surfactant contact is modelled using

$$F_{rep,ij} = -\nabla U_s = \frac{16\pi N_s k_b T}{\delta_i+\delta_j}\frac{R_{p,i}^2 R_{p,j}^2}{(R_{p,i}+R_{p,j})^2}\ln\left(\frac{R_{p,i}+R_{p,j}}{r_{ij}}\right), \quad (6)$$

where R_p, δ and N_s are, respectively, the radius of particle, the thickness of the surfactant layer and the surface density of surfactant molecules [19].

2.5 Viscous drag

The viscous drag on a particle is computed using Stokes' formula

$$F_{D,i} = D\frac{dx_i}{dt}, \quad (7)$$

where $D = 6\pi\eta R_{hyd,p}$ is the drag coefficient, η is the fluid viscosity and $R_{hyd,p}$ is the hydrodynamic radius of the particle.

2.6 Interparticle hydrodynamics interactions

Interparticle hydrodynamic interactions are taken into account using lubrication theory, with a force of the form [18],

$$F_{lub,ij} = \frac{6\pi\mu_f V_{r,i,j}}{h_{ij}}\frac{R_{p,i}^2 R_{p,j}^2}{(R_{p,i}+R_{p,j})^2}, \quad (8)$$

where h_{ij} is the separation bewteen the surfaces and $V_{r,i,j}$ is the relative velocity between the particles.

2.7 Brownian diffusion

The Brownian force was modeled as a Gaussian white noise process. The magnitde of this force in one dimension is

$$F_{B,i}(t) = \xi\sqrt{\frac{2Dk_BT}{\Delta t}} \quad (9)$$

where is Boltzmann's constant, D is is the Stokes' drag coefficient as described above and ξ is a random number with a Gaussian distribution.

2.8 Equations of Motion

Particle motion during assembly is predicted by solving Langevin's equation,

$$m_i\frac{d^2x_i}{dt^2} + D\frac{dx_i}{dt} = F_{sum,i}, \quad (10)$$

where

$$F_{sum,i} = F_{mag,i} + F_{B,i}(t) + \sum_{\substack{j=1\\ j\neq i}}^{N}\left(F_{dd,ij}+F_{vdw,ij}+F_{rep,ij}+F_{lub,ij}\right), \quad (11)$$

Eq. (10) is solved after reduction to coupled first-order system of equations,

$$\Delta x_i = \frac{F_{sum,i}}{D}\tau + \frac{m_i}{D}\left(v_{i,0}-\frac{F_{sum,i}}{D}\right)\left(1\text{-}e^{-\frac{D}{m_i}\tau}\right), \quad (12)$$

$$v_{i,f} = \frac{F_{sum,i}}{D} + \left(v_{i,0}-\frac{F_{sum,i}}{D}\right)e^{-\frac{D}{m_i}\tau}, \quad (13)$$

that are integrated using a numerical time-stepping scheme.

3 RESULTS

We have used the model to study the system that consists of soft-magnetic cylinders that are embedded in a non-magnetic substrate **Fig. 1**. A uniform bias field H_{bias} is applied perpendicular to the substrate to magnetize the elements. We study the self-assembly of Fe_3O_4-SiO_2 core-shell particles with a core diameter of 17 nm ($R_{core} = 8.5\ nm$) and a shell thickness of 11.5 nm. The template elements are permalloy (78% Ni, 22% Fe), which has a saturation magnetization $M_{e,s} = 8.6\times10^5\ A/m$. The bias field is taken to be $H_{bias} = 3.9\times10^5\ A/m$ ($B_{bias} = 5000\ Gauss$), which is sufficiently strong to saturate both the nanoparticles and the template elements. Their hydrodynamic radius is assumed to be the same as their physical radius $R_{hyd,p} = 40\,nm$. The carrier fluid is assumed to be nonmagnetic with a viscosity and density equal to that of water.

We use field and force equations from our previous work [20-21] to compute the radial and axial magnetic field and force components as shown in **Fig.2**. We choose the radius and height of soft-magnetic cylindrical temple to be $R_m = 200\,nm$ and $h = 300\,nm$, which has a demagnetization factor as 0.46. Thus, a uniform external bias field (0.5 T) will be applied to saturate the soft-magnetic elements, which ensures that $\boldsymbol{H}_{bias} \geq N_d \boldsymbol{M}_{es}$. The spacing between the elements is taken to be 2 μm center-to-center so that there is negligible overlap of their fields. The radial and axial force components $F_{mag,r}$ and $F_{mag,z}$ are computed at a distance $z = 100\,nm$ above the template geometry as shown in **Fig. 2c** and **d**. The analysis shows that there is a relatively strong attractive axial force over the center of the element, $r/R_m < 1$, which promotes assembly in this region. There is also a relatively weak repulsive (upward-directed) axial force above near the exterior edge of cylinder ($r = R_m$) that prevents particles from assembling there. Note from **Fig. 2c** that this force component peaks (in a negative sense) above the radial edge of the element and therefore acts to move the particles inward towards the center of the element.Thus, the magnetic force directs particles towards the center of the template during assembly, which promotes the formation of a tapered 3D pyramid structure. The ability to produce regions of attractive and repulsive magnetic force is a key feature of the proposed assembly method that enables nanoscale precision of particle placement.

Next, we use the computational model to study the dynamics of the assembly process. The bias field and template dimensions are as above. We use a computational domain centered with respect to a single element. The domain spans a unit cell, i.e. 1 μm along both the x and y axes and 1 μm in the z-direction as shown in **Figs. 3a**. The base of the domain is at $z = 0$, which coincides with the top surface of both the substrate and the template element. Periodic boundary conditions for particle transport are imposed at the lateral sides of the domain to account for a 2D array of template elements. We simulate the assembly of 40 nm Fe_3O_4-SiO_2 particles with a 17 nm Fe_3O_4 core. Note that the assembled structure shown in **Fig 3c** has a uniform hcp crystalline structure that is tapered due to focusing by the magnetic force as described above. This structure is very similar to those observed by Yellen et al. as shown in **Fig. 3d**[22]. The individual colored-coded layers in the

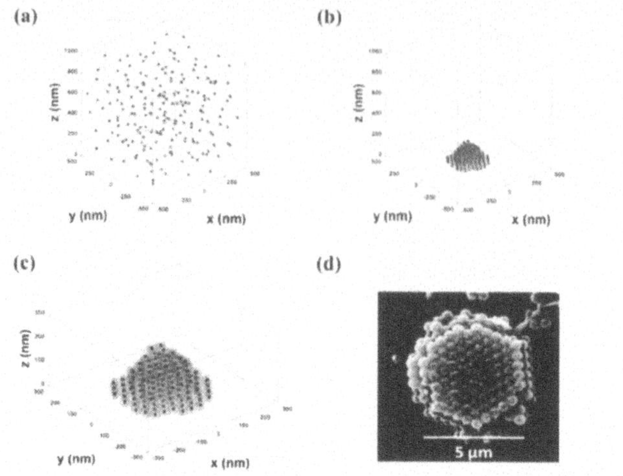

Fig. 3 Initial and final particle distributions for a cylindrical template element: (a) initial random particle distribution, (b) perspective of final assembled particle structure, (c) lateral view of assembled particle structure, (d) field-directed assembly of non-magnetic beads in a ferrofluid using cylindrical traps [22].

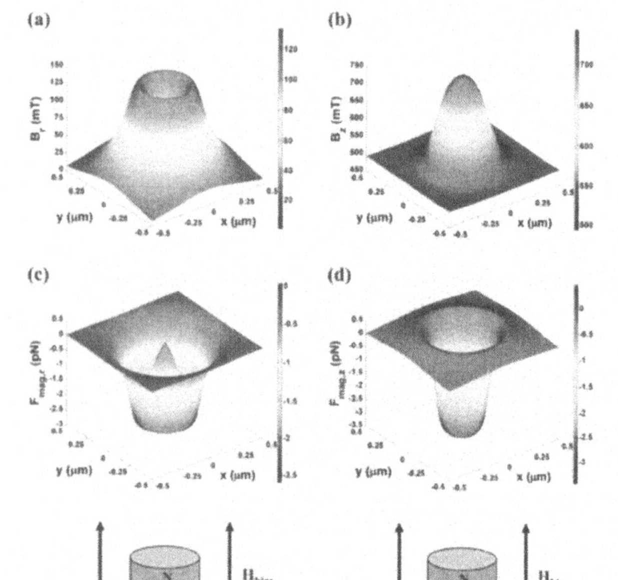

Fig.2 Axisymmetric magnetic field and force components at z=100 nm above the template element: (a) Br, (b) Bz, (c) Fmag,r, and (d) Fmag,z

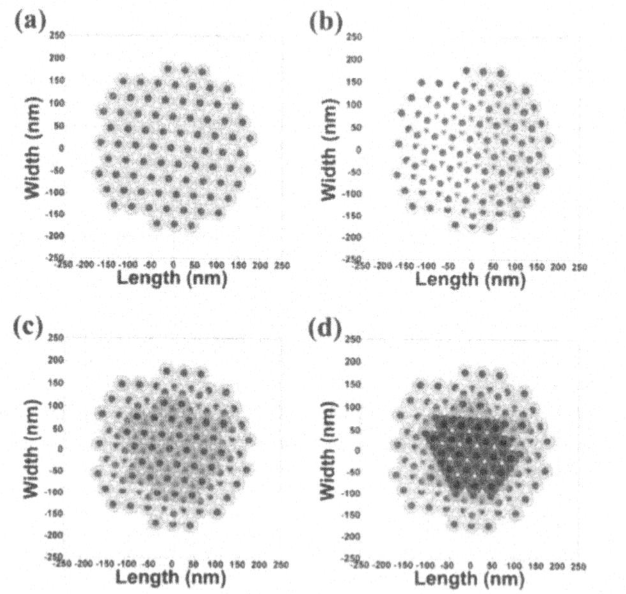

Fig. 4 Multi-layer hexagonal packed structure. (a), (b), (c) and (d) are indicated as 1st, 2nd, 3rd and 4th layer of the particles' structure after assembly. The particle core is shown as a solid sphere and the shell is

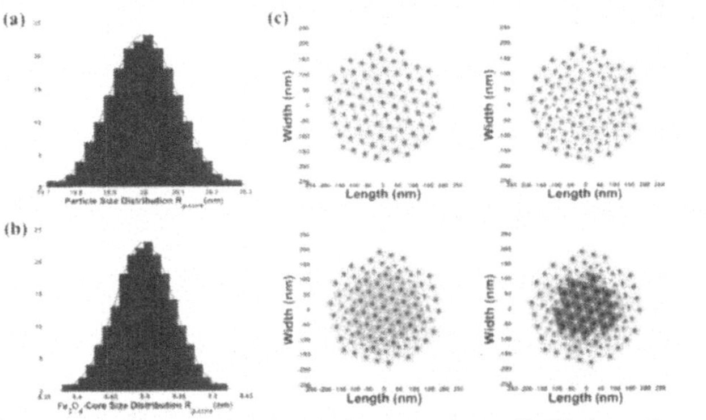

Fig. 5 Multi-layer hexagonal packed structure for a cylindrical template element with particle size distribution: (a) particle size distribution, (b) Fe3O4 core size distribution, (c) 4 layers' hexagonal packed structure after assembly.

assembled structure are shown in **Fig. 4**. Note that there are no defects (i.e. interparticle gaps) in these layers. However, as the number of partice layers increases, defects in crystalline structure can occur, which are due to a decrease in effect of the gradient field force that tends to pack the particles.

Next, we study the effcet of a size distribution on the self-assembly process. We consider a polydisperse system of core-shell particles that is based on a Gaussian size distribution as shown in **Fig.5a** and **b**. The oveall particle size varies between 19.7 nm and 20.3 nm, while the core size varies between 8.3 nm and 8.7 nm, as shown in **Fig.5a** and **b**. As shown in **Fig 5c** the assembled structure has a uniform hcp crystalline structure that is similar to the monodisperse system. However, our works show that as the size distribution increases, defects occur in the crystalline structure.

4 CONCLUSIONS

We have described a method for directing the assembly of colloidal magnetic-dielectric core-shell nanoparticles into extended 3D crystalline structures with nanoscale precision. In this method, sub-micron soft-magnetic template elements are used to guide the assembly in the presence of a uniform bias field. The use of a uniform bias field combined with localized high gradient fields produced by the template elements is a key feature of the method that enables nanoscale precision of particle placement, which translates to nanoscale feature resolution in the assembled structure. We have demonstrated proof-of-concept of the method using a computational model that takes into account the dominant assembly mechanisms. The method broadly applies to arbitrary template geometries and multi-layered particles with at least one magnetic component. The ability to produce such materials opens up opportunities for the scalable high-throughput fabrication of multifunctional nanostructured materials for a broad range of applications and our computational model enables the rational design of such media.

ACKNOWLEDGEMENTS

The authors acknowledge financial support from the U.S. National Science Foundation, through award number CBET-1337860.

REFERENCES

[1] Pankhurst, Q.A.; Thanh, N.K.T.; Jones, S.K.; Dobson, J. Progress in applications of magnetic nanoparticles in biomedicine. In J. Phys. D Appl. Phys., 2009, 42(22), 224001/224001-224001/224015.

[2] Pamme, N. Magnetism and microfluidics. In Lab Chip, 2006, 6(1), 24-38.

[3] Berry, C.C. Progress in functionalization of magnetic nanoparticles for applications in biomedicine. In J. Phys. D Appl. Phys., 2009, 42(22), 224003/224001-224003/224009.

[4] Furlani, E.P. Magnetic biotransport: analysis and applications. In Materials, 2010, 3, 2412-2446.

[5] Dobson, J. Gene therapy progress and prospects: magnetic nanoparticle-based gene delivery. In Gene Ther., 2006, 13(4), 283-287.

[6] Furlani, E.P.; Xue, X. Field, force and transport analysis for magnetic particle-based gene delivery. In Microfluid. Nanofluid., 2012, 13(4), 589-602.

[7] Furlani, E.P.; Xue, X. A Model for Predicting Field-Directed Particle Transport in the Magnetofection Process. In Pharm. Res., 2012, 29(5), 1366-1379.

[8] Arruebo, M; Fernández-Pacheco, R; Ibarra, M.R; Santamaría, J. Magnetic nanoparticles for drug delivery. In Nanotoday, 2007, 2(3):22–32.

[9] Ganguly, R.; Puri, I.K. Microfluidic transport in magnetic MEMS and bioMEMS. In Wiley Interdiscip. Rev. Nanomed. Nanobiotechnol., 2010, 2(4), 382-399.

[10] Zborowski, M; Chalmers J.J. Magnetic cell separation, vol 32 (Laboratory techniques in biochemistry and molecular biology). In Elsevier Science, 2007, New York.

[11] Yethiraj, A.; Thijssen, J.H.J.; Wouterse, A.; Van Blaaderen, A. Large-area electric-field-induced colloidal single crystals for photonic applications. In Adv. Mater. (Weinheim, Ger.), 2004, 16(7), 596-600.

[12] Xia, Y.; Gates, B.; Li, Z.-Y. Self-assembly approaches to three-dimensional photonic crystals. In Adv. Mater. (Weinheim, Ger.), 2001, 13(6), 409-413.

[13] Lu, Y.; Yin, Y.; Xia, Y. A self-assembly approach to the fabrication of patterned, two-dimensional arrays of microlenses of organic polymers. In Adv. Mater. (Weinheim, Ger.), 2001, 13(1), 34-37.

[14] Shipway, A.N.; Katz, E.; Willner, I. Nanoparticle arrays on surfaces for electronic, optical, and sensor applications. In ChemPhysChem, 2000, 1(1), 18-52.

[15] Hochbaum, A.I.; Fan, R.; He, R.; Yang, P. Controlled growth of Si nanowire arrays for device integration. In Nano Lett., 2005, 5(3), 457-460.

[16] Furlani, E.P. Analysis of particle transport in a magnetophoretic microsystem. In J. Appl. Phys., 2006, 99(2), 024912/024911-024912/024911.

[17] Furlani, E.P.; Ng, K.C. Nanoscale magnetic biotransport with application to magnetofection. In Phys. Rev. E Stat., Nonlinear, Soft Matter Phys., 2008, 77(6-1), 061914/061911-061914/061918.

[18] Russel, W.B.; Saville, D.A.; Schowalter, W.R., Colloidal Dispersions, Cambridge University Press, 1989.

[19] Rosensweig, R.E. Fluid dynamics and science of magnetic liquids. In Advances in Electronics and Electron Physics. 1979, Vol. 48: Martin M (ed.) Academic Press, New York, pp. 103-99.

[20] Xue, X.; Furlani, E.P. Template-assisted nano-patterning of magnetic core–shell particles in gradient fields. Physical Chemistry Chemical Physics, 2014, 16(26), 13306-13317.

[21] Xue, X.; Furlani, E.P. Analysis of the Dynamics of Magnetic Core-shell Nanoparticles and Self-assembly of Crystalline Superstructures in Gradient Fields. The Journal of Physical Chemistry C, 2015.

[22] Yellen, B.B.; Hovorka, O.; Friedman, G. Arranging matter by magnetic nanoparticle assemblers. Proceedings of the National Academy of Sciences of the United States of America, 2005, 102(25), 8860-8864.

Low-cost spin-coatable, transferable, and high-κ ion gel dielectric for flexible electrowetting

V. Narasimhan, S. Vijayavenkataraman and S.-Y. Park

National University of Singapore, Singapore, mpeps@nus.edu.sg

ABSTRACT

A novel ion gel material was demonstrated as a high-capacitance dielectric for flexible electrowetting-on-dielectric (EWOD) applications. Unlike conventional dielectric materials such as SiO_2 and Al_2O_3 commonly used for EWOD, an ion gel offers two or three order higher capacitance than them, while being fabricated through a simple low-cost spin-coating process without complex and expensive fabrication setups such as high vaccume facilities. In this paper, a brief introduction into the constituents of the ion gel and its electrical modeling for EWOD are discussed, followed by fabrication methodologies. With the ion gel films consisting of a [P(VDF-HFP)] copolymer and a [EMIM][TFSI] ionic liquid, we compared EWOD performance of the ion gel samples with other conventional dielectric materials and showed the potential of the ion gel material capable of low-cost, spin-coatable, transferable and high-capacitanec dielectric for flexible EWOD applications.

Keywords: electrowetting-on-dielectric (EWOD), ion gel, high-κ dielectric

1 INTRODUCTION

Electrowetting-on-dielectric (EWOD) is an emerging liquid-handling technology enabling the surface tension modulation of a liquid with an applied electric field. When an electric potential is applied between a liquid and a solid electrode, the charge re-distribution modifies the surface tension at the liquid-solid interface where the like-charges repulsion decreases the work by expanding the surface area. Due to its force dominance over body forces in small scale, the technology has been widely applied. In recent years, EWOD has been attractively used for 3D flexible applications such as electronic display [1], lab on a chip [2], and biochemical sensors [3].

An EWOD phenomenon can be mathematically explained by the Young-Lippmann equation, showing the relationship between the applied electric potential (V) and is the contact angle (θ) of the droplet, given as [4]:

$$\cos\theta = \cos\theta_0 + \frac{1}{2\gamma_{LG}} cV^2 \quad (1)$$

where θ_0 is the contact angle of the droplet with zero potential application, γ_{LG} is the surface tension between two immiscible fluids, and c is the capacitance per unit area. According to this equation, the EWOD performance can be improved by using high-capacitance dielectric materials. To meet such requirements, extensive studies have been conducted on dielectric materials that can offer a high capacitance to lower operating voltage. Several high-κ materials such as Si_3N_4, Al_2O_3, and Ta_2O_5 are typically proposed to provide a high-capacitance benefit. However, these materials need expensive and complex vacuum facilities like CDV, PECVD, and sputtering for layer deposition. Particularly for flexible EWOD applications, those conventional fabrication processes can't practically provide the rigidity of the materials. Alternatively, polymer-based materials such as PDMS, SU-8, and Parylene C have been used to permit deformability. However, a typical spin-coating method for deposition of those material yields a thick dielectric layer in the range of μm. Consequently, they offer very low capacitance requiring operational voltages as high as 700 V [3].

Here, we present, for the first time, the use of an ion gel material as a high-capacitance dielectric for flexible EWOD applications. It offers two or three order higher capacitance than conventional dielectrics such as such as Si_3N_4, Al_2O_3, and Ta_2O_5, while being fabricated through a simple low-cost spin-coating process without any expensive vacuum facilities. We successfully fabricated the ion gel films, which consist of a copolymer poly(vinylidene uoride-co-hexauoropropylene) [P(VDF-HFP)] and an ionic liquid, 1-ethyl-3-methylimidazolium bis(triuoromethylsulfonyl)imide [EMIM][TFSI], through a simple spin-coating process. Our study for EWOD performance indicates that ion gel was capable of providing an improved high-capacitance dielectric and lowering operational voltage without bubble generation by electrolysis. The deformability of ion gel was demonstrated by transferring to a flexible substrate and subsequently bending it.

2 THEORETICAL MODELING

The ion gel is a recently developed material in the form of a thin-film made of an ionic liquid and a structuring polymer. Ionic liquids (also known as room temperature molten salts) themselves have been very attractive due to their favorable properties such as a wide electrochemical

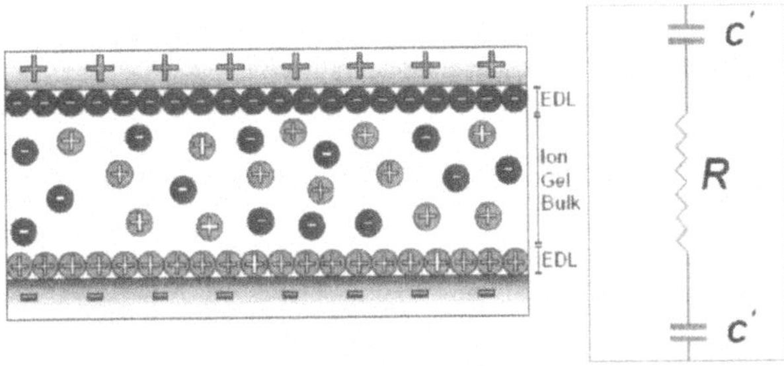

Figure 1. Schematic of the ion gel layer sandwiched between two electrified surfaces of opposite polarity (left) and its equivalent circuit model (right). Counter-ions from the gel pack themselves against their respective electrode surfaces, forming thin EDL capacitors, *c'*. In the bulk of the ion gel layer, a resistance *R* exists due to electrolyte resistance emerging from the constituent ionic liquid. At DC or a very low frequencies typically used for EWOD operation, the polymer resistance sandwiched by two capacitors can be simply negligible.

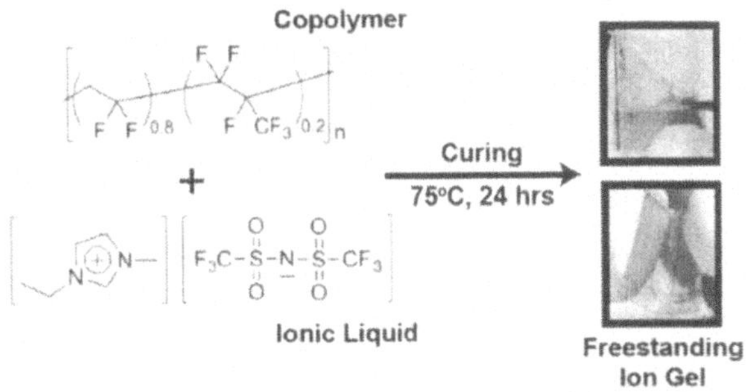

Figure 2. Chemical structures of the P(VDF-HFP) copolymer (top) and the [EMIM][TFSI] ionic liquid (bottom) which are cured to obtain free standing ion gels. The images show layers of an ion gel and Teflon fabricated on a glass substrate (right top) and then transferred on a flexible ITO substrate, which can be bent with ease (right bootom).

window, high thermal stability, almost negligible vapor pressure, and large ionic conductivity and capacitance. For several applications of the ionic liquid, it is desirable to utilize them in the form of a solid film. Such a thin-film form known as ion gels is attained through the addition of a structuring polymer to the ionic liquid thereby inducing either physical or chemical crosslinking.

For a fundamental study of the ion gel used for EWOD as a dielectric layer, it is first necessary to understand the nature of capacitance within ionic liquids, because they are an essential building block of the ion gel. At electrified interfaces, free counter-ions inside ionic liquids accumulate at the liquid-electrode interface and form an EDL capacitor. One of the accurate forms of measurement of this EDL capacitance of ionic liquids is obtained through the measurement of electrostatic forces normal to the ionic liquid-electrode interfaces. This approach provides an approximate EDL capacitance value in the range of 3 ~ 11 $\mu F/cm^2$ for most commonly used ionic liquids [5]. Some more studies reported that the structuring polymers used to provide a network around the ionic liquids thereby forming a solid thin-film do not contribute to the capacitance of the gel [6]. This also suggested that the high capacitance reported earlier is as a result of the EDL formation that is not influenced by the type of structuring polymer used.

Based on the studies of ion gels previously reported, the ion gel dielectric for EWOD can be modelled with an equivalent circuit as presented in Figure 1. Following its constituent ionic liquid, the ion gel also forms compact EDL layers at electrified interfaces while at the same time there is a bulk electrolyte resistance also emerging from the constituent ionic liquid. The effective capacitance of the ion gel is entirely dependent on its EDL capacitance making it overall layer thickness independent. At DC or a very low frequencies typically used for EWOD operation, the voltage consumption across the polymer resistance, which is sandwiched by two capacitors, can be simply neglected.

3 EXPERIMENTAL METHODS

We successfully obtained a spin-coatable ion gel layer for flexible EWOD applications. P(VDF-HFP) polymer was selectively dissolved in [EMIM][TFSI] ionic liquid and diluted in acetone. It was then spin-coated on an ITO substrate and cured in an oven to completely remove the solvent. The gelation process occurs when these crystals are brought together by polymer chains dissolved in the solution. Figure 2 illustrates the chemical structures of the copolymer and the ionic liquid. The layer initially fabricated on a glass wafer was transferred with the help of tweezers onto a flexible ITO substrate indicating ease of transferability.

Figure 3 graphically represents some of the different gel layers that were fabricated under various conditions. The thickness was characterized with the dry film after curing. By varying spin speed as well as the wt % ratio of the copolymer, ionic liquid and solvent, we were able to vary the ion gel thickness from 4 μm to 18 μm. Subsequently, a 6 % Teflon AF as well as 1 % Teflon AF solution were spin-coated and cured over the ion gel layer to provide a hydrophobic surface.

After the layer deposition, we conducted both static and dynamic EWOD tests to investigate the performance of the new ion gel layer. Figure 4 shows experimental setup for the static EWOD studies. A water droplet was placed on the spin-coated ion gel and Teflon layers. The thickness of the gel layer and the Teflon layer were individually varied to study their impact on the EWOD performance. A platinum (Pt) probe was inserted into the droplet while the buried ITO electrode was grounded. Voltage was then applied between the probe and the ground using a low voltage DC source controlled by LabVIEW. The contact angle was then measured at incremental voltages in steps of 5 V.

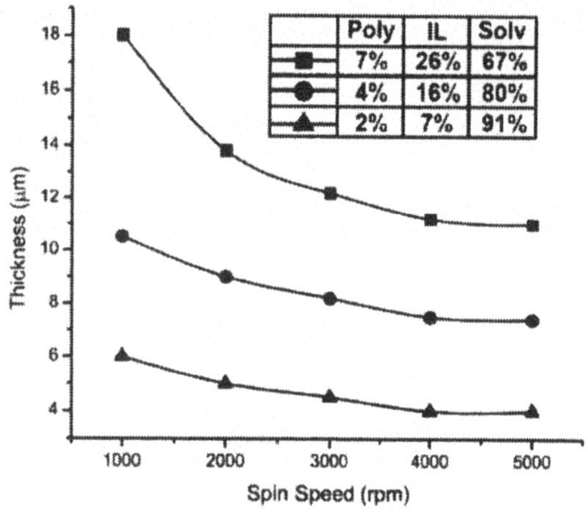

Figure 3. Various thicknesses of an ion gel layer fabricated by varying spin speed and relative wt % of the mixtures. The legend indicates the relative wt % ratio of the copolymer P(VDF-HFP), the ionic liquid [EMIM][TFSI], and the acetone solvent.

4 RESULTS AND DISCUSSION

First, static EWOD tests were performed to benchmark the performance of the ion gel against certain standard dielectrics commonly used in EWOD. This would help gauge the practicality of its usage in various EWOD applications. Al_2O_3 was chosen to represent standard thin-film dielectric materials deposited through atomic layer deposition (ALD) processes that typically require expensive high-vacuum facilities. In contrast, PDMS was chosen to represent dielectric materials fabricated by a simple spin-coating procedure specifically used for flexible EWOD applications. The thickness of the ion gel was 10 μm, while that of Al_2O_3 and PDMS were 100 nm and 10 μm respectively. Teflon layer of the same thickness (170 nm) was spin-coated on the above mentioned dielectrics for comparison.

Figure 5 shows that the gel layer performs much better than other benchmarked dielectrics with a given thickness of a Teflon layer. As expected, EWOD performances are shown in the order of Al_2O_3 and PDMS, which is the same order of

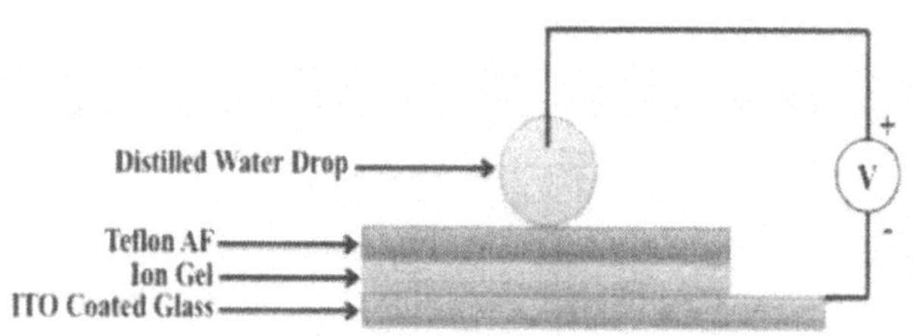

Figure 4. Experimental setup for static EWOD study. A 5 μL distilled water droplet is loaded on top of the hydrophobic (Teflon AF) and dielectric (ion gel) layers.

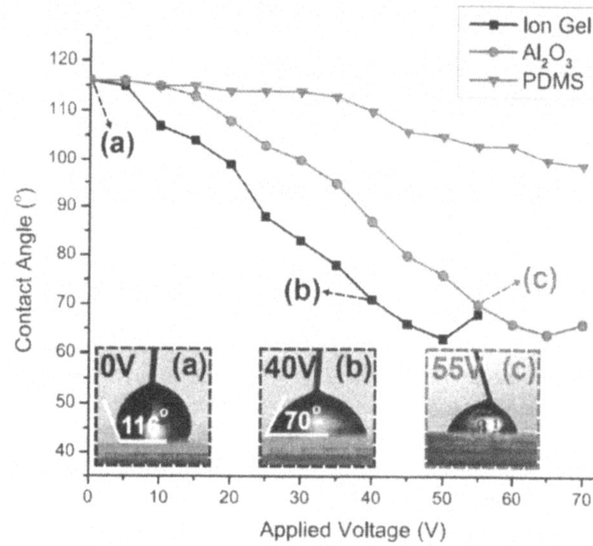

Figure 5. A comparative study of the EWOD performance of the ion gel versus Al_2O_3 and PDMS. Al_2O_3 was chosen to represent standard thin-film dielectric materials used in EWOD while PDMS was chosen to represent dielectric materials requiring a simple fabrication procedure. Insets (a) and (b) display images of a 5 μL water droplet over the Teflon / ion gel stack at 0 V and 40 V respectively. Inset (c) displays an image of the water droplet over the Teflon / Al_2O_3 stack at 55 V where electrolysis is clearly observed (circled in red).

their capacitance magnitudes. One interesting observation was that electrolysis occurs on the Al_2O_3 sample at 55 V (see the inset (c)), while no such phenomena occurs in the case of the ion gel (see the insets (a) and (b)) even at the saturation voltage around 50 V. This implies that dielectric layers deposited at such small thicknesses may not be very practical for use in microfluidic devices that require higher voltages, while the ion gel which offers a thickness independent capacitance can be simply spin-coated on patterned electrodes for robust usage in microfluidic applications. Additionally, being a high capacitance material, the ion gel induces large electrostatic storage energy within the hydrophobic and dielectric stack which is in turn crucial for large surface tension modulation. This again is evident from the theoretical study in Eq. (1), which is verified by our experimental results in Figure 5.

We also studied the droplet actuation phenomena using the novel ion gel dielectric. For the experiments, an electrode pattern was created through standard photolithography processes using AZ P4260 photoresist spin-coated on an ITO coated substrate. The pattern was then wet etched using 37% concentrated HCl for 4 minutes. The photoresist was then stripped using acetone to realize the electrode pattern. Over the pattern, a 15 μm ion gel layer and an 880 nm Teflon AF layer were deposited and cured to complete the fabrication of the device. Figure 6 shows the movement of a 10 μL water droplet at a speed of 17.5 cm/sec on the Teflon / ion gel

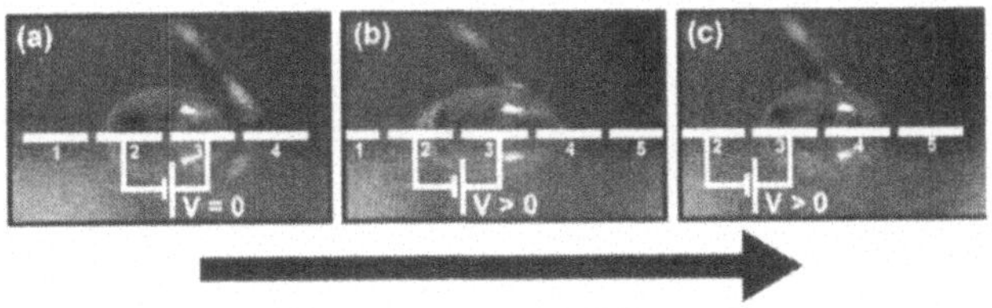

Figure 6. EWOD based droplet movement was demonstrated using the novel ion gel dielectric. Some of the electrodes in the series are labeled from 1 to 5. The speed of the droplet was measured as 17.5 cm/sec

coated substrate. Some of the electrodes in the series are labeled from 1 to 5. The droplet was initially over electrode 2 and 3 and there was no applied potential (Figure 6a). DC voltage of 100 V is applied between electrode 2 and 3 realizing droplet actuation (Figure 6b). When the voltage is released at Figure 6 (c), the droplet has moved forward and now covers electrodes 3 and 4. This process is repeated iteratively to achieve droplet movement.

5 CONCLUSION

A novel ion gel capable of spin-coatable, transferable and high-capacitance dielectric was studies for flexible EWOD applications. At electrified interfaces, a large number of free counter-ions inside ionic liquids tightly accumulate at the liquid-electrode interface and form an EDL. Such thin but highly-densified ion distributions allow to hold large electrostatic energy across the EDL capacitor and hence large surface tension modification necessary for the improved EWOD performance. It was reported that the gel layer can provide a few order higher capacitance than that of conventional dielectric materials commonly used for EWOD. The fundamental study of the ion gel material interestingly indicates that the capacitance is thickness independent. This suggests that a relatively thick ion gel over 10 μm can be used to provide a stable high-capacitance dielectric property without the layer breakdown (i.e. electrolysis) for wide EWOD applications.

The ion gel films, which consist of a copolymer poly(vinylidene uoride-co-hexauoropropylene) [P(VDF-HFP)] and an ionic liquid [EMIM][TFSI], 1-ethyl-3-methylimidazolium bis(triuoromethylsulfonyl)imide, was succefully fabricated through a simple spin-coating process without expensive high-vacuum facilities that are typically required conventional IC fabrication processes. It was further transferred to flexible ITO substrates and demonstrated the deformability of the ion gel dielectric by subsequently bending it. The ion gel's EWOD performance was compared with certain commonly used dielectric materials for traditional as well as flexible EWOD and experimentally showed the lower operation voltage at a given contact angle modulation on ion gel samples. Finally, we also demonstrated droplet movement at a speed of 17.5 cm/sec on the Teflon / ion gel coated substrates.

The paper informs that an ion gel is capable of lowering operational voltage and providing an improved high-capacitance dielectric without bubble generation by electrolysis. With its relatively large capacitance and easy fabrication, the ion gel is seen as a material with great potential for 3D flexible EWOD applications.

REFERENCES

[1] D. Y. Kim and A. J. Steckl, “Electrowetting on paper for electronic paper display," ACS Appl. Mater. Interfaces, 2, 3318-3323, 2010.

[2] S.-K. Fan, H. Yang, and W. Hsu, “Droplet-on-a-wristband: Chip-to-chip digital microfuidic inter- faces between replaceable and exible electrowetting modules," Lab Chip, 11, 343-347, 2010.

[3] M. Abdelgawad, S. L. S. Freire, H. Yang, and A. R. Wheeler, “All-terrain droplet actuation," Lab Chip, 8, 672-677, 2008.

[4] M. G. Lippmann, "Relations entre les phénomènes electriques et capillaires," Annales de Chimie et de Physique, vol. 5, pp. 494-549, 1875.

[5] J. H. Cho, J. Lee, Y. Xia, B. Kim, Y. He, M. J. Renn, T. P. Lodge, and C. D. Frisbie, “Printable ion-gel gate dielectrics for low-voltage polymer thin-film transistors on plastic," Nat. Mater., 7, 900-906, 2008.

[6] S. Zhang, K. H. Lee, C. D. Frisbie, and T. P. Lodge, “Ionic conductivity, capacitance, and viscoelastic properties of block copolymer-based ion gels, Macromolecules, 44, 940949, 2011.

[7] K. H. Lee, M. S. Kang, S. Zhang, Y. Gu, T. P. Lodge, and C. D. Frisbie, “Cut and stick" rubbery ion gels as high capacitance gate dielectrics," Adv. Mater., 24, 44574462, 2012.

An Integrated Biometric Platform for Evaluation of Targeted Drug Delivery

Antony Thomas[1], Yaling Liu[1,2,]*

[1]Bioengineering program
[2]Department of Mechanical Engineering and Mechanics
Lehigh University, Bethlehem, PA 18015, USA, yal310@lehigh.edu

ABSTRACT

Endothelial cells (ECs) form the inner lining of blood vessels and are exposed to various factors like hemodynamic conditions, biochemical signals and interaction with other cell types. Blood vessel functions are regulated by interactions among these factors. The occurrence of pathological conditions lead to localized upregulation of cell adhesion molecules on the endothelial lining of blood vessels. This process is promoted by cytokines such as tumor necrosis factor-alpha (TNF-α), which leads to expression of intercellular adhesion molecule-1 (ICAM-1) on the EC surface among others. We developed a novel biomimetic blood vessel model by culturing confluent, flow aligned, ECs in a microfluidic platform, and performed real-time *in situ* characterization of localized pro-inflammatory endothelial activation. The model mimics the physiological phenomenon of cytokine activation of endothelium from tissue side and studies the heterogeneity in localized surface ICAM-1 expression and F-actin arrangement. Fluorescent antibody coated particles were used as imaging probes for identifying cell surface ICAM-1 expression. The binding properties of particles were evaluated under flow for different particle sizes and antibody coating densities. This allowed the investigation of spatial resolution and accessibility of ICAM-1 molecules on ECs, along with their sensitivity in receptor-ligand recognition and binding. This work has developed an *in vitro* blood vessel model that integrate various heterogeneous factors to effectively mimic a complex and physiologically relevant endothelial microenvironment.

Keywords: Vascular pathology, Microfluidics, Receptor-ligand recognition, Endothelial microenvironment, Real-time study

1 INTRODUCTION

The local vascular dynamics of ECs under disease conditions these changes and their induction remain largely enigmatic, in part due to the difficulties of modeling. *In vivo* studies most adequately reflect the pathophysiological context, but the system complexity is beyond our current means to accurately dissect specific aspects. Parameters of the heterogeneous physical, chemical, and biological pathways influencing the blood vessel add to the complexity [1]. On the other hand, the biological relevance of conventional cell culture models is limited. Therefore, both understanding of pathophysiology and designing adequate interventions demand alternative bio-mimetic model platforms.

This study aims to develop a bio-mimetic blood vessel to study dynamics of endothelial activation induced by local action of pro-inflammatory cytokines. To achieve this goal, we combined endothelial microfluidics/flow adaptation approach with local cytokine application via a semi-permeable "sub-endothelial compartment" and employed antibody-coated nanoparticles as imaging probes. This approach allowed us to define flow-mediated local heterogeneity of endothelial activation by cytokines that appears to reflect the pathophysiology of the process. The platform facilitated real-time microscopic analysis of cellular characteristics and particle binding as well as differentiation between localized differences in cell responses to treatment. Furthermore, the results of this study support design of drug delivery systems based on affinity nanocarriers targeted to the pathological endothelium.

2 MATERIALS AND METHODS

2.1 Fabrication and EC culture on bio-mimetic blood vessel model

Blood vessel mimicking channels are photolithographically fabricated and BAOEC culture is attained as explained prior [2]. The upper and lower channels are casted out of Sylgard 184 polydimethylsiloxane (PDMS) (Dow Corning Corp.). The upper channel is 20 mm long, 350 μm wide and 100 μm tall, and the lower channel is 5 mm long, 1000 μm wide and 100 μm tall. A polycarbonate (PC), track-etched thin clear membrane (Whatman, GE Healthcare) with 1 μm diameter pores and an average calculated pore density of $1.5*10^7$ pores/cm^2 is embedded between two PDMS channels. The bottom PDMS channel layer is bonded on to a thin glass slide by exposing the sides in contact to oxygen plasma. Upper PDMS channels is bonded to the membrane by using a thin PDMS mortar film (10:1 ratio of base and curing agent with toluene in equal proportion). The upper channel bonded to the membrane is then bonded to lower channel after making sure the channels are aligned. After each step the components are placed in an oven at 60oC for the appropriate time to enhance bonding. Inlet and outlet ports were punched to provide access to upper and lower channels.

The device was sterilized in UV light overnight and the upper channel of the device was coated with 50 μg/ml fibronectin solution (Sigma-Aldrich) overnight at 37oC. BAOECs cells were seeded in the upper channel at a density of 2*107 cells/m and let to attach on the semi-permeable membrane. Cell seeded devices were placed in an incubator under standard culture conditions (37oC and 5% CO2) overnight to allow cell attachment and spreading on the membrane. On reaching confluence, the BAOECs were subjected to a FSS of 12 dyne/cm^2. The flow was brought about using a high precision and extremely low pulsation peristaltic pump (ISMATEC, IPC-N series) and the entire setup was placed in standard culture conditions. Recirculation of media was not permitted while thrombin and tracer molecule was flowed through the channels. To calculate the volumetric flow rate that correspond to the required maximum FSS experienced by the ECs, the following equation was used [3]. $\tau_{cell} = \frac{6\mu Q}{wh^2}$

2.2 TNF- α treatment

The dynamics of endothelial activation and ICAM-1 expression by local action of TNF-α were studied. BAOECs were subjected to 24 hrs or more of flow at 12 dyne/cm^2 and were finally activated with TNF-α to study surface ICAM-1 expression. TNF-α was mixed in media at a concentration of 10 Units (U)/ml and the treatment was performed for 2 hrs on a confluent EC layer (>80%). BAOECs were locally activated by introducing TNF-α in the lower channel of the device which has direct access to a section of the upper channel. TNF-α diffused from the lower to the upper channel through the pores of the membrane and locally activated the endothelial layer from the basal side.

2.3 Particle binding study

The particles were introduced to the flow in the upper channel of the device and after their designated flow time unbound particles were removed by flushing with a buffer solution. The wash buffer contains a plasma membrane stain (CellMask™, Life technologies) to fluorescently tag the BAOECs. The particle bound BAOECs were then fixed in paraformaldehyde (3.7%). Particle binding was analyzed by phase contrast and fluorescence microscopy (FV1000-IX81, Olympus) and image analysis was performed using ImageJ software.

To demonstrate the applicability of our platform towards real time *in situ* studies we characterized dynamic binding of anti-ICAM-1 coated NPs to BAOECs being subjected to TNF-α treatment. The enclosed nature of the biomimetic blood vessel platform allowed maintenance of sterile conditions. We performed a continuous particle binding study over time where the flow set-up was assembled on a microscope table. BAOECs were subjected to 12 dyne/cm^2 of FSS for 6 hrs before localized TNF-α treatment from the lower channel while a steady flow is continued in the upper channel. Real time binding of anti-ICAM-1 coated 210 nm particles (232.5 ± 25 anti-ICAM-1/particle) was analyzed to characterize the dynamic nature of ICAM-1 expression by BAOECs. The cells were cultured in HEPES buffered media and such cultures do not require a controlled gaseous atmosphere [4]. The pump controlled the flow rate of the media and an in-line solution heater (Warner Instruments, SH-27B) maintained the media temperature at 37°C.

3 RESULTS AND DISCUSSION

3.1 On-chip cell culture

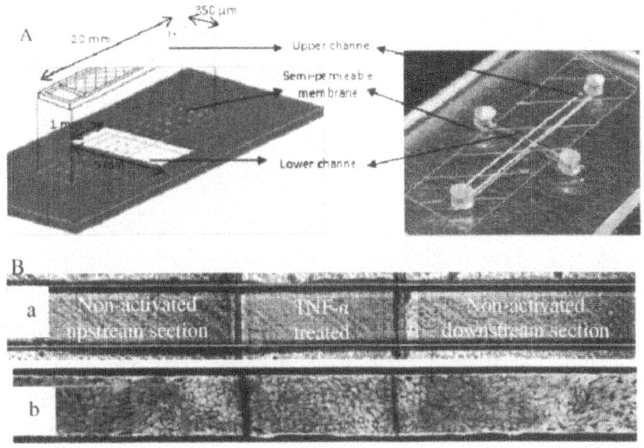

Figure 1: A, Schematic and photograph of the bi-layer device showing the upper (yellow lines) and lower (red lines) channel separated by the porous membrane; B (a) Bright field image of the upper (yellow lines) and lower (red lines) channel defining the upstream, TNF-α activated and downstream sections, (b) Fluorescence labeled actin cytoskeleton (FITC phalloidin) images of confluent BAOEC layer aligned to flow (12 dyne/cm^2 FSS for 24 hrs). The cells grow on the semi-permeable membrane in the upper channel of the device. Arrow shows flow direction

Pro-inflammatory and proliferative signaling pathways of vascular ECs become down-regulated when the EC layer is exposed to unidirectional sustained FSS [5]. Because the quiescent phenotype of ECs is critical to the success of a biomimetic blood vessel model and subjecting the ECs to physiologically relevant FSS is important [6]. A confluent and flow-aligned BAOEC monolayer was maintained for 5 days and the F-actin arrangement pattern was comparable to that of BAOECs subjected to 12 dynes/cm^2 FSS for 24 hrs.

3.2 BAOEC stress fiber alignment under FSS

To further examine the EC layer subjected to flow in our device, their actin assembly characteristics were studied. ECs exposed to FSS change shape and microfilament network to align with the direction of flow [7]. We examined this by staining F-actin stress fibers of BAOECs growing in the upper channel. The exposure of cells to 12 dynes/cm^2 FSS resulted in alteration of cell shape from the typical cobblestone pattern to fusiform as observed in ECs *in vivo* [8-9]. The cells and their actin stress fibers were uniformly aligned in the direction of flow (Fig. 2 B). The control (no flow) case had the cortical actin and F-actin fibers arranged in a radial pattern (Fig. 2 A). To better depict the relationship between cell orientation and flow direction, the angle (α) between the orientation of stress fibers and the width (short-axis) of the microchannel was plotted. It is observed that for the control case the stress fibers align at an average angle of 51.4°, while the cells exposed to flow show an average angle of 90.1° (Fig. 2 C).

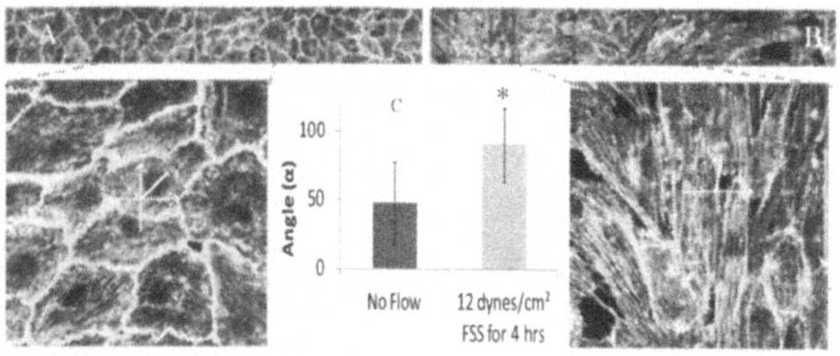

Figure 2: Organization of F-actin stress fibers (green) in BAOECs assessed by confocal microscopy (A) Static case; Cell stress fibers tend to align parallel to the width of the channel. (B) After 4 hrs of flow at 12 dynes/cm2 FSS; Cell stress fibers tend to align parallel to the flow direction (C) Quantitative measurement of stress fiber alignment angle (α) to the width of the channel under static and flow shear stress case (*$p<0.001$, by Student's t test)

3.3 Characterization of flow-mediated localized BAOEC activation using fluorescent probes

ICAM-1 expression was evaluated by studying specific binding of anti-ICAM-1 coated particles to ECs in the upper channel. Saturated anti-ICAM-1 coating on 210 nm particles produced a high antibody coating density of 232.5 ± 25 anti-ICAM-1/particle. This was brought down to a low antibody coating density of 112.9 ± 19 anti-ICAM-1/particle using control IgG antibody. 210 nm nanoparticles coated with anti-ICAM-1 or control IgG were perfused at 6 dyne/cm^2 FSS.

Having both direct cytokine-activated and non-activated ECs in the same channel allowed investigation of the heterogeneous nature of ICAM-1 expression and the hemodynamic control of the marginal zone between inflammation foci and relatively normal vasculature. As the ECs were fluorescently tagged, the boundaries of individual cells were marked and the number of particles binding per cell was counted using ImageJ software. The binding of fluorescent particles to ICAM-1 in the upstream, TNF-α treated and downstream areas of the channel were studied using antibody coated 210 nm and 1 μm particles. Flow rates of 6, 12 and 18 dyne/cm^2 were employed after the BAOECs were treated with TNF-α for 2 hrs under a steady and sustained FSS of 12 dyne/cm^2.

BAOECs in the TNF-α treated section of the upper channel have around 4-5 times higher particle binding density compared to the upstream section for 210 nm particle with a coating density of 232.5 ± 25 anti-ICAM-1/particle (Fig. 3 B) ($p<0.01$). A similar trend was observed for the lower antibody coating density case (Fig. 3 A) and this was consistent for all the flow cases for both particle antibody coating densities. This clearly showed a significant increase in surface ICAM-1 expression in BAOECs at the TNF-α treated section. The p value is <0.01 by one way ANOVA test for particle binding data compared between TNF-α treated, upstream and downstream regions for both 210 nm particle antibody coating density cases. The downstream section of the channel also showed significantly higher (around 2 times) particle binding density compared to the upstream section ($p<0.01$).

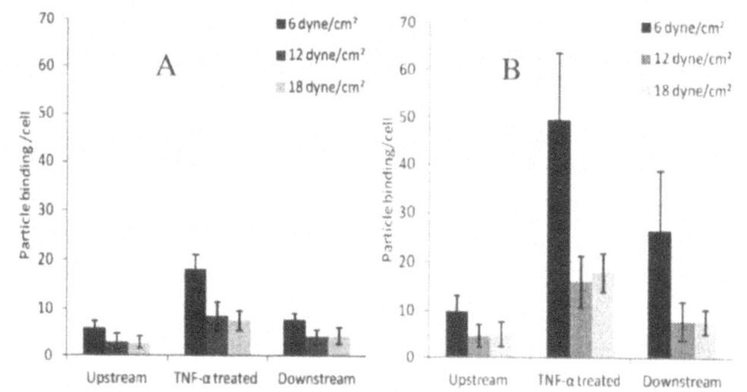

Figure 3: Targeted binding of anti-ICAM-1 coated particles of different antibody coating densities on BAOECs in the upstream, TNF-α treated and downstream sections of the channel at different shear rates. Quantification of particle binding density per cell at 6, 12 and 18 dyne/cm^2 for 210 nm particles of; (A), Anti-ICAM-1 coating density of 112.9 anti-ICAM-1/particle and; (B), 232.5 anti-ICAM-1/particle. $p<0.01$ by one way ANOVA test for particle binding data compared between upstream, TNF-α treated and downstream regions for both 112.9 and 232.5 anti-ICAM-1/particle cases. Data are shown as mean ± S.D. (n=100 cells from three independent experiments for each case).

3.4 Advantages and improvements over existing methods, devices or materials

Gold standards platforms currently used to model blood vessels *in vitro* are on a petridish and flow chambers. Competitors in today's market for microfluidic technology similar to ours are "Cellix LTD", "Ibidi GmbH" and "CFD Research Corporation". These companies offer microfluidic platforms on which cells can be cultured to study different phenomena/processes. The microfluidic chips they offer have some features that match our technology. But our product has several other key features not offered by our competitors like, the ability to closely mimic *in vivo* healthy and diseased blood vessel conditions in the same channel, ability to grow multiple cell lines, tunable geometry etc.

The introduction of a bi-layer system allows us to locally trigger receptor expression on cells that mimic vascular disease, grow two types of cells in the same system, and test targeted drug particle delivery and uptake. Our product allows control and target cells to be grown on the same device, integrated *in situ* imaging and testing, patient specific vascular geometry and flow condition, and fast parametric evaluation with minimal sample volume. The introduction of the bi-layer design supported by the semi-permeable membrane allows for an advantage in studies related to exchange, uptake and delivery, above and beyond what the competitors offer making the platform well suited for studying targeted drug delivery and other phenomena.

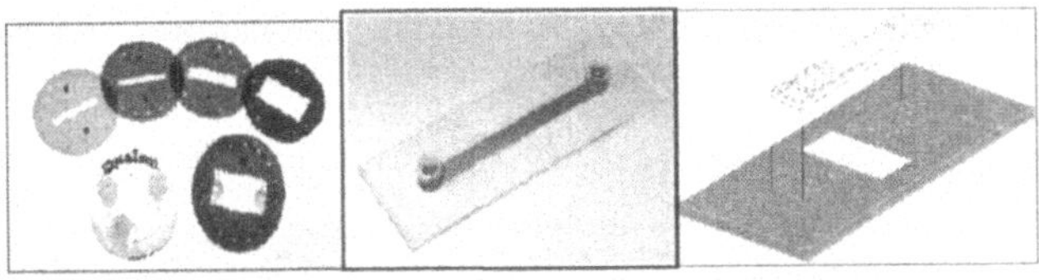

Flow channel with preset dimensions in flow chambers Courtesy: Glycotech | Current single layer platforms Courtesy: Ibidi | Our platform with the bi-layer advantage and tunable geometry

Flow chambers	Single layer channel	Our bi-layer channel
~ 1000 $ Reusable Tedious to set-up Not a lot of options Higher running cost: Need more sample volumes	~30 $/device Not reusable Single layer cell culture Customer specific channels not available Can only do one cell line	~10 $/device Not reusable Bi- layer cell culture Customer specific channels available Capable of doing multiple cell lines (top & bottom channel)

Table 1: Comparison of existing *in vitro* blood vessel models to our bi-layer platform

4 CONCLUSION

Our invention is a bio-mimetic microfluidic platform that can prototype a blood vessel outside human body. This platform integrates an EC layer in a microfluidic channel, which can be subjected to specific flow and chemokine stimulations. It consists of a top and bottom channel separated by a semi-permeable, porous, cell culture friendly membrane. ECs that form the inner lining of blood vessels are coated on the semi-permeable membrane in the top channel. Our platform has the capability to access specific sections of the top channel from the bottom channel, which allows spatially controlled simulations of these ECs from the basal side. This *in vitro* blood vessel model has the capacity to be applied as a generic platform to study various topics on, targeted drug delivery (ligand-receptor specificity, drug carrier features, hemorheology factors), cardiovascular conditions (atherosclerosis, drug eluting stents), immunology (inflammation, leukocyte adhesion and migration), as well as perform explicit studies on patient specific blood vessels (design channel geometries specific to patient blood vessel; integrate ECs from patient) to understand the best treatment strategy for a disease condition or study various factors that culminated in the onset of a disease condition, etc. This model understands the current/upcoming need of a blood vessel model that can cater to the demands for patient specific therapeutics, as well as provide a more realistic platform to enhance current conventional research studies.

REFERENCES

1. Staton, C.A., et al., *Current methods for assaying angiogenesis in vitro and in vivo.* International Journal of Experimental Pathology, 2004. **85**(5): p. 233-248.
2. Antony Thomas, H.D.O.-Y., Linda Lowe-Krentz, Vladimir R. Muzykantov, Yaling Liu,, *Biomimetic channel modeling local vascular dynamics of pro-inflammatory endothelial changes.* Lab on a chip: Under Review, 2015.
3. Khan, O. and M. Sefton, *Endothelial cell behaviour within a microfluidic mimic of the flow channels of a modular tissue engineered construct.* Biomedical Microdevices, 2011. **13**(1): p. 69-87.
4. Shipman, C., *Evaluation of 4-(2-Hydroxyethyl)-1-piperazineëthanesulfonic Acid (HEPES) as a Tissue Culture Buffer.* Experimental Biology and Medicine, 1969. **130**(1): p. 305-310.
5. Chien, S., *Effects of Disturbed Flow on Endothelial Cells.* Annals of Biomedical Engineering, 2008. **36**(4): p. 554-562.
6. Wasserman, S.M. and J.N. Topper, *Adaptation of the endothelium to fluid flow: in vitro analyses of gene expression and in vivo implications.* Vascular Medicine, 2004. **9**(1): p. 35-45.
7. Malek, A.M. and S. Izumo, *Mechanism of endothelial cell shape change and cytoskeletal remodeling in response to fluid shear stress.* Journal of Cell Science, 1996. **109**(4): p. 713-726.
8. Wong, A., T. Pollard, and I. Herman, *Actin filament stress fibers in vascular endothelial cells in vivo.* Science, 1983. **219**(4586): p. 867-869.
9. Kim, D.W., A.I. Gotlieb, and B.L. Langille, *In vivo modulation of endothelial F-actin microfilaments by experimental alterations in shear stress.* Arteriosclerosis, Thrombosis, and Vascular Biology, 1989. **9**(4): p. 439-45.
10. Goldblum SE, D.X., Campbell-Washington J., *TNF-alpha induces endothelial cell F-actin depolymerization, new actin synthesis, and barrier dysfunction.* Am J Physiol., 1993. **264**(4 Pt 1): p. 894-905.
11. Peng, J., et al., *Protein kinase C-alpha signals P115RhoGEF phosphorylation and RhoA activation in TNF-alpha-induced mouse brain microvascular endothelial cell barrier dysfunction.* Journal of Neuroinflammation, 2011. **8**(1): p. 28.
12. McKenzie, J.A.G. and A.J. Ridley, *Roles of Rho/ROCK and MLCK in TNF-α-induced changes in endothelial morphology and permeability.* Journal of Cellular Physiology, 2007. **213**(1): p. 221-228.

Guidelines for Choosing an Appropriate Turbulence Model when Simulating Blood Flows in Cardiovascular Devices

B.A.F. Simpson* and G.D. Tansley*

*School of Engineering, Griffith University, Parklands Drive, Southport, Queensland, 4222, Australia, ben.simpson@griffith.edu.au

ABSTRACT

Results are presented from a study to determine which turbulence models are the most appropriate to use in the simulation of cardiovascular flows. The research team undertook a systematic and detailed comparison of the performance of different turbulence models to predict typical velocity fields and shear forces experienced in blood flows. The computational results were validated against experimental data, and guidelines established that will help engineers choose the best models in future studies. The range of turbulence models tested included those specifically developed to predict laminar-to-turbulent transitional flow.

Keywords: computational fluid dynamics, cardiovascular, simulation, turbulence, microfluidics

1 INTRODUCTION

1.1 Computational Fluid Dynamics

Computational Fluid Dynamics (CFD) is a simulation technology that is commonly used in product development and to investigate 'what if...' design scenarios in a controlled, safe, virtual environment. In the last 10 years CFD has helped drive the development of cardiovascular devices which are used to diagnose and treat heart disease and related health problems. The devices are designed to have minimal haemolytic properties and to avoid thrombogenicity and other damage to formed elements in the blood (e.g. von Willebrands monomers). The detrimental effects of high shear stresses on blood have been known for many years. Simulations can be used to measure, assess and compare the performance of different devices across a wide range of operating conditions. Then only the optimal designs need to be fabricated and tested *in vitro*. Although it is generally agreed that simulations are very useful in analysing cardiovascular devices, engineers face significant challenges in generating them.

1.2 Turbulence Modelling Overview

It is quite difficult to define turbulence in either words or mathematically. The flow is so irregular that many researchers have tended to use the word random to describe the fluid motion, and for practical simulation this concept of randomness has prompted the application of statistical tools rather than a deterministic approach to the subject. However, today it is mostly accepted that a turbulent flow has structure and is chaotic rather than random. At any rate it is a phenomenon that is highly complex and stochastic.

It is widely accepted that for incompressible Newtonian fluids, the Navier-Stokes equations coupled with the continuity equation correctly describe the behaviour of flow under laminar and turbulent conditions. However, it is very computationally expensive to calculate all the structures present in a fully turbulent flow. Furthermore, small changes in the initial conditions can lead to large, difficult to predict changes in the long term evolution of the flow variables. So a more practical approach is to predict the overall flow behaviour and mean flow variable quantities. Traditionally there have been four approaches to solving the governing fluid equations. They may either be solved directly - Direct Numerical Simulation (DNS), temporally averaged - Reynolds Averaged Numerical Simulations (RANS), spatially averaged - Large Eddy Simulation (LES) or both spatially and temporally averaged - Detached Eddy Simulation (DES).

Increasing Computational Cost ↓

One-Equation models	• Spalart-Allmaras
Two-Equation models	• Standard K-ε • RNG K-ε • Realizable K-ε • Standard K-ω • BSL K-ω • SST K-ω
Reynolds Stress Model (RSM)	• Linear Pressure-Strain • Quadratic Pressure Strain • Stress-Omega
Transition Models	• K-Kl-ω Transition Model • SST Transition Model
DES	• Spalart-Allmaras • Relizable K-epsilon • SST k-ω
LES	• Scale-Adaptive Simulation
DNS	Directly solving the governing equations

Table 1: The different approaches to turbulence modelling available in the ANSYS CFD (FLUENT) code v.16.0 [1].

Modern commercial CFD packages have a large range of turbulence models available. A detailed description of the models listed in Table 1 can be found in the ANSYS FLUENT Theory Manual [1]. The wide range of competing models developed over more than 40 years of global research and development can often seem daunting. The most common question asked by researchers and industry practitioners alike is therefore 'which turbulence model is best for my application?'

1.3 Blood flows in Cardiovascular Devices

The rheology of blood makes it a difficult fluid to model. Furthermore, cardiovascular devices may involve complex rotating geometry, the flow fields may have transient features such as moving points of separation and flapping jets, the boundary conditions may be difficult to define and the flow field in the device is likely to contain laminar, turbulent and transitional flow regions.

Despite a considerable amount of research effort being spent on simulating cardiovascular flows [2, 3,], there has been little agreement as to the best approach to modelling the turbulence. As a result there has been a lot of confusion and in recent years authorities such as the U.S. Food Drug Administration (FDA) have called for further investigation and model validation [2]. Furthermore, blind tests on an idealized medical device geometry performed by several experienced CFD groups produced a wide variation in results [2]. This demonstrated that CFD for these applications cannot be considered "push button" technology and emphasised the need for better simulation procedural guidelines.

1.4 Selecting a Turbulence Model

When performing a CFD simulation many assumptions have to be made (modelling goals, geometry, physics, boundary conditions etc.) and there are many potential sources of error (round off, iteration, solution, model, user etc.). It is therefore essential for the CFD Engineer to review the analysis performed and consider the limitations in the context of the modelling goals.

Turbulence modelling can potentially be one of the greatest sources of error in an analysis. Assumptions and simplifications lie at the heart of all the models. The significance of these assumptions should be considered in the context of the flow being analysed.

When selecting a turbulence model a CFD engineer is advised to take several steps:

I. Decide clear modelling goals with a focus on the accuracy required. Furthermore, identify the fluid phenomena that the simulation is seeking to predict.
II. Determine if the flow to be analysed is laminar, turbulent or transitional. Although, in many fluid flows there may be localised areas of laminar and turbulent flows.
III. Review the literature and then select a model that best suites the physics of the problem and the available computational resources.
IV. Validate the selected models for the problem at hand against either experimental data or analytical data when it is available.
V. Perform a sensitivity study on the mesh, turbulence model, wall approach, turbulence model parameters and the turbulent boundary conditions.
VI. Review the limitations of the models and the potential sources of error in the analysis.

The undertaking of these steps should help the CFD engineer reduce the error in the analysis and gain confidence in the results of the simulation performed.

2 SIMULATION RESULTS

2.1 Idealized Medical Device Benchmark

In 2012 Stewart *et al* reported the results from the first computational inter-laboratory study that assessed CFD performance in simulations of an idealized medical device. A group of participants from universities, industry and the U.S. FDA defined the test geometry illustrated in Figure 1. Three independent laboratories generated experimental validation data using particle image velocimetry. This experimental data makes a useful dataset against which CFD simulations can be compared.

The flow rates were chosen to cover all flow regimes – laminar, transition and turbulent [2]. The geometry is orientated so that the flow enters a conical collector, passes through a throat and exits through a sudden expansion.

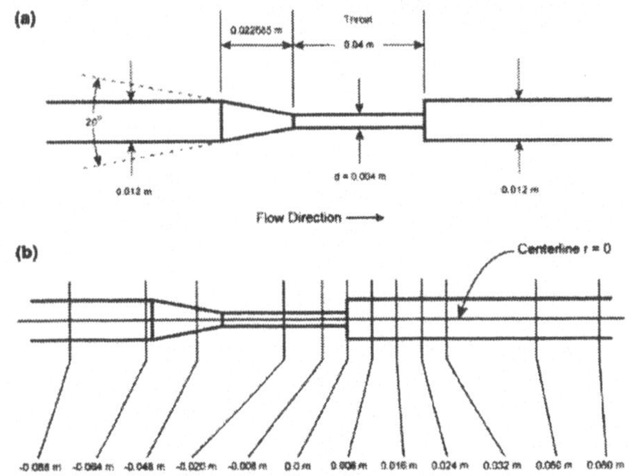

Figure 1: Schematic presentation of the computational domain for the idealized FDA benchmark nozzle geometry [2]

The idealized geometry has many of the characteristics that are found in microfluidic or cardiovascular devices; contractions (streamline curvature, localized acceleration of the flow), expansions (wall separations, recirculating fluid zones, jetting fluid and possibly jet break up) and channel flows (wall shear dominated zones). The present flow has a Reynolds No. of 6,500 based on the throat diameter suggesting that the flow will be mostly turbulent. Consequently the SST k-ω model was employed in the current studies, unless otherwise stated.

2.2 Mesh Sensitivity Study

One of the critical components of any simulation is the sensitivity of the solution to the mesh. The general rule is to have enough mesh to resolve the flow gradients. Various rules and predictive calculations exist to calculate the size of appropriate cells near walls and in free shear layers. However, one good approach is to carry out a grid sensitivity study. Figure 2 illustrates the results of a grid sensitivity study performed on an axisymmetric model of the idealized medical device. The course mesh had 3,350 cells, the medium mesh 11,000 cells and the fine mesh 44,000 cells.

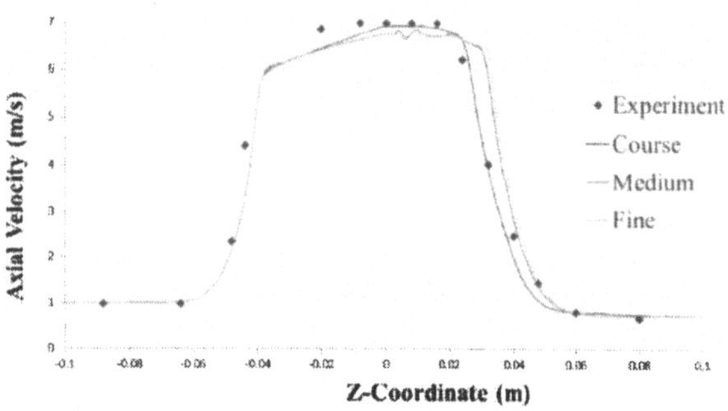

Figure 2: Time averaged axial velocity along the axis line comparing three simulations with different mesh sizes.

2.3 Turbulence Model Comparison

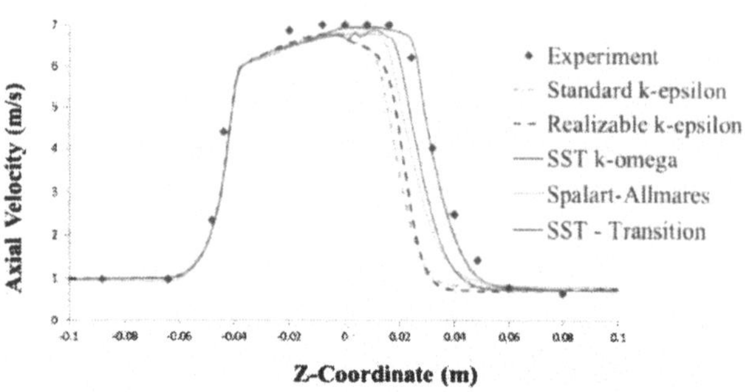

Figure 3: Time averaged axial velocity along the axis line comparing different turbulence models.

Only the SST K-ω model was able to predict the sudden expansion within the confidence levels of the experimental data.

2.4 Comparing Axisymmetric / 3D Results

If the geometry and the flow field are both symmetric around the centerline then an axisymmetric simulation is a valid approach. Axisymmetric simulations significantly reduce the computational effort.

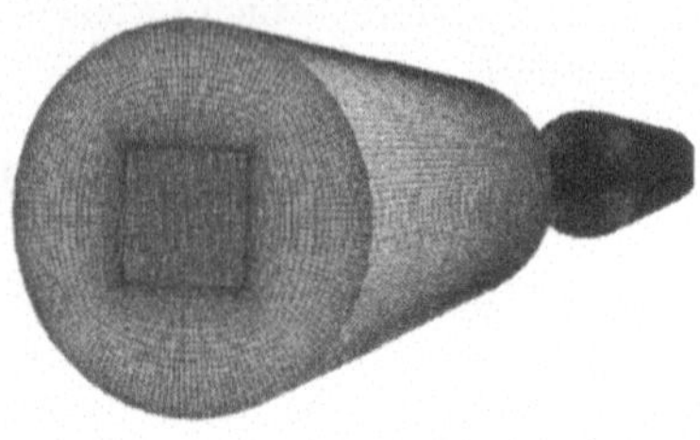

Figure 4: 3D mesh with 1,210,000 cells.

No significant difference was observed between the flow fields calculated using a large 3D simulation and the simpler axisymmetric simulation as illustrated in Figure 4.

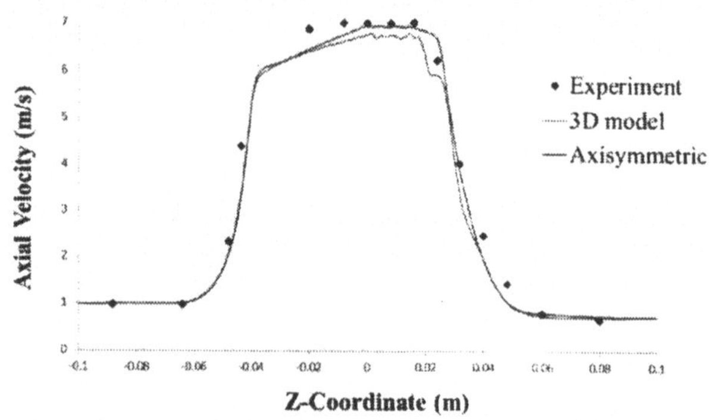

Figure 5: Time averaged axial velocity along the axis line comparing 3D and Axisymmetric models.

2.5 Sensitivity to the Inlet Boundary Conditions

The profile of the boundary condition only affected the flow in the inlet channel and contractor (Figure 6).

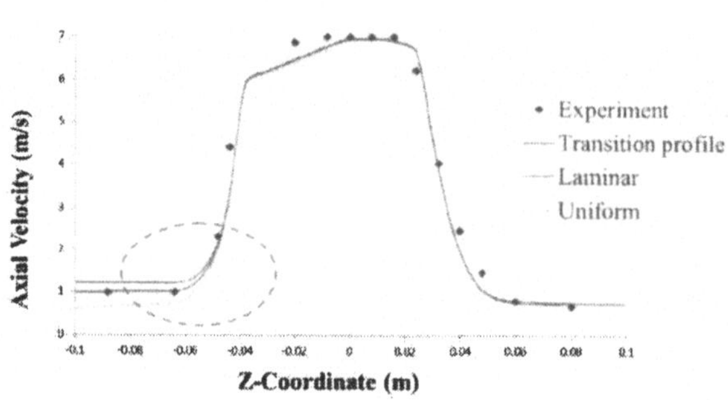

Figure 6: Time averaged axial velocity along the axis line comparing different inlet boundary conditions.

2.6 Transient Features

In some of the simulations which used more refined meshes, vortices were observed forming in the free shear layers after the sudden expansion. These structures appeared to be transient in nature.

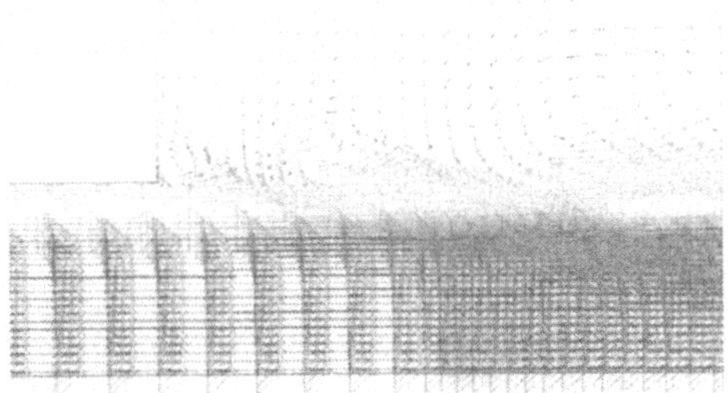

Figure 7: Vortices forming in the free shear layer.

3 SIMULATION CONCLUSIONS

The presented simulations of the idealized medical device benchmark geometry have demonstrated significant sensitivity to the mesh and the turbulence model choice. In addition, the more refined meshes resolved transient structures in the free shear layers, making convergence sometimes difficult to achieve. It is therefore perhaps not that surprising that the blind tests in the original study delivered such a wide range of results.

4 GUIDELINES FOR SELECTING AN APPROPRIATE TURBULENCE MODEL

Considering the modelling complexities associated with blood flows in cardiovascular devices, great care and good practice should be adopted to ensure accurate results. The current paper focuses specifically on guidelines for successfully modelling the effects of turbulence. CFD engineers undertaking a simulation of a cardiovascular flow are advised to:

I. Identify the flow behaviors that need to be captured to correctly predict the shear stresses. Determine if the flow is predominantly turbulent, laminar or transitional. This key information will help inform the CFD engineer which turbulence approach should be adopted.
II. Consider employing a transitional model that may include experimental correlations to trigger the transition onset, when the flow is determined to be in the transitional regime. An alternative approach which is computationally more expensive and less tested is to investigate using DES or LES.
III. Employ a k-ω formulation when the blood flow is determined to be predominantly turbulent. The complex geometry of cardiovascular devices means a turbulence model must be able to predict flow separation from a smooth surface under an adverse pressure gradient, such as in the diffuser section of the idealized geometry modelled, rather than simply from a sharp edge. Models that can integrate to the wall rather than depend on wall functions will therefore likely produce better results. The more sophisticated commercial CFD codes feature models that automatically adapt to the near-wall resolution by blending between wall functions and viscous sublayer models [4]. The present study demonstrated that the SST K-ω models performed the best.
IV. Undertake at least one transient simulation, since it is very likely that the blood flow in cardiovascular devices exhibits transient behavior. The simulation will help determine the source of the transient behavior and determine the effect on the shear stresses in the flow, although static simulations may still yield useful design results.
V. Specify realistic turbulent quantities and profiles at the inflow boundaries. It is reasonable to assume that the overriding flow mechanism in the self-similar region along the walls and areas of separation will be the shear layers. However, the specification of unrealistic boundary conditions such as a uniform velocity will mean the turbulence specified at the inflow boundary will be quickly dissipated and the shear layer development will be delayed [5].

5 FUTURE WORK

Turbulence modelling continues to be one of the driving technologies in CFD [4]. RANS modelling is likely to remain the preferred option for simulating blood flows in cardiovascular and microfluidic devices due to the balance of accuracy and computational resource. However, as computer power continues to grow year on year, the benefits of the DES and LES turbulence approaches will become more attainable. Modelling transitional flows will continue to be a challenge and the topic of continued research.

REFERENCES

[1] Ansys FLUENT Theory Guide – Chapter 4: Turbulence, 2013.

[2] S.F.C. Stewart, E.G. Paterson, G.W. Burgreen, P. Hariharan, M. Giarra, V. Reddy, S.W. Day, K.B. Manning, S. Deutsch, M.R. Berman, M.R. Myers and R.A. Malinauskas, Cardiovascular Engineering Technology, Vol. 3, No. 2, p408-426, 2012.

[3] K.H. Fraser, M.E. Taskin, B.P. Griffith, Z.J. Wu, Medical Engineering & Physics, 33, p. 263-280, 2011.

[4] F. Menter, Turbulence Modeling for Engineering Flows, Technical Paper from ANSYS Inc, 2011.

[5] B.A.F. Simpson and A.E. Holdo, Applied Mechanics and Engineering. Vol.4, No. 4 p801-815, 1997.

Plasmonic Nanocages as Photothermal Transducers for Laser Induced Heating and Bubble Generation

Ioannis H. Karampelas[1], Fatema Alali[2] and Edward P. Furlani[1,2]

[1] Dept. of Chemical and Biological Engineering, [2] Dept. of Electrical Engineering,
University at Buffalo SUNY, NY 14260, Office: (716) 645-1194, Fax: (716) 645-3822, efurlani@buffalo.edu

ABSTRACT

Plasmonics is emerging from among the most promising means for generating and controlling thermal energy at the nanoscale. In this approach, metallic nanoparticles are laser heated at plasmon resonant wavelengths that depend on the size, shape and properties of the particles. Key attributes of this method include remote optical activation, nanoscale resolution and efficient photothermal transduction. In this presentation we discuss photothermal transduction using laser-pulsed colloidal gold nanocage structures. We explore fundamental aspects of this process using computational models that predict the absorption spectra of the nanoparticles, photothermal transduction at plasmon resonance, heat transfer to the fluid and the dynamics of bubble generation under conditions of superheating. We demonstrate that process variables can be tuned to achieve optimal photothermal performance. We also discuss applications involving drug delivery and therapy of malignant tumors.

Keywords: Localized surface plasmon resonance (LSPR), nanocages, plasmonic nanocages photothermal energy conversion, plasmonic-enhanced photothermal energy transfer, LSPR-induced optical absorption, pulsed-laser photothermal heating, photothermal therapy, plasmonic nanobubble cancer treatment.

1 INTRODUCTION

In recent years, there has been a proliferation of research on gold nanocages. Representing a fairly recent advance in the field of nanoparticle synthesis, such nanostructures can now be readily fabricated using bottom-up techniques [1-3]. One major advantage of this geometry is its porosity. This property makes nanocages a great nanoparticle candidate for drug encapsulation and optically controlled photothermal drug delivery [4, 5]. Moreover, nanocages have also shown increased potential for a wealth of applications including photothermal treatment [6], catalysis [7], optical imaging [8], cancer diagnosis [9] and photodynamic therapy [10] among others. Some of the most promising areas of interest involves the use of gold nanocages for cancer detection and subsequent photothermal destruction of cancer cells [11, 12]; a process where an increase in the temperature of a colloid of gold nanoparticles, preferably uptaken by cancer cells, is achieved remotely through laser irradiation, thus leading to cell apoptosis or necrosis. However, if such temperature rise is not sufficiently localized there may be unwanted damage to nearby healthy cells. Hence, the need arises to better study the diffusion and control thermal energy at the nanoscale or investigate alternative methods of treatment such as cell membrane lysis through explosive nanobubble nucleation.

Laser-based plasmonic photothermal energy conversion is of particular interest since it allows for highly localized efficient heating. In this approach, illuminating the targeted nanostructures with a pulsed laser at their plasmon resonance frequency can result in optimum energy absorption because of the coherent and collective oscillation of electrons within the nanostructures, resulting in a maximized colloidal heating effect. The plasmon resonance frequency can, in turn, be tuned from the ultraviolet (UV) through the near-infrared (NIR) spectrum by controlling their size and especially their shape during the synthesis process. [1, 2, 11-13].

Thus far, we have used computational models to elucidate the fundamental physics of plasmon-enhanced photothermal transduction for various nanoparticle geometries that exhibit some form of axial symmetry, such as nanorods, nanotori and nanorings [1, 2]. It suffices to use a 2D analysis for these structures because of their axial symmetry. In this presentation, we study the photothermal physics of laser heated plasmonic nanocages, which is more challenging since it requires a full 3D analysis. Through the use of computational models, we are able to investigate the plasmon resonance behavior of gold nanocages as a function of their geometric aspect ratios and orientation with respect to the incident polarization. We also study their thermofluidic behavior, i.e. heat transfer to the fluid and the dynamics of bubble generation under conditions of superheating. For all cases studied, we consider nanosecond-pulsed laser illumination of the nanoparticles in fluid leading to bubble nucleation, wherein the pulse duration exceeds the characteristic time constants for initial non-equilibrium photothermal transient effects. Our results show that randomly oriented nanocages are best suited for colloidal photothermal manifestation due to their higher absorption efficiency. We demonstrate that various nanocage geometries also exhibit a resonance peak in the NIR, which in combination with their high absorption in random orientations, qualify them as exceptional photothermal transducers for deep tumor penetration and photoactivated drug delivery.

2 RESULTS AND DISCUSSION

We model laser-induced plasmon-enhanced photothermal effects using continuum-level photonic and fluidic analysis. We use computational electromagnetics to predict photothermal energy conversion within the nanoparticles, i.e. the time-averaged power absorbed by a particle as a function of the wavelength and polarization of the incident light. Subsequently, the absorbed power is converted into heat and the particle becomes a heat source within the fluid. As the particle heats up, thermal energy is transferred to the fluid and, under illumination of sufficient intensity, a vapor bubble can nucleate at the particle-fluid interface if the critical vaporization temperature of the fluid is reached. Once nucleated, a bubble will exhibit a dynamic behavior (expansion and collapse) that is a complex function of the heat and mass transfer at the bubble-fluid interface as well as the temperature and flow in the surrounding fluid. This simulation is achieved using computational fluid dynamic (CFD) analysis.

2.1 Photonic Simulations

For the photonic analysis we use 3D full-wave time-harmonic field theory. We perform the numerical simulations using the finite element-based COMSOL RF solver (wwww.comsol.com).

The first nanocage structure we consider, shown in **Fig.1,** is modeled as a hollow cubical structure with sides that are 5nm thick and 50nm in length. The nanocage is located at the origin of the computational domain and the incident field is generated by a surface current (not shown) in the xy-plane at the top of the computational domain,

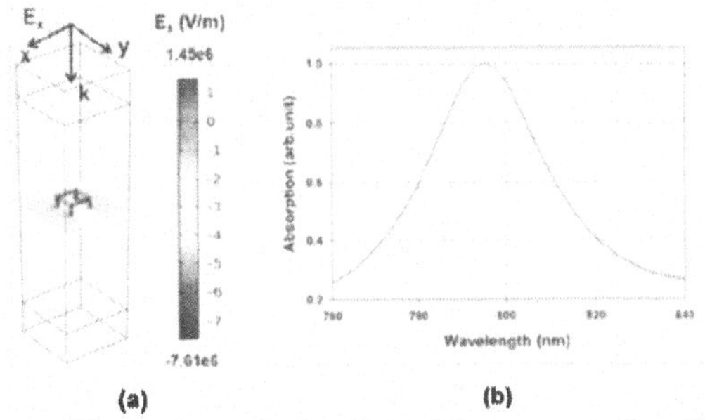

Figure 1. Photonic analysis of a gold nanocage (a= 50 nm and t = 5 nm) with parallel alignment to the incident polarization: (a) Computational domain and plot of E_x through a cross section of the domain, (b) absorbed power vs. wavelength at parallel orientation.

which results in a TEM plane wave with the electric field polarized in the x-direction and the wave propagating downwards along the z- axis as indicated by the **k** vector as shown in **Fig. 1a**. The permittivity of gold nanoparticles is predicted using an analytical expression that is based on a Drude-Lorentz model as described in our previous work [1, 2]. The parameters for this equation can be found in the literature [14-16]. We performed a parametric analysis of the power absorbed by the nanocage as a function of wavelength for λ = 760-840 nm. Plasmon resonance was found to occur at 795 nm (**Fig. 1b**).

Following our initial results, we proceeded with a parametric analysis of the absorption of a nanocage with the same dimensions at different orientations relative to the incident polarization. In **Fig. 2**, a comparison is shown between the absorption of a nanocage, of same dimensions as our initial analysis, and a nanorod with a length of 60nm and a diameter of 17nm. The absorption spectra vs. orientation of the nanocage is radically different from that of the nanorod. In the case of the nanocage, a very high

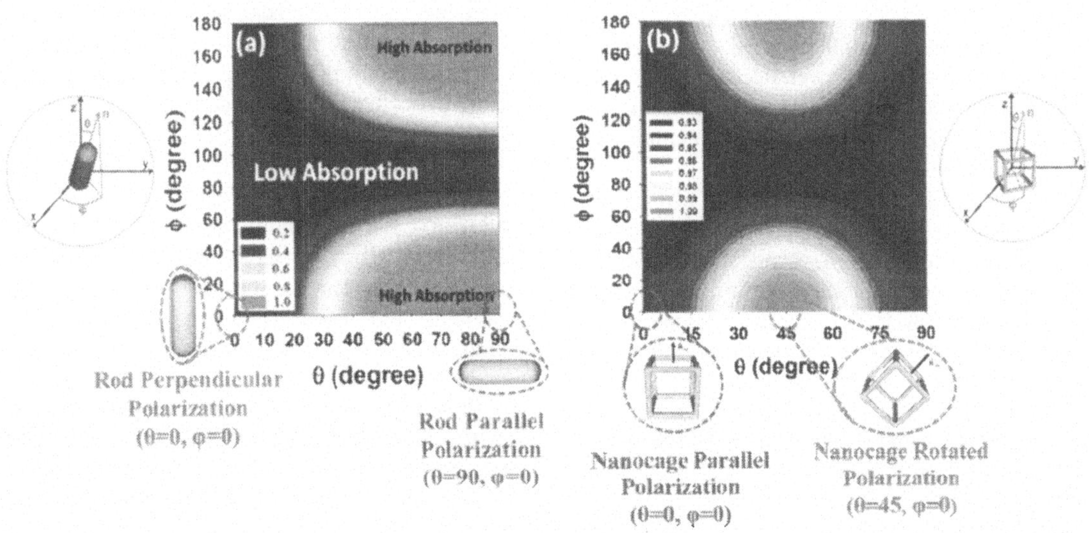

Figure 2. Normalized peak absorption at respective fixed LSPR wavelengths as a function of particle orientation (θ, φ): (a) nanorod at λ = 770 nm, (b) nanocage at λ = 795 nm [2]

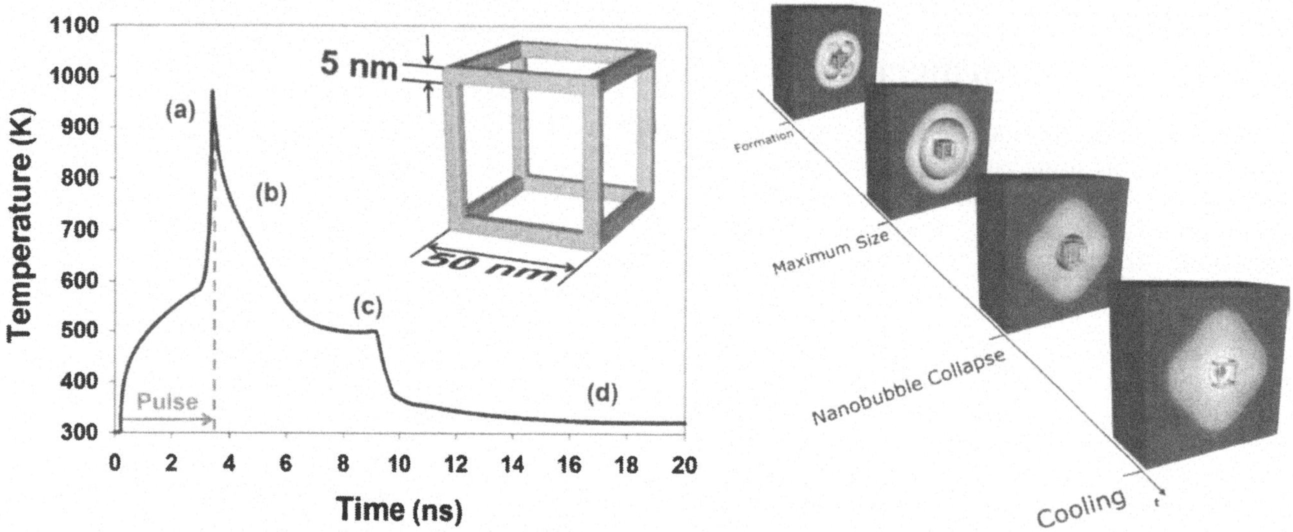

Figure 3. Photothermal heat cycle of a nanocage (a=50nm, t=5nm) (perspective 1/2 view): plot of nanocage temperature vs. time, pulse duration indicated by red arrow and dashed line and inset plots showing various phases of the thermal cycle; (a) nanobubble formation, (b) nanobubble (maximum size), (c) nanobubble collapse, (d) cooling.

absorption (more than 90%) is achieved throughout a wide range of possible orientations while in the case of the nanorod the absorption remains high only in certain favorable directions. Moreover, our results show that peak absorption for the nanocage occurs at orientations offset by 45° in an angular sense from parallel alignment with the incident field. Hence, we find that colloidal Au nanocages could be excellent as photothermal transducers for deep tumor photothermal treatment or drug delivery as they have a high transduction efficiency. This is in contrast to the nanorod where, for NIR resonance, high absorption is restricted to (near) parallel polarized orientations [2].

2.2 Fluidic Simulations

A 3D CFD analysis was implemented in order to ascertain the pulse duration and power necessary for the nucleation of an explosive nanobubble without causing potential damage to the nanostructure. Due to its symmetry, it sufficed to model one octant of the nanocage.

The fluidic analysis can be divided into two phases. In the first phase, a thermal analysis was performed to predict the power level required to heat a nanocage with an edge length of 50nm and an edge thickness of 5nm from an ambient temperature of 300K to the superheat temperature of water i.e. 580K. This depends on the pulse intensity and duration, which was constrained to be between 3 to 5 ns. In the second phase of the analysis, we applied the power levels obtained in phase one and slightly increased the pulse duration so that the nanoparticles were heated beyond the superheat temperature, which caused vaporization of the surrounding fluid and bubble nucleation. The pulse duration was tuned so that the nanoparticle achieved a temperature that was sufficiently high to generate a sustained nanobubble, but low enough (< 1100K) to avoid melting the nanoparticle. Based on our analysis, it was found that a power of 180 μW with pulse duration of 3.18 ns is adequate for nucleation. The results of the thermo-fluidic analysis are shown in **Fig. 3**.

The nanocage and the fluid are initialized at a temperature of 300K. After 0.2 ns, a heat pulse caused by laser illumination result in a sudden rise in the temperature of the nanoparticle, which is indicated in the temperature vs. time plot of **Fig. 3**. When the nanocage reaches the superheat temperature, a homogeneous water vapor bubble is nucleated around it. Upon nucleation, the temperature of the nanoparticle rises rapidly since it is being insulated by a thin sheath of vapor while still being heated. This continues until the end of the heat pulse, which coincides with the bubble formation and maximum temperature as shown in plot segment and inset figure (a). The pulse duration is 3.18 ns as indicated by the red dashed line. After it is nucleated the nanobubble expands because of its higher pressure compared to the surrounding fluid. The nanobubble reaches a maximum size as shown in inset figure (b). For this geometry, the maximum bubble radius achieved was 120 nm. Eventually, the nanobubble collapses bringing fluid back in contact with the nanoparticle as shown in plot segment and inset figure (c), gradually reducing its temperature to the ambient temperature (d). An interesting observation of this process is the formation of a hot droplet of fluid in the middle of the nanoparticle, which partially evaporates as the nanobubble expands. The CFD simulation was performed on 12-core workstation with 64 GB of RAM. The time for the simulation was approximately 100 hours.

3 CONCLUSIONS

We have used a combination of computational electromagnetic and fluid dynamic analysis to study the effects of nanosecond-pulsed laser heating of Au nanocages. Our results demonstrate that nanocages exhibit a much higher absorption (photothermal transduction) over a wide range of orientations with respect to the incident polarization as compared to other nanoparticles with less 3D symmetry such as nanorods. Nanocages have a more than 90% average absorption for random orientations when illuminated with the appropriate plasmon resonance frequency. Hence, Au nanocage structures are well-suited as photothermal transducers and hold promise for photothermal therapy. Our modeling approach can be used to explore fundamental behavior of nanoscale photothermal processes and is useful for the rational design of novel plasmonic structures for a wide range of applications.

REFERENCES

[1] Furlani, E. P., Karampelas I. H., et al. (2012), Lab on a Chip **12**(19): 3707-3719
[2] Alali, F., Karampelas, I.H., et al. (2013), Journal of Physical Chemistry C **117**: 20178-20185
[3] Skrabalak, S.E., et al. (2008), Accounts of Chemical Research, 41(12): 1587-1595
[4] Khlebtsov, N., et al. (2013) Theranostics, 3(3): 167-180
[5] Yavuz, MS, Cheng, YY et al. (2009) Nature Materials 8(12): 935-939
[6] Chen, JY, Glaus C et al. (2010) SMALL **6**(7):811-817
[7] Zeng, J, Zhang, Q et al. (2010) Nano Letters **10**(1):30-35
[8] Cang, H, Sun, T et al. (2005) Optics Letters **30**(22):3048-3050
[9] Kim, C, Cho, EC et al. (2010) ACS Nano **4**(8):4559-4564
[10] Gao, L, Fei, JB et al. (2012) ACS Nano **6**(9):8030-8040
[11] Skrabalak, SE, Au, L et al. (2007) Nanomedicine **2**(5):657-668
[12] Chen, JY, Wang, DL et al. (2007) Nano Letters **7**(5):1328-1322
[11] West, J. L. and Halas N. J. (2000). Current Opinion in Biotechnology **11**(2): 215-217
[12] Pitsillides, C.M., et al. (2003). Biophysics Journal. **84:** 4023-4032
[13] Roper, D.K., Ahn W. et al. (2007). J. Phys. Chem. C **111**(9): 3636-3641
[14] Jain P. K., Huang X. H., et al. (2008), Acc. Chem. Res. **41**: 1578–1586
[15] Kreibig U. and Vollmer M. (1995), Optical Properties of Metal Clusters, Springer, Berlin
[16] Lewinski N., Colvin V., et al. (2008), SMALL. **4**: 26–49.

Bioabsorbable Anti-inflammatory Material: A Novel Protein-Conjugated Polymer with Applications in the Implantable Medical Device Field

Crystal Rapier[1*], Esther Chen[1,2], Wendy F. Liu[1,2,3], Abraham P. Lee [1]

1 Department of Biomedical Engineering, University of California, Irvine, CA, USA
2 The Edwards Lifesciences Center for Advanced Cardiovascular Technology, University of California, Irvine, CA, USA
3 Department of Chemical Engineering and Materials Science, University of California, Irvine, CA, USA
*E-mail: crapier@uci.edu

ABSTRACT

In this paper, we present a microfluidic method of creating hybrid protein-PLGA particles. We are combining the bioabsorbable property of PLGA and the anti-inflammatory property of an immunomodulatory protein. We aim to assess whether this protein can inhibit inflammatory response to PLGA's acidic subunits as it degrades. Particles are made using a double emulsion and solvent extraction process. To date, our microfluidic system has been optimized to create single or multi-cavity particles, thin shell particles, and solid particles using a non-toxic solvent. In the near future we will characterize and compare the release profiles of the hybrid protein-PLGA particles vs. conventional protein encapsulated particles.

Keywords: PLGA, stent, particles, microfluidics, anti-inflammatory, immunomodulatory

INTRODUCTION

The human immune system is designed to recognize and fight "foreign" objects, pathogens, or domestic diseased cells. Part of the immune system's defenses includes an inflammatory response in which the area surrounding damaged or infected tissue becomes red and swollen (inflamed). This is in part due to white blood cell activation from the presence of a foreign object or laceration. Inflammation poses a major problem to the successful function of essentially "foreign" implanted medical devices such as vascular stents, heart valves, hip or knee replacements, etc. Conventional implants are permanent and made from surgical grade stainless steel metal which does not provoke an immune response. However, there is a need for non-permanent implant solutions. Biodegradable/bioabsorbable materials are a relatively new movement in the multibillion dollar implantable medical device field. These materials break down into smaller absorbable subunits in the body after it has served its purpose. Despite the promise of non-permanent implants, subunits of degraded bioabsorbable materials can induce a localized inflammatory immune response. This can be seen in a commonly used bioabsorbable material called Poly(lactic-co-glycolic acid) (PLGA), which is currently used as degradable stent and suture material among other things. PLGA degrades into nontoxic lactic and glycolic acid subunits. Although PLGA is a safe material for implantation into the body, accumulation of its acidic subunits in a localized area can cause irritation thereby invoking an inflammatory response. To overcome this limitation, one must design a degradable polymer that is capable of reducing acute inflammation while promoting wound healing. In previous studies, an immunomodulatory protein coated on polystyrene surfaces has been found to reduce the activation of white blood cells and their subsequent secretion of inflammatory factors [1]. Our work aims to combine the bioabsorbable property of PLGA and the anti-inflammatory property of an immunomodulatory protein described by [1]. We are using a microfluidic system designed to conjugate functionalized proteins to PLGA, forming protein-polymer hybrid particles. Our lab has previously reported on two microfluidic approaches for the synthesis of PLGA micro and nanoparticles using a non-toxic organic solvent [2]. We are assessing whether an immunomodulatory protein can be conjugated to PLGA microfluidically, displayed throughout the particle lifetime, and retain its functionality. For the purpose of this study, bovine serum albumin (BSA) is used as a model protein to test the feasibility of protein-polymer conjugation and to fine-tune the parameters of microfluidic particle production. The next step is to conjugate PLGA with an immunomodulatory protein as described by [1] and observe its functionality. In the future, we hope to create vascular stents with this anti-inflammatory PLGA material which would provide a non-permanent, biodegradable alternative to conventional stent materials.

APPROACH

Microfluidic devices are fabricated by the poly(dimethyl) siloxane (PDMS) replica molding process using standard soft lithography techniques [3]. Device design includes two flow focusing junctions, the first junction (Fig. 1)creates discrete droplets of protein immersed in a PLGA-solvent solution (1-2.5wt% PLGA 50:50 or 2wt% PLGA 85:15 in DMC (dimethyl carbonate)). The second junction (Fig. 2) creates water/oil/water double or multi emulsions with 1-2.5wt% PVA (polyvinyl alcohol) solution as the continuous phase. The device outlet is connected to a mixing well where the emulsions (Fig. 3) are stirred to help the solvent evaporation process and the subsequent formation of particles (Fig. 4). Particles are formed after the non-toxic solvent, DMC, is evaporated from the emulsion while in aqueous solution. Particles are collected, centrifuged, and washed several times before further testing.

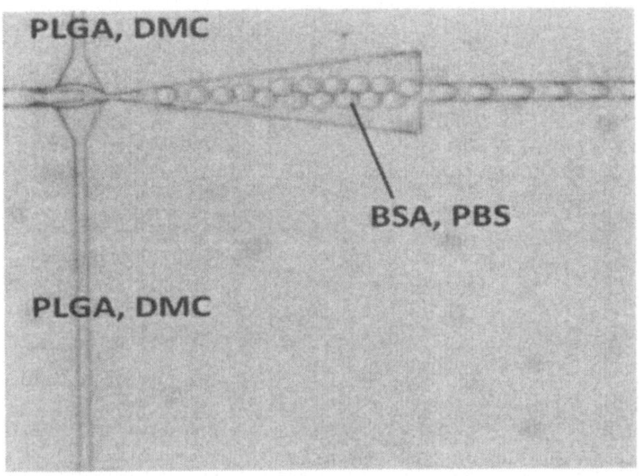

Figure 1: Microfluidic device, primary flow focusing junction.

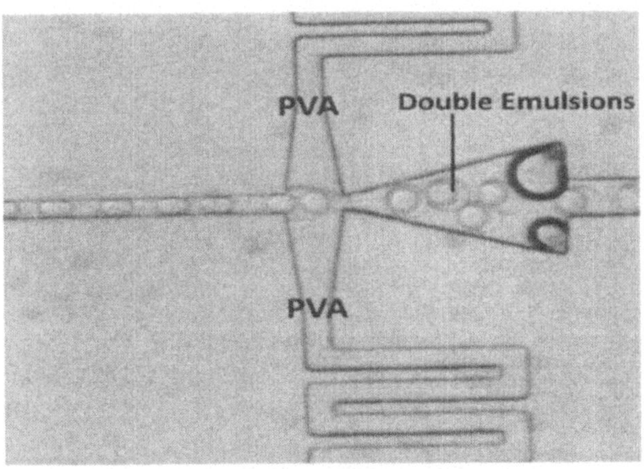

Figure 2: Microfluidic device, secondary flow focusing junction.

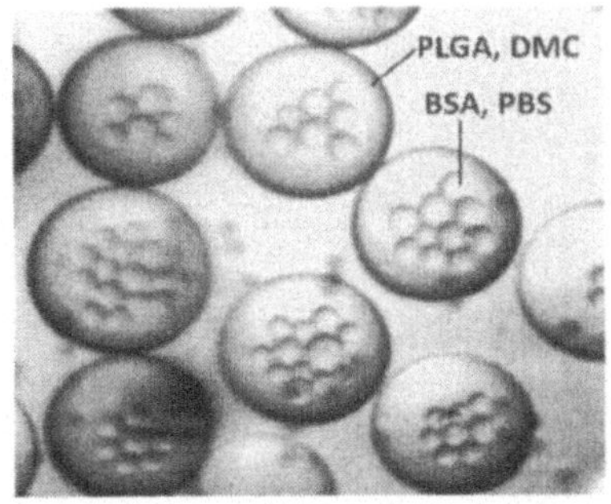

Figure 3: Multi-emulsion PLGA droplets before solvent evaporation (4X magnification).

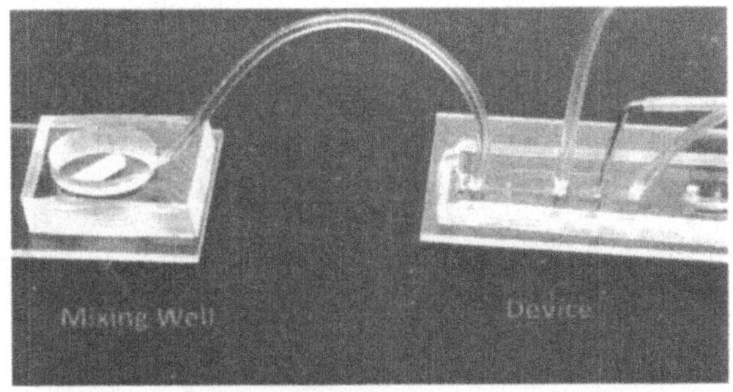

Figure 4: PDMS microfluidic device and PDMS mixing well

RESULTS & DISCUSSION

To date, our microfluidic system has been optimized to create single or multi-cavity particles, thin shell particles, and solid particles using a non-toxic solvent **(Fig. 5)**. To monitor and characterize the release profiles of the particles, we have encapsulated fluorescent BSA **(Fig. 6)**. We are currently comparing two approaches for the conjugation of protein to functionalized PLGA. In the near future we will characterize and compare the release profiles of the hybrid protein-PLGA particles vs. conventional protein encapsulated particles. Once the BSA release profiles are characterized, CD200 will be substituted in place of BSA. Particles will be created with different dosages of CD200, cultured with macrophage, and extensively observed for biological response to its degradation. This includes observing CD200-PLGA particle influence on cell activation and secretion of pro-inflammatory factors.

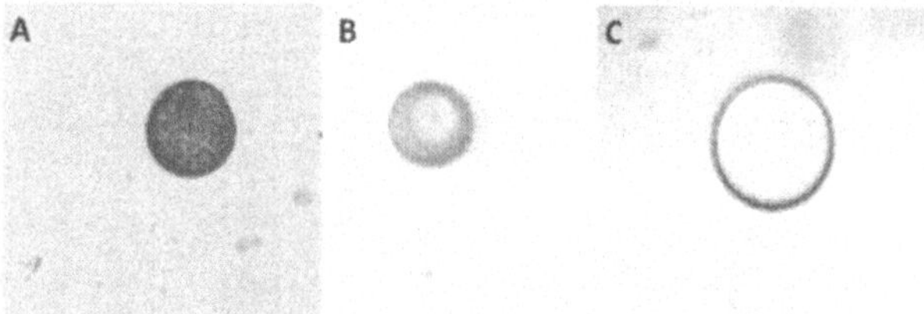

Figure 5: Inverted microscope image of solid PLGA particle at 20X magnification (A), Single cavity particle at 20X magnification (B), and a thin shell particle at 10X magnification (C).

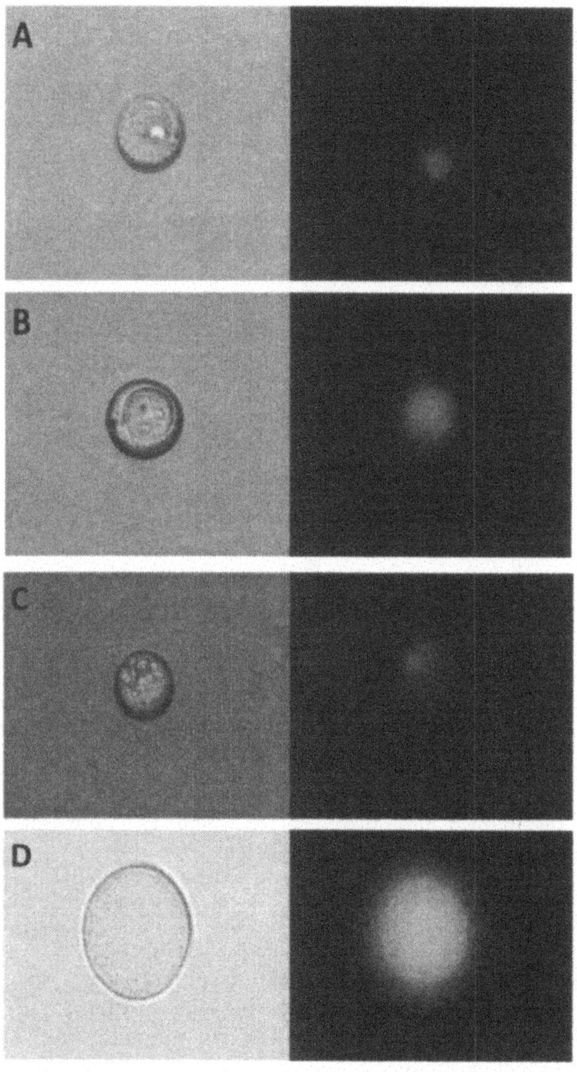

Figure 6: Inverted microscope bright field and FITC filtered images of PLGA particles encapsulating fluorescent BSA at 40X magnification. Images of a single small cavity particle (A), a large single cavity (B), a double cavity (C), and a large thin shell particle (D).

REFERENCES

[1] Kim, Y.K., Que, R., Wang, S.-W. and Liu, W. F. "Modification of Biomaterials with a Self-Protein Inhibits the Macrophage Response". Advanced Healthcare Materials, 3: 989–994. (2014)

[2] Hung L.-H., The S.-Y., Jester J., and Lee A.P. "PLGA Micro/Nanosphere Synthesis by Droplet Microfluidic Solvent Evaporation and Extraction Approaches". Lab Chip **10**, 1820-1825. (2010).

[3] Xia Y.N. and Whitesides G.M. "Soft lithography". Angewandte Chemie-International Edition, vol. 37, pp. 551-575. (1998).

Adsorption of Cd (II) Ions from Aqueous Solution Using Activated Carbon Prepared from *Vitellaria paradoxa* Shell

G.B Adebayo*, F. A Adekola*, A. A Jimoh**, F. O Alao* and H. I Adegoke***

*Department of Industrial Chemistry, University of Ilorin, P.M.B 1515, Ilorin, Nigeria, adebayochem@yahoo.com,

**Department of Chemical, Geological and Physical Sciences, Chemistry Unit, Kwara State University, Malete Ilorin, Nigeria.

***Department of Chemistry, University of Ilorin, P.M.B 1515, Ilorin, Nigeria

ABSTRACT

Activated carbon prepared from *vitellaria paradoxa* shell using ortho-phosphoric acid as activating agent was used for the removal of Cd (II) ions from aqueous solution. The adsorbent was characterized by some physicochemical and spectroscopic methods such as pH, point of zero charge (PZC), moisture content, iodine number, loss on ignition, bulk density, XRF, SEM and TEM. The pH and PZC of the sample were found to be 4.0 and 5.8 respectively. The adsorbent sample has moisture content of 5.22 ± 0.1%, iodine number of 256.50 ± 59%, loss on ignition of 10.71 ± 0.18% and bulk density of 0.84 ± 0.09%. The XRF analysis indicated that Ca, Si and Fe were major constituents in the adsorbent sample. The morphology of the adsorbents as revealed by SEM and TEM analysis results indicated good adsorptive characteristics of the adsorbent. The equilibrium adsorption of Cd (II) ions data were well fitted with Langmuir and Freundlich isotherms with correlation coefficient of 0.996 and 0.998 respectively. The kinetics study revealed that the experimental data fitted pseudo-second order kinetic better when compared to zero-order kinetics. Thermodynamics experiment indicated that the adsorption process was endothermic with ΔH value of +27.84 kJ/mol and ΔS of +0.993 kJ/molK and ΔG is negative at all temperature indicating the feasibility of the adsorption process.

Keywords: *Vitellaria paradoxa,* adsorbent, adsorption, SEM and TEM, heavy metals, kinetics, thermodynamics

1.0 INTRODUCTION

Cadmium is released by various natural and anthropogenic sources to the atmosphere, aquatic environments (fresh and salt water environments) and terrestrial environments.

In the environment, cadmium is toxic to plants, animals and micro-organisms. Being an element, cadmium is persistent – it cannot be broken down into less toxic substances in the environment. The degree of bioavailability and potential for effects varies depending on the form of cadmium.

Because of its toxicity and bioaccumulation, Cd (II) is considered as a priority pollutant by the U S Environmental Protection Agency. The permissible limit for Cd(II) as described by WHO is 0.01 mg/dm^3. The main anthropogenic pathway through which Cd(II) enters the water bodies is via wastes from industrial processes such as electroplating, plastic manufacturing, metallurgical processes and industries of pigments and Cd/Ni batteries (Cheremisinoff, 1995).

The shea tree (*Vitellaria paradoxa*) is found in areas with 400-1800 mm rainfall per year. It is a multi-purpose tree daily used by rural African communities. The fruit when ripe is either eaten as a snack and also as a famine food. It can be eaten raw or slightly cooked. The pulp can be processed into juice. According to McAllan *et al.* (1996), the pulp could also be removed by fermentation. The nuts are laid out to dry in the sun. The kernels are extracted usually before the butter making starts, by cracking open the nuts with stones or gently pounding in a mortar and the powder is made into butter.

This study investigated the potential of the *Vitellaria paradoxa* shell which are agricultural waste as adsorbent for the removal of Cd(II) ions from aqueous solution.

2.0 MATERIALS AND METHODS

2.1 Activated Carbon preparation

The Shea nut shells (*vitellaria paradoxa*) were collected from a small scale Shea butter producing factory at Budo oja at Ifelodun L.G.A. area in Kwara state. The sample was washed with plenty of water to remove surface impurities and sundried. Sample was later dried in an oven at 105°C overnight (Fan *et al.*, 2003, Mozammel *et al.*, 2002, Zahangir *et al.*, 2008, Itodo *et al.*, 2009 and Omonhenle *et al.*, 2006). A 100g of pre-treated sample (< 2mm mesh size) was introduced into a clean and pre weighed crucibles and carbonized in a furnace at 500°C for 5 minutes and washed with cool water and 10% HCl to remove surface ash. Accurately weighed 3g carbonized sample was mixed with $2cm^3$ of each 1M activating agent (H_3PO_4). The sample was introduced into a furnace, heated at 800°C for 15 minutes. The activated sample was cooled with ice cold water. Excess water was drained and sample was allowed to dry at room temperature (Gimba *et al.*, 2004).

2.2 Sample preparation

The stock solution of $CdSO_4$ was prepared and diluted into the required concentration using de-ionised water whenever necessary.The adsorption to equilibrium time was determined using 0.1 g of activated vitellaria paradoxa, 50 mL of metal solution with a concentration of 200 mg/L. The mixture was shaken (120 rpm) for 2 hours at ambient temperature. The amount of heavy metal adsorption per unit weight of activated *vitellaria paradoxa* at time t (q_t) in (mg/g), was calculated. Adsorption isotherms of the heavy metals were determined by varying the initial concentration of the metal solutions in range of 10-200 mg/L at constant and time.

3.0 RESULTS AND DISCUSSION

3.1 Physico-chemical properties of adsorbents

The moisture content of the activated carbon sample was 20.20% which suggested extensive porosity in the structure of adsorbent (Suguvadevi *et al.*, 2002).

The percentage ash content was 10.71% for activated carbon sample which indicated a decrease in percentage volatile matter (Gan *et al.*, 2004). The pH of the activated adsorbent after mixing with de-ionised water was 4.0 which might be the kind of treatment to which the adsorbent was subjected. The result of the iodine number of the activated carbon sample was 256.2 indicating good adsorptive capacity and better development of the microporous structure (Aziz *et al.*, 2009; Baccar *et al.*, 2009).The SEM of the adsorbent showed rough areas of surface of the carbon and very minute micropores (figure 1).

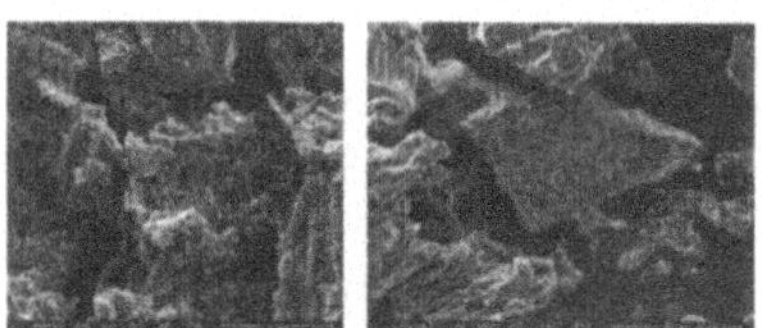

A B

Figure 1. SEM Micrograph of activated carbon sample (A) and SEM Micrograph of activated sample after adsorption (B)

Transmission electron microscope (TEM) was used to determine the internal morphology of the adsorbent. It was observed (figure 2) that the adsorbent was effective on Cd adsorption, since the internal pore sizes was reduced in Cd-embedded adsorbent.

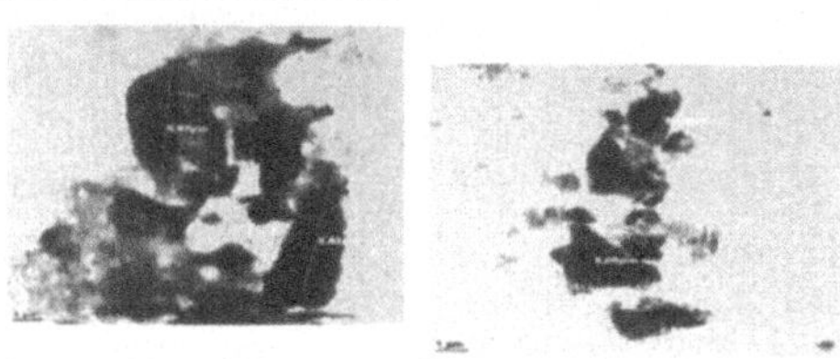

A B

Figure 2.TEM Micrograph of activated sample(A) and TEM Micrograph of activated sample after adsorption(B)

The elemental analysis by XRF of the adsorbent result showed that Ca, Si and Fe are present as major constituents (1.528- 12.18%); Na, Ti and Zn are minor constituents (0.36-0.6775%) while Mg, Al, P, Mn, Co, Ni, Cu and Cd are in traces (0.0018-0924%).

The pH_{pzc} was found to be 5.8 for the activated carbon while its pH was 4.0 indicating that the adsorption of anions could be enhanced (Hamadaoui, 2007; Almeida *et al.*, 2009 and Farahani *et al.*, 2011

3.2 Batch Adsorption Studies

The degree of metal ions adsorption onto the adsorbent increased from 50.48mg/L to maximum of 57.37mg/L of Cd(II) when the solution pH was increased from 2-6(Sari and Tuzen, 2009). The maximum percentage removal of 87.37% of Cd (II) was observed at 0.1g. The results were subjected to Langmuir and Freundlich adsorption isotherm models. The correlation coefficient of 0.996 and 0.998 were obtained for Cd(II) indicating monolayered multilayer adsorption nature of Cd (II) ions. The pseudo-first order plot is linear with correlation coefficient of 0.361. The pseudo-second order kinetic plot is of better linearity with correlation coefficient 0.97. The ΔG value of Cd (II) was negative, confirming the thermodynamic feasibility of adsorption of Cd (II) on *Vitellaria paradoxa* (Chowdhury *et al.*, 2011).

4.0 CONCLUSION

The equilibrium adsorption data for the removal of Cd(II) ion from aqueous solution showed satisfactory correlation with the Langmuir adsorption and Freundlich. The kinetic study revealed that pseudo-second order kinetic was better fitted compared to first-order kinetic. Thermodynamic study revealed a spontaneousity of the process.

REFERENCES

Annadurai, G., Juang, R.S., Lee, D.J. (2002). Adsorption of heavy metals from water using banana and orange peels. *Water Science and Technology*, 47(1), 185-190.

Aziz. A, Ovali. M.S, Elanduloussi E.H 2009; Chemically modified olive stone a low cost sorbent for heavy metals and basic dyes removal from aqueous solution. J. Hazard materials, 163;441-447.

Baccar R, Bouzid J, Feki M, Montel A 2009; Preparation of activated carbon from Tunisian olive waste cakes and its application for adsorption of heavy metal ions. J. Hazard materials, 162; 1522-1529

Cheremisinoff P.N. and Angelo C.M., (1980). Carbon adsorption applications, in: Carbon Adsorption Handbook, Ann Arbor Science Publishers, Inc., Ann Arbor, MI, pp. 1–54.

Rapid prototyping bacterial spore hygro-actuators for soft robotics and adaptive materials

A.-H. Cavusoglu*, X. Chen**, K. Tatem***, and O. Sahin****

*Department of Chemical Engineering, ac3236@columbia.edu
**Department of Biological Sciences, xc2239@columbia.edu
***Department of Physics, kvt2103@columbia.edu
****Department of Biological Sciences and Physics, Columbia University, New York, NY, USA, sahin@columbia.edu

ABSTRACT

Over the past decade, rapid prototyping has migrated from research and industrial labs into the homes of consumers. However, these solid models are limited in demonstrating dynamic actuation. A large range of soft materials are readily available for sensing and structural applications, but soft actuators are still slow to manufacture and inefficient. One method of introducing strong, controllable actuation to devices with gentle stimulation is through hygroscopic actuators. Here, we present a platform that demonstrates the potential of printable soft actuators. We develop a spore suspension that resists sedimentation, improves ink jet performance, and optimizes surface wetting. We demonstrate control of the direction, speed, and magnitude of actuation. Extended work by the researchers demonstrate potential for both contractile and expansive actuation. Finally, stimuli-responsive folding architectures are presented as a possible tool for hierarchical design of humidity-responsive soft robots.

Keywords: water, bacteria, soft robotics, actuators, adaptive materials

1 INTRODUCTION

Additive manufacturing (3-D printing) is growing in popularity and are entering the homes of consumers. Many of these platforms allow users to manufacture solid models constructed from plastic parts and computer aided design (CAD) data. Some of these platforms have been modified for 4-D printing [1]. However, these models are limited in dynamic actuation options. A large range of soft materials are readily available for sensing and structural applications, but soft actuators are still relatively weak and inefficient [2]. The high voltages usually required to operate electroactive polymers can degrade some materials and can prove to be a user hazard, whereas lower voltage iconic polymer metal composites (IPMCs) are currently inefficient, slow, and weak. Pneumatic actuation requires extensive additional pressure infrastructure which can make these systems bulky. Finally, shape-memory wires have significant power requirements that make them unreasonable for smaller devices.

One method of introducing strong, controllable actuation to devices with gentle stimulation is through hygroscopic actuators [3, 4]. Hygroscopic actuators exhibit significant, and reversible, volume expansion resulting from the absorption/desorption of water vapor. One potential hygroscopic material is B. subtilis spores [5]. These spores have demonstrated high energy density (~10 MJ/m^3) due to the relatively gentle stimulus of water vapor. These spores have been imbedded in the semi-crystalline adhesive poly-vinyl alcohol to create composite materials that change shape due to changes in local relative humidity (Figure 1).

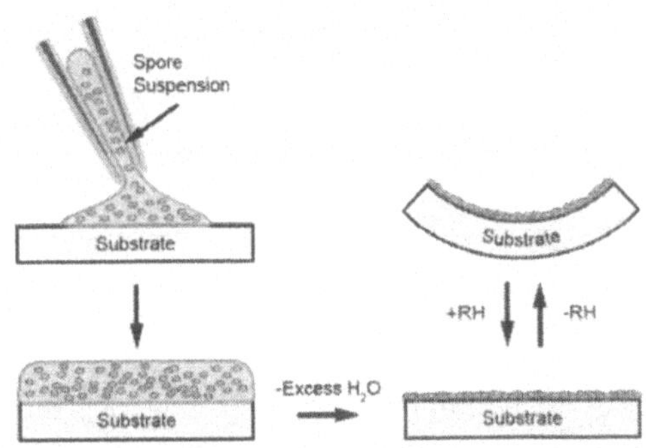

Figure1: Schematic of the micropipette drop coating process and the shape changes due to local relative humidity levels.

Inspired by the energy of these spores and the simplicity of fabricating these bio-composite materials, we manufacture a computer numerical control (CNC) platform combined with modified consumer inkjet cartridges to deposit a solution of spores onto flexible substrates in various geometries to achieve controlled shape actuation.

Here, we present the CNC platform and the potential toward developing printable actuators for soft robotic applications. We develop an improved spore suspension that resists sedimentation and improves ink jetting and surface wetting. We demonstrate that the vector of print direction can control the direction, speed, and magnitude of actuation. Extended work demonstrates the potential for both contractile and expansive actuation. Finally, stimuli-responsive folding architectures are presented as a possible tool for hierarchical design of humidity-responsive soft robots.

2 THE RAPID PROTOTYPE PLATFORM

Previously, these bio-composite materials were manufactured via manual, micropipette dispensed drop coating. To eliminate the manual labor involved, as well as enable higher resolution and unique geometries, we assembled a rapid prototype platform that merges the operation of a CNC router stage (ZenTool Works) with an open source inkjet controller (InkShield) that enables us to modify and control the operation of a commercial inkjet cartridge (HP C6602) (Figure 2) [6].

Figure 2: The Rapid Prototype Platform.

2.1 Hardware Setup

The CNC stage can move at a maximum of 2 mm/sec in any axial direction before the onset of mechanical instabilities. To maintain safe mechanical operation, the maximum operation speed was set at 1.67 mm/sec. The InkShield can fire the 12 nozzles of the HP C6602 cartridge at a maximum liquid feed rate of 199 nL per second per nozzle before burning out the heating elements that drive the thermal inkjet process.

2.2 Software Setup & Vector Design

After constructing and installing the necessary hardware components, several software packages need to be installed to render the system operational. First, the Arduino compiler needs to be installed along with additional InkShield and Grbl packages to facilitate the operation of the InkShield and CNC stage, respectively. Second, the appropriate parameters need to be calibrated for the CNC stage to allow for an accurate conversation of vector print image to G-Code to final printed product. Finally, vector graphics are converted to G-code via the online software package MakerCam.

3 INK: SEDIMENTATION & WETTING

A critical component of this rapid prototyping platform is the ink composition. The final composition will have an impact on its ability to jet (print), to remain homogeneous (resist sedimentation), and to appropriately coat the surface (wetting) while maintaining functionality.

3.1 Ink Preparation

While previous experiments have demonstrated the appropriate ink compositions for micropipette drop coating of bio-composite materials, inkjet technology has a limited operating window [7]. This can be illustrated by a relationship of two dimensionless parameters: the Reynolds Number and the Ohnesorge Number (Figure 3). To adjust our ink to be in the printable fluid regime, we introduced thixotropic methylcellulose and an ethanol co-solvent.

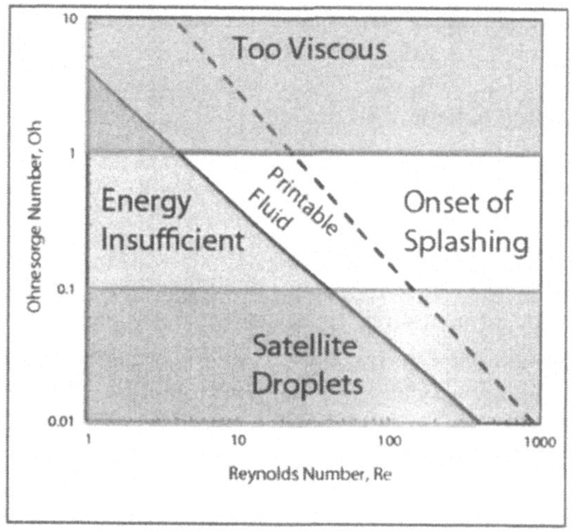

Figure 3: Illustration of the regimes encountered in inkjet systems at different Re and Oh. Adapted from [7]

3.2 Sedimentation Control

One concern during an extended print run is the need to limit inhomogeneity in the ink suspension. One phenomena that can be controlled to limit homogeneity is sedimentation. The force balance of sedimentation can be simplified as Equation 1.

$$v_t = \frac{F_{net}}{6\pi r_0 \eta} \tag{1}$$

The terminal velocity can be controlled by appropriately modifying the viscosity of the ink in the cartridge, as long as the size of the spores and specific gravity of the solution is not significantly altered. By introducing methylcellulose into the ink solution, we reduced sedimentation by nearly 50%, enabling print sequences that last up to 3 hours with minimal ink composition changes (Figure 4).

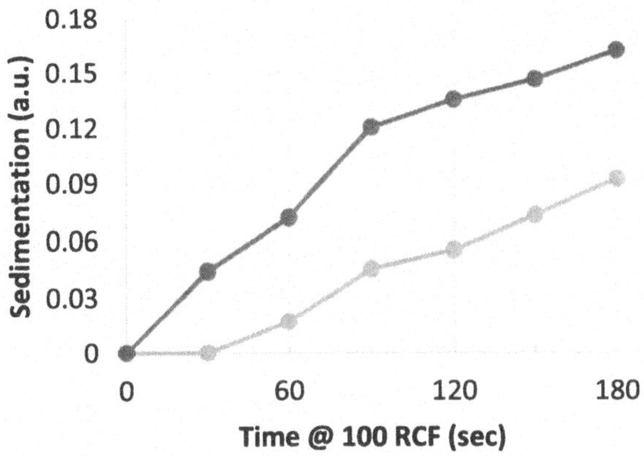

Figure 4: Sedimentation of the original spore/PVA solution (blue) vs. the spore/PVA/methylcellulose solution (green) sampled at various times of centrifugation.

3.3 12 Nozzle Horizontal Sweeps

After formulating our ink appropriately, we tested the ability of our new ink composition to wet a 9 by 3 cm area (27 square mm) on the surface of three different plastic substrates, Kapton (polyimide), Mylar (PET), and Polypropylene. Our results are illustrated in Figure 5. We are able to observe the desired print area on the Kapton and Mylar substrates with acceptable variance, while the polypropylene substrate exhibits significant dewetting, resulting in smaller contact patches. Due to this effect, our following work focused on using Kapton and Mylar.

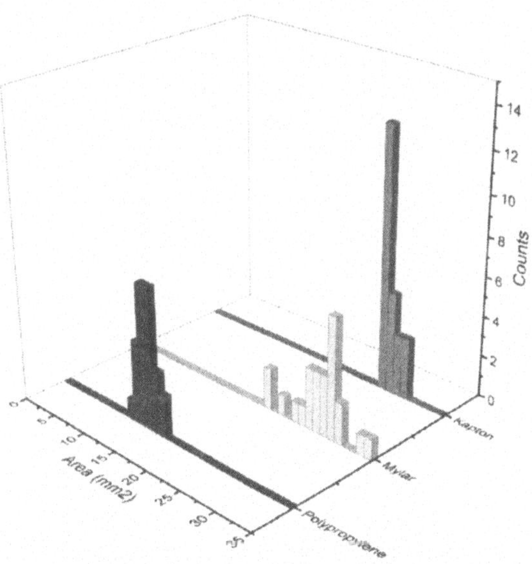

Figure 5: Histograms of 12 Nozzle Test Results on Kapton (red), Mylar (green), and Polypropylene (blue). Note the Kapton and Mylar distribution around 27 mm^2, while Polypropylene is around 15 mm^2.

3.4 Single Nozzle Resolution

After observing that Kapton and Mylar are appropriate substrates for large scale actuator printing, we continued our tests to determine the narrowest lines we can produce as a function of inkjet nozzle frequency at a fixed transverse velocity of 1.67 mm/sec along with recoating to dispense the same volume of ink on each test. Figure 6 illustrates the trend that lowering our printing frequency results in smaller feature sizes (~300 micrometers) along with lower variability in line width.

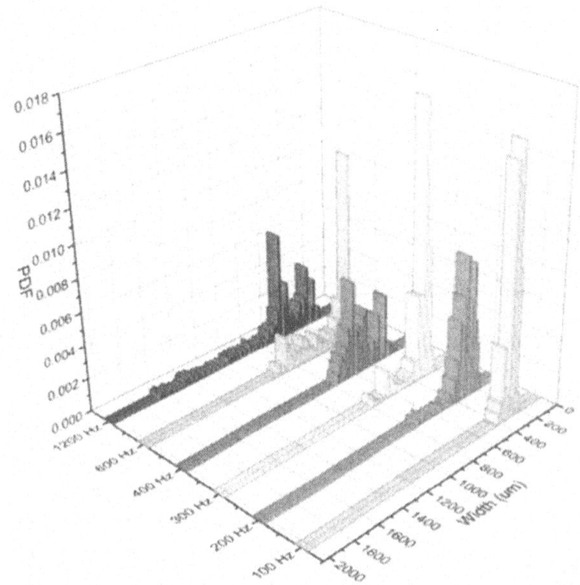

Figure 6: Histogram of our single nozzle test on Kapton for a range print frequencies of 100 – 1200 Hz. Note the narrow width distribution for samples from 100 – 300 Hz.

4 DESIGN & APPLICATIONS

This rapid prototyping platform now enables controlled fabrication of bio-composite actuators with features below 400 microns in size. The following section now details the results of various actuators designed and controlled by the method of fabrication

4.1 Curling Actuators

Previous experiments with this bio-composite actuator focused on curling, or contractile, composites. These actuators are flat at high relative humidity and curl as the relative humidity decreases.

One difficulty with this class of actuator is in controlling the direction of bending. Due to the complicated energy landscape of the relatively large surfaces being coated by the manual micropipette method, the curling actuators would occasionally bend in directions that were not beneficial for device fabrication because of lower energy barriers for these bending modes.

To impose control on our system, we fabricate our curling actuators on Mylar via our single nozzle printing method. On a 60 mm by 20 mm surface of Mylar, we printed 10 parallel lines, each 60 mm long and 0.3 mm wide (a 15% surface coverage). As the relative humidity decreases, we observe that the direction of curling is orthogonal to the direction of printing and that these actuators curl to a width 75% of the starting width (Figure 7). This is confirmed on several different Mylar samples with various orientations of rolling stress.

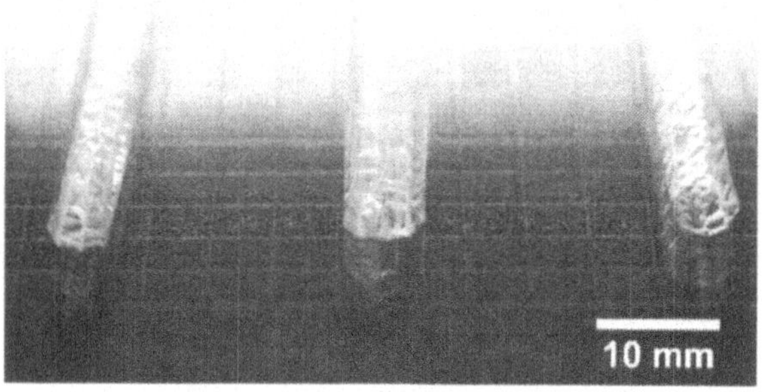

Figure 7: Dry Mylar Based Curling Actuators (25% RH).

4.2 Expanding Actuators

We further explored using this method of actuation to develop expanding actuators that remain flat at low relative humidity and expand at higher humidity levels. We repeated our printing method of section 4.1 on a Kapton substrate and observed that this structure exhibits both curling and expanding modes of actuation (Figure 8).

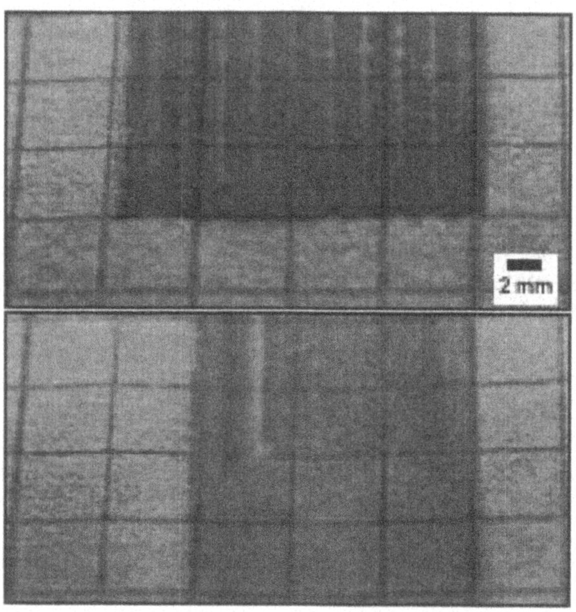

Figure 8: Expanding Actuator. (T: 25% RH, B: 90% RH)

5 CONCLUSION

Here, we presented a CNC platform and the potential toward developing printable actuators for soft robotic applications. We developed an improved spore suspension that resists sedimentation and improves ink jetting and surface wetting and demonstrate that ink vectors control the direction, speed, and magnitude of actuation, both contractile and expansive. We believe that these stimuli-responsive folding architectures will be a possible tool for hierarchical design of humidity-responsive soft robots.

REFERENCES

1. Raviv, D., et al., *Active Printed Materials for Complex Self-Evolving Deformations.* Scientific Reports, 2014. **4**: p. 7422.
2. Lipson, H., *Challenges and Opportunities for Design, Simulation, and Fabrication of Soft Robots.* Soft Robotics, 2014. **1**(1): p. 21-27.
3. Taccola, S., et al., *Toward a New Generation of Electrically Controllable Hygromorphic Soft Actuators.* Advanced Materials, 2015. **27**(10): p. 1668-1675.
4. Reichert, S., A. Menges, and D. Correa, *Meteorosensitive architecture: Biomimetic building skins based on materially embedded and hygroscopically enabled responsiveness.* Computer-Aided Design, 2015. **60**: p. 50-69.
5. Chen, X., et al., *Bacillus spores as building blocks for stimuli-responsive materials and nanogenerators.* Nat Nanotechnol, 2014. **9**(2): p. 137-41.
6. Ou, J., et al., *bioPrint.* 2014: p. 121-122.
7. Derby, B., *Inkjet Printing of Functional and Structural Materials: Fluid Property Requirements, Feature Stability, and Resolution.* Annual Review of Materials Research, 2010. **40**(1): p. 395-414.

Microbial-mediated Au@biolayer Core-shell Nanoparticles: Synthesis and Application

W. Liu and C. Jing*

*State Key Laboratory of Environmental Chemistry and Ecotoxicology, Research Center of Eco-Environmental Sciences, Chinese Academy of Science, Beijing 100085, China, cyjing@rcees.ac.cn

ABSTRACT

Gold nanoparticles (Au NPs) bio-synthesized through microbial activities create biocompatible and non-toxic nanomaterials that may have great applications. Previous studies have hypothesized that such biogenic Au NPs are covered by a layer of proteins. However, because of the high electron density of the Au core, the electron-opaque leaflet biolayers are hard to discern. Herein, we demonstrated that *Pantoea* sp. IMH can generate Au@biolayer core-shell NPs. Multiple complementary identification and characterization results suggest that the ~3 nm thin layer was coated on the spinous Au NPs. This unique composition enables the application of the biogenic Au@biolayer NPs in the removal and detection of dyes using surface enhanced Raman scattering (SERS).

Keywords: Au@biolayer NPs, biosynesize, *Pantoea* sp. IMH, dye removal, SERS

1 INTRODUCTION

Microorganisms have adapted to the presence of toxic heavy metals by developing a variety resistance mechanisms including reducing soluble toxic inorganic ions to insoluble metal nanoparticles (NPs). Gold (Au) NPs produced through microbial activities create biocompatible and non-toxic nanomaterials that have great applications in analysis, catalysis, and biomedicine.[1-3] The biogenic Au NPs are hypothesized to be covered by a layer of proteins.[4-6] However, because of the high electron density of the Au core, the electron-opaque leaflet biolayers are hard to discern. Investigating the existence of biolayers is important to understand the composition of microbial-mediated Au NPs, to study the biomineralization mechanism, and to explore its application.

Core-shell nanoparticles greatly break application restrictions of metallic nanoparticles which are easily and rapidly oxidized and dissolved. Organic shell coated on the Au NPs could not only protect the Au core, but also offer modification sites for target molecules, resulting in adsorption and detection of pollutants and biomolecules. Au NPs coating with surface materials has been reported to be synthesized through a variety of chemical fabrication methods. However, its widespread use is also restricted due to the chemical pollution and complicated procedure. Thus, the eco-friendly and simple method to synthesis Au@organic molecules core-shell NPs also motivated our research.

The objectives of this study were 1) to confirm the existence of biolayer on the Au nanoparticles synthesized by *Pantoea* sp. IMH (JX861130), with the formation of microbial–mediated Au@biolayer core-shell nanoparticles, and 2) to explore the application potential of microbial-mediated Au@biolayer core-shell NPs in adsorption and detection of dyes with surface enhanced Raman scattering (SERS) technique.

2 EXPERIMENTAL SECTION

2.1 Materials

Chloroauric acid ($HAuCl_4 \cdot 4H_2O$) and all chemicals were purchased from Sinopharm Chemical Reagent Co., Ltd (China). Microbiological media and ingredients were procured from Oxoid Ltd. Basingstoke Hampshire (England). Ultrapure Millipore water (18.2 MΩ) was used in all experiments. The stock solution of $HAuCl_4$ (100 mM) was prepared by dissolving $HAuCl_4 \cdot 4H_2O$ in Milli-Q water and stored in a brown bottle at 4 °C.

2.2 Synthesis of biogenic Au@protein NPs by Pantoea sp.IMH

Pantoea sp.IMH (JX861130) from our laboratory was maintained in Luria–Bertani (LB) broth. The organism was cultivated by inoculating freshly grown bacteria in a liquid medium with LB broth. The cultures were aerobically grown to late-log phase at 30 °C for 150 rpm. Cells were then harvested by centrifugation at 10,000 g (5424, Eppendorf, Hamburg, Germany) for 3 min, washed with PBS buffer solution (pH 7.4) for three times, and used for Au@protein nanoparticles synthesis. The harvested cells were resuspended in phosphate buffer (pH 7.4). Then, Au (III) was added to 10 mL of the above cells suspension and the final concentration of Au (III) reached to 1.8 mM. The mixture was incubated at 30 °C under shaking (150 rpm). Control experiments without cells were also performed. After 8h, the cells and nanoparticles were harvested by centrifugation at 10,000 rpm for 10 min. All of the experiments were performed in triplicate.

The distribution, morphology, crystal structure, and chemical composition of biogenic Au@protein NPs by the *Pantoea sp.* IMH were characterized through HRTEM-

SAED-EDS (Tecnai G^2 F20 S-Twin, FEI), FESEM-EDS (SU 8020, Hitachi), and XRD (X'Pert PRO MPD, PANalytical) analysis. Gold nanoparticles were analyzed using XANES at beamline14W1 at the Shanghai Synchrotron Radiation Facility (SSRF), China. Infrared spectra of bio-Au NPs were recorded on Thermo-Nicolet Nexus 6700 FTIR spectrometer equipped with a liquid-nitrogen-cooled mercury−cadmium−telluride (MCT) detector.

2.3 Adsorption of dyes by biogenic Au@protein NPs

The biogenic Au@protein NPs (30 mg) were added to 100 mL of crystal violet, malachite green, rhodamine 6G, and congo red solutions (0.05 mM), each taken in separate 250 mL Erlenmeyer flasks. The mixture was stirred at room temperature for 20 min with the magnetic stirrer. The samples were collected at regular intervals up to 20 s in the first 2 min and then at 30 s interval during the rest of adsorption experiment. Then the solution was filtered through a 0.22 μm membrane and the dye concentration was measured using a spectrophotometer (UV-2550 Shimadzu). Each experiment was performed five times.

2.4 SERS detection of dyes on biogenic Au@protein NPs

To obtain the SERS spectrum of capping protein on Au NPs, biogenic Au@protein NPs were transferred whole to silicon wafer, and exposed to the laser for 10 s to collect the SERS signal. The samples of crystal violet, malachite green, rhodamin 6G and congo red were prepared by diluting the stock solution in DI water to reach a final concentration of twice detection concentration. The samples (0.5 mL) were mixed with the biogenic Au@protein NPs (0.5 mL). Then about 10 μL above mixture was placed on the silicon wafer, and exposed to the laser for 5 s to collect the SERS signal. Raman spectra were obtained using a portable Raman spectrometer (Enwave Optronics, Inc. USA) with a 4 cm^{-1} resolution at the excitation energy of 785 nm.

3 RESULTS AND DISCUSSION

3.1 Microbial-mediated synthesis of Au@biolayer core-shell NPs

Incubation of *Pantoea* sp. IMH with $HAuCl_4$ solution resulted in color change of the bacterial suspension with time from pale white to purple. After 12 h, a peak in UV-vis spectrum of the purple solution was at ~541 nm due to the surface plasmon resonance (SPR) of Au NPs, indicating the formation of Au NPs. The formation of Au NPs was confirmed by characterizing thin sections of strain IMH incubated with Au (III) using high-resolution transmission electron microscope (HRTEM).

The energy-dispersive X-ray analysis (EDXA) shows the appearance of Au peaks (Figure 1). Meanwhile, the dominant feature of particles in X-ray absorption near edge structure (XANES) spectra (Figure 1) had an absorbance edge (11,919 eV) in accord with zero valent Au. Both selected area electron diffraction (SAED) and XRD analysis (Figure 1) shows the Scherrer ring patterns characteristic of Au NPs at (111) and (220).

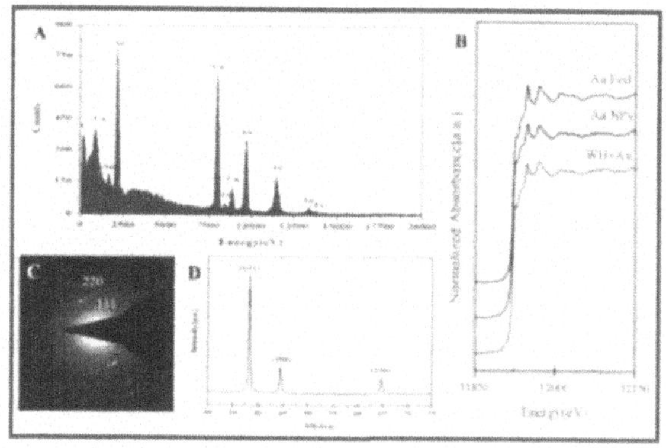

Figure 1: (A) EDS spectrum of nanoparticles observed in HRTEM images. (B) Au L-edge XANES spectra of nanoparticles in microbial environment (red line), Au NPs synthesized with chemical method (blue line), and Au foil as reference (black line). SAED (C) and XRD patterns (D) of Au NPs.

To confirm the presence of biolayer on the biogenic Au NPs, biogenic Au NPs in ordinary sample was measured through HRTEM. Figure 2 shows that a thin layer (~2.5 nm) constituted by amorphous substance was coated on the biogenic Au NPs. Au NPs were ensured through the lattice fringe spacing of 0.225 nm, corresponding to the distance between the Au (1 1 1) crystal planes. The SEM results show that clusters of biogenic Au NPs wires on the cell surface were coated by organic substance due to the sample conductivity. The EDS spectrum confirmed capping substance on the Au NPs was organic matter as a result of the appearance of C and O peaks. And the organic matter should be biomolecules owing to the biological system.

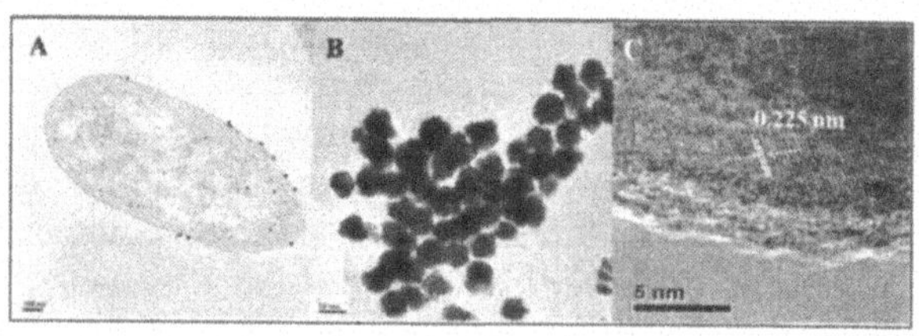

Figure 2: (A and B) HRTEM micrographs of thin section of cells with Au(III) treatment.(C) HRTEM image of nanoparticles formed in microbial medium containing 1.8 mM Au(III).

Therefore, Au NPs synthesized by bacteria were directly proved to be coated by the thin biolayer, with the formation of Au@biolayer core-shell NPs by the strain sp. IMH. Amine group and carboxyl group of nanoparticles as detected by FTIR could form peptide bond between adjacent NPs. As a result, nanoparticles would tend to form wirelike assemblies driven by Brownian motion and short-range interaction. The structure of biogenic Au@biolayer NPs was similar with that of magnetosome synthesized by several magnetotactic bacteria under iron ion stress. Thus, capping biolayer on Au NPs would be generated in the bioreduction and pumping out process by the bacteria under Au(III) stress. Biosynthesis Ag nanoparticles using *Shewanella oneidensis* were reported to be stable and speculated to be coated by protein/peptide. According to the phenomenon of biosynthesis Fe_3O_4, Au, and Ag nanoparticles, the existence of biolayer including lipid, proteins and other biomolecules would be prevalent in the biosynthesis nanoparticles by bacteria under metal stress.

3.2 Adsorption of dyes by biogenic Au@biolayer NP

Our adsorption experimental results (Figure 3) show that the removal efficiency of crystal violet, malachite green, and rhodamine 6G on biogenic Au@biolayer NPs rapidly reached equilibrium within 2 min at 98.3%, 59.2%, and 97.7%, respectively. The fast adsorption rate reflects good accessibility of the binding sites of the biogenic Au@biolayer NPs to dye molecules. While the adsorption of congo red was only 3.6% nearly could be ignorable. According to the structure of dyes, crystal violet, malachite green, and rhodamine 6G are cation-type dye, while congo red is anion-type dye. The adsorption of crystal violet, malachite green and rhodamine 6G on biogenic Au@biolayer NPs was much more than that of congo red on Au@biolayer NPs, consistent with the dyes structure. Therefore, biogenic Au@biolayer NPs were indicated to be negatively charged because of electrostatic adsorption interaction. Moreover, FTIR results that organic groups coated on Au NPs with negatively charge further testified our conclusion. Carboxylic group was reported to combine with anionic amino groups on the dyes. It was indicated that crystal violet, malachite green and rhodamine 6G were rapidly removed through adsorption mechanism involving the reaction between carboxyl group on the proteins and amino group on the dyes.

In addition, biogenic Au@layer NPs had higher adsorption ability to crystal violet than to malachite green even these two dyes are categorized as triphenylmethane dye, resulting from the structure difference of crystal violet and malachite green. The hydrophilic property of crystal violet is better than that of malachite green due to the molecular structure and biogenic Au@biolayer NPs is also hydrophilic due to the capping amine group, carboxyl group and other hydrophilic groups. Thus the hydrophilic-hydrophilic interaction may cause the adsorption difference between crystal violet and malachite green on biogenic Au@biolayer NPs.

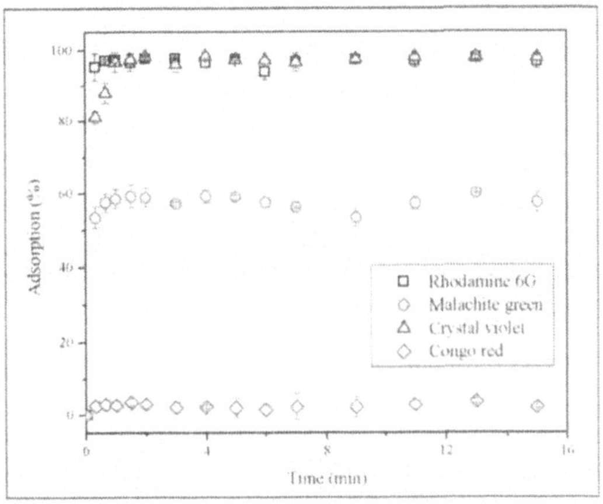

Figure 3: Adsorption kinetic of crystal violet, malachite green, rhodamine 6G, and congo red on biogenic Au@protein nanoparticles.

The adsorption of crystal violet, malachite green, and rhodamine 6G on biogenic Au@biolayer NPs were significantly higher than those on other three adsorbents as shown in Figure 4, while the adsorption of congo red on four adsorbents had no obvious difference. High adsorption capacity to dyes suggested that biogenic Au@biolayer NPs could be applied in the removal of dyes due to core-shell structure—spinous Au NPs with capping proteins

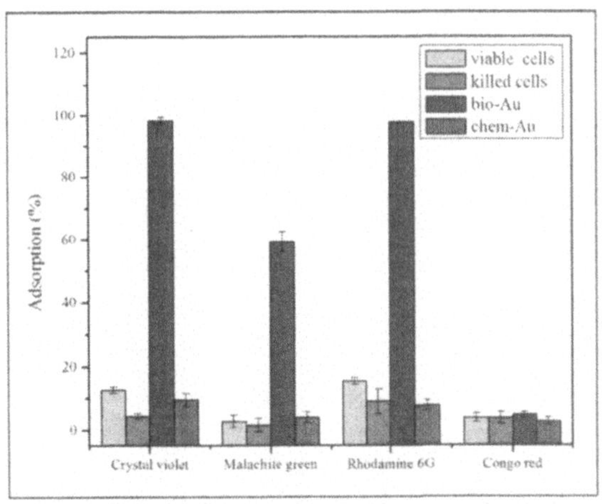

Figure 4: Adsorption of four dyes on the visible cells, dead cells, biogenic Au@protein NPs, and Au NPs synthesized with chemical method.

3.3 SERS detection of dyes on biogenic Au@biolayer NPs

Surface-enhanced Raman scattering (SERS) provides an alternative method for environmental analyses mainly due to its substantial electromagnetic enhancement induced by local surface plasmon resonance. Biogenic Au@biolayer

NPs were confirmed to be SERS-active and the capping biolayer was constituted by amine, carboxyl, phosphate, and other organic groups according to peak assignments, in accordance with FTIR results. Biogenic Au NPs by *Aspergillus nidulans* colonies are determined to be SERS-active through spectra from an extended scan map and only one genuine SERS spectrum was obtained when examined with Raman microscopy. SERS spectra based on our biogenic Au@biolayer NPs as SERS substrate are stable and could be easily acquired by a portable Raman spectrometer. Therefore, biogenic Au@biolayer NPs have the great application potential for SERS detection.

To justify our hypothesis that biogenic Au@biolayer NPs could be applied in SERS detection as Au NPs synthesized by chemical method, four dyes, including crystal violet, malachite green, rhodamine 6G, and congo red, were analyzed with SERS. Their spectra are shown in Figure 5. For comparison purposes, spectra of the dyes standards and blank substrate were also collected. Nevertheless, the background peaks would not interfere with the detection of the four dyes. In line with previous reports and our DFT calculations, the observed SERS spectra agreed well with the corresponding Raman spectra of dye standards. The lack of change in SERS peaks from the Raman spectra implies that the enhancement may be mainly due to the electromagnetic mechanism. The biogenic Au@biolayer NPs, with the spinous structure, could form "hot spots" between adjacent nanoparticle, resulting in pronounced SERS enhancement for molecules for them.

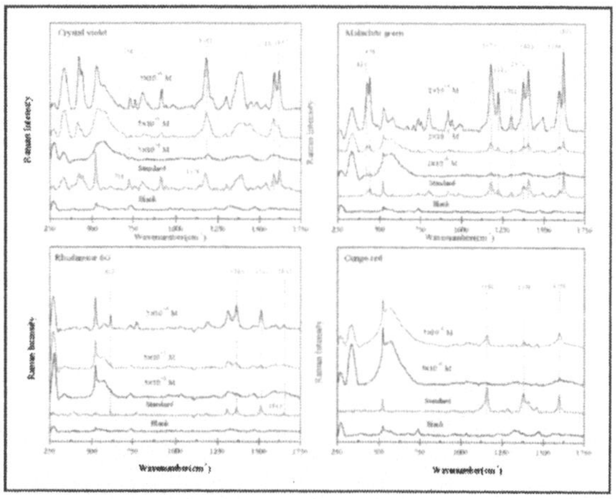

Figure 5: SERS spectra of four dyes with different concentrations at 10^{-4} M, 10^{-5} M, and 10^{-6} M based on biogenic Au@protein NPs as SERS substrate.

Figure 5 shows that the detection limit of crystal violet, malachite green and rhodamine 6G was about 10^{-6} M, while the lowest detection concentration of congo red was only 5×10^{-5} M. Therefore, it was revealed that biogenic Au@protein NPs tend to adsorb more molecules of crystal violet, malachite green and rhodamin 6G than that of congo red, according with the adsorption results of dyes on biogenic Au@biolayer NPs (Figure 8). SERS experiment demonstrated that biogenic Au@biolayer NPs exhibit surface plasmon resonance and could be used as substrate in SERS detection of organic pollutants such as dyes.

Upon successful adsorption removal and SERS detection of dyes, biogenic Au@biolayer NPs, with the special structure-spinous shape and capping biolayer, could be used in dye contaminated water treatment. And it has the possibility that fast adsorption removal of dyes, with the simultaneous monitoring dye concentration, provides a nanotechnology-based one-step green approach for water purification and improvement of environmental sustainability.

4 CONCLUSION

In conclusion, we have demonstrated that *Pantoea* sp.IMH can generate Au@biolayer core-shell nanoparticles. Examination of *Pantoea* sp.IMH incubated in $HAuCl_4$, using UV-Vis spectroscopy, HRTEM-ED-EDS, XRD, XANES, SEM-EDS, and FTIR, provided complementary identification and characterization. Microbial-mediated Au@biolayer NPs were one-step synthesized core-shell nanoparticles that thin (~2.5 nm) layer was coated on the spinous Au nanoparticles. In biogenic Au@biolayer NPs — dyes system, capping biomolecules could be combined with dye molecules, resulting in high adsorption efficiency and good SERS enhancement to dyes. With the high adsorption capacity, the biogenic Au@biolayer NPs could be applied for the removal of dyes due to the special structure of nanoparticles. Biogenic Au@biolayer NPs exhibit surface plasmon resonance and can be used as substrate for SERS detection of dyes. Our results will help to understand the transformation of metal ions to nanoparticles in nature, the gold biosynthesis mechanism, and the development of green approach for dyes-contaminating water treatment.

REFERENCES

[1] F. Reith, S.L. Rogers, D. C. Mcphail, D. Webb, Science, 313, 233-236, 2006.

[2] M. Lengke, G. Southam, Geochimica Et Cosmochimica Acta, 70, 3646-3661, 2006.

[3] F. Reith et al., Proceedings of the National Academy of Sciences of the United States of America, 106, 17757-17762, 2009.

[4] S. K. Das, J. Liang, M. Schmidt, F. Laffir, E. Marsili, ACS Nano, 6, 6165-6173, 2012.

[5] T. Ahmad, I. A. Wani, N. Manzoor, J. Ahmed, A. M. Asiri, Colloids and Surfaces B-Biointerfaces, 107, 227-234, 2013.

[6] K. Kalishwaralal, V. Deepak, S. R. K. Pandian, M. Kottaisamy, S. BarathManiKanth, B. Kartikeyan, S. Gurunathan, Colloids and Surfaces B-Biointerfaces, 77, 257-262, 2010.

Bio-inspired Polydopamine Treatment and Nanosilver Formation for Antibacterial Uses

Y.-H. Chiu*, M.-C. Li*, K.-M. Wu*, and Y.-C. Chung*, **

*Research Center of Biomimetics and Medicare Technology, National University of Kaohsiung
Kaohsiung, 811 Taiwan R.O.C. ericyhchiu@gmail.com
**Department of Chemical and Materials Engineeing, National University of Kaohsiung
Kaohsiung, 811 Taiwan R.O.C. ycchung@nuk.edu.tw

ABSTRACT

We employed mussels' adhesive secretion, dopamine, and its reducing ability toward silver ions, to create a novel antibacterial effect on a facial mask surface eliminating the need for harmful antiseptics and preservatives. Most past methods for depositing silver particles onto facial masks have lacked suitable processing conditions, and they require additives which limit their uses. Dopamine was adsorbed onto the nonwoven surface and then underwent a self-polymerization reaction to have silver nanoparticles (AgNPs) reduced on the surface. Surface characterization was carried out using microscopic analysis. Photospectrometers were employed to analyze size, deposition, and distribution of silver nanoparticles. For a facial mask design, the common substrates are nonwoven fabrics, which usually accumulate a lot of bacteria when immersed into nutrient-rich extracts and essential oils. Nanosilver for antibacterial uses has attracted much attention owing to its release of silver ions. With this simple strategy, we confirmed the antimicrobial performance of the facial mask surfaces against bacteria.

Keywords: mussels, polydopamine, silver nanoparticles, antibacterial, nonwoven fabric

1 INTRODUCTION

Mechanically, silver-based materials which account for their antibacterial properties are thought to release silver ions (Ag+) and attach to specific thiol (–SH) groups in a variety of structural and functional bacterial proteins [1-3]. Another possible mechanism involves an accumulation of reactive oxygen species (ROS), inducing a bacterial apoptosis-like response [4]. Therefore, AgNPs are being increasingly utilized in the medical industry due to their antibacterial properties. AgNPs can be synthesized in a variety of different ways; the most common methods of AgNP synthesis employ chemical reduction of silver salt (e.g., $AgNO_3$) using a reducing agent (e.g., sodium borohydride) [5]. However, reducing agents and organic solvents used in chemical reduction, such as N,N-dimethylformamide (DMF), hydrazine hydrate (N_2H_4), and sodium borohydride ($NaBH_4$), are highly reactive and pose potential environmental and biological risks [6]. So we focused on developing new green chemistry methods of AgNP synthesis with the advantage of using natural products and avoiding toxic reducing agents, organic solvents, and wasteful purifications with high cytotoxic residuals. With these methods, molecules produced by living organisms such as mussels can replace the reducing and capping agents. Polydopamine (PDA) is a biomimetic polymer that is based on the mussel's adhesive protein that is rich in L-3,4-dihydroxyphenylalanine (L-DOPA) and L-lysine and enables these animals to tightly attach themselves to wet surfaces. Dopamine, containing two hydroxyl groups at the orthopositions to one another, could reduce metal ions to metal nanoparticles, such as Au^+ and Ag^+. Under oxidative and alkaline conditions, dopamine provides a facile approach to surface modification in which self-polymerization of dopamine produces an adhesive PDA coating on a variety of substrates [7]. In this study, dopamine was coated onto nonwoven fabric. It performed a self-polymerization reaction to reduce silver nanoparticles and created a novel antibacterial effect on the facial mask surface, eliminating the need for harmful antiseptics and preservatives.

2 MATERIALS AND METHODS

2.1 Materials

Nonwoven fabric was ultrasonically cleaned in ethanol and deionized water for 15 min and dried in an oven at 60 °C before use. 3-Hydroxytyramine hydrochloride (dopamine/ HCl, 99%, Acros Co.), silver nitrate (99.9%, Alfa Aesar) and Tris hydrochloride (ultra-pure, from MP Biomedicals), were used as purchased. All chemical reagents were used as received and without further purification.

2.2 Dopamine Oxide Polymerization on Nonwoven Surface

3.152 g Tris–HCl was dissolved in 150 ml deionized water, mixed with 10 wt% NaOH aqueous solution to control pH value to 8.5, and then diluted to 200 ml with water for use as a 0.1 M alkaline Tris–HCl buffer solution. The nonwoven fabric was immersed in an alkaline Tris–HCl buffer solution of dopamine (1 g/L) for 15-240 min at room temperature. After a predetermined reaction time, the resulting nonwoven fabric was filtered and rinsed thoroughly with an ultrasonic cleaner for 3 h and dried in an oven at 60 °C. The obtained sample was denoted as nonwoven-PDA.

2.3 Electroless Deposition of Silver Coating on the Nonwoven-PDA Surface

The nonwoven-PDA substrate was incubated in aqueous solution of $AgNO_3$ (1-100 mM). The reduction was allowed to continue for 30 min at 70 °C. The resulting sample was separated by filtration, washed thoroughly with an ultrasonic cleaner for 3h and dried in an oven at 60 °C. The obtained sample is denoted as nonwoven-PDA/Ag in the discussion below. To systematically demonstrate the use of polydopamine in antibacterial applications, we developed a technique for producing a polydopamine-based coating containing nanosilver as seen in Figure 1.

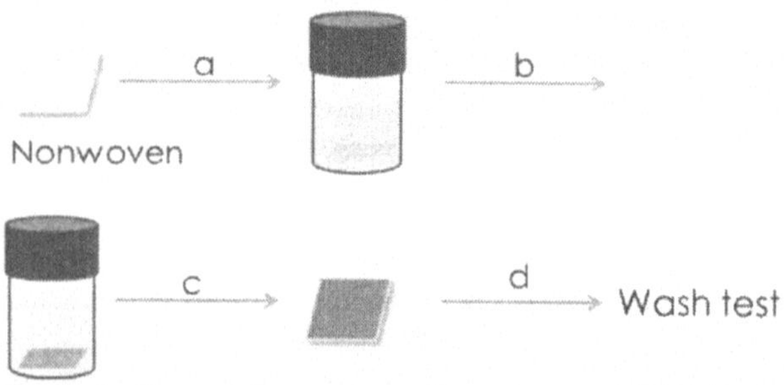

Figure 1. Preparation process for nonwoven-PDA/Ag. (a) Tris-HCl buffer solution of dopamine (1 g/L) (pH=8.5), room temperature, 15-240 min. (b) Aqueous solution of $AgNO_3$ (1-100 mM), 70 °C, 30 min. (c) Dried in oven, 60 °C. (d) Washed thoroughly with ultrasonic cleaner, 3h.

2.4 Microbiological evaluations

Antimicrobial behavior of the nonwoven surfaces was evaluated with aerobic plate count and fungi, yeast, and mold plate count. 1 g sterile nonwoven fabric was placed into a screw-cap test tube containing 1 ml sterile Tween 80 and the product was dispersed with a sterile spatula. 8 ml sterile modified letheen broth was added and mixed thoroughly to 10^{-1} dilution. The spread plate technique was used to facilitate recognition of different colony types. Duplicate sets of petri dishes, containing modified letheen agar or potato dextrose agar, were prepared and labeled. 0.1 ml of each dilution was mixed and pipetted onto surfaces of solid media in petri dishes, and inoculum was spread over the entire surface with a bent glass rod that was first sterilized by being dipped in 95% ethanol and quickly flamed to remove the ethanol. The medium was allowed to absorb the inoculum before the plates were inverted and incubated. All colonies in the plates were counted and average colony counts were obtained.

3 RESULTS AND DISCUSSION

3.1 Dopamine modified nonwoven fabric (nonwoven-PDA)

The dopamine solution was colorless and transparent. During the dopamine's oxidative self-polymerization, the color of the solution turned to deep brown. When the nonwoven fabric was immersed into dopamine solution for a longer period of time, the color of became darker (as shown in Figure 2). We characterized the surface morphology of nonwoven-PDA using SEM, and the results are shown in Figure 2. The longer the treatment time, the more uneven the fibers on the surface of the nonwoven fabric became. The elemental composition of the nonwoven fabric was investigated via EDX. Figure 3 shows the EDX patterns of the samples before and after dopamine modifications. It can be seen from Figure 2(b) that the nonwoven-PDA sample is composed of carbon, nitrogen, and oxygen elements. The results suggest the polydopamine was successfully deposited onto the nonwoven fabric.

Figure 2. SEM images of polydopamine deposited on nonwoven polypropylene fabric with various reacting times: (a) 0 min, (b) 120 min, (c) 240 min.

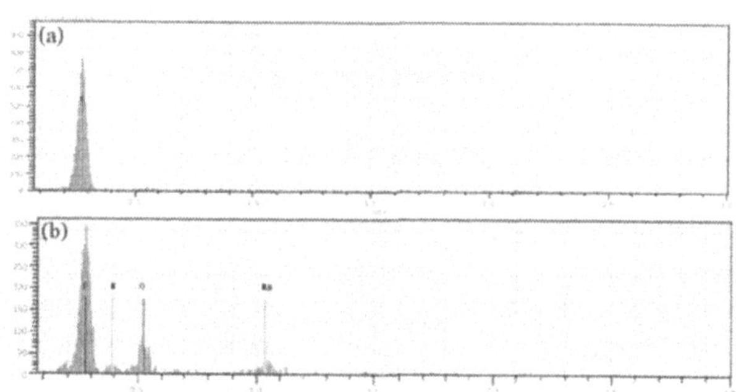

Figure 3. EDX patterns of (a) nonwoven fabric, (b) nonwoven-PDA fabric.

3.2 Nano-silver coated on the nonwoven-PDA surface (nonwoven-PDA/Ag)

The polydopamine on the nonwoven surface has been shown to be an effective functional layer for reduction of ionic silver. The silver-binding ability of the catechol and N-containing groups present in the PDA coating can make

TechConnect Briefs 2015, TechConnect.org, ISBN 978-1-4987-4729-5

the reduced AgNPs attach on the surface of nonwoven fabric, as shown in Figure 4. The color becomes dark yellow after silver is coated on the nonwoven fabric. According to SEM analysis, the diameter of AgNPs were below 200 nm. We further confirmed the formation of AgNPs through emergence of a silver plasmon resonance, evident by an increase in absorbance around 420 nm by UV-vis spectrometer. The number and size of the AgNPs could be tailored via variation of immersion time with dopamine, as shown in Figure 5. In order to confirm that the AgNPs can bind well with the nonwoven-PDA, the nonwoven-PDA/Ag samples were washed with an ultrasonic cleaner for 3 h and investigated with a UV-vis spectrometer.The result shows 98 % of AgNPs can still be attached on the nonwoven-PDA.

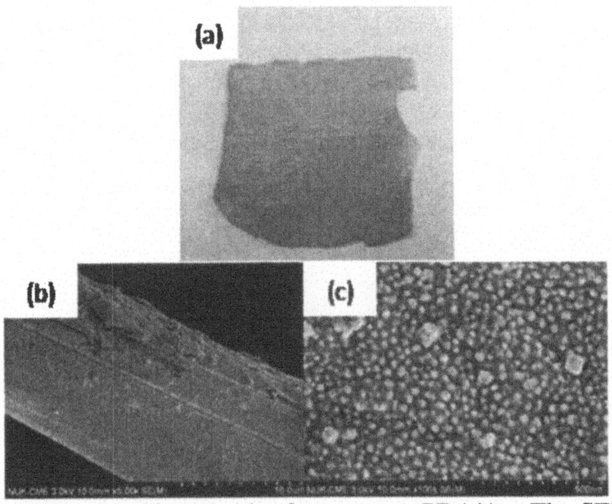

Figure 4. (a)Photograph of nonwoven-PDA/Ag. The SEM images show (b) the formation of nanoparticles on the fabric surfaces and its close-up look. (c) Using high magnification.

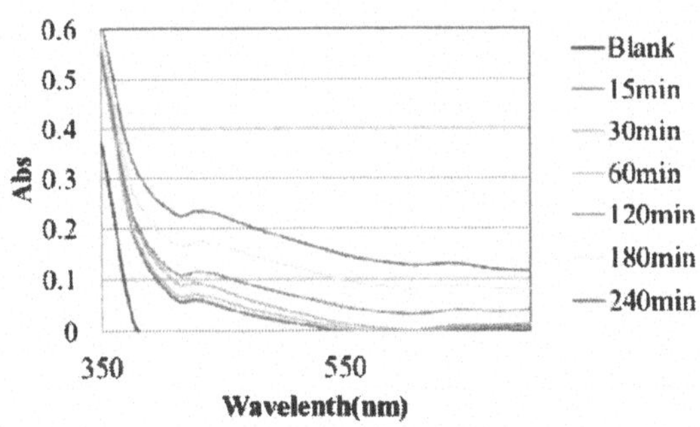

Figure 5. UV-vis spectrum of nonwoven fabric via variation of immersion time with dopamine solution. All samples were immersed in the same silver coating.

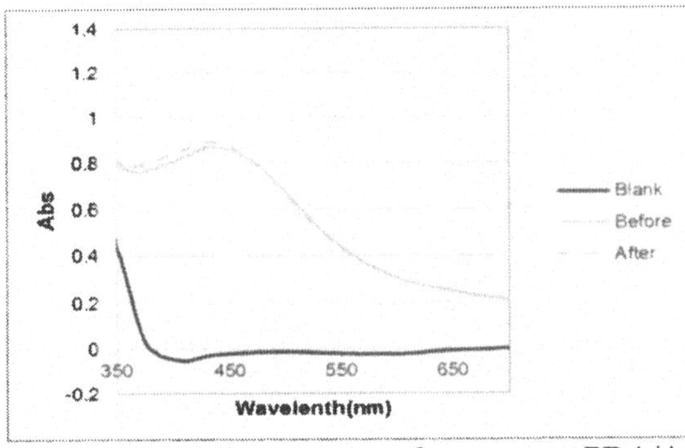

Figure 6. UV-vis spectrum of nonwoven-PDA/Ag sample before and after ultrasonic washing.

3.3 Bacterial Attachment and Death Assays

Antimicrobial behavior of the generated surfaces wa evaluated using the colony-forming unit (CFU) counting method. The results of antibacterial measurement are shown in Table 1. No CFUs were observed on the total bacterial counts or total fungi count tests (Figure 7b-c). Furthermore, we used chromogenic *E.coli*-coliform agar and chromogenic MRSA agar to detect *E. coli*, coli form and *Staphylococcus aureus* strains, and no CFUs were observed even after 72 h (Figure 7d-e). The result indicated that the *in situ* formation of AgNPs via exposure of PDA membranes to $AgNO_3$ solutions can ensure the complete inactivation of bacterial cells.

Table 1. Antibacterial measurement

	Detection items	Methods	Results
1	total bacterial counts	Letheen Agar, Modified (MLA)	**0CFU/cm^2**
2	total fungi counts	Potato Dextrose Agar (PDA)	**0CFU/cm^2**
3	*E. coli*, Coli form	Chromogenic E. Coli /Coliform Agar	**N.D.**
4	*Staphylococcus aureus*	Chromogenic MRSA Agar	**N.D.**
Note	total bacterial counts：35 oC for 48~72 h total fungi counts：25 °C for 72~96 h test temperature：24.0 °C relative humidity：54 %		

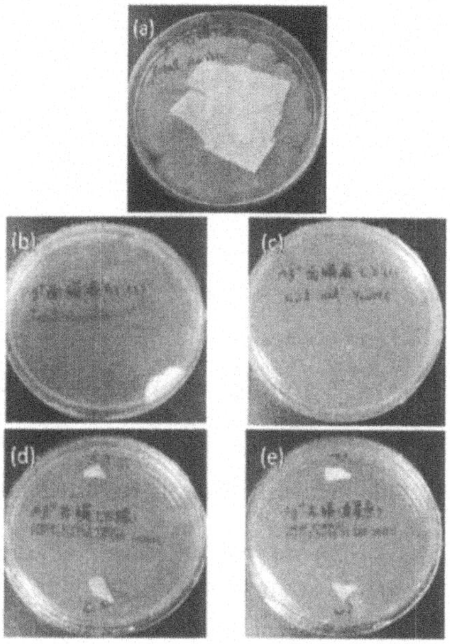

Figure 7. Photographs of antibacterial measurement. (a) total aerobic incubation with nonwoven fabric; nonwoven-PDA/Ag were incubated with (b) total aerobic incubation, (c) mold and yeasts, (d) *E. coli* and coliform, (e) *Staphylococcus aureus.*

4 CONCLUSION

The popular treatment of substrates for silver nanoparticle coating are physical adsorption of silver nanoparticles via vapor deposition methods, or electroless plating of silver salts, which still cause nanoparticle depletion from the surfaces and induce some toxic reactions. We demonstrated a novel immersion strategy for nonwoven fabric coatings that actively kill bacteria through silver ion release. When nanosilver comes into contact with bacteria, the silver ions can pass through the cell walls to react with RSH groups of proteins and therefore destroy the bacteria. The strategy is enabled by a polydopamine layer reducing silver nanoparticles from a silver salt solution onto nonwoven fabric. Finally, we confirmed the antimicrobial performance of the facial mask surfaces against both gram-positive and gram-negative bacterial strains. Thus, our approach represents a simple method that avoids toxic reducing agents and harmful antiseptics and preservatives; the results suggest significant potential for expansion of this concept to other coatings in the future.

REFERENCES

[1] P. Sanpui, A. Murugadoss, P. V. D. Prasad, S. S. Ghosh and A. Chattopadhyay, "The antibacterial properties of a novel chitosan-Ag-nanoparticle composite," Int. J. Food Microbiol. 124, 142, 2008.

[2] J. R. Morones, J. L. Elechiguerra, A. Camacho, K. Holt, J. B. Kouri, J. T. Ramírez and M. J. Yacaman, "The bactericidal effect of silver nanoparticles," Nanotechnology, 16, 2346, 2005.

[3] S. Shrivastava, T. Bera, S. K. Singh, G. Singh, P. Ramachandrarao and D. Dash, "Characterization of antiplatelet properties of silver nanoparticles," ACS Nano, 3, 1357, 2009.

[4] W. Lee, K. J. Kim, and D. G. Lee, "A novel mechanism for the antibacterial effect of silver nanoparticles on Escherichia coli," Biometals, 27, 1191, 2014.

[5] R. Sato-Ber'ru, R. Red'on, A. V'azquez-Olmos and J. M. Saniger, "Silver nanoparticles synthesized by direct photoreduction of metal salts. Application in surface-enhanced Raman spectroscopy," J. Raman Spectrosc., 40, 376, 2009.

[6] N. Vigneshwaran, R. P. Nachane, R. H. Balasubramanya and P. V. Varadarajan, "A novel one-pot "green" synthesis of stable silver nanoparticles using soluble starch," Carbohyd. Res. 341, 2012, 2006.

[7] H. Lee, S. M. Dellatore, W. M. Miller and P. B. Messersmith, "Mussel-inspired surface chemistry for multifunctional coatings" Science 318, 426, 2007.

Three-Dimensional Macroporous Carbon Nanotube Scaffolds for Stem Cell Expansion and Maintenance

Gaurav Lalwani, Anu Gopalan, Michael D'Agati, Sunny Patel, Yahfi Talukdar and Balaji Sitharaman

Department of Biomedical Engineering, Stony Brook University, Stony Brook, NY 11794-5281

1. Introduction

Human mesenchymal stem cells (MSCs) show promise for several therapeutic applications including regenerative medicine, drug discovery, cellular therapy and disease modeling. Expansion of stem cells and maintenance of their self-renewal capacity *in vitro* requires specialized robust cell culture systems. Conventional approaches consisting of animal derived matrices and a cocktail of growth factors have several limitations such as consistency, scalability, and pathogenicity (risk of infection). Furthermore, to achieve high cell densities for practical therapeutic applications, 3D culture systems have been recommended over conventional 2D substrates. To overcome the above limitations, multifunctional 3D porous scaffold, fabricated using synthetic materials that permit stem cell expansion and maintenance *in vitro* would be a significant advancement.

Carbon nanotubes (CNTs) exhibit excellent physiochemical properties such as high mechanical strength, electrical conductivity as well as unique opto-acoustic and electromagnetic response. Therefore, they have been investigated for several therapeutic and diagnostic biomedical applications. Indeed, there has been a growing interest in assembling carbon nanomaterials into various two- and three-dimensional architectures for the fabrication of next-generation of energy storage, electronic, super-capacitor, photovoltaic and biomedical devices and implants.[1] For biomedical applications, carbon nanotubes and graphene have been assembled into two-dimensional films (using vacuum filtration and chemical vapor deposition (CVD) methods) and 3D foams (using CVD and sacrificial template-transfer methods) and reported as cytocompatible substrates for cellular function (proliferation and differentiation of stem cells for applications in bone, neuron and cardiac tissue engineering).[2-5] These methods have several limitations. CVD method requires very specific substrates capable of withstanding high temperatures and pressure. Vacuum filtration and spray coating methods can produce 2D substrates that may not be suitable for tissue engineering of larger organs that demand three-dimensional scaffolds. Furthermore, a general limitation of these methods is that in the absence of strong chemical bonds between the individual nanomaterials, the structural integrity of architectures assembled relies mainly on weak Van der Wall forces or on physical entanglement of the nanoparticles, leaving them prone to dissociation under physiological shear forces experienced by *in vivo* biomedical devices and implants. Furthermore, methods such as sacrificial template transfer do not allow control over the porosity of the assembled 3D scaffold, which depends on the template architecture. Therefore, the assembly of carbon nanomaterials into three-dimensional (especially with >1 mm in all three dimensions) macroporous tissue engineering scaffolds with tunable porosity across various length scales (macro, micro and nanoscopic) would constitute a significant advancement. We have recently reported a novel, cheap and easy method to fabricate macroscopic, 3D, free standing, all-carbon scaffolds (macroporous architectures) by radical initiated thermal crosslinking and annealing of CNTs.[1] This method can fabricate macroporous (20 nm – 300 μm) all-carbon structures with high porosities (~80-85%). Furthermore, porosity of scaffolds can be controlled by the amount of radical initiator used in the crosslinking process, thereby allowing the fabrication of porous all-carbon scaffolds tailored towards specific applications.

Towards the development as multifunctional 3D CNT scaffolds for tissue engineering applications, we report the cell-material interactions of two type of all-carbon scaffolds – prepared using single- and multi- walled carbon nanotubes (SWCNTs and MWCNTs). In this study, we characterize the cytocompatibility of 3D SWCNT and MWCNT scaffolds using human adipose derived mesenchymal stem cells (ADSCs). We study the cell viability, adhesion, proliferation and infiltration of ADSCs on 3D MWCNT and SWCNT scaffolds. Furthermore, we assess the plasticity of ADSCs after long-term expansion on 3D SWCNT and MWCNT scaffolds by inducing tri-lineage (osteogenic, adipogenic and chondrogenic) differentiation according to International Society for Cellular Therapy guidelines.[6]

2. Materials and Methods:

Preparation of SWCNT and MWCNT scaffolds and ADSCs cell cultured has been described previously.[1, 7] 3D porous poly(lactic-co-glycolic) acid (PLGA) scaffolds were used as positive controls. Cell death of ADSCs after 1, 3, and 5 days of culture on SWCNT and MWCNT scaffolds was analyzed by lactate dehydrogenase (LDH) assay. Calcein-AM was used for LIVE cell imaging after day 5. Immunofluorescence imaging was used to determine the presence of vinculin and Ki-67. To determine MSC phenotype, surface markers were analyzed as described previously. Furthermore, to confirm the plasticity of MSCs after 15 days of culture on SWCNT and MWCNT scaffolds, cells were harvested from SWCNT and MWCNT scaffolds and cultured in adipogenic, osteogenic and chondrogenic media (Lonza, MD, USA) in 24 well-plates. Cells were then assessed for tri-lineage differentiation (adipogenic: Oil-red-O staining and elution; osteogenic: Alizarin red staining, ALP assay, calcium deposition; and chondrogenic: Alcian Blue staining and collagen-II immunohistochemistry.

3. Results and Discussions:

Detailed chemical, physical, mechanical and electrical characterization of MWCNT scaffolds has been reported previously.[1] The scaffolds used in this study were macroscopic (>1 mm in all three directions) architectures. The crosslinking of SWCNTs or MWCNTs to form 3D SWCNT or MWCNT scaffolds was confirmed by high-resolution SEM and TEM imaging (results not shown). The porosity values of MWCNT and SWCNT scaffolds were 45.82 ± 3.59% and 38.62 ± 2.91%, respectively, quantified by microCT and SEM image processing. The presence of macro-, micro-, and nano-porosity is critical for the exchange of nutrients and waste metabolites during cell culture. ADSCs cultured on SWCNT and MWCNT scaffolds show negligible cell death quantified by LDH assay after 1, 3, and 5 days of culture compared to poly(lactic-co-glycolic) acid (PLGA) scaffolds (control group). Calcein-

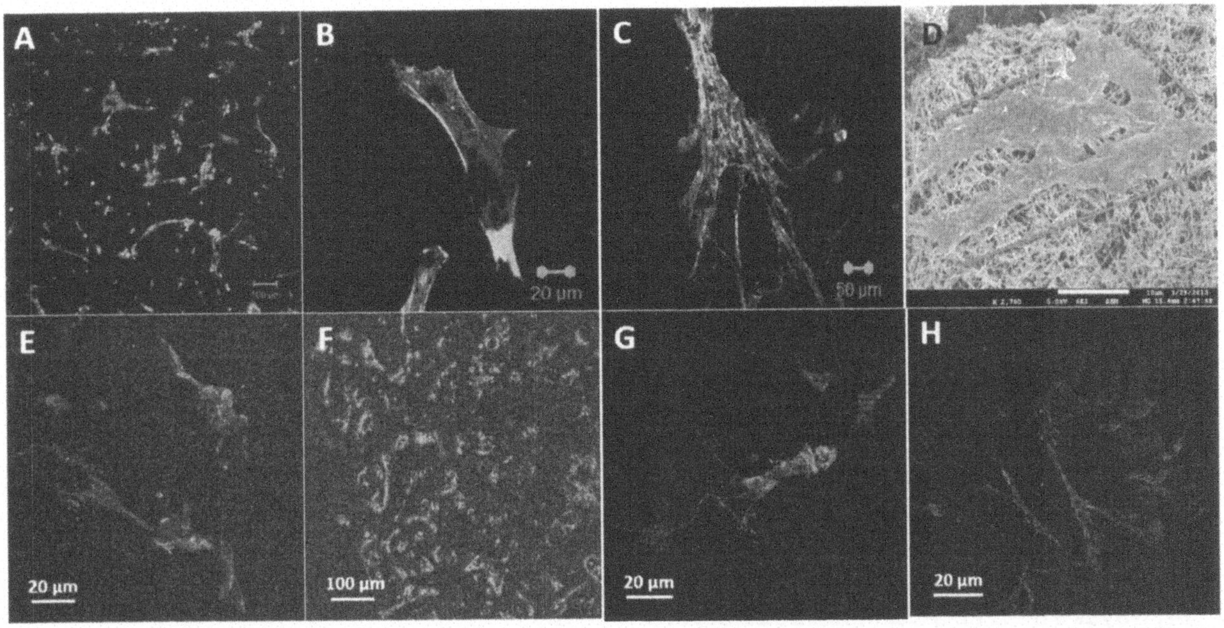

Figure 1: (A) Calcein-AM stained image showing live cells. Scale bar is 100um. (B and C) Immunofluorescence images showing expression of Ki-67 and vinculin (both in red). Cytoplasm is stained green. (D) SEM image showing cell spreading and attachment. Scale bar is 10µm (E-H) immunofluorescence images showing the expression of CD105, CD90, and absence of CD45 and CD19, respectively. Surface antigens are green in color and cytoplasm is stained red for actin.

AM staining show the presence of live ADSCs on SWCNT and MWCNT scaffolds, corroborating LDH results (Figure 1A). Immunofluorescence results confirm the expression of cell attachment vinculin and proliferation marker Ki-67 and indicating that ADSCs attach and proliferate on SWCNT and MWCNT scaffolds (Figure 1B and C). Furthermore, SEM of ADSCs showed the cytoplasmic extensions attach to the underlying nanotube network (Figure 1D). ADSCs cultured on SWCNT and MWCNT scaffolds express stem cell phenotypic surface markers (ADSCs were positive for CD105, CD73 and CD90, and negative for CD45, CD34, CD14 and CD19) after 15 days (Figure 1 E-H). Additionally, SWCNT and MWCNT scaffolds can be de-assembled into smaller components to allow the harvest of ADSCs throughout the scaffold; hitherto not possible with polymeric scaffolds such as PLGA. The harvested ADSCs are capable of *in-vitro* expansion. Tri-lineage differentiation studies (results not shown) confirm that the ADSCs differentiate towards adipocytes (Oil-red-O staining and elution), osteoblasts (Alizarin red staining, ALP assay, calcium deposition) and chondrocytes (Alcian Blue staining) and indicate that the scaffolds do not affect differentiation capabilities of the ADSCs.

To the best of our knowledge, this is the first study showing that 3D, macroscopic, porous SWCNT and MWCNT scaffolds with tunable porosities can be used for the expansion and maintenance of human ADSC. The ADSCs cultured on these 3D CNT scaffolds retain plasticity and can be harvested for lineage specific differentiation depending on patient-specific demand. The results of this study open avenues towards future *in vivo* safety and efficacy studies on the biocompatibility of 3D SWCNT and MWCNT scaffolds. In addition to stem cell expansion and maintenance, due to multifunctional properties of CNTs, one can envision the development these scaffolds to induce/guide specific cellular processes (such as differentiation, specific protein expression), support non-invasive longitudinal diagnosis/monitoring of tissue regeneration, and permit stimulus based drug and gene delivery, a significant advancement towards next-generation tissue engineering and regenerative medicine applications.

4.References

[1] Lalwani G et. al., Carbon. 2013;53:90 100.

[2] Li N et. al., Scientific reports. 2013;3:1604.

[3] Crowder SW et. al., Nanoscale. 2013;5:4171-6.

[4] Nayak TR et. al. ,ACS nano. 2011;5:4670-8.

[5] Mooney E et. al., Biomaterials. 2012;33:6132-9.

[6] Dominici M et. al., Cytotherapy. 2006;8:315-7.

[7] Talukdar Y et. al., Biomaterials. 2014;35:4863-77.

One and Two Dimensional Inorganic Nanoparticles' Interactions with Fibroblasts and Stem Cells

J. T. Rashkow[*], Y. Talukdar[*] G. Lalwani[*] and B. Sitharaman[*]

[*]Department of Biomedical Engineering, Stony Brook University
Bioengineering Bldg, Room 115, Stony Brook University, Stony Brook, NY, USA,
balaji.sitharaman@stonybrook.edu

ABSTRACT

We report the effects of one and two dimensional transition metal dichalcogenide nanoparticles, tungsten disulfide nanotubes (WSNTs) and molybdenum disulfide nanoplatelets (MSNPs), on fibroblastic cell (mouse fibroblasts (NIH-3T3) and human adipose derived stem cell (MSC)) viability and MSC differentiation potential. Cytotoxicity of MSNPs and WSNTs dispersed in 1,2-Distearoyl-sn-Glycero-3-Phosphoethanolamine conjugated polyethylene glycol (DSPE-PEG) was assessed by Presto Blue and lactate dehydrogenase (LDH) viability assays at concentrations ranging from 5 to 300 μg/ml for 6, 12 or 24 h. Viability assays showed no dose or time dependent increase in toxicity for NIH-3T3 cells treated with WSNTs or MSCs treated with MSNPs or WSNTs. Viability of NIH-3T3 cells treated with MSNPs at concentrations above 50 μg/ml was significantly lower compared to untreated cells. Analysis of MSCs differentiation to adipocytes and osteoblasts showed that MSC differentiation potential was not significantly affected after treatment with potentially safe low (10 μg/ml) and high (50 μg/ml) doses of MSNPs and WSNTs for 24 h. TEM analysis showed that both MSNPs and WSNTs are internalized into the cytoplasm by MSCs. The results lay the foundation to further explore the potential of these nanoparticles at potentially safe doses as multifunctional agents for biomedical applications.

Keywords: inorganic nanoparticles, stem cells, cytotoxicity, differentiaiton, tissue engineering

1 INTRODUCTION

Molybdenum disulfide nanoplatelets (MSNPs) and tungsten disulfide nanotubes (WSNTs) are layered inorganic compounds whose interesting physiochemical properties have been investigated for tribological and electronic applications [1]. Biomedical applications of MSNPs and WSNTs include investigation as lubricants for orthodontic wires and catheters [2]. Recently, we have found that these inorganic nanoparticles are able to reinforce polymeric nanocomposites for tissue engineering better than carbon nanotubes and graphene [3,4]. In this application, fibroblastic cell types, including mesenchymal stem cells (MSCs), would be exposed to these particles as the scaffolds degrade, necessitating investigation into the cytotoxicity of these particles. The few studies performed on the cytotoxicity of these nanoparticles have focused on cell types that would be affected during environmental exposure to the particles through inhalation or ingestion and not on stem cells. This study investigates the cytotoxicity of MSNPs and WSNTs to fibroblasts and human MSCs, the particles effect on differentiation potential of MSCs, and the uptake of these nanoparticles to determine potentially safe doses for biomedical applications.

2 MATERIALS AND METHODS

2.1 Synthesis and Characterization

MSNPs were synthesized using a high temperature reaction of MoO_3 and sulfur powder as described previously [5] while WSNTs were purchased from APNano (NY, USA). Particles were characterized by transmission electron microscopy (TEM) and Raman spectroscopy. The nanoparticles were water-solubilized with 1,2-Distearoyl-sn-Glycero-3-Phosphoethanolamine conjugated polyethylene glycol (DSPE-PEG).

2.2 Cytotoxicity

For cytotoxicity assays, cells were treated with nanoparticle concentrations of 5, 10, 50, 100, or 300 μg/ml for 6, 12, or 24 hours. The two positive controls were untreated or treated with the DSPE-PEG, and the negative control was treated with 70% ethanol to kill the cells. Cytotoxicity was tested by Presto Blue and lactate dehydrogenase (LDH) assays.

2.3 Differentiation

For differentiation studies, cells were treated with 10 or 50 μg/ml of nanoparticles for 24 hours and then osteo- or adipogenic differentiation media was used to induce differentiation. Alizarin Red S staining and alkaline phosphatase and calcium assays were used to assess osteogenesis while Oil Red O staining and elution were used to assess adipogenesis against a control of DSPE-PEG treated cells.

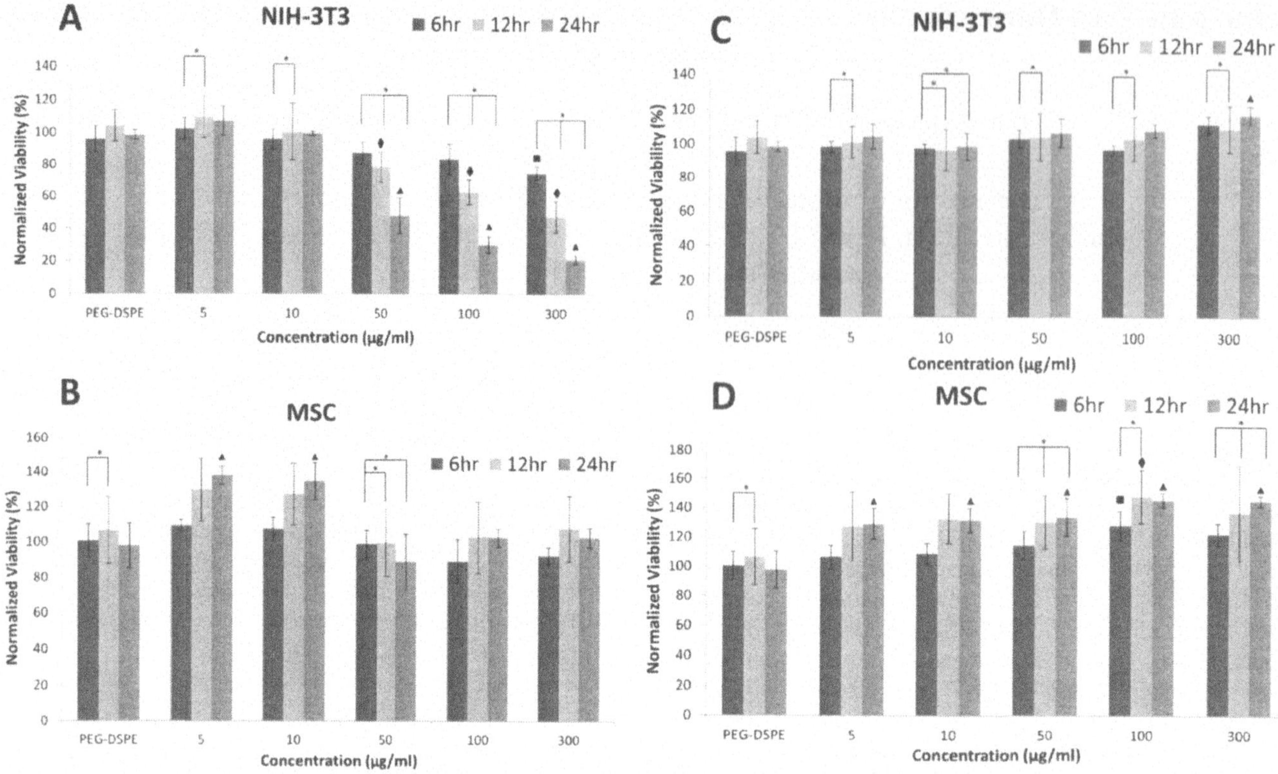

Figure 1. Presto Blue assay results at 6, 12, and 24 hours after treatment. For each nanoparticle, cells were treated with PEG-DSPE, 5 μg/ml, 10 μg/ml, 50 μg/ml, 100 μg/ml, and 300 μg/ml concentrations. MSNPs (A) and WSNTs (C) for NIH-3T3 fibroblasts; MSNPs (B) and WSNTs (D) for MSCs. Data are presented as mean ± standard deviation of percentage viability compared to untreated cells ($n = 4$). Statistical significance ($p = 0.05$) with respect to untreated groups at 6, 12 and 24 hours are denoted by (■), (◊), (▲) respectively. Statistical significance between time points within groups is denoted by (*).

2.4 Uptake

MSCs were seeded on ACLAR film and treated with 50 μg/ml of either MSNPs or WSNTs for 24 hours. After treatment, cells were fixed and processed for TEM.

2.5 Statistics

A sample size of four was used for all groups and Kruskal-Wallis with a Dunn post hoc was used to compare within the groups and with the controls.

3 RESULTS AND DISCUSSION

3.1 Characterization

MSNPs were round plates with diameters in the range of 60-90 nm and thickness of about 8 nm [3]. WSNTs were tube-like structures with diameters in the range of 50-100 nm and lengths in the range of 1-15 μm. Raman spectra of the nanoparticles showed the expected peaks for hexagonal molybenum disulfide and tungsten disulfide (data not shown).

3.2 Cytotoxicity

Only treatment of NIH-3T3 cells with MSNPs showed a decrease in viability with increasing concentration as compared to DSPE-PEG. Treatment of fibroblasts with MSNPs and treatment of MSCs with MSNPs or WSNTs showed little to no cytotoxicity at any of the concentrations (Figure 1). An increase in MSC proliferation was observed for all WSNT treatment concentrations compared to the DSPE-PEG control with the highest increase of 45% at 24 hours.

3.3 Differentiation

Figure 2 shows the results of the alkaline phosphatase and calcium assays. MSCs treated with low (10 μg/ml) and high (50 μg/ml) concentrations of MSNPs and WSNTs for 24 hours maintained their differentiation potential to adipocytes and osteoblasts.

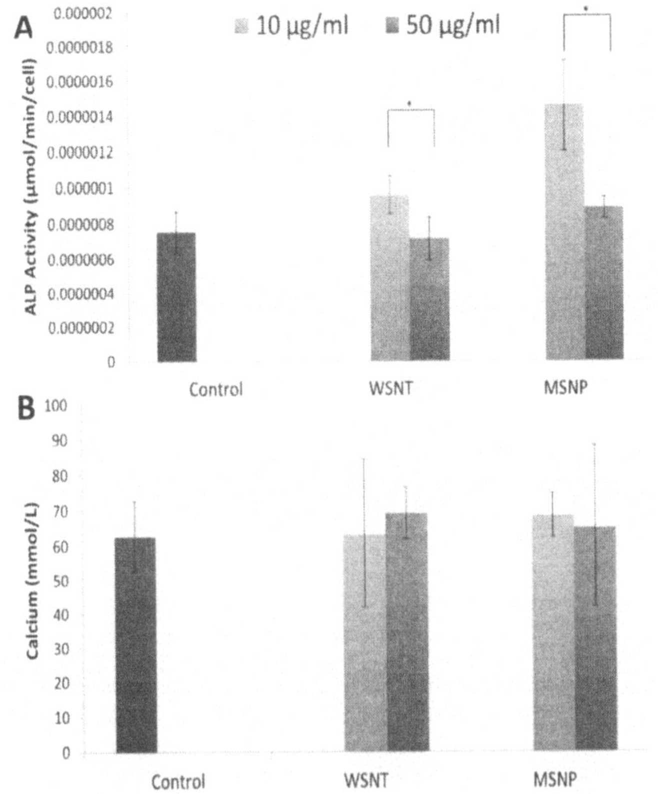

Figure 2. Osteogenesis results for MSCs after treatment for 24 h with either 10 or 50 µg/ml of MSNPs or WSNTs respectively, followed by 14 days incubation with osteogenic differentiation media. (A) ALP activity, (B) Calcium content. Data are presented as mean ± standard deviation ($n = 3$). Statistical significance ($p < 0.05$) was determined by the Kruskal-Wallis test with Dunn's post hoc. (*) indicates significance within groups.

3.4 Uptake

MSNPs are internalized in vesicles in the cells while WSNTs are internalized in vesicles as well as cytoplasmic matrix (data not shown).

4 CONCLUSIONS

The results indicate that concentrations less than 50 µl/ml of MSNPs or WSNTs do not significantly affect viability of NIH-3T3 cells or MSCs and do not affect the differentiation potential of MSCs. The results provide preliminary safety guidelines to further explore the potential of these nanoparticles at potentially safe doses as multifunctional agents for biomedical applications.

5 REFERENCES

[1] R. Tenne, "Inorganic nanotubes and fullerene-like nanoparticles," Nat. Nanotechnol. 1, 103-111, 2006.

[2] A. Katz, M. Redlich, H.D. Wagner, and R. Tenne, "Self-lubricating coatings containing fullerene-like WS_2; nanoparticles for orthodontic wires and other possible medical applications," Tribol. Lett. 21, 135-139, 2006.

[3] G. Lalwani, A.M. Henslee, B. Farshid, L. Lin, F. K. Kasper, Y-X. Qin, A.G. Mikos and B. Sitharaman, "Two-dimensional nanostructure-reinforced biodegradable polymeric nanocomposites for bone tissue engineering," Biomacromolecules, 14, 900-909, 2003.

[4] G. Lalwani, A.M. Henslee, B. Farshid, P. Parmar, L. Lin, Y-X. Qin, F.K. Kasper, A.G. Mikos, and B. Sitharaman. Acta Biomater. 9, 8365-8373, 2013.

[5] C.F. Castro-Guerrero, F.L. Deepak, A. Ponce, J. Cruz-Reyes, M.D. Valle-Granados, S. Fuentes-Moyado, D.H. Galván, M. José, "Structure and catalytic properties of hexagonal molybdenum disulfide nanoplates," Catal. Sci. Technol. 1, 1024-1031, 2011.

Preparation and investigation of nano-carbonated hydroxyapatite/polymer composite spongy scaffolds for biomedical applications

S.A. Siddiqi[1], F. Manzoor[1], U. Azhar[1], A. Jamal[1], M. Tariq[2], M.S. Anwar[3], M. Saleem[3], A.A. Chaudhry[1], I.U. Rehman[1,4]

1. Interdisciplinary Research Centre in Biomedical Materials (IRCBM), COMSATS IIT. Defence Road, Off Raiwind Road, Lahore
2. Department of Biology, Syed Babar Ali School of Science and Engineering, LUMS, DHA, Lahore 54792, Pakistan
3. Department of Physics, Syed Babar Ali School of Science and Engineering, LUMS, DHA, Lahore 54792, Pakistan
4. Department of Material Science and Engineering, The Kroto Research Institute, University of Sheffield, Broad Lane, S3 7HQ, Sheffield, UK

ABSTRACT

Spongy scaffolds containing nano-carbonated hydroxyapatite (nCHA), with composition $Ca_{10}(PO4)_{6-y}(CO3)_y(OH)_2$,with y=2,4, were prepared by the lyophilization method. Different polymers, natural gelatin (G), chitosan (CS) and polyvinyl alcohol (PVA), alone or in combination, were used to make their composites with nCHA. All the samples were characterized by (XRD), (SEM), (FT-IR) techniques. Particle size and surface area were determined by BET surface area analysis. The nCHA containing scaffolds mediated cytotoxicity (MTT assay) and cell attachment studies (SEM) were carried out on rat bone marrow derived mesenchymal stem cell (MSC). SEM studies revealed varying microstructures for pore sizes, shapes and pore interconnectivity. All the scaffolds were found to be highly porous, porosities beyond 70%. It is observed that the scaffold morphology changes significantly for one nCHA/polymer combination to the other. Furthermore we investigated the biocompatibility of these scaffolds by performing cell viability and cell attachment tests. Our results indicated evident cell attachment on all the materials and also no significant cytotoxicity was observed in cells grown with these materials compared with control. The preparation of spongy scaffolds and their biocompatibility augment their potential use in tissue engineering and biomedical applications.

Keywords: *carbonated hydroxyapatite, cell attachment, freeze drying, spongy scaffolds, polymer scaffolds, lyophilization*

1 INTRODUCTION

Human bone contains up to 8wt% carbonate ions that occupy phosphate and hydroxide positions in the apatite lattice. It has been shown that the presence of carbonate in synthetic hydroxyapatite (HA) leads to structural disorder and a higher solubility. However, this disorder is useful because it gives rise to higher HA bonding to surrounding tissue. Carbonate-substituted HA is therefore, considered as an optimized biomaterial, which leads to faster bonding between an inserted implant surface and human bone[1].

In recent past, considerable attention has been given to the fabrication of porous ceramic scaffolds for osseous tissue regeneration by different preparation methods. Freeze casting is a simple technique to produce complex-shaped highly porous ceramic or polymeric scaffolds. An ideal scaffold should have the following characteristics: non-toxicity, biocompatibility, suitable mechanical strength, biodegradability with a rate that can match the rate of tissue regeneration, and should not have negative effects on the surrounding tissues and organs. Meanwhile, it should have high porosity and interconnected porous structure to provide adequate space for the cells' seeding, growth and proliferation. With these requirements in mind,we report here thesynthesis of porous scaffolds of nCHA with different biologically accepted polymers (G, PVA, CS) and explore them for their biomedical applications [1,2].

2 EXPERIMENTAL METHODS

2.1 Preparation of Freeze dried spongy scaffolds

The scaffolds are prepared by the procedure given below

2.1.1 Preparation of Gelatin/ Carbonated-HA scaffold

The analytical grade Gelatin powder (Fluka, Germany), 1g, was dissolved in 40 ml distilled water at 40°C with constant stirring for 6h and 1g each of hydrothermally pre-prepared [3], nano-CHA, $Ca_{10}(PO4)_{6-y}(CO3)_y(OH)_2$with y=2,4, were added to transparent solution with constant stirring to obtain a milky slurry of nCHA and G. The slurry was poured into 24-well plate template and kept at 40°C in a freezer for 6 h. Subsequently, the samples were lyophilized for 24 h.

2.1.2 Preparation of PVA/ Carbonated-HA scaffold

Similarly PVA (Merck M_w=72,000) powder, 1g, was dissolved in 40ml distilled water at 90°C with constant stirring for 1h. Pre-prepared nano-CHA, $Ca_{10}(PO4)_{6-y}(CO3)_y(OH)_2$ with y=2,4, 1g each, were added, one by one, to PVA transparent solution with constant stirring to obtain a milky slurry of nCHA and PVA. The slurry was poured into 24-well plate template and kept at -40°C for 6 h in a freezer. Subsequently, the samples were lyophilized for 24 h.

2.1.3 Preparation of Chitosan/ Carbonated-HA scaffold

Similarly chitosan powder, extracted from shrimpshells, 81% Degree of Deacetylation (DD), (M_w=88kDa), was dissolved in distilled water at 50°C with constant stirring for 6h. A few drops of acetic acid were added to enhance CS dissolution. Similarly, pre-prepared nano-CHA, with y=2,4, 1g each, were added to CS solution with constant stirring to obtain milky slurries of nCHA and CS. The slurries were poured into 24-well plate template and kept at -40°C for 6 h. Subsequently, the samples were lyophilized for 24 h.

2.1.4 Preparation of PVA/Chitosan/ Carbonated-HA scaffold

Pre-prepared 20ml distilled water solutions, with each of 0.5g PVA solution, at 90°C, and CS, at 40°C, were mixed together at room temperature with constant stirring for 6h. Pre-prepared 1g nano-CHA, powders, with y=2,4, were added to PVA and CS solutions with constant stirring to obtain a milky slurry of nCHA, PVA and CS. The slurries were poured into 24-well plate template and freezed at 40°C for 6 h. Subsequently, the samples were lyophilized for 24 h [4].

2.2 In vitro culture of rat mesenchymal stem cells (MSC)

Rat mesenchymal stem cells (rMSC) were isolated from femur of 4-5 weeks rats using direct adherence method. The femur was isolated under sterile conditions. A disposable aseptic syringe was used to draw antibiotic supplemented L-DMEM medium and to repeatedly fill bone marrow cavity to collect cells in a sterile petri dish. The obtained cell suspension was centrifuged at 250xg for 5 minutes. The cell pellet was resuspended in DMEM (GIbco) containing 10% FBS (Gibco) and 0.1% penicillin and streptomycin (GIbco) and transferred to T25 tissue culture flask. The flasks were incubated at 37 deg. C in a 5 % CO2 incubator. Cells isolated from one rat were cultured in one flask. The first medium was changed after 4 days. Later on the medium was changed on alternative days until the cells become 70–80 % confluent. MSC were sub-cultured at 70–80 % confluence. The cells were trypsinized, counted (dead cells excluded by trypan blue assay) and passaged in T-75 flasks. Second- or third-passage MSC were used for cytotoxicity and SEM analysis.

2.3 Cytotoxicity Assay

Cellular toxicity was determined by 3-(4,5-Dimethylthiazol-2-yl)-2,5-diphenyltetrasodium bromide (MTT) assay. Prior to cell culture, all the scaffolds were sterilized under UV light for 6 hrs/autoclave. Immediately before cell seeding the scaffolds were washed 2-3 times with PBS and pre-conditioned in DMEM medium for an hour. MSC were seeded in 24-well cell culture plate with 5*104 cells per well with or without scaffolds. Cells seeded in 24-plate wells without scaffold were used as positive control. Post day 7 the medium was discarded and cells/scaffolds were washed with 1 ml PBS. 1 ml (0.5 mg/ml) MTT solution was added to each well and the plate was incubated at 37°C for 3 hrs. The MTT solution was discarded and the cells/scaffolds were washed once with 1 ml PBS. To solubilize the formazan crystals 0.5 ml dimethyl sulfoxide (DMSO) was added to each well and the plate was kept under shaking for 15-20 minutes. The optical density (OD) of the dissolved crystals was measured by using microplate reader at 590 nm. The assay was set up in triplicate with MSC derived from 3 different rats for each sample. % Viability is represented as mean ± SD of 3 independent experiments.

X-ray diffraction (XRD) was performed on XPERT-PRO diffractometer, operated at 40 kV and 40 mA using Cu Karadiation. The detector was scanned over a range of 2θangles from 20°to 80°at a step size of 0.02°. FT-IR (Thermo-Nicolet 6700 P Spectrometer (USA)) was taken in the wavenumber range of 500 cm^{-1} to 4000 cm^{-1} in the photo-acoustic mode in the resolution of 8 cm^{-1} with 256 numbers of consecutive scans.Scanning electron microscopy was performed on Nova NanoSEM-450. SEM was operated on low vacuum mode at 50 Pa. Low vacuum detector was attached with its cone under pole piece. All other parameters are mentioned on bar of the images. The pore size and surface was determined by BET (Tristar II 3020analyzer) using liquid nitrogen.

3 RESULTS AND DISCUSSION

3.1 XRD of Carbonated Hydroxyapatite

XRD patterns of nCHA, y=2, with gelatin (G), polyvinyl alcohol (PVA) and chitosan (CS) are shown in Fig 1a.Hydroxyapatite peaks are appeared when 2θ is 26.1°, 31.8°, 46.2° and 51.3°. These patterns resemble the JCPDScard number 9-0432 which is the evidence of hydroxyapatite. However, these spectra are different somewhat which is due to the presence of the polymers in CHA. In Fig. 2 XRD pattern of CHA y=4 with polymers has been shown. The spectra clearly show the presence of CHA.

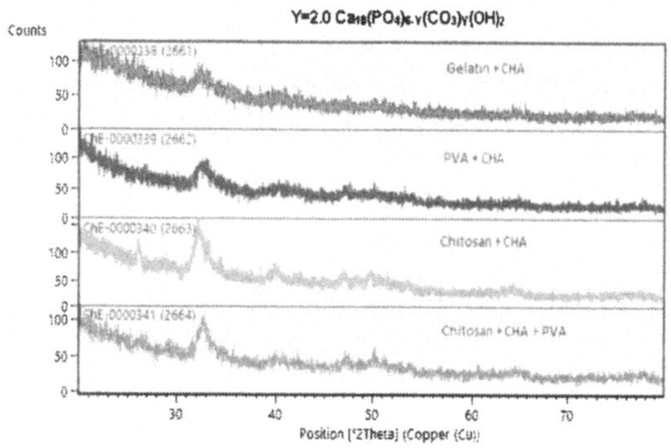

Figure 1 XRD pattern of spongy scaffolds y=2

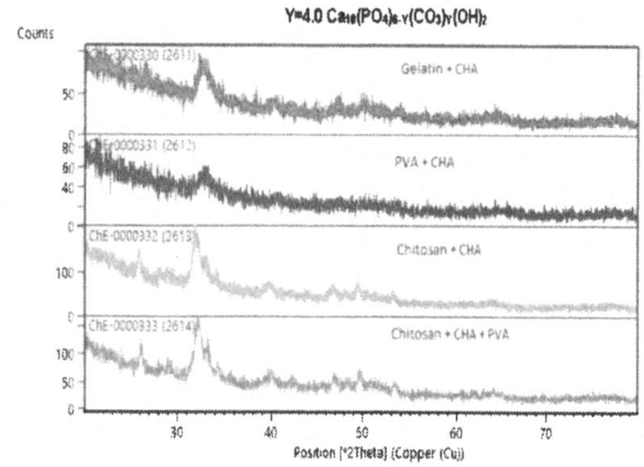

Figure 2 XRD pattern of spongy scaffolds y=4.

3.2 Fourier Transform InfraredSpectroscopy (FT-IR)

The FT-IR spectra of CHA with polymers have shown in the Fig. 3. These are the spectra of CHA y=2 with different polymers. The PO_4^{-3} appeared in the spectra at the peaks 562cm^{-1} and 1014cm^{-1} while CO_3^{-2}peaks appeared in the range of 745cm^{-1} to 755cm^{-1} which is due to the presence of carbonate in the CHA. OH^{-1} stretch can be seen between 3480 cm^{-1} and 3560 cm^{-1}.

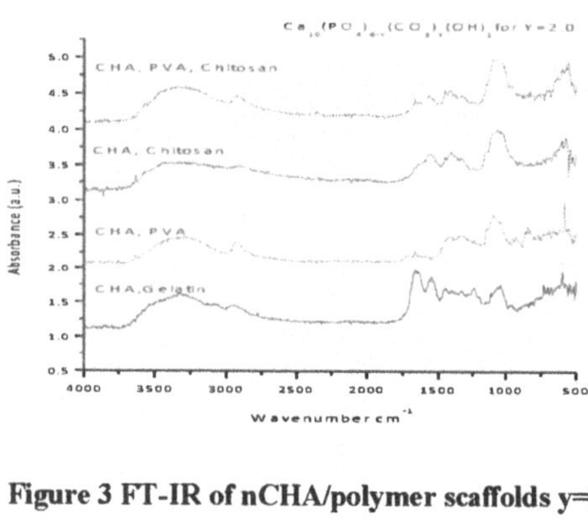

Figure 3 FT-IR of nCHA/polymer scaffolds y=2

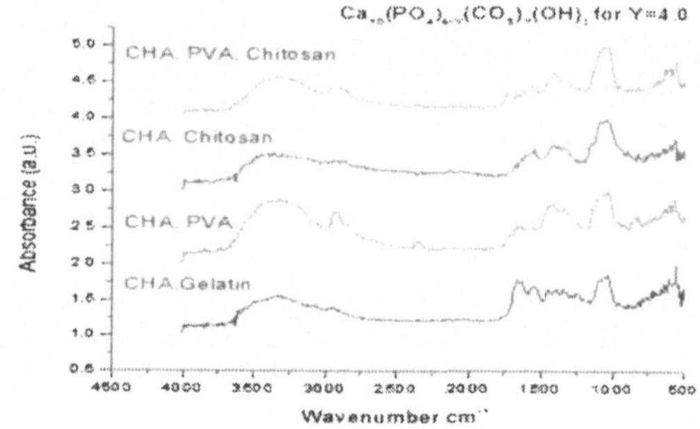

Figure 4 FT-IR of nCHA/polymer spongy scaffolds y=4

3.3 Scanning Electron Microscopy (SEM)

To study the surface morphology of the freeze dryer and MSC cell attached samples, electron microscopy was performed on all the samples and their micrographs has shown below in the fig. SEM images show that the polymer matrix has uniformly dispersed in the nCHA matrix. The interconnected porousstructure can also be seen in the micrographs. It has noted that the pore size and porosity increased from y=2 to y=4

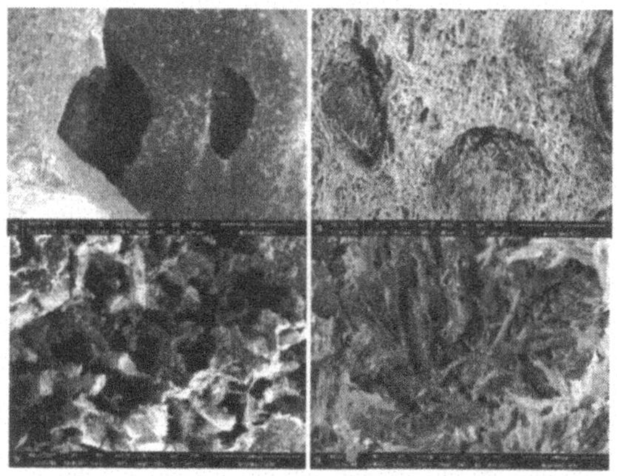

Figure 5 Freeze dried scaffolds of CHA y=4 with A) Gelatin, B) PVA, C) Chitosan, D) PVA and Chitosan

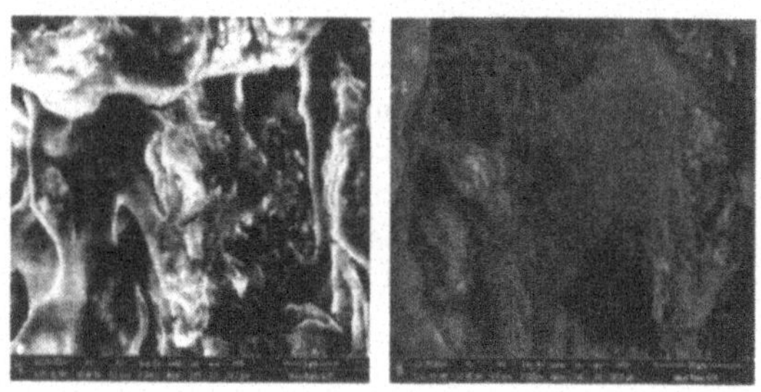

Figure 6 SEm showing MSC cell attachment on the spongy scaffolds

Figure 7 Freeze dried scaffolds of CHA y=2 with A) Gelatin, B) PVA, C) Chitosan, D) PVA and Chitosan

Scaffolds of nCHA with	Y	Surface Area (m^2/g)	Pore size (nm)	Pore volume (cm^3/g)
G	4.0	3.6889	9.59.02	0.884
PVA	4.0	16.4998	57.49	0.237
CS	4.0	27.6406	107.92	0.746
PVA+CS	4.0	18.2690	21.78	0.0994
G	2.0	5.6739	22.66	0.032
PVA	2.0	40.7931	12.04	0.123
CS	2.0	23.5861	19.54	0.115
PVA+CS	2.0	21.0259	13.61	0.071

Table 1 BET data of spongy scaffolds

3.4 Surface area and pore size

Surface area plays an important role in the cell attachment. It is shown in the Figs 8,9 that our scaffolds are mesoporous. The pore size is maximum, as shown in the Table 1.In the case of gelatin for nCHAboth for y=2 and 4 that may be due to the dispersion of polymer in the nCHA matrix.

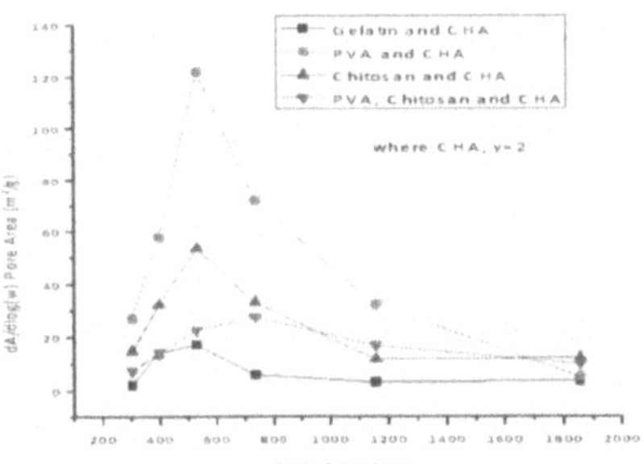

Figure 8 BET of spongy scaffolds for y=2

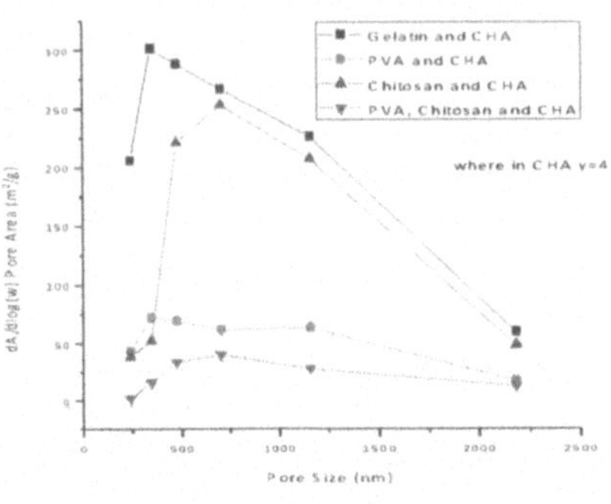

Figure 9 BET of spongy scaffolds for y=4

3.5 Cytotoxicity

Cytotoxicity tests showed that the samples are non-toxic for tissue engineering applications.

Figure 10 Cytotoxicity of spongy scaffolds

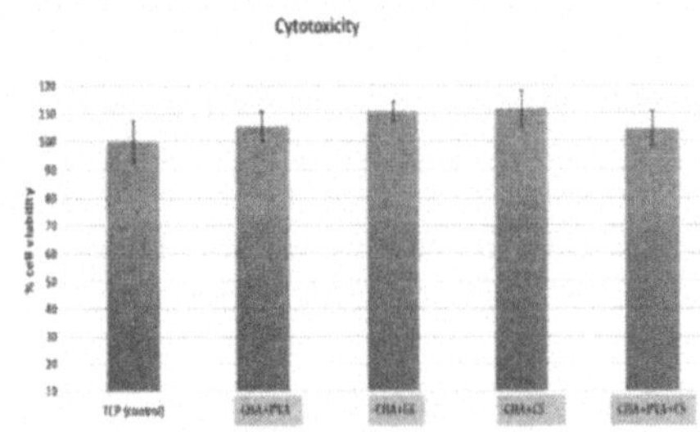

4 CONCLUSION

The studies showed that the spongy scaffolds resemble the natural bone and the biological studies suggest that the scaffolds are biocompatible and can be used in bone tissue engineering.

Acknowledgements:

The authors are grateful to HEC, Govt. of Pakistan, for providing funds under NRPU project No. 1835.

References

1. Lijun Kong, Yuan Gao, Guangyuan Lu, Yandao Gong,Nanming Zhao, Xiufang Zhang,European Polymer Journal 42 3171–3179,2006.

2. Hae-Won Kima, Hyoun-EeKima, Vehid Salih, Biomaterials, 26 5221–5230,2005.

3. J S Earl, D J Wood and S J Milne, Journal of Physics: Conference Series, 26, 268–2712006.

4. Sylvain Deville, Eduardo Saiz, Antoni P. Tomsia, Biomaterials, 27 5480–5489, 2006.

Electrophoretic Coating of Hydroxyapatite-Boron Nitride Nanocomposites on Titanium Implants

F. Bakan*, M. Sezen*, M. Gecgin***, Y. Goncu**, N. Ay***

*Sabanci University Nanotechnology Research and Application Center (SUNUM), Istanbul, Turkey, feraybakan@sabanciuniv.edu
**BORTEK Boron Technologies &Mechatronics Inc., Eskisehir, Turkey, yyuceturk@yahoo.com
*** Anadolu University, Materials Science and Engineering, Eskisehir, Turkey, nay@anadolu.edu.tr

ABSTRACT

Titanium (Ti) and its alloys have long been used as implant materials in dental and orthopedic applications. However, they can suffer certain disadvantages, such as poor osteoinductive properties and low corrosive-wear resistance. The potential of hydroxyapatite (HAp) coatings to enhance the long-term stability of the bond between metallic implants and the surrounding bone structure has been well documented. The HAp-coated implant can combine the high mechanical strength of the metal with the excellent biocompatibility and bioactivity of the ceramic and is therefore suitable for implants in high load-bearing applications. Recently, in order to improve the mechanical properties of the HAp coating itself, HAp composite coatings can used. Accordingly, nano-sized hexagonal boron nitride (hBN)-HAp composites coated on Ti by Electrophoretic deposition (EPD) process and structural and morphological properties of coatings were investigated. For the structural characterization, XRD, Raman Spectroscopy and EDS were used, and for the identification of morphology and surface features, SEM and FIB analyseswere conducted.

Keywords: hydroxyapatite, boron nitride, nanocomposite, electrophoretic deposition (EPD), characterization

1. EXPERIMENTAL

The starting powders used in the present work were nano sized hydroxyapatite (nHAp, Nanotech Advance Technology and Electric Electronic Systems Co. Ltd., Eskişehir, Turkey) and nano sized hexagonal boron nitride (nhBN; 120nm; Boron Technologies and Mechatronics Inc., Eskişehir, Turkey) powders. Commercial pure titanium (Ti-Gr4) plates used as substrate were provided from TIMET Metal&Medical, Izmit, Turkey.

Each powder was dispersed in absolute ethanol (Sigma-Aldrich, USA) under magnetic stirring and sonicaton. The suspensions were prepared by magnetically stirring (Heidolph MR Hei-Standart) of nHap-nhBN dispersions certain ratios in absolute ethanol. Each suspension contains 1% nHAp. The concentration of nhBN was chosen 0-2.0-10.0 and 25.0 % (w) by the percentage of nHAp.

Commercial pure titanium sheet was used and cut in a way to have the same surface area (10x20x1.5 mm). Each sample was polished using SiC papers. Subsequently the samples were washed with detergent in ultrasonic bath and then ultrasonically cleaned with acetone. Substrates were washed with distilled water and kept in ethanol.

Titanium plates were used as working and counter electrodes. The distance between the electrodes was chosen 10mm. Experiments were performed at 100V using a D.C volt regulator (BioRAD Power Pac Basic). Voltages were applied on Ti plates for 5 seconds. After deposition, all of the samples were heat treated respectively in tube furnace 800°C, for 2h in argon atmosphere at a heating rate 1°C/min and cooling rate 1°C/min. Sample abbreviations are given in Table 1.

Notation	nhBN amount (w%)
HB0.0	0
HB5.0	5
HB10.0	10
HB25.0	25

Table 1: Description of the coated samples

For the examination of structural and morphological characteristics of the composites various complimentary analysis methods were performed. For the structural characterization, XRD and Raman Spectroscopy were used while, SEM and FIB techniques were carried out for revealing the morphological characterization.

To investigate the phase composition, XRD analyses were performed at a Bruker AXS Advance D8 using a CuKα radiation with the step scanning mode, applying a tube voltage of 40 kV and a tube current of 40 mA. The XRD patterns of BN/HAp composites sintered at 1100°C for 2 h are shown for different BN amounts. Phase analyeses were performed using PDF card no 009-0432 for HAp, 09-169 for Ca_3PO_4 (β-TCP) and 034-421 for BN.

Raman Spectroscopy measurements were taken by Renishaw Raman inVia Spectrometer couoled with a laser source of 532 nm and a Leica microscope (measurements performed using 50x objective).

Scanning electron microscopy (SEM) examinations were done for investigating the morphology of the composites using a JEOL 4601F MultiBeam Platform with FEG-SEM, equipped with an energy dispersive spectrometer (EDS-Xmax ^{N} SDD Oxford Instruments). An electron probe at low acceleration voltages (3-5 kV) were applied. Prior to morphologic characterization, all of the samples were coated with a thin layer of Pt/Pd by sputtering (Cressignton, USA). For the gallium ion milling, 30 keV ion energy was used at intermediate milling currents (0.1-1 nA), while the SE images were acquired by the FEG-SEM column of the platform at low electron energies vaying from 3 to 5 keV.

2. RESULTS AND DISCUSSION

2.1 X-ray diffractrometry (XRD)

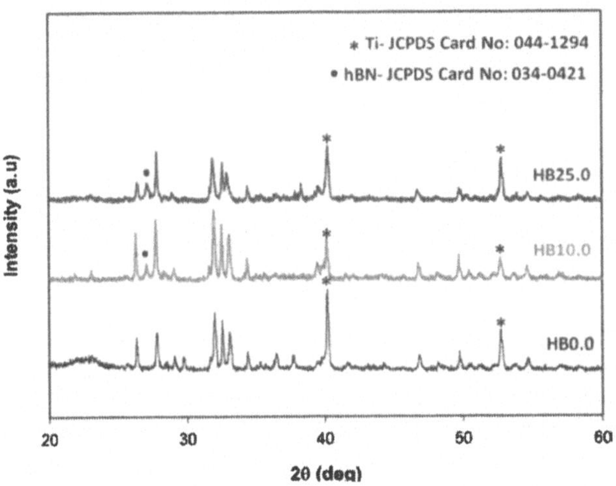

Figure 1: XRD patterns of HAp-hBN coatings on Ti-Gr4.

XRD patterns of samples sintered at 800°C under Ar atmosphere is shown in Figure 1. Phase analysis was performed using PDF card no. 009-0432 for nHAp, 034-0421 for hBN and 034-0421 for Ti, providing information in the 2Θ range of 20°–60°. Phase analysis revealed that all major peaks of HAp were present for all the samples.

2.2 Raman Spectroscopy

Figure 2 shows the Raman spectra of the samples. All bands have been assigned to internal vibrational modes of the phosphate groups. The strong peak at 960 cm^{-1} is assigned to the totally symmetric stretching mode (υ1) of the tetrahedral PO_4 group (P–O bond) [1-3]. Triply degenerate bending mode (υ4) of the PO_4 group (O–P–O bond) was observed at 590 cm^{-1} and 578 cm^{-1}. Triply degenerate asymmetric stretching mode (υ3) of the PO_4 group (P–O bond) was observed at 1046 cm^{-1} [4-5]. For structural characterization of hBN in the composites, the peak observed at 1368 cm^{-1} can be attributed to the characteristic adsorption of in-plane B–N stretching vibration mode of hBN [6-8].

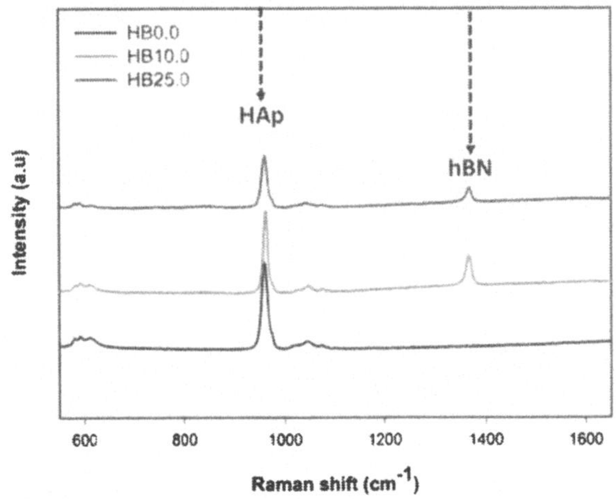

Figure 2: Raman spectra of the HAp-hBN coatings on Ti.

2.3 Morphology

Figure 3 shows the secondary electron (SE) images of the coatings. Increasing the BN amount in the composite caused structural change from spherical-like to plate-like form. Also the highest porosity was observed for HB10.0 compared to other structures.

In addition, cross-sections from each sample were acquired by means of gallium ion milling at the FIB/SEM platforms. The cross-sections revealed the nanocomposite coating layer as well as the Ti substrate, while the interfaces between those layers could also be observed. The related SE images from the cross-sections are given in Fig. 4, Fig.5 and Fig. 6 respectively.

Here it should be noted that, the vertical lines are formed due to the uneven topography of the sample surface, resulting in milling lines caused by curtain effect of ion milling.

Figure 3: FEG-SEM images of HAp-BN composite coatings on Ti substrate.

According to FIB cross-section analyses, it can be observed that the coatings have different adhesion behaviour on to the substrates as well as the altered morphologies, such as porosity and packing. Moreover, with the increasing amount of hBN, there seems to be more integration with the Ti matrix as the interface between the substrate and the coating becomes less in thickness. This can be obviosly tracked via the EDS line profile analysis which will be given in Fig.7.

Figure 4: SE images of FIB cross-section which belongs to HB 0.0 sample.

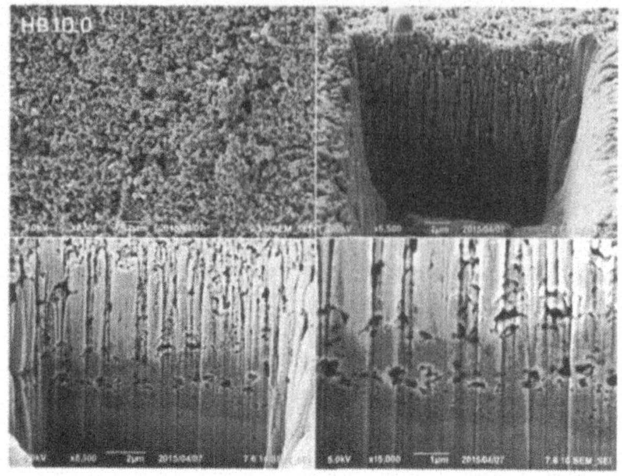

Figure 5: SE images of FIB cross-section which belongs to HB 10.0 sample.

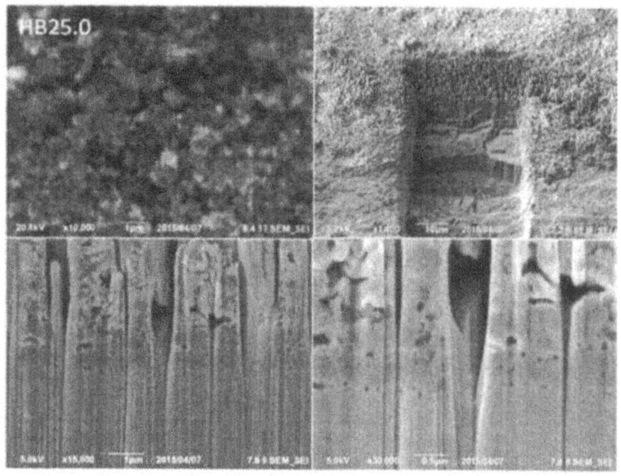

Figure 6: SE images of FIB cross-section which belongs to HB 25.0 sample.

EDS was used in order to track the changes in the chemistry throuh the cross-sections of 3 different materials. Figure 7 shows the elementel distrubution within the line profiles acqired at the transition zones within FIB cross-sections of the samples that contains different percentages of hBN. The interface layer has found to be rich in oxygen when compared to other layers which can be due to native oxide layer formed by Ti substrate. However the thickness of this layer and the detected compositional percentage of O_2 varies by the hBN concentration of the nanocomposite coating. This can be attributed to the adhision behaviour of the coatings on the Ti matrices.

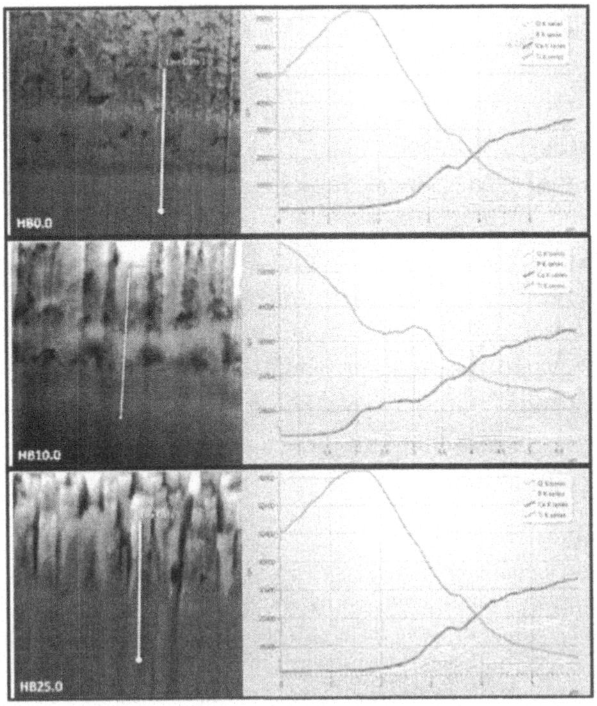

Figure 7: EDS line profiles through the FIB crossections of the samples

3. CONCLUSION

Titanium (Ti) and its alloys have long been used as implant materials in dental and orthopedic applications. However, they can suffer certain disadvantages, such as poor osteoinductive properties and low corrosive-wear resistance. In this study, ceramic composites with HAp matrix and nano-sized hexagonal boron nitride (hBN) having varying compositional percentages were coated on commercial pure titanium (Ti-Gr4) for improved properties and the effect of hBN amount on microstructure and morphological properties of the coatings were investigated. For the structural characterization, XRD, Raman Spectroscopy and EDS were used, whilst for the identification of morphology and surface features, comprehensive SEM and FIB analyses were perfomed for a complimentary analysis. According to XRD results, all major peaks of HAp were present for all samples. hBN and Ti peaks could also be tracked. For structural characterization of hBN in the composites, Raman measurments were done and the peak observed at 1368 cm^{-1} was attributed to the characteristic adsorption of in-plane B–N stretching vibration mode of hBN.

The FIB cross-sections revealed the nanocomposite coating layer as well as the Ti substrate, while the interfaces between those layers could also be observed. Via FIB-SEM and EDS analysis it was observed that the coatings have different adhesion behaviour on to the substrates as well as the altered morphologies, such as porosity and packing. With the increasing amount of hBN, there seems to be more integration with the Ti matrix as the interface between the substrate and the coating becomes less in thickness, which was tracked via the EDS line profiles. The interface layer has found to be rich in oxygen when compared to other layers which can be due to native oxide layer formed by Ti substrate. However the thickness of this layer and the detected compositional percentage of O_2 varies by the hBN concentration of the nanocomposite coating. This can be attributed to the adhision behaviour of the coatings on the Ti matrices.

Acknowledgements

112M592, 112M590 TUBITAK Primary Subjects R&D Funding Program projects are gratefully acknowledged for the financial support.

REFERENCES

[1] O'Shea DC, Bartlett ML, Young RA, "Compositional analysis of apatites with laser-Raman spectroscopy: (OH, F, Cl) apatites", Arch Oral Biol,19, 995–1006, 1974.

[2] Griffith WP, "Raman studies on rock-forming minerals. Part II. Minerals containing MO3, MO4, and MO6 groups", J Chem Soc (A) ,286–291, 1970.

[3] De Aza PN, Guitian F, Santos C, De Aza S, Cusco R, Artus L, "Vibrational investigation of calcium phosphate compounds 2. Comparison between hydroxyapatite and β-tricalcium phosphate", Chem Mater,9, 916–922 1997.

[4] Tsuda H, Arends J, "Raman spectra of human dental calculus", J Dent Res,72,1609–1613, 1993.

[5] Sauer GR, Zunic WB, Durig JR, Wuthier RE, "Fourier transform Raman spectroscopy of synthetic and biological calcium phosphates", Calcif Tissue Int, 54, 414–420, 1994.

[6] M. V. Ugarov, V. P. Ageev, and V. I. Konov, "Chemical vapour deposition of boron nitride films stimulated by ultraviolet radiation pulses from a KrF excimer laser,"Quantum Electronics, 25, 7,. 679–683, 1995.

[7] M. V. Ugarov, V. P. Ageev, A. V. Karabutov et al., "UV laser induced interfacial synthesis of CN-BCN layers on diamond films in borazine and ammonia,"Applied Surface Science, 138-139, 1–4, 359–363, 1999.

[8] G. Lian, X. Zhang,L. Zhu, M. Tan, D. Cui, Q. Wang, "A facile solid state reaction route towards nearly monodisperse hexagonal boron nitride nanoparticles", J. Mater. Chem., 20, 3736–3742, 2010.

Nonclinical evaluation of heparin binding copolymer – HBC1 for restoring of blood coagulation after heparin treatment

B. Kalaska*, E. Sokolowska*, K. Kaminski**, J. Miklosz*, Shin-ichi Yusa***, A. Blazejczyk****, J. Wietrzyk****, I. Kasacka*****, K. Szczubialka**, D. Pawlak*, M. Nowakowska** and A. Mogielnicki*

* Department of Pharmacodynamics, Medical University of Bialystok, Poland, amogiel@umb.edu.pl
** Faculty of Chemistry, Jagiellonian University, Krakow, Poland, szczubia@chemia.uj.edu.pl
*** Department of Materials Science and Chemistry, University of Hyogo, Japan, yusa@eng.u-hyogo.ac.jp
**** Department of Experimental Oncology, Polish Academy of Sciences, Wroclaw, Poland, wietrzyk@iitd.pan.wroc.pl
***** Department of Histology and Cytophysiology, Medical University of Bialystok, Poland, kasacka@umb.edu.pl

ABSTRACT

Protamine sulfate – a protein isolated from salmons is registered antidote of heparin. It is widely used to restore blood coagulation in case of severe bleedings related to heparin therapy, but may also induce serious side effects. We synthesized compounds able to bind heparin in blood. One of the Heparin-binding Copolymers (HBC) efficiently neutralized heparin activity in the living animal. HBC1 injected intravenously 3 minutes after heparin stopped bleeding and restored clotting to normal values. When administered in the therapeutic doses, HBC1 did not influence blood count and chemistry, platelets activity, cardiovascular and respiratory functions of rat. Intravital imaging revealed rapid elimination of polymer with urine. HBC1 is efficient, safe and easy to synthesize polymer that could replace protamine for the treatment of heparin associated bleeding. Our technology presents a pharmaceutical companies a unique opportunity to optimize a lead compound into a marketable therapeutic.

Keywords: heparin, antidote, polymer, thrombosis, coagulation, animal model

1 BACKGROUND

The incredible progress and wide use of various medical devices during intravascular or cardiac interventions increased the need for administration of anticoagulants, such as heparins. Protamine sulfate – a protein isolated from sperm of salmon fished around Japan is the only one registered antidote of unfractionated heparin (UFH). It is used more than 600,000 times each year in US alone to restore blood coagulation in case of severe bleeding related to heparin therapy after open-heart procedures. However, around one thousand deaths a year could be attributed to complications after protamine injection; mostly because of its allergenic properties. The quest to replace protamine has a long history, and was recently described [1]. Our group took a challenge 5 years ago. We have already shown some promising natural and synthetic polymers able to bind UFH [2-5]. Some of them have patents as heparin binding agents [6-8] with possible application in anemia [9]. Our present aims were to: synthesize new heparin-binding copolymers, select the most potent and safe one in vitro and show the efficacy and safety of most promising lead candidate for human medicine in preclinical studies.

2 MATERIAL AND METHODS

We synthesized a group of heparin binding copolymers (HBC). To estimate pharmacokinetics fluorescently labeled with Alexa Fluor 750 variants of the polymers were also obtained. Physicochemical characteristics of the polymers was performed to find out their molecular weight and dispersity of molecular weight. These polymers were tested for the ability to bind UFH in vitro. The amount of each polymer needed to bind 1 mg of UFH was determined using Azure A assay. The size of the particles formed in solution due to complexation of UFH with the polymers was found using dynamic light scattering (DLS) experiments. For the most promising polymers their interaction with serum proteins was also studied using gel electrophoresis and DLS. Blood count and aggregation in whole blood (Aggregometer Model 700, Chrono-log, USA) were performed to check potential blood toxicity. From several HBC polymers we selected *in vitro* the one – HBC1 with the best efficacy/toxicity index, and this lead polymer was further tested in 80 Wistar male rats, weighing 180-220 g and 30 NMRI-Foxn1nu/Foxn1nu mice weighing 20-30 g. All the procedures involving animals and their care were approved by Local Ethical Committees. They were randomly divided into 3 groups treated intravenously with PBS solution of: vehicle, UFH (300 UI/kg) and UFH+HBC1 (UFH 300 UI/kg followed by HBC1 1.95 mg/kg). We examined the efficacy of HBC1 in the model of arterial thrombosis electrically induced in the carotid artery of 36 rats according to experimental plan: UFH-3 min-

HBC1-7 min-electrical stimulus (1 mA)-10 min-thrombosis development-45 min-end. At the end tail bleeding time was measured, formed thrombus was removed (it was weighed after 24 h incubation) and blood from the heart for further measurement was collected. All the technical details were previously published by our team [3-4]. Additionally, 10 rats were injected with enoxaparin or enoxaparin followed by 5-minute intravenous infusion of HBC1. Changes in activated clotting time (ACT) were measured in blood samples collected from tail at the following time points: baseline, 5, 10, 30 and 60 minutes after anticoagulant administration. HBC1 infusion started at 5 minute of experiment and was continued for 5 minutes. The body distribution of HBC1 was measured in 0, 5, 30, 60 and 120 minute after the injection of UFH followed by AlexaFluor®750- labelled HBC1 to 30 mice (5 animals for each time point) using In-vivo MS FX PRO system (Carestream Health INC., USA). Then, brain, lungs, kidneys, heart, liver and spleen were isolated from mice and their fluorescence was measured ex vivo. Blood count and blood chemistry was monitored 1 hour or once a week in 28 rats for over 1 month. All organs were macroscopically examined for gross pathology. Samples of the lungs and liver were immediately fixed in formalin and routinely embedded in paraffin. Specimens were stained with hematoxylin and eosin for general histological examination. Additionally, we examined a possible acute toxicity of HBC1 injected intravenously to 15 heparinized rats in 3 times higher than therapeutic dose by monitoring blood pressure for over 1 hour (Plugsys, Transonics System, USA), heart rate, tissue perfusion, blood oxygen saturation, respiration rate, peak exhaled CO_2 and body temperature (PhysioSuite, USA).

3 RESULTS

3.1 In vitro tests

From a series of polymers obtained we selected the HBC1 polymer which exhibited the strongest ability to bind UFH (measured as the weight required to bind a unit mass of UFH), while the size of its complexes with UFH was the smallest. In vitro assays showed that only few polymers, including HBC1, did not change RBC count, platelet count and aggregation. Additionally, in vitro studies revealed the most suitable ratio of HBC1/heparin for in vivo studies.

3.1 Animal tests

UFH administered intravenously decreased thrombus weight, prolonged aPTT, bleeding time and increased anti-factor Xa activity. HBC1 reversed UFH effects to the control values (Fig. 1). Following the administration of the enoxaparin we observed significant rise in the ACT. Infusion of HBC1 rapidly shortened ACT value to normal levels and the neutralization was maintained throughout the course of the experiment (Fig. 2).

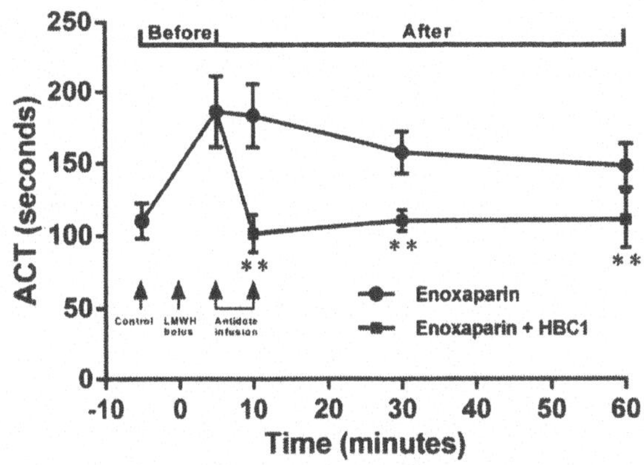

Figure 2: Reversing of ACT prolongation after enoxaparin by HBC1 in rats. ** P<0.01 vs. enoxaparin; Mann-Whitney test. Results are shown as mean ± SD, n = 6.

The signal of overall fluorescence was the highest 15 minutes after injection of Alexa Fluor®750-labeled HBC1 and could be detected in urine in all mice (Fig. 3). Ex vivo visualization of the fluorescent probe localized in isolated internal organs revealed that HBC1 is transiently distributed to liver (41% of the total fluorescence) and kidneys (18% of total fluorescence recorded). We did not observe any signal over 120 minutes in brain, lungs, heart and spleen. Although we found small pulmonary congestion in microscopic examination, none of the respiratory parameters changed within 1 hour after UFH+HBC1 administration.

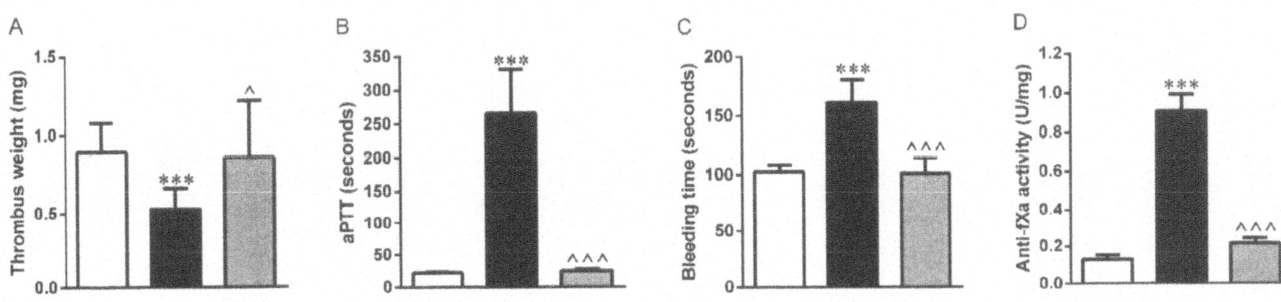

Figure 1. Reversing of UFH effects by HBC1 in rats developing arterial thrombosis. *** P<0.001 vs. vehicle; ^ P<0.05, ^^^ P<0.001 vs. UFH 300 UI/kg, Mann-Whitney test. Results are shown as mean ± SD, n = 9-14.

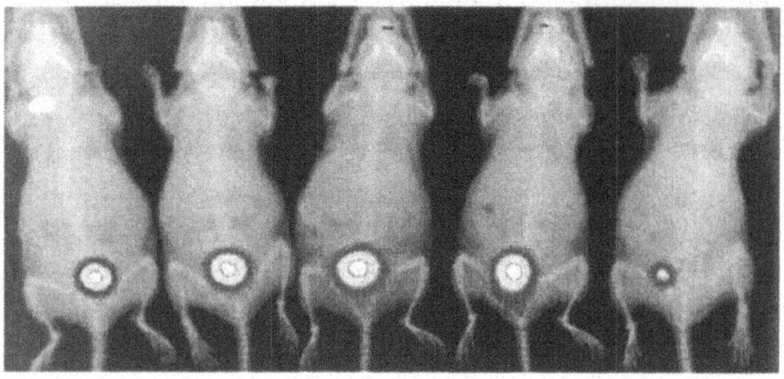

Figure 3: Distribution of HBC1 15 minutes after injection.

Routine histopathological examinations showed slight enlargement of hepatic cells 1 month after HBC1 injection. However, measurements in blood collected from rats once a week during 1 month did not reveal any hepatic, nephrotoxicity or blood abnormalities. HBC1 didn't change significantly blood pressure, heart rate and respiratory parameters, when infused into rats in efficient neutralizing dose.

4 CONCLUSION

HBC1 is easy to synthesize, efficient and safe heparin-binding copolymer that could replace protamine for the treatment and prevention of heparin associated bleeding. Significant restoring of Anti-factor Xa activity in heparinized animals and neutralization of anticoagulant effects of enoxaparin suggest good efficacy against low molecular weight heparins and fondaparinux, in case of which there are currently no specific antidotes. This property would extend the indications for use of HBC1, but needs financial support and further investigation. Our technology presents pharmaceutical companies unique opportunity to optimize a lead compound into marketable therapeutic.

FUNDING

The study was supported by grant no. DEC-2011/03/B/NZ7/00755 and additionally by grant no. UMO-2013/09/D/ST5/03864 from National Science Centre in Poland. BK was supported by funds from Leading National Research Center in Bialystok (31/KNOW/2013).

REFERENCES

[1] Bromfield SM, Wilde E, Smith DK (2013), Chem Soc Rev 42: 9184-9195.
[2] Kaminski K, Szczubialka K, Zazakowny K, Lach R, Nowakowska M (2010), J Med Chem 53: 4141-4147.
[3] Kaminski K, Plonka M, Ciejka J, Szczubialka K, Nowakowska M, Lorkowska B, Korbut R, Lach R (2011), J Med Chem 54: 6586-6596.
[4] Kalaska B, Sokolowska E, Kaminski K, Szczubialka K, Kramkowski K, Mogielnicki A, Nowakowska M, Buczko W (2012), Eur J Pharmacol 5: 81-89.
[5] Kaminski K, Kalaska B, Koczurkiewicz P, Michalik M, Szczubialka K, Mogielnicki A, Buczko W, Nowakowska M (2014), MedChemComm 5: 489-495.
[6] WO 2010002281 A1: Kaminski K, Zazakowny K, Szczubialka K, Nowakowska M (2009)
[7] US 20110301120 A1: Nowakowska M, Szczubialka K, Kaminski K, Zazakowny K (2011),
[8] US 20130034516 A1: Nowakowska M, Szczubialka K, Kaminski K (2013),
[9] WO 2013157967 A1: Nowakowska M, Szczubialka K, Kaminski K, Buczko W, Mogielnicki A, Kalaska B (2013)

Dex40-GTMAC3 modified dextran as a novel heparin antagonist

B. Kalaska*, K. Kaminski**, E. Sokolowska*, D. Czaplicki***, M. Kujdowicz**, K. Stalinska***, K. Szczubialka**, J. Bereta***, D. Pawlak*, M. Nowakowska** and A. Mogielnicki*

*Department of Pharmacodynamics, Medical University of Bialystok, Poland, amogiel@umb.edu.pl
**Faculty of Chemistry, Jagiellonian University, Krakow, Poland, szczubia@chemia.uj.edu.pl
***Department of Cell Biochemistry, Biophysics and Biotechnology, Jagiellonian University, Krakow, Poland, joanna.bereta@uj.edu.pl

ABSTRACT

Protamine is the only registered antidote preventing bleeding in patients treated with unfractionated heparin (UFH). However, protamine may induce anaphylactic shock or serious hypotension. We aimed to develop an alternative UFH antidote as efficient as protamine, but safer and easier to produce. As a starting material, we have chosen generally non-toxic, biocompatible, widely available, inexpensive, and easy to functionalize polymer - dextran. Our approach was to synthesize and purify dextran substituted with glycidyltrimethylammonium chloride (GTMAC) groups at a ratio of 0.65 per glucose unit (Dex40-GTMAC3), then to check it for potential heparin-reversal activity in vitro, and finally to examine efficacy and safety of the modified dextran in Wistar rats and BALB/c mouse models of experimentally induced arterial and venous thrombosis. Efficacy studies included the measurement of thrombus formation and coagulation parameters; safety studies included the measurement of pharmacokinetics, hemodynamic, hematologic, biochemical and immunogenic endpoints. Dex40-GTMAC3 is immediately eliminated from blood. It is as efficient as protamine but less immunogenic UFH inhibitor. Basing on our non-clinical evaluation of Dex40-GTMAC3, we believe that the novel polymer has very promising properties and can potentially substitute protamine as the UFH antidote.

Keywords: heparin, protamine, antidote, dextran, polysaccharides

1 BACKGROUND

Protamine is the only registered antidote preventing bleeding of patients treated with unfractionated heparin (UFH). However protamine, isolated from salmon sperm, may produce a number of adverse effects, such as anaphylactic shock or serious hypotension. We aimed to develop an alternative, safer than protamine, UFH antidote. This issue is continuously studied over many years but remains unsolved [1]. As a starting material, we have chosen generally non-toxic, biocompatible, widely available, inexpensive, and easy to functionalize polymer – dextran. We have already shown the potential of modified dextran to reverse heparin activity *in vitro* [2] and *in vivo* [3]. Our material and its indication for use to neutralize heparin activity or prevent anemia is worldwide patented technology [4,5].

2 METHODS

Basing on *in vitro* heparin-reversal activity assay of several polysaccharide cationic polymers we chose a high molecular weight dextran substituted with glycidyltrimethylammonium chloride groups at a ratio of 0.65 per glucose unit (Dex40-GTMAC3). The ability of Dex40-GTMAC3 to bind 1 mg of UFH was determined using Azure A assay. The size of the particles formed in solution due to complexation of UFH with polymer was found using dynamic light scattering (DLS) experiments. Interaction with serum proteins was also studied using gel electrophoresis and DLS. The potential direct blood toxicity of Dex40-GTMAC3 was estimated after incubation of polymer with whole blood by measuring: RBC, hemoglobin, hematocrit, mean corpuscular volume, hemoglobin and platelets in ABC Vet counter (Horiba, Germany) and osmotic resistance. The efficacy of Dex40-GTMAC3 was tested in 68 male Wistar rats weighing 217.7 ± 10.5 g, developing electrically induced arterial thrombosis [6], and 25 male BALB/c mice weighing 24.2 ± 0.2 g, developing electrolytic inferior vena cava thrombosis [7]. The pharmacokinetics, acute and chronic toxicity were estimated in 45 healthy male Wistar rats, weighing 220-300 g. The immunization experiment was performed in 20 female BALB/c mice weighing 26.3 ± 0.5 g. Animals were randomly divided into treatment groups. They were treated intravenously with UFH (in doses of 150 or 300 U/kg), Dex40-GTMAC3 (depending on the experiment in doses of 2.5, 3.75, 7.5 or 22.5 mg/kg b.w.) or protamine (1.5, 3 or 9 mg/kg b.w.). The control group received vehicle (phosphate buffer solution, PBS). All the procedures involving animals and their care were approved by Local Ethical Committees in Bialystok and Krakow (Permit Numbers 28/2012 and 92/2012, respectively) and conducted in accordance with ARRIVE guidelines, directive 2010/63/EU and the national laws. The efficacy endpoints in animal models of thrombosis were: arterial (rats) or venous (mice) thrombus weight, tail bleeding time, activated partial thromboplastin time (aPTT) (automatically determined in coagulometer – Coag-Chrom 3003, Bio-ksel, Poland), anti-factor Xa activity (measured in microplate reader – Dynex Tech., USA according to instructions of Sekisui Diagnostics, USA kit). All the procedural details were previously published by our team [3,8]. Plasma concentration of Dex40-GTMAC3 labelled with fluorescein was fluorometrically measured, and pharmacokinetic parameters were calculated by using PK Solver Excel add-in program [9]. The blood pressure, heart rate, blood count and blood chemistry was measured 1 hour after administration of studied compounds to rats. In chronic setting body weight, blood count and blood chemistry were monitored once a week over 1 month. Brain, lungs, kidneys, heart, liver, spleen, bladder, the large and small intestine were collected from rats 1 hour or 28 days after single dose injection. All organs were macroscopically examined for gross pathology. To compare immune response of Dex40-GTMAC3 and protamine, the heparinization/neutralization regimen was repeated 5 times, once every week (days 1, 8, 15, 22, and 29). Blood samples were collected from the tail vein of each mouse one day prior UFH administration and serum was isolated by centrifugation. One week after the last injection (day 36) all mice were sacrificed; final blood samples were collected and spleens of the animals were isolated for evaluation. The levels of antibodies specific to protamine and Dex40-GTMAC3 were evaluated using standard indirect ELISA.

3 RESULTS

3.1 In vitro UFH binding by Dex40-GTMAC3

The concentration of free UFH in the solution decreased with increasing concentration of Dex40-GTMAC3. The amount of polymer necessary for binding 90% of 1 mg of UFH was used as a measure of the binding efficiency. This value was used for the preliminary assessment of the dose of a cationic polymer required in biological tests.

3.2 Reversal of the effects of UFH on the arterial thrombosis in rats

UFH decreased the weight of arterial thrombus in rats. Dex40-GTMAC3 reversed antithrombotic effect of UFH (Figure 1). The polymer was administered in doses corresponding with the ratio of polymer/UFH estimated in the UFH binding assay. When administered alone Dex40-GTMAC3 did not influence thrombus weight (0.98 ± 0.25 mg) in comparison to control group (0.92 ± 0.17 mg).

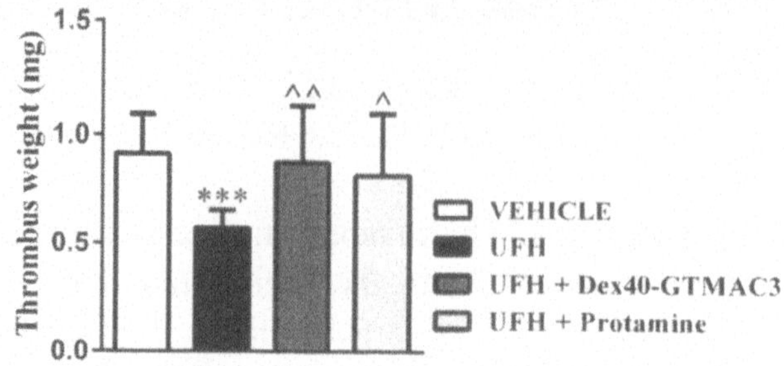

Figure 1: Reversal of the effects of UFH on the arterial thrombosis in rats. Dry thrombus weight in Wistar rats treated with vehicle (PBS), UFH (300 U· kg^{-1}) alone or followed by Dex40-GTMAC3 (7.5 mg·kg^{-1}), and protamine (3.0 mg·kg^{-1}), ***P<0.001 vs. vehicle; ^P<0.05, ^^P<0.01 vs. UFH 300 U·kg^{-1}, Mann-Whitney test. Results are shown as mean ± SD, n = 8-10.

3.3 Influence of Dex40-GTMAC3 on aPTT in rats

Dex40-GTMAC3 significantly shortened aPTT which was prolonged by earlier administration of UFH (Figure 2).

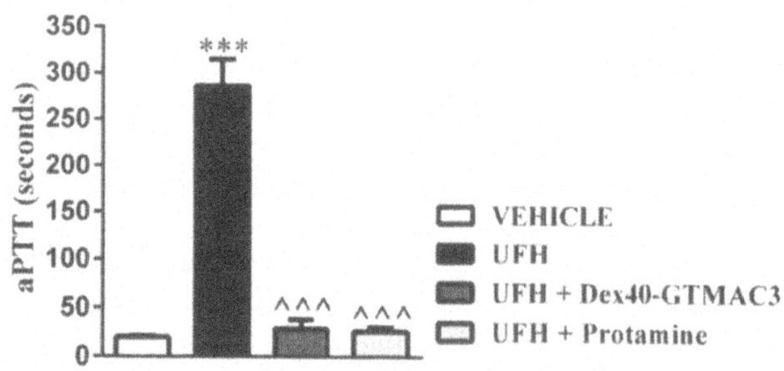

Figure 2: Influence of the cationic polymers on aPTT in rats. aPTT in Wistar rats treated with vehicle (PBS), UFH (300 U·kg^{-1}) alone or followed by Dex40-GTMAC3 (7.5 mg·kg^{-1}), and protamine (3.0 mg·kg^{-1}), ***P<0.001 vs. vehicle; ^^^P<0.001 vs. UFH 300 U·kg^{-1}, Mann-Whitney test. Results are shown as mean ± SD, n = 8-10.

3.4 Reversal of tail bleeding time by Dex40-GTMAC3

Similarly to aPTT measurement, UFH prolonged the tail transection bleeding time. When administered in the ratio of 2.5 mg for every 100 U·kg^{-1} b.w of UFH, Dex40-GTMAC3 completely reversed tail bleeding time to the control value (Figure 3). Dex40-GTMAC3 administered alone did influence bleeding time (111.3 ± 18.9 sec.) in comparison to vehicle (104.3 ± 4.6 sec.).

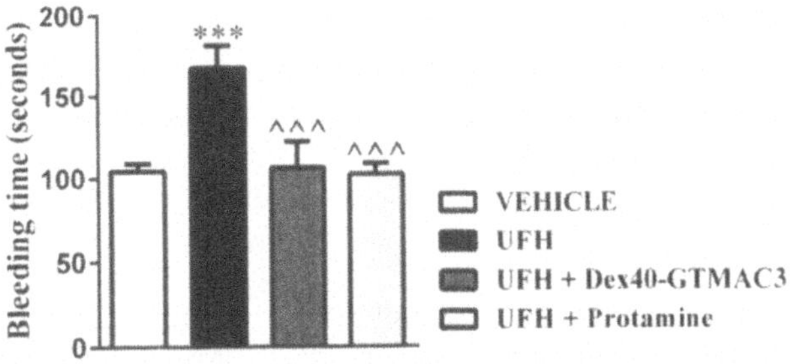

Figure 3: Reversal of tail bleeding time by the cationic polymers. Tail bleeding time in Wistar rats treated with vehicle (PBS), UFH (300 U·kg^{-1}) alone or followed by Dex40-GTMAC3 (7.5 mg·kg^{-1}), and protamine (3.0 mg·kg^{-1}), ***P<0.001 vs. vehicle; ^^^P<0.001 vs. UFH 300 U·kg^{-1}, Mann-Whitney test. Results are shown as mean ± SD, n = 8-10.

3.5 Reversal of plasma anti-Xa activity by Dex40-GTMAC3

Dex40-GTMAC3 significantly reversed increase of anti-Xa activity in UFH-treated animals (Figure 4).

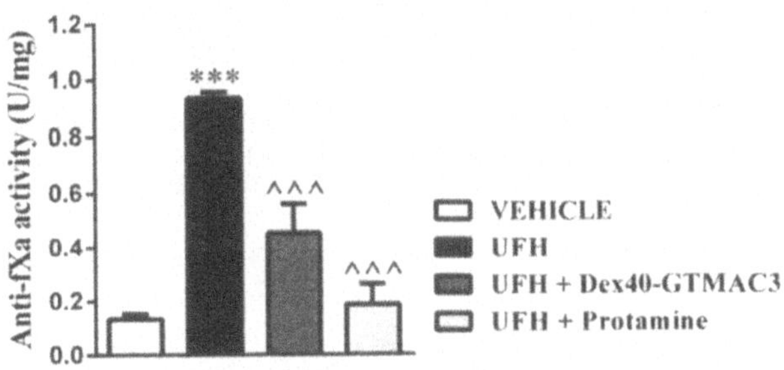

Figure 4: Reversal of plasma anti-fXa activity by the cationic polymers. Anti-fXa activity in Wistar rats treated with vehicle (PBS), UFH (300 U·kg^{-1}) alone or followed by Dex40-GTMAC3 (7.5 mg·kg^{-1}), and protamine (3.0 mg·kg^{-1}), ***P<0.001 vs. vehicle; ^^^P<0.001 vs. UFH 300 U·kg^{-1}, Mann-Whitney test. Results are shown as mean ± SD, n = 8-10.

3.6 Reversal of the effects of UFH on the venous thrombosis in mice

Similarly to the results in the model of the arterial thrombosis in rats, UFH significantly decreased thrombus weight and prolonged aPTT in mice developing electrically induced venous thrombosis. We confirmed the ability of Dex40-GTMAC3 to reverse antithrombotic (Figure 5) and anticoagulant (Figure 6) effects of UFH in mice.

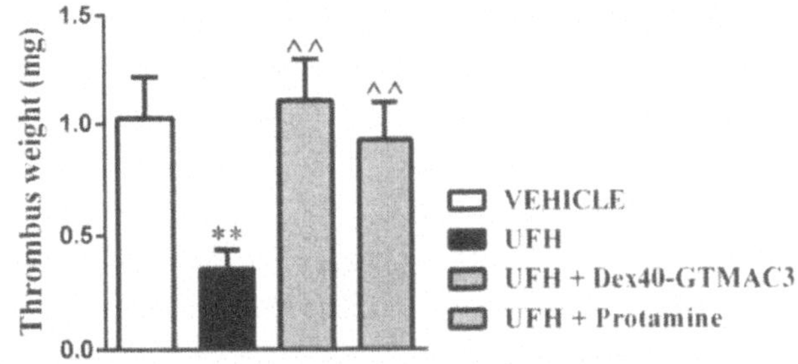

Figure 5: Reversal of the effects of UFH on the venous thrombosis in mice. Dry thrombus weight in mice treated with vehicle (PBS), UFH (300 U·kg^{-1}) alone or followed by Dex40-GTMAC3 (7.5 mg·kg^{-1}), and protamine (3.0 mg·kg^{-1}), **P<0.01 vs. vehicle; ^^P<0.01 vs. UFH 300 U·kg^{-1}, Mann-Whitney test. Results are shown as mean ± SD, n = 5-7.

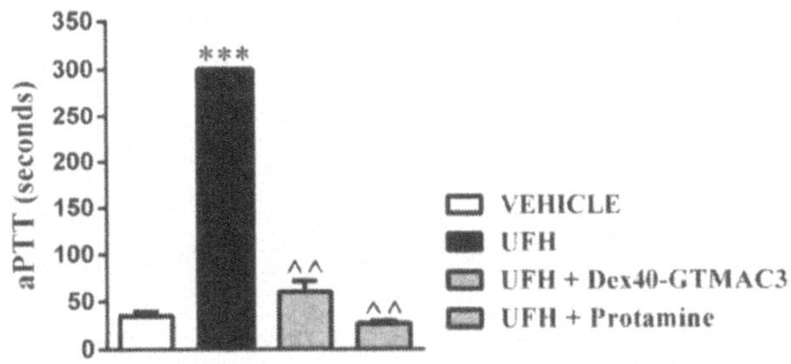

Figure 6: Reversal of the effects of UFH on aPTT in mice. aPTT in mice treated with vehicle (PBS), UFH (300 U·kg^{-1}) alone or followed by Dex40-GTMAC3 (7.5 mg·kg^{-1}), and protamine (3.0 mg·kg^{-1}), ***P<0.001 vs. vehicle; ^^P<0.01 vs. UFH 300 U·kg^{-1}, Mann-Whitney test. Results are shown as mean ± SD, n = 5-7.

3.7 Effects of Dex40-GTMAC3 on blood pressure in rats

We found significant, but much weaker and temporary change of mean blood pressure (MBP) shortly after administration of Dex40-GTMAC3 (decrease in MBP). If UFH was injected 3 minutes before the administration of Dex40-GTMAC3, the cationic polymer did not decrease MBP.

3.8 Evaluation of the immune response to Dex40-GTMAC3

IgG-class response was detectable in all mice in the protamine group beginning from day 21st on. On day 36 the mean protamine-specific IgG signal in the protamine group was significantly higher than the background signal in the UFH group (P<0.01). In contrast to this result, no significant levels of IgG specific to Dex40-GTMAC3 were detected (Figure 7).

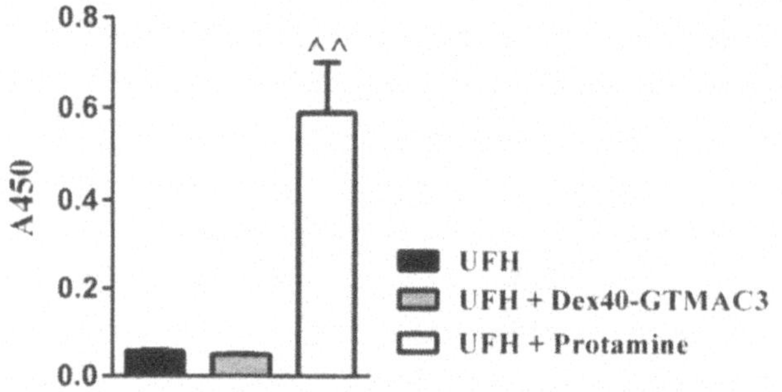

Figure 7: The levels of IgG specific toward Dex40-GTMAC3 or protamine in mice in 36 day of therapy presented as values of absorbance in ELISA test. ^^ P<0.01 vs. UFH, Mann-Whitney test. Results are shown as mean ± SD, n = 5.

3.9 Pharmacokinetic parameters, blood chemistry and toxicity studies

The concentration of Dex40-GTMAC3 in plasma after drug administration rapidly declined with $T_{1/2}$ of 29.4 ± 0.47 min. After 1 hour the amount of the circulating polymer was 7.7 % of the amount measured after 5 minutes. The toxicity studies did not reveal any pathological signs of acute or chronic hepato- or nephrotoxicity. The body weight, gross changes of internal organs, and blood chemistry remained the same as control values.

3.10 Blood count

UFH significantly increased white blood cells (WBC). Dex40-GTMAC3 partially, but significantly, inhibited increase of WBC in UFH treated animals. Dex40-GTMAC3 did not change any other blood count.

4 CONCLUSIONS

Documented efficacy without hepato-, nephro-, immunotoxicity or blood toxicity of Dex40-GTMAC3 make this agent advantageous over protamine. Dex40-GTMAC is easy to synthesize, efficient and safe heparin-binding polysaccharide polymer with a potential as marketable therapeutic.

FUNDING

The study was supported by grant no. DEC-2011/03/B/NZ7/00755 and additionally by grant no. UMO-2013/09/D/ST5/03864 from National Science Centre in Poland. BK was supported by funds from Leading National Research Center in Bialystok (31/KNOW/2013).

REFERENCES

[1] SM. Bromfield, E. Wilde, DK Smith, Chem Soc Rev. 42, 9184-9195, 2013.
[2] K Kaminski, K Szczubialka, K Zazakowny, R Lach, M Nowakowska, J Med Chem. 53, 4141-4147, 2010.
[3] B Kalaska, E Sokolowska, K Kaminski, K Szczubialka, K Kramkowski, A Mogielnicki, M Nowakowska, W Buczko, Eur J Pharmacol. 5, 81-89, 2012.
[4] M Nowakowska, K Szczubialka, K Kaminski, US 20130034516 A1, 2013.
[5] M Nowakowska, K Szczubialka, K Kaminski, W Buczko, A Mogielnicki, B Kalaska, WO 2013157967 A1, 2013.
[6] S Guarini, J Pharmacol Toxicol Meth. 35, 101-105, 1996.
[7] JA Diaz, SK Wrobleski, AE Hawley, BR Lucchesi, TW Wakefield, DD Myers Jr., J Vis Exp. 53, 27-37, 2011.
[8] K Kaminski, B Kalaska, P Koczurkiewicz, M Michalik, K Szczubialka, A Mogielnicki, W Buczko, M Nowakowska, MedChemComm 5, 489-495, 2014.
[9] Y Zhang, M Huo, J Zhou, S Xie, Comput Methods Programs Biomed. 99, 306-14, 2010.

Design of a novel melatonin Pickering emulsion sunscreen

H. Ribeiro[*], J Marto[*,**], A Ascenso[*], P Manteigas[*], CV Eleutério[*], L Gouveia[*], SR Loreau[*], E Oliveira[**], A Almeida[*],

[*] Research Institute for Medicines (iMed.ULisboa), Faculty of Pharmacy, University of Lisbon, Lisbon, Portugal;

[**] Laboratórios Atral S.A., Castanheira do Ribatejo, Portugal

INTRODUCTION

Melatonin, the main secretory product of the pineal gland, was recently found to have a protection effect against photocarcinogenesis based on its free radical scavenger and antioxidant activity. On the other hand, the daily use of sunscreens is particularly recommended for patients treated with immunosuppressive drugs and submitted to surgeries, chemotherapy and radiotherapy. Therefore, melatonin associated with sunscreens may be the key strategy in order to prevent UV-radiation induced skin damages, such as pre-cancerous and cancerous skin lesions. Given the increase in skin cancer incidence and the existence of an emerging group of high risk patients treated with immunosuppressive drugs submitted to oncological therapy the application of sunscreens with a high sun protection factor is one of the main strategies strongly recommended to prevent UV-induced cell damage. This work arose from the necessity to fill the gap of photoprotection market, by developing a formulation with high SPF and a lower content of substances of synthetic origin, i.e., less aggressively to the skin.

The aims of this study were the design and stabilization of melatonin in a novel sunscreen Pickering emulsion and its morphological characterization, droplet size distribution, contact angle and sun protection factor.

Keywords: Melatonin, Pickering Emulsion System, UV protection, Physical filters.

1 MATERIALS AND METHODS

1.1 Materials

Melatonin was purchased from Alfa Aesar (Ward Hill, USA). Ethanol was obtained from Merck® (Kenilworth, USA). Green coffee oil was supplied by COOXUPE´ – Cooperativa de Cafeicultores de Gauxupé, (Minas Gerais, Brazil). Titanium dioxide was a gift from Sensient (Milwaukee, USA). Liquid paraffin was obtained from José Vaz Pereira, S.A. (Lisbon, Portugal). All other reagents were HPLC grade and used as received. Purified water was obtained by inverse osmosis (Millipore, Elix® 3).

1.2 Methods

Solid Particles

Wettability measurements

The measurement of contact angles of water and liquid paraffin on TiO_2 in air was performed at room temperature by using ConAnXL - a Microsoft Excel based workbook and Add-in software. TiO_2 surface were prepared by compressing suitable amounts of particles size under a pressure of 10 metric tons applied by a KBr press. Briefly, a drop (approximately 10 µL) of water, paraffin and green coffee oil was carefully deposited on the solid surface. The contact angle was defined as the angle between the PS surface and the tangent to the surface of the drop. All measurements were performed in triplicate.

Particle size distribution

Particle size distribution was determined using a Malvern Mastersizer 2000 (Malvern Instruments, UK), coupled with a Hydro S accessory. The refractive index used was 1.52 (default). For a correct turbidity, about 0.1-0.5 g of TiO_2, corresponding to an obscuration between 1-5%, was added to the sample chamber containing 150 mL of purified water. The readings were performed under mechanical stirring (3500 rpm). Data was expressed in terms of relative distribution of volume of particles in the range of size classes, and given as diameter values corresponding to percentiles of 10, 50 and 90. The Span value is a useful statistical parameter to characterize the particle size distribution, and it was calculated by using the **(1.**

$$Span = \frac{d(90)-d(10)}{d(50)} \quad (1)$$

Preparation of the Melatonin SunScreen (MSS)

According to the pre-formulation studies, four final formulations were selected (**Erro! A origem da referência não foi encontrada.**) based on macroscopic appearance, stability and SPF value. The continuous oil phase (Phase A) consisted of green coffee oil, and the aqueous phase (Phase B) was composed by purified water, ethanol and melatonin. TiO_2 was firstly dispersed in the oil phase. The oil and aqueous phases were then mixed using an UltraTurrax® homogenizer (IKA-Werke GmbH & Co. KG, Germany) at room temperature (cold process).

Ingredients	Quantitative Composition (%)			
	MSS 1	MSS 2	MSS 3	MSS 4
Phase A (external)				
TiO2	7.5	10	15	25
Green Coffee Oil	62,5	60	55	45
Phase B (internal)				
Melatonin	1	1	1	1
Ethanol	6	6	6	6
H2O	23	23	23	23

Table 1: **Qualitative and Quantitative composition of the final formulations**

Characterization Studies of MSS

Emulsion Type

The emulsion type was identified by the dilution method and staining technique.

In the first, a drop of emulsion was added to a certain volume of pure water and another to pure oil (liquid paraffin). If it disperses in one of them, then that liquid is the continuous or external phase of the emulsion. The liquid in which it remains as drops, is the discontinuous or internal phase.

In the second, a drop of methylene blue (hydrophilic dye) was added to the emulsion and then the emulsion was inspected in an optical microscope to detect whether or not the water phase was the continuous one.

Droplet Size

After one day of storage at room temperature, the emulsions were investigated using an optical microscope (Olympus CX40, Japan), which was equipped with a video camera. One drop of the emulsion was added on a glass plate without cover glass and diluted with two drops of paraffin for microscope observation and photographed for posterior size analysis. The droplet size was determined using the image analysis software Olympus Stream Essentials®. The size data was expressed in terms of relative size distribution of particles, and given as diameter values corresponding to percentiles of 10%, 50%, 90%. The Span value was also taken into consideration since it is a statistical parameter useful to characterize the particle size distribution, and is calculated using the equation 1.

Efficacy of the Sunscreen Formulations

Theoretical SPF

The algorithm indicated in the BASF sunscreen simulator was used in order to find a sunscreen with a sun protection factor (SPF) of 50+ that protects the skin against both UVA and UVB radiation. The European Legislation (CE) N.° 1223/2009 - Annex VI was also considered.

The BASF Sunscreen Simulator is a tool that estimates the SPF as well as various common UVA-Metrics. This tool helps to plan the composition of active ingredients in sunscreen formulations. However, it does not replace the formulation study and *in vivo* SPF determination of the final product.

2 RESULTS AND DISCUSSION

2.1 Solid Particles

Wettability Studies

In this type of emulsions, one of the liquids will probably wet the solid more than the other liquid, with the more poorly wetting liquid becoming the disperse phase. The importance of the wettability of the particles at the oil-water interface is quantified by the contact angle, θ, that the particle makes with it, which will determine the type of emulsion. If the contact angle, measured through the aqueous phase, is inferior to 90° the emulsion will be o/w and, by contrast, if the contact angle is greater than 90° the emulsion will be w/o [1, 2, 3].

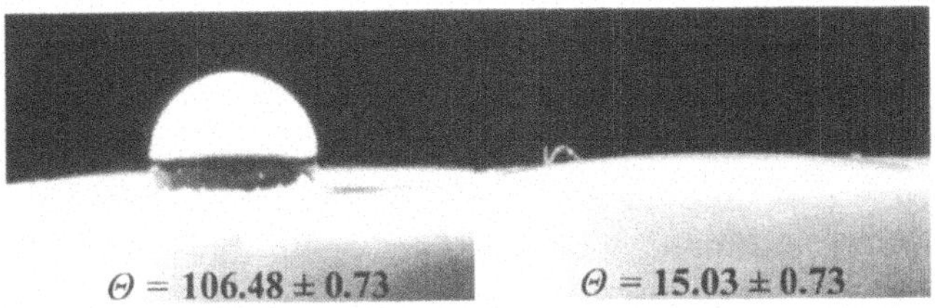

Figure 1: Contact angles of Titanium dioxide in (a) water and (b) liquid paraffin.

By analogy, TiO_2 will stabilize more effectively w/o emulsions, because the contact angle of water greater than 90° and simultaneously, a contact angle of paraffin smaller than 90° (Figure 1).

Particle Size

In this assay it is important to prove that the particles have a size greater than 100 nm [4], in order to avoid regulatory issues, including the needed to perfome cito- and genotoxicity studies [5,6] and to prevent nano-ethical [7] problems.

This premise was thoroughly, as evidenced by the particle size distribution: d(0.5) > 100 nm.

High values of Span were recorded for TiO_2, which can be explained by hydrophobic character of this type and probably associated with an aggregation phenomenon.

Particle size distribution (μm)			
Span	**d (0.1)**	**d (0.5)**	**d (0.9)**
37.36 ± 1.38	0.14 ± 0.01	0.19 ± 0.01	7.12 ± 0.30

Table 2: Particle size distribution of TiO2 (mean ± SD,n=2).

2.2 Characterization Studies of MSS

Emulsion Type

Theoretically the type of emulsion was already defined - w / o, however to assure its veracity further studies concerning the determination of the type of the emulsion were performed.

The emulsions were inspected visually and a drop of emulsion was added to water and to liquid paraffin (oil).

A drop of oil mixed very fast within the emulsion sample, unlike the water drop, showing that oil corresponded to the continuous phase as expected. Finally, the microscopic observation of the emulsion stained with methylene blue (hydrophilic dye) confirmed the presence of a w/o emulsion system (Figure 2).

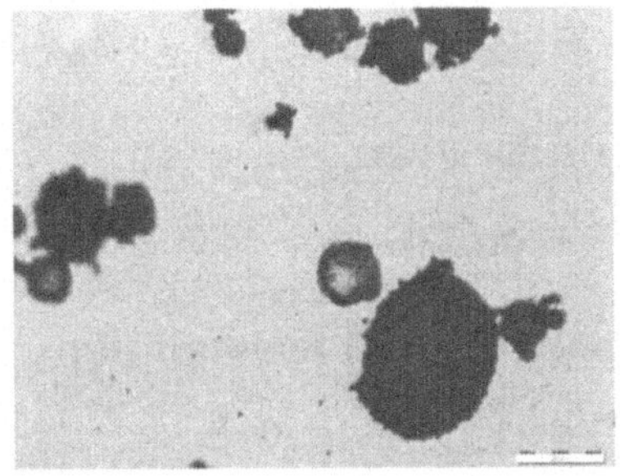

Figure 2: Emulsion dyed with methylene blue visualized by optical microscopy. Scale bar: 20μm.

Droplet Size

In this assay we evaluated the influence of the TiO_2 percentage on Pickering emulsions droplet size. Assuming that by getting smaller droplets it was possible to obtain a more stable emulsion and that an increase in the content of the solid particles leads to a dense coating of the droplet, the excess of TiO_2 will allow forming smaller droplets with a larger surface area.

Based in these results it can be inferred that an increase in the TiO2 will produce an emulsion with smaller droplets (Table 3 and Figure 3).

Similar results have already been published by other authors, such as Li et al. [8], who determined in their experiments that a a higher concentration of solid particles can stabilize a larger total interfacial area, which in turn means that a smaller average droplet size can be achieved.

	***Droplet size* distribution (μm)**			
MSS	**Span**	**d (0.1)**	**d (0.5)**	**d (0.9)**
1	0.402	8.365	12.209	22.861
2	0.849	2.161	4.343	7.517
3	1.578	1.273	2.227	4.506
4	2.756	1.006	1.931	6.167

Table 3: Influence of the TiO_2 percentage on Pickering emulsions droplet size – d(0.1), d(0.5), d(0.9) and Span. (mean; n=625).

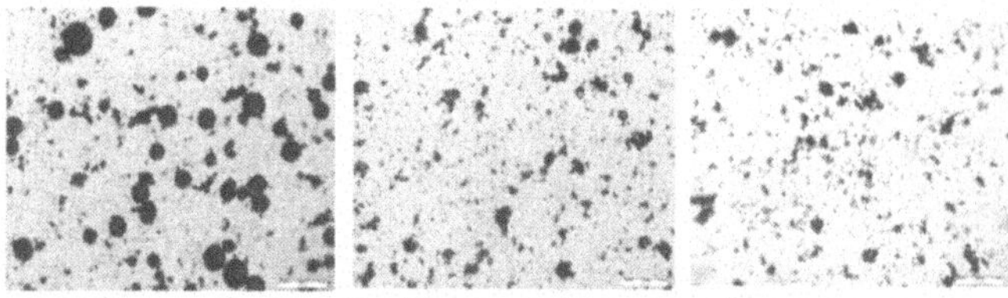

Figure 3: Emulsion samples with different TiO_2 percentage visualized by optical microscopy: left- 7.5%; center- 15% and right- 25%. Scale bar=100 µm.

Efficacy of the Sunscreen Formulations

Theoretical SPF

According to the desired SPF and the limits imposed by the UE, the limit of 25% for TiO_2 could not be exceeded. Thus, it was necessary to combine UV-filters according to its SPF to achieve the maximum protection as possible. Using the BASF simulator and considering the droplet size assay, we tried to establish some formulations that meet the criteria of SPF 50+. Regarding TiO_2 content, we tried not exceed 20% in order to avoid values very close of the maximum allowed concentration and to ensure that this limit wasn't exceeded. Only with 25% of TiO_2 (physical filter) is sufficient to achieve the desired protection factor – 50.

	MSS			
	1	2	3	4
SPF	14	19	30	51.1

Table 4: SPF obtained with different % of TiO_2

Regarding the Pickering Emulsions, they proved to be a good solution, allowing the combination of stabilizing and photoprotective properties of TiO2 with the melatoninergic action. The result obtained was a simple emulsion that offers an excellent compromise between stability, UV protection and cosmeticity and that could allow an easy scale-up.

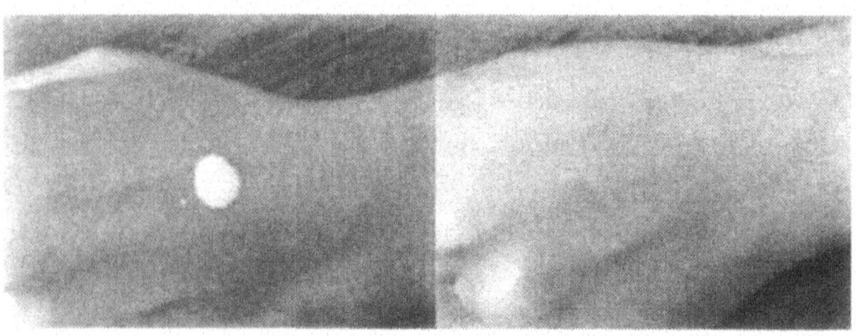

Figure 4: Topical application of MSS.

3 CONCLUSIONS

The formulation obtained showed good cosmeticity and to be of a low irritant nature. Surprisingly, a TiO_2-based Pickering emulsion containing suitable excipients provided good results in terms of physical stability and high UV protection. Consequently, it offers an excellent compromise between stability, UV protection and cosmeticity.

REFERENCES

[1] Binks BP. Particles as surfactants – similarities and differences. Current Opinion in Colloid & Interface Science, 7: 21-41, 2002.

[2] Marku D, Wahlgren M, Rayner M, Sjöö M, Timgren A. Characterization of starch Pickering emulsions for potential applications in topical formulations. International Journal of Pharmaceutics, 428: 1-7, 2012.

[3] Schmitt V, Ravaine V. Surface compaction versus stretching in Pickering emulsions stabilised. Current Opinion in Coloid & Interface Science, 18: 532-541, 2013..

[4] Smijs TG, Pavel S. Titanium dioxide and zinc oxide nanoparticles in sunscreens: focus on their safety and effectiveness. Nanotechnology, Science and Applications, 12; 95-112, 2011.

[5] Smijs TGM, Bouwstra JA. Focus on skin as a possible port of entry for solid nanoparticles and the toxicological impact. J Biomed Nanotech. 6:469–484, 2010.

[6] Menard A, Drobne D, Jemec A. Ecotoxicity of nanosized TiO2. Review of in vivo data. Environ Pollut, 159:677–684, 2011.

[7] Jacobs JF, van den Poel I, Osseweijer P. Sunscreens with Titanium Dioxide (TiO2) Nano-Particles: A Societal Experiment. Nanoethics, 4: 103-113, 2010.

[8] Li C, Li Y, Sun P, Yang C. Pickering emulsions stabilized by native starch granules. Colloids and Surfaces A: Physicochemical and Engineering Aspects, 431: 142-149, 2013.

The use of Ag_5IO_6 as an antibiofilm agent in wound dressings

V. Incani*, A. Omar*, G. Prosperi-Porta*,** and P. Nadworny*

*Innovotech, Inc., Suite 101, 2011 94 St., Edmonton, AB, Canada
**Department of Chemical and Materials Engineering, University of Alberta, 7th Floor Electrical and Computer Engineering Research Facility, 9107 116 St., Edmonton, AB, Canada
E-mails: vanessa.incani@innovotech.ca, amin.omar@innovotech.ca, prospeg@mcmaster.ca
Corresponding author: patricia.nadworny@innovotech.ca

ABSTRACT

The novel antibiofilm activity of Ag_5IO_6 was explored. Wound dressings impregnated with Ag_5IO_6 were compared to commercially available silver, chlorhexidine, and PHMB dressings. The dressings were tested against *P. aeruginosa*, *S. aureus*, and *C. albicans* for their ability to prevent microorganism adherence, eliminate planktonic species, and eliminate mature biofilms within 24h. Only the Ag_5IO_6 dressings were able to prevent adherence and eliminate surrounding planktonic microorganisms for all species tested for ≥28 days of elution with log reductions >4. Moreover, only these dressings were able to generate >4 log reductions against all mature biofilms. Thus, Ag_5IO_6 has superior activity to a number of antimicrobials, with long-term prevention of microbial adherence, rapid kill of planktonic microorganisms, and the ability to eliminate mature biofilms. Ag_5IO_6 may be a valuable antimicrobial agent for use in a number of medical device applications, including catheters, and implants.

Keywords: Silver periodate, antimicrobial, wound dressing, infection, biofilm

1 INTRODUCTION

Silver agents as topical antimicrobials have been used for hundreds of years in wound care [1]. Silver has been used to prevent or manage infection in its solid elemental form, as solutions of silver salts used to cleanse wounds, and more recently as creams or ointments containing a silver-antibiotic compound. However, these applications of silver have come with considerable challenges that include complex methods for incorporation/coating of silver into/onto materials, lack of efficacy against planktonic microorganisms and biofilms in part due to inactivation by bodily fluid components [2, 3], cost effectiveness, and questions about safety [1]. Recently, numerous new silver compounds/products have been developed and added to medical devices in efforts to overcome these obstacles, mostly as antibiotic alternatives in the face of limited antimicrobial spectrum and growing microbial resistance. Bacteria exist predominantly in biofilms [4]. Biofilms are responsible for 80% of human infections [5], and are typically 100-1000x more resistant to treatment than planktonic microorganisms [6]. Therefore, when testing new silvers for potential clinical relevance, they need to demonstrate efficacy against biofilm phenotypes.

Ag_5IO_6 (pentasilver hexaoxoiodate) possesses promising properties for efficacy against biofilms, and simplicity for coating onto/incorporating into medical devices with approriate release characteristics. These properties include the presence of silver in both cation ($[Ag_3]^{3+}$), and anion-complexed with highly oxidized iodine ($[Ag_2IO_6]^{3-}$) [7]. This may enhance penetration of the biofilm relative to Ag^+, which binds to negatively charged surface components of biofilms. The periodate structure around the anionic silver may slow its inactivation by bodily fluid components. Combining silver and iodine may increase antimicrobial mechanisms of action, reducing the likelihood of developing resistance to Ag_5IO_6. The synthesis method used here combines small grain size with large particle size, resulting in polycrystallinity on a large surface area which may improve its antimicrobial activity relative to other Ag-compounds. Ag_5IO_6 has demonstrated excellent stability (storage; thermal; and in the presence of light, water, saline, organic solvents, autoclaving, ethylene oxide, etc.), and can be easily synthesized with high purity in a form that is simple to deposit onto metals or incorporate into wound dressings, gels, polymers, etc. [8]

This study examined whether Ag_5IO_6 could prevent biofilm formation on surfaces and eliminate mature biofilms, with comparison to commercially available antimicrobials to determine whether its activity was novel.

2 EXPERIMENTAL SECTION

Materials. Ag_5IO_6 was synthesized in-house [8]. Ag_5IO_6 was coated onto two bandages - a dressing with two outer layers of high-density polyethylene (HDPE) and a rayon/polyester core (3ply-Ag_5IO_6); and a rayon/cotton elastic adhesive bandage (elastic-Ag_5IO_6) [9]. Commercial dressings included in testing: nanoAg, alginate-Ag, NaCMC-Ag, oxysilver, chlorhexidine, and PHMB (polyhexamethylene biguanide). **Microorganisms.** *Pseudomonas aeruginosa* ATCC 27853, *Staphylococcus aureus* ATCC 6538, *S. aureus* ATCC 29213, and *Candida albicans* ATCC 18804. **Materials for tests.** 0.9% NaCl was used in pre-treatments. Human plasma (Human Male AB; SeraCare Life Sciences Inc., Gaithersburg, MD) was used in challenge medium. The recovery/neutralization buffers used as appropriate were STS [0.1% (w/v) sodium thioglycolate (Fisher Scientific, Ottawa, ON, Canada) and 0.5% (w/v) Tween 80 (Fisher Scientific) in phosphate-buffered saline), and Difco Dey/Engley (DE) neutralizing broth (Becton Dickinson, Mississauga, ON, Canada) plus 5 g/L each of L-cysteine and L-glutathione. Tryptic soy agar and broth were used for plating and medium, respectively, unless otherwise indicated. For *C. albicans*, Sabouraud dextrose broth supplemented with 500 mM galactose was used as needed. **Method.** The methodology followed for this study comprises a) Anti-adherence: pre-treatment of the wound dressing, challenge, planktonic recovery, and adherent microorganism recovery [10]. b) Mature biofilm: biofilm growth, challenge, and recovery [10, 11]. c) Planktonic log reduction: organism growth, challenge, and recovery [3, 12].

3 RESULTS AND DISCUSSION

The anti-adherence method simulated a scenario in which an antimicrobial wound dressing is used to prevent a biofilm infection. Dressings were exposed to a wound-like environment for different time periods (1, 7, 14, and 28 days) followed by exposure to microorganisms under conditions that would promote biofilm formation, allowing for determination of use-times for devices. Control dressings were used to account for material effects. Overall, only the Ag_5IO_6 dressings were active up to 28 days against all three microorganisms (Fig.1). The alginate-Ag and NaCMC-Ag reached 28 days with *P. aeruginosa* only, likely because *P. aeruginosa* is more sensitive to silver than the other micro-organisms tested [12], and PHMB reached 28 days with *C. albicans* only. It was interesting that nanoAg had a shorter life, given its superior anti-planktonic kinetics compared with alginate-Ag and NaCMC-Ag [10, 12]. Possible reasons include that the active agent is coated onto the outer surface rather than impregnated in the dressing, meaning that uncoated layers may act as surfaces for biofilm growth. As well, the active agent was likely inactivated by the saline pre-soaks [13], given the precipitate observed. Furthermore, the manufacturer's recommended dressing usage is 3 days. An observed change in color of the oxysilver suggested deactivation/depletion of the active agent resulting in its use time of <14 days. The chlorhexidine showed virtually no activity despite manufacturers' claims of 3-4 days activity [14].

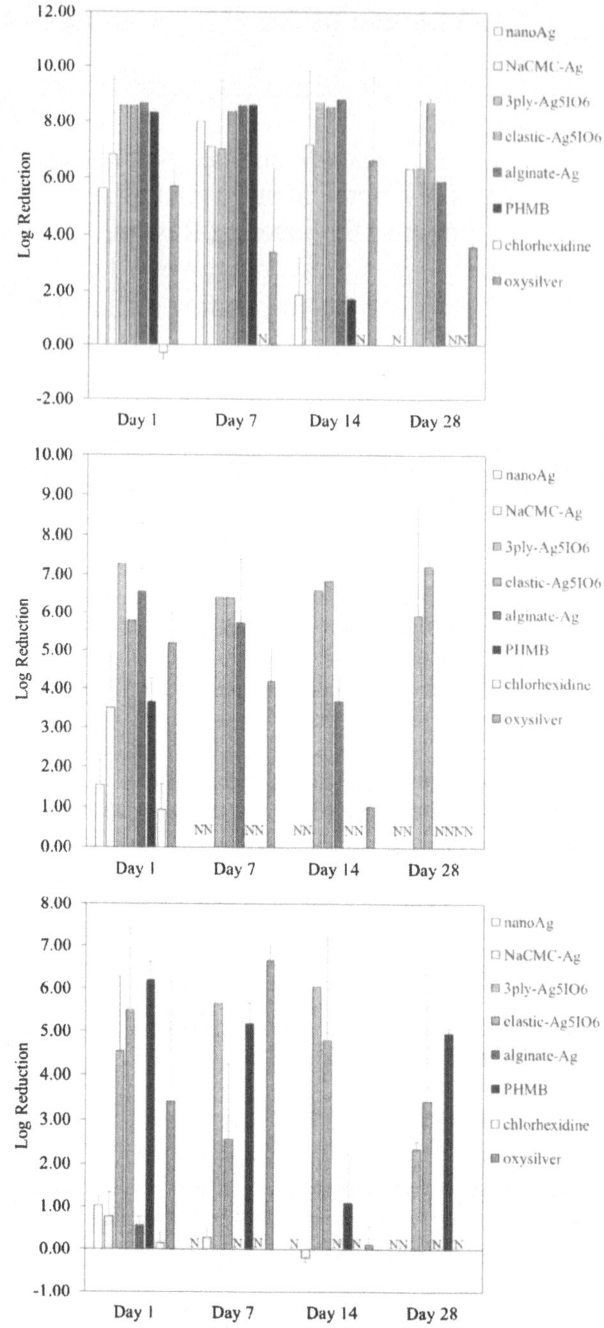

(a)

(b)

(c)

(Fig. 1 continued on next page.)

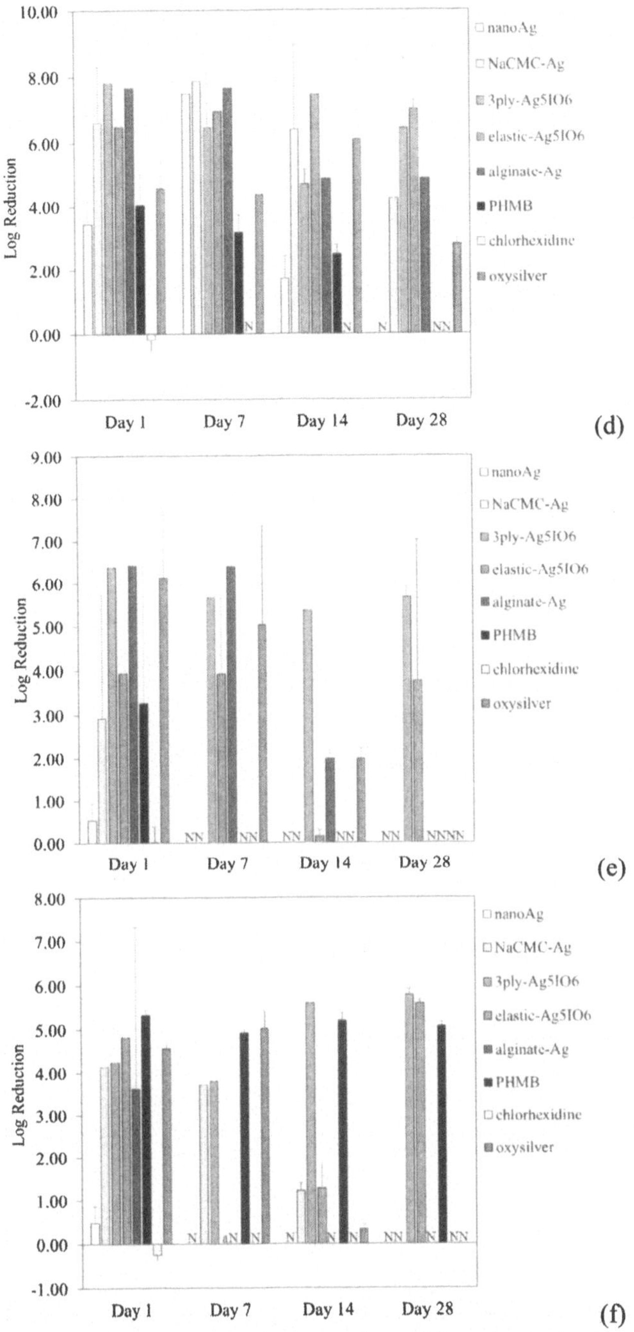

Figure 1. 28-day anti-adherence data. Mean log reductions on each test day, calculated relative to control dressings, are shown along with the standard deviations (SD). SD of zero indicate total kill of the microorganisms. When a dressing generated a log reduction <4 both for planktonic microorganisms and adhered biomass, the dressing was eliminated from the study after that time point (indicated by 'N'). (a) Planktonic *P. aeruginosa*, (b) planktonic *S. aureus*, (c) planktonic *C. albicans*, (d) adhered *P. aeruginosa*, (e) adhered *S. aureus*, (f) adhered *C. albicans*. For data point 'a', there was no growth of the microorganism on the test or control dressings.

Comparing commercially intended concentrations (ca. 1 punch for most dressings, ca. 3 punches for PHMB, Fig. 2), the relative activity of the dressings against mature biofilms was as follows: for *P. aeruginosa*, elastic-Ag_5IO_6 > PHMB > 3ply-Ag_5IO_6 > alginate-Ag > nanoAg > NaCMC > chlorhexidine > oxysilver; for *S. aureus*, elastic-Ag_5IO_6 > PHMB > 3ply-Ag_5IO_6 > chlorhexidine > NaCMC-Ag > nanoAg > oxysilver > alginate-Ag; and for *C. albicans*, 3ply-Ag_5IO_6 > elastic-Ag_5IO_6 > NaCMC-Ag > nanoAg > chlorhexidine > oxysilver > alginate-Ag > PHMB.

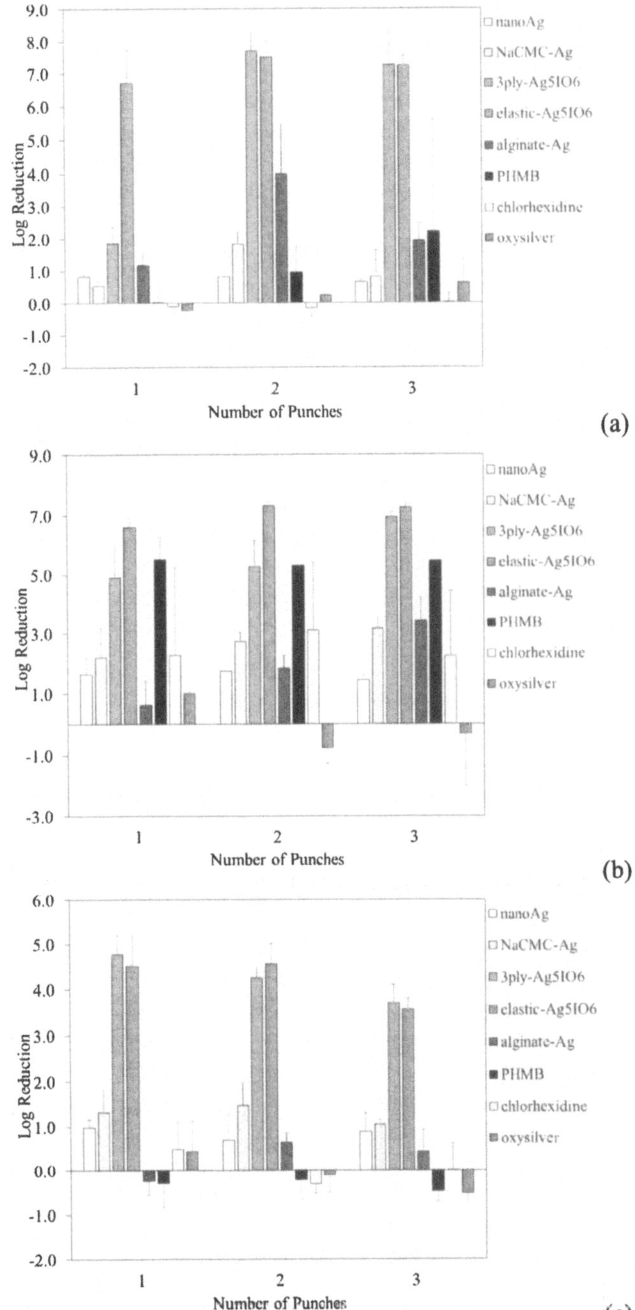

Figure 2. Ability of dressings to eliminate mature biofilm in 24h. One, two, or three dressing punches were exposed to mature biofilm for 24h. Mean log reductions, calculated relative to control dressings, are shown along with the SD. SD of zero indicate total kill of the microorganisms. (a) *P. aeruginosa*, (b) *S. aureus*, (c) *C. albicans*.

None of the commercial silver and chlorhexidine dressings disrupted/killed mature biofilms. The charge (Ag^+) and structure (chlorhexidine) of these materials may cause them to bind to extracellular biofilm components prior to interaction with microorganisms. Some species-specific activity was seen with PHMB.

In the standardized planktonic log reduction assay, rapid kill of planktonic microorganisms (total kill in 30 min) was achieved with Ag_5IO_6. These results were comparable to nanoAg and PHMB, and significantly better than NaCMC-Ag and alginate-Ag (0 log reduction). Rate of kill at medical device surfaces is important both for preventing biofilm formation by eliminating planktonic microorganisms before biofilms are started, and microorganism translocation and subsequent systemic infection.

4 CONCLUSION

The application of Ag_5IO_6 in wound dressings has shown efficacy with combined abilities to rapidly kill planktonic micro-organisms, prevent adherence of micro-organisms for ≥28 days in-use, and disrupt/eliminate mature biofilms. This makes Ag_5IO_6 an excellent candidate for use in a variety of medical devices, including those that need a coating to prevent microbial adherence both for short- and longer-term applications, and in treatment of mature biofilms. These applications could include use in wound dressings and gels, catheters, needles, connectors, ports, and implants.

REFERENCES

[1] Lansdown, Br J Nurs, 13, S6-S19, 2004.

[2] Phillips, Yang, Davis, Sampson, Azeke, Hamad, and Schultz, Int Wound J, published online:1-15, Sept 2013.

[3] Nadworny and Burrell, J Wound Technol, 2, 6-13, 2008.

[4] Costerton, Cheng, Geesey et al, Ann Rev Microbiol, 41, 435-465, 1987.

[5] Cowan, J Wound Care, 20, 220-226, 2011.

[6] Gilbert and McBain, Am J Infect Control, 29, 252-255, 2001.

[7] Kovalevskiy and Jansen, Z Anorg Allg Chem, 632, 577-581, 2006.

[8] Nadworny, Prosperi-Porta, and Unrau , Application No. PCT/CA2013/000735, 16 pages, 2014

[9] Olson, Anderson, Nadworny, and Omar, Application No. US 13/214,286, 14 pages, 2012.

[10] Incani, Omar, Prosperi-Porta, and Nadworny, Int J Antimicrob Agents, 8 pages, 2015.

[11] ASTM International, E2799-11, 2011.

[12] Taylor PL, Ussher AL, Burrell RE, Biomaterials, 26, 7221-7229, 2005.

[13] Smith&Nephew, Inc., http://www.smith-nephew.com/key-products/advanced-wound-management/acticoat [accessed 08.04.15].

[14] Smith&Nephew, Inc., http://www.smith-nephew.com/canada/products/advanced-wound-management/bactigras/,[accessed 07.04.15].

Electrospun Functional Poly(ether amide) Composite Nanofibres

Xingmao Jiang*, Shuai Liang*, Fei Wang*, Wangchang Geng**

* Changzhou University, School of Petrochemical Engineering, 213164, PR China

** Northwestern Polytechnical University, School of Science, Xi'an 710072, P.R. China

Corresponding author: E-mail: jxm@cczu.edu.cn; Fax: +86 519 86330251; Tel: +86 519 86330253

ABSTRACT

Due to its unique physical and chemical properties, poly (ether amide) (Pebax) has been widely applied in membrane separation, chemical sensing, and medical applications. Pebax composite nanofibres offer advantages such as high contact area and fast transport for ions and water molecules. (Pebax) nanofibers containing Ag nanoparticles were prepared by an electrospinning method. The Ag/Pebax nanofibers with ultra-low Ag loadings demonstrated superior antimicrobial properties against Escherichia coli and Staphylococcus aureus. Moreover, LiCl has been uniformly distributed in polyether domains of Pebax, and Pebax 2533 composite nanofibers with different amount of $LiCl{\cdot}H_2O$ were successfully fabricated by electrospinning. The electrospun nanofibres possessed high humidity sensitivity and displayed quick response and recovery.

Keywords: Pebax, Composite nanofibers, Electrospinning, antimicrobial properties, humidity sensitivity

Pebax 2533 is a kind of poly (ether amide) (Pebax) copolymer, and its structure unit consists of dodecyl amide and ether components, which offers a perfect combination of mechanical strength, breathability, flexibility, chemical resistance, no skinallergic and ease of processing. Therefore, it can be applied in many fields, such as medical materials, sports market, daily life, etc [1–3]. As medical material, it can be used to make urinary catheter, arterial catheter, PTCA (percutaneous transluminal coronary angioplasty) catheter device, infusion bags, and so forth [4–6]. If the antibacterial ability can be rendered to the Pebax by embedding Ag nanoparticles, the performance of Pebax will be dramatically improved with certainty. However, to the best of our knowledge, the preparation of Ag/Pebax composite has not been reported so far.

On the other hand, conducting polymers can be used as humidity sensor materials. However, current commonly used conducting polymers are still not sufficiently sensitive to changes in humidity, especiallyat high relative humidity and their stability and antijamming capability is poor [7]. Thus, how to overcome these shortcomings is a great challenge for humidity sensor research. Now, it has been proposed that the Pebax 2533 will form hydrophobic/hydrophilic phase-separated mesostructure under solvothermal conditions due to its composition of a hydrophobic polyamide (PA) hard domain and a soft hydrophilic polyether (PE) domain, in which PE was periodically distributed within the hydrophobic rigid PA domains [8]. It is expected that water soluble $LiCl{\cdot}H_2O$ will be located in the hydrophilic domains. However, to the best of our knowledge, Pebax-based humidity sensingmaterial has not been reported so far.

Herein, we present excellent antimicrobial material, Ag/Pebax 2533 composites and a highly efficient humidity sensing material, LiCl-doped Pebax 2533 composite nanofibers through a novel electrospinning technique, respectively. Ag/Pebax presented excellent antimicrobial properties at an ultralow $AgNO_3$ content in the precursor (0.15‰), which was much lower than that of conventional antimicrobial Ag/polymer composites reported in the literature [9-13], making Ag/Pebax composite a promising antimicrobial material with low-cost and excellent antimicrobial properties, which could be widely used in many fields in the near future. And sensors with high sensitivity among the whole relative humidity (RH) range have been obtained by appropriate loading of $LiCl{\cdot}H_2O$. Humidity sensing mechanism was analyzed by complex impedance spectra and the corresponding equivalent circuit.

We fabricated Pebax composites nanofibers with different amounts of $AgNO_3$ or $LiCl{\cdot}H_2O$ by an electrospinning method. The process started with an anhydrous Pebax solution in isopropanol and the mixture was magnetically stirred for 5 h at 60 °C to obtain a Pebax solution with a concentration of 66.7 mg mL^{-1}. Then a given amount of $AgNO_3$ or $LiCl{\cdot}H_2O$ solution was added into the Pebax solution and the content of them in Pebax composites nanofibers were controlled from 0.05 ‰ to 0.25 ‰ and 16.7 to 44.4 wt%, respectively. After sonication, the homogeneous solution was fed into a positively charged spinneret with a diameter of 0.6 mm attached to an electrospinning apparatus [14]. Finally, the Ag/Pebax composite nanofibers and LiCl-doped Pebax composite nanofibers can be obtained by the electrospinning device in the specified conditions. The composite nanofibers were collected on a filter paper or TEM grids.

The SEM and TEM images (Figure 1) indicated that the electrospun Pebax was mesh-like on a micro level formed by the connection between Pebax composite nanofibers. Especially, the interlaced $LiCl{\cdot}H_2O$–Pebax 2533 electrospun nanofibers were deposited on the pure interdigital gold electrodes as shown in Figure 1c. And the nanofibers with the diameter of about 200 nm were obviously observed in both SEM and TEM images. These results prove that the electrospinning technique can be applied successfully to Pebax.

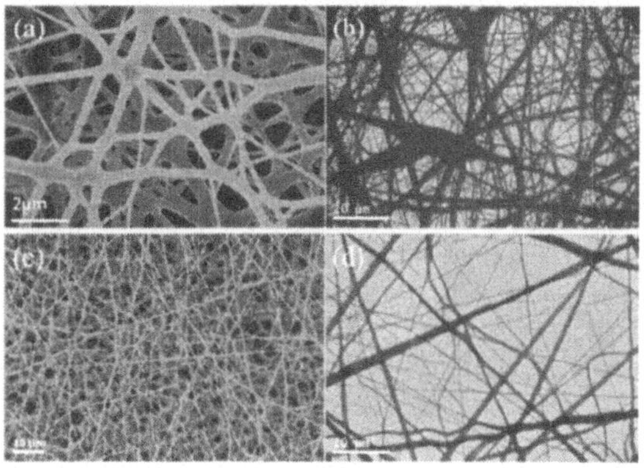

Figure 1: the SEM and TEM images of Ag/Pebax composite nanofibers (a,b) and LiCl-doped Pebax composite nanofibers(c,d).

The potential use of Ag/Pebax composite as antibacterial material was assessed by observing their antibacterial activity against two kinds of common bacteria, E. coli and S. aureus. The electrospun Pebax without Ag was evaluated in a control experiment. The bacteria were incubated in a growth medium containing various nanofibers and the antibacterial properties of samples were tested by the membrane adhering method, following the antibacterial testing standard (JISZ 2801:2010). The changes in the content of Ag had a significant effect on the antibacterial properties of Ag/Pebax composite. The number density of bacterial colony for pure Pebax was high (Figure 2a), and both of the antibacterial rate (1) and antimicrobial activity (2) were nearly zero (Figure 3); that is, there was no antibacterial activity observed for Ag free pure Pebax.

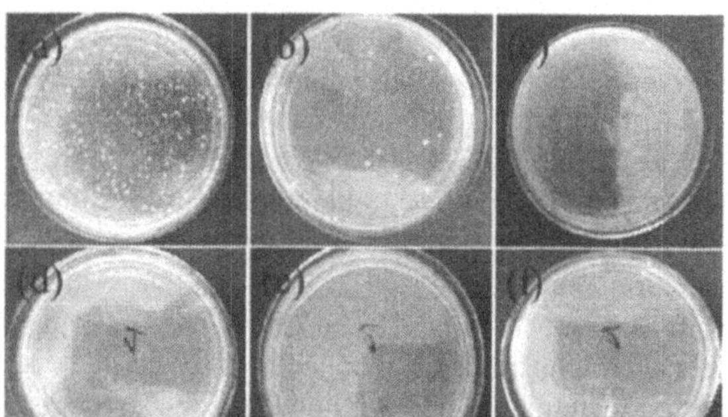

Figure 2: Comparison of the antibacterial ability of (a) Pebax, (b) Ag/Pebax-1, (c) Ag/Pebax-2, (d) Ag/Pebax-3, (e) Ag/Pebax-4, and (f) Ag/Pebax-5 against E. coli (the mass ratio of $AgNO_3$ to Pebax at 0.05 ‰, 0.10 ‰, 0.15 ‰, 0.20 ‰, and 0.25 ‰, respectively).

In contrast, the samples added with small amount of $AgNO_3$ in the Pebax precursor solution showed an inhibitory effect on E. coli and S. aureus. Although the content of $AgNO_3$ was only 0.05 ‰ (Figures 2b and 3b), the antibacterial rate against E. coli and S. aureus can reach 88.2 % and 76.1 %, respectively.

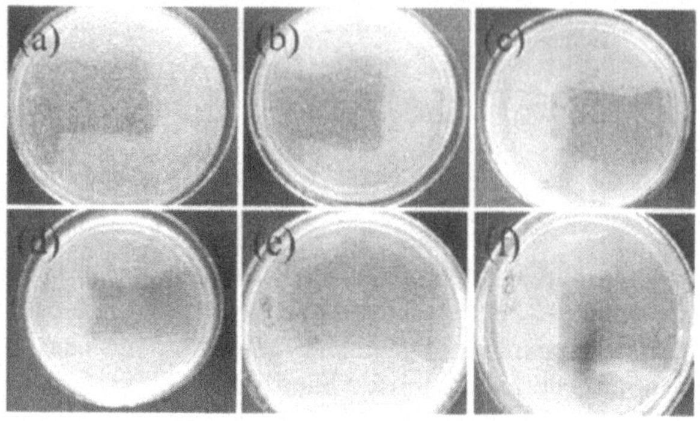

Figure 3: Comparison of the antibacterial ability of (a) Pebax, (b) Ag/Pebax-1, (c) Ag/Pebax-2, (d) Ag/Pebax-3, (e) Ag/Pebax-4, and (f) Ag/Pebax-5 against S. aureus (the mass ratio of $AgNO_3$ to Pebax at 0.05 ‰, 0.10 ‰, 0.15 ‰, 0.20 ‰, and 0.25 ‰, respectively).

Furthermore, the antibacterial properties of Ag/Pebax composite were enhanced gradually with the increase of $AgNO_3$ concentration (Figures 2b–2d, 3b–3d and 4). When the $AgNO_3$ content reached 0.15 ‰, no bacterial colony can be observed (Figures 2d and 3d) and the inhibition rates all larger than 99.9 %, and the value of antimicrobial activity against E. coli and S. aureus was 5.8 and 5.6, respectively (Figure 4). Further increasing the content of Ag failed to enhance the antibacterial properties of Ag/Pebax any more (Figures 2e–2f, 3e–3f and 4). Because the Pebax did not show antibacterial ability, Ag nanoparticles were responsible for the antibacterial ability of Ag/Pebax composites. It was reported that the antimicrobial property of material containing silver is mainly determined by the total release rate of silver ions [15]. In this work, owing to the stable particle size, the increase of Ag content will lead to the enlargement of the number and the total surface area of Ag nanoparticles; thus the release rate of silver ions was accelerated, giving rise to the improvement of antibacterial ability of Ag/Pebax composites (Figure 4). According to the antimicrobial testing standard JISZ 2801:2010, the sample whose antimicrobial activity is greater than 2 ($R > 2$) passes the antibacterial test of the strain ($R > 2$ means the antibacterial efficiency can reach 99 % or more). According to the experimental results, it can be concluded that the Ag/Pebax composite with $AgNO_3$ content larger than 0.15 ‰ presented excellent antibacterial properties, and this Ag content was much lower than that of Ag/polymer composites reported in the literature (usually larger than 1 wt.%) [9,11,12,16-19], making Ag/Pebax a promising antibacterial material with potential application in a number of fields, such as biomedical materials, sports apparatus, and laminating films.

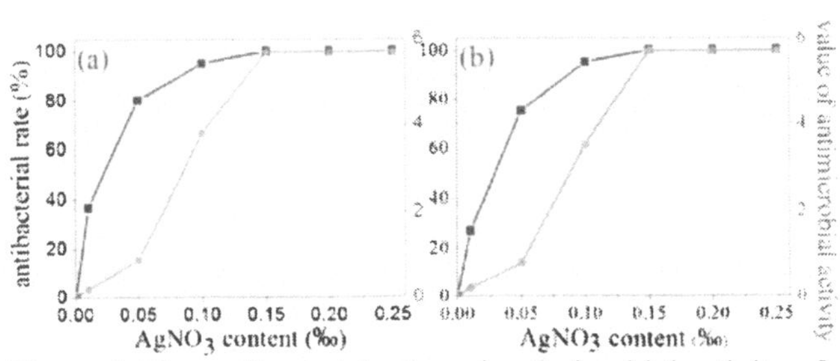
Figure 4: The antibacterial rate and antimicrobial activity of Ag/Pebax against (a) E. coli and (b) S. aureus as a function of $AgNO_3$ content in the precursor.

Figure 5a reveals the hysteresis data of $LiCl \cdot H_2O$–Pebax 2533 composite with $LiCl \cdot H_2O$ content at 44.4 wt%. As we can see from Figure 5a this as-synthesized sample has a negligible hysteresis, indicating that this material is a kind of promising humidity sensing material. Response and recovery time are also very important to a humidity sensor. Figure 5b provides the response and recovery time curve of $LiCl \cdot H_2O$–Pebax 2533 composite with $LiCl \cdot H_2O$ content at 44.4 wt%. According to the literature report [20], the time taken by a sensor to achieve 90 % of the total impedance change is defined as the response time or recovery time. For the sample of $LiCl \cdot H_2O$–Pebax 2533 composite with $LiCl \cdot H_2O$ content at 44.4 wt%, the response time was about 30 s with relative humidity increasing from 11% RH to 95% RH, and the recovery time was 80 s with relative humidity decreasing from 95% RH to 11% RH.

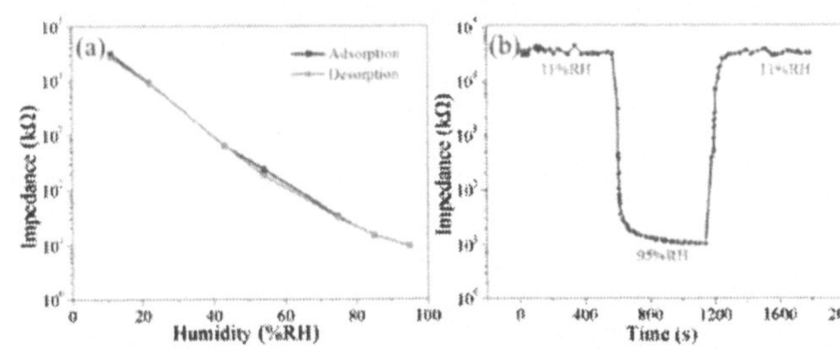
Figure 5: (a) Hysteresis, (b) Response and recovery curve of $LiCl \cdot H_2O$-Pebax 2533 composite with $LiCl \cdot H_2O$ content at 44.4 wt% (measuring frequency: 100 Hz, voltage: 1 V).

In order to explain the humidity sensing mechanism of this $LiCl \cdot H_2O$–Pebax 2533 composite material, complex impedance plots and the corresponding equivalent circuits (inset) of $LiCl \cdot H_2O$–Pebax 2533 composite with $LiCl \cdot H_2O$ content at 44.4 wt% under different relative humidities were provided as shown in Figure 6. The measuring frequency was from 50 Hz to 100 kHz, the horizontal and vertical axis value corresponding to the real and imaginary part of the complex impedance of the sensing material, respectively. When the relative humidity was low (22% RH), few water molecules were adsorbed on the surface of sensing material, and proton migrates only by hopping from site to site across the sample surface [21], so the resistance of the sensor under this relative humidity mainly comes from the intrinsic high resistance of Pebax 2533, and its impedance is mainly determined by the capacitance of sensing film. The corresponding complex impedance plot looks like a straight line and the equivalent circuit can be modeled by a constant phase element (CPE) [22] as shown in inset of Figure 6a, the film capacitance calculated from the constant phase element was 1.6×10^{-11} F. When RH increased to the middle region (43% and 54% RH), more water molecules are adsorbed and serial water layer is formed. According to the ion transfer mechanism of Grotthuss [23], the serial water layer accelerates the transfer of H^+ and H_3^+O dissociated from the adsorbed water layer. Therefore, the resistance of the sensing film decreased, the film resistance decreased from 2.87×10^7 Ω to 7.39×10^6 Ω. In addition, the film capacitance value increased gradually because that the adsorbed water increased the dielectric constant of the sensing film. The corresponding complex impedance plot is a semicircle as shown in Figure 6b and c and the equivalent circuit can bemodeled by the parallel combination of a resistor (R) and a capacitor (C) as shown in inset of Figure 6b and c. When RH further increased to 75% RH, more water molecules are adsorbed and serial water layer can reach three layers according to literature [24]. According to the model of ion transport mechanism reported by Casalbore Miceli [25], Li^+ dissociated from the dissolved LiCl become free ions after getting rid of the interaction of the opposite charges Cl^- and the resistance of the sensing film decreased sharply due to the quick transfer of the free ions. In addition, under this RH, water molecules also accumulated at the interface between sensing filmand electrode. Hence, the interface resistance and capacitance also contributed to the impedance of the sensor under this RH. We can see that the complex impedance plot includes a semicircle in left side (high frequency region) and a short line in right side (low frequency region) as shown in Figure 6d. The semicircle is caused by the resistance R_f and capacitance C_f of $LiCl \cdot H_2O$–Pebax 2533 film. And the short line in low frequency side is caused by the diffusion process of ions or charge carriers (mainly includes H^+ and free Li^+) at the sensing film/electrode interface [22]. Under this RH, equivalent circuit shown in inset of Figure 6d can account for the sensing mechanism. When the relative humidity increased to higher RH region (85% RH, 95% RH), due to the hygroscopicity of LiCl, water multilayers formed both in sensing film and in sensing film/electrode interface. More and more LiCl dissolved in these water layers to form a solution, the dissolved Li^+ and Cl^- have a big contribution to the conduction of the sensing film, so the sensing film resistanceand interface resistance further decreased. Under this RH range (85–95% RH), the LiCl solution resistance also contributed to the conduction of the sensor and it decreased with the increasing RH. Therefore, we can use the equivalent circuit shown in inset of Figure 6e and f to describe sensing mechanism under this RH range. Interestingly, the film capacitance C_f increased nearly two orders of magnitude (from 3.7×10^{-10}F to 3.5×10^{-8} F) when RH increased from 85% RH to 95% RH. Our opinion is that under 95% RH, the internal of the sensing material are fully filled by adsorbed water which will greatly increase the dielectric constant of the sensing material, so the capacitance also increased sharply.

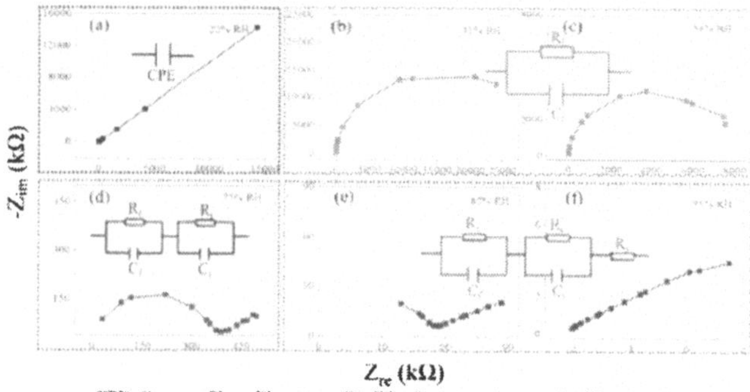

CPE: Constant Phase Element R_f: Film Resistance C_f: Film Capacitance
R_i: Interface Resistance C_i: Interface Capacitance R_s: Solution Resistance

Figure 6: Complex impedance plots and the corresponding equivalent circuits (inset) of $LiCl \cdot H_2O$-Pebax 2533 composite with $LiCl \cdot H_2O$ content at 44.4 wt% under different relative humidities (measuring frequency: 16 frequency points from 50 Hz to 100 kHz, voltage: 1 V).

In summary, this study has described a simple route to prepare Ag/Pebax composite nanofibers and LiCl-doped Pebax composite nanofibers via an electrospinning process. Ag nanoparticles with an average diameter of ca.15 nm were dispersed homogeneously in Pebax nanofibers. The antimicrobial activities against E. coli and S. aureus were evaluated by membrane adhering method. Study results indicated that the Ag/Pebax composites presented outstanding antimicrobial properties at ultralow Ag content; for example, the inhibition rate can reach >99.9% and antimicrobial activity against E. coli and S. aureus was 5.8 and 5.6 at the critical value of $AgNO_3$ content (0.15 ‰), respectively, which suggested that the Ag/Pebax composite can be used as a promising antimicrobial material. And study results of humidity sensing performance indicated that $LiCl \cdot H_2O$–Pebax 2533 composite with LiCl·H2O content at 44.4 wt% exhibited the best sensing property. Free Li ion dissociated from dissolved LiCl at higher relative humidity region play an important role in improving the sensing propertyof this material. The present work proved that electrospinning technology is a convenience and effective way to prepare LiCl/block polymer composite-based humidity sensing materials with high sensitivity.

This work was financially supported by the National Natural Science Foundation of China under Grant (no. 21373034); Jiangsu Key Laboratory of Advanced Catalytic Material and Technology; and Key Laboratory of Fine Petrochemical Engineering and PAPD of Jiangsu Higher Education Institutions.

REFERENCES

[1] C. L. Restuccia and C. LoFaro, U.S. Patent. No. 0170746 A1. 2010.
[2] N.L. Le, Y. Wang, T.S. Chung, J. Membr. Sci. 379,174–183, 2011.
[3] C. Cheng, R. Tang, F. Xi, Rapid Commun. 26,744–749, 2005.
[4] T. H. Gilman, U.S. Patent. no. 0015192 A1, 2012.
[5] D. J. Buckley, D. G.Wildes,W. Lee, andW. B. Griffin, U.S. Patent 0256502 A1, 2010.
[6] N. L. Le, Y.Wang, and T. S. Chung, J Membrane Sci. 379(1-2) 174–183, 2011.
[7] K. Jiang, T. Fei, T. Zhang, Sens. Actuators B. 199, 1–6, 2014.
[8] X.M. Jiang, C.J. Brinker, Chem. Commun. 466, 123–6125, 2010.
[9] M. S. Islam and J. H. Yeum, Colloid Surf. A-Physicochem. Eng. Asp. 436, 279–286, 2013.
[10] H. H. Chae, B. Kim, K. S. Yang, and J. I. Rhee, Synthetic Met. 161, 2124–2128, 2011.
[11] C. D. Saquing, J. L. Manasco, and S. A. Khan, Small, 5, 944–951, 2009.
[12] F. A. Sheikh, N. A. M. Barakat, M. A. Kanjwal et al., Macromol. RES. 17, 688–696, 2009.
[13] K. A. Khalil, H. Fouad, T. Elsarnagawy, and F. N. Almajhdi, Int. J. Electrochem. Sci. 8, 3483–3493, 2013.
[14] Q. Shi, N. Vitchuli, J. Nowak et al., Eur. Polym. J. 47, 1402–1409, 2011.
[15] S. Zhang, F. Peng, H. Wang, H. Yu, J. Yang, and H. Zhao, Catal. Commun. 12, 689–693, 2011.
[16] Y. Matsumura, K. Yoshikata, S. Kunisaki, and T. Tsuchido, Appl. Environ. Microb. 69, 4278–4281, 2003.
[17] D. P. Dowling, K. Donnelly, M. L. McConnell, R. Eloy, and M. N. Arnaud, Thin Solid Films, 398-399, 602–606, 2001.
[18] Q. Shi, N. Vitchuli, J. Nowak et al., Journal of Materials Chemistry, 21, 10330–10335, 2011.
[19] A. Celebioglu, Z. Aytac, O. C. Umu, A. Dana, T. Tekinay, and T. Uyar, Carbohydrate Polymers, 99, 808–816, 2014.
[20] S. Agarwal, G.L. Sharma, Sens. Actuators B. 85, 205–211, 2002.
[21] J.H. Anderson, G.A. Parks, J. Phys. Chem. 72, 3662–3668, 1968.
[22] W.C. Geng, Q. Yuan, X.M. Jiang, J.C. Tu, L.B. Duan, J.W. Gu, Q.Y. Zhang , Sens. Actuators B. 174, 513–520, 2012.
[23] F.M. Ernsberger, J. Am. Ceram. Soc. 11, 747–750, 1983.
[24] C.L. Li, Y.T. Ma, Y. Li, F.H. Wang, Corros. Surf. 52, 3677–368, 2010.
[25] G. Casalbore-Miceli, M.J. Yang, N. Camaioni, C.M. Mari, Y. Li, H. Sun, M. Ling, Solid State Ionics 131, 311–321, 2000.

Transformative Design of Medical Oxygen Concentrator by Rapid Pressure Swing Adsorption

Rama Rao Vemula*, Mayuresh V. Kothare**[1] and Shivaji Sircar***

Department of Chemical and Biomolecular Engineering,
C329, Iacocca Hall, 111 Research Drive,
Lehigh University, Bethlehem, PA, USA 18015
*rav212@lehigh.edu, **mvk2@lehigh.edu, ***sircar@aol.com

ABSTRACT

A novel, single-bed, 4-step, Rapid Pressure Swing Adsorption (RPSA) process is designed for producing 90% O_2 from compressed air for medical use by patients with Chronic Obstructive Pulmonary Disease (COPD) and other lung disorders. The proprietary design integrated with product storage tank is capable of continuously producing 1 - 3 LPM of ~90±% O_2 by employing a commercial sample of LiLSX zeolite as an adsorbent for air separation. An optimum cycle time of 4-6 s was identified from the experimental results at which a lower Bed Size factor (BSF) of ~ 41 kgs/TPDO2 and corresponding oxygen recovery (R) of ~29 % were obtained. An optimum adsorbent particle size range was also identified by simulating the RPSA process. Thus, the transformative design of medical oxygen concentrator (MOC) using RPSA Technology has a potential to reduce the adsorber size by a factor of 2 to 3 while offering similar or ~10% higher O_2 recovery.

Keywords: medical oxygen concentrator, rapid pressure swing adsorption, charonic obstructive pulmonary disease, air separation, litheium exhanged zeolite

1 INTRODUCTION

Medical oxygen concentrators(MOC) directly produce 1-10 SLPM of ~ 90 % O_2 from ambient air by employing a rapid pressure swing adsorption (RPSA) technology using a N_2 selective zeolitic adsorbent. The O_2 is used to help breathing of patients suffering from non-curable lung disorders like COPD, fibrosis, etc. Nearly 5 % of world population suffer from these conditions which are one of the leading causes of deaths [1-2]. Several MOCs designs have been marketed by various manufacturers during the past five to ten years. The potential U.S and Global market for MOCs is growing rapidly and it is estimated to reach $ 2 billion by 2017 from $ 0.4 billion in 2014 [3].

The commercial stationary MOCs are bulky and heavy, whereas the portable MOCs deliver oxygen in a pulse mode. The battery life is also an issue for the portable units. Consequently, A light-weight and compact MOC delivering a continuous supply of O_2 is a very highly desirable product for this market.

Therefore, the objective of this research work was the design of next generation MOC, which is a smaller, lighter, more energy efficient unit that can deliver continuous (as opposed to pulsed boluses of) ~90% oxygen at 1-3 slpm using a novel rapid pressure swing adsorption. Furthermore, the critical resistances effecting the RPSA process performance were identified from the modeling and simulation study.

2 MODELING AND SIMULATION OF RPSA PROCESS

The Cyclic Skarastrom type rapid pressure swing adsorption (RPSA) technology is commonly used for the design of a commercial MOC unit [4]. The cycles consist of various combinations of adsorption, desorption and complimentary steps and a nitrogen selective lithium exchanged X zeolite is employed for the air separation. The key performance variables for an MOC are bed size factor (BSF = kg adsorbent / ton of product/ day), and oxygen recovery (R = amount of product produced/ amount of oxygen in feed air). The perfomance of RPSA process is primanrly governed by the individal step durations, feed pressure, purchage rate and adsorbent particle size. In addition, the nonidealities like momentum, heat and mass transfer resistance impadiment the performance. A mathematical model of RPSA process was developed and simulated to study the cyclic adsorption and desorption of N_2 from a zeolite bed. The effect of individual process variables and non-idealities were thoroughly investigated in the simulation study. A similar situation exisits in air separation. A N_2- He was selected as adsorbates for this study. The mathamatical model equations and boundary conditions were reported elsewhere[5]. The Comsol Multiphysics software with MATLAB, which uses finitite element method for discritizing the model equations, was used to solve the model equatins. All the nonidelaities present in the RPSA process were considered in these simulations.

---

[1] Corresponding author: Dr. Mayuresh V.Kothare at mvk2@lehigh.edu

The effect of RPSA cycle time (t_c) on Bed Size Factor (BSF) and He Recovery (R) is shown in Figure 1 at cyclic steady state conditions. The BSF decreases with decreasing t_c up to 5 s and then increase with further drease in t_c. Thus, a minimum of BSF was observed for an optimum cycle time (t_c) of ~3- 4 s. The increase in BSF with increasing of t_c for longer cycle times is due to lower frequency of operation of the RPSA process and it lowers the specific productivity of helium from the PSA system. However, the derecease in BSF for shorter cycle times (t_c< 3s) is due to the detrimental effects of the mass, heat and momentum transfer resistances in the system. The He recovery (R) from the feed gas is nearly constant for longer t_c and it derecses with decrease of tc below 5 s as shown in Figure 1.

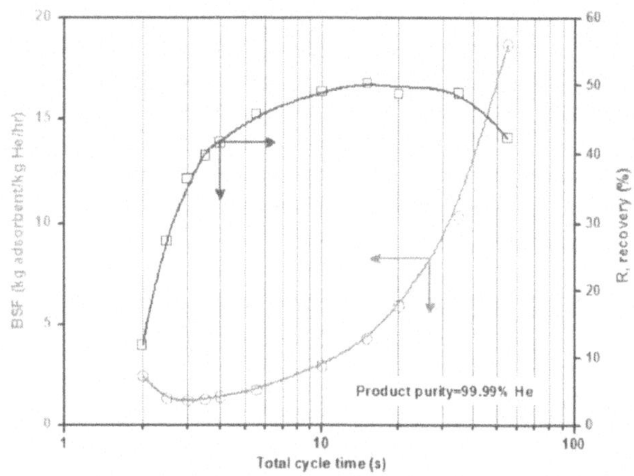

Figure 1: Effect of cycle time on BSF and R for 99.99% He product purity

In another study, the effect of adsorbent particle size (d_p) on rapid pressurization and depressurization of RPSA process was studied by simulating the individual pressurization and depressurization steps of the RPSA cycle [6]. The N_2 was selected as an adsorent and 5A zeolite material was chosen as an adsorbent. An optimum adsorbent particle size (d_p) range between 200- 350 μm was identified to lower the pressurization and de-pressurization times below 1s from the simulation study as shown in Figure 2. The decrease in adsorbent particle size below the optimum size range leads to increase in column pressure drop and also increases the mass transfer coefficicent. On the other hand, the increase of particle size leads to drease in mass transfer coefficient and lowers the column pressure drop. Thus, an optimum particle size range exists as shown in Figure 2 with in which both the resistances are counter balanced, and the pressurization and depressrization step durations were less than 1 s.

Therefore, it is evident from the above simulation results that the BSF of a RPSA process cannot be reduced indefinitely by lowering the process cycle time due to impediments created by finite the mass and heat transfer resistances whose detrimental effects on the process performance become pronounced at smaller cycle times, which was also experimentally demonstated in our study reported below [7].

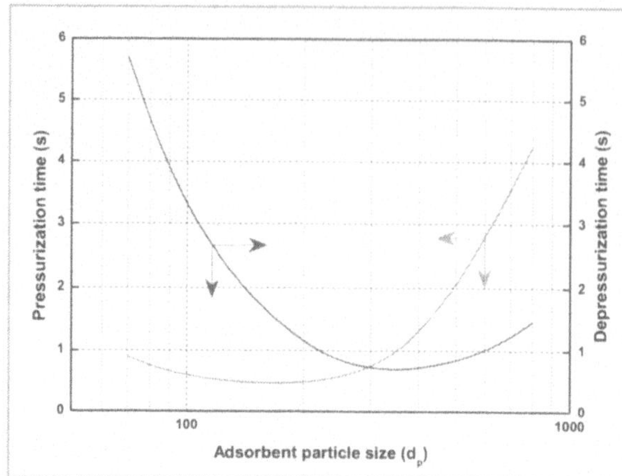

Figure 2: Simulated base case pressurization-depressurization times as functions of adsorbent particle size.

3 ACTUAL EXPERIMENTAL OPERATION OF A RPSA-MOC PROCESS

A bench scale unit consisted of a single adsorber vessel (diameter = 0.0498 m, Length = 0.127 m) was designed and constucted to study the miniaturization of MOC using RPSA Technology. The novel design has an intergated-coaxial product storage tank for continuous supply of product O_2 as well as to back purge the oxygen in purge step of RPSA cycle. The schematic and photograph of experimental set is shown elsewhere in our earlier publication[10]. It is capable of producing a continuous stream of 90 % O_2 at a flow rate of 1 – 3 SLPM from a compressed air gas as feed employing a total cycle time of 2.5 seconds or more.

The cyclic steady performance of the RPSA process using a commercial sample of LiLSX zeolite obtained from Zeochem Corporation was extensively studied. The synthetic air (21% O_2 + 78% N_2+ 1% Ar) was used as feed gas for the continuous production of 90 ± 1.0% O_2. The adsorbent sample was used as-received after thermal regeneration at ~ 350° C under dry N_2 flow for 8 hrs. A modified four-step Skarstrom-like PSA cycle was employed, which consisted of (i) column pressurization using a part of the O_2 enriched product gas from storage tank, (ii). Adsorption by flowing compressed air into the adsorber and collecting the product O2 in the tank, (iii) blowdown to ambient pressure for partial desorption of N_2 and (iv) back purge with a part of enriched O_2 from the tank for further desorption of N_2. The effect of cyclic time (tc) on RPSA process was extensively studied in the bench experimental unit. Cyclic steady state was usually reached after 50 cycles of operation. Only those runs where the over-all and component (O_2) mass balances between inlet and outlet flow streams closed within ± 5.0% were accepted. The effect of cycle time on BSF and R was shown in figure 3. A minimum BSF of is ~ 40.8 kgs/TPDO2 was observed at a total cycle time of ~ 5.5s from the experimental results and the correspond oxygen recovert is 24%.

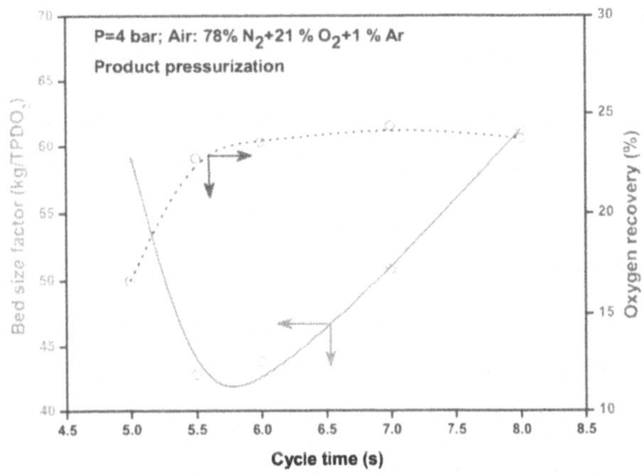

Figure 3: Experimental plots of BSF and O_2 recovery vs total cycle time at P_A=4 bar, P_D ~ 1 bar. Adsorbent particle size (dp)=400-800 μm.

Therefore, our propritory design of RPSA can potentially lower the adsorbent inventory of a RPSA-MOC by a factor of 2–3 and improving the O2 recovery by ~10% compared with the performance of a commercial MOC unit.

Salient Features of Lehigh Patented RPSA Design for MOC.

- The RPSA MOC design delivers a continuous stream of product O_2, which is preferred by 80% of the patient suffers with COPD and other lung diseases.
- It can be used to design both portable and stationary units of various sizes because of its ease of scalability.
- A unique single-bed adsorber enclosed in a product gas storage tank allows compact, light - weight design.
- The single adsorber RPSA permits easy process control vis a vis a conventional two or multi-adsorber RPSA design.
- The optimum total and individual -step cycle times of RPSA to obtain minimum adsorbent inventory and the maximum O_2 recovery were identified through our R& D efforts.
- The R & D led to the use of an optimum adsorbent particle size to maximize the RPSA performance, which was also previously unknown.

4 CONCLUSIONS

A novel, RPSA system based MOC design was constructed and successfully tested for continuous production of 1–3 lpm of 90% O_2 for medical use of COPD patients using a commercial sample of pelletized LiLSX zeolite (0.15 kg) as the air separation sorbent and a total cycle time (tc) of only 3–9 s. A lower BSF of 41 kgs/TPDO2 and at a corresponding recovery of 24 % were achieved using LiLSX zeolite at cycle time of 5–6 s with the current design. An optimum adsorbent particle size range of 200- 350 μm was identified by modeling and simulation of RPSA cycle. Thus, our transformative RPSA design of MOC is capable of reducing the adsorbent inventory by a factor of 2–3 and improving the O_2 recovery by ~10% compared with the performance of a commercial MOC unit.

REFERENCES

[1] Kim. V, Benditt. J. O, Wise. R. A and Sharafkhaneh. A, "Oxygen therapy in chronic obstacle pulmonary disease," Proc. Amer. Thoracic Soc, 5, 513-518, 2008.

[2] Chatburn. R. L and Williams. T. J, "Performance comparison of four portable oxygen concentrators," Respiratory Care, 55, 433- 442, 2010.

[3] Winter Green Market Research Inc, "Oxygen Concentrator Market Opportunities, Market Forecasts and Market Strategies: 2011-2017," www.wintergreenresearch.com, Lexington, MA, March 2011.

[4] Skarstrom. C. W, "Method and apparatus for fractionating gaseous mixtures by adsorption," US Patent 2, 944, 627, 1960.

[5] Rama Rao. V, Kothare. M. V and Sircar. S, "Anatomy of a Rapid Pressure Swing Adsorption (RPSA) Process Performance," AIChE J, 2015 (available online).

[6] Rama Rao. V, Kothare. M. V and Sircar. S, "Numerical simulation of rapid pressurization and depressurization of a zeolite column using nitrogen," Adsorption, 20, 53-60, 2014.

[7] Rama Rao. V, Kothare. M. V and Sircar. S, "Novel Design and Performance of a Medical Oxygen Concentrator using a Rapid Pressure Swing Adsorption Concept," AIChE, 60, 9, 3330-3335, 2014.

Magnetic Nanofluids in Ultrasonic Testing

V. V. Sokolov and M. I. Osipov

Moscow State University of Instrument Engineering and Computer Science
ul. Stromynka 20, Moscow 107996, Russia, vvs1953@rambler.ru

ABSTRACT

The use of the magnetic nanofluid as the bonding medium find an ever increasing application in ultrasonic non-destructive testing method.A thin layer of magnetic nanofluid which is adhered on the transducers face with a permanent magnet enhance the ability of movement and provide good bonding for transducers. Using the results of the hydrodynamical analysis and these experimental data we evaluated the magnetic fluid losses when the transducer is moving along the tested surface. Measurements of the temperature dependence of the ultrasonic velocity, density, viscosity and surface tension of water-based, dodecane-based, kerosene-based, antifreeze-based magnetic nanofluids have been made. The obtained results can be use for the choice of an optimum type of magnetic nanofluid in the design of ultrasonic transducers.

Keywords: magnetic nanofluid, ultrasound, acoustic impedance, ultrasonic testing

1 INTRODUCTION

Ferroflud, also called magnetic nanofluid is synthesized colloidal mixture of non-magnetic carrier liquid containing single domain permanently magnetized particles, typically magnetite, with diameters of order 10 nm. To prevent particles agglomeration due to dipole– dipole interaction, they are coated with a stabilizing layer, for instance, a surfactant, the choice of which is determined by the type of carrier liquid [1] .

Ultrasonic defectoscopy is one of the non-destructive testing methods. When setting the tasks for calculating and designing ultrasonic defectoscopes, one also considers the problem of how to ensure maximal stability of the acoustic contact of the ultrasonic sensor with the control subject. Water, alcohol, mineral oils, glycerine and various special multicomponent liquids serve as a coupling medium [2]. The quality of the acoustic contact, which to a greater extent determines the sensitivity of an ultrasonic defectoscope, depends on the following parameters and physical properties of the coupling fluid: the thickness of the contact layer, acoustic resistance of the coupling fluid, the ultrasound absorption ratio in the coupling fluid, the wettability of the working surfaces by the coupling fluid and continuity of the contact layer. The authors of patent [3] enjoy the privilege of first using magnetic nanofluids to ensure acoustic contact in ultrasonic spectroscopy. The main advantage of this contact is associated with an opportunity to hold magnetic nanofluid in the working gap with the help of a magnetic field specific configuration created by constant magnets. Besides, there is an opportunity to scan the surfaces controlled, e.g. those of a pipeline, with the given advance speed of ultrasonic sensors. It is also a matter of great import to study the influence of a magnetic field on the velocity of ultrasound propagation in a magnetic nanofluid. However, it is historical that in one of the first experimental works on measuring the impact of a magnetic field on the ultrasound propagation velocity in a water-based magnetic nanofluid [4] there crept in a procedure error. According to [4], the change of ultrasonic velocity on the exposure to relatively weak fields (of several hundreds of gausses) amounted to 30 – 50 %. These data served the basis for a number of inventions, e.g. various delay lines; they also provided grounds for critique of theoretical research on ultrasound propagation in magnetic nanofluids. Paper [5] pointed at the procedure error committed by the authors of work [4] when the latter were processing the experimental data. In paper [4] in order to determine the changes of phase ultrasonic velocity in a magnetic nanofluid when exposed to a magnetic field, they used a phase method, which essentially means measuring the phase shift between two harmonic signals, one of which passes through an acoustic cell with the studied magnetic nanofluid, whereas the other passes through a delay line. However, while processing the experimental data, the authors of [4] did not take into account the phase advance at the length equalling the distance between the piezoelectric converters, which led to as is shown in [5] an upward bias of the relative change in ultrasonic velocity by two orders. Thereafter, this result was proven experimentally and theoretically [6].

2 EXPERIMENTAL DATA

One of the major requirements for the coupling fluid is for it to provide a reliable acoustic contact when carry-

Table 1: Experimental data

Sample	$\rho, kg/m^3$	$c, m/s$	$\eta \cdot 10^3$, $Pa \cdot s$
A1	1103	1718	7.73
A2	1119	1695	8.17
A3	1259	1572	25.91
K1	956	1228	1.78
K2	1064	1165	3.23
K3	1325	1132	7.60
K4	1565	1094	19.85
H	1101	1186	3.41
D	1135	1112	5.07
W1	1022	1473	1.11
W2	1058	1464	1.23
W3	1100	1440	1.39
W4	1135	1420	2.12
W5	1197	1416	2.43

ing out ultrasonic control in a wide temperature range,as well as to minimise the magnetic nanofluid waste when scanning the object at a fixed speed. In this regard, there have been studied magnetic nanofluids based on: the antifreeze (A1,A2,A3), the kerosene (K1,K2,K3,K4), the heptane (H), the dodecane (D), and the water (W1, W2, W3, W4). The antifreeze was the mixtures of ethylene glycol and water with ratio 2/3 and 1/3 of volume parts respectively.

The physical properties of samples, corresponding to temperature $T = 293\,K$, are given in Table 1.For the water-based magnetic nanofluids, the measurements were taken in the temperature range of $273 - 313\,K$, whereas for the antifreeze-based ones the measurements were conducted in the range of $233 - 313\,K$. Temperature dependence of the ultrasonic velocity c in water-based magnetic nanofluids qualitatively repeats a similar dependence for pure water. In antifreeze-based fluids, the ultrasonic velocity decreases linearly with the temperature growth. Temperature dependence of density ρ in all samples studied is close to linear one.

With the temperature increase, the shear viscosity η decreases nonlinearly, so, for instance, for A2 sample the shear viscosity reduced by more than two orders.

2.1 ACOUSTIC IMPEDANCE

The acoustic impedance Z of a material is defined as the product of its density ρ and sound velocity c,

$$Z = \rho c \tag{1}$$

The attenuation does not contribute significantly to the acoustic impedance. Even in highly concentrated colloids at high frequency, when the attenuation becomes about $1000\,dB/cm$, it still contributes only 2% to the acoustic impedance. This allows us to approximate the acoustic impedance as a real number. Acoustic impedance is important in determination of acoustic transmission and reflection at the boundary of two materials having different acoustic impedance [7].

Basing on the obtained experimental results, we have calculated the acoustic impedance of samples. Figure 1 features the dependence of acoustic impedance on the antifreeze-based,kerose, magnetic nanofluids temperature.

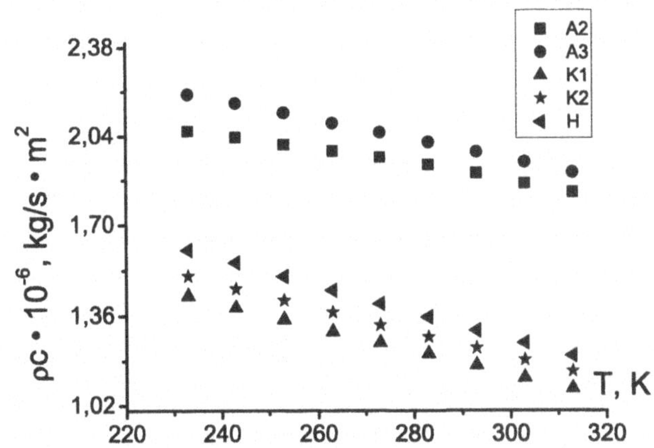

Figure 1: Acoustic impedance of different magnetic nanofluids *versus* temperature.

In the water-based magnetic nanofluids with the increase of solid-phase particles concentration, the temperature dependence of wave resistance is less pronounced (Figure 2), and is practically non-existent in sample W5.

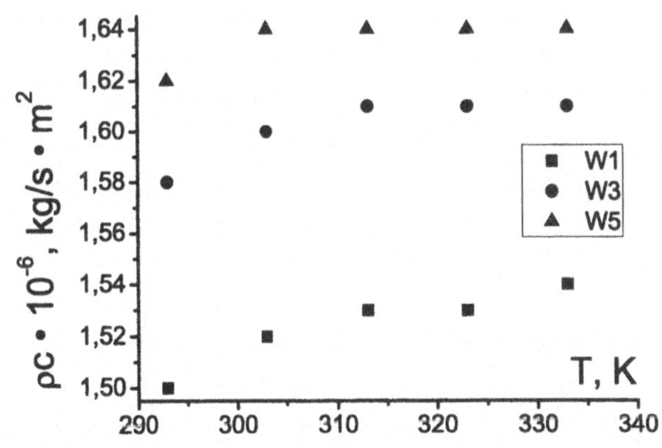

Figure 2: Acoustic impedance of water-based *versus* temperature.

2.2 SURFACE TENSION OF MAGNETIC NANOFLUIDS

When performing optimization calculations of magnetic fluid contact, alongside with the afore-listed values we must know the temperature dependence of surface tension coefficient. In order to define this dependence, there has been elaborated a new method taking into account the specific properties of magnetic nanofluids [8]. The very pinch of the method is to generate gravitational capillary waves on the magnetic nanofluid surface with the help of static and varying magnetic fields, orthogonal to the fluid surface. Gravitational capillary waves excited in such a manner have been recorded by holographic interferometry in the colliding beams with the time averaging. He -Ne laser with wavelength of $0.63mk$ was used as a source of coherent radiation. The physical reason for forming standing waves on the magnetic nanofluid surface is a parametric swing of the free surface by a varying magnetic field. A layer of magnetic nanofluid was submerged in homogeneous, orthogonal to the surface static and varying magnetic fields, i.e.

$$H = H_0 + h\cos\omega t \tag{2}$$

with the amplitude of varying magnetic field h being considerably smaller than the value of the static magnetic field H_0. When carrying out this method, we took into consideration the restriction for the upper limit of H_0 change in order to prevent appearing static monotonic instability. The fields mentioned were created by two pairs of Helmholtz coils that were placed on the alignment table, which allowed setting the magnetic nanofluid surface orthogonally to the magnetic field intensity vector every time prior to conducting the measurements. The frequency of varying magnetic field varied within $10 - 60\,Hz$ range. Magnetic nanofluid was placed in Petri dish with inner diameter of $61\,mm$, the fluid layer thickness was $20\,mm$. With the help of stroboscopic illumination, we discovered that the frequency of standing waves oscillations coincides with the frequency of variable component of a magnetic field. The recording of holographic interferograms was done on a photo plate, the exposition time was 1 2.5 minutes. Under these conditions, we registered high-quality holographic interferograms which confirmed the stability of surface waves.

The dispersion equation of gravitational capillary waves of a magnetic nanofluid layer resting between the poles of a magnet has the form [9]

$$\omega^2(k) = \frac{\sigma k^3}{\rho} + gk + \frac{(\mu-1)^2 H_0^2 k^3}{4\pi\rho(\mu+1)}, \tag{3}$$

where k is the wave number, σ is the required surface tension coefficient, g is the gravitational acceleration, μ is the permeability of magnetic nanofluid. The last term of this equation is negligibly small.Therefore we calculated the surface tension using the following expression

$$\sigma = \frac{\rho\omega^2}{k^3} - \frac{\rho g}{k^2} \tag{4}$$

To ensure usability of the data obtained in engineering designs, the latter were approximated by the following dependences. For water-based magnetic nanofluids the dependence of surface tension coefficient can be satisfactorily described by the following expression:

$$\sigma(T) = 18.8\left(2.73 - \frac{T}{273}\right)10^{-3}$$

in the temperature range of $288 - 313\,K$. Similar relations are true for antifreeze-based magnetic nanofluids:

$$\sigma(T) = 16.3\left(3.32 - \frac{T}{273}\right)10^{-3}$$

in the temperature range of $273 - 303\,K$.

3 CONCLUSION

During magnetic fluid acoustic contact, there arises a problem how to minimize the irreversible loss of magnetic nanofluid that occurs when the defectoscope ultrasonic head moves on the product surface. Theoretical studies connected with calculating the thickness of the layer left on the surface as a function of surface movement speed and the parameters characterising the properties of the fluid are based on the findings of paper [10] . The peculiarity of such thin films flow is a significant role of surface tension and fluid shear viscosity. According to [11], a customary mode of liquid viscous flow with the constant thickness of the film is realized with Reynolds numbers not exceeding $20-30$. With $Re > 30$ there occurs a wave flow in the fluid film, which becomes an additional source of disturbances for acoustic signal passage. Besides, with small thickness of the coupling liquid layer, it can decompose into single drops due to a prevailing role of capillary forces. Paper [10] singles out two limit cases covering high and low velocities. The latter case is more common for a magnetic fluid contact, which corresponds to low velocity of fluid flow v:

$$v \ll \frac{\sigma}{d} \tag{5}$$

In this approximation, paper obtained an expression defining the rate of magnetic nanofluid consumption from the drop held by a system of magnets in a flat gap between two mutually shifting surfaces. The analysis was conducted for a thin layer, when viscous and capillary forces play the main role in forming a fluid film carried by a moving border. Thereat, it was experimentally discovered that the thickness of the film drifting in a wide range of cross sectional dimensions turned out to be considerably less than the thickness of the contact layer. According to the authors of [11], the consumption of magnetic nanofluid per a unit of length due to the movement of one of the layers borders at velocity v is defined by the following expression:

$$Q = 0.66vd\left(\frac{\eta v}{\sigma}\right)^{\frac{3}{2}} \qquad (6)$$

This expression shows that consumption of magnetic nanofluid entrained by the moving boundary of the layer increase with an increase in velocity $\backsim v^{5/3}$ and viscosity $\backsim \eta^{2/3}$ and decreases with increases in surface tension $\backsim \sigma^{-2/3)}$. We chose the contact layer thickness equal to half the length of an ultrasonic wave λ with frequency of $f_0 = 3MHz$. Therefore, Eq. 6 takes the form

$$Q = 0.66v\frac{c}{2f_0}\left(\frac{\eta v}{\sigma}\right)^{\frac{3}{2}} \qquad (7)$$

The obtained array of experimental data enabled calculating irreversible losses with the help of this relation. The product scanning speed varied from 5 mm/s to 10 cm/s. The outcomes of the calculations performed by the formula stated above allowed making the following conclusions. Irreversible losses sharply increase with the increase of the scanning speed in low temperatures range of $233 - 283\,K$. The antifreeze-based magneto-acoustic contact has the biggest losses.

For instance, at the scanning speed of $10cm/s$ the permanent loss amounts to , $10^{-4}m^2/s$, which is two orders more than in other types of magnetic nanofluids. The temperature dependence of irreversible losses is close to linear one at the aforementioned scanning speeds. An increase in magnetite concentration results in the increase of irreversible losses, moreover, this dependence most sharply manifests itself in the antifreeze-based magnetic fluid, which first and foremost is stipulated by high viscosity of the carrier fluid. Among the magnetic fluid samples studied, there is no universal working temperature range. In the temperature range of $233-273\,K$ it is optimal to use antifreeze-based magnetic fluid but with a small volume of magnetite content, not exceeding 0.5 %. The main advantage of water-based magnetic nanofluids is connected with the independence of the wave resistance from the temperature, thus such fluids are reasonable to use when we control the products under conditions of varying surface temperature. It goes without saying, all the aforementioned presupposes the fluid to be aggregately stable. It is quite a tough requirement as the magnetic fluid is being held under the piezoelectric converter with the help of a nonhomogeneous magnetic field. In a production environment, one can evaluate the magnetic fluid quality with the help of an ultrasonic defectoscope. For this purpose, it is enough to record temperature dependence of the signal level at a fixed position of the converter. If the change of absorption meets the inequality $\Delta\alpha \leqslant dB/cm$, the magnetic nanofluid can be used as a coupling fluid.

Acknowledgements

Authors want to thank Prof. P. A. Eminov for the discussion of this paper results.

REFERENCES

[1] R. Rosensweig, "Ferrohydrodynamics," Cambridge University Press, 8-24, 1985.

[2] Charles J. Hellier "Handbook of nondestructive evaluation, " McGRAW-HILL, New York, 7.13, 2003.

[3] P. P. Prokhorenko and A. R. Baev "Method of creating acoustic contact for for ultrasonic measurements", Patent 697916 (USSR), 1979.

[4] D.Y.Chung and W.E. Isler, "Ultrasonic velocity anisotropy in ferrofluids under the influence of a magnetic field," J. Appl. Phys., 43, 1809-1811, 1978.

[5] V. V. Sokolov "Comments to the results of Chung and Isler,devoted magnetic field effect on the sound velocity in a magnetic fluid," Magnetogydrodynamics, 4, 136, 1986.

[6] V. V. Sokolov, "Wave Propogation in Magnetic Nanofluids (A review)", Acoustical Physics, 56, 972, 2010.

[7] Andrei S. Dukhin and Philip J. Goetz, "Characterization of Liquids, Nano- and Microparticulates, and Porous Bodies Using Ultrasound," second ed., Elsevier, New York, 95-97, 2010.

[8] V.M. Kondrachov, V.V. Sokolov "The Golographic Study of Surface Properties of Magnetic Fluids, " Abst. Six Int. conf. on Magn. Fluids, Paris, 288-289, 1992.

[9] R. E. Zelazo and Melcher, "Dynamics and stability of ferrofluds: surface interactions," ,J. Fluid Mech., 39, 1-24, 1969.

[10] L. D. Landau and V. G. Levich, "Entrainment of liquid by moving plate, " in L. D. Landau "Collected Works" , Gordon and Breach, Science Publishers, New York , 355-363,1965.

[11] V. M. Korovin and Yu. L. Raiker, "Entrainment of a magnetic fluid by the moving boundary of a plane- parallel layer", Magnetohydrodynamics, 23, 41-44, 1987.

Linseed oil miniemulsions stabilized by a polyelectrolyte complex (gum Arabic-chitosan)

M.G. Michel Barba and H. Espinosa Andrews*

Food Technology, Centro de Investigación y Asistencia en Tecnología y Diseño del Estado de Jalisco A.C. AV. Normalistas 800, Colinas de la Normal, Guadalajara, Jalisco, México, *hespinosa@ciatej.mx

ABSTRACT

An linseed oil miniemulsion was improved using the microfludization process and stabilizing by a polyelectrolyte complex. Variables such as gum Arabic/oil ratio (GA/O), oil volume fraction (ϕ), homogenization pressure and steps of homogenization were studied. The results show that microfluidization is an effective process to produce primary emulsions characterized by monomodal distribution between 440 and 826 nm and ζ potential values of −12.1 and −44.6 mV. Afterwards, a selected emulsion was titrated with a chitosan (Ch) solution. The droplet size and charge of the secondary emulsion (multilayer) changed from 585 to 428 nm and −34 to +56 mV with the increase of Ch concentration. A bridging flocculation process was observed in a biopolymer mass ratio ($R_{GA/Ch}$) of R_5, because the net charge was close to zero. This study provides useful information on the formation of miniemulsion with droplet size, and charge specific which may increase the rate of lipid digestion.

Keywords: miniemulsions, polyelectrolyte complex, gum Arabic, chitosan.

1 INTRODUCTION

Today we know that fatty acids are essential for normal growth and development and may play an important role in the prevention and treatment of coronary artery disease, hypertension, diabetes, arthritis, other inflammatory and autoimmune disorders, and cancer [1]. Linseed oil contains 55% of α-linolenic acid (Omega-3) and 16% of linoleic acid (Omega-6) [2]. Commonly, these fatty acids are marketed as emulsions, and their therapeutic efficacy depends on their appropriated design. An emulsion is a mixture of two immiscible liquids, in which one liquid is dispersed in the form of droplets in another liquid that forms the continuous phase. Based on the emulsion droplet size, emulsions can be divided into micro- (10–100 nm), mini (nano)- (100–1000 nm) and macro-emulsions (0.5–100 μm) [3]. Recently, developments in droplet microfluidics enable the generation of monodisperse and size-controlled single/multiple emulsion droplets [3-5]. These are frequently used as a precursor to fabricate particles or capsules for industrial applications [5]. Actually, polyelectrolyte complexes (PEC) membranes are widely used as potential carrier systems in the development of new products (emulsions, capsules, or gels), because these provide specific flow behaviors, textures, and appearances, tactile and oral properties [6]. PEC formation is observed when the two oppositely charged polyelectrolyte are mixed in aqueous solution under accurate conditions [6, 7].

Gum Arabic (GA) is one of the world's most common gums, widely used in the production of beverage and flavor emulsions and meal replacers. Recently, it was reported the PEC formation between GA and chitosan (Ch) [8]. The aim of this study was to develop and characterize linseed oil emulsions stabilized by a polyelectrolyte complex (gum Arabic/chitosan).

2 METHODOLOGY

2.1 Materials

Gum Arabic (GA) and linseed oil were purchased in Natural Products of Mexico SA de CV (Morelos, Mexico) and chitosan (Ch, degree of deacetylation 92.2%) in Sigma Aldrich (State of Mexico, Mexico).

2.2 Emulsions

To obtain a fine emulsion with desirable properties (lower particle sizes and higher electric charge), twenty emulsions were produced using an M110 PS microfluidizer (Microfluidics, USA) according with a 2^4 factorial design with 4 central points: Pressure: 34.5 to 206.8 MPa, steps (cycles) 1 to 5, oil volume fraction (ϕ) 0.05 to 0.1, and GA/O ratio 1 to 3. Coarse emulsions were prepared using a high shear dispersion equipment (Silverson, England) at 5000 RPM for 10 minutes. The charge and size of the emulsions were evaluated.

2.3 ζ potential

The ζ-potential was determined using a light scattering equipment Zetasizer Nano ZS90 (Malvern Instrument, UK). The emulsions were diluted in desionized water (1/100) to make the appropriate measurements. The ζ potential was calculated using the Smoluchowsky mathematical model measuring the direction and velocity of the emulsions as they moved along the applied electric field at 20°C.

2.4 Average particle size and distribution

The hydrodynamic size (Z-average diameter, d_h) and distribution of the emulsions was estimated using a Zetasizer Nano ZS90. The d_h was calculated by the Stokes–Einstein equation: $d_h = k_B T / 3\pi\eta_s D$; where k_B is the Boltzmann constant, T is the absolute temperature, η_s is the dynamic viscosity of the solvent and D is the z-average translational diffusion coefficient.

2.5 Colloidal titration

One selected emulsion was used to obtain the multilayer emulsions. A titration experiment was performed adding a Ch solution (10 mg/mL) into a 10 mL of the emulsion (1/100) above the isoelectric point of the system [8]. The charge and the d_h of the system was determined. The titration measurements are reported as the average ± standard deviation of measurements made on three independent samples with three measurements made per run at 20 °C.

2.6 Formation of multilayer emulsions

Electrodeposition technique layer by layer [9] was used for the formation of multilayer emulsions. Based on the charge of the emulsions (cationic, neutral, and anionic), three selected biopolymers mass ratios ($R_{GA/Ch}$) were prepared. Similarly, the charge and d_h of the emulsions were monitored.

2.7 Statistics

All data were reported as mean values and standard deviations of three measurements. Statgraphics Centurion XV version 2.15.06 was used for analysis of variance with a confidence level of 95%.

3 RESULTS

3.1 Microfluidization

All factors (ϕ, GA/O, pressure and steps) showed a significant effect ($p > 0.05$) on the d_h and charge of the emulsions.

3.1.1. Particle size and distribution

The pre-emulsions showed a d_h up to 1500 nm. In general, the d_h of the emulsions decreased and a mono-modal distributions with hydrodynamic droplet sizes between 440 and 720 nm was observed. These emulsions are classified as "miniemulsions" and showed a better stability to gravitational separation and aggregation than conventional emulsions. Into the microfluidizer chamber, the emulsion droplets are subjected to high shear and collision that reduce the average size of the emulsion allowing a narrow droplet size distribution [10].

A d_h decrement was observed with the increment GA/O ratio and with the decrement of the ϕ (Figure 1). Higher emulsifier concentrations were required to produce a kinetically stable emulsion against coalescence and flocculation [10].

When the pressure of microfluidization increased from 34.5 to 206.8 MPa, an increased of the d_h and distribution of particle sizes were observed. This is known as over-processing of the emulsion, at it is related to the phenomena of coalescence and rupture of some droplets during extreme conditions inside the microfluidizer chamber. Similar results were reported by Mahdi Jafari, He [11]. They found that the optimum conditions to produce fine emulsions of *d*-limonene with whey protein isolated by microfluidization were moderate pressure (42-63 MPa) and 2 cycles.

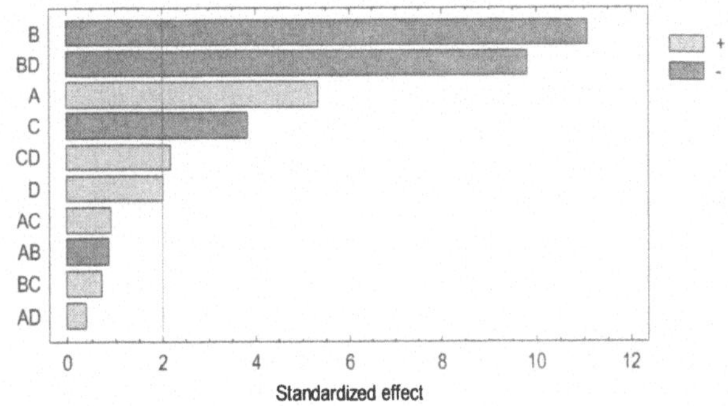

Figure 1. Standardized pareto chart of d_h ($p>0.05$): A (ϕ), B (GA/O), C (cycle) & D (pressure).

3.1.2. ζ potential

In general, droplets with ζ potential outside the range -30 mV to +30 mV are considered stable to the flocculation [12]. The charge of the pre-emulsion droplets was around -14.6 mV. After microfluidization steps, the ζ potential of the emulsions decreases in some cases significantly showing values between −12.1 and −44.6 mV. A positive effect of the GA/O ratio was observed, an increase of the amount of emulsifier, the particle size decreases, increasing the electrostatic repulsion between the emulsion droplets (Figure 2). Meanwhile, an increase of the ϕ, increase the particle size of the emulsion, decreasing the repulsive force between oil droplets. This behavior may favor the floculation and coalescence of droplets [13]. Finally, an increase in the pressure and cycle causes a decrease in the net charge of the system, possibly due to an increase of the collisions between particles. Similar results were obtained by Salvia-Trujillo, Rojas-Graü [14] in the formation of microfluidized emulsions of essential oil of lemongrass with alginate.

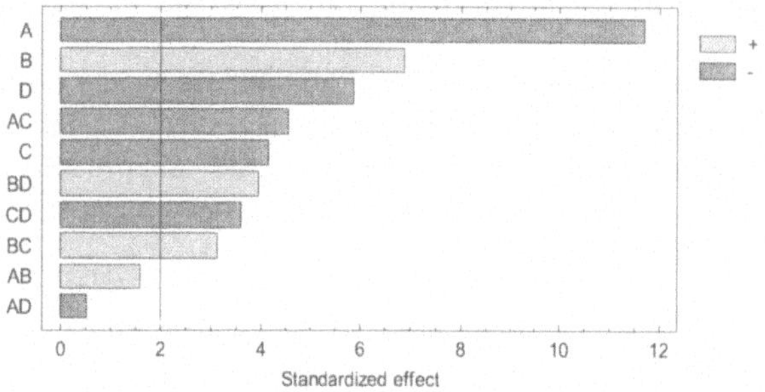

Figure 2. Standardized pareto chart of ζ potential (p>0.05): A (ϕ), B (GA/O), C (cycle) & D (pressure).

3.2 Formation of multilayer emulsions

In the electrodeposition technique layer by layer, anionic emulsifier is used to produce a "primary" emulsion containing small droplets (anionic), then an oppositely charged polyelectrolyte is added to the system that adsorbs to the droplet surfaces and produces a "secondary" emulsion containing droplets coated with a two-layer interface (cationic) [9].

Based on the results, an emulsion with 5% disperse phase, GA/O 3:1, 2 cycle at 172 MPa was produced. Figure 3 shows the evolution of the ζ potential and d_h as the primary miniemulsion (anionic) was titrated with the Ch (cationic) dispersion.

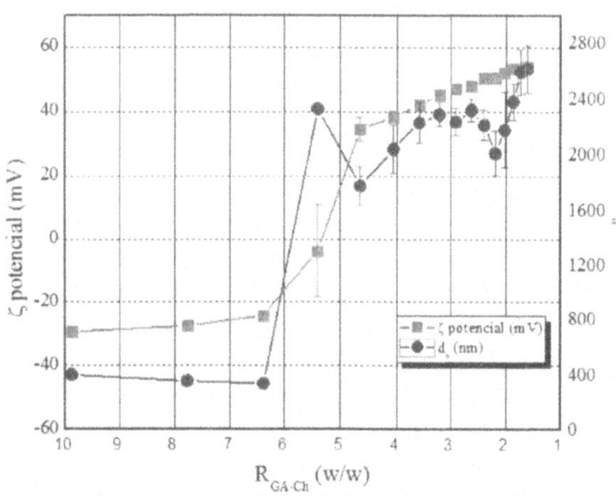

Figure 3. Evolution of the ζ potential (■) and d_h (●) as the emulsion was titrated with the chitosan.

In biopolymers ratio greater than R_{10}, the miniemulsions were decorated with Ch molecules, the electric charge of the systems keep the droplets without floculation, but the d_h decrease significantly from 573.2±11.8 to 432.6±4.6 nm with the increase of Ch concentration.

At a $R_{GA/Q}$ ~ 5 a stronger electrostatic interaction between biopolymers favors that, the Ch segments are adsorbed on multiple adjacent miniemulsions, bridging them together to form a complex coacervation miniemulsion. As a consequence, the d_h increase abruptly and the net charge close to zero. At a relatively high biopolymers ratio (>$R_{6.5}$), the net charge of the miniemulsion trends to +50 mV. However, through the flocculation process a highly viscous coacervate emulsion was formed inside the cuvette, therefore the d_h values are not trustworthy at all. Similar behavior was observed in the biopolymer alone by Espinosa-Andrews, Enríquez-Ramírez [8].

Based on these results, three miniemulsions with different charge were prepared (Table 1): cationic (R_{20}), neutral (R_5), and anionic (R_1).

Biopolymer matrix	ζ potential (mV)	d_h (nm)
GA	−33.5±0.2^a	573.2±11.7^a
R_{20}	−21.3±0.4^b	554.5±6.2^b
R_5	0.4±0.4^c	–
R_1	55.0±1.5^d	432.6±4.5^c

$^{a, b, c, d}$ The mean values (± SD, n = 3) that share the same letter in the same column are homogeneous groups (p> 0.05) according to Fisher LSD test.

Table 1. Physicochemical parameters of multilayer miniemulsions.

The ζ potential results agree with the titration results. In R_{20}, Ch addition decrease significantly the charge and d_h of the miniemulsion, without macroscopic phase separation. In R_5, a complex coacervation miniemulsion was formed, the net charge was close to zero and a high flocculate viscous emulsion was observed. Finally, R_1 did not show phase separation, it was characterized by a cationic emulsion droplet, where the d_h decreased as a consequence of PEC formation.

Droplet charge approached to electroneutrality at the biopolymers coacervation, negatively increased with a GA excess and positively with a Ch excess that inhibiting bridging flocculation by keeping the droplets emulsion separated by electrostatic effects. If the droplets in an emulsion have a positive charge in the small intestine, they may also become trapped within the mucous layer (which consists of anionic biopolymers) [15]. Therefore, positive miniemulsion may improve bioavailability of lipophilic components. In general terms, the average droplet size of the secondary miniemulsion decreased with increasing the Ch content. It is known that the bioavailability of lipophilic functional components encapsulated is increased when the droplet size is decreased. Some possible reasons discussed in the literature are that small droplets have a large surface area and may therefore be digested faster by digestive enzymes so that their contents can be released and absorbed easily. The small droplets may penetrate into the mucous

 TechConnect Briefs 2015, TechConnect.org, ISBN 978-1-4987-4729-5

layer, thereby increasing their residence time and bringing them closer to the site of absorption. The water-solubility of highly lipophilic components increases as the droplet size decreases, which may enhance absorption [15].

CONCLUSIONS

Microfluidization technique was useful for producing linseed oil miniemulsion. Gum Arabic/oil ratio, oil volume fraction, homogenization pressure and cycles of homogenization showed a significant effect ($p > 0.05$) on the Z-average diameter and charge of the miniemulsions.

The biopolymers mass ratio between gum Arabic/chitosan showed significant influence ($p > 0.05$) on the hydrodynamic size and ζ potential due conformational changes caused by the attractive interaction between the macromolecules. Polyelectrolyte complexes of gum Arabic and chitosan were attributed special features to miniemulsions that can be manipulated for specific applications as carriers for bioactive lipid.

REFERENCES

1. Simopoulos, A.P., *Essential fatty acids in health and chronic disease.* The American journal of clinical nutrition, 1999. **70**(3): p. 560s-569s.
2. Kumar, S., et al., *QTL for fatty acid composition and yield in linseed (Linum usitatissimum L.).* Theoretical and Applied Genetics, 2015: p. 1-20.
3. Jafari, S.M., Y. He, and B. Bhandari, *Production of sub-micron emulsions by ultrasound and microfluidization techniques.* Journal of Food Engineering, 2007. **82**(4): p. 478-488.
4. Kong, T., et al., *Droplet based microfluidic fabrication of designer microparticles for encapsulation applications.* Biomicrofluidics, 2012. **6**(3): p. 034104.
5. Keng-Shiang, H., et al., *Calcium alginate microcapsule generation on a microfluidic system fabricated using the optical disk process.* Journal of Micromechanics and Microengineering, 2007. **17**(8): p. 1428.
6. Thünemann, A.F., et al., *Polyelectrolyte Complexes*, in *Advances in Polymer Science*. 2004. p. 113-171.
7. Espinosa-Andrews, H., et al., *Gum Arabic−Chitosan Complex Coacervation.* Biomacromolecules, 2007. **8**(4): p. 1313-1318.
8. Espinosa-Andrews, H., et al., *Interrelationship between the zeta potential and viscoelastic properties in coacervates complexes.* Carbohydrate Polymers, 2013. **95**(1): p. 161-166.
9. Guzey, D. and D.J. McClements, *Formation, stability and properties of multilayer emulsions for application in the food industry.* Advances in Colloid and Interface Science, 2006. **128–130**(0): p. 227-248.
10. Jafari, S.M., Y. He, and B. Bhandari, *Optimization of nano-emulsions production by microfluidization.* European Food Research and Technology, 2007. **225**(5-6): p. 733-741.
11. Mahdi Jafari, S., Y. He, and B. Bhandari, *Nano-Emulsion Production by Sonication and Microfluidization—A Comparison.* International Journal of Food Properties, 2006. **9**(3): p. 475-485.
12. Heurtault, B., et al., *Physico-chemical stability of colloidal lipid particles.* Biomaterials, 2003. **24**(23): p. 4283-4300.
13. Qian, C. and D.J. McClements, *Formation of nanoemulsions stabilized by model food-grade emulsifiers using high-pressure homogenization: Factors affecting particle size.* Food Hydrocolloids, 2011. **25**(5): p. 1000-1008.
14. Salvia-Trujillo, L., et al., *Effect of processing parameters on physicochemical characteristics of microfluidized lemongrass essential oil-alginate nanoemulsions.* Food Hydrocolloids, 2013. **30**(1): p. 401-407.
15. McClements, D.J. and J. Rao, *Food-grade nanoemulsions: formulation, fabrication, properties, performance, biological fate, and potential toxicity.* Critical reviews in food science and nutrition, 2011. **51**(4): p. 285-330.

Polymer Gels: Fracture, Friction, Healing and Superabsorbency

K. A. Erk*^, T. L. Thornell*, Q. Zhu*, and M. J. Krafcik*

* School of Materials Engineering, Purdue University, West Lafayette, IN, USA
^ erk@purdue.edu

ABSTRACT

This proceeding decribes a series of experimental investigations performed to better understand and ultimately control the nonlinear mechanical behavior of soft polymeric gels. In the first part, shear rheophysical experiments are described in which the shear-induced flow instability formation of fracture planes and shear bands is studied for a model physically associating polymer gel. When the fractured gel was allowed to rest for a given time period, the mechanical strength of the gel progressively increased as the internal macromolecular network structure healed. In the second part, a specific application of polymer gels as internal curing agents in high performance concrete is described. The swelling performance of model chemically crosslinked polymer gels was found to be strongly affected by the ionic nature of the aqueous environment, ultimately leading to reductions in absorption capacity that could be detrimental to the internal curing of concrete.

Keywords: self-healing, injectable, infrastructure, concrete, soft materials

1 INTRODUCTION

Polymeric gels are used in a variety of industrial and biological applications. Taking advantange of their tunable chemical structures and physical properties, polymer gels have been widely employed from drug delivery aids in the human body [1–3] to internal curing agents for pavement concrete. [4–6] More recently, many research studies have focused on optimizing their properties for use as injectable materials for regenerative medical applications. [7–9]

In these applications, it is important to understand the mechanical properties of the gels, including their fracture toughness, self-healing ability, and swelling response. However, characterizing the properties of these mechanically soft materials using traditional techniques is difficult as the materials will deform or deswell with only slight changes in applied pressure or environment. For example, many soft materials can be nearly impossible to grip during traditional tensile tests and solvent evaporation can be significant during compression testing. Some soft materials are also viscoelastic and can thus deform and relax due to gravity and are thus impossibleto mold into a well-defined, isolated shape for mechanical testing.

One solution to these characterization limitations is to employ *shear rheometry* to probe the fracture, friction, and healing properties of soft materials. By using shear rheometry, experiments have been designed to successfully quantify the strain stiffening behavior of gels [10], the onset and evolution of fracture and shear banding flow instabilities [11], the sliding friction response [12] and, more recently, the healing kinetics of fractured gels. [13] Additionally, traditional swelling studies have also been performed on hydrogels for specific application as internal curing agents in high performance concrete. [14] In this proceeding, the results of these studies on the properties and applications of polymer gels are summarized. For more details, refer to Refs. 10-14.

2 EXPERIMENTAL METHODS

The model gels described here are polyacrylic-based macromolecular networks immersed in solvent. For the stiffening, fracture, friction, and healing experiments, a triblock copolymer consisting of 8.9 kg/mol poly(methyl methacrylate) endblocks connected by a 53 kg/mol poly(n-butyl acrylate) midblock was dissolved at different concentrations in 2-ethyl hexanol. At the temperature range of interest (20-28 °C), the copolymer self assembles to form a macromolecular network composed of physically associated endblock aggregates interconnected by flexible midblock bridges, resulting in a physically crosslinked gel. Rheological characterization was performed in a shear rheometer (Anton Paar MCR 302) using a concentric cylinder fixture. Rheophysical experiments were performed to correlate rheometry data with the macroscale response of the sample, as quantified by optical tracking of micron-scale inert tracer particles within the gel during the rheometry experiments. [12,13] For the hydrogel swelling experiments, random poly(acrylic acid-acrylamide) copolymer with varying acrylic acid and acrylamide ratios were custom synthesized and chemically crosslinked to form a strong hydrogel for use as a concrete internal curing agent. Gels were immersed in different aqeuous environments and the tea bag test was used to quantify the overall swelling capacity and swelling kinetics. [14]

3 RESULTS AND DISCUSSION

3.1 Shear-Induced Strain Stiffening

When the model physically associating polymer gels are deformed in shear in a rheometer cell, strain-stiffening behavior is observed at relatively low values of strain (see upturn in stress-strain data of Figure 1A). The observed stiffening was accurately described by a constitutive model with a single fitting parameter related to the dimensions of the internal polymer network:

$$\tau = G_0 \gamma exp\left[\left(\frac{\gamma}{\gamma^*}\right)^2\right], \qquad (1)$$

where τ is the measured shear stress, G_0 is the small-strain (elastic) shear modulus, γ is the applied shear strain, and γ^* is the critical value of shear strain at which stiffening becomes significant. The same model is also effective for describing the stiffening of actin, collagen, and other biopolymer networks. [10]

3.2 Flow Instabilities and Sliding Friction

If shear deformation is applied at relatively high shear rates, the physical crosslinks within the gel's network structure are disrupted, and the gel displays a corresponding overshoot in the measured stress response (see overshoot in Figure 1A). This behavior was found to be due to cohesive fracture, which was confirmed by particle tracking flow visualization techniques. [11,12] Once shear-induced fracture or shear banding interfaces were created in the gel, the resulting stress response was found to be related to sliding friction forces at that interface by applying theories from polymer physics. [12] The rheological response following the instability formation within the gel was power-law shear thinning with a fixed slope, regardless of the type of instability – *i.e.*, high-rate fracture or low-rate shear banding. Thus, at the molecular level, fracture can be considered to be a special case of shear banding within these physical gels. [12]

3.3 Fracture and Healing Behavior

When the fractured gel was heated above a critical temperature (T_{crit} ~ 33-36 °C) and then cooled to $T < T_{crit}$, the physically crosslinked network was re-formed and the gel was found to recover its original elasticity. A shear rheometry protocol was designed to study the fracture-healing behavior in a contolled way. [13] First, the shear stress response measured from applying a fixed shear rate (e.g., 1 s^{-1}) to the gel was recorded as the 'primary fracture' (see Figure 1A). Fracture was confirmed by simultaneous optical particle tracking analysis. Second, the shear rate was reduced to zero and the fractured gel was allowed to wait in the rheometer undisturbed for a desired 'resting time' of 5 minutes to several hours. Third, the shear rate was re-applied at the same magnitude (e.g., 1 s^{-1}) and the gel was subsequently re-fractured to determine the 'recovery fracture' stress response (see Figure 1A). The ratio of the maximum stress of the 'recovery fracture' to the maximum stress of the 'primary fracture' was used to determine the percentage of strength recovery for each resting time.

Figure 1 displays examples of the collected results. Figure 1A represents the primary and recovery shear stress responses of a 25 °C, 5.5 vol.% gel for various resting times. The maximum stress of the recovery fracture increased at greater resting times as shown in Figure 1B. At a resting time of 5 minutes, 45% of the primary strength was recovered while after a resting time of 45 minutes, the strength recovered was 79%. The magnitude of strength recovered for a given resting time was also dependent on the temperature of the gel, as shown in Figure 1B.

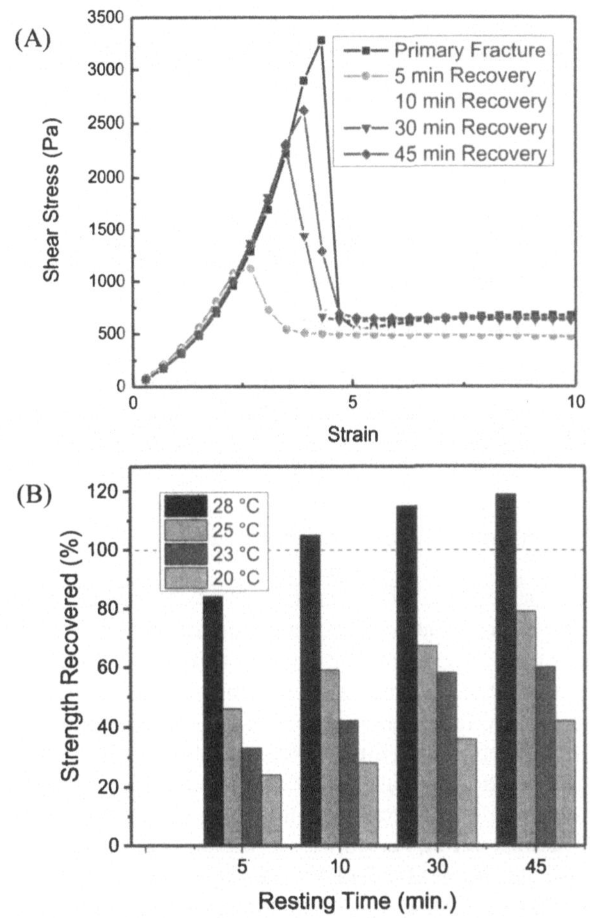

Figure 1. (A) Primary and 5-45 min. recovery stress-strain curves of 5.5 vol.% gel at 25 °C. (B) Recovery kinetics for a 5.5 vol.% gel at different temperatures. Adapted from Ref. [13].

The time to achieve 100% strength recovery was determined from power law fitting of experimental data. These 'full recovery' times were determined for concentrations of 5, 5.5, and 6 vol.% gel and temperatures of 20, 23, 25, and 28 °C. These values are presented in Figure 2 in comparison with the small-strain, characteristic relaxation time values of the different gels as determined from stress relaxation experiments. In addition to the similar Arrhenius behavior and activations energies in the range of 300-500 kJ/mol, the most important observation from the experimental data was the complete separation of timescales, with the characteristic relaxation time of the gels ranging from a few seconds to a few minutes and the full recovery times ranging from tens of minutes to tens of hours. The mechanism by which the gel heals and recovers its elastic strength was endblock re-association and re-formation of the endbock aggregates, which act as the physical crosslinks of the macromolecular network.

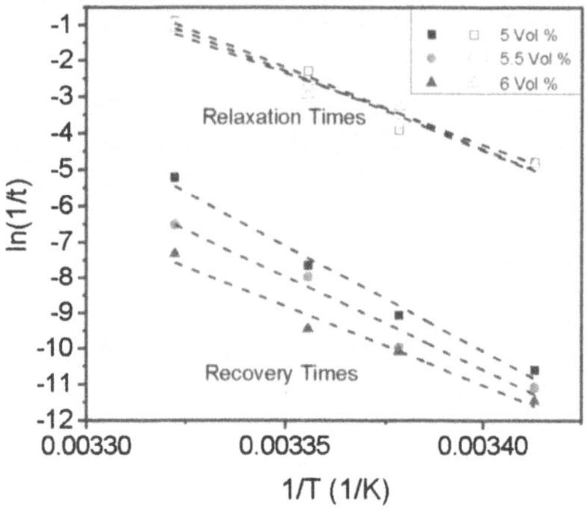

Figure 2. Arrhenius relationships of small-strain characteristic relaxation times and full recovery times at various temperatures and gel concentrations. Adapted from Ref. [13].

3.4 Hydrogel Superabsorbency

When superabsorbent polymer (SAP) hydrogels are used as internal curing agents in concrete, their role is to adsorb water during cement mixing and subsequently release the water within the concrete after placement. [15,16] As the concrete cures, the mixture water is consummed during the hydration reaction with unreacted portland cement, forming the hardened aggregate-cement composite material commonly known as concrete. Incorporation of hydrogel-based internal curing agents within the mixture ensures that water is available for the hydration reaction and is dispersed evenly throughout the volume of the concrete, promoting even curing. The incorporation of SAP also leads to a reduction in autogenous shrinkage of the cement and the reduced formation of shrinkage-induced microcracks within the system, thus creating a stronger, more durable final product. [4] Therefore, similar to how injectable gels can be used as drug delievery agents in medicine, hydrogels used as internal curing agents can be thought of as similarly beneficial transport agents in concrete, delivering the water that is required to fuel the curing reaction.

SAP hydrogels are comprised of a chemically crosslinked polyelectrolyte network that is swollen with water during immersion. The overall anion concentration within the network, controlled by the chemical composition of the polymer, directly controls the driving force for swelling. In applications involving concrete, the absorbed water will typically contain a variety of ions derived from the cement hydration reaction. Unfortunately, these ions can deleteriously affect the swelling behavior of the polymer hydrogel. [17] The research summarized here employed model SAP hydrogels with varying anionic concentrations to systemmatically quantify the magnitude of SAP-ion interactions. [14]

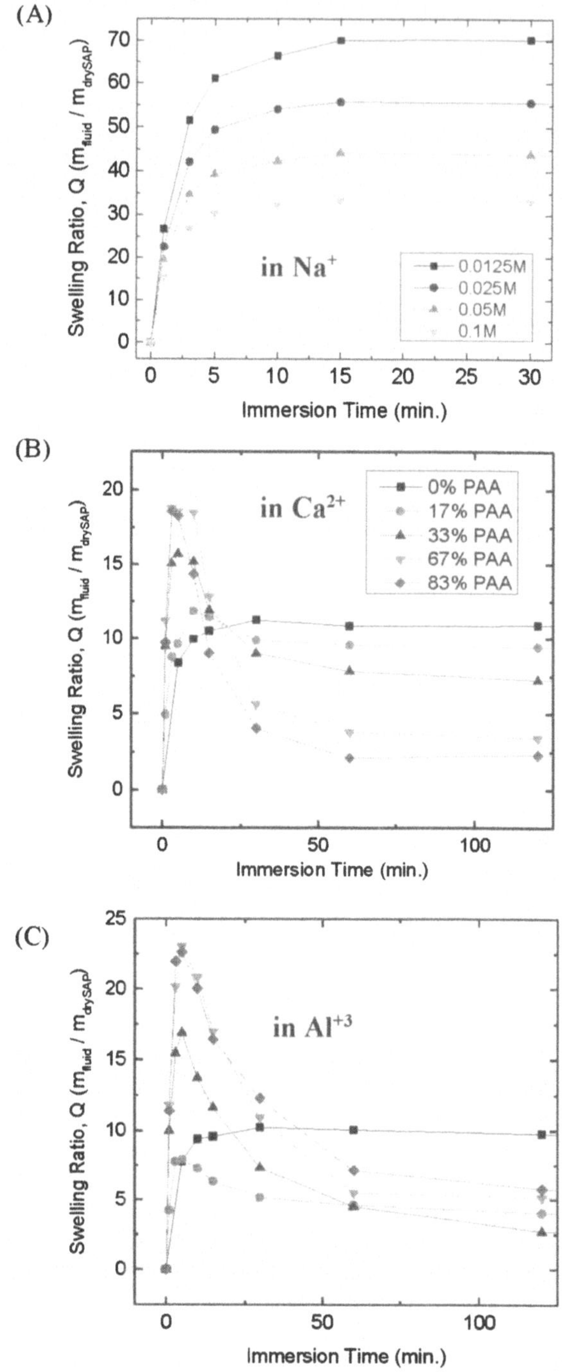

Figure 3. (A) Swelling response of 33% PAA SAP hydrogels immersed in different concentration solutions of Na^{+}. Swelling and rapid deswelling response of SAP hydrogels immersed in (B) Ca^{2+} solution and (C) Al^{3+} solutions. Adapted from Ref. [14]

TechConnect Briefs 2015, TechConnect.org, ISBN 978-1-4987-4729-5

In the experiments, chemically crosslinked gels created from neutralized poly(acrylic acid-polyacrylamide) copolymer were swollen in different aqueous environments. As shown in Figure 3, the overall equilibrium swelling capacity and the swelling kinetics were strongly dependent on the concentration and type of salt ions in solution. In the presence of multivalent cations, such as calcium (Ca^{2+}) and aluminum (Al^{3+}), the swelling capacity was strongly reduced and a strong desweling behavior was observed after only 5-10 minutes of immersion. Additionally, the presence of Al^{3+} resulted in the unexpected formation of a mechanically stiff barrier layer at the hydrogel's surface, which hindered the release of fluid and caused the overall elastic modulus of the hydrogel to increase from ~10 kPa for hydrogels immersed in Ca^{2+} solutions to ~100 kPa for hydrogels immersed in Al^{3+} solutions.

The impact of ions on the SAP behavior was found to be stronger for SAP hydrogels which contained the highest concentraion of anions in their polymer network (which resulted from deprotonation of the acrylic acid monomer during immersion). [14] It was believed that the reduced swelling capacity and fast deswelling was due to ionic crosslinking taking place within the hydrogels due to complexation of neighboring anionic groups in the polymer network by the multivalent cations in the absorbed fluid. This observed ionic sensitivity of SAP would most likely lead to an overall performance reduction when SAP-based hydrogels are used as internal curing agents in concrete. Current work at Purdue is focusing on the bottom-up synthetic design of SAP hydrogel systems that are more ionically resistant and incorporate additional inorganic phases within the network that will act to further enhance the cement curing reaction and lead to the formation of stronger, more durable concrete.

4 CONCLUSIONS

Controlled rheometry and swelling experiments of model physically and chemically crosslinked polymeric gels are useful to accurately and consistently quantify the nonlinear properties of these soft materials, including the gel's strain-stiffening response, fracture toughness, internal sliding friction, self-healing abilities, mechanical strength, and equilibrium swelling capacity and kinetics. By studying gels that are custom synthesized in house, complete control can be had over the physical and chemical parameters of the polymeric network, which can allow for the gel's properties and performance in specific applications (e.g., cement curing) to be directly linked to the synthetic structure of the gel. In the future, this will allow for the bottom-up design and development of synthetic strategies to create gel systems with properties that are fully optimized for their required applications.

REFERENCES

[1] A. S. Hoffman, Adv. Drug Deliv. Rev. **64**, 18 (2012).

[2] N. A. Peppas, K. B. Keys, M. Torres-Lugo, and a M. Lowman, J. Control. Release **62**, 81 (1999).

[3] L. Zha, B. Banik, and F. Alexis, Soft Matter **7**, 5908 (2011).

[4] V. Mechtcherine, M. Gorges, C. Schroefl, A. Assmann, W. Brameshuber, A. B. Ribeiro, D. Cusson, J. Custódio, E. F. Silva, K. Ichimiya, S. Igarashi, A. Klemm, K. Kovler, A. N. Mendonça Lopes, P. Lura, V. T. Nguyen, H.-W. Reinhardt, R. D. T. Filho, J. Weiss, M. Wyrzykowski, G. Ye, and S. Zhutovsky, Mater. Struct. **47**, 541 (2013).

[5] O. M. Jensen and P. F. Hansen, Cem. Concr. Res. **32**, 973 (2002).

[6] O. M. Jensen, Concr. Int. **35**, 48 (2013).

[7] D. J. Overstreet, D. Dutta, S. E. Stabenfeldt, and B. L. Vernon, J. Polym. Sci. Part B Polym. Phys. **50**, 881 (2012).

[8] C. Yan, A. Altunbas, T. Yucel, R. P. Nagarkar, and P. Schneiderk, Soft Matter **6**, 5143 (2010).

[9] B. D. Olsen, J. A. Kornfield, and D. A. Tirrell, Macromolecules **43**, 9094 (2010).

[10] K. A. Erk, K. J. Henderson, and K. R. Shull, Biomacromolecules **11**, 1358 (2010).

[11] K. A. Erk and K. R. Shull, Macromolecules **44**, 932 (2011).

[12] K. A. Erk, J. D. Martin, Y. T. Hu, and K. R. Shull, Langmuir **28**, 4472 (2012).

[13] T. Thornell, B. Helfrect, S. Mullen, A. Bawiskar, and K. A. Erk, ACS Macro Lett. **3**, 1069 (2014).

[14] Q. Zhu, C. W. Barney, and K. A. Erk, Mater. Struct. **In Press**, doi: 10.1617/s11527 (2014).

[15] S. Friedrich, in *Appl. Superabsorbent Polym. Concr. Constr.*, edited by V. Mechtcherine and H.-W. Reinhardt (RILEM State of the Art Reports 2, 2012), pp. 13–19.

[16] H. W. Reinhardt, A. Assman, and S. Monning, in *Int. Conf. Microstruct. Relat. Durab. Cem. Compos.*, edited by W. Sun, K. van Breugel, C. Miao, G. Ye, and H. Chen (2008), pp. 313–322.

[17] C. Schröfl, V. Mechtcherine, and M. Gorges, Cem. Concr. Res. **42**, 865 (2012).

Simulation of the suspension structure for the improvement of colloidal processes

M. Cerbelaud*, A. Videcoq*, B. Lestriez**, D. Guyomard**, R. Ferrando***

* SPCTS, UMR 7315, ENSCI, CNRS, CEC, Limoges, France, manuella.cerbelaud@unilim.fr
** Institut des Matériaux Jean Rouxel (IMN), Université de Nantes, CNRS, Nantes, France
Réseau sur le Stockage Electrochimique de l'Energie (RS2E), FR CNRS 3459, France
*** Dipartimento di Fisica and CNR-IMEM, Genoa, Italy

ABSTRACT

Colloidal processes are used in many fields of applications to produce multi-material components with specific microstructures. Here, controlling the structure of colloidal suspensions is crucial in order to reliably adapt the suspension rheology to the process and to get the desired microstructure in the final part. We propose to show that numerical simulations, which take into account the discrete nature of the colloidal components, may be of help in this goal. Indeed, we develop and use simulation codes that can be employed to control the suspension stability and the suspension structure: particle ordering, compactness of the aggregate network, percolation properties, etc. We will first explain how a colloidal suspension can be modeled and then illustrate the topic with the specific case of a model alumina-silica system, in which simulation results will be compared to experimental characterizations. We will show that the simulations can well describe the experimental structures of suspensions. Finally, we will present how this kind of computational study can be applied to the complex suspensions of composite electrodes for Li-ion batteries.

Keywords: simulations, colloidal suspensions

1 INTRODUCTION

The colloidal route is commonly used in technological processes. A lot of ceramics shaping processes use suspensions. This is the case for example of the processes of casting, of stereolithography, of inkjet printing... For all these shaping processes, the microstructure of the final shaped materials, which determines the final properties of the body (mechanical, electrical, optical...), depends on the suspension behavior used in the process. It is indeed crucial to understand in details this behavior to improve the process and by consequence the final properties of the shaped materials. To achieve this goal, we develop and use Brownian dynamics simulations, which take into account the discrete nature of colloids. These simulations can be employed to understand and to control the suspension stability and the suspension structure. In the following, this method of simulation will be described and applications to two different kinds of suspensions will be presented.

2 BROWNIAN DYNAMICS SIMULATIONS

Brownian dynamics simulations are used to simulate the behavior of colloidal suspensions. In these simulations, the fluid is considered as a continuum medium and its effects on the colloids are taken into account through frictional and random forces. The motion of a colloid i is then described by the Langevin equation:

$$m_i \frac{dv_i(t)}{dt} = -\zeta_i v_i(t) + \sum_j F_{ij}(r_{ij}(t)) + \Gamma_i(t) \quad (1)$$

where m_i and v_i are respectively the mass and the velocity of the colloid i. ζ_i is the friction coefficient which depends on the viscosity of the solvent η and the radius of the colloid a_i ($\zeta_i = 6\pi\eta a_i$). $F_{ij}(r_{ij}(t))$ denotes the pair interaction force between the colloids i and j separated by the center-to-center distance r_{ij}. These forces derive generally from pair interaction potentials ($F_{ij}(r_{ij}(t)) = -\nabla U_{ij}(r_{ij}(t))$). $\Gamma_i(t)$ is the random force. Brownian dynamics simulations consist in integrating the Langevin equation numerically and therefore they describe the evolution of the colloid positions as a function of time.

3 APPLICATIONS

3.1 Suspensions of oxide particles for a wet granulation process

We have developed and used Brownian dynamics simulations in the past to study the aggregation process in aqueous suspensions composed of alumina and silica particles, where heteroaggregation occurs. Indeed, the silica is negatively charged and the alumina is positively charged. This study was based on a real system used in a wet granulation shaping process. A more detailed description of this study can be found in Ref. [1, 2, 3]. Here we just focus on one of the important results. In the simulations, the interactions between the particles were modeled by DLVO potentials whose parameters were extracted from experimental measurements. By

comparing simulation results with experimental characterizations, we have shown that the Brownian dynamics simulations can well describe the structure of suspensions in terms of adsorption of small silica particles onto larger alumina particles; aggregation of alumina particles with small silica particles making 'bridges' in between adjacent alumina particles; spatial distribution of the two types of particles as a function of mass and size ratio between the two components. Thus we have shown that the simulations help predicting the structure appearing in such suspensions.
Simulations also allowed us to understand the final structure of the granules obtained from different suspensions. It is indeed observed that when granulation is performed with suspensions where silica and alumina particles have similar sizes, after drying and sintering steps, the granules have a hollow core [3]. However, when silica and alumina particles have a high size ratio, the granules are full and compact. According to the simulation results, this behavior is attributed to the difference of porosity in the aggregates. When silica and alumina particles have similar sizes, the aggregates are made of an intercalation between alumina and silica particles, because similar particles are repelling each other. This guarantees highly porous structures. However, when silica particles are much smaller than the alumina ones, they are responsible of the aggregation of the alumina particles by bridging, but there is much less porosity inside the aggregates. Brownian dynamics simulations appear as a useful tool to understand and predict the qualitative porosity of the granules obtained by this granulation process.
Because of the sphericity of colloids and the well known interactions between the components used in this study, the system can be considered as model. In the following, we will present how Brownian dynamics simulations can be used in much complex suspensions of electrodes for Li-ion batteries.

3.2 Suspensions of electrodes for Li-ion batteries

Electrodes of Li-ion batteries are composite materials generally obtained by tape-casting or extrusion. They are often composed of active materials, conductive additives and polymers. During the last years, it has been shown that the electrochemical performances of Lithium batteries are intimately linked to the structure of the electrodes which depends on the behavior of the suspensions used during the shaping process [4]. It is thus crucial to well understand this behavior to better control the process and the structure of electrodes in order to improve the efficiency of the battery. To achieve good electrochemical performances, different conditions on the structure of electrodes have to be fulfilled:
- a percolating network of the conductive additives has to be built up in order to ensure the transport of the electrons within the electrodes if the active material is not conductive. If the active material is conductive, the percolating network is not mandatory however it remains generally helpful to facilitate the electronic conductivity.
- a high number of contacts between the active material and the conductive network is generally preferable to better ensure the transfer of electrons between the active mass and the conductive additives.

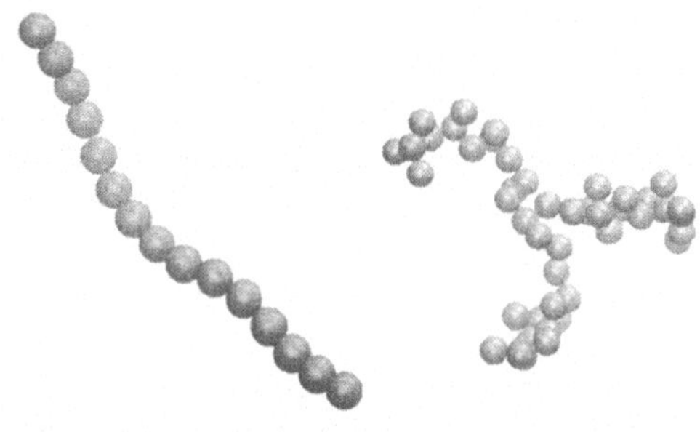

(a) (b)

Figure 1: (a) Example of one carbon fiber used in the simulations and (b) example of one non-deformable carbon black aggregate used in simulations. For visibility the images are not represented at the same scale.

A lot of parameters can influence the structure of suspensions for electrodes (pH, size of components, temperature, additives...) and by consequence on the structure of the final electrodes. Simulations that take into account the discrete nature of particles in suspensions, appear in this case as a very useful tool to understand the role of each parameter. However, these simulations are actually few developed for these systems because of the complexity of the suspension modeling. In the following, we will present how simple Brownian dynamics simulations can help to understand the role of the conductive additive shape on the structure of the suspension.
Description of the simulations: Simulations are performed for suspensions composed of lithium titanate particles (LTO) of 500 nm ($\rho_L = 4000\,\mathrm{kg\,m^{-3}}$) and carbon conductive additives ($\rho_c = 2000\,\mathrm{kg\,m^{-3}}$). Two kinds of non spherical additives are considered : carbon black aggregates (CB) and carbon fibers (CF). Carbon fibers are elongated entities and carbon blacks are fractal aggregates composed of fused elementary particles. In the simulations, LTO particles will be represented as spherical particles. Carbon fibers will be modeled as a chain of 15 spherical particles of 100 nm linked together by an elastic potential (see Figure 1a). In order to avoid large deformations of the fibers which would not be realistic, a linear repulsive potential is added between non-

adjacent spheres when the angle between adjacent pairs of spheres exceeds 10°. Carbon blacks are modeled as non-deformable fractal aggregates composed of elementary particles of 20 nm, which do not interact together (see Figure 1b). The carbon aggregates are moved as a whole. For that, the Langevin equation is modified to introduce the diffusion coefficient of the global aggregate. Because of its non-deformable shape, rotation is also introduced for carbon black aggregates. More details on this modeling can be found in Ref [5, 6]. The polymeric components are implicitly included in the solvent and in the interaction forces. In this study, all the interactions between the components are considered as attractive and are modeled by a generalized short-range Lennard-Jones potential:

$$U_{ij} = 4\epsilon_{ij} k_B T \left[\left(\frac{a_i + a_j}{r_{ij}} \right)^{36} - \left(\frac{a_i + a_j}{r_{ij}} \right)^{18} \right] \quad (2)$$

where k_B is the constant of Boltzmann, T the temperature and ϵ_{ij} is the potential well depth in units of $k_B T$. ϵ_{ij} is fixed at 7 for the interactions between two LTO particles and for the interactions between two carbon components (fibers or carbon blacks). ϵ_{ij} is fixed at 14, for the interactions between LTO and a carbon component. Simulations are performed at 298 K for a solvent of viscosity $\eta = 10^{-3}$ Pa s. A cubic box with periodic boundary conditions is used. The box size depends on the solid volume fraction ϕ present in the system (here $\phi = 0.2$). The time step is fixed at $\delta t = 5.10^{-9}$ s. All the simulations are performed with 14 LTO particles. Results are analyzed at $t = 0.05$ s.

Results and discussion: First, simulations are performed without carbon fibers. Two compositions are considered: one with 5w% of CB and a second with 10w% of CB. For each composition, 10 independent simulations are carried out and in the following the results are averaged over all the independent simulations. A snapshot of simulation for 5w% of carbon black is shown in Figure 2. At the end of simulations, all the components are mixed in the simulation box and aggregated. As already mentioned, two structural properties appear as primordial for the application. First, the carbon network must percolate and then the number of contacts between the carbon network and the active materials must be high. Since the Brownian dynamics simulations give the positions of the components in the box, they allow to study these properties. For each simulation, the number of contacts between the carbon network and the LTO particles is computed and the dimensionality of percolation of carbon network is determined as described in Ref. [7]. The dimensionality of percolation determines how the carbon network spans in the simulation box. A dimensionality of percolation equal to 1 means that the carbon network spans in the box from end to end in 1 dimension and a dimensionality of

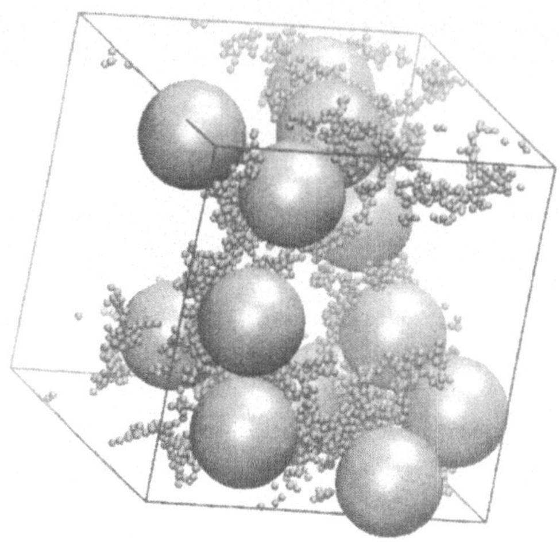

Figure 2: Simulation snapshot at $t = 0.05$ s for a suspension composed of 14 LTO particles and 5w% of carbon black aggregates ($\phi = 0.2$). CB is represented in orange and LTO in cyan.

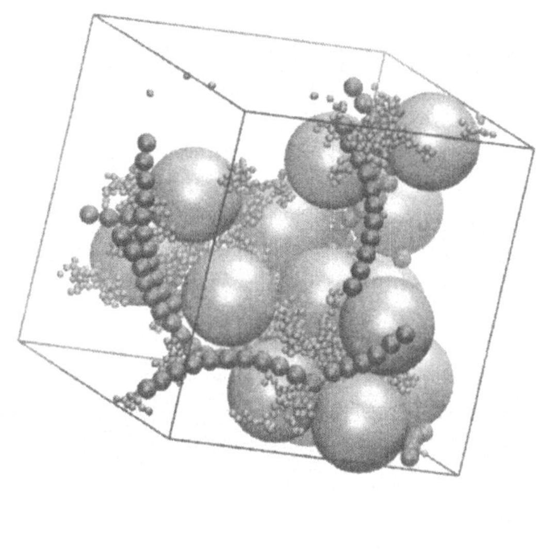

Figure 3: Simulation snapshot at $t = 0.05$ s for a suspension composed of 14 LTO particles, 2.5w% of carbon black aggregates and 2.5w% of carbon fibers ($\phi = 0.2$). CB is represented in orange, CF in red and LTO in cyan.

Table 1: Results of simulations at $t = 0.05$ s. ns is the averaged number of contacts between LTO and carbon additives and D_{perc} is the averaged dimensionality of percolation. (Averages obtained on 10 simulations)

Compositions	D_{perc}	ns
5w% of CB	0.6	1067
10w% of CB	2.9	1187
2.5w% of CB + 2.5%w of CF	1.6	689

percolation of 2 or 3 means that the network spans in 2 or 3 dimensions in the box respectively. Results obtained for the simulations at $t = 0.05$ s are reported in Table 1. An increase of the carbon content in the suspension increases the dimensionality of percolation and also the number of contacts between the LTO particles and the carbon network, which shows that increasing the content of CB can improve the structural properties. Nevertheless, it is noteworthy that the mass of additives must remain low in order to not decrease too much the mass loading of the electrodes.

Actually, formulation of electrodes for Lithium battery use carbon additives with a higher shape anisotropy than CB (carbon nanotubes, carbon fibers...) [8]. In a second part of the numerical study, carbon fibers are thus added in Brownian dynamics simulations. Simulations are performed with a total amount of 5w% of carbon, with half of the carbon blacks replaced by carbon fibers (2.5w% of CB + 2.5w% of CF). A snapshot of simulation at $t = 0.05$ s is shown in Figure 3. As previously, all the components are aggregated. The number of contacts between the LTO particles and the carbon network (ie network composed of CF and CB) are reported in Table 1 as well as the dimensionality of percolation of the carbon network. The presence of CF increases the dimensionality of percolation, which should result in a better electronic wiring in the suspension. However the number of contacts between LTO particles and the carbon network decreases. These results are in agreement with previous experimental observations, which have shown that introduction of fibers in the formulation of a carbon coated Li-$FePO_4$ based composite electrodes leads to a better electronic wiring when the porosity is not optimized (case of electrodes before calendaring step) [8].

This example shows that even if the Brownian dynamics simulations used here are quite simple, they can be used to understand the role of the shape of conductive additives on the suspension structure. They highlight that the introduction of fibers in the suspensions enhanced the dimensionality of percolation of the carbon network.

4 CONCLUSION

Brownian dynamics simulations have been developed and applied to two different kinds of suspensions. First simulations have been applied to a model suspension composed of oxide spherical particles and then to more complex suspensions of electrodes for Li-ion batteries where some colloids are not spherical. They have shown that even if they are simple, they can be helpful to understand the structure of the suspensions used during the shaping process and therefore to understand the microstructure of the final object (granules or electrodes respectively). To conclude, such simulations appear as a very interesting tool to improve the colloidal processes.

Acknowledgements

The authors thank CALI and its team for computing facility. Figures 1, 2 and 3 have been obtained by VMD, a molecular graphics program developed by the Theoretical Biophysics Group in the Beckman Institute for Advanced Science and Technology at the University of Illinois at Urbana-Champaign [9].

REFERENCES

[1] M. Cerbelaud, A. Videcoq, P. Abélard, C. Pagnoux, F. Rossignol and R. Ferrando, Langmuir, 24, 3001, 2008.

[2] M. Cerbelaud, A. Videcoq, P. Abélard and R. Ferrando, J. Colloid Interface Sci., 332, 360, 2009.

[3] M. Cerbelaud, A. Videcoq, P. Abélard, C. Pagnoux, F. Rossignol and R. Ferrando, Soft Matter, 6, 370, 2010.

[4] D. Guy, B. Lestriez and D. Guyomard, Advanced Materials 2004, 16, 554-557

[5] M. Cerbelaud, B. Lestriez, R. Ferrando, A. Videcoq, M. Richard-Plouet, M.T. Caldes and D. Guyomard, Langmuir, 30, 2660, 2014

[6] M. Cerbelaud, B. Lestriez, A. Videcoq, R. Ferrando and D. Guyomard, 'Understanding the structure of electrodes suspensions for Li-ion batteries: a numerical study', submitted.

[7] M. Piechowiak, A. Videcoq, R. Ferrando, D. Bochicchio, C. Pagnoux and F. Rossignol, Phys. Chem. Chem. Phys., 14, 1431, 2012

[8] C. Fongy, S. Jouanneau, D. Guyomard and B. Lestriez, J. Power Sources, 196, 8494, 2011

[9] W. Humphrey, A. Dalke and K. Schulten, J. Mol. Graphics, 14, 33, 1996.

SRD-MD modelling of hydrodynamic interactions in colloidal systems: playing between computational cost and resolution

A. M. K. Laganapan*, A. Videcoq*, M. Cerbelaud*, M. Mouas*, M. Bienia*, and R. Ferrando**

* SPCTS, UMR 7315, ENSCI, CNRS, Centre Européen de la Céramique
12 rue Atlantis, 87068 Limoges cedex France
** Dipartimento di Fisica and CNR-IMEM, via Dodecaneso 33, Genoa I-16146, Italy
aleena.laganapan@etu.unilim.fr

ABSTRACT

A ubiquitous problem that has always been a subject of intensive research is the role played by hydrodynamic interactions (HIs) in several fundamental problems in colloid rheology and aggregation kinetics. However, the high computational demand that is often associated with HIs modelling presents a huge challenge. Even with the computer power that is available today, a full molecular dynamics simulation is still unsustainable. Meanwhile, there is a diverse number of technological applications that call for a means to address this limitation. To that end, we use a hybrid Stochastic Rotation Dynamics - Molecular Dynamics (SRD-MD) method, which can provide a compromise between computational efficiency and molecular dynamics resolution.

We present our SRD-MD simulation results in two parts. The first one is modelling of the rheological behaviour of colloidal suspensions. We successfully reproduce, both qualitatively and quantitatively, the divergence of the suspension viscosity from linear behaviour when HIs are properly included. The second part deals with the aggregation kinetics of colloidal suspensions. We show that HIs speed up the aggregation process of colloids and slow down the reorganization of the clusters formed. Consequently, we demonstrate that the probability of percolation is higher for systems with HIs in comparison with systems without.

Keywords: simulations, rheology, shear viscosity, percolation,colloidal suspension

1 INTRODUCTION

The significant progress in numerical methods over the recent years has provided a better understanding of the various applications of colloids in natural and industrial settings [1]. Nevertheless, due to the increasing complexity of materials that are being discovered everyday, there is a considerable effort devoted to the development of advanced numerical methods that can keep up with technology's rising demands.

A straightforward technique that can be used to obtain the dynamic properties of many-particle system is Molecular Dynamics (MD). However, there is an extensive class of complex fluids where the timescale separation, between the motion of the embedded solid and the underlying solvent, presents a challenging task. Since this can be a serious problem with MD, several coarse-grained models have been proposed to resolve this issue. Examples are Brownian Dynamics (BD), BD with Yamakawa-Rotne-Prager tensors (BD-YRP), Dissipative Particle Dynamics (DPD) and Lattice Boltzmann (LB) technique. Among them, the most popular method is BD, mostly because of its simplicity. This is achieved by representing the effects of the solvent particles on the solute by a combination of random forces and frictional terms [2]. In reality however, hydrodynamic forces are omnipresent in complex fluids and while the extension to BD-YRP incorporates hydrodynamic interactions (HIs), its validity is constrained to relatively dilute suspensions. This is because the computational cost of BD-YRP increases drastically with the number of colloid particles N i.e. it requires a tensor evaluation that scales as $\mathcal{O}(N^2)$ and diagonalization that scales as $\mathcal{O}(N^3)$ [3, 4]. Alternatively, HIs can be solved using particle-based or lattice-based methods. Among the widely-celebrated models are DPD and LB. In DPD, the system is completely defined by specifying all the positions and momenta of the solvent particles making it almost similar to MD. The difference is that instead of using an explicit force term, it uses a weight function to define the collision rule [5]. In LB, a linearized and preaveraged Boltzmann equation is discretized and solved on a lattice. Here, the position and velocity probability density functions of each of the nodes are calculated every time step [6, 7].

A versatile model should include all the relevant aspects of the dynamics of the system, i.e. thermal fluctuations and HIs, and still be computationally reasonable to be of value in predicting transport properties, testing physical theories and offering insights to experiments. To this end, we use a hybrid Stochastic Rotation Dynamics - Molecular Dynamics (SRD-MD) method that is first introduced by Malevanets and Kapral over a decade ago [8]. While it can be argued that the cost of computation still depends on the system being modeled, the main advantage of SRD-MD over the previously mentioned methods is that it is generally simpler and faster. Ref. [9] referred to it as the "Ising model" of hydrodynamics. Moreover, its simplified dynamics has allowed the analytic calculation of several transport coefficients [10, 11]. In comparison to the other methods mentioned above, SRD-MD is a relatively new technique so that most of the work are still in the development stage. But due to its promising potential as a simulation model for complex fluids, it has been gaining a lot of attention and is continuously being developed. In this study, we show that despite the simplicity of SRD-MD, the fundamental rheological and structural behaviors of colloidal systems are still reproduced.

2 SRD-MD

Simply put, the fluid dynamics is treated using SRD and the colloid dynamics is treated using MD. SRD consists of a two-step process that involves the streaming and collision of fluid particles. In the streaming step, the positions of the particles evolve according to [8]:

$$\mathbf{r}_i(t+\Delta t_{\mathrm{SRD}}) = \mathbf{r}_i(t) + \mathbf{v}_i(t)\Delta t_{\mathrm{SRD}}, \qquad (1)$$

where $\mathbf{r}_i$ is the particle position, $\mathbf{v}_i$ is the particle velocity and Δt_{SRD} is the SRD time step. In the collision step, the coarse graining is done by dividing the simulation box of side length L into manageable collision cells, where all the particles in the same cell undergoes the same rotation:

$$\mathbf{v}_i(t+\Delta t_{\mathrm{SRD}}) = \mathbf{v}_{\mathrm{cm}} + \mathbf{R}(\alpha)\left[\mathbf{v}_i(t) - \mathbf{v}_{\mathrm{cm}}\right], \qquad (2)$$

where $\mathbf{v}_{\mathrm{cm}}$ is the average velocity of the cell and $\mathbf{R}(\alpha)$ is a rotation matrix. The SRD parameters that define the dynamics of the fluid are the cell size a_0, which we choose to be half the radius $a = 300$ nm of the colloid that is embedded in the fluid, the rotation angle $\alpha = 90°$, the dimensionless mean-free path $\lambda = 0.1$, the time step $\Delta t_{\mathrm{SRD}} = 7.37 \times 10^{-4}$ s and the mass of the individual fluid particles $m_f = 6.75 \times 10^{-19}$ kg. The SRD parameters used in this study are the same as in Refs. [12, 13], where a more detailed discussion can be found.

The coupling of embedded colloids in the fluid can be done in two ways. The first is to couple by SRD, where the colloids are treated as point particles and are rotated along with the fluids in the SRD collision steps [14]. The second one is to couple by MD, where the the colloids have a defined interaction radius σ and are evolved along with the colloids during MD steps [15]. In the MD procedure, the colloid-fluid and colloid-colloid interactions obey Newton's equations of motion. The potentials used are of the form:

$$V_{cf} = \begin{cases} \varepsilon\left(\frac{\sigma}{r}\right)^{12} & (r < r_c) \\ 0 & (r > r_c), \end{cases} \qquad (3)$$

for the colloid-fluid interaction; and

$$V_{cc} = \begin{cases} \varepsilon\left(\frac{\sigma'}{r}\right)^{12} & (r < r_c') \\ 0 & (r > r_c'), \end{cases} \qquad (4)$$

for the colloid-colloid interaction. In Eqs. 3 and 4, $\sigma = 0.8a$, $\sigma' = 2a$, $\varepsilon = 2.5k_BT$, $r_c = 2.5\sigma$ and $r_c' = 2.5\sigma'$.

3 SHEAR VISCOSITY

When a system is undergoing a linear perturbation, the viscosity can be obtained by using $\eta = \sigma_{yx}/\dot{\gamma}$. To introduce shear, we use Lees-Edwards boundary conditions (LEBC) so that the particles in the simulation box follow a planar Couette flow [1]. Here, we demonstrate the method at a shear rate of $\dot{\gamma} = 50\ \mathrm{s}^{-1}$ (Pe ≈ 6), where a linear velocity gradient is produced after 3 s of waiting time. For the calculation of σ_{yx}, we adapt the stress tensor provided in Ref. [16] for our system:

$$\sigma^{\mathrm{fluid}}_{\mathrm{kin},yx} = -\frac{m_f}{V_f}\sum_{i=1}^{N_f}\hat{v}'_{iy}\hat{v}_{ix} - \frac{m_f\dot{\gamma}\Delta t_{\mathrm{SRD}}}{2V_f}\sum_{i=1}^{N_f}v_{ix}^2, \qquad (5)$$

$$\sigma^{\mathrm{fluid}}_{\mathrm{col},yx} = -\frac{1}{V_f\Delta t_{\mathrm{SRD}}}\sum_{i=1}^{N_f}\Delta p_{iy}r'_{ix}, \qquad (6)$$

$$\sigma^{\mathrm{colloid}}_{yx} = -\frac{M_c}{V}\sum_{i=1}^{N_c}\hat{v}'_{iy}\hat{v}_{ix} - \frac{M_c\dot{\gamma}\Delta t_{\mathrm{SRD}}}{2V}\sum_{i=1}^{N_c}v_{ix}^2 - \frac{1}{V}\sum_{i=1}^{N_c}F_{iy}r'_{ix}, \qquad (7)$$

where $V = L^3$, $N_f = 163840$ is the number of fluid particles, $M_c = 2.49 \times 10^{-16}$ kg is the mass of the colloid and the shorthand notations $\hat{v} = v(t - \Delta t)$ and $\Delta p_{iy} = m_f(v_{iy} - \hat{v}_{iy})$ are used [12]. F_{iy} is obtained from $\partial\left(V_{cc,i} + V_{cf,i}\right)/\partial r_y$. The apostrophe symbol (′) is for quantities measured after LEBC are imposed.

Because of the coupling used, the free volume accessible to the fluid is now defined as $V_f = L^3 - N_c\frac{4}{3}\pi\sigma^3$, where $L = 32a_0$ and N_c is the number of embedded colloids. Consequently, we define the volume fraction occupied by the colloids as $\phi = (N_c/L)4\pi\sigma^3/3$. Note that because of our definition of ϕ, we use σ as our best approximation of the hydrodynamic radius of the colloids.

The continuous pumping of energy from shearing increases the kinetic energy of the system. To ensure that the kinetic energy is conserved, a Monte-Carlo thermostat [17] is also implemented.

Finally, the total stress tensor of the suspension is calculated from the sum of the contributions from the embedded colloids and the suspending fluid:

$$\sigma_{yx} = \sigma^{\mathrm{fluid}}_{\mathrm{kin},yx} + \sigma^{\mathrm{fluid}}_{\mathrm{col},yx} + \sigma^{\mathrm{colloid}}_{yx}. \qquad (8)$$

The results are shown in Fig. 3 and are superimposed with the results of Refs. [18, 19]. We are able to reproduce the divergence from Einstein's linear equation $\eta_R = 1 + 2.5\phi$, which is only valid for dilute suspensions.

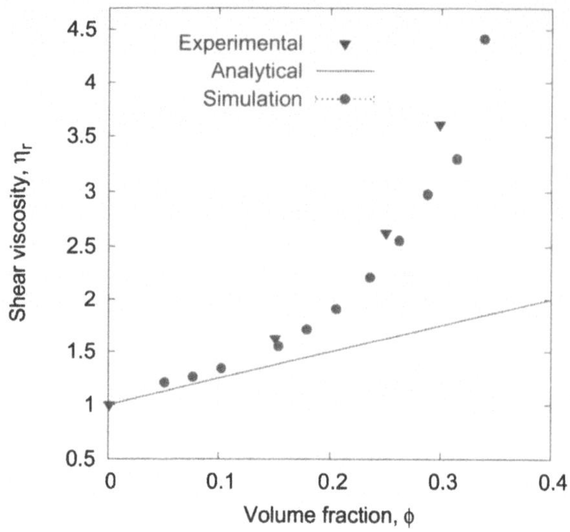

Figure 1: Shear viscosity as a function of volume fraction for $\dot{\gamma} = 50\ \mathrm{s}^{-1}$ superimposed with the experimental results of Ref. [19] and Einstein's equation $\eta_R = 1{+}2.5\phi$ [18].

Moreover, we are able to demonstrate the SRD-MD calculation of shear viscosities even for the concentrated cases.

4 AGGREGATION KINETICS

To compare the differences in the aggregation kinetics of colloidal systems, we performed two sets of simulations: BD for systems without HIs and SRD-MD for systems with HIs. For the colloids to be attractive, we substitute the V_{cc} in Eq. 4 by a generalized Lennard Jones potential of the form:

$$V_{cc} = \begin{cases} 4\varepsilon_{\mathrm{LJ}}\left[\left(\frac{\sigma'}{r}\right)^{36} - \left(\frac{\sigma'}{r}\right)^{18}\right] & (r < r'_{\mathrm{LJ}}) \\ 0 & (r > r'_{\mathrm{LJ}}), \end{cases} \tag{9}$$

where $\varepsilon_{\mathrm{LJ}} = 14k_BT$ and $r'_{\mathrm{LJ}} = 1.6\sigma'$.

Fig. 4 shows the aggregation kinetics of a 500-particle colloidal system in a simulation box of side length $L = 55a_0$ with a volume fraction of 10%. By considering an aggregate as an assembly of at least two colloids, we can quantify the aggregation kinetics by observing the temporal evolution of the number of aggregates (N_A) in the system. We divide the aggregation process in two stages. Stage I is determined by the increase in N_A due to the formation of small clusters from individual colloids. From Fig. 4, we can observe that the increase in N_A is faster for the system without HIs. This is attributed to the slow dimer formation observed in the SRD-MD simulation [20]. A maximum number of clusters is then achieved which marks the beginning of Stage II. N_A starts to decrease because the clusters coalesce

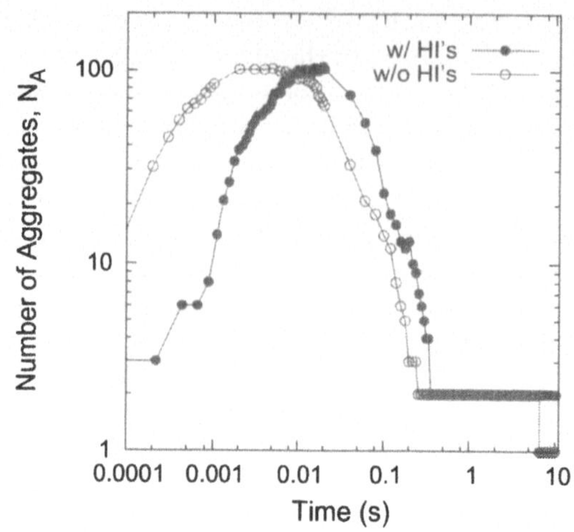

Figure 2: Comparison of the aggregation kinetics between the system with HIs simulated using SRD-MD and the system without HIs simulated using BD. The volume fraction used is 10%.

to form bigger aggregates. In Ref. [20], it was found that the diffusivity increases when HIs are taken into account. This therefore explains the faster coalescence of aggregates in Stage II for the case with HIs.

The two stages of aggregation kinetics both contribute to the final structure of the colloids. Hence from our knowledge of the aggregation process, we turn our attention to percolating networks. Several studies [21, 22] have suggested that the probability of percolation may increase due to the mere presence of hydrodynamic forces. Our results are shown in Fig. 3. This is a snapshot of the colloids at ≈ 0.5 s, where the biggest aggregate is shown in red. For the case with HIs, the red aggregate is connected in all 3 directions; while for the case without HIs, the red aggregate is connected in only 2 directions. The difference observed may be due to the rapid formation of big aggregates in Stage II, which does not provide enough time for the colloids to reorganize. Therefore, the reorganization process is another stage in the aggregation kinetics of colloids that needs to be considered. Because of this, we also note that the structures presented here can be considered as transient structures and may change given a longer time scale.

5 CONCLUSIONS

We have demonstrated that SRD-MD is an effective simulation technique for the study of the rheology and structure of colloidal suspensions. The calculation of shear viscosity as a function of volume fraction is one of the benchmarks for the effectiveness of the method in modelling hydrodynamics.

In our simulations, BD remains to be the fastest method for the study of aggregation kinetics. However

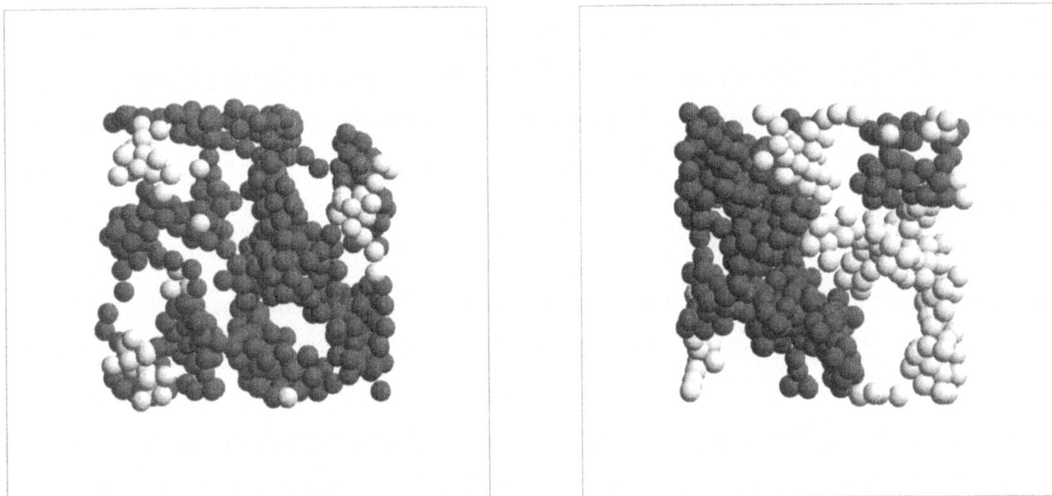

Figure 3: Snapshots at 0.5 s of colloids simulated using SRD-MD (left) and BD (right). The colored spheres corresponds to the biggest cluster present in the system.

in the system described here, HIs also affect the different stages of aggregation of colloids. Thus when HIs cannot be neglected, SRD-MD is an effective approach and can be potentially useful in the study of percolation.

While there is not a single simulation model that can encompass all kinds of systems, we were able to demonstrate that the relevant aspects of the dynamics of colloidal suspensions, especially HIs, are reproduced using SRD-MD despite the approximated fluid and colloid dynamics employed in the model.

Acknowledgements

A. K. L. acknowledges support from the Programme VINCI 2013 of the French-Italian University on the subject "Structuration of ceramic suspensions under constraints." The authors also acknowledge CALI and its team, where the calculations in this paper were carried out.

REFERENCES

[1] Allen and Tildesley, "Computer Simulation of Liquids," Oxford University Press, 1987.

[2] D. L. Ermak and A. J. McCammon, J. Chem. Phys. 69, 1352, 1978.

[3] J. Rotne and S. Prager, J. Chem. Phys. 50, 1969.

[4] H. Yamakawa, J. Chem. Phys. 53, 1970.

[5] P. J. Hoogerbrugge and J. M. V. A. Koelman, Europhys. Lett. 19, 1992.

[6] A. Ladd and R. Verberg, J. Stat. Phys. 104, 1191, 2001

[7] S. T. T. Ollila, C. Denniston, M. Karttunen, and T. Ala-Nissila, J. Chem. Phys. 134, 2011.

[8] A. Malevanets and R. Kapral, J. Chem. Phys. 110, 1999; 112 ,2000.

[9] G. Gompper, T. Ihle, D. M. Kroll, and R. G. Winkler, "Advance computer simulation ap- proaches for soft matter sciences iii," Springer, pp. 1-88 , 2009.

[10] J. T. Padding and A. A. Louis, Phys. Rev. E. 74, 2006.

[11] T. Ihle and D. M. Kroll, Phys. Rev. E. 67, 2003.

[12] A. M. K. Laganapan, A. Videcoq, M. Bienia, T. Ala-Nissila, D. Bochicchio and R. Ferrando, J. Chem. Phys. 142, 2015.

[13] A. Tomilov, A. Videcoq, T. Chartier, T. Ala-Nissila, and I. Vattulainen, J. Chem. Phys. 137, 2012.

[14] M. Hecht, J. Harting, M. Bier, J. Reinshagen, and H. Herrmann, Phys. Rev. E. 74, 2006.

[15] G. Batôt, V. Dahirel, G. Mériguet, A. Louis, and M. Jardat, Phys. Rev. E. 88, 2013.

[16] R. G. Winkler and C. C. Huang, J. Chem. Phys. 130, 2009.

[17] C. C. Huang, A. Varghese, G. Gompper, and R. G. Winkler, Phys. Rev. E. 91, 2015.

[18] A. Einstein, Ann. Phys. 19, 1906; 34, 1911.

[19] P. N. Segrè, S. P. Meeker, P. N. Pusey, and W. C. K. Poon, Phys. Rev. Lett. 75, 1995.

[20] A. Tomilov, A. Videcoq, M. Cerbelaud, M. A. Piechowiak, T. Chartier, T. Ala-Nissila, D. Bochicchio, and R. Ferrando, J. Phys. Chem. B 117, 2013.

[21] H. Tanaka and T. Araki, Phys. Rev. Lett. 85, 2000.

[22] J. Whitmer and E. Luijten, J. Phys. Chem. B. 115, 2011.

New Insights into Microstructure, Rheology & Molecular Level Structural Changes During Self-Assembly & Gelation in Complex Fluids Through Combined DLS-Optical Microrheology-Raman Spectroscopy

S. Amin, S. Blake, S. Kenyon & E. N. Lewis

Malvern Instruments, 7221 Lee Deforest Drive, Suite 300 Columbia Maryland, USA

Contact email: samiul.amin@malvern.com

Summary: Soft matter / complex fluid systems are ubiquitous across a range of industrial and consumer sectors with common examples being inks, paints, drilling fluids, cosmetics, personal care products and foodstuffs. In many instances, the final product format of these complex fluids are gels or soft solids and the processing and product functionality attributes of such materials are often dependent on their rheological response and viscosity. The rheology evolution in such complex fluid systems as a function of formulation parameters (e.g. pH, ionic strength) is intimately connected to corresponding changes in the micro- and mesostructure and intermolecular and intramolecular associations and interactions. Most insights developed into understanding the self-assembly and rheology evolution process in such systems has primarily focused on elucidating the associated micro- and mesostructural changes through various scattering (light, x-ray, neutron) and imaging techniques (cryo-TEM, SEM, AFM). Furthermore detailed insights into the associated chemical conformational/structural changes and various non-covalent interactions (e.g. H-bonds, hydrophobic interactions) leading to the self-assembly process have been very limited. An understanding of the changes in the molecular level structural changes as self-assembly and gelation progresses will provide new mechanistic insights that will allow better optimization and control over performance controlling formulation design rules in industrial complex fluids. In this talk, we provide new structural/interaction insights into the self-assembly and gelation mechanism of complex fluids through combination of a number of well-established analytical techniques namely dynamic light scattering (DLS), Raman spectroscopy and optical microrheology. The talk will illustrate the utility of the combination of mesoscale structure-property elucidation techniques such as DLS/microrheology with the high resolution chemical structure/conformation elucidation techniques such as Raman spectroscopy in generating novel mechanistic insights that will allow the performance engineering of complex fluids and soft matter systems. This will be exemplified through studies into the self-assembly/gelation mechanisms in three very different complex fluids-mixed anionic/zwitterionic surfactant wormlike micelles, polyvinyl alcohol and a thermo-reversible, gel-forming agarose biopolymer.

Introduction: Technique

A combined DLS-Raman technique (Helix, Malvern Instruments Ltd) has been used to obtain DLS (nanoscale particle size) and Raman (structural information) data sequentially on a single sample. Microrheological measurements (viscosity and viscoelasticity) were performed with the same instrument. The Helix system uses a proprietary non-invasive backscatter (NIBS) detector with dynamic (DLS), static (SLS) and electrophoretic (ELS) light scattering to measure the hydrodynamic radius of particles from 0.15 nm to 5 µm. Raman spectra are collected using 785 nm excitation (~280 mW) from 150 to 1925 cm^{-1} at 4 cm^{-1} resolution. To perform the microrheological measurements, 900 nm polystyrene particles were added to obtain a final concentration of 0.1% w/w in the complex fluid systems.

Results: DLS, Raman Spectroscopy and DLS-based optical microrheology were carried out on three different self-assembling systems-wormlike micelle forming SLES/CAPB (investigated as function of NaCl addition), polyvinyl alcohol (PVA) (investigated as function of extent of hydrolysis) and agarose (investigated as function of temperature). Microrheology data (complex viscosity, elastic modulus G') was obtained and the results were compared with low frequency Raman data and DLS data (size) to derive new insights into the link between rheology-microstructure-molecular structural changes as these systems undergo self-assembly. Some of the comparative data (highlights) on the worm-like micelles, agarose and PVA are shown below:

System: SLES/CAPB wormlike micelles

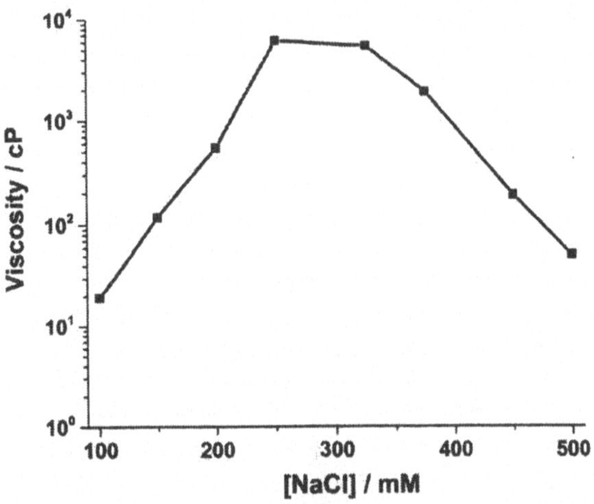

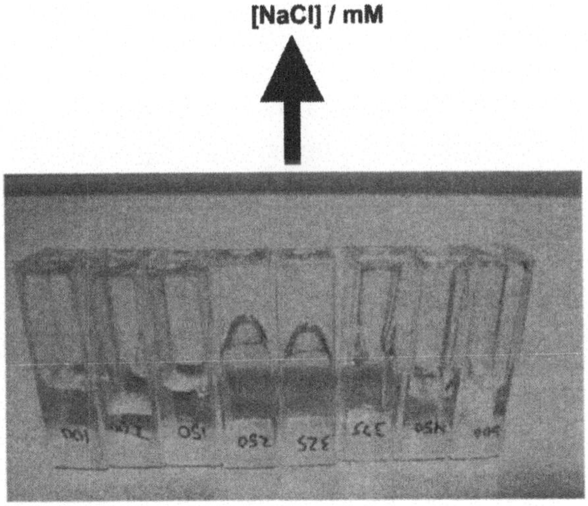

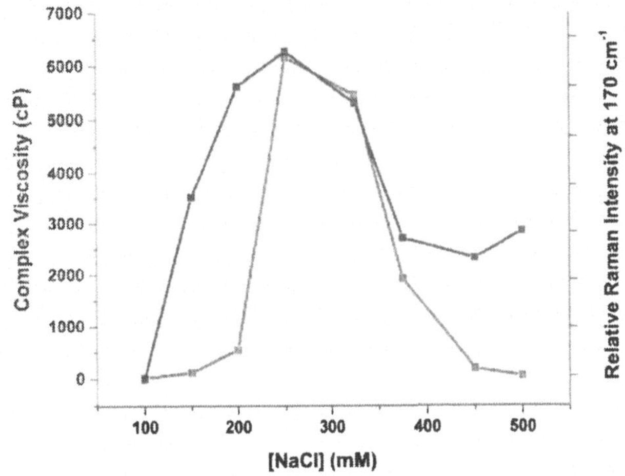

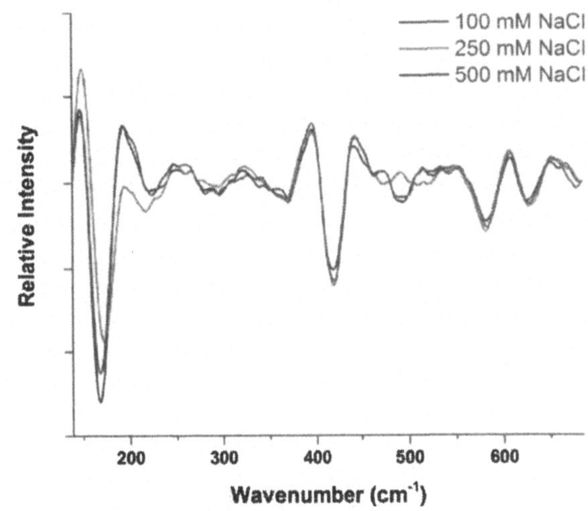

***Figure:** SLES/CAPB: Viscosity enhancement and then decrease with NaCl addition. Interesting correlation with low frequency Raman marker. Raman low frequency intensity marker behavior may be sensitive to microstructural differences (e.g. branching) at high and low concentrations of NaCl due to differences in water structuring.*

System: Agarose

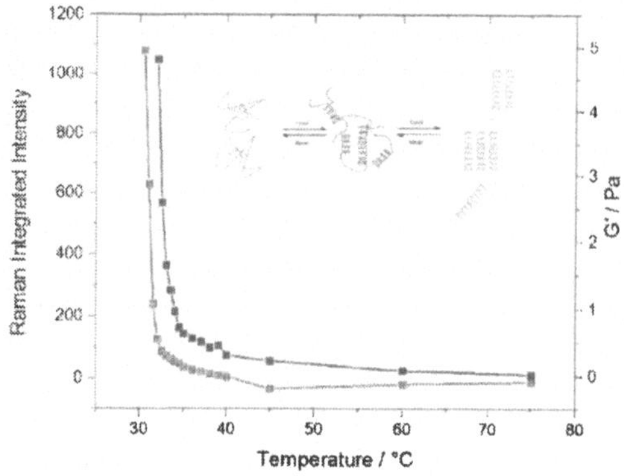

Figure: Agarose: Enhancement of both G' and low frequency Raman marker with decreasing temperature. Low frequency Raman exhibits good correlation with elastic modulus G' for agarose gelation. A two-step process is seen in both G' and low frequency Raman data. The low frequency Raman data seems to be a good indicator of molecular nature of gelation/self-assembly in range of complex fluid systems

System:Polyvinyl Alcohol:

LOW FREQUENCY RAMAN

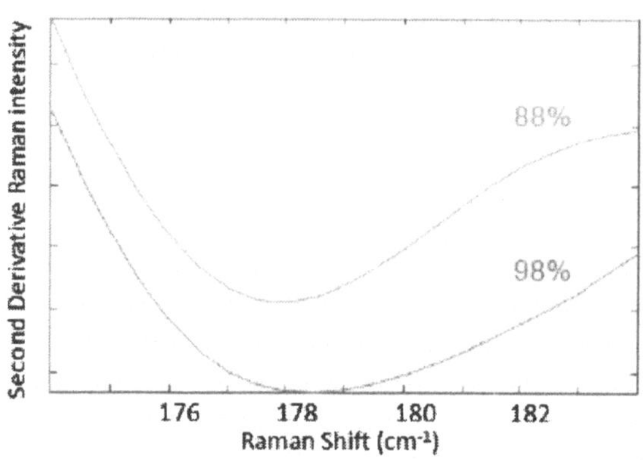

MICRORHEOLOGY

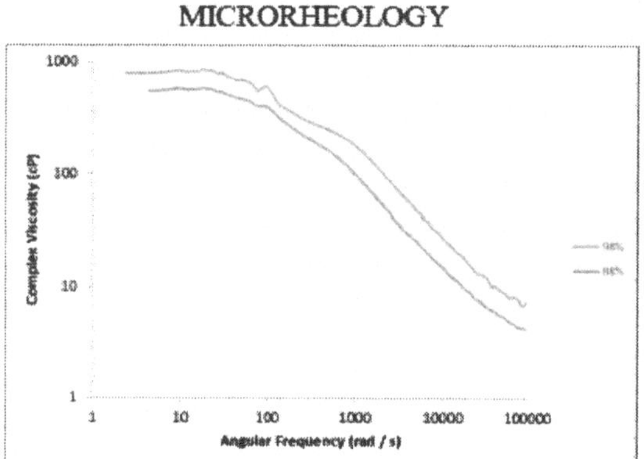

Hyrdolysis	Viscosity (cP)	Intensity
88%	542.7	-4.18
98%	947.2	-4.50

Figure: *PVA viscosity enhancement with extent of hydrolysis correlates well with low frequency Raman intensity.*

Conclusions:

The connectivity between rheology and molecular level structural changes is a new unexplored area that has been opened up by combined- DLS-DLS optical microrheology-Raman Spectroscopy and should help to develop new insights that will allow the optimization of performance enhancing formulation design rules for industrial complex fluids.

References:

1. S.Amin, S.Blake, S.M.Kenyon, R.C.Kennel , E.N.Lewis, A novel combination of DLS-optical microrheology and low frequency Raman spectroscopy to reveal underlying biopolymer self-assembly and gelation mechanisms. J. Chem. Phys. 141, 234201 (2014)

2. S. Amin, C. A. Rega and H. Jankevics, Detection of viscoelasticity in aggregating dilute protein solutions through dynamic light scattering-based optical microrheology. Rheol. Acta 51 (4), 329-342 (2012).

Non-destructive imaging analysis of unstable foam structures using 3D X-ray computed tomography

A. Eggert*, M. Supper*, F. Nachtrab*, J. Dombrowski**, A. Rack*** and S. Zabler* ****

* Fraunhofer Development Center X-ray Technology EZRT, Flugplatzstraße 75, 90768 Fürth, Germany, anja.eggert@iis.fraunhofer.de

** Chair for Food Process Engineering and Dairy Technology, TU München, Germany, jannika.dombrowski@tum.de

*** European Synchrotron Radiation Facility, France, alexander.rack@esrf.fr

**** Chair of X-ray Microscopy, Physics and Astronomy, University Würzburg, Germany, simon.zabler@physik.uni-wuerzburg.de

ABSTRACT

For the engineering of foamed products, knowledge about the foam structure as well as about its dynamics and stability are of critical importance. Using X-ray computed tomography (CT) accurate 3D information about the foams structure is obtained, but the structural characterization of liquid foams, which feature structural changes, is still challanging.

In this contribution a fast laboratory CT setup is presented allowing the acquisition of 3D data sets almost instantaneously.

Reached limits for the CT system:

- 15 s measurement time for a complete 3D recording
- 3.5 µm resolution for small objects (I.D. 10 mm)
- time-resolved structure characterization
- constant climate conditions (temperature, humidity)
- non-destructiv

The suitability and possibilities of such a system for non-solidified foams and other porous materials are shown.

Keywords: foam, computed tomography, fast CT, X-ray imaging,

1 INTRODUCTION

Foamed products play an essential role in modern food industry. The contained air changes the physical properties and influences product attributes. Their fluffy structure fulfils the consumer's desire for light fare and has a special impact on mouth feel and flavour perception. However, the characterization of sensitive porous structures is difficult, because of the foams sensitivity against mechanical influences. Still, appropriate quantitative methods are not readily available today [1].

X-ray micro computed tomography (µCT) offers a non-destructive 3D imaging technique with a resolution of few micrometres [2]. From the 3D data the microstructure can be measured directly to analyse the air fraction of the product regarding air volume and bubble size distribution, accurately and with an extensive set of statistics [3]. µCT is suited well for foam investigation, because it is based on the different absorption of X-rays through various materials. Foams have a great amount of air, which don't absorb X-rays, and the continuous phase, which absorb X-rays depending on the used material. This resulted in a high image contrast between air and foaming material in the received images.

A challenge for CT is non-solidified foam, because of the time dependency. The recording time for a normal µCT scan is between 10 and 100 min, during this time the object should not change, but non-solidified foams change their structure within few seconds [3]. Therefore we optimized a laboratory CT system concerning recording speed and image quality to realize CT scans within a few seconds, allowing for dynamic, time-resolved measurements. Structure changes and foam development can be studied in place without interference.

2 MATERIAL AND METHODS

2.1 Foam

For the laboratory X-ray µCT scans a β-lactoglobulin (BLG) protein-foam was chosen for its good foaming properties. It is a model system for the investigation of unstable foams. The foam is produced on site seconds before the experiment by a foaming apparatus.

BLG powder was dissolved in deionized water to yield a protein concentration of 0.25 % (w/w). For the present study the pH of the solution was set to 5.2 by adding 1 M hydrochloric acid (HCl). The temperature of the solutions was kept constant around 20 °C.

For the foaming process, a DFA100 Dynamic Foam Analyzer unit from Krüss GmbH Germany was used. The system is based on the aeration method introduced by Waniska and Kinsella [4]. Afterwards the pillar with the foam is taken out of the apparatus and directly placed in the CT system.

2.2 Computed tomography

For this foaming experiment we employed a µ-CT setup at the Fraunhofer Development Center X-ray Technology (EZRT) which has already been described by Zabler et al. [5]. The µCT setup was used with a Varian PaxScan 2520D/CL detector, which is a Si-TFT panel and operates at a maximum frame rate of 10 images s^{-1} in 14-bits full frame mode (1920 × 1536 pixels) and 30 images s^{-1} in 2 × 2 binning mode. The tube was operated in 'high-power' mode at 60 kV acceleration voltage and 240 µA tube current (approx. 14.4 W power) allowing for exposure times as short as 34 ms.

The foam was generated in a 20 mm wide plexiglass column and images were recorded within 30 s at three time points: 3 min, 5 min and 7 min after the initial foaming process. The column was scanned with continuous sample rotation (fly-by) over 200°. The foam structure was thereby recorded with an effective voxel sampling of 21 µm and a spatial resolution of approx. 43 µm. Depending on the rotation speed, CT volumes can be recorded in scan times down to 15 s for a volume of 512^3 voxels (400 projection images).

2.3 Data processing and analysis

Prior to tomographic reconstruction the radiographic images were filtered with a Median kernel. The radiographs were then processed by a filtered back-projection (FBP). Figure 2a shows a typical slice from the CT-volume data. The structural analysis of the foam is detailed in Figure 2. First the volume images are binarized by applying an Otsu threshold (see Figure 2b) [6]. This method avoids the erroneous selection of non-foam particles during the first step of the binarization. For retrieving the complete foam borders, the resulting binary 3D images are segmented using a Euclidean distance transformation (see Figure 2c) and an adaptive water-shedding algorithm (see Figure 2d) from the software ToolIP (Tool for Image Processing, Fraunhofer ITWM, Kaiserslautern, Germany). Single bubbles down to 54 µm diameter are thus segmented, labelled and analyzed in terms of position, size and volume. Based on this data histometric bubble size distributions can be calculated.

For the CT series of further foam studies published by Eggert et al. [3] individual growth and shrinkage of a selected bubble cluster was analyzed, which coalesced at the latest time point into one single bubble. Starting from the latter a 3D mask was created based on this the coalesced bubbles were identified in the previous time step, and so on until the initial bubble ensemble was determined at the first time step. Prior to this 3D masking-and-finding procedure it was necessary to register all datasets in polar coordinates so that central coordinates of corresponding bubbles would match. This registration is performed manually using the software ImageJ.

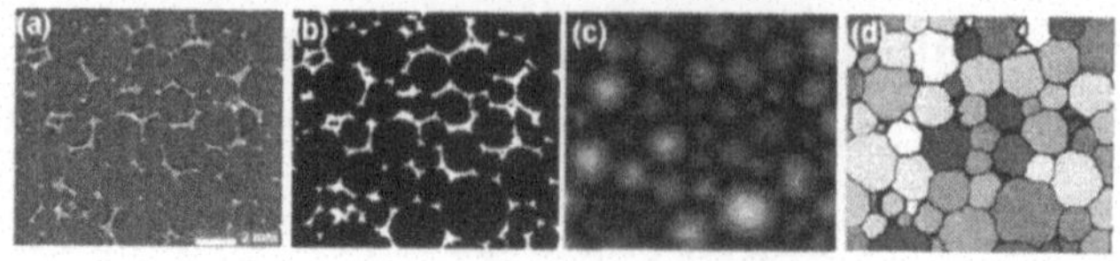

Figure 2: Pivotal steps in the image processing chain shown for the central slice: (a) binarization of the original image, (b) the binary image serves as input for the distance transformation and (c) the watershed image masked with (a) to obtain the resulting label image (d). [3]

2.4 Climatic chamber

The climatic chamber, schematically shown in Figure 3, guarantees temperature and humidity of microstructures during radiographical investigation. The specimen is placed into an EPS foam cup (max. volume: 250 ml) to allow transmission of the X-rays. Cup base is cut off and watertight fastened to a copperplate. An experimentally determined amount of water is pipetted into the cup in order to ensure certain humidity. Control of temperature is possible through a regulated TEC1-12706 Peltier element situated between copperplate and cooling element. The 127 couples 40 x 40 mm size module is supplied by a DC voltage using the NTS 150 W multi USB EuP 8.5A power supply (Goobay, Germany) and linked to a QC-PC-CO-CH1 controller (Quick-Ohm Küpper & Co. GmbH, Germany). The temperature can be varied within the range of -40 to 100°C. A NTC 10 Ω (β=3977K) temperature sensor, that is linked to the controller, is situated at the height of the specimen to provide the feedback signal for the temperature control.

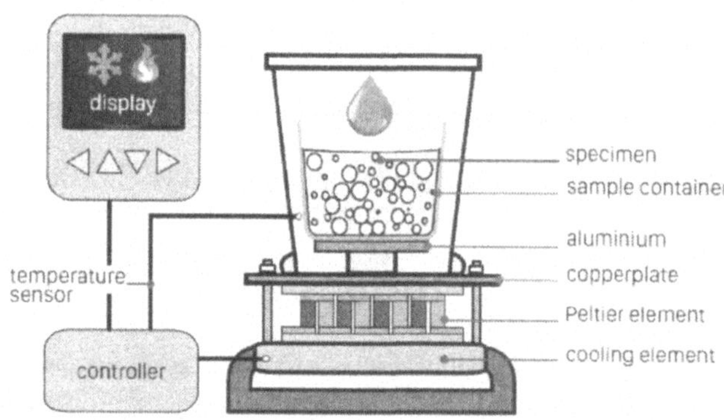

Figure 3: Schematic setup of the cooling/heating chamber with certain humidity.

3 RESULTS AND DISCUSSION

The aim of the work presented here was to generate a comprehensive knowledge of foam forming and stabilizing using non-destructive, fast and high-resolution techniques.

3.1 Foaming

As a first step, a method for the gentle production of foams had to be found, which allowed an in-situ characterization of the aerated structures.

Therefor the DFA100 Dynamic Foam Analyzer unit from Krüss GmbH Germany was used. The system is based on the aeration method introduced by Waniska and Kinsella [4]. The foam is produced on site seconds before the experiment. Afterwards the pillar with the foam is taken out of the apparatus and directly placed in the CT system, see Figure 1.

3.2 CT-system "Fly-by"

To describe foam decay bubble size distributions and further bubble properties were examined using 3D-μCT. Therefore we improved a CT-system concerning recording speed and image acquisition suited for unstable foams.

The measurement time for standard CT recordings is about 10 min at best. This is far too long for unstable foams, which constantly change their structure. The movement of the foam causes image artefacts, that's why the recording time has to be reduced. A standard CT-system works after stop-and-go principle, meaning that a projection is recorded and saved for each angle step. In fact that's quit time consuming.

Using the fly-by method the CT-recording is restricted to the essential steps only. The image acquisition occurs "on the fly" and the image seqeuence is processed afterwards for 3D reconstruction. With this method the recording time is only linked to the rotation speed of the specimen during the CT-scan. An important condition for fly-by is a fast detector with a frame rate above 15 images s^{-1}. For a further speed up the CT-scan can be cut down from a whole 360° rotation to a limited angel of about 180°. The recording time can thus be reduced by factor two down to 15 s at all. This allows for new and detailed insights into the dynamics of instable foams, i.e. to assess foam properties during foam decay.

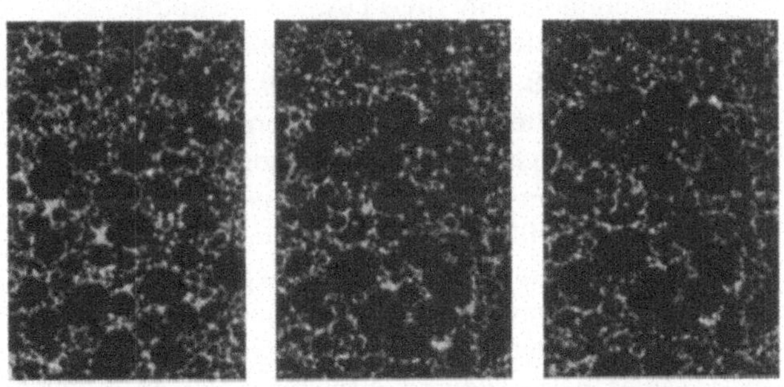

Figure 4: Sequence of binary axial CT slices of BLG foam at pH 5.2 after 3, 5 and 7 min.

3.3 Foam analysis

For foam analysis during foam decay the example of BLGs foaming behavior is used here. Figure 4 shows an image sequence of BLG foam recorded with fly-by CT in 15 s.

The time development of the BLG foam is regarded at the isoelectric point (pH 5.2). The image sequence in Figure 4 shows a fast structure change at the beginning of the recording. After 5 min the shapes in figure 4 seem to have stabilized. In the next minutes only slight changes in the microstructure can be observed. The foam moves up in the pillar because of the drainage collects below and pushes the foam up. The liquid loss because of drainage can also be investigated by the decreasing thickness of the lamellas between the foam cells. The borders between the foam cells become less visible, because the thinner lamellas absorb less X-rays.

In Figure 5 the bubble size distribution is shown to visualize the foams change over time. The bubble size distribution shifts in the area of bigger bubble diameters with the time. The number of bubbles decreases, especially the number of small bubbles reduces strongly. The mean diameter rises from 469 to 549 μm accordingly.

3.4 Foam development

Furthermore, an image processing chain was created and subsequently automated using the software ToolIP (Tool for Image Processing) from the Fraunhofer Institute ITWM. Thus, a quantitative analysis of the microstructure even of instable foams was possible, providing time-resolved information about bubble size, diameter, shape and position. The obtained results were supplemented by an additional series of measurements at the European Synchrotron Radiation Facility (ESRF) in Grenoble, which

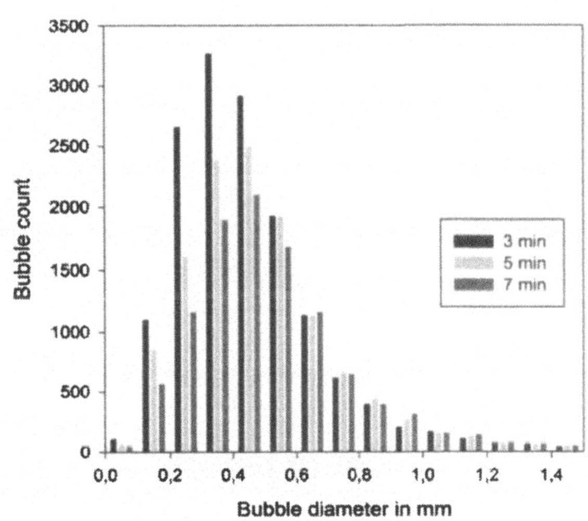

Figure 5: Bubble size distribution of BLG foam at pH 5.2 after 3, 5 and 7 min.

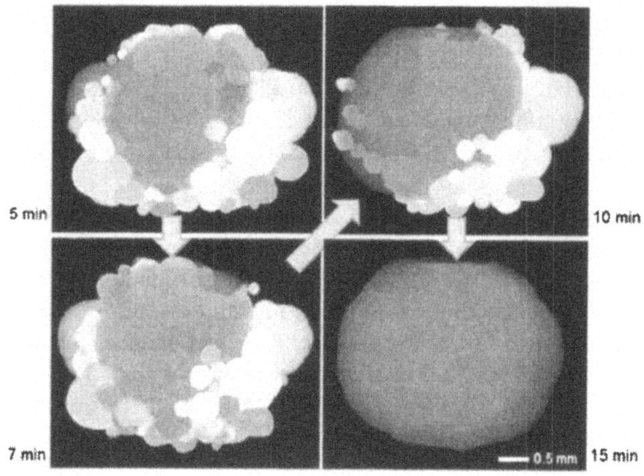

Figure 6: Development of a foam cluster over 15 min [3].

allowed the quantitative description of phenomenological characterized destabilization mechanisms like coalescence and disproportionation presented by Eggert et al. [3]. Thereby bubble growth, shrinkage and fusion can be observed or followed (see Figure 6).

The results of these investigations can be used for developing new micro structured food products and for a better understanding of the relation between the production and the structure of foamed material.

3.5 Temperature and humidity

The climatic chamber has been designed for controlling temperature and humidity of microstructures during radiographical investigation to guarantee a native investigation. The climatic chamber is placed within the X-ray beam and images of microstructures are recorded in a temperature and humidity controlled environment, thus long term X-ray scans of sensitive materials are possible (see Figure 7).

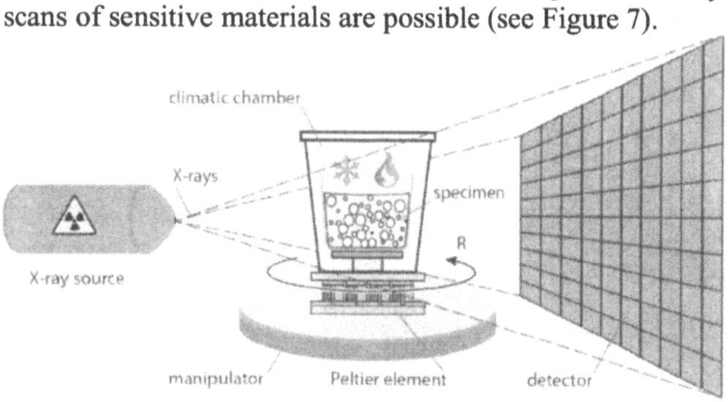

Figure 7: Schematic measurement setup with climatic chamber between X-ray source and detector.

4 CONCLUSION

In this contribution a laboratory µCT setup was presented allowing the acquisition of 3D data sets in down to 15 s. The suitability and possibilities of such a system for non-solidified foams was shown.

Reached limits for the CT system:

- 15 s measurement time for a complete 3D recording
- 3.5 µm resolution for small objects (I.D. 10 mm)
- time-resolved structure characterization
- constant climate conditions (temperature, humidity)
- non-destructive

From the 3D data the microstructure of the foam can be measured directly to analyze the air fraction of the product regarding air volume and bubble size distribution, accurately and with an extensive set of statistics. Apart from the foams structure, its dynamic changes over time can be studied in place without interference. Thereby bubble growth, shrinkage and fusion can be observed or followed.

The results of these investigations can be used for developing new micro structured food products and for a better understanding of the relation between the production and the structure of foamed material.

5 ACKNOWLEDGEMENTS

We thank Dr. Katja Schladitz (ITWM) for her generous support and assistance with the ToolIP software. The research project AiF 17124 N was supported by the German Ministry of Economics and Technology (via AiF) and the FEI (Forschungskreis der Ernährungsindustrie e.V., Bonn). The authors gratefully acknowledge the funding of the German Research Council (DFG), who, in the framework of the "Excellence Initiative" supports the Cluster of Excellence ''Engineering of Advanced Materials'' at the University of Erlangen-Nürnberg. These studies were further co-financed by the free state of Bavaria. The authors thank the ESRF for granting beamtime at beamline ID19 in frame of proposal MA1548.

REFERENCES

[1] Lim and Barigou, "X-ray micro-computed tomography of cellular food products", Food Research International, 37, 1001–1012, 2004.

[2] Bonse and Busch, "X-ray computed microtomography (µCT) using synchrotron radiation (SR)", Progress in Biophysics and Molecular Biology, 65, 133–169, 1996.

[3] Eggert, Müller, Nachtrab, Dombrowski, Rack and Zabler, "High-speed in-situ tomography of liquid protein foams", International Journal of Materials Research, 105, 632-639, 2014.

[4] Waniska, Kinsella, "Foaming properties of proteins:evaluation of a column aeration apparatus using ovalbumin", J. Food Sci. 44, 1398, 1979.

[5] Zabler, Fella, Dietrich, Nachtrab, Salamon, Voland, Ebensperger, Oeckel, Hanke, Uhlmann, "High-resolution and high-speed CT in industry and research", Proc. SPIE 8506, 850617, 2012.

[6] Ohser and Schladitz, "3d Images of Materials Structures – Processing and Analysis", 124, 2009

Preparation of ion group functionalized diblock copolymers and examination of their thermogelling properties

Bo Keun Lee*, Jin Seon Kwon*, Ji Hoon Park*, Doo Yeon Kwon*, Jae Ho Kim*, Se Heang Oh**, Sang Jin Lee***, Moon Suk Kim*,1

* Department of Molecular Science and Technology, Ajou University, Suwon 443-749, Korea, moonskim@ajou.ac.kr
** Department of Nanobiomedical Science, Dankook University, 330-714, Korea, seheangoh@dankook.ac.kr
*** Wake Forest Institute for Regenerative Medicine, NC, USA, sjlee@wfubmc.edu

ABSTRACT

Recently several thermogelling materials have been developed for biomedical applications. To investigate thermogelling behavior, in this study, methoxy polyethylene glycol-b-[poly(ε-caprolactone)-ran-poly(6-chloride-ε-caprolactone) (PCL-ran-PfCL)] (PCL-Cl) diblock copolymers were designed such that the PfCL content in the PCL segment was varied. The PCL-Cl diblock copolymers were derivatized by introducing a pendant amine group at the PfCL segment (PCL-NH_2). When the temperature was increased above room temperature, PCL-NH_2 diblock copolymer solutions exhibited sol-to-gel phase transitions. PCL-NH_2 diblock copolymer solutions formed gels immediately after injection into animal. The in vivo formed gel integrity persisted up to ten weeks under physiological conditions. We confirmed that that PCL-NH_2 diblock copolymer solution may potentially be used as thermogelling materials into the body in a minimally invasive manner.

Keywords: thermogelling, block copolymer, pendant amine group, in vivo injection

1 INTRODUCTION

Over the past few decades, several thermogelling materials have been focused extensively on a broad range of biomedical applications, such as drug or cell delivery carriers [1]. The thermogelling materials can absorb water and swell in aqueous environments that allow them to retain large amounts of fluids.

The fully hydrated thermogelling materials can incorporate various therapeutic agents, such as drugs, growth factors, gene products, or cells simply by mixing. The thermogelling materials can be injected and solidified at the specific target tissue, and can exhibit biological characteristics similar to those of live tissues.

Various thermogelling block copolymers have been developed to apply in biomedical field. Thermogelling block copolymers consisting of (methoxy) polyethylene glycol [(M)PEG], poly(propylene oxide) (PPO), and biodegradable polyesters, such as poly(L-lactic acid) (PLLA), poly(glycolic acid) (PGA), their copolyesters (PLGA), or poly(ε-caprolactone) (PCL), have been prepared as potential candidates [2]. Previously, we depeloped MPEG–PCL (MC) and MPEG–polyester diblock copolymers, [MPEG-b-(PCL-co-PLLA), MPEG-b-(PCL-co-PTMC), and MPEG-b-(PCL-co-PDO)], as thermogelling polymers [3-6]. Among them, MC material generates favorable thermogelling properties of long gel-persistence time and biodegradation.

In a later study, we examined the thermogelling properties of MC diblock copolymers derivatized with carboxylic acid group, tertiary amine group, and zwitterionic sulfobetaine in the end position in MC chain [7]. The functional groups on MC diblock copolymers engaged in intra- and intermolecular interactions to stabilize or destabilize polymer chain aggregates. The end group derivatization of MC diblock copolymers can lead to macroscopic gelation through the formation of intra and intermolecular interactions.

Scheme 1

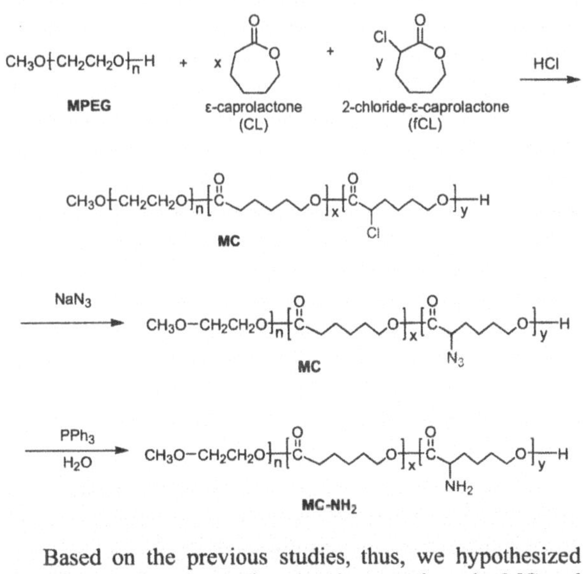

Based on the previous studies, thus, we hypothesized that introducing various functional groups into the MC end position could alter the thermogelling behavior of the MC

copolymer in aqueous solutions. This approach is based on the hypothesis that the formation and destruction of a structured network of PCL hydrophobic blocks of MC depends on the functional groups in the PCL chain end positions and induces the thermal phase transition behavior of the MC copolymer [8].

No	[CL]/[fCL] molar ratio	M_n calcd MPEG-polyester	M_n NMR [a] MPEG-polyester	Yield (%)	M_w/M_n [b]	Phase Transition
MC-Cl	97/3	750-2410	750-2380	91	1.4	O
MC-N_3		750-2420	750-2410	81	1.4	X
MC-NH_2		750-2380	750-2400	72	1.8	O
MC-Cl	95/5	750-2430	750-2540	94	1.4	O
MC-N_3		750-2570	750-2550	87	1.6	X
MC-NH_2		750-2540	750-2540	68	1.7	O
MC-Cl	90/10	750-2480	750-2610	94	1.3	X
MC-N_3		750-2660	750-2630	84	1.3	X
MC-NH_2		750-2600	750-2670	74	1.5	O
MC-Cl	85/15	750-2420	750-2430	96	1.3	X
MC-N_3		750-2515	750-2520	89	1.3	X
MC-NH_2		750-2440	750-2440	80	1.6	X

Table 1: Synthesis of MC-Cl, MC-N_3 and MC-NH_2, and thermogelling properties

Chloride, azide and amine functional pendant groups were tested for their effects on intra- and intermolecular interactions in this work as shown in Scheme 1. Caprolactone (CL) and 2-chloride-ε-caprolactone (fCL) monomers were randomly copolymerized with varying feed ratios to prepare MPEG-b-[poly(ε-caprolactone)-ran-poly(2-chloride-ε-caprolactone) (PCL-ran-PfCL)] (MC-Cl) diblock copolymers and then followed by the reaction with sodium azide. Then subsequent derivatization using amine group provide MPEG-b-[poly(ε-caprolactone)-ran-poly(2-amine-ε-caprolactone) (PCL-ran-PfCL-NH_2)] (MC-NH_2). Then we examined the sol-to-gel phase transitions as a function of temperature using MC-Cl, MC-N_3 and MC-NH_2.

2 RESULT AND DISCUSSION

MC-Cl was synthesized via the ring-opening polymerization of the monomer CL and fCL using the terminal alcohol of MPEG as the initiator. The colorless MC-Cl diblock copolymers were obtained in almost quantitative yield. MC-NH_2 was prepared by introducing a pendant azide and amine group at the chloride group on MC-Cl.

Table 1 summarizes the synthetic results of MC-Cl, MC-N_3 and MC-NH_2 prepared from PfCL content of 0–15 mol% in the PCL segment.

All diblock copolymers exhibited ^{1}H- and ^{13}C-NMR peaks characteristic of PCL, PfCL, MPEG, or MC-NH_2 (Figure 1).

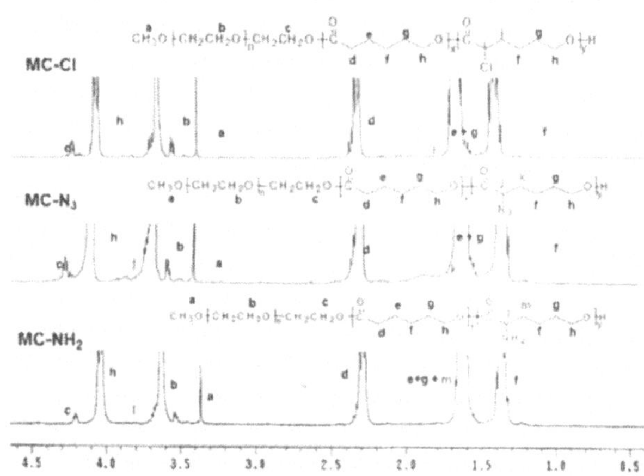

Figure 1: ^{1}H NMR spectra of MC-Cl, MC-N_3 and MC-NH_2.

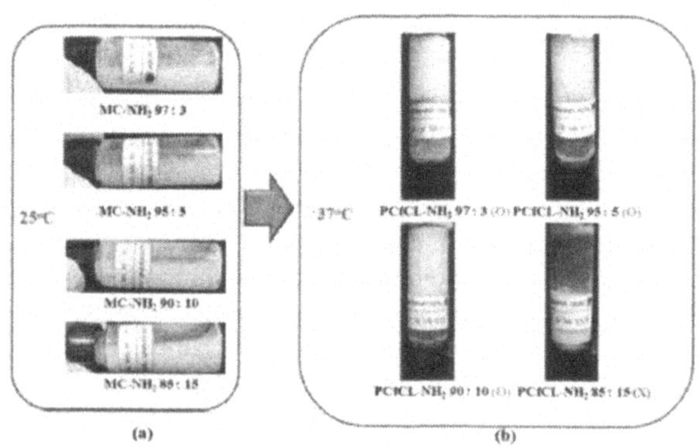

Figure 2: Phase transition of MC-NH_2 with amine content of 3–15 mol% in the PCL segment.

Aqueous solutions of MC-NH_2 with a PfCL content of 3–15 mol% in the PCL segment were prepared by dissolving diblock copolymers with 20 wt% concentration at 80°C in deionized water. At room temperature, MC-NH_2 solutions formed a translucent liquid that flowed when tilted (Figure 2a). The MC-NH_2 copolymer solution with 10 wt% of PfCL in the PCL segment became hydrogel indicating exhibition of distinct sol-to-gel phase transitions at 37°C (Figure 2b).

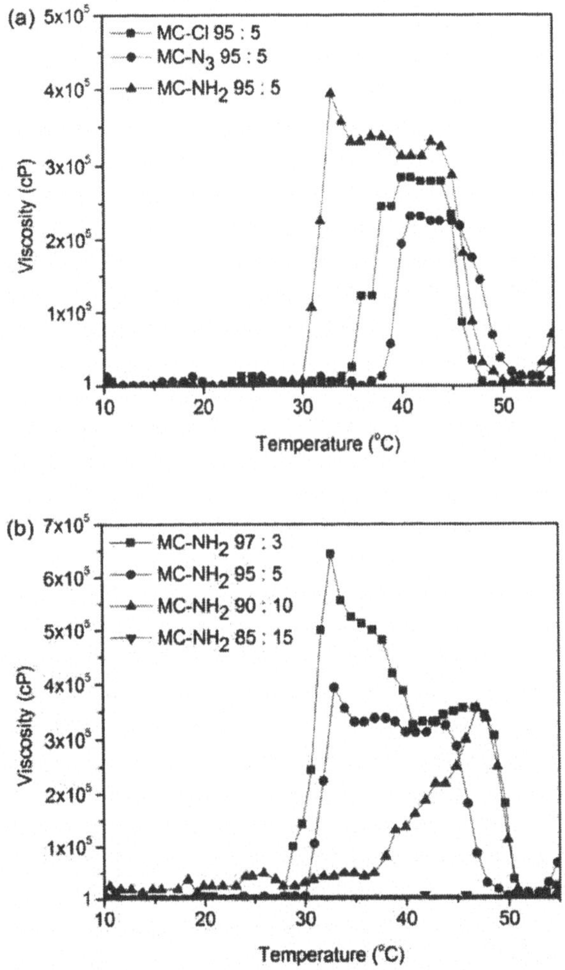

Figure 3: Viscosity versus temperature curves of (a) MC-Cl, MC-N_3 and MC-NH_2 with amine content of 5 mol (b) MC-NH_2 with amine contents of 3, 5, 10, and 15 mol% in the PCL segment.

The phase transition behavior was monitored by measuring the solution viscosity as a function of temperature from 10°C to 60°C (only MC-NH_2 viscosities in Figrue 3). At low temperatures, MC-Cl, MC-N_3 and MC-NH_2 solution were a homogeneous solution with a viscosity of 1 cP. As the temperature was increased, the MC-Cl, MC-N_3 and MC-NH_2 solutions exhibited gel-like behavior at an onset temperature marked by an increase in the viscosity. As the PfCL content increased, MC-NH_2 copolymers showed increased the onset temperatures and decreased maximum viscosities.

The sol-to-gel phase transition depended on the PfCL content in the PCL segment and the identity of the functional group in the pendant position. The sol-to-gel phase transition was not observed at 15 mol% PfCL content in the PCL segment. The MC-NH_2 solutions exhibited viscosities of 3.6 x 10^5–6.6 x 10^5 cP. The viscosity may be correlated with the strength of the gel formed by MC-NH_2 copolymer. This indicated that the amine groups on MC-NH_2 affect aggregation between polyester chains.

3 CONCLUSION

These results indicate that a sol-to-gel phase transition of MC-NH_2 copolymers depended on both the type of functional pendant amine group and the composition of PfCL in the PCL segment. We conclude that the MC-NH_2 copolymers featuring pendant amine group designed in this study can be used as potential thermogelling materials for drug and cell.

Acknowledgments

This study was supported by a grant from a Research Program NRF-2014M3A9E5073700 and Priority Research Centers Program (2011-0022978) through the National Research Foundation of Korea (NRF) funded by the Ministry of Education.

REFERENCES

[1] D. Y. Kim, D. Y. Kwon, J. S. Kwon, J. H. Kim, B. H. Min and M. S. Kim, "Stimuli-responsive injectable in situ-forming hydrogels for regenerative medicines" Polymer Reviews, DOI: 10.1080 / 15583724.2014.983244.

[2] M. S. Kim, H. Hyun, K. S. Seo, Y. H. Cho, G. Khang, H. B. Lee, "Preparation and characterization of MPEG-PCL diblock copolymers with thermo-responsive sol-gel-sol behavior" Journal of Polymer Science Part A: Polymer Chemistry, 44, 5413-5423, 2006.

[3] H. Hyun, Y. H. Kim, I. B. Song, J. W. Lee, Moon Suk Kim, G. Khang, K. Park and H. B. Lee, "In vitro and in vivo release of albumin using biodegradable MPEG-PCL diblock copolymer as an in situ gel forming carrier" Biomacromolecules, 8, 1093-1100, 2007.

[4] Y. M. Kang, S. H. Lee, J. Y. Lee, J. S. Son, B. S. Kim, B. Lee, H. J. Chun, B. H. Min, J. H. Kim and

M. S. Kim, "A biodegradable, injectable, gel system based on MPEG-b-(PCL-ran-PLLA) diblock copolymers with an adjustable therapeutic window" Biomaterials, 31, 2453-2460, 2010.

[5] J. S. Kwon, S. M. Yoon, D. Y. Kwon, D. Y. Kim, G. Z. Tai, L. M. Jin, B. Song, B. Lee, J. H. Kim, D. K. Han, B. H. Min and M. S. Kim, "Injectable in situ-forming hydrogel for cartilage tissue engineering" Journal of Materials Chemistry B, 1, 3314-3321, 2013.

[6]. J. S. Kwon, S. W. Kim, D. Y. Kwon, S. H. Park, A R. Son, J. H. Kim and M. S. Kim, "In vivo osteogenic differentiation of human turbinate mesenchymal stem cells in an injectable in situ-forming hydrogel" Biomaterials, 35, 5327-5346, 2014.

[7] S. W. Shim, D. Y. Kwon, J. H. Park, H. J. Chun, Y. J. Koh and M. S. Kim, "Preparation of zwitterionic sulfobetaine end-functionalized poly(ethylene glycol)-b-poly(caprolactone) diblock copolymers and examination of their thermogelling properties" Journal of Polymer Science Part A: Polymer Chemistry, 52, 2185-2191, 2014.

[8] J. I. Kim, D. Y. Kim, D. Y. Kwon, H. J. Kang, J. H. Kim, B. H. Min and M. S. Kim, "An Injectable Biodegradable Temperature-Responsive Gel with an Adjustable Persistence Window" Biomaterials, 33, 2823-2834, 2012.

[1] Department of Molecular Science and Technology, Ajou University, Suwon 443-749, Korea Ph: (82) 31-219-2608, Fax: (82) 31-219-3931, moonskim@ajou.ac.kr

A new quality and process control technology for real-time testing of viscoelastic changes in dairy and food products

S. Strandman (Principal Scientist), A. Hadj Henni (CEO) and C. Schmitt (CTO)
Rheolution Inc.
Montreal, QC, H2S 2X3, Canada
satu.strandman@rheolution.com, +1 (514) 586-2006

ABSTRACT

Measuring the viscoelastic properties of soft food materials with precision for industrial QC and production process control is a continuing challenge. Research laboratory equipments are typically not adapted for industrial use and are costly because of required capital investment and high operational costs (only highly skilled professionals can effeciently operate rheometers). On the other hand, texturometers that are widely used in the industry lack precision and do not allow controlling the production process.

Rheolution Inc. has developed new, simple to use and cost-effective instrument series, ElastoSens™ and CoaguSens™, for real-time quality control and production process control of food products and ingredients. The technological principle of these instruments is purely mechanical and its day-to-day use is much simpler than the use of traditional instruments, such as rheometers. The capabilities of ElastoSens™ will be presented through the results of viscoelastic testing of gelation and melting processes of various ingredients and food products. CoaguSens™, the processing hall version of the technology, was used in cheese production plants for at-line mechanical monitoring of milk coagulation processes and cutting time determination on the basis of curd elasticity. The results showed reduced variability (-45%) in yield after adopting CoaguSens™ as part of cheesemaking process. This allows refining the process and leads to considerable savings upon improved yield and quality. Rapid quantitative measurement of curd firmness facilitates further automation of the cutting step.

Keywords: viscoelasticity, quality control, real-time process monitoring, gelation, coagulation, soft food products

1 THE CHALLENGE: MEASURING THE ELASTICITY OF SOFT MATERIALS

Viscoelasticity testing during and after the structure formation or structure loss allows elucidating structure-texture property relationships of food and dairy ingredients and monitoring and controlling processes which involve changes in mechanical properties. Examples of such processes include coagulation, gelation, melting, or degradation of products and formulations, such as dairy products, protein- and polysaccharide-based gels, and food emulsions. Currently, it is still a challenge to monitor the mechanical properties of food products in production environments to control quality and processes.

Because of the continuous need to control and modify food ingredients to meet specific textural requirements, or to control industrial processes such as coagulation of milk, Rheolution Inc. has developed a new analytical technology to measure elasticity of food materials both in laboratory and at-line (in plant). The aim of this simple-to-use technology is to measure in real-time the elasticity of soft materials during and after the structure formation without perturbing the process, without destroying the sample and in a cost effective way.

1.1. ElastoSens™: Lab QC applications

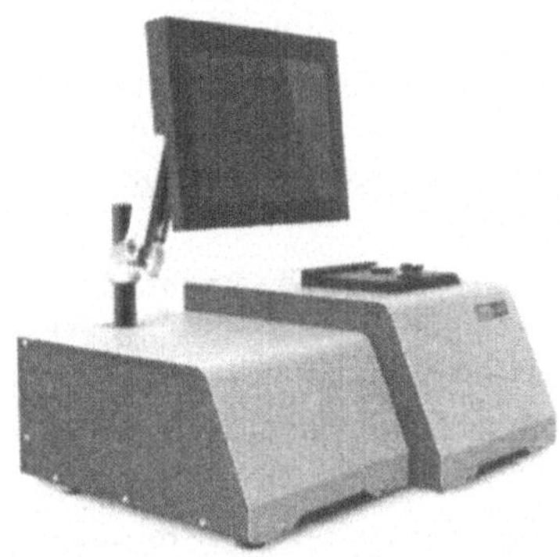

Most available non-destructive instruments are based on optical methods, focusing on molecular level characteristics (at nanometer-range distances), and few techniques with the exception of certain rheological methods (such as small-amplitude oscillatory shear, SAOS) allow monitoring larger-scale material properties, such as viscoelasticity. However, following the same sample over long periods of time is seldom possible without repeated sample loading and altering the sample properties.

For this purpose, Rheolution Inc. has developed ElastoSens™ X3, the first lab mechanical testing instrument allowing:

- To characterize in real time and non-destructively the viscoelasticity of food ingredients and formulations.
- To measure the mechanical properties without contact and without perturbing the structure formation or loss.
- To use removable and sterilizable sample holders to follow the evolution of the same sample over long periods of time.
- To measure the mechanical impact of physical stimuli such as temperature and pH on food products.

1.2. CoaguSens^{TM}: Automatic & at-line process control of milk coagulation in the cheese industry

The technology has also been developed as a processing hall-adapted version, **CoaguSens^{TM}**, which is used in cheese plants for at-line monitoring of the coagulation of milk during the cheesemaking process. In cheese production, milk gel is usually cut at predetermined time or according to cheesemaker's subjective evaluation of gel texture and appearance in order to separate whey from the cheese curd. These can lead to considerable variation in yield, because many factors affecting the gel strength do not remain constant.

There is a pressing need for objective and quantitative real-time milk gelation and cutting time determination that help refine the cheesemaking process, improve process automation, reduce the variations, and maximize yield. In cheese production, CoaguSens^{TM} allows:

- Measuring milk gel elasticity during coagulation.
- At-line monitoring of milk gel formation in real time.
- Precise and automatic monitoring of cutting time.

2 THE TECHNOLOGY

Contactless and non-destructive ElastoSens^{TM}/ CoaguSens^{TM} technology measures in real time the viscoelastic properties of gels as a function of time and temperature or other stimuli. Both heating and cooling ramps can be applied to the sample. The patented technological principle behind this instrument is purely mechanical. The sample is contained in a cylindrical disposable and detachable sample cup having a flexible bottom. Small and gentle vibration of a few micrometers is applied to the sample and the response is measured using a laser probe without contact (Fig. 1).

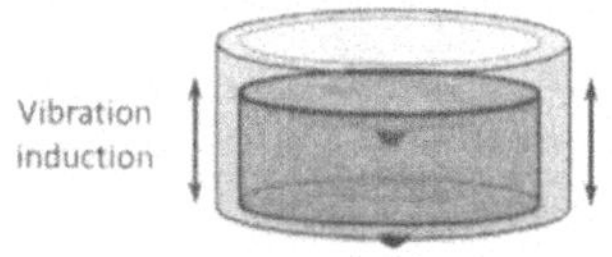

Fig. 1 Contactless and non-destructive viscoelasticity measurements by ElastoSens^{TM} and CoaguSens^{TM} using a few micrometer-amplitude vibrations.

The material response is processed to obtain the viscoelastic properties (i.e., shear storage and loss moduli). This process is repeated in desired intervals to characterize the mechanical evolution of the studied material.

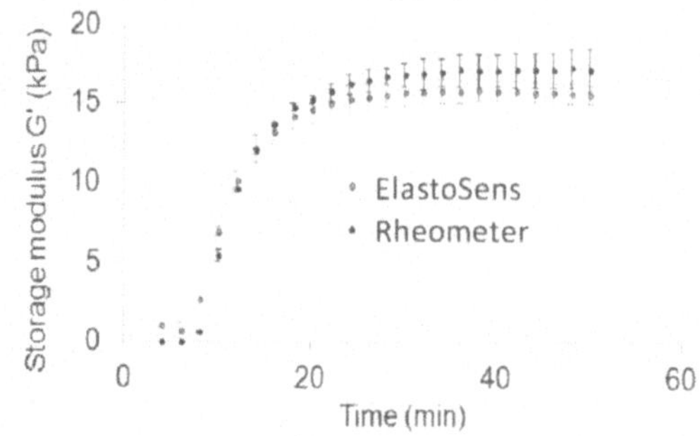

Fig. 2 Comparison of ElastoSens^{TM} X3 with a rotational rheometer (n = 3).

The technology was validated by a comparison with a commercial rotational rheometer that provides identical viscoelastic information. Fig. 2 shows the elasticity results on the same sample (1.4 % w/w solution of cell culture-grade agar), measured by ElastoSens^{TM} X3 and a rotational rheometer (f = 1 Hz, strain = 0.1 %).

3 VISCOELASTICITY OF FOOD INGREDIENTS

3.1 Gelation at constant temperature

Hydrocolloids are polymers that form viscous dispersions or gels in water. In food industry, they are used as thickeners, gelling agents, emulsifiers, stabilizers and coating ingredients, owing to their ability to modify the rheology of food systems. Common hydrocolloids include agar, alginate, carrageenan, various cellulose derivatives, chitosan, gums (gum arabic, guar, xanthan), pectin, starches and gelatin. The gelation of hydrocolloids can occur via ionic interactions, cold-setting or heat-setting.

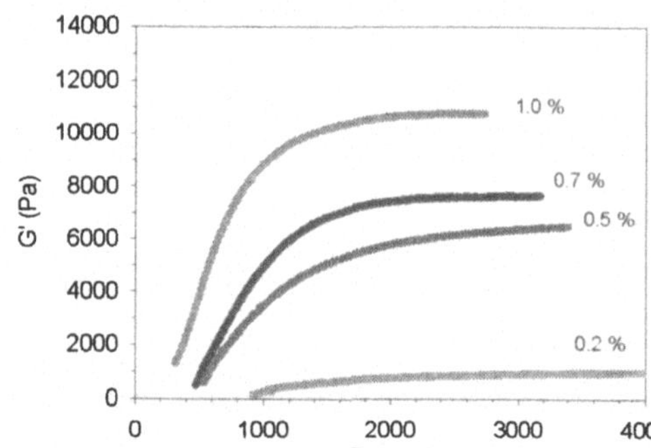

Fig. 3 Concentration dependence of agar gelation kinetics at 25 °C, as measured by ElastoSens^{TM} X3(n = 3).

Agar is a cold-setting gelling agent than can be used for both hot and cold foods owing to its high dissolution temperature. ElastoSens^{TM} X3 was used for non-destructive monitoring of food-grade agar gelation upon cooling the solution to room temperature without perturbing the gelation process. As the polysaccharide chains of agar are prone to spontaneous or acid-induced degradation, testing the batch-to-batch variation of gelling properties or the stability of formulations is an important part of quality control and R&D processes.

The gelation of agar was monitored using ElastoSens™ X3 after injecting warm solution (40 °C) in a sample cup at 25 °C. The pH of the solution was 7.4. At low concentrations (Fig. 3), the probability of chain association is lower and thus, the gelation occurs slowly and the gels are weak.

3.2 Temperature-induced changes

As described above, thermal gelation of hydrocolloids can occur either via heating or cooling. In addition to structure formation, structure loss is important for textural attributes, usable temperature range, and storage stability. For example, meltability and corresponding elasticity of cheese is important for both processing and use in cooking, and depends not only on the type and composition of cheese but also on the milk processing conditions. Fig. 4 shows an example of the evolution of viscoelastic properties of processed cheese upon heating, as monitored using ElastoSens™ X3. The instrument can be used for testing the quality of formulations or the effect of composition on the viscoelastic properties and thermal transitions without perturbing the processes.

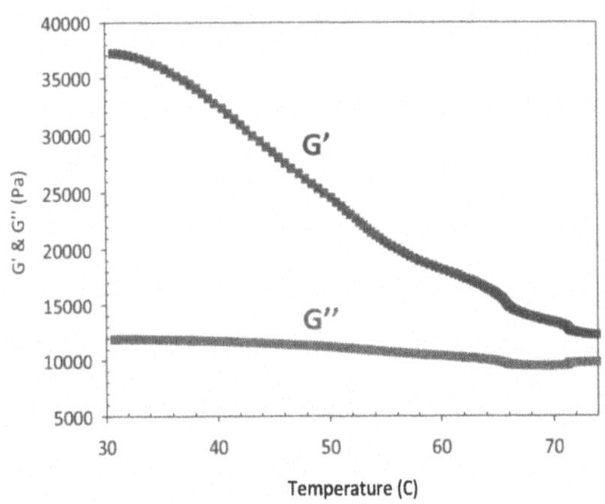

Fig. 4 Gradual melting of processed cheese upon heating, as monitored using ElastoSens™ X3.

4 REAL-TIME MONITORING OF MILK COAGULATION: IN-PLANT STUDY

4.1 On CoaguSens™

Cutting time selection of milk gels greatly affects the yield, moisture and quality of cheese. Cutting the gel too soon when the curd is not firm enough leads to lower cheese yield through increased curd fines and whey fat loss, whereas delayed cutting results in higher curd moisture content due to reduced collapse of the gel. This may further alter the ripening process and final product quality. Optimizing the process and reducing the variability in cheese yield converts into interesting savings, as the milk price makes most of the cheese production costs.

CoaguSens™ instrument has been developed for precise quantitative monitoring of the coagulation and gelation processes at food processing hall conditions. The instrument comes with CoaguTouch™, a modular, user-friendly and connected touchscreen-based user interface designed to configure the instrument. CoaguTouch™ provides user-oriented tools and functions to manage, analyse, display, store and transfer data and may be used in two different operating modes: 1) Process calibration mode or 2) Process control mode (communication with PLCs).

An example of how CoaguSens™ monitors the milk coagulation is shown in Fig. 5. In this case, the coagulation kinetics and milk gel elasticity depended strongly on rennet (90 % chymosin solution) concentration.

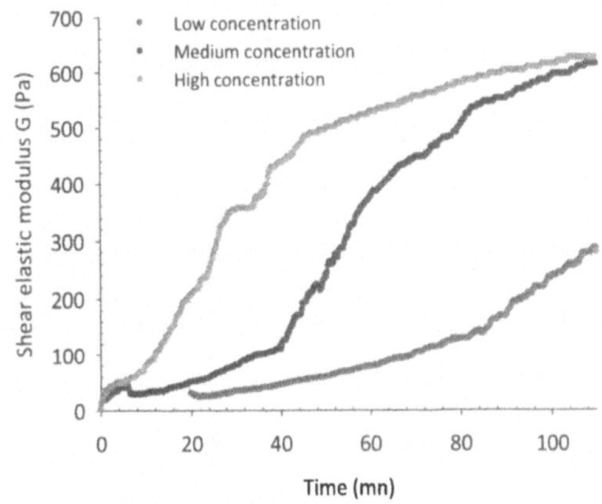

Fig. 5 Gelation of milk at 41 °C after adding rennet solution at different concentrations (0.05, 0.42 and 3.33 % v/v).

4.2 In-plant study

The study conducted in a medium-sized cheese plant was divided into two phases to determine the impact of using CoaguSens™ on cheese yield: (I) Database development, and (II) Cutting time control.

Phase I: cutting time was determined according to the cheesemakers' subjective evaluation of milk gel texture and appearance while simultaneously monitoring the coagulation process and the evolution of milk gel elasticity (curd firmness) at-line by CoaguSens™. The mean values and variability of milk gel elasticity and cutting time were analyzed.

Phase II: the milk gel formation was followed by CoaguSens™, as above, but now the cutting step was manually started when the gel elasticity reached the mean elasticity obtained in Phase I.

Seasonal variation in milk composition was observed between the two phases. In short, milk in phase II was less rich than in phase I. Consequently, the yield was expected to decrease in phase II.

4.3 Results

In Phase II, the cutting of curd was started when the elasticity reached the mean value measured in Phase I. This lead to greater variability (+ 51 %) in cutting times due to batch-to-batch variation of coagulation kinetics, but reduced the variability in cutting elasticity by -51 % (Fig. 6).

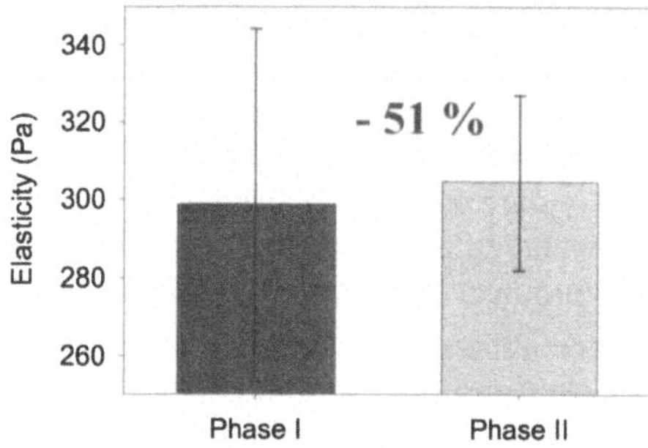

Fig. 6 Mean values and the variability of milk gel elasticity at cutting in Phase I (red) and Phase II (green), without and with CoaguSens™, respectively.

Importantly, reduced variability in curring elasticity was reflected in cheese yield: the variability in actual yield reduced from ± 0.47 % to ± 0.25 %, that is, by 45 % (Fig. 7). The mean yield in Phase II remained stable compared to Phase I, because the mean cutting elasticity was identical in both phases. The cutting time monitoring by CoaguSens™ prevented the decrease in yield expected due to the seasonal change in milk composition.

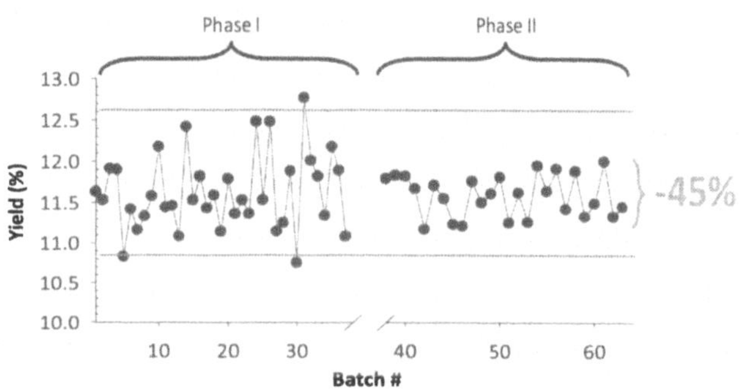

Fig. 7 Cheese yield and its variability in phases I (without CoaguSens™) and II (with CoaguSens™).

CoaguSens™ has a proven capacity to control and help minimizing the variability of the yield of industrial cheese making process. High variability in production process can lead to increase raw material and operating costs, lower plant capacity, and product quality issues. In cheese industry, reduced variability allows improving the manufacturing process and optimizing the yield within the confinements of product regulations (composition, humidity). This has important economic and environmental benefits.

More importantly, CoaguSens™ may be used by manufacturers as a routine process control instrument to set the best cutting elasticity that ensures high yield and optimal moisture content. Using a simple calculation based on milk price in the US and Europe, 0.2 % increase in cheddar cheese yield when using CoaguSens™ in a plant processing 600 tons of milk daily would lead to nearly 2 M USD annual savings. In countries such as Canada or Switzerland, where milk prices are high, the annual savings are even higher.

5 CONCLUSIONS

Among the most interesting applications of the new ElastoSens™/CoaguSens™ technology is non-destructive real-time measurement of mechanical evolution of various food ingredients and formulations during and after the structure formation or structure loss without perturbing the process. Measurements are conducted without contact and the samples remain sterile. This affordable and simple to use technology has an innovative feature of removable and sterilizable sample holder that can be stored in a controlled environment to study the long-term stability and viscoelastic changes in different conditions. The measuring accessory with three parallel sample holders can greatly improve the productivity of routine testing and the statistical significance of measurements.

The advantages in QC and production process control are:

- Real-time control of product quality.
- Fast correction of the recipe to reach optimal quality.
- Adjusting process conditions such as formulation and temperature.
- Real-time testing of the impact of additives.
- Stopping gel formation in time.
- Avoiding technical accidents and losses in production.
- Standardizing the product quality.

All these capabilities are made possible by the new ElastoSens™ technology designed to overcome challenges in food and dairy industry QC and process control.

The industrial processing hall version of the technology, CoaguSens™, allows at-line testing of products and controlling production processes through rapid individual tests or continuous monitoring. In cheese manufacturing plants, the data can be used for automated control of processes such as cutting of milk gels at desired curd firmness. Reduced variability in yield and the possibility for process optimization convert into interesting savings.

The Secret of Chia Gel Revealed - Nanoscale 3D Network Retards and Prolongs Food Release

N. Pottackal[1,3], P. Das[2,3], **M. Samateh**[3] and G. John[3*]

[1]Edison High School (EHS), Edison, NJ, USA
[2]South Brunswick High School, Monmouth Junction, NJ, USA
[3]Dept. of Chemistry, The City College of New York, New York, NY, USA
E-Mail: john@sci.ccny.cuny.edu

ABSTRACT

Chia gels have been associated with numerous health benefits over centuries, like curbing diabetes and obesity by retarding digestion rate and quenching craving for more food. However, the exact mode of gelation has always been quite ambiguous; the usual explanation often taken as the mucilage's swelling in or coagulation of water. Gel making is our passion and expertise. Our working hypothesis is that chia seed mucilage forms extended fibrous structures in water to ultimately form a 3D network at nanoscale. We are interested to probe the mechanism of gelation in many ways including microscopic techniques. Our preliminary studies entailed gelation tests, which revealed the polar nature of the mucilage. Subsequently, we systematically performed microscopic studies using optical microscope and scanning electron microscope. The optical micrographs revealed fiber bundles at microscale level while the SEM micrograph revealed a 3D network of nanoscale fibers with diameters around 50 nm, confirming our hypothesis. This breakthrough would surely revolutionize how we envision gelation process of Chia and other gel forming seeds and materials.

Keywords: chia gel, chia gel 3D network, chia nano fibers, chia digestion, chia slow food release

1. INTRODUCTION

Chia (Salvia hispanica) seed was native to the valleys of Mexico and northern Guatemala. Its uses included offering to the gods and painting of the face and body. Ancient people like the Mayans consumed it to prolong duration of fullness during war and other tasks. Chia is regarded as "super food" and contains nutritional and health benefits like antioxidants, omega-3 and -6, fibers, vitamins, and minerals.[1-4]

Unlike most other seeds, it has the unique property of gelling water. According to Munoz [5], its mucilage has water retaining and gelling properties with applications such as food thickening and emulsion stabilization. The mucilage, comprising mainly polysaccharides on the seed surface, swells and extends in water, trapping food particles (like sugars) and, thus, retarding digestion in the body.[5-6]

We aimed to understand the fundamental science underlying how chia seeds form 3D network that leads to gel formation. Despite the current notion that gel formation is merely due to the swelling and coagulation in water by the seed's mucilage, our hyposthesis is that the mucilage of chia contains nanoscale fibers that extend in water to form 3D network, which entraps and "solidifies" water to form a gel.

We began by deducing the effect of various parameters like solvent polarity on gelation in order to discern conditions that influence network formation by the mucilage. Optical microscopy zoomed in and elucidated how the fibers grew around the seed in water. Finally, scanning electron microscopy was used to further zoom in and study the nature of fiber configuration and size in the network at nanoscale. This knowledge will be vital for future advancements in materials design and medicine, especially to provide the general public with convincing information to best adhere to effective health styles and to prevent conditions like diabetes and obesity.[7-9]

2. MATERIALS

Chia seeds were obtained from Costco, NJ. Unless otherwise stated, all the solvents including hexane, ethanol and distilled water were obtained from Sigma-Aldrich. Canola oil was obtained from Costco, NJ. Congo red dye and Methylene blue dye obtained from Sigma-Aldrich.

3. METHODS

3.1 Effects of Different Parameters on Gelling Efficiency

NOTE: Gelation was confirmed by observing no flow of the water/seed mixture due to gravity upon inverting vial upside down; swelling was confirmed when the seeds swelled instead of remaining the same or shrinking.

(A) EFFECT OF SEED CONCENTRATION: Different concentrations (4, 5, 6, 8, 10, 15 & 20% g/mL) of chia seed in deionized water were prepared. Each mixture was allowed time to gel and the minimum gelation concentration (MGC) capable of gelling water determined.

(B) EFFECT OF SOLVENT POLARITY: The MGC was used to prepare mixtures of chia in 100, 50, and 0 % water in water/ethanol mixture respectively.

(C) EFFECT OF MODE OF ADDITION OF COMPONENTS OF THE 50:50 SOLVENT MIXTURE: The MGC was used to prepare mixtures by changing the order of adding components in 50:50 water/ethanol.

(D) EFFECT OF HYDROPHOBIC SOLVENTS: Each of hexane and canola oil was added to chia based on MGC.

3.2 Imaging Fiber Growth

Prior to each imaging, the fibers were grown by gelling or swelling in water. For SEM, the water was completely removed while the fibers were extended by lyophilization.

(A) OPTICAL MICROSCOPY: The extruding fibers of chia seeds in water were viewed using the Nickon TiE optical microscope. The samples were prepared by adding Congo red or methylene blue dye to the seeds in water in 6x4 multiwell plate wells. The plate was loaded onto the stage, and the images focused at different magnifications and satisfactory images captured by an attached camera.

(B) SCANNING ELECTRON MICROSCOPY (SEM): The lyophilized, dry fibers were used for SEM imaging at statndard conditions using a scanning electron microscope (ZEISS Supra 55 VP). The SEM was operated at various magnifications to capture satisfactory SEM images.

4. RESULTS AND DISCUSSION

4.1 Results of Effects of Different Parameters

The results of the effects of various parameters on gelation are shown in Tables 1 - 4.

Table 1. Effect of chia seed concentration on gelation

Concentration (% g/mL)	4	5	6	8	10	15	20
Gelation Result	N	N	N	N	N	G	G

G = gel formed; N = gel not formed.

Table 2. Effect of water/ethanol ratio on gelation

% of water in mixture	100	50	0
Gelation Result	G	N	N

G = gel formed; N = gel not formed.

Table 3. Effect of mode of adding 50:50 water/ethanol on gelation or swelling

Mode of addition	H_2O 1st	H_2O last	together
Gelation Result	SW	N	N

SW = seed swollen in solution; N = not swollen.

Table 4. Effect of hydrophobicity on gelation or swelling

Hydrophobic solvent	hexane	canola oil
Gelation Result	N	N

SW = seed swollen in solution; N = not swollen.

Based on the above results, the ideal conditions for chia seed gelation are a minimum of 15% g/mL seed concentration (Table 1) and a polar, aqueous solvent, preferably 100% water (Tables 2 to 4). This observation may be attributed to the gelling agent on the seed's surface being polar, thus, resulting in a more favorable interaction with water compared to with less polar solvents systems.

4.2 Results of Imaging Fiber Growth

Optical Microscopy

A photograph of chia seed gel and optical micrographs of grown chia seed fibers are shown in Figure 1. Chia seeds formed a gel which was verified by inverting the vial containing the gel upside down (Figure 1A); no flow despite the pull of gravity proved a gel was formed. The evenly spaced out seeds in the gel led us to believe that there must some sort of fibers responsible for separating the seeds apart as well as entrapping the water molecules.

Figure 1. (A) Image of chia seed gel, inverted to verify a gel formation. (B & C) Optical micrograph of chia seed fiber bundles in water - dyed with Methylene blue (B) & Congo red (C)

To some extent, optical microscope gave some evidence that there were some kind of fibers covering the seed surface (Figure 1B & C). These figures show bundles of extended fibers of part of a chia seed soaked in water and dyed with Methylene Blue (Figure 1B) and Congo Red (Figure 1C). However, the difficulty to see distinct individual fibers still remained.

Electron Microscopy

Figure 2 shows SEM micrographs obtained using the ZEISS Supra 55 VP electron microscope. The SEM analyses resulted in an enhanced means of zooming in more to deduce the nature of the fibers covering the surface of the seeds. The samples were prepared by forming chia seed hydrogel, which was then lyophilized in order to carefully void the gel of water content while leaving the fibers in their extended state. The extremely dried, lyophilized samples were viewed under the SEM. The SEM imaging corroborated our hypothesis; observed was an extensive network of nanoscale fibers extended from proximate seeds to all directions (Figure 2). The magnification was ~1,000X. Furthermore, higher magnification (~30,000X) revealed nanoscale fibers with a diameters around 50 nm.

This discovery is a breakthrough towards understanding the fundamental science and the hypothesis that chia seed must form some sort of 3D network of fibers at nanoscale to induce gel formation in water.

This study has a major implication to science. Gelation has always been associated with the coagulation or solidification of a liquid solvent. The mechanism of gelation using small synthesized molecules, or even refined natural macromolecules, has been studied and somewhat quite understood to proceed via the interaction of nanoscale fibers to form a 3D network. [10-11] However, on the contrary, the mode of gelation using unrefined natural materials such as whole seeds like chia has been simply taken for granted as undergoing "coagulation", "swelling", etc. without a thorough investigation or understanding of the mechanism. Consequently, for the first time, our discovery that these whole seeds form such elegant 3D structures at nanoscale level will inevitably shed more light on the process of 3D network formation and gelation. This would surely provide the opportunity to study and hopefully mimic the efficiency of nature (biomimicry) in order to develop more efficient gels and nano softmaterials, as well as pave way to discovery more gel-forming seeds.

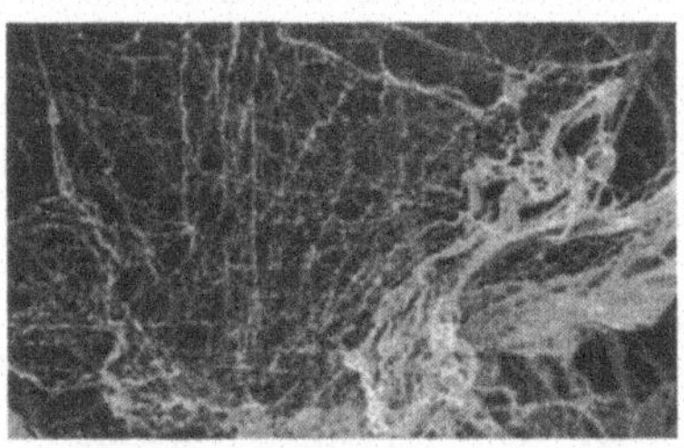

Figure 2. SEM micrograph of Chia seed fibers: 1000X magnification.

5. FUTURE DIRECTION

We plan to perform do more microscopic studies using transmission electron microscopy (TEM) to further elucidate the nature of the nanoscale network. We are also plan to study and confirm the effect of the nanoscale 3D network seed fibers on the hydrolysis of starch by salivary amylase (α-amylase).

6. CONCLUSION

Our studies have revealed that the gelation of chia seed is due to nanoscale fibers with diameters around 50 nm that formed a nanoscale 3D network. Based on the results, these fibers, covering seed surface, are most likely polar moieties that extends in polar water. These results would surely revolutionize how we envision gelation process of Chia and other gel forming seeds and materials, and potentially provide the opportunity to study and mimic chemistry in nature to develop more efficient gels and softmaterials.

REFERENCES

[1] Adams, S.M.; Standridge, J.B. What Should We Eat? Evidence from Observational Studies. *Southern Medical Journal*, **2006**, *99*(7), 744-748.

[2] Lien, N.; Lytle, L.A.; Klepp, K.I. Brief Report - Stability in Consumption of Fruit, Vegetables, and Sugary Foods in a Cohort from Age 14 to Age 21. *Preventive Medicine*, **33,** 217–226.

[3] Rozin, P. The Meaning of Food in Our Lives: A Cross-Cultural Perspective on Eating and Well-Being. *J Nutr Educ Behav.* **2005**, *37* Suppl 2:S107-12.

[4] Anderson, W. The Ancient Food of the Future. http://www.chiabill.com/articles/ancient-food-of-the-future.html Date Acessed: 11/17/2014.

[5] Munoz, L.A.; Cobos, A.; Diaz, O.; Aquilera, J.M. Chia seeds: Microstructure, mucilage extraction and hydration. *Journal of Food Engineering*, **2012**, 108, 216–224.

[6] Cummins, S.; Petticrew, M.; Higgins, C.; Findlay, A.; Sparks, L. Large scale food retailing as an intervention for diet and health: quasi-experimental evaluation of a natural experiment. *J. Epidemiol Community Health*, **2005**, *59*, 1035–1040.

[7] American Diabetes Association. Foot Complications. http://www.diabetes.org/living-with-diabetes/complications/foot-complications/ Date Accessed: 11/17/2014.

[8] Mayo Clinic. Amputation and diabetes: How to protect your feet. http://www.mayoclinic.org/diseases-conditions/diabetes/in-depth/amputation-and-diabetes/art-20048262 Date Accessed: 11/17/2014.

[9] Vuksan, V.; Whitham, D.; Sievenpiper, J. L.; Jenkins, A. L.; Rogovik, A. L.; Bazinet, R. P.; Vidgen,

[10] Sangeetha, N. M.; Maitra, U., Supramolecular gels: functions and uses. *Chemical Society reviews* **2005,** *34* (10), 821-36.

[11] Maity, G. C., Low Molecular Mass Gelators of Organic Liquids. *Journal of Physical Sciences* **2007,** *11*, 156-171.

Evaluation of Biodegraded and Undegraded Plantain Peels as Replacement to Wheat Offal in Broiler Production

F. Aderemi * O.Alabi and A. Awe
*Department of Animal Science and Fisheries management, Bowen University
Iwo Osun State Nigeria.
*Corresponding Author Email faaderemi@yahoo.co.uk

ABSTRACT

This study evaluated replacement of wheat offal with biodegraded and un-degraded plantain peel in raising broilers plantain peels was biodegraded using fungus. Three diets were formulated diet I with 12% wheat offal was the control and diets II and III wheat offal was replaced with undegraded and biodegraded plantain peels. Ninety day old broilers were randomly assigned to these diets replicated thrice and lasted for eight weeks Results revealed that body weight gain was similar across the diets at both starter and finisher phases Those on dietI had a similar feed intake to those on biodegraded diet, but higher than those on the undegraded plantain peel at the starter phase, while at the finisher phase the intake was similar across the diets Feed conversion ratio of those fed control was higher than others at both phases mortality was not be linked to dietary treatments. Serum metabolites were within normal range Morphological analysis of the birds revealed significant effect of treatments. In the main plantain peels could replace wheat offal without adverse effect on the broilers.

Keywords: Plantain peels, broilers, performance characteristics.

1 INTRODUCTION

The sub-optimal consumption of animal protein by a large percentage of Nigerian population has posed a challenge to livestock farmers, researchers and the government. The high cost of feed ingredients in most tropical countries clearly indicates there is need for utilization of unconventional materials [1]. Pressure from consumers for cheaper animal products have necessitated the assessment of local feed ingredients as sole diets in replacement of conventional ingredients during the periods of scarcity and soaring costs[2]. Plantain peels are agro-industrial-byproducts left behind after the edible portion of plantain has been processed into various food items by cooking, roasting or milling into flour Plantain peels, a waste from plantain is observed to have some nutritional values as it contains about 12% crude protein, 16% crude fibre and 1300kcal/kg energy on dry matter basis [3]. Utilization of micro-organisms in enhancing the bio-availability of nutrients in plant by-products has been in use for long[4].This study was embarked upon to evaluate the bioconversion of plantain peels and its utilization by broilers in order to solve problem of inadequate animal protein consumption.

2 MATERIALS AND METHODS

The study was carried out at the poultry section of the Teaching and Research farm Bowen University Iwo with the environmental temperature range of 15-28°C during the harmattan season and 20-36°C in the hot season. It receives an annual rainfall of 1400 mm.

2.1 Experimental diet

Solid state fermentation has been used tremendously to improve nutritive value of feed ingredients which otherwise was considered as a waste. *Rhizopus stolonifer* was *collected* from Microbiology and used for the biodegradation of Plantain peels. *Rhizopus* was introduced into the grated ripe plantain peel in conical flasks then sugar syrup was added to supply enough carbon for the growth of the Rhizopus. The conical flask was covered with cotton wool and foil paper after which the conical flasks were later incubated in a chamber. The larger quantity of grated plantain was poured in a sterilized polythene bag mixed with substrate from the flask and tied to avoid housefly contamination. It was left to biodegrade with frequent turning for ten (10) days after which it was oven dried at a temperature of 60°C. The dried sample is tagged biodegraded plantain peels other grated sample was oven dried and tagged undegraded. Three diets were formulated such that diet I with 12% wheat offal served as the control and in diets II and III the wheat offal was replaced with undegraded and biodegraded plantain peels respectively. This feeding trial lasted for eight weeks during which data on performance characteristic and blood parameters of the birds were taken.

2.2 Birds and their management

Ninety Anak day old broiler chicks were weighed and randomly allotted to three dietary treatments replicated thrice in a completely randomized design. Each replicate had ten chicks. Birds were brooded and reared in a deep litter experimental poultry house. Routine medication and vaccination were carried out to ensure proper health of the birds.. Feed and water were given ad libitum. Blood samples were also taken at the eighth week for analysis all data generated were subjected to analysis.

3 RESULTS AND DISCUSSION

The gross composition of the experimental diets is as shown in Table 1 the diets were formulated to meet the nutrient requirement of the birds. Table 2 shows the performance characteristics of the broilers. The feed intake of chicks on control was similar to those on diet II but higher than undegraded peels at starter phase.The feed intake value of the chicks fed experimental diets may have to do with characteristics nature of the ingredients,its physic-chemical properties as observed by [1] It has been reported that at higher levels of inclusion of unconventional feedstuffs several properties of the feed may be altered the texture, colour, taste and odour[5].While at finisher phase the intake was similar across the diets. This perhaps suggest similarities between plantain peels(degraded or undegraded) and wheat offal. This could also be due to increased digestive capacity and gut motility of older birds. It has been reported that feed intake is a function of the dietary energy [6] Mortality that was recorded could not be traced to dietary treatment. The feed conversion ratio of those fed diet III were better than others.i.e convert feed to flesh at a minimum costs at starter phase whereas at finisher phase broilers on diet II had the best feed conversion ratio. The Carcass characteristics and organ weights of broilers fed experimental diets observed for this study is as shown in Table 3. The observed values for organ weight of broilers were all normal. Serum metabolites of broilers fed experimental diets is as shown in Table 4.The cholesterol of broilers fed diets I and II were significantly lower than diet III. This could be that recycling of bile acid is diminished thereby synthesis of more bile acids from cholesterol is stimulated. This then result again in lowering of serum cholesterol levels this is in agreement with the findings of [7].Total protein albumin and globulin values in this study indicated efficient protein utilization by the broilers similar to the findings of Tewe (1994).It also implied that the birds were not dehydrated. It could be observed that at the finisher phase plantain peels promoted muscle wastage as higher levels of creatinine were observed with birds fed these rations. The glucose level does not have a particular statistical trend. The observed enzymes transaminase (AST) and Aspartate transferase (ALT) were within normal range indicative of healthy broilers. This confirmed absence of liver diseases in the experimental birds similar to the findings of Aderemi et al [8].

4 CONCLUSION

In the main this study shows that biodegraded and /or undegraded plantain peels could replace wheat offal in the diet of broilers without adversely affecting the performance, the blood serum metabolites or even the morphological parameters of the broilers. These in turn promote the consumption of animal protein among the populace.

Table 1: Gross composition of experimental diets for Broiler.

Ingredients	Starter Phase I	II	III	Finisher Phase I	II	III
Maize	57.0	57.0	57.0	54.0	54.0	54.0
Soybean meal	11.0	11.0	11.0	11.0	11.0	11.0
Wheat offal	12.0	0.0	0.0	16.0	0.0	0.0
Un-degraded plantain peels	0.0	12.0	0.0	0.0	16.0	0.0
Degraded plantain peels	0.0	0.0	12.0	0.0	0.0	16.0
Palm kernel cake	3.5	3.5	3.5	3.5	3.5	3.5
Groundnut cake	9.0	9.0	9.0	4.0	4.0	4.0
Fish meal	3.0	3.0	3.0	3.0	3.0	3.0
Bone meal	2.0	2.0	2.0	3.0	3.0	3.0
Oyster shell	2.0	2.0	2.0	5.0	5.0	5.0
*Premix	.25	.25	.25	.25	.25	.25
Salt	.25	.25	.25	.25	.25	.25
Total	100	100	100	100	100	100
Calculated analysis						
Crude protein (%)	20.76	20.13	20.32	18.60	18.56	18.21
ME kcal/kg	3022	3045	3137	3018	3023	3029

*Vitamin/mineral premix provided the following vitamin and minerals per kg of diet: A, 10,000 I.U.; D_3, 300 I.U.; E. 8.0 I.U.; K, 2.0mg; B_1, 2.0mg; B_6, 1.2mg; B_{12}, 0.12mg; Nianin 1.0mg; Panthothenic acid, 7.0mg; Folic acid, 0.6mg; Cholic, 500mg; C, 10.0mg; Fe, 60mg; Mn, 80mg; Mg, 100mg; Cu, 8.0mg; Zn, 50mg; Co, 0.45mg; I, 2.0mg and Se, 0.1mg.ANUPCO, Anglican Nutrition Products Company, England

Table 2: Performance Characteristics of broilers fed experimental diets.

Starter phase	I	II	III	SEM
Weekly feed intake/bird(kg)	0.67^a	0.69^a	0.64^b	0.05
Final body weight gain/bird(kg)	0.37^a	0.39^a	0.38^a	0.001
Feed conversion ratio	2.03	1.97	1.88	-
Mortality (%)	3.0	3.0	2.5	0.01
Finisher phase				
Weekly feed intake/bird(kg)	2.21	2.16	2.16	0.01
Weekly weight gain/bird(kg)	0.82	0.84	0.82	0.01
Feed conversion ratio	2.70	2.57	2.63	-
Mortality (%)	0	0	0	-

abc Means with the same superscript along a row are not significantly different (P>0.05)

Table 3: Carcass characteristics and organ weights of broilers fed experimental diets.

Parameters	I	II	III	SEM
Liveweight (kg)	1.33	1.33	1.32	0.02
Shank (g)	12.38^{a}	10.90^{c}	11.41^{a}	0.40
Wing (g)	5.82^{a}	5.64^{b}	3.14^{c}	0.10
Back (g)	16.39^{b}	18.03^{a}	13.08^{c}	0.05
Breast (g)	13.71^{c}	18.86^{a}	14.99^{b}	0.50
Kidney (%)	0.54	0.56	0.53	0.21
Liver (%)	2.57^{a}	2.00^{c}	2.17^{b}	0.10
Heart (%)	0.69^{a}	0.49^{c}	0.58^{b}	0.08

abc Means with the same superscript along a row are not significantly different (P>0.05)

Table 4: Serum metabolites of broilers fed experimental diets.

Parameters	I	II	III	SEM
Glucose (mg/dl)	156.68^{a}	144.03^{a}	166.04^{a}	5.00
AST (i.µ/l)	1290^{a}	1268^{a}	1257^{a}	2.11
ALT (i.µ/l)	76.33^{a}	75.52^{a}	74.73^{a}	0.66
T.P (mg/dl)	4.48^{b}	4.86^{a}	4.32^{a}	0.04
Cholesterol (mg/dl)	89.35^{c}	107.73^{b}	123.71^{a}	5.00
Urea (mg/dl)	3.12^{b}	3.43^{a}	1.49^{c}	0.20
Creatinine (mg)	0.34^{b}	0.46^{a}	0.48^{a}	0.33

REFERENCES

[1] Aderemi Foluke Utilization of graded levels of biodegraded cassava peels in broiler ration. Electronic Journal of Environmental,Agricultural and Food Chemistry (EJEAFChe) 9(4)672-678.2010.

[2] Oluokun J.A. performance off growing Rabbitsfed a low protein diet supplemented with urea or roasted soybean meal.Journal a Animal production Research 17:89-99. 2001.

[3] Ajasin F.O. Omole A.J. Oluokun J.A. Obi O.O. and Owosibo A. performance characteristics of weaned rabbits fed plantain peel as replacement for maize (review);Emerald group publishing Ltd. 2008.

[4] Iyayi E.A. Changes in the cellulose, sugar and crude protein contents of agro- industrial by-products fermented with aspergillus niger,Aspergillus flavus and penicillin sp. Afr.J. Biotechnol 3 :186-188. 2004.

[5] Umoren U.E, Isika M.A., Asanga E.P. and Ezeigwe P.N. Effect of replacement of erath ball (Icacinia manni) meal on the performance of broiler chickens.Pakistan Journal of Biological Science 10:2368-2373. 2007.

[6] Omole A.J. Ayodeji I.O and Raji M.A. The Potential of peels of mango, plantain, cocoyam and pawpaw as diets for growing snail(Archahatina marginata). Livestock Research for Rural development.16 (12): 1- 5 2004.

[7] Ezeagu I.E. Fetzke K.J. Ologhobo .A.D. Metges C.C. and Akinsoyinu A.O. Compositional evaluation and feed potential of the African kudzu (Pueraria phaseoloides,Roxb.Benth) seed meal Journal of Food ScienceTechnol 37 191-195. 2000.

[8] Aderemi F.A ,T.E. Lawal, O.M. Alabi, O.A. Ladokun ,G.O. Adeyemo. Effect of enzyme supplemented cassava root sieviate on egg quality gut morphology and performance of egg type chicken.International Journal of Poultry science 5(6): 526-529.2006.

Index of Authors

Index of Keywords

For Product Safety Concerns and Information please contact our EU representative GPSR@taylorandfrancis.com Taylor & Francis Verlag GmbH, Kaufingerstraße 24, 80331 München, Germany

Printed and bound by CPI Group (UK) Ltd, Croydon, CR0 4YY
14/09/2026
02216251-0001